NON-SELF-ADJOINT BOUNDARY EIGENVALUE PROBLEMS

NORTH-HOLLAND MATHEMATICS STUDIES 192

(Continuation of the Notas de Matemática)

Editor: Jan van Mill
Faculteit der Exacte Wetenschappen
Amsterdam, The Netherlands

2003
ELSEVIER
Amsterdam – Boston – Heidelberg – London – New York – Oxford
Paris – San Diego – San Francisco – Singapore – Sydney – Tokyo

NON-SELF-ADJOINT BOUNDARY EIGENVALUE PROBLEMS

Reinhard MENNICKEN
University of Regensburg
Regensburg, Germany

and

Manfred MÖLLER
University of the Witwatersrand
Johannesburg, South Africa

2003
ELSEVIER
Amsterdam – Boston – Heidelberg – London – New York – Oxford
Paris – San Diego – San Francisco – Singapore – Sydney – Tokyo

ELSEVIER SCIENCE B.V.
Sara Burgerhartstraat 25
P.O. Box 211, 1000 AE Amsterdam, The Netherlands

First edition 2003

Library of Congress Cataloging in Publication Data
A catalog record from the Library of Congress has been applied for.

British Library Cataloguing in Publication Data
A catalogue record from the British Library has been applied for.

ISBN: 0-444-51447-3
ISSN: 0304-0208

♾ The paper used in this publication meets the requirements of ANSI/NISO Z39.48-1992 (Permanence of Paper).
Printed in Hungary.

PREFACE

The purpose of this book is the study of non-self-adjoint boundary eigenvalue problems for first order systems of ordinary differential equations and n-th order scalar differential equations. The coefficients of the differential equations as well as the boundary conditions are allowed to depend polynomially, holomorphically or asymptotically on the eigenvalue parameter. The boundary conditions may contain infinitely many interior points and an integral term. With the boundary eigenvalue problem a bounded operator function is associated which consists of two components, the differential operator function and the boundary operator function. These operator functions depend in general nonlinearly on the eigenvalue parameter.

Various eigenfunction expansions are proved by the contour integral method under regularity conditions which originally were introduced by BIRKHOFF and STONE in case of λ-independent boundary condtitions. The calculation of the Fourier coefficients of these expansions is based on the theory of the inverses of holomorphic Fredholm operator valued functions which for the sake of completeness is included in this book. An important aspect of this theory is the representation of the principal parts of the inverses of these functions at their poles by root functions (eigenvectors and associated vectors) of the given operator functions and their adjoints. The proofs of the eigenfunction expansions are based on sharp asymptotic estimates of the resolvents (Green's functions) for large values of the eigenvalue parameter.

Our approach is based on functional analytic methods. The reader should be familiar with basic concepts of Banach spaces and Lebesgue integration and should have some knowledge about distributions. Whenever we use these basic results we give references so that the reader unfamiliar with these concepts can easily find them. Our main references to the basic topics are the monograph [KA] of T. KATO for Banach spaces, the monograph [HS] of E. HEWITT and K. STROMBERG for the theory of Lebesgue integration, and the monograph [HÖ2] of L. HÖRMANDER for the theory of distributions.

Each chapter ends with a short section containing historical notes.

Chapters I and II are concerned with preparations from functional analysis and Sobolev space theory. In Chapters III–V first order systems are considered, followed by n-th order equations in Chapters VI–IX. Since n-th order equations are reduced to first order systems, some of the results of Chapters III–V are needed in Chapters VI–IX. Chapter X contains applications to problems from physics and engineering.

The literature for n-th order linear differential equations and first order systems is vast, and the bibliography is only a selection of publications in this field. The list of notations and the index should help the reader to navigate through the text.

CONTENTS

CHAPTER IV **Birkhoff regular and Stone regular boundary eigenvalue problems**

CHAPTER V **Expansion theorems for regular boundary eigenvalue problems for first order systems**

CHAPTER VI ***n*-th order differential equations**

CHAPTER VII **Regular boundary eigenvalue problems for *n*-th order equations**

APPENDIX A Exponential sums

INTRODUCTION

In this monograph we consider first order systems of ordinary differential equations

$$y' = (\lambda A_1 + A_0)y$$

on a bounded interval $[a,b]$ together with boundary conditions of the form

$$\sum_{j=0}^{\infty} W^{(j)}(\lambda)y(a_j) + \int_a^b W(x,\lambda)y(x)\,\mathrm{d}x = 0,$$

where $a_0 = a$, $a_1 = b$ and the other points a_j lie in the interior of $[a,b]$ and are mutually distinct. The $n \times n$ matrices $W^{(j)}(\lambda)$ and $W(x,\lambda)$ depend polynomially on λ, A_0 and A_1 are $n \times n$ matrix functions depending on the independent variable $x \in [a,b]$, and $A_1(x)$ is a not necessarily invertible diagonal matrix, but satisfies certain other restrictions.

Apart from λ-nonlinear boundary eigenvalue problems for first order systems such problems for n-th order scalar differential equations will be investigated, more precisely,

$$y^{(n)} + \sum_{i=0}^{n} p_i(\cdot,\lambda)y^{(i)} = 0$$

together with boundary conditions

$$W^{(0)}(\lambda)(y(0),\ldots,y^{(n-1)}(0))^{\mathsf{T}} + W^{(1)}(\lambda)(y(1),\ldots,y^{(n-1)}(1))^{\mathsf{T}} = 0.$$

The functions p_i are polynomials in λ whose coefficients are, for example, continuous functions on $[a,b]$, and $W^{(0)}(\lambda)$, $W^{(1)}(\lambda)$ are $n \times n$-matrices depending polynomially on λ. For simplicity we have here only written down two-point boundary conditions, but in the general treatment we will also have multipoint and integral boundary terms as in the case of first order systems.

Boundary eigenvalue problems of these types occur in various branches of engineering and physics, e. g. in elasticity theory, control theory, hydrodynamics, magnetohydrodynamics, and even in meteorology. For more details we refer to Chapter X of this monograph.

Since every n-th order differential equation can be transformed into a first order system, we will first develop a spectral theory for first order systems. Statements about n-th order differential equations will then be proved by using results for first order systems, with possible special attention to improvements due to the special structure of n-th order differential equations.

With a boundary eigenvalue problem for a first order systems we will associate the operator pencil

$$T(\lambda) = \begin{pmatrix} T^D(\lambda) \\ T^R(\lambda) \end{pmatrix} : (H_1(a,b))^n \to (L_2(a,b))^n \times \mathbb{C}^n,$$

where $H_1(a,b)$ is the Sobolev space of order 1, and

$$T^D(\lambda)y = y' - (\lambda A_1 + A_0)y,$$

$$T^R(\lambda)y = \sum_{j=0}^{\infty} W^{(j)}(\lambda)y(a_j) + \int_a^b W(x,\lambda)y(x)\,\mathrm{d}x.$$

Similarly, for an n-th order scalar differential equation we introduce the operator pencil

$$L(\lambda) = \begin{pmatrix} L^D(\lambda) \\ L^R(\lambda) \end{pmatrix} : H_n(a,b) \to L_2(a,b) \times \mathbb{C}^n,$$

where $H_n(a,b)$ denotes the Sobolev space of order n and

$$L^D(\lambda)y = y^{(n)} + \sum_{i=0}^{n} p_i(\cdot,\lambda)y^{(i)},$$

$$L^R(\lambda)y = W^{(0)}(\lambda)(y(0),\ldots,y^{(n-1)}(0))^\mathsf{T} + W^{(1)}(\lambda)(y(1),\ldots,y^{(n-1)}(1))^\mathsf{T}.$$

It is the particular goal of this monograph to prove eigenfunction expansions for such λ-nonlinear boundary eigenvalue problems, both for first order systems and for n-th order differential equations. In the following paragraphs of this introduction we will restrict our considerations to boundary eigenvalue problems for first order systems and consequently to operator functions $T(\lambda)$; nearly analogous statements hold in the case of differential equations and for the operator function $L(\lambda)$.

The realization of eigenfunction expansions consists of two parts: first explicit formulas for the Fourier coefficient of these expansion have to be found, and secondly the convergence of these expansions for a suitable class of functions has to be proved. The solution of both problems is highly nontrivial. For a satisfactory solution of the first problem a functional analytic setting is appropriate. For the convergence proof we apply the powerful contour integral method.

Such a functional analytic setting has already been put into place in the definition of the operator function $T(\lambda)$. For fixed $\lambda \in \mathbb{C}$, $T(\lambda)$ is a bounded Fredholm operator defined on the whole domain $(H_1(a,b))^n$. With respect to λ it is a holomorphic function in $\mathbb{C}$. For such operator functions a well-established spectral theory is available: The notions *resolvent set*, *spectrum*, *discrete spectrum*, *eigenvector* and *associated vectors* or *root functions* are well-known. The adjoint operator function $T^*(\lambda)$ is also a holomorphic bounded Fredholm operator valued function in $\mathbb{C}$. Thus the same notions as for $T(\lambda)$ are available.

If the resolvent set of such an operator function is non-empty, i. e., there is a point $\lambda_0 \in \mathbb{C}$ such that $T(\lambda_0)$ has a bounded inverse, then the spectrum of this operator function is discrete and its resolvent is a finitely meromorphic operator function in $\mathbb{C}$. According to a theorem of I. C. GOHBERG and E. I. SIGAL [GS], in a special case due to M. V. KELDYSH [KE1], [KE2], the finite rank operators in the principal part of $T(\lambda)^{-1}$ of the Laurent expansion in a neighbourhood of an eigenvalue λ_0 can be represented by biorthogonal canonical systems the eigenvectors and associated vectors of $T(\lambda)$ and $T^*(\lambda)$ at $\lambda = \lambda_0$.

To explain how this representation yields the Fourier coefficients in the eigenfunction expansions, let us assume for simplicity that the operator function $T(\lambda)$ is linear in λ, i. e., $T(\lambda) = T_0 + \lambda T_1$. Then for λ in the resolvent set of T, the identity

$$\frac{1}{\lambda} f = T^{-1}(\lambda) T_1 f + \frac{1}{\lambda} T^{-1}(\lambda) T_0 f$$

holds. If we integrate this identity along a contour around 0 in the resolvent set, say, a circle with centre 0 and radius r, then we obtain

$$f = \sum \operatorname{res}_{\lambda=\mu} (T^{-1}(\lambda) T_1 f) + \frac{1}{2\pi i} \int_{|\lambda|=r} \frac{1}{\lambda} T^{-1}(\lambda) T_0 f \, \mathrm{d}\lambda,$$

where the summation is taken over all finitely many eigenvalues of T with $|\mu| < r$. An application of the above mentioned representation theorem yields the corresponding partial sum of the eigenfunction expansion for f with explicit formula for the Fourier coefficients. If we extend the summation in this partial sum over all eigenvalues of T and the question of convergence of this series is not addressed, then it is referred to as the formal eigenfunction expansion of f.

In order to prove the convergence of the formal eigenfunction expansion, we have to impose so-called regularity conditions, originally introduced by G. D. BIRKHOFF and M. H. STONE, respectively, on the given boundary eigenvalue problem, i. e., on $T(\lambda)$, and certain boundary conditions on the function f which is to be expanded. By a tedious estimate of the resolvent $T^{-1}(\lambda)$, Birkhoff regularity yields that this resolvent is bounded on a certain sequence of circles Γ_n with centres 0 and radii r_n tending to infinity. The Stone regularity condition is weaker and means that the resolvent divided by some power of λ, say λ^s, is bounded on such a sequence of circles. If such regularity conditions assure that in a suitable function space the integral in the foregoing equation tends to zero as $r = r_n$ tends to infinity, then in this function space the convergence of the eigenfunction expansion holds. Otherwise conditions have to be imposed on the function f.

We will explain how this leads to boundary conditions on f in a quite natural way. Choose a function $f^{[1]}$ in $(H_1(a,b))^n$ such that

$$T_0 f + T_1 f^{[1]} = 0$$

and substitute $T_1 f^{[1]}$ for $T_0 f$ in the above identity for $T^{-1}(\lambda)T_1 f$. By this procedure we obtain

$$T^{-1}(\lambda)T_1 f = \frac{1}{\lambda} f + \frac{1}{\lambda} T^{-1}(\lambda)T_1 f^{[1]}$$

and further by iteration

$$\frac{1}{\lambda} + \frac{1}{\lambda^2} f^{[1]} = T^{-1}(\lambda)T_1 f + \frac{1}{\lambda^2} T^{-1}(\lambda)T_0 f^{[1]},$$

whence

$$f = \sum \mathrm{res}_{\lambda=\mu}(T^{-1}(\lambda)T_1 f) + \frac{1}{2\pi i}\int_{|\lambda|=r} \frac{1}{\lambda^2} T^{-1}(\lambda)T_0 f^{[1]}\, \mathrm{d}\lambda,$$

where the summation is again taken over all eigenvalues μ of T with $|\mu| < r$. If now the second term on the right-hand side of this equation converges to 0 in the considered function space, then we stop this procedure. Otherwise, we repeat it. Since by the regularity conditions the function $\lambda^{-s}T(\lambda)$ is bounded on the circles Γ_n as n tends to infinity, the procedure will stop after at most $s+1$ iterations. Next we will illustrate the meaning of the above condition on $f^{[1]}$. Recall that the operator function $T(\lambda)$ consists of two components: namely the differential operator part

$$T^D(\lambda)y = y' - (\lambda A_1 + A_0)y$$

and the boundary part

$$T^R(\lambda)y = T_0^R y + \lambda T_1^R y.$$

Let us assume for simplicity that the matrix $A_1(x)$ is invertible for all $x \in [a,b]$. Then the differential part of the equation for $f^{[1]}$ means that

$$f^{[1]} = -A_1^{-1}(f' - A_0 f),$$

and the boundary part yields the boundary condition

$$T_0^R f + T_1^R f^{[1]} = 0$$

which the function f has to fulfil. If the boundary condition is independent of λ, then this condition reduces to the usual boundary condition

$$T_0^R f = 0.$$

If the matrix $A_1(x)$ is not invertible or the boundary conditions are polynomial in λ, the foregoing procedure has to be modified, but works in priciple in the same way.

Some hard analysis consists in the derivation of practical criteria for Birkhoff or Stone regularity and in the proof of asymptotic estimates of the resolvent, i. e., of the Green's matrix function. The proofs of these criteria and estimates rely on the construction of a suitable asymptotic fundamental system of solutions of the first order system. With respect to the n-th order scalar differential equation it is advantageous to transform it into a first order system. The transformed system is then asymptotically linearized with respect to the eigenvalue parameter. For this reason we consider a slightly more general first order system than written down above, namely

$$y' = \left(\lambda A_1 + A_0 + \frac{1}{\lambda} A^0(\cdot,\lambda)\right) y,$$

where the $n \times n$ matrix function $A^0(\cdot,\lambda)$ is uniformly bounded as λ tends to infinity, together with the above given boundary conditions in an appropriate asymptotically constant form. For the asymptotic system a suitable fundamental system is constructed. With respect to this fundamental system the determinant of the characteristic matrix associated with the asymptotic boundary conditions becomes an exponential sum whose coefficients are λ-asymptotic polynomials. This representation yields the regularity criteria. Birkhoff regularity is characterized by the leading term of the asymptotic fundamental system and the constant terms of the asymptotic boundary conditions at the endpoints of the interval $[a,b]$. A characterization of Stone regularity is much more involved. We present a practical geometric criterion due to R. H. COLE [CO3] in a slightly generalized form.

The estimates of the Green's matrix function and the resolvent of the λ-asymptotic boundary eigenvalue problem are established in various function spaces. They require a careful analysis of the Green's matrix function and skilful arrangements of its components. The resulting estimates for the λ-asymptotic boundary eigenvalue problem which are valuable for themselves lead to a variety of eigenfunction expansion theorems concerning Birkhoff and Stone regular boundary eigenvalue problems both for first order systems and n-th order scalar differential equations as stated at the beginning of this introduction.

In this monograph we only deal with the problem of expandability in terms of eigenfunctions and associated functions. Expandability implies completeness, but does not imply minimality or basisness of these functions. Only recently, the problems of minimality and basisness for Birkhoff regular boundary eigenvalue problems defined by systems of differential equations $y' = (\lambda A_1 + A_0)y$ and n-th order scalar differential equations $\mathbb{K}\eta = \lambda\mathbb{H}\eta$ both with λ-polynomial boundary conditions were successfully approached by C. TRETTER in a series of papers [TR7], [TR8], [TR9], and [TR10]. The key point is a new linearization method of the λ-polynomial boundary conditions which leaves the structure of the diagonal matrix unchanged by blowing it up with additional zeros. Boundary eigenvalue problems with such noninvertible diagonal matrices were first investigated in the authors' publications [MM4], [MM5].

Purely functional analytic methods, not considering the Green's function, are applied to λ-polynomial boundary eigenvalue problems of different types in numerous publications. A. DIJKSMA [DIJ] and A. DIJKSMA, H. LANGER and H. DE SNOO [DLS1], [DLS2], [DLS3] treated self-adjoint λ-polynomial boundary eigenvalue problems and transformed them to common spectral problems for self-adjoint operators in Hilbert or Krein spaces by problem oriented linearization methods. M. V. KELDYSH [KE1], [KE2], I. C. GOHBERG and M. G. KREIN [GK], V. M. VIZITEĬ and A. S. MARKUS [VM], M. FAIERMAN, A. S. MARKUS, V. MATSAEV and M. MÖLLER [FMMM], for numerous further references see the substantial monograph of A. S. MARKUS [MA4], developed a comprehensive spectral theory for operator polynomials in Hilbert and Banach spaces using abstract linearization procedures and perturbation arguments to prove so-called n-fold completeness, minimality and basisness for the corresponding eigenvectors and associated vectors. These publications include various applications to λ-polynomial ordinary and partial differential equations, however, nearly exclusively with λ-independent boundary conditions. An exception is A. A. SHKALIKOV'S paper [SH5] which treats boundary eigenvalue problems for n-th order scalar differential equations where the differential equations as well as the boundary conditions depend polynomially on the eigenvalue parameter.

The functional analytic approach of combining the differential operator and the boundary operator to a two-component operator defined on a fixed space, not depending on the eigenvalue parameter, was introduced for S-Hermitean boundary eigenvalue problems by F. W. SCHÄFKE and A. SCHNEIDER in [SCHSCH1], [SCHSCH2], [SCHSCH3], see also H.-D. NIESSEN [NI]. For non-self-adjoint boundary eigenvalue problems this setting is due to R. Mennicken and M. Möller in [MM4], [MM5]. One advantage of this approach is the fact that the boundary conditions are considered inhomogeneously in a natural way, see also the monograph [STV] of Š. SCHWABIK, M. TVRDÝ and O. VEJVODA. Another advantage is that, in the case of a non-self-adjoint problem, the adjoint boundary value problem including all spectral data, as for example associated vectors, is defined without any additional restrictions.

Because of the diversity of the subject and the enormous number of publications we abstain here from a further historical excursion, but refer the reader to special sections with historical remarks at the end of each chapter.

Chapter I deals with spectral theory for holomorphic Fredholm valued operator functions, in particular, the principal parts of their inverses at the poles are investigated. It is shown that these principal parts can be written in terms of eigenvectors and associated vectors of the operator function and its adjoint. One-to-one connections between biorthogonal systems of eigenvectors and associated vectors and the principal parts of the inverse operator functions are established. Special attention is paid to the case of λ-linear problems.

Chapter II contains the preriquisites for the study of differential operators. Sobolev spaces on intervals are introduced and their properties are investigated. These results are essentially well-known but, in general, are stated and proved for subsets of $\mathbb{R}^n$. The one dimensional case is easier and gives some additional properties. Therefore, and to make the monograph more self-contained, this chapter is included. Also, some basic results for differential equations are stated.

Chapter III starts with the definition of boundary eigenvalue problems for first order systems. The adjoint and the inverse are calculated, and their relations to the "classical" adjoint and inverse for the differential operator considered in L_p spaces are discussed. Some examples show the difficulties which arise if one considers the classical adjoint. The inverse is an extension of the classical inverse, which is an integral operator whose kernel is the GREEN'S matrix function.

Chapter IV is devoted to the estimate of the GREEN'S matrix function. To this end, Birkhoff regularity is introduced for systems which are asymptotically linear in the eigenvalue parameter, and necessary and sufficient conditions for Birkhoff regularity are given. The characteristic determinant is estimated below away from its zeros, which are the eigenvalues of the given boundary eigenvalue problem. Then the GREEN'S matrix function and finally the resolvent of the boundary eigenvalue operator functions are estimated on suitable circles in the complex plane tending to infinity.

In Chapter V the estimates of the previous chapter are used to prove expansion theorems for first order systems which are linear in the parameter; the boundary conditions are still allowed to depend polynomially on the parameter. Not only Birkhoff regularity is considered but also Stone regularity. Whereas all functions in $L_p(a,b)$, $1 < p < \infty$, are expandable in eigenfunctions and associated functions if the problem is Birkhoff regular, the expandable functions must be sufficiently smooth and must satisfy certain auxiliary boundary conditions if the problem is Stone regular. Also uniform convergence is investigated, where even for Birkhoff regular problems the expandable functions have to satisfy certain regularity conditions and some boundary conditions.

Chapter VI is concerned with n-th order differential equations. Here the corresponding results of Chapter III are obtained for an n-th order differential equation, where also the equivalence of this problem to one for a first order system is established.

Chapter VII deals with boundary value problems for differential equations whose equivalent first order system can be linearized asymptotically. Using the estimates of Chapter IV, expansion theorems are proved.

Chapter VIII is concerned with regular two-point boundary eigenvalue problems for the differential equation $\mathbf{K}\eta = \lambda\mathbf{H}\eta$, where $\mathbf{K}$ and $\mathbf{H}$ are differential operators such that $\mathbf{K}$ is of higher order than $\mathbf{H}$. The structures of the fundamental system, its adjoint, and the Green's function are investigated more thoroughly. However, the estimates of Chapter IV can still be used. Some applications to problems from mechanics are given, to which the results of Chapter VII are not directly applicable.

In Chapter IX problems depending polynomially on the eigenvalue parameter λ are linearized with respect to λ. The corresponding convergence theorems for first order systems lead to n-fold expansions for the original problem. Completeness and minimality for these problems are considered.

Chapter X contains further examples dealing with problems from mechanics like elastic bars and control of beams, fluid mechanics, magnetohydrodynamics, and meteorology.

Appendix A deals with estimates of exponential sums. They are needed for the estimate of the characteristic determinant in Chapter IV.

Chapter I

OPERATOR FUNCTIONS IN BANACH SPACES

This chapter deals with holomorphic Fredholm operator valued functions in Banach spaces. The structure of the resolvent of such an operator function is discussed in detail. It is shown that, on a domain in $\mathbb{C}$, its resolvent is finitely meromorphic if its resolvent set is non-empty (Theorem 1.3.1). The operators in the principal parts of the resolvent are expressed in terms of biorthogonal canonical systems of root functions (eigenvectors and associated vectors) of the corresponding operator function and its adjoint operator function (Theorems 1.5.4, 1.5.9 and 1.6.5, 1.6.7). This representation can be understood as the formal eigenfunction expansion with respect to the given operator function since it yields explicit formulæ for the corresponding Fourier coefficients. As a consequence of this representation theorem concerning the principal parts of the resolvent we obtain a local factorization theorem for the operator function itself (Theorem 1.8.4). Particular properties, such as minimality of the system of eigenvectors and associated vectors, are discussed for Fredholm operator valued functions which are linear in the eigenvalue parameter.

Boundary eigenvalue problems as considered in later chapters have an underlying abstract operator theoretic structure, which is investigated in Section 1.11. For these abstract boundary eigenvalue problems the notions fundamental matrix function and characteristic matrix function are introduced, generalizing the concepts of fundamental matrix and characteristic matrix, which is well-known for boundary value problems for ordinary linear differential equations and systems. It is shown that the operator function defined by such an abstract boundary eigenvalue problem, shortly called boundary eigenvalue operator function, is globally holomorphically equivalent to a canonical extension of its characteristic matrix function (Theorem 1.11.1). The characteristic matrix function is defined on a smaller (finite-dimensional) Banach space than the corresponding boundary eigenvalue operator function, but nevertheless contains all the information about its spectral data.

1.1. Banach spaces

Let E be a vector space. We always assume that E is a vector space over the field of complex numbers $\mathbb{C}$. A map $|\ |: E \to \mathbb{R}$ is called a *norm* if

$$\begin{aligned} &|y| \geq 0 \quad (y \in E), \\ &|\alpha y| = |\alpha|\,|y| \quad (\alpha \in \mathbb{C},\ y \in E), \\ &|x+y| \leq |x| + |y| \quad (x, y \in E), \\ &|y| = 0 \Rightarrow y = 0 \quad (y \in E). \end{aligned}$$

If $|\ |$ is a norm on E, then $E = (E, |\ |)$ is called a *normed space*.

Let E be a normed space, $y \in E$ and $(y_n)_{n\in\mathbb{N}}$ be a sequence in E. Then $(y_n)_{n\in\mathbb{N}}$ is said to *converge* to y if

$$|y_n - y| \to 0 \text{ as } n \to \infty,$$

and we write

$$y_n \to y \text{ as } n \to \infty \quad \text{or} \quad \lim_{n\to\infty} y_n = y.$$

The sequence $(y_n)_{n\in\mathbb{N}}$ is said to be a *Cauchy sequence* in E if for each $\varepsilon > 0$ there is a number $n_0 \in \mathbb{N}$ such that $|y_n - y_m| \leq \varepsilon$ for all $n, m \geq n_0$. A normed space E is called a Banach space if each Cauchy sequence in E converges to some y in E. A Banach space is a complete metric space, see e. g. [DI, p. 88].

Let E and F be Banach spaces. Then $L(E,F)$ denotes the space of all continuous linear operators on E to F, i. e., $T \in L(E,F)$ if and only if $T: E \to F$ is linear and

$$|T| := \sup\{|Ty|_2 : y \in E,\ |y|_1 \leq 1\} < \infty,$$

where $|\ |_1$ and $|\ |_2$ are the norms on E and F, respectively. It is well-known that $|\ |$ is a norm on $L(E,F)$ which makes $L(E,F)$ a Banach space, see e. g. [KA, p. 150]. We write $L(E)$ instead of $L(E,E)$. By id_E we denote the identity on E, i. e., $\mathrm{id}_E\, x = x$ for $x \in E$. For an operator $T \in L(E,F)$, $N(T) := \{x \in E : Tx = 0\}$ denotes the *null space* and $R(T) := \{Tx : x \in E\}$ the *range* of T. $T \in L(E,F)$ is called a *Fredholm operator* if both its *nullity* $\operatorname{nul} T := \dim N(T)$ and its *deficiency* $\operatorname{def} T := \operatorname{codim} R(T)$ are finite. $\Phi(E,F)$ denotes the set of all Fredholm operators on E to F. If $T \in \Phi(E,F)$, then $\operatorname{ind} T := \operatorname{nul} T - \operatorname{def} T$ is well-defined and called the *index* of T.

The operator $T \in L(E,F)$ is called *invertible* if there is an operator $T^{-1} \in L(F,E)$ such that $TT^{-1} = \mathrm{id}_F$ and $T^{-1}T = \mathrm{id}_E$. If T is invertible, then T^{-1} is unique. From the closed graph theorem, see e. g. [KA, p.166], we know that T is invertible if and only if T is bijective.

$E' := L(E, \mathbb{C})$ is called the *dual space* of E. The bilinear form $\langle\ ,\ \rangle$ on $E \times E'$ is defined by $\langle y, u \rangle = u(y)$, where $y \in E$ and $u \in E'$. With respect to the norm

$$|u| := \sup\{|\langle y, u\rangle| : |y| \leq 1\},$$

which is the operator norm on $L(E,\mathbb{C})$, E' is a Banach space. For T in $L(E,F)$ there is a unique $T^* \in L(F',E')$ such that

$$\langle Ty, v\rangle = \langle y, T^*v\rangle \quad (y \in E, v \in F').$$

The operator T^* is called the *adjoint* of T. Note that $|T| = |T^*|$, cf. e. g. [KA, p. 154]. If T is a Fredholm operator, then T^* is also a Fredholm operator, cf. e. g. [KA, p. 234]. T is invertible if and only if T^* is invertible, and in this case $(T^{-1})^* = (T^*)^{-1}$.

Note that E' is the set of continuous linear functionals on E and not the set of continuous complex conjugate linear functionals. This is advantageous since we are often dealing with holomorphic operator functions, in which case the adjoint will be holomomorphic, see Proposition 1.2.6. Otherwise, the adjoint would be anti-holomorphic, which would generate unnecesssary complications. Although T' might seem to be more appropriate as a notation for the adjoint of T, we have chosen T^* since T' is reserved for the derivative.

For a set G we denote the set of $k\times n$ matrices with entries from G by $M_{k,n}(G)$. We write G^k if $n = 1$, and $M_n(G)$ if $n = k$. For

$$y = (y_{ij})_{i=1,j=1}^{k,n} \in M_{k,n}(E) \tag{1.1.1}$$

we set

$$|y| := \sum_{i=1}^{k} \max_{j=1}^{n} |y_{ij}|. \tag{1.1.2}$$

Then $M_{k,n}(E)$ is a Banach space with respect to this norm.

If y is given by (1.1.1), then

$$y^{\mathsf{T}} := y^* = (y_{ij})_{j=1,i=1}^{n,k}.$$

For $y \in E$ and $v \in F'$ we define the *tensor products* $y\otimes v$ and $v\otimes y$ by

$$(y\otimes v)(w) := \langle w, v\rangle y \quad (w \in F),$$
$$(v\otimes y)(u) := \langle y, u\rangle v \quad (u \in E').$$

PROPOSITION 1.1.1. *Let $y\in E$ and $v \in F'$. Then $y\otimes v \in L(F,E)$, $v\otimes y \in L(E',F')$, $(y\otimes v)^* = v\otimes y$, and $|y\otimes v| = |v\otimes y| = |y|\,|v|$.*

Proof. Obviously, $y\otimes v$ and $v\otimes y$ are linear. We have

$$\begin{aligned}|y\otimes v| &= \sup\{|(y\otimes v)(w)| : w \in F,\ |w| \le 1\}\\ &= \sup\{|\langle w, v\rangle|\,|y| : w \in F,\ |w| \le 1\}\\ &= |y|\,|v|,\end{aligned}$$

which implies $y\otimes v\in L(F,E)$. Let $w\in F$ and $u\in E'$. Then

$$\begin{aligned}\langle (y\otimes v)(w),u\rangle &= \langle \langle w,v\rangle y,u\rangle = \langle w,\langle y,u\rangle v\rangle\\ &= \langle w,(v\otimes y)(u)\rangle.\end{aligned}$$

This proves $(y\otimes v)^* = v\otimes y$. □

PROPOSITION 1.1.2. *Let E, F and G be Banach spaces, $y\in E$, $v\in F'$, $S_1\in L(E,G)$ and $S_2\in L(G,F)$. Then*

$$S_1(y\otimes v) = (S_1 y)\otimes v \quad \textit{and} \quad (y\otimes v)S_2 = y\otimes (S_2^* v).$$

Proof. For $w\in F$ and $u\in G$ we have

$$(S_1(y\otimes v))(w) = \langle w,v\rangle S_1 y = ((S_1 y)\otimes v)(w)$$

and

$$\begin{aligned}((y\otimes v)S_2)(u) &= (y\otimes v)(S_2 u) = \langle S_2 u,v\rangle y\\ &= \langle u,S_2^* v\rangle y = (y\otimes (S_2^* v))(u). \quad \Box\end{aligned}$$

A set $A\subset E$ is called *bounded* if $\sup\{|y| : y\in A\} < \infty$.

Let $T\in L(E,F)$. Since the Banach space F is a complete metric space, the following three conditions are equivalent:

i) for each bounded subset B of E, the set $T(B)$ is relatively compact;

ii) for each bounded subset B of E, the set $T(B)$ is precompact;

iii) for each bounded sequence $(y_n)_{n\in\mathbb{N}}$ in E there is a convergent subsequence of $(Ty_n)_{n\in\mathbb{N}}$;

see e. g. [DI, 3.16.1]. The operator T is called *compact* if it fulfils one of these equivalent conditions.

DEFINITION 1.1.3. Let E and F be Banach spaces, G be a nonempty open subset of E, and $f: G\to F$. Then f is called *differentiable* if for each $y_0\in G$ there are a linear operator $f'(y_0)\in L(E,F)$ and a map $\varepsilon_{y_0}: G\to F$ such that $\varepsilon_{y_0}(y)\to 0$ as $y\to y_0$ and

$$f(y) = f(y_0) + f'(y_0)(y-y_0) + \varepsilon_{y_0}(y)|y-y_0|$$

for all $y\in G$. The operator $f'(y_0)$ is called the *derivative* of f at y_0.

PROPOSITION 1.1.4. *Let E and F be Banach spaces and assume that the set $\mathscr{I}(E,F)$ of invertible continuous linear operators from E to F is nonempty.*

i) Let $T\in\mathscr{I}(E,F)$ and $B\in L(E,F)$ such that $|B| < |T^{-1}|^{-1}$.
Then $T+B\in\mathscr{I}(E,F)$ and

$$|(T+B)^{-1}| \le \frac{|T^{-1}|}{1-\delta}, \tag{1.1.3}$$

$$|(T+B)^{-1} - T^{-1}| \le \frac{\delta}{1-\delta}|T^{-1}|, \tag{1.1.4}$$

where $\delta := |BT^{-1}| < 1$.

ii) *The set* $\mathscr{I}(E,F)$ *is open, the map* $T \mapsto T^{-1}$ *from* $\mathscr{I}(E,F)$ *to* $\mathscr{I}(F,E)$ *is differentiable, and its derivative at* $S_0 \in \mathscr{I}(E,F)$ *is given by* $T \mapsto -S_0^{-1}TS_0^{-1}$ $(T \in L(E,F))$.

Proof. i): From $\delta = |BT^{-1}| \le |B|\,|T^{-1}| < 1$ we infer

$$\sum_{\nu=0}^{\infty} \left|T^{-1}\left(BT^{-1}\right)^{\nu}\right| \le |T|^{-1} \sum_{\nu=0}^{\infty} |BT^{-1}|^{\nu} \le \frac{|T^{-1}|}{1-\delta}. \tag{1.1.5}$$

Hence the series $\sum_{\nu=0}^{\infty} (-1)^{\nu} T^{-1}\left(BT^{-1}\right)^{\nu}$ converges in $L(E,F)$ and thus defines an element in $L(E,F)$. The equations

$$\begin{aligned}
&(T+B)\sum_{\nu=0}^{\infty}(-1)^{\nu}T^{-1}\left(BT^{-1}\right)^{\nu} \\
&\quad = \sum_{\nu=0}^{\infty}(-1)^{\nu}\left(BT^{-1}\right)^{\nu} + \sum_{\nu=0}^{\infty}(-1)^{\nu}\left(BT^{-1}\right)^{\nu+1} = \mathrm{id}_F
\end{aligned}$$

and

$$\begin{aligned}
&\left\{\sum_{\nu=0}^{\infty}(-1)^{\nu}T^{-1}\left(BT^{-1}\right)^{\nu}\right\}(T+B) \\
&\quad = \sum_{\nu=0}^{\infty}(-1)^{\nu}T^{-1}\left(BT^{-1}\right)^{\nu}T + \sum_{\nu=0}^{\infty}(-1)^{\nu}T^{-1}\left(BT^{-1}\right)^{\nu+1}T = \mathrm{id}_E
\end{aligned}$$

yield that $T+B$ is invertible and

$$(T+B)^{-1} = \sum_{\nu=0}^{\infty}(-1)^{\nu}T^{-1}\left(BT^{-1}\right)^{\nu}. \tag{1.1.6}$$

From (1.1.5) we infer (1.1.3). Finally,

$$\left|(T+B)^{-1} - T^{-1}\right| \le \sum_{\nu=1}^{\infty}\left|T^{-1}\left(BT^{-1}\right)^{\nu}\right| \le \frac{\delta}{1-\delta}|T^{-1}|$$

completes the proof of part i).
ii): Let $S_0 \in \mathscr{I}(E,F)$. For each $S \in L(E,F)$ with $|S-S_0| < |S_0^{-1}|^{-1}$, part i) yields $S \in \mathscr{I}(E,F)$, which proves the openness of $\mathscr{I}(E,F)$. For $S \in \mathscr{I}(E,F)$ we set

$$\varepsilon_{S_0}(S) := \begin{cases} |S-S_0|^{-1}\left(S^{-1} - S_0^{-1} + S_0^{-1}(S-S_0)S_0^{-1}\right) & \text{if } S \neq S_0, \\ 0 & \text{if } S = S_0. \end{cases}$$

Then

$$S^{-1} = S_0^{-1} - S_0^{-1}(S-S_0)S_0^{-1} + |S-S_0|\varepsilon_{S_0}(S)$$

holds for $S \in \mathscr{I}(E,F)$. For $|S-S_0| < |S_0^{-1}|^{-1}$ and $S \neq S_0$ we conclude from (1.1.6) with $T = S_0$ and $B = S - S_0$ that

$$\left|\varepsilon_{S_0}(S)\right| = |S-S_0|^{-1}\left|\sum_{\nu=2}^{\infty}(-1)^{\nu} S_0^{-1}((S-S_0)S_0^{-1})^{\nu}\right| \leq \frac{|S-S_0|\,|S_0^{-1}|^3}{1-|S-S_0|\,|S_0^{-1}|}\,.$$

This proves $\varepsilon_{S_0}(S) \to 0$ as $S \to S_0$. Obviously, $T \mapsto -S_0^{-1}TS_0^{-1}$ is a linear map from $L(E,F)$ to $L(F,E)$, and $\left|-S_0^{-1}TS_0^{-1}\right| \leq \left|S_0^{-1}\right|^2|T|$ proves its continuity. □

1.2. Holomorphic vector valued functions

Let Ω be an open nonempty subset of $\mathbb{C}$.

DEFINITION 1.2.1. Let E be a Banach space, $y : \Omega \to E$, and $\lambda_0 \in \Omega$. The vector function y is called *holomorphic at* λ_0 if there are a number $r > 0$ and $y_n \in E$ $(n \in \mathbb{N})$ such that $K_r(\lambda_0) := \{\lambda \in \mathbb{C} : |\lambda - \lambda_0| < r\} \subset \Omega$,

$$\sum_{n=0}^{\infty} r^n |y_n| < \infty, \tag{1.2.1}$$

and

$$y(\lambda) = \sum_{n=0}^{\infty} (\lambda - \lambda_0)^n y_n \tag{1.2.2}$$

for all $\lambda \in K_r(\lambda_0)$. Because of (1.2.1) the series (1.2.2) is absolutely convergent, and thus $y(\lambda)$ is well-defined. The vector function y is called *holomorphic in* Ω if it is holomorphic at each $\lambda \in \Omega$.

$H(\Omega,E)$ denotes the set of all holomorphic functions from Ω to E; if E_0 is any subset of E, then

$$H(\Omega,E_0) := \{f \in H(\Omega,E) : f(\Omega) \subset E_0\}.$$

REMARK 1.2.2. A function $f : \Omega \to E$ is holomorphic if and only if it is (continuously) differentiable, see e. g. [DI, (9.3.6) and 9.10, Problem 1)].

PROPOSITION 1.2.3. *Let E_1, E_2, E_3 be Banach spaces and $\cdot : E_1 \times E_2 \to E_3$ be a bilinear and continuous mapping, i. e., $|x \cdot y| \leq C|x|\,|y|$ for some $C > 0$ and all $x \in E_1$ and $y \in E_2$. If $u \in H(\Omega,E_1)$ and $v \in H(\Omega,E_2)$, then $u \cdot v \in H(\Omega,E_3)$, where $(u \cdot v)(\lambda) := u(\lambda) \cdot v(\lambda)$ $(\lambda \in \Omega)$.*

Proof. Let $\lambda_0 \in \Omega$. There are $r > 0$, $u_n \in E_1$ and $v_n \in E_2$ for $n \in \mathbb{N}$ such that $K_r(\lambda_0) \subset \Omega$,

$$\sum_{n=0}^{\infty} r^n |u_n| < \infty, \quad \sum_{n=0}^{\infty} r^n |v_n| < \infty,$$

and

$$u(\lambda) = \sum_{n=0}^{\infty} (\lambda - \lambda_0)^n u_n, \quad v(\lambda) = \sum_{n=0}^{\infty} (\lambda - \lambda_0)^n v_n$$

for all $\lambda \in K_r(\lambda_0)$. Then

$$\sum_{n=0}^{\infty} r^n \sum_{k=0}^{n} |u_k||v_{n-k}| = \Big(\sum_{n=0}^{\infty} r^n |u_n|\Big)\Big(\sum_{n=0}^{\infty} r^n |v_n|\Big) < \infty$$

and

$$u(\lambda) \cdot v(\lambda) = \sum_{n=0}^{\infty} (\lambda - \lambda_0)^n \sum_{k=0}^{n} u_k \cdot v_{n-k}$$

for $\lambda \in K_r(\lambda_0)$. □

We immediately infer

COROLLARY 1.2.4. *Let E, F, and G be Banach spaces, $T \in H(\Omega, L(E,F))$, $S \in H(\Omega, L(F,G))$, $y \in H(\Omega, E)$ and $v \in H(\Omega, F')$.*
Then $ST \in H(\Omega, L(E,G))$, $Ty \in H(\Omega, F)$ and $y \otimes v \in H(\Omega, L(F,E))$.

For $T \in H(\Omega, L(E,F))$,

$$\rho(T) := \{\lambda \in \Omega : T(\lambda) \text{ is invertible}\}$$

is called the *resolvent set* of T and $\sigma(T) := \Omega \setminus \rho(T)$ the *spectrum* of T. We set $T^{-1}(\lambda) := T(\lambda)^{-1}$ for $\lambda \in \rho(T)$. The operator function T^{-1} is called the *resolvent* of T.

PROPOSITION 1.2.5. *Let E and F be Banach spaces and $T \in H(\Omega, L(E,F))$. Then $\rho(T)$ is open and $T^{-1} \in H(\rho(T), L(F,E))$.*

Proof. Let $\lambda_0 \in \rho(T)$. From the holomorphy of T and the openness of $\mathscr{I}(E,F)$, see Proposition 1.1.4 ii), we infer that $\lambda \in \rho(T)$ for all λ in some neighbourhood of λ_0. This proves the openness of $\rho(T)$.

T^{-1} is the composition of $\lambda \mapsto T(\lambda)$ ($\lambda \in \rho(T)$) and $S \mapsto S^{-1}$ ($S \in \mathscr{I}(E,F)$). The first map is differentiable by assumption, cf. Remark 1.2.2. Since the second map is also differentiable, see Proposition 1.1.4 ii), the map T^{-1} is differentiable as the composition of differentiable maps is differentiable, see e. g. [DI, 8.2.1]. This proves $T^{-1} \in H(\rho(T), L(F,E))$, once more because of Remark 1.2.2. □

Proposition 1.2.5 can also be proved by calculating the power series expansion of T^{-1} in a neighbourhood of $\lambda \in \Omega$. Conversely, Proposition 1.2.3 can be proved as Proposition 1.2.5 by using the differentiability of the bilinear map, see e. g. [DI, 8.1.4].

PROPOSITION 1.2.6. *Let E and F be Banach spaces and $T \in H(\Omega, L(E,F))$. Set $T^*(\lambda) := (T(\lambda))^*$ for $\lambda \in \Omega$. Then $T^* \in H(\Omega, L(F',E'))$.*

Proof. Let $\lambda_0 \in \Omega$. Then there are a number $r > 0$ and, for $n \in \mathbb{N}$, $T_n \in L(E,F)$ such that $K_r(\lambda_0) \subset \Omega$,

$$\sum_{n=0}^{\infty} r^n |T_n| < \infty,$$

and

$$T(\lambda) = \sum_{n=0}^{\infty} (\lambda - \lambda_0)^n T_n \quad (\lambda \in K_r(\lambda_0)).$$

Since $|T_n^*| = |T_n|$, we obtain

$$\sum_{n=0}^{\infty} r^n |T_n^*| < \infty,$$

whence

$$\widehat{T}(\lambda) := \sum_{n=0}^{\infty} (\lambda - \lambda_0)^n T_n^* \quad (\lambda \in K_r(\lambda_0))$$

is well-defined. For $y \in E$, $v \in F'$ and $\lambda \in K_r(\lambda_0)$ we obtain

$$\begin{aligned}\langle y, T^*(\lambda)v\rangle = \langle T(\lambda)y, v\rangle &= \sum_{n=0}^{\infty} (\lambda - \lambda_0)^n \langle T_n y, v\rangle \\ &= \sum_{n=0}^{\infty} (\lambda - \lambda_0)^n \langle y, T_n^* v\rangle \\ &= \langle y, \widehat{T}(\lambda)v\rangle,\end{aligned}$$

which proves that

$$T^*(\lambda) = \widehat{T}(\lambda) = \sum_{n=0}^{\infty} (\lambda - \lambda_0)^n T_n^*.$$

Thus T^* is holomorphic. □

Let $\mu \in \mathbb{C}$ and $y \in H(U' \setminus \{\mu\}, E)$ for some open neighbourhood U' of μ. We say that y is *meromorphic* at μ if there is a nonnegative integer s such that $(\cdot - \mu)^s y$ has a holomorphic continuation to all of U'. The smallest such number s is called the *pole order* of y at μ, and μ is called a *pole* of y if this number is positive. Note that y has a holomorphic extension to μ if and only if the pole order is zero. Since $(\cdot - \mu)^s y$ is holomorphic at μ, the Laurent series expansion

$$y = \sum_{j=-s}^{\infty} (\cdot - \mu)^j y_j$$

holds in some punctured neighbourhood of μ. We call

$$\sum_{j=-s}^{-1} (\cdot - \mu)^j y_j$$

the *principal part* of y at μ.

Let U be an open subset of Ω and $y \in H(U,E)$. We say that y is *meromorphic in* Ω if $\Omega \setminus U$ is a discrete subset of Ω and y is meromorphic at each point μ in $\Omega \setminus U$.

1.3. The inverse of a Fredholm operator valued function

Let Ω be a domain in $\mathbb{C}$ and E and F be Banach spaces.

THEOREM 1.3.1. *Let $T \in H(\Omega, \Phi(E,F))$ and assume that $\rho(T) \neq \emptyset$. Then $\sigma(T)$ is a discrete subset of Ω and T^{-1} is a meromorphic operator function in Ω. If $\mu \in \sigma(T)$, then T^{-1} has a pole at μ, i. e.,*

$$T^{-1} = \sum_{j=-s_\mu}^{\infty} (\cdot - \mu)^j S_{j,\mu} \tag{1.3.1}$$

in some punctured neighbourhood of μ, where $s_\mu \in \mathbb{N} \setminus \{0\}$ and $S_{-s_\mu,\mu} \neq 0$. In addition, for $-s_\mu \leq j \leq -1$ the operators $S_{j,\mu}$ are degenerate operators, i. e., $\dim R(S_{j,\mu}) < \infty$, *and $S_{0,\mu} \in \Phi(F,E)$ with* $\operatorname{ind} S_{0,\mu} = 0$. *$T^{-1}$ is called a finitely meromorphic operator function, cf.* GOHBERG *and* SIGAL [GS].

Proof. Let Ω_1 be the set of all $\mu \in \Omega$ such that $U \setminus \{\mu\} \subset \rho(T)$ for some neighbourhood U of μ. Define Ω_2 to be the set of all $\mu \in \Omega$ having a neighbourhood which is contained in $\sigma(T)$. We assert that $\Omega = \Omega_1 \cup \Omega_2$. Obviously, $\rho(T) \subset \Omega_1$ since $\rho(T)$ is open by Proposition 1.2.5. Now let $\mu \in \sigma(T)$. Since $T(\mu)$ is a Fredholm operator, there are a finite-codimensional subspace $M \subset E$ and a finite-dimensional subspace $N \subset F$ such that

$$E = M \dotplus N(T(\mu)), \quad F = R(T(\mu)) \dotplus N. \tag{1.3.2}$$

We may assume that M is a closed subspace of E, see e. g. [KA, p. 135]. $T(\mu)$ is a Fredholm operator and thus $R(T(\mu))$ is a closed subspace of F, see e. g. [GO, Corollary IV.1.13]. Hence the direct sums in (1.3.2) are also topologically direct, see e. g [TL, p. 247]. This yields the operator matrix representation

$$T(\lambda) = \begin{pmatrix} T_{11}(\lambda) & T_{12}(\lambda) \\ T_{21}(\lambda) & T_{22}(\lambda) \end{pmatrix} : M \dotplus N(T(\mu)) \to R(T(\mu)) \dotplus N \tag{1.3.3}$$

for $\lambda \in \Omega$. The operators T_{ij} $(i,j = 1,2)$ are holomorphic in Ω. For example, $T_{11}(\lambda) = QT(\lambda)P$, where P is the continuous embedding of M into $M \dotplus N(T(\mu))$ and Q is the continuous projection of $R(T(\mu)) \dotplus N$ onto $R(T(\mu))$. For $m \in M$ and $n \in N(T(\mu))$ we have

$$T_{11}(\mu)m = T(\mu)m = T(\mu)(m+n).$$

This immediately shows that $N(T_{11}(\mu)) = \{0\}$ and $R(T_{11}(\mu)) = R(T(\mu))$. Hence $T_{11}(\mu)$ is invertible. Thus the operator $T_{11}(\lambda) : M \to R(T(\mu))$ is invertible for each λ in some neighbourhood U of μ since $\rho(T_{11})$ is open by Proposition 1.2.5. In U we consider the SCHUR factorization

$$\begin{pmatrix} T_{11} & T_{12} \\ T_{21} & T_{22} \end{pmatrix} \tag{1.3.4}$$
$$= \begin{pmatrix} \mathrm{id}_{R(T(\mu))} & 0 \\ T_{21}T_{11}^{-1} & \mathrm{id}_N \end{pmatrix} \begin{pmatrix} T_{11} & 0 \\ 0 & T_{22} - T_{21}T_{11}^{-1}T_{12} \end{pmatrix} \begin{pmatrix} \mathrm{id}_M & T_{11}^{-1}T_{12} \\ 0 & \mathrm{id}_{N(T(\mu))} \end{pmatrix}.$$

It is easy to see that the right-hand factor and the left-hand factor on the right-hand side of (1.3.4) are invertible on U, e. g.

$$\begin{pmatrix} \mathrm{id}_M & T_{11}^{-1}T_{12} \\ 0 & \mathrm{id}_{N(T(\mu))} \end{pmatrix}^{-1} = \begin{pmatrix} \mathrm{id}_M & -T_{11}^{-1}T_{12} \\ 0 & \mathrm{id}_{N(T(\mu))} \end{pmatrix}$$

on U. We set

$$S(\lambda) := T_{22}(\lambda) - T_{21}(\lambda)T_{11}^{-1}(\lambda)T_{12}(\lambda) \quad (\lambda \in U).$$

The representation (1.3.4) yields that $T(\lambda)$ is invertible for $\lambda \in U$ if and only if

$$\begin{pmatrix} T_{11}(\lambda) & 0 \\ 0 & S(\lambda) \end{pmatrix}$$

is invertible. Since T_{11} is invertible on U, we infer for $\lambda \in U$ that $T(\lambda)$ is invertible if and only if $S(\lambda)$ has this property. $S(\lambda) : N(T(\mu)) \to N$ is a linear operator in finite-dimensional spaces. If $\dim N \neq \dim N(T(\mu))$, then $S(\lambda)$ is not invertible for any $\lambda \in U$, and μ belongs to Ω_2. If $\dim N = \dim N(T(\mu))$, then $S(\lambda)$ is invertible if and only if $\det S(\lambda) \neq 0$, where the determinant is taken relative to some bases of N and $N(T(\mu))$. Since the holomorphic function $\det S$ is zero in some neighbourhood of μ or does not have a zero in some punctured neighbourhood of μ, we infer that μ belongs to $\Omega_1 \cup \Omega_2$. The openness of the sets Ω_1 and Ω_2 and $\Omega_1 \cap \Omega_2 = \emptyset$ are obvious by definition of Ω_1 and Ω_2. Since Ω is connected and $\Omega_1 \supset \rho(T) \neq \emptyset$, the set Ω_2 is empty, which proves the discreteness of $\sigma(T)$.

Now let $\mu \in \sigma(T)$ and S be defined as above. By CRAMER'S rule, applied to the matrix corresponding to $S(\lambda)$ with respect to some bases of N and $N(T(\mu))$, we obtain a holomorphic operator function $\widetilde{S} : U \to L(N, N(T(\mu)))$ such that

$$\det S(\lambda)\, \mathrm{id}_N = S(\lambda)\widetilde{S}(\lambda).$$

The SCHUR factorization (1.3.4) yields

$$T^{-1} = \begin{pmatrix} \mathrm{id}_M & -T_{11}^{-1}T_{12} \\ 0 & \mathrm{id}_{N(T(\mu))} \end{pmatrix} \begin{pmatrix} T_{11}^{-1} & 0 \\ 0 & S^{-1} \end{pmatrix} \begin{pmatrix} \mathrm{id}_{R(T(\mu))} & 0 \\ -T_{21}T_{11}^{-1} & \mathrm{id}_N \end{pmatrix}$$
$$= \begin{pmatrix} T_{11}^{-1} & 0 \\ 0 & 0 \end{pmatrix} + (\det S)^{-1} \begin{pmatrix} T_{11}^{-1}T_{12}\widetilde{S}T_{21}T_{11}^{-1} & -T_{11}^{-1}T_{12}\widetilde{S} \\ -\widetilde{S}T_{21}T_{11}^{-1} & \widetilde{S} \end{pmatrix}$$

in some punctured neighbourhood of μ. Since $\det S$ is a holomorphic function which is not identically zero and $\widetilde{S}$ is an operator function in finite-dimensional spaces, we obtain

$$(\det S)^{-1}\begin{pmatrix} T_{11}^{-1}T_{12}\widetilde{S}T_{21}T_{11}^{-1} & -T_{11}^{-1}T_{12}\widetilde{S} \\ -\widetilde{S}T_{21}T_{11}^{-1} & \widetilde{S} \end{pmatrix} = \sum_{j=-\bar{s}_\mu}^{\infty} (\cdot-\mu)^j A_{j,\mu}$$

in some punctured neighbourhood of μ, where $\bar{s}_\mu$ is the order of the zero of $\det S$ at μ and all $A_{j,\mu}$ are degenerate operators. Suppose that T^{-1} is holomorphic at μ. Since $TT^{-1} = \mathrm{id}_F$ and $T^{-1}T = \mathrm{id}_E$ in some punctured neighbourhood of μ, this would imply $T(\mu)T^{-1}(\mu) = \mathrm{id}_F$ and $T^{-1}(\mu)T(\mu) = \mathrm{id}_E$, which contradicts $\mu \in \sigma(T)$. Hence T^{-1} has a pole of order $s_\mu \le \bar{s}_\mu$ at μ, the representation (1.3.1) holds with $S_{j,\mu} = A_{j,\mu}$ for $-s_\mu \le j \le -1$,

$$S_{0,\mu} = \begin{pmatrix} T_{11}(\mu)^{-1} & 0 \\ 0 & 0 \end{pmatrix} + A_{0,\mu} \in \Phi(F,E)$$

since the sum of a Fredholm operator and a degenerate operator is a Fredholm operator, and

$$\operatorname{ind} S_{0,\mu} = \operatorname{ind}\begin{pmatrix} T_{11}(\mu)^{-1} & 0 \\ 0 & 0 \end{pmatrix} = \operatorname{ind} T_{11}(\mu)^{-1} + \dim N - \dim N(T(\mu)) = 0.$$

For the the arguments used in the proof of the properties of $S_{0,\mu}$ see e. g. [KA, Theorem IV.5.26]. □

PROPOSITION 1.3.2. *Let $T \in H(\Omega, \Phi(E,F))$ such that $\rho(T) \neq \emptyset$ and let $\mu \in \sigma(T)$. The holomorphic part in the Laurent series expansion* (1.3.1) *is called the reduced resolvent of T with respect to μ and denoted by S_μ. In a neighbourhood of μ we have*

$$S_\mu = \sum_{j=0}^{\infty}(\cdot-\mu)^j S_{j,\mu} = T^{-1} - \sum_{j=-s_\mu}^{-1}(\cdot-\mu)^j S_{j,\mu}, \tag{1.3.5}$$

which shows that $S_\mu \in H(\rho(T)\cup\{\mu\}, L(F,E))$. If the pole order of T^{-1} at μ is 1, *then*

$$T(\mu)S_\mu(\mu)T(\mu) = T(\mu). \tag{1.3.6}$$

If T is a polynomial of degree 1, *then*

$$S_\mu(\mu)T(\mu)S_\mu(\mu) = S_\mu(\mu). \tag{1.3.7}$$

Proof. Let

$$T = \sum_{j=0}^{\infty}(\cdot-\mu)^j T_j$$

be the power series expansion of T at μ. First we consider the case that T^{-1} has a pole of order 1 at μ. From $T^{-1}T = \mathrm{id}_E$ we infer

$$S_{-1,\mu}T_0 = 0,$$

and $TT^{-1} = \mathrm{id}_F$ yields

$$T_0 S_{0,\mu} + T_1 S_{-1,\mu} = \mathrm{id}_F \,. \tag{1.3.8}$$

Thus

$$T(\mu) = T_0 S_{0,\mu} T_0 + T_1 S_{-1,\mu} T_0 = T(\mu) S_\mu(\mu) T(\mu).$$

Now let T be a polynomial of degree 1. In this case, (1.3.8) also holds. Hence

$$S_\mu(\mu) T(\mu) S_\mu(\mu) = S_{0,\mu} - S_{0,\mu} T_1 S_{-1,\mu} \,.$$

From $T^{-1}T = \mathrm{id}_E$ and $TT^{-1} = \mathrm{id}_F$ we infer

$$S_{k,\mu} T_1 = -S_{k+1,\mu} T_0 \quad \text{and} \quad T_0 S_{-k-1,\mu} = -T_1 S_{-k-2,\mu}$$

for $k \in \mathbb{N}$, where $S_{-j,\mu} := 0$ for $j > s_\mu$. These identities yield

$$\begin{aligned} S_{0,\mu} T_1 S_{-1,\mu} &= -S_{1,\mu} T_0 S_{-1,\mu} = S_{1,\mu} T_1 S_{-2,\mu} \\ &= \cdots = S_{s_\mu,\mu} T_1 S_{-s_\mu-1,\mu} = 0. \end{aligned}$$

Hence (1.3.7) is proved. □

In the following examples we shall see that (1.3.6) and (1.3.7) are not necessarily true without the restrictions imposed in Proposition 1.3.2.

EXAMPLE 1.3.3. Let $T \in H(\mathbb{C}, L(\mathbb{C}^2))$ be given by

$$T(\lambda) = \begin{pmatrix} \lambda & 1 \\ 0 & \lambda \end{pmatrix} \quad (\lambda \in \mathbb{C}).$$

T is a polynomial of degree 1 and

$$T^{-1}(\lambda) = \begin{pmatrix} \lambda^{-1} & -\lambda^{-2} \\ 0 & \lambda^{-1} \end{pmatrix} \quad (\lambda \in \mathbb{C} \setminus \{0\}),$$

i. e., 0 is a pole of T of order 2. Obviously,

$$T(0) = \begin{pmatrix} 0 & 1 \\ 0 & 0 \end{pmatrix}, \quad S_0(0) = 0.$$

Hence (1.3.6) does not hold. □

EXAMPLE 1.3.4. Let $T \in H(\mathbb{C}, L(\mathbb{C}^2))$ be given by

$$T(\lambda) = \begin{pmatrix} 1 & 1 \\ -\lambda & \lambda^2 \end{pmatrix} \quad (\lambda \in \mathbb{C}).$$

T is a polynomial of degree 2 and

$$T^{-1}(\lambda) = \begin{pmatrix} \frac{\lambda}{1+\lambda} & -\frac{1}{\lambda+\lambda^2} \\ \frac{1}{1+\lambda} & \frac{1}{\lambda+\lambda^2} \end{pmatrix} \quad (\lambda \in \mathbb{C} \setminus \{0, -1\}),$$

i. e., 0 is a pole of T of order 1. Obviously,

$$T(0)=\begin{pmatrix}1&1\\0&0\end{pmatrix},\quad S_0(0)=\begin{pmatrix}0&-1\\1&1\end{pmatrix},$$

$$S_0(0)T(0)S_0(0)=\begin{pmatrix}0&0\\1&0\end{pmatrix}.$$

This shows that (1.3.7) does not hold. □

EXAMPLE 1.3.5. Let $T\in H(\mathbb{C},L(\mathbb{C}^2))$ be given by

$$T(\lambda)=\begin{pmatrix}1&\lambda\\-\lambda&\lambda^2\end{pmatrix}\quad(\lambda\in\mathbb{C}).$$

T is a polynomial of degree 2 and

$$T^{-1}(\lambda)=\begin{pmatrix}\frac{1}{2}&-\frac{1}{2\lambda}\\\frac{1}{2\lambda}&\frac{1}{2\lambda^2}\end{pmatrix}\quad(\lambda\in\mathbb{C}\setminus\{0\}),$$

i. e., 0 is a pole of T of order 2. Obviously,

$$T(0)=\begin{pmatrix}1&0\\0&0\end{pmatrix},\quad S_0(0)=\begin{pmatrix}\frac{1}{2}&0\\0&0\end{pmatrix}.$$

Hence neither (1.3.6) nor (1.3.7) holds. □

1.4. Root functions of holomorphic operator functions

Let Ω be a domain in $\mathbb{C}$ and E and F be Banach spaces.

DEFINITION 1.4.1. Let $T\in H(\Omega,\Phi(E,F))$ and $\mu\in\Omega$. The vector function y in $H(\Omega,E)$ is called a *root function* of T at μ if $y(\mu)\neq 0$ and $(Ty)(\mu)=0$. The number $\nu(y)$ denotes the order of the zero of Ty at μ and is called the *multiplicity of* y (with respect to T at μ).

Since, by Theorem 1.3.1, the inverse of a holomorphic Fredholm operator valued function on a domain is meromorphic if its resolvent set is nonempty, we obtain

LEMMA 1.4.2. *Let $T\in H(\Omega,\Phi(E,F))$ such that $\rho(T)\neq\emptyset$. Let $\mu\in\sigma(T)$ and denote the pole order of T^{-1} at μ by s_μ. Then*

$$s_\mu=\max\{\nu(y): y \textit{ root function of } T \textit{ at } \mu\}.$$

Proof. Let y be a root function of T at μ. Since $(\cdot-\mu)^{s_\mu}T^{-1}$ is holomorphic at μ,

$$(\cdot-\mu)^{s_\mu}y=(\cdot-\mu)^{s_\mu}T^{-1}Ty$$

has a zero of order $\geq\nu(y)$ there. From $y(\mu)\neq 0$ we infer $\nu(y)\leq s_\mu$. This proves

$$s_\mu\geq\max\{\nu(y): y \text{ root function of } T \text{ at } \mu\}.$$

For the proof of the reverse inequality, we use the Laurent series expansion

$$T^{-1} = \sum_{j=-s_\mu}^{\infty} (\cdot - \mu)^j S_{j,\mu}$$

in a punctured neighbourhood of μ, where the operators $S_{j,\mu}$ belong to $L(F,E)$ for $j = -s_\mu, -s_\mu + 1, \dots$ and $S_{-s_\mu,\mu} \neq 0$. Choose $x_0 \in F$ such that $S_{-s_\mu,\mu}x_0 \neq 0$. Set $z := (\cdot - \mu)^{s_\mu} T^{-1} x_0$. The function z is holomorphic at μ, $z(\mu) = S_{-s_\mu,\mu}x_0 \neq 0$, and $Tz = (\cdot - \mu)^{s_\mu} x_0$ has a zero of order s_μ at μ. Then the Taylor polynomial

$$y = \sum_{j=-s_\mu}^{-1} (\cdot - \mu)^{s_\mu + j} S_{j,\mu} x_0$$

of z at μ of order $s_\mu - 1$ is a root function of T at μ with $\nu(y) \geq s_\mu$. □

Though it is useful to have a root function defined as a holomorphic function, in general we only need the "principal part"

$$\sum_{l=0}^{\nu(y)-1} (\cdot - \mu)^l y_l$$

of a root function y at μ, where

$$\sum_{l=0}^{\infty} (\cdot - \mu)^l y_l$$

is the power series expansion of y at μ. Thus we can deal with polynomials, if necessary. On the other hand, we often only need a power series expansion in a neighbourhood of μ; i. e., it is sufficient to have a root function of T at μ defined in a neighbourhood of μ.

Let $T \in H(\Omega, \Phi(E,F))$ such that $\rho(T) \neq \emptyset$. Let $\mu \in \sigma(T)$ and $n \in \mathbb{N} \setminus \{0\}$. Then $\widetilde{L}_n$ denotes the set of all $y_0 \in N(T(\mu))$ such that there is a root function y with $y(\mu) = y_0$ and $\nu(y) \geq n$. Obviously,

$$L_n := \widetilde{L}_n \cup \{0\} \tag{1.4.1}$$

is a subspace of $N(T(\mu))$. For $j \in \mathbb{N}$ with $0 < j \leq \operatorname{nul} T(\mu)$ we define

$$m_j := \max\{n \in \mathbb{N} \setminus \{0\} : \dim L_n \geq j\}. \tag{1.4.2}$$

The numbers m_j are called the *partial multiplicities* of T at μ. They are well-defined since $L_1 = N(T(\mu))$ and $L_n = \{0\}$ if n is larger than the pole order of T^{-1} at μ. Obviously, $m_j \geq m_{j+1}$.

The number $r = \dim N(T(\mu))$ is called the *geometric multiplicity* of T at μ, and the number

$$m = \sum_{j=1}^{r} m_j$$

is called the *algebraic multiplicity* of T at μ.

REMARK 1.4.3. Let $0 < j \le \operatorname{nul} T(\mu)$. Then $\dim L_{m_j+1} < j \le \dim L_{m_j}$.

PROPOSITION 1.4.4. *Let $T \in H(\Omega, \Phi(E,F))$, assume that $\rho(T) \neq \emptyset$ and let $\mu \in \sigma(T)$. Let $0 < r \le \operatorname{nul} T(\mu)$ and let $y_1, \ldots, y_r$ be root functions of T at μ such that $y_1(\mu), \ldots, y_r(\mu)$ are linearly independent.*
The following conditions are equivalent:

i) $\nu(y_j) = \max\{\nu(y) : y$ *is a root function of* T *at* μ *and* $y(\mu) \notin \operatorname{span}\{y_1(\mu), \ldots, y_{j-1}(\mu)\}\} \quad (j = 1, \ldots, r)$,

ii) $\nu(y_j) = m_j \quad (j = 1, \ldots, r)$,

iii) $\nu(y_j) \ge m_j \quad (j = 1, \ldots, r)$.

Proof. i) $\Rightarrow$ ii). The condition i) implies $\nu(y_k) \ge \nu(y_{k+1})$ $(k = 1, \ldots, r-1)$. Let $j \in \{1, \ldots, r\}$. Then

$$\operatorname{span}\{y_1(\mu), \ldots, y_j(\mu)\} \subset L_{\nu(y_j)}.$$

We infer $\dim L_{\nu(y_j)} \ge j$. Hence $m_j \ge \nu(y_j)$ by (1.4.2). We know that there is a vector $y_0 \in L_{m_j} \setminus \operatorname{span}\{y_1(\mu), \ldots, y_{j-1}(\mu)\}$ since $j - 1 < \dim L_{m_j}$, see Remark 1.4.3. By the definition of L_{m_j} there is a root function y with $y(\mu) = y_0$ and $\nu(y) \ge m_j$. Hence the number on the right-hand side of i) is greater or equal m_j, i. e., $\nu(y_j) \ge m_j$. Thus we have proved $\nu(y_j) = m_j$.

The conclusion ii) $\Rightarrow$ iii) is obvious.

iii) $\Rightarrow$ i). Let $j \in \{1, \ldots, r\}$. Since $y_j(\mu) \notin \operatorname{span}\{y_1(\mu), \ldots, y_{j-1}(\mu)\}$, we see that $\nu(y_j)$ does not exceed the number on the right-hand side of i). We set $i := \nu(y_j) + 1$. For $k = 1, \ldots, \dim L_i$ we have $i \le m_k$ and thus $i \le \nu(y_k)$ by assumption iii). This proves that $y_k(\mu) \in L_i$ for $k = 1, \ldots, \dim L_i$. Hence

$$L_i = \operatorname{span}\{y_1(\mu), \ldots, y_{\dim L_i}(\mu)\}. \tag{1.4.3}$$

From $\nu(y_j) = i - 1$ and the assumption $\nu(y_j) \ge m_j$ we conclude $m_j < i$ and thus $j > \dim L_i$. In view of (1.4.3) we infer

$$L_i \subset \operatorname{span}\{y_1(\mu), \ldots, y_{j-1}(\mu)\}. \tag{1.4.4}$$

Now let y be a root function of T at μ with $y(\mu) \notin \operatorname{span}\{y_1(\mu), \ldots, y_{j-1}(\mu)\}$. Because of (1.4.4), $y(\mu) \notin L_i$, which implies $\nu(y) < i$. Therefore $\nu(y) \le i - 1 = \nu(y_j)$. This proves that i) holds. □

DEFINITION 1.4.5. Let $T \in H(\Omega, \Phi(E,F))$, $\rho(T) \neq \emptyset$ and $\mu \in \sigma(T)$. A system $\{y_1, \ldots, y_r\}$ of root functions of T at μ is called a *canonical system of root functions* (CSRF) if $\{y_1(\mu), \ldots, y_r(\mu)\}$ is a basis of $N(T(\mu))$ and one of the equivalent conditions i), ii) or iii) in Proposition 1.4.4 is fulfilled.

PROPOSITION 1.4.6. *Let $T \in H(\Omega, \Phi(E,F))$, $\rho(T) \neq \emptyset$ and $\mu \in \sigma(T)$. We set $r := \operatorname{nul} T(\mu)$ and let $k \in \{0,\dots,r-1\}$. Let $y_1,\dots,y_k$ be root functions of T at μ with $\nu(y_j) \geq m_j$ for $j = 1,\dots,k$ such that $y_1(\mu),\dots,y_k(\mu)$ are linearly independent, where the numbers m_j are the partial multiplicities defined in (1.4.2). Then there are root functions $y_{k+1},\dots,y_r$ of T at μ such that $\{y_1,\dots,y_r\}$ is a canonical system of root functions of T at μ.*

Proof. With $r_n := \dim L_n$ we have $r = r_1$. If $m_0 := m_1 + 1$, then $L_{m_0} = \{0\}$ by definition of m_1. From $\nu(y_j) \geq m_j \geq m_k$ for $j = 1,\dots,k$ we infer

$$y_1(\mu),\dots,y_k(\mu) \in L_{m_k}.$$

Choose root functions $y_{k+1},\dots,y_{r_{m_k}}$ such that

$$\operatorname{span}\{y_1(\mu),\dots,y_{r_{m_k}}(\mu)\} = L_{m_k}$$

and $\nu(y_j) \geq m_k (\geq m_j)$ for $j = k+1,\dots,r_{m_k}$. For $n = m_k - 1, m_k - 2,\dots,1$ there are root functions $y_{r_{n+1}+1},\dots,y_{r_n}$ of T at μ such that

$$L_{n+1} \dot{+} \operatorname{span}\{y_{r_{n+1}+1}(\mu),\dots,y_{r_n}(\mu)\} = L_n$$

and $\nu(y_j) \geq n$ for $j = r_{n+1}+1,\dots,r_n$. For these j we have $j > r_{n+1} = \dim L_{n+1}$ and hence $n+1 > m_j$. Thus $m_j \leq n \leq \nu(y_j)$. This proves that condition iii) of Proposition 1.4.4 holds for all $j \in \{1,\dots,r\}$. From

$$\begin{aligned} N(T(\mu)) &= L_2 + \operatorname{span}\{y_{r_2+1}(\mu),\dots,y_r(\mu)\} \\ &= \cdots = L_{m_k} + \operatorname{span}\{y_{r_{m_k}+1}(\mu),\dots,y_r(\mu)\} \\ &= \operatorname{span}\{y_1(\mu),\dots,y_r(\mu)\} \end{aligned}$$

we infer that $\{y_1(\mu),\dots,y_r(\mu)\}$ is a basis of $N(T(\mu))$. □

Taking $k = 0$ in the above proposition we obtain

PROPOSITION 1.4.7. *Let $T \in H(\Omega, \Phi(E,F))$, $\rho(T) \neq \emptyset$ and $\mu \in \sigma(T)$. Set $r := \operatorname{nul} T(\mu)$. Then there is a canonical system of root functions $\{y_1,\dots,y_r\}$ of T at μ.*

Another result on completing incomplete systems of root functions is the following one:

PROPOSITION 1.4.8. *Let $T \in H(\Omega, \Phi(E,F))$, $\rho(T) \neq \emptyset$ and $\mu \in \sigma(T)$. Set $r := \operatorname{nul} T(\mu)$. Let $x_1,\dots,x_k$ $(0 < k < r)$ be root functions of T at μ such that $x_1(\mu),\dots,x_k(\mu)$ are linearly independent and such that for each set $\Lambda \subset \{1,\dots,k\}$ and for each root function y of T at μ with $y(\mu) \in \operatorname{span}\{x_j(\mu) : j \in \Lambda\}$ the estimate*

$$\nu(y) \leq \max\{\nu(x_j) : j \in \Lambda\} \tag{1.4.5}$$

holds. Then there is a canonical system of root functions $y_1,\dots,y_r$ of T at μ such that $\{x_1,\dots,x_k\} \subset \{y_1,\dots,y_r\}$.

Proof. We first take those $x_{j_1}(\mu),\dots,x_{j_s}(\mu)$ which belong to L_{m_1}. Then we choose $z_{s+1},\dots,z_{r_{m_1}} \in L_{m_1}$ such that $\{x_{j_1}(\mu),\dots,x_{j_s}(\mu),z_{s+1},\dots,z_{r_{m_1}}\}$ is a basis of L_{m_1}. Here $r_m := \dim L_m$. Now we set $y_i := x_{j_i}$ for $i = 1,\dots,s$, and we choose root functions $y_{s+1},\dots,y_{r_{m_1}}$ of T at μ of multiplicity m_1 such that $y_j(\mu) = z_j$ for $j = s+1,\dots,r_{m_1}$.

Next we take those $x_{k_1}(\mu),\dots,x_{k_{\tilde{s}}}(\mu)$ which belong to $L_{m_1-1} \setminus L_{m_1}$. The assumption (1.4.5) says that

$$L_{m_1} \cap \operatorname{span}\{x_{k_1}(\mu),\dots,x_{k_{\tilde{s}}}(\mu)\} = \{0\}.$$

Then we choose $w_{\tilde{s}+1},\dots,w_{s'}$, where $s' := \dim L_{m_1-1} - \dim L_{m_1}$, such that

$$L_{m_1-1} = L_{m_1} \dotplus \{x_{k_1}(\mu),\dots,x_{k_{\tilde{s}}}(\mu),w_{\tilde{s}+1},\dots,w_{s'}\}.$$

We set $y_{i+r_{m_1}} := x_{k_i}$ for $i = 1,\dots,\tilde{s}$ and for $j = \tilde{s}+1,\dots,s$ we choose root functions $y_{j+r_{m_1}}$ of T at μ of multiplicity $m_1 - 1$ such that $y_{j+r_{m_1}}(\mu) = w_j$. Proceeding in this way, we obtain a system $\{y_1,\dots,y_r\}$ of root functions with $\{x_1,\dots,x_k\} \subset \{y_1,\dots,y_r\}$. By construction, $y_1(\mu),\dots,y_r(\mu)$ are linearly independent. From $j \le \dim L_{m_j}$ we infer $y_j(\mu) \in L_{m_j}$. Therefore $\nu(y_j) \ge m_j$ follows if we show that $\nu(y) \le \nu(y_j)$ holds for each root function y of T at μ with $y(\mu) = y_j(\mu)$. If $y_j = x_{j'}$ for some j', this follows from (1.4.5) with $\Lambda = j'$. And if $y_j \notin \{x_1,\dots,x_k\}$, it holds by construction. □

PROPOSITION 1.4.9. *Let $T \in H(\Omega,\Phi(E,F))$, $\rho(T) \neq \emptyset$ and $\mu \in \sigma(T)$. Let $\{y_1,\dots,y_r\}$ be a canonical system of root functions of T at μ. Then*

$$L_n = \operatorname{span}\{y_j(\mu) : j \le \dim L_n\} = \operatorname{span}\{y_j(\mu) : m_j \ge n\}$$

holds for each $n \in \mathbb{N}\setminus\{0\}$.

Proof. Since there are no root functions y of T at μ with $\nu(y) > m_1$, we have

$$L_n = \{0\} = \operatorname{span}\{y_j(\mu) : j \le \dim L_n\} \text{ for } n > m_1.$$

Now let $n \in \mathbb{N}\setminus\{0\}$ with $n \le m_1$ and set $r_n := \dim L_n \le r$. The definition of m_j in (1.4.2) yields $n \le m_{r_n}$, and hence $\nu(y_1) \ge \cdots \ge \nu(y_{r_n}) \ge n$ follows. This proves that $L_n \supset \operatorname{span}\{y_j(\mu) : j \le r_n\}$. Since $y_1(\mu),\dots,y_{r_n}(\mu)$ are linearly independent, we obtain that $\operatorname{span}\{y_j(\mu) : j \le r_n\}$ is an r_n-dimensional subspace of L_n. From $r_n = \dim L_n$ we infer $L_n = \operatorname{span}\{y_j(\mu) : j \le \dim L_n\}$.

The second equality follows since, by definiton of m_j, $\dim L_n \ge j$ holds if and only if $m_j \ge n$. □

1.5. Representation of the principal part of a finitely meromorphic operator function

Let Ω be a domain in $\mathbb{C}$ and E and F be Banach spaces.

PROPOSITION 1.5.1. *Let N be a finite-dimensional subspace of E and $\{y_1,\dots,y_k\}$ be a basis of N. Assume that $A \in L(F,E)$ and $R(A) \subset N$. Then there are $v_1,\dots,v_k \in F'$ such that*

$$A = \sum_{i=1}^{k} y_i \otimes v_i.$$

Proof. The HAHN-BANACH theorem yields a family $(u_i)_{i=1}^k$ in E' which is biorthogonal to $(y_i)_{i=1}^k$, i. e., $\langle y_j, u_i\rangle = \delta_{ij}$ holds for $i, j = 1,\dots,k$; see e. g. [HO, p. 51]. It follows that

$$y = \sum_{i=1}^{k} \langle y, u_i\rangle y_i \quad (y \in N).$$

For $i = 1,\dots,k$ we set $v_i := A^* u_i$. Let $w \in F$. Since $Aw \in N$, the above equation yields

$$Aw = \sum_{i=1}^{k} \langle Aw, u_i\rangle y_i = \sum_{i=1}^{k} \langle w, v_i\rangle y_i = \sum_{i=1}^{k} (y_i \otimes v_i)(w). \quad \square$$

REMARK 1.5.2. Let N be a finite-dimensional subspace of F' and $\{v_1,\dots,v_k\}$ be a basis of N. Assume that $A \in L(F,E)$ and $R(A^*) \subset N$. Then there are $y_1,\dots,y_k \in E$ such that

$$A = \sum_{i=1}^{k} y_i \otimes v_i.$$

Proof. The linear independence of the $v_1,\dots,v_k$ implies that there is a family $(x_i)_{i=1}^k$ in F which is biorthogonal to $(v_i)_{i=1}^k$; see e. g. [RR, p. 32]. Setting $y_i := Ax_i$, we obtain as in the proof of Proposition 1.5.1 that

$$A^* = \sum_{i=1}^{k} v_i \otimes y_i. \quad \square$$

PROPOSITION 1.5.3. *Let E_1 and E_2 be Banach spaces and $\mu \in \Omega$. Assume that $x_1,\dots,x_r \in H(\Omega,E_1)$ and $z_1,\dots,z_r \in H(\Omega\setminus\{\mu\},E_2')$. Assume that the vectors $x_1(\mu),\dots,x_r(\mu)$ are linearly independent and that the functions $z_1,\dots,z_r$ are meromorphic at μ. If $\sum_{j=1}^{r} x_j \otimes z_j$ is holomorphic at μ, then all the z_j $(j = 1,\dots,r)$ are holomorphic at μ.*

Proof. There are a neighbourhood $U \subset \Omega$ of μ and for each $j \in \{1,\dots,r\}$ an integer $s_j \in \mathbb{N}$ such that

$$z_j(\lambda) = \sum_{i=-s_j}^{\infty} (\lambda-\mu)^i z_j^{(i)} \qquad (\lambda \in U \setminus \{\mu\}).$$

Suppose that the assertion of the proposition does not hold. Then it follows that $s := \max\{i \in \mathbb{N} : \exists\, j \in \{1,\dots,r\}\ z_j^{(-i)} \neq 0\} > 0$. From the holomorphy of the x_i and of $\sum_{i=1}^{r} x_i \otimes z_i$ at μ we obtain

$$\sum_{i=1}^{r} z_i^{(-s)} \otimes x_i(\mu) = \Big((\cdot-\mu)^s \Big(\sum_{i=1}^{r} x_i \otimes z_i\Big)^*\Big)(\mu) = 0. \tag{1.5.1}$$

We fix some $j \in \{1,\dots,r\}$ such that $z_j^{(-s)} \neq 0$. Since the $x_i(\mu)$ are linearly independent, the HAHN-BANACH theorem yields an element $w \in E_1'$ such that

$$\langle x_i(\mu), w\rangle = \delta_{ij} \quad (i=1,\dots,r),$$

see e. g. [HO, p. 51]. Then (1.5.1) leads to the contradiction

$$z_j^{(-s)} = \sum_{i=1}^{r} \langle x_i(\mu), w\rangle z_i^{(-s)} = 0. \qquad \square$$

THEOREM 1.5.4. *Let $T \in H(\Omega, \Phi(E,F))$, $\rho(T) \neq \emptyset$, and let $\mu \in \sigma(T)$. Let $\{y_1,\dots,y_r\}$ be a CSRF of T at μ. Then there are polynomials $v_j : \mathbb{C} \to F'$ of degree less than m_j and a function $D \in H(U, L(F,E))$, where $U \subset \Omega$ is a suitable neighbourhood of μ, such that*

$$T^{-1}(\lambda) = \sum_{j=1}^{r} (\lambda-\mu)^{-m_j} y_j(\lambda) \otimes v_j(\lambda) + D(\lambda) \tag{1.5.2}$$

for all $\lambda \in U \setminus \{\mu\}$. The polynomials v_j are uniquely determined by the system $\{y_1,\dots,y_r\}$,

$$\{v_1,\dots,v_r\} \text{ is a CSRF of } T^* \text{ at } \mu, \tag{1.5.3}$$

$\nu(v_j) = m_j$ $(j=1,\dots,r)$, where $m_1,\dots,m_r$ are the partial multiplicities of T at μ, and the biorthogonal relationships

$$\frac{1}{l!}\frac{d^l}{d\lambda^l}\langle \eta_{i,h}, v_j\rangle(\mu) = \delta_{ij}\delta_{m_i-h,l} \tag{1.5.4}$$
$$(1 \le i \le r,\ 1 \le h \le m_i,\ 1 \le j \le r,\ 0 \le l \le m_j - 1)$$

are fulfilled, where

$$\eta_{i,h}(\lambda) := (\lambda-\mu)^{-h}(Ty_i)(\lambda).$$

Proof. Let s be the pole order of T^{-1} at μ. First we prove the following statement by induction:
(1.5.5) For $\kappa = 0,\dots,s$ and $j = 1,\dots,r$ there are polynomials $v_j^\kappa : \mathbb{C} \to F'$ of degree less than m_j such that the pole order of

$$T^{-1} - \sum_{j=1}^{r} (\cdot - \mu)^{-m_j} y_j \otimes v_j^\kappa \tag{1.5.6}$$

at μ does not exceed $s-\kappa$.

For $\kappa = 0$, (1.5.5) holds with $v_j^0 = 0$ $(j = 1,\dots,r)$. Suppose that the statement (1.5.5) is fulfilled for some $0 \le \kappa < s$. We set

$$\begin{aligned} A(\lambda) :&= \sum_{l=-s+\kappa}^{\infty} (\lambda-\mu)^l A_l \\ :&= T^{-1}(\lambda) - \sum_{j=1}^{r} (\lambda-\mu)^{-m_j} y_j(\lambda) \otimes v_j^\kappa(\lambda) \end{aligned} \tag{1.5.7}$$

for λ in some punctured neighbourhood of μ (where $A_l \in L(F,E)$). For $j \in \{1,\dots,r\}$, $(\cdot-\mu)^{-m_j} T y_j$ is holomorphic at μ because of $m_j = \nu(y_j)$. Hence, by (1.5.7), Proposition 1.1.2 and Corollary 1.2.4, TA is holomorphic at μ. For $y_0 \in R(A_{-s+\kappa}) \setminus \{0\}$ choose some $x \in F$ such that $y_0 = A_{-s+\kappa} x$ and define $y(\lambda) := (\lambda-\mu)^{s-\kappa} A(\lambda) x$. y is a root function of T at μ with $y(\mu) = y_0$ and $\nu(y) \ge s-\kappa$. This proves $R(A_{-s+\kappa}) \subset L_{s-\kappa}$, where $L_{s-\kappa}$ is defined by (1.4.1). By Proposition 1.5.1 with $N = L_{s-\kappa}$ and Proposition 1.4.9 we find $z_j \in F'$ $(j = 1,\dots,r)$ such that

$$A_{-s+\kappa} = \sum_{j=1}^{\dim L_{s-\kappa}} y_j(\mu) \otimes z_j = \sum_{j=1}^{r} y_j(\mu) \otimes z_j, \tag{1.5.8}$$

where $z_j := 0$ if $m_j < s-\kappa$. We set

$$v_j^{\kappa+1}(\lambda) := v_j^\kappa(\lambda) + (\lambda-\mu)^{m_j-s+\kappa} z_j \quad (j = 1,\dots,r).$$

The $v_j^{\kappa+1} : \mathbb{C} \to F'$ are polynomials of degree less than m_j. From (1.5.7) and (1.5.8) we conclude that

$$\begin{aligned} &T^{-1}(\lambda) - \sum_{j=1}^{r} (\lambda-\mu)^{-m_j} y_j(\lambda) \otimes v_j^{\kappa+1}(\lambda) \\ &\quad = A(\lambda) - \sum_{j=1}^{r} (\lambda-\mu)^{-s+\kappa} y_j(\lambda) \otimes z_j \\ &\quad = \sum_{l=-s+\kappa+1}^{\infty} (\lambda-\mu)^l A_l - \sum_{j=1}^{r} (\lambda-\mu)^{-s+\kappa} (y_j(\lambda) - y_j(\mu)) \otimes z_j, \end{aligned}$$

whence (1.5.5) holds for $\kappa+1$ since the pole order of $(\cdot-\mu)^{-s+\kappa}(y_j - y_j(\mu))$ at μ is less than $s-\kappa$.

The assertion (1.5.2) follows if we set $v_j := v_j^s$ $(j = 1, \ldots, r)$.

To prove uniqueness let

$$T^{-1}(\lambda) = \sum_{j=1}^{r} (\lambda - \mu)^{-m_j} y_j(\lambda) \otimes \tilde{v}_j(\lambda) + \widetilde{D}(\lambda)$$

for λ in a punctured neighbourhood of μ, where $\tilde{v}_j : \mathbb{C} \to F'$ $(j = 1, \ldots, r)$ are polynomials of degree less than m_j and $\widetilde{D}$ is holomorphic in a neighbourhood of μ. Then

$$\sum_{j=1}^{r} y_j \otimes ((\cdot - \mu)^{-m_j}(v_j - \tilde{v}_j)) = \widetilde{D} - D$$

is holomorphic in a neighbourhood of μ. Since $y_1(\mu), \ldots, y_r(\mu)$ are linearly independent, the functions $(\cdot - \mu)^{-m_j}(v_j - \widetilde{v}_j)$ are holomorphic at μ by Proposition 1.5.3. As v_j and $\tilde{v}_j$ are polynomials of degree less than m_j, we obtain $v_j - \tilde{v}_j = 0$.

We still have to prove the assertions (1.5.3) and (1.5.4). Multiplying (1.5.2) with $T(\lambda)$ from the right-hand side yields

$$\mathrm{id}_E = \sum_{j=1}^{r} (\lambda - \mu)^{-m_j} y_j(\lambda) \otimes (T^* v_j)(\lambda) + (DT)(\lambda) \tag{1.5.9}$$

for $\lambda \in U \setminus \{\mu\}$. Hence the sum

$$\sum_{j=1}^{r} y_j \otimes (\cdot - \mu)^{-m_j} T^* v_j$$

is holomorphic in U. Since $y_1(\mu), \ldots, y_r(\mu)$ are linearly independent, the functions

$$(\cdot - \mu)^{-m_j} T^* v_j \quad (j = 1, \ldots, r) \tag{1.5.10}$$

are holomorphic at μ by Proposition 1.5.3. We apply (1.5.9) to $y_i(\lambda)$ and obtain

$$y_i(\lambda) = \sum_{j=1}^{r} (\lambda - \mu)^{-m_j} \langle y_i(\lambda), (T^* v_j)(\lambda) \rangle y_j(\lambda) + (DTy_i)(\lambda)$$

for $\lambda \in U \setminus \{\mu\}$. At μ the vector function DTy_i has a zero of order at least m_i, whence

$$\sum_{j=1}^{r} y_j \{ (\cdot - \mu)^{-m_i} (\delta_{ij} - \langle (\cdot - \mu)^{-m_j} Ty_i, v_j \rangle) \}$$

is holomorphic there. Thus, by Proposition 1.5.3 with $E_2 = \mathbb{C}$, the functions

$$(\cdot - \mu)^{-m_i} (\delta_{ij} - \langle (\cdot - \mu)^{-m_j} Ty_i, v_j \rangle) = (\cdot - \mu)^{-m_j} (\delta_{ij} - \langle (\cdot - \mu)^{-m_i} Ty_i, v_j \rangle)$$

for $i, j = 1, \ldots, r$ are holomorphic at μ. We infer for $i, j = 1, \ldots, r$ and $h = 1, \ldots, m_i$ that the function

$$\delta_{ij} (\cdot - \mu)^{m_i - h} - \langle (\cdot - \mu)^{-h} Ty_i, v_j \rangle$$

has a zero of order not less than m_j at μ, whence the equations (1.5.4) are proved.

In (1.5.4) we set $h = m_i$ and $l = 0$ and obtain

$$\langle \eta_{i,m_i}(\mu), v_j(\mu) \rangle = \delta_{ij} \quad (i, j = 1, \dots, r). \tag{1.5.11}$$

These equations show that the vectors $v_1(\mu), \dots, v_r(\mu)$ are linearly independent. Since the vector functions in (1.5.10) are holomorphic at μ, the v_j $(j = 1, \dots, r)$ are root functions of T^* at μ with $\nu(v_j) \geq m_j$.

We multiply the equation (1.5.2) with $T(\lambda)$ from the left-hand side and obtain

$$\mathrm{id}_F = \sum_{j=1}^{r} (\lambda - \mu)^{-m_j} (Ty_j)(\lambda) \otimes v_j(\lambda) + (TD)(\lambda)$$

for $\lambda \in U \setminus \{\mu\}$. Forming the adjoint yields

$$\mathrm{id}_{F'} = \sum_{j=1}^{r} (\lambda - \mu)^{-m_j} v_j(\lambda) \otimes (Ty_j)(\lambda) + (D^* T^*)(\lambda)$$

for $\lambda \in U \setminus \{\mu\}$. Let v be a root function of T^* at μ. The above equation yields

$$v(\lambda) = \sum_{j=1}^{r} (\lambda - \mu)^{-m_j} \langle (Ty_j)(\lambda), v(\lambda) \rangle v_j(\lambda) + (D^* T^* v)(\lambda)$$

for $\lambda \in U \setminus \{\mu\}$. As μ is a zero of $T^* v$, it follows that

$$v(\mu) = \sum_{j=1}^{r} \langle ((\cdot - \mu)^{-m_j} Ty_j)(\mu), v(\mu) \rangle v_j(\mu), \tag{1.5.12}$$

which proves that $v(\mu) \in \operatorname{span}\{v_1(\mu), \dots, v_r(\mu)\}$. Hence $\{v_1(\mu), \dots, v_r(\mu)\}$ is a basis of $N(T^*(\mu))$. If $1 \leq k \leq r$ and

$$v(\mu) \notin \operatorname{span}\{v_1(\mu), \dots, v_{k-1}(\mu)\},$$

then, by (1.5.12), there is an integer $j \geq k$ such that the function

$$\langle (\cdot - \mu)^{-m_j} Ty_j, v \rangle$$

has a non-zero value at μ. But this function is equal to

$$\langle y_j, (\cdot - \mu)^{-m_j} T^* v \rangle,$$

whence $\nu(v) \leq m_j$. Since $j \geq k$, the inequality $m_j \leq m_k$ holds. Furthermore, $m_k \leq \nu(v_k)$ as was shown above. Thus $\nu(v) \leq m_k \leq \nu(v_k)$. We infer that

$$\nu(v_k) = \max\big\{ \nu(v) : v \text{ is a root function of } T^* \text{ at } \mu \text{ and } v(\mu) \notin \operatorname{span}\{v_1(\mu), \dots, v_{k-1}(\mu)\} \big\}$$

and $\nu(v_k) = m_k$. This proves the assertion (1.5.3). □

An immediate consequence of the properties of the CSRF of T^* is

COROLLARY 1.5.5. *Let $T \in H(\Omega, \Phi(E,F))$, $\rho(T) \neq \emptyset$ and $\mu \in \sigma(T)$. Then* $\operatorname{nul} T(\mu) = \operatorname{nul} T^*(\mu)$, *and the partial multiplicities of T and T^* at μ coincide.*

COROLLARY 1.5.6. *Let $T \in H(\Omega, \Phi(E,F))$, $\rho(T) \neq \emptyset$, and $\mu \in \sigma(T)$. We set $r = \operatorname{nul} T(\mu)$. Then there are root functions $y_1, \ldots, y_r$ of T at μ and $v_1, \ldots, v_r$ of T^* at μ such that the following properties hold:*

$$\nu(y_j) = \nu(v_j) = m_j \quad (j = 1, \ldots, r);$$

(1.5.13)
$$\frac{d^l}{d\lambda^l} \langle \eta_i, v_j \rangle(\mu) = \delta_{ij}\, \delta_{0l} \quad (1 \le i \le r, 1 \le j \le r, 0 \le l \le m_j - 1),$$

where

$$\eta_i := (\cdot - \mu)^{-m_i} T y_i \qquad (i = 1, \ldots, r);$$

(1.5.14)
$$\frac{d^l}{d\lambda^l} \langle y_i, \zeta_j \rangle(\mu) = \delta_{ij}\, \delta_{0l} \quad (1 \le i \le r, 1 \le j \le r, 0 \le l \le m_i - 1),$$

where

$$\zeta_j := (\cdot - \mu)^{-m_j} T^* v_j \qquad (j = 1, \ldots, r);$$

(1.5.15)
$$D := T^{-1} - \sum_{j=1}^{r} (\cdot - \mu)^{-m_j} y_j \otimes v_j$$

is holomorphic at μ. The systems $\{y_1, \ldots, y_r\}$ and $\{v_1, \ldots, v_r\}$ fulfilling the above properties are called biorthogonal CSRFs of T and T^* at μ.

Proof. Choose a CSRF $\{y_1, \ldots, y_r\}$ of T at μ according to Proposition 1.4.7. Let $\{v_1, \ldots, v_r\}$ be a CSRF of T^* at μ according to Theorem 1.5.4. Then (1.5.13) and (1.5.15) immediately follow from (1.5.4) and (1.5.2). The relationships (1.5.13) say that each of the functions $\delta_{ij} - \langle \eta_i, v_j \rangle$ $(i, j = 1, \ldots, r)$ has a zero of order $\ge m_j$ at μ. The definition of the ζ_j and the η_i immediately yields

$$\delta_{ij} - \langle y_i, \zeta_j \rangle = (\cdot - \mu)^{m_i - m_j} \big(\delta_{ij} - \langle \eta_i, v_j \rangle\big) \quad (i, j = 1, \ldots, r).$$

Hence each of these functions has a zero of order $\ge m_i$ at μ. This proves (1.5.14). □

REMARK 1.5.7. The biorthogonal relationships in (1.5.13) are formally weaker than the biorthogonal relationships (1.5.4). But they are easily seen to be equal if we observe that (1.5.4) means that, for $i, j = 1, \ldots, r$ and $h = 1, \ldots, m_i$,

$$(\cdot - \mu)^{m_i - h} (\delta_{ij} - \langle \eta_i, v_j \rangle)$$

has a zero of order $\ge m_j$ at μ, whereas (1.5.13) means that this holds for $h = m_i$.

REMARK 1.5.8. Assume that $T \in H(\Omega, \Phi(E,F))$, $\rho(T) \neq \emptyset$ and $\mu \in \sigma(T)$. Let $\{v_1, \ldots, v_r\}$ be a CSRF of T^* at μ. Note that the numbers $m_j = \nu(v_j)$ $(j = 1, \ldots, r)$ are the partial multiplicities of T and T^* at μ. Then there exist polynomials $y_j : \mathbb{C} \to E$ of degree less than m_j and a function $D \in H(U, L(F,E))$, where $U \subset \Omega$ is a suitable neighbourhood of μ, such that $\{y_1, \ldots, y_r\}$ is a CSRF of T at μ and the representation (1.5.2) as well as the biorthogonal relationships (1.5.4) hold, where the polynomials $y_1,\dots,y_r$ are uniquely determined by $v_1,\dots,v_r$.

Proof. This is analogous to the proof of Theorem 1.5.4. We only have to use Remark 1.5.2 instead of Proposition 1.5.1 in order to obtain a representation

$$A_{-s+\kappa} = \sum_{j=1}^{r} x_j \otimes v_j(\mu)$$

with $x_j \in E$. Now we proceed as in the proof of Theorem 1.5.4, considering T^* instead of T. □

Often it is not easy to find the partial multiplicities or to check condition i) of Proposition 1.4.4. But the following theorem says that an estimate of the multiplicities of root functions and the validity of biorthogonal relationships is sufficient to find biorthogonal CSRFs.

THEOREM 1.5.9. *Let $T \in H(\Omega, \Phi(E,F))$, $\rho(T) \neq \emptyset$, $\mu \in \sigma(T)$ and $r := \operatorname{nul} T(\mu)$. Let $k_1 \geq \cdots \geq k_r$ be positive integers, $y_1,\dots,y_r$ be root functions of T at μ, and $v_1,\dots,v_r$ be root functions of T^* at μ. Assume that $\nu(y_j) \geq k_j$ and $\nu(v_j) \geq k_j$ for $j = 1,\dots,r$. Set*

$$\eta_i := (\cdot - \mu)^{-k_i} T y_i \qquad (i = 1,\dots,r).$$

Then the following properties are equivalent:
i) *the systems $\{y_1,\dots,y_r\}$ and $\{v_1,\dots,v_r\}$ are biorthogonal, i. e.,*

$$\text{(1.5.16)} \qquad \frac{\mathrm{d}^l}{\mathrm{d}\lambda^l}\langle \eta_i, v_j\rangle(\mu) = \delta_{ij}\,\delta_{0l} \qquad (1 \leq i \leq r,\ 1 \leq j \leq r,\ 0 \leq l \leq k_j - 1);$$

ii) *the operator function*

$$\text{(1.5.17)} \qquad D := T^{-1} - \sum_{j=1}^{r} (\cdot - \mu)^{-k_j} y_j \otimes v_j$$

is holomorphic at μ.
If one of these properties holds, then $\{y_1,\dots,y_r\}$ is a CSRF of T at μ, $\{v_1,\dots,v_r\}$ is a CSRF of T^* at μ, and $\nu(y_j) = \nu(v_j) = k_j$ for $j = 1,\dots,r$.

Proof. i) ⇒ ii): Set

$$\zeta_j := (\cdot - \mu)^{-k_j} T^* v_j \quad (j = 1,\dots,r).$$

These functions are holomorphic at μ as $\nu(v_j) \geq k_j$. The definition of ζ_j and η_i immediately yields

$$\text{(1.5.18)} \qquad \delta_{ij} - \langle y_i, \zeta_j\rangle = (\cdot - \mu)^{k_i - k_j}(\delta_{ij} - \langle \eta_i, v_j\rangle) \quad (i,j = 1,\dots,r).$$

By (1.5.16), the function $\delta_{ij} - \langle \eta_i, v_j\rangle$ has a zero of order $\geq k_j$ at μ. Since $k_i > 0$ we infer that $(\cdot - \mu)^{k_i - k_j}(\delta_{ij} - \langle \eta_i, v_j\rangle)$ has a zero at μ. This proves

$$\text{(1.5.19)} \qquad \langle y_i, \zeta_j\rangle(\mu) = \delta_{ij} \quad (i,j = 1,\dots,r).$$

Thus the linear independence of the vectors $y_1(\mu),\ldots,y_r(\mu)$ is proved. From $r = \operatorname{nul} T(\mu)$ we infer that $\{y_1(\mu),\ldots,y_r(\mu)\}$ is a basis of $N(T(\mu))$. Since T^{-1} has a pole at μ, D also has a pole at μ or is holomorphic there. Hence there is an integer κ_0 such that

$$D(\lambda) = \sum_{\kappa=\kappa_0}^{\infty} (\lambda-\mu)^{\kappa} D_{\kappa} \tag{1.5.20}$$

in a punctured neighbourhood of μ. Suppose that $\kappa_0 < 0$ and $D_{\kappa_0} \neq 0$. From the definition of D in (1.5.17) and the definition of η_j we obtain

$$TD = \mathrm{id}_F - \sum_{j=1}^{r} \eta_j \otimes v_j. \tag{1.5.21}$$

This proves that TD is holomorphic at μ as the vector functions η_j and v_j are holomorphic there. Since $T(\mu)D_{\kappa_0}$ is the coefficient of $(\cdot-\mu)^{\kappa_0}$ in the Laurent series expansion of TD at μ, we have $T(\mu)D_{\kappa_0} = 0$. Hence

$$R(D_{\kappa_0}) \subset \operatorname{span}\{y_1(\mu),\ldots,y_r(\mu)\}.$$

Choose some $x \in F$ such that $D_{\kappa_0}x \neq 0$. The biorthogonal relationships (1.5.19) imply that

$$\langle D_{\kappa_0}x, \zeta_{j_0}(\mu)\rangle \neq 0 \quad \text{for some } j_0 \in \{1,\ldots,r\}. \tag{1.5.22}$$

The function $\langle Dx, T^*v_{j_0}\rangle = \langle TDx, v_{j_0}\rangle$ is holomorphic at μ. From (1.5.20) we infer the expansion

$$\langle Dx, T^*v_{j_0}\rangle = (\cdot-\mu)^{\kappa_0+k_{j_0}}\langle D_{\kappa_0}x, \zeta_{j_0}(\mu)\rangle + \text{ terms of higher order.}$$

In view of (1.5.22) and $\kappa_0 < 0$ this shows that $\langle Dx, T^*v_{j_0}\rangle$ does not have a zero of order $\geq k_{j_0}$. For $j = 1,\ldots,r$ the biorthogonal relationships (1.5.16) yield that the function $\delta_{j,j_0} - \langle \eta_j, v_{j_0}\rangle$ has a zero of order $\geq k_{j_0}$ at μ. In view of (1.5.21) we obtain that

$$\begin{aligned}\langle Dx, T^*v_{j_0}\rangle &= \langle x, v_{j_0}\rangle - \sum_{j=1}^{r}\langle \eta_j, v_{j_0}\rangle\langle x, v_j\rangle \\ &= \sum_{j=1}^{r}\big(\delta_{j,j_0} - \langle \eta_j, v_{j_0}\rangle\big)\langle x, v_j\rangle\end{aligned}$$

has a zero of order $\geq k_{j_0}$. This contradiction proves the holomorphy of D at μ.

Let i) be fulfilled. We shall prove that $\{y_1,\ldots,y_r\}$ is a CSRF of T at μ, $\{v_1,\ldots,v_r\}$ is a CSRF of T^* at μ and $\nu(y_j) = \nu(v_j) = k_j$ for $j = 1,\ldots,r$. We have shown that $\{y_1(\mu),\ldots,y_r(\mu)\}$ is a basis of $N(T(\mu))$. From (1.5.16) with $l = 0$ we infer that the vectors $v_1(\mu),\ldots,v_r(\mu)$ are linearly independent. This shows that $\{v_1(\mu),\ldots,v_r(\mu)\}$ is a basis of $N(T^*(\mu))$ as $\operatorname{nul} T(\mu) = \operatorname{nul} T^*(\mu)$, see Corollary 1.5.5. By Definition 1.4.5 and Corollary 1.5.5 we know that $\{y_1,\ldots,y_r\}$ and

$\{v_1,\dots,v_r\}$ are CSRFs if we prove that $k_j \ge m_j$ for $1 \le j \le r$, where the numbers m_j are the partial multiplicities of T at μ. Suppose that there is a $j_0 \in \{1,\dots,r\}$ such that $k_j \ge m_j$ for $1 \le j \le j_0 - 1$ and $k_{j_0} < m_{j_0}$. Set $\tilde{y}_j = y_j$ for $j = 1,\dots,j_0 - 1$. By Proposition 1.4.6 there are root functions $\tilde{y}_{j_0},\dots,\tilde{y}_r$ such that $\{\tilde{y}_1,\dots,\tilde{y}_r\}$ is a CSRF of T at μ. We choose the CSRF $\{\tilde{v}_1,\dots,\tilde{v}_r\}$ of T^* at μ according to Theorem 1.5.4. By (1.5.2) the operator function

$$T^{-1} - \sum_{j=1}^{r} (\cdot - \mu)^{-m_j} \tilde{y}_j \otimes \tilde{v}_j$$

is holomorphic at μ. Since the property ii) holds, we infer that the operator function

$$\sum_{j=1}^{r} (\cdot - \mu)^{-m_j} \tilde{y}_j \otimes \tilde{v}_j - \sum_{j=1}^{r} (\cdot - \mu)^{-k_j} y_j \otimes v_j$$

is holomorphic at μ. We multiply the above operator function by the holomorphic function $(\cdot - \mu)^{m_{j_0}-1}$ and obtain that

$$\sum_{j=1}^{j_0-1} \tilde{y}_j \otimes \left((\cdot - \mu)^{-m_j+m_{j_0}-1} \tilde{v}_j - (\cdot - \mu)^{-k_j+m_{j_0}-1} v_j\right)$$
$$+ \sum_{j=j_0}^{r} \tilde{y}_j \otimes (\cdot - \mu)^{-m_j+m_{j_0}-1} \tilde{v}_j$$

is holomorphic at μ since $(\cdot - \mu)^{-k_j+m_{j_0}-1}$ is holomorphic for $j = j_0,\dots,r$ as $k_j \le k_{j_0} < m_{j_0}$. Since the $\tilde{y}_j(\mu)$ are linearly independent, Proposition 1.5.3 yields that $(\cdot - \mu)^{-1}\tilde{v}_{j_0}$ is holomorphic at μ, i. e., we have $\tilde{v}_{j_0}(\mu) = 0$. But this is impossible since $\tilde{v}_{j_0}$ is a root function. Finally, $k_j \ge m_j$ implies $m_j \le k_j \le \nu(y_j) = m_j$, whence $k_j = \nu(y_j)$ and, similarly, $k_j = \nu(v_j)$ for $j = 1,\dots,r$.

ii)$\Rightarrow$i): Let $y \in N(T(\mu))$. Property ii) yields

$$0 = \left(T^{-1} - \sum_{j=1}^{r} (\cdot - \mu)^{-k_j} y_j \otimes v_j\right)(\mu) T(\mu) y$$
$$= y - \sum_{j=1}^{r} \langle y, ((\cdot - \mu)^{-k_j} T^* v_j)(\mu) \rangle y_j(\mu),$$

whence $N(T(\mu)) \subset \operatorname{span}\{y_1(\mu),\dots,y_r(\mu)\}$. Here we have used that the vector function $(\cdot - \mu)^{-k_j} T^* v_j$ is holomorphic at μ since $\nu(v_j) \ge k_j$. From the assumption that $\dim N(T(\mu)) = r$ we infer that $y_1(\mu),\dots,y_r(\mu)$ are linearly independent. By property ii), for each $i \in \{1,\dots,r\}$ the function

$$\sum_{j=1}^{r} (\delta_{ij} - \langle y_i, \zeta_j \rangle) y_j = \left(T^{-1} - \sum_{j=1}^{r} (\cdot - \mu)^{-k_j} y_j \otimes v_j\right) T y_i$$

has a zero of order $\geq k_i$ at μ since $\nu(y_i) \geq k_i$. From Proposition 1.5.3 with $E_2 = \mathbb{C}$ we infer that for $i, j = 1, \dots, r$ the functions

$$(\cdot - \mu)^{-k_j}(\delta_{ij} - \langle \eta_i, v_j \rangle) = (\cdot - \mu)^{-k_i}(\delta_{ij} - \langle y_i, \zeta_j \rangle)$$

are holomorphic at μ, which yields the biorthogonal relationships (1.5.16). □

1.6. Eigenvectors and associated vectors

Let Ω be a domain in $\mathbb{C}$ and E and F be Banach spaces.

DEFINITION 1.6.1. Let $T \in H(\Omega, \Phi(E,F))$, $\rho(T) \neq \emptyset$ and $\mu \in \sigma(T)$.
i) An ordered set $\{y_0, y_1, \dots, y_h\}$ in E is called a *chain of an eigenvector and associated vectors* (CEAV) of T at μ if

$$y := \sum_{l=0}^{h} (\cdot - \mu)^l y_l$$

is a root function of T at μ with $\nu(y) \geq h+1$.
ii) Let $y_0 \in N(T(\mu)) \setminus \{0\}$. Then $\overline{\nu}(y_0)$ denotes the maximum of all multiplicities $\nu(y)$, where y is a root function of T at μ with $y(\mu) = y_0$. $\overline{\nu}(y_0)$ is called the *rank of the eigenvector* y_0.
iii) A system $\{y_l^{(j)} : 1 \leq j \leq r, 0 \leq l \leq \overline{m}_j - 1\}$ is called a *canonical system of eigenvectors and associated vectors* (CSEAV) of T at μ if

$$\{y_0^{(1)}, \dots, y_0^{(r)}\} \text{ is a basis of } N(T(\mu)),$$
$$\{y_0^{(j)}, y_1^{(j)}, \dots, y_{\overline{m}_j-1}^{(j)}\} \text{ is a CEAV of } T \text{ at } \mu \quad (j = 1, \dots, r),$$
$$\overline{m}_j = \max\{\overline{\nu}(y) : y \in N(T(\mu)) \setminus \operatorname{span}\{y_0^{(k)} : k < j\}\} \quad (j = 1, \dots, r).$$

PROPOSITION 1.6.2. *Let $T \in H(\Omega, \Phi(E,F))$, $\rho(T) \neq \emptyset$ and let $\mu \in \sigma(T)$. Let $\{y_l^{(j)} : 1 \leq j \leq r, 0 \leq l \leq \overline{m}_j - 1\}$ be a CSEAV of T at μ and set*

$$y_j := \sum_{l=0}^{\overline{m}_j - 1} (\cdot - \mu)^l y_l^{(j)} \quad (j = 1, \dots, r).$$

Then $\{y_1, \dots, y_r\}$ is a CSRF of T at μ, and the numbers $\overline{m}_j$ are the partial multiplicities of T at μ.

Proof. By assumption, $\{y_1(\mu), \dots, y_r(\mu)\} = \{y_0^{(1)}, \dots, y_0^{(r)}\}$ is a basis of $N(T(\mu))$. The definition of a CEAV yields $\nu(y_j) \geq \overline{m}_j$ for $j = 1, \dots, r$. Hence

$$\begin{aligned} \nu(y_j) &\geq \max\{\overline{\nu}(y) : y \in N(T(\mu)) \setminus \operatorname{span}\{y_0^{(k)} : k < j\}\} \\ &= \max\{\nu(y) : y \text{ is a root function of } T \text{ at } \mu \text{ and} \\ &\qquad y(\mu) \notin \operatorname{span}\{y_1(\mu), \dots, y_{j-1}(\mu)\}\}. \end{aligned}$$

The above inequality is an equality since the estimate $\leq$ is obvious. So $\{y_1,\dots,y_r\}$ is a CSRF of T at μ with $\nu(y_j)=\overline{m}_j$ $(j=1,\dots,r)$, see Definition 1.4.5. □

PROPOSITION 1.6.3. *Let $T\in H(\Omega,\Phi(E,F))$, $\rho(T)\neq\emptyset$ and $\mu\in\sigma(T)$. Let $\{y_1,\dots,y_r\}$ be a CSRF of T at μ. Set*

$$y_l^{(j)}:=\frac{1}{l!}\frac{\mathrm{d}^l y_j}{\mathrm{d}\lambda^l}(\mu)\quad (j=1,\dots,r;\, l=0,\dots,m_j-1),$$

where the numbers m_j are the partial multiplicities of T at μ. Then the set of vectors $\{y_l^{(j)}:1\leq j\leq r, 0\leq l\leq m_j-1\}$ is a CSEAV of T at μ, and $\overline{\nu}(y_0^{(j)})=m_j$ for $j=1,\dots,r$.

Proof. The set $\{y_0^{(1)},\dots,y_0^{(r)}\}$ is a basis of $N(T(\mu))$ as $y_0^{(j)}=y_j(\mu)$ for $j=1,\dots,r$. Since

$$\sum_{l=0}^{m_j-1}(\cdot-\mu)^l\frac{1}{l!}\frac{\mathrm{d}^l y_j}{\mathrm{d}\lambda^l}(\mu)$$

is the Taylor polynomial of y of order m_j-1 at μ, it is a root function of order $\geq\nu(y_j)$ at μ. Hence $\{y_0^{(j)},y_1^{(j)},\dots,y_{m_j-1}^{(j)}\}$ is a CEAV of T at μ for $j=1,\dots,r$. Finally, for $j=1,\dots,r$,

$$\begin{aligned}m_j&=\max\{\nu(y): y \text{ is a root function of } T \text{ at } \mu \text{ and}\\ &\qquad y(\mu)\notin\operatorname{span}\{y_1(\mu),\dots,y_{j-1}(\mu)\}\}\\ &=\max\{\overline{\nu}(y): y\in N(T(\mu))\setminus\operatorname{span}\{y_0^{(k)}:k<j\}\}.\end{aligned}$$ □

PROPOSITION 1.6.4. *Let $T\in H(\Omega,\Phi(E,F))$, $\rho(T)\neq\emptyset$ and $\mu\in\sigma(T)$. We set $r:=\operatorname{nul}T(\mu)$ and let m_j, $j=1,\dots,r$, be the partial multiplicities of T at μ. Then there is a CSEAV $\{y_l^{(j)}:1\leq j\leq r, 0\leq l\leq m_j-1\}$ of T at μ.*

Proof. The result immediately follows from Propositions 1.4.7 and 1.6.3. □

THEOREM 1.6.5. *Assume that $T\in H(\Omega,\Phi(E,F))$, $\rho(T)\neq\emptyset$ and $\mu\in\sigma(T)$. Let $\{y_l^{(j)}:1\leq j\leq r, 0\leq l\leq m_j-1\}$ be a CSEAV of T at μ. Then there are vectors $v_l^{(j)}\in F'$ $(j=1,\dots,r;\, l=0,\dots,m_j-1)$ and a function $\widetilde{D}\in H(U,L(F,E))$, where $U\subset\Omega$ is a suitable neighbourhood of μ, such that*

$$T^{-1}(\lambda)=\sum_{j=1}^{r}\sum_{l=1}^{m_j}(\lambda-\mu)^{-l}\sum_{h=0}^{m_j-l}y_h^{(j)}\otimes v_{m_j-l-h}^{(j)}+\widetilde{D}(\lambda)\tag{1.6.1}$$

for all $\lambda\in U\setminus\{\mu\}$. The vectors $v_l^{(j)}$ are uniquely determined by the given system $\{y_l^{(j)}:1\leq j\leq r, 0\leq l\leq m_j-1\}$,

$$\{v_l^{(j)}:1\leq j\leq r, 0\leq l\leq m_j-1\} \textit{ is a CSEAV of } T^* \textit{ at } \mu,$$

$\overline{\nu}(v_j^{(0)}) = m_j$ $(j = 1, \ldots, r)$, *and the biorthogonal relationships*

$$(1.6.2) \quad \sum_{k=0}^{l} \sum_{q=1}^{m_i} \frac{1}{(k+q)!} \left\langle \frac{\mathrm{d}^{k+q} T}{\mathrm{d}\lambda^{k+q}}(\mu) y_{m_i-q}^{(i)}, v_{l-k}^{(j)} \right\rangle = \delta_{ij}\delta_{0l}$$

$$(1 \le i \le r,\ 1 \le j \le r, 0 \le l \le m_j - 1)$$

are fulfilled.

Proof. Define the CSRF $\{y_1, \ldots, y_r\}$ of T at μ according to Proposition 1.6.2. By Theorem 1.5.4 there are unique root functions

$$v_j = \sum_{l=0}^{m_j-1} (\cdot - \mu)^l v_l^{(j)} \quad (j = 1, \ldots, r)$$

such that the representation (1.5.2) of T^{-1} holds. Since $\{v_1, \ldots, v_r\}$ is a CSRF of T^* at μ by (1.5.3), Proposition 1.6.3 yields that $\{v_l^{(j)} : 1 \le j \le r, 0 \le l \le m_j - 1\}$ is a CSEAV of T^* at μ. An easy calculation gives that

$$\sum_{j=1}^{r} \sum_{l=1}^{m_j} (\cdot - \mu)^{-l} \sum_{h=0}^{m_j-l} y_h^{(j)} \otimes v_{m_j-l-h}^{(j)}$$

is the principal part of

$$\sum_{j=1}^{r} (\cdot - \mu)^{-m_j} y_j \otimes v_j$$

at μ. This proves the representation (1.6.1), and the uniqueness of the v_j implies the uniqueness of the $v_l^{(j)}$.

For $i = 1, \ldots, r$ and $h = 1, \ldots, m_i$ let $\eta_{i,h}$ be defined as in Theorem 1.5.4. Since Ty_i has a zero of order $\ge m_i$, we infer

$$\eta_{i,h} = \sum_{k=0}^{\infty} (\cdot - \mu)^{-h+m_i+k} \sum_{q=1}^{m_i} \frac{1}{(k+q)!} \frac{\mathrm{d}^{k+q} T}{\mathrm{d}\lambda^{k+q}}(\mu) y_{m_i-q}^{(i)}.$$

Hence

$$\langle \eta_{i,h}, v_j \rangle = \sum_{l=0}^{\infty} (\cdot - \mu)^{-h+m_i+l} \sum_{k=\max\{0, l-m_j+1\}}^{l} \sum_{q=1}^{m_i} \frac{1}{(k+q)!} \left\langle \frac{\mathrm{d}^{k+q} T}{\mathrm{d}\lambda^{k+q}}(\mu) y_{m_i-q}^{(i)}, v_{l-k}^{(j)} \right\rangle.$$

For $i = 1, \ldots, r$, $j = 1, \ldots, r$, $h = m_i$, and $l = 0, \ldots, m_j - 1$ the biorthogonal relationships (1.5.4) yield

$$\sum_{k=0}^{l} \sum_{q=1}^{m_i} \frac{1}{(k+q)!} \left\langle \frac{\mathrm{d}^{k+q} T}{\mathrm{d}\lambda^{k+q}}(\mu) y_{m_i-q}^{(i)}, v_{l-k}^{(j)} \right\rangle = \delta_{ij}\, \delta_{0l}. \quad \square$$

COROLLARY 1.6.6. *Let $T \in H(\Omega, \Phi(E,F))$ such that $\rho(T) \neq \emptyset$. Let $\mu \in \sigma(T)$, $r = \operatorname{nul} T(\mu)$, and m_j $(j = 1, \ldots, r)$ be the partial multiplicities of T at μ. Then there are CEAVs $y_0^{(j)}, \ldots, y_{m_j-1}^{(j)}$ $(j = 1, \ldots, r)$ of T at μ and $v_0^{(j)}, \ldots, v_{m_j-1}^{(j)}$ $(j = 1, \ldots, r)$ of T^* at μ such that the following properties hold:*

$$\overline{\nu}(y_0^{(j)}) = \overline{\nu}(v_0^{(j)}) = m_j \quad (j = 1, \ldots, r),$$

$$(1.6.3) \quad \sum_{k=0}^{l} \sum_{q=1}^{m_i} \frac{1}{(k+q)!} \left\langle \frac{d^{k+q}T}{d\lambda^{k+q}}(\mu) y_{m_i-q}^{(i)}, v_{l-k}^{(j)} \right\rangle = \delta_{ij} \delta_{0l}$$

$$(1 \leq i \leq r,\ 1 \leq j \leq r,\ 0 \leq l \leq m_j - 1),$$

and the operator function

$$(1.6.4) \qquad \widetilde{D} := T^{-1} - \sum_{j=1}^{r} \sum_{l=1}^{m_j} (\cdot - \mu)^{-l} \sum_{h=0}^{m_j-l} y_h^{(j)} \otimes v_{m_j-l-h}^{(j)}$$

is holomorphic at μ. We call the systems $\{y_l^{(j)} : 1 \leq j \leq r, 0 \leq l \leq m_j - 1\}$ and $\{v_l^{(j)} : 1 \leq j \leq r, 0 \leq l \leq m_j - 1\}$ biorthogonal CSEAVs of T and T^* at μ.

The above corollary follows from Corollary 1.5.6 in the same way as we obtained Theorem 1.6.5 from Theorem 1.5.4. Similarly, Theorem 1.5.9 yields

THEOREM 1.6.7. *Assume that $T \in H(\Omega, \Phi(E,F))$, $\rho(T) \neq \emptyset$ and $\mu \in \sigma(T)$. Let $r = \operatorname{nul} T(\mu)$ and $k_1 \geq \cdots \geq k_r$ be positive integers. Let $y_0^{(j)}, \ldots, y_{k_j-1}^{(j)}$ $(j = 1, \ldots, r)$ be CEAVs of T at μ and $v_0^{(j)}, \ldots, v_{k_j-1}^{(j)}$ $(1, \ldots, r)$ be CEAVs of T^* at μ. Assume that $\overline{\nu}(y_0^{(j)}) \geq k_j$ and $\overline{\nu}(v_0^{(j)}) \geq k_j$ for $j = 1, \ldots, r$. Then the following properties are equivalent:*
i) *$\{y_l^{(j)} : 1 \leq j \leq r, 0 \leq l \leq m_j - 1\}$ and $\{v_l^{(j)} : 1 \leq j \leq r, 0 \leq l \leq m_j - 1\}$ are biorthogonal, i.e.*

$$(1.6.5) \quad \sum_{k=0}^{l} \sum_{q=1}^{m_i} \frac{1}{(k+q)!} \left\langle \frac{d^{k+q}T}{d\lambda^{k+q}}(\mu) y_{m_i-q}^{(i)}, v_{l-k}^{(j)} \right\rangle = \delta_{ij} \delta_{0l}$$

$$(1 \leq i \leq r,\ 1 \leq j \leq r,\ 0 \leq l \leq m_j - 1);$$

ii) *the operator function*

$$(1.6.6) \qquad \widetilde{D} := T^{-1} - \sum_{j=1}^{r} \sum_{l=1}^{m_j} (\cdot - \mu)^{-l} \sum_{h=0}^{m_j-l} y_h^{(j)} \otimes v_{m_j-l-h}^{(j)}$$

is holomorphic at μ.
If one of these properties holds, then $\{y_l^{(j)} : 1 \leq j \leq r, 0 \leq l \leq m_j - 1\}$ is a CSEAV of T at μ, $\{v_l^{(j)} : 1 \leq j \leq r, 0 \leq l \leq m_j - 1\}$ is a CSEAV of T^* at μ, and $\overline{\nu}(y_0^{(j)}) = \overline{\nu}(v_0^{(j)}) = k_j$ for $j = 1, \ldots, r$.

1.7. Semi-simple eigenvalues

Let Ω be a domain in $\mathbb{C}$ and E and F be Banach spaces.

DEFINITION 1.7.1. Let $T \in H(\Omega, \Phi(E,F))$ and $\mu \in \sigma(T)$. Then μ is called a *semi-simple eigenvalue* of T if for each $y \in N(T(\mu)) \setminus \{0\}$ there is a $v \in N(T^*(\mu))$ such that

$$\left\langle \left(\frac{\mathrm{d}}{\mathrm{d}\lambda} T\right)(\mu) y, v \right\rangle \neq 0.$$

If $\mu \in \sigma(T)$ is semi-simple and $\operatorname{nul} T(\mu) = 1$, then μ is called a *simple eigenvalue*.

PROPOSITION 1.7.2. *Let $T \in H(\Omega, \Phi(E,F))$, $\rho(T) \neq \emptyset$ and $\mu \in \sigma(T)$. The following properties are equivalent:*
i) *μ is a semi-simple eigenvalue of T;*
ii) *there are CSRFs $\{y_1, \ldots, y_r\}$ of T at μ and $\{v_1, \ldots, v_r\}$ of T^* at μ such that*

$$\left\langle \left(\frac{\mathrm{d}}{\mathrm{d}\lambda} T\right)(\mu) y_i(\mu), v_j(\mu) \right\rangle = \delta_{ij} \quad (i, j = 1, \ldots, r);$$

iii) *for each root function y of T at μ we have $\nu(y) = 1$;*
iv) *the pole order of T^{-1} at μ is 1.*

Proof. iii) $\Leftrightarrow$ iv) is obvious because of Lemma 1.4.2.
i) $\Rightarrow$ iii): Suppose that there is a root function y of T at μ with $\nu(y) \geq 2$. Then

$$\begin{aligned}\left(\frac{\mathrm{d}}{\mathrm{d}\lambda} T\right)(\mu) y(\mu) &= \left(\frac{\mathrm{d}}{\mathrm{d}\lambda} Ty\right)(\mu) - T(\mu)\left(\frac{\mathrm{d}}{\mathrm{d}\lambda} y\right)(\mu) \\ &= -T(\mu)\left(\frac{\mathrm{d}}{\mathrm{d}\lambda} y\right)(\mu).\end{aligned}$$

Now let $v \in N(T^*(\mu))$. Then

$$\begin{aligned}\left\langle \left(\frac{\mathrm{d}}{\mathrm{d}\lambda} T\right)(\mu) y(\mu), v \right\rangle &= -\left\langle T(\mu)\left(\frac{\mathrm{d}}{\mathrm{d}\lambda} y\right)(\mu), v \right\rangle \\ &= -\left\langle \left(\frac{\mathrm{d}}{\mathrm{d}\lambda} y\right)(\mu), T^*(\mu) v \right\rangle = 0,\end{aligned}$$

which contradicts i) since $y(\mu) \in N(T(\mu)) \setminus \{0\}$.
iii) $\Rightarrow$ ii): By Corollary 1.5.6 there are biorthogonal CSRFs $\{y_1, \ldots, y_r\}$ of T at μ and $\{v_1, \ldots, v_r\}$ of T^* at μ. The assumption iii) implies that $m_j = 1$ for all $j \in \{1, \ldots, r\}$. For $i, j \in \{1, \ldots, r\}$, (1.5.13) yields

$$\langle ((\cdot - \mu)^{-1} T y_i)(\mu), v_j(\mu) \rangle = \delta_{ij}.$$

Since

$$\begin{aligned}((\cdot - \mu)^{-1} T y_i)(\mu) &= \frac{\mathrm{d}}{\mathrm{d}\lambda}\left(T y_i\right)(\mu) \\ &= \left(\frac{\mathrm{d}}{\mathrm{d}\lambda} T\right)(\mu) y_i(\mu) + T(\mu)\left(\frac{\mathrm{d}}{\mathrm{d}\lambda} y_i\right)(\mu),\end{aligned}$$

we infer

$$\begin{aligned}\Big\langle\Big(\frac{\mathrm{d}}{\mathrm{d}\lambda}T\Big)(\mu)y_i(\mu),v_j(\mu)\Big\rangle &= \langle((\cdot-\mu)^{-1}Ty_i)(\mu),v_j(\mu)\rangle - \Big\langle\Big(\frac{\mathrm{d}}{\mathrm{d}\lambda}y_i\Big)(\mu),T^*(\mu)v_j(\mu)\Big\rangle \\ &= \delta_{ij}.\end{aligned}$$

ii) $\Rightarrow$ i): Let $y \in N(T(\mu)) \setminus \{0\}$. Then there are $\alpha_i \in \mathbb{C}$ $(i = 1,\dots,r)$ such that

$$y = \sum_{i=1}^{r} \alpha_i y_i(\mu),$$

where $\alpha_{i_0} \neq 0$ for some $i_0 \in \{1,\dots,r\}$. We infer

$$\Big\langle\Big(\frac{\mathrm{d}}{\mathrm{d}\lambda}T\Big)(\mu)y,v_{i_0}(\mu)\Big\rangle = \sum_{i=1}^{r}\alpha_i\Big\langle\Big(\frac{\mathrm{d}}{\mathrm{d}\lambda}T\Big)(\mu)y_i(\mu),v_{i_0}(\mu)\Big\rangle = \alpha_{i_0},$$

which proves i) since $v_{i_0}(\mu) \in N(T(\mu))$. □

Using eigenvectors the above proposition reads

PROPOSITION 1.7.3. *Let $T \in H(\Omega,\Phi(E,F))$, $\rho(T) \neq \emptyset$ and $\mu \in \sigma(T)$. The following properties are equivalent:*
i) *μ is a semi-simple eigenvalue of T;*
ii) *there are bases $\{y_1,\dots,y_r\}$ of $N(T(\mu))$ and $\{v_1,\dots,v_r\}$ of $N(T^*(\mu))$ such that*

$$\Big\langle\Big(T\Big)(\mu)y_i,v_j\Big\rangle = \delta_{ij} \quad (i,j = 1,\dots,r);$$

iii) *each eigenvector of T at μ has rank* 1;
iv) *the pole order of T^{-1} at μ is* 1.

The following result shows that we can easily create operators with non-semi-simple eigenvalues.

LEMMA 1.7.4. *Let $T \in H(\Omega,\Phi(E,F))$, $\rho(T) \neq \emptyset$, $\mu \in \sigma(T)$, and consider the operator $S = \begin{pmatrix} T & 0 \\ \frac{d}{d\lambda}T & T\end{pmatrix} \in H(\Omega,\Phi(E\times E,F\times F))$. Then $\rho(S) = \rho(T)$. If μ is a semi-simple eigenvalue of T, then $\mu \in \sigma(S)$ is a non-semi-simple eigenvalue of S. In particular, if $\sigma(T)$ is infinite, then T or S has infinitely many eigenvalues which are not semi-simple.*

Proof. For $\lambda \in \rho(T)$ we have

$$S(\lambda)^{-1} = \begin{pmatrix} T(\lambda)^{-1} & 0 \\ -T(\lambda)^{-1}\Big(\frac{\mathrm{d}}{\mathrm{d}\lambda}T\Big)(\lambda)T(\lambda)^{-1} & T(\lambda)^{-1}\end{pmatrix},$$

which shows that $S(\lambda)^{-1}$ has a pole at each eigenvalue of T, i. e., $\sigma(S) = \sigma(T)$. Since μ is a simple pole of T, there are biorthogonal systems of eigenvectors of T and T^* at μ such that

$$T(\lambda)^{-1} - \frac{1}{\lambda - \mu} \sum_{j=1}^{r} y^{(j)} \otimes v^{(j)}$$

is holomorphic at μ, see (1.6.1). Therefore the proof of the lemma is complete if we show that the coefficient of $(\lambda - \mu)^{-2}$ in $T(\lambda)^{-1}T'(\lambda)T(\lambda)^{-1}$ is nonzero. But this coefficient is

$$\begin{aligned}
&\Big(\sum_{j=1}^{r} y^{(j)} \otimes v^{(j)}\Big)\Big(\frac{\mathrm{d}}{\mathrm{d}\lambda}T\Big)(\mu)\Big(\sum_{k=1}^{r} y^{(k)} \otimes v^{(k)}\Big) \\
&= \sum_{j=1}^{r}\sum_{k=1}^{r} (y^{(j)} \otimes v^{(j)})\Big(\Big(\frac{\mathrm{d}}{\mathrm{d}\lambda}T\Big)(\mu)y^{(k)} \otimes v^{(k)}\Big) \\
&= \sum_{j=1}^{r}\sum_{k=1}^{r} \Big\langle\Big(\frac{\mathrm{d}}{\mathrm{d}\lambda}T\Big)(\mu)y^{(k)}, v^{(j)}\Big\rangle y^{(j)} \otimes v^{(k)} = \sum_{j=1}^{r} y^{(j)} \otimes v^{(j)},
\end{aligned}$$

where we have used the biorthogonal relationships (1.6.2). □

1.8. Local factorizations

Let Ω be a domain in $\mathbb{C}$ and E and F be Banach spaces.

PROPOSITION 1.8.1. *Let $T \in H(\Omega, \Phi(E,F))$, $\rho(T) \neq \emptyset$ and $\mu \in \sigma(T)$. Let $\{y_1,\dots,y_r\}$ and $\{v_1,\dots,v_r\}$ be biorthogonal CSRFs of T and T^* at μ. Let D, η_j and ζ_j $(j = 1,\dots,r)$ be as defined in Corollary 1.5.6. Then:*

i) *For all λ in some neighbourhood of μ we have*

$$T(\lambda)D(\lambda) = \mathrm{id}_F - \sum_{j=1}^{r} \eta_j(\lambda) \otimes v_j(\lambda),$$

$$D(\lambda)T(\lambda) = \mathrm{id}_E - \sum_{j=1}^{r} y_j(\lambda) \otimes \zeta_j(\lambda).$$

ii)
$$T(\mu)D(\mu)T(\mu) = T(\mu).$$

iii) *For all $i, j = 1,\dots,r$ and each w in some Banach space G we have*

$$(w \otimes y_j(\mu))\zeta_i(\mu) = \delta_{ij}w = (w \otimes \zeta_i(\mu))y_j(\mu)$$

and

$$(w \otimes v_j(\mu))\eta_i(\mu) = \delta_{ij}w = (w \otimes \eta_i(\mu))v_j(\mu).$$

iv) *For all $i = 1,\dots,r$ we have*

$$(TD\eta_i)(\mu) = 0,\ (T^*D^*\zeta_i)(\mu) = 0.$$

Proof. i) immediately follows from the definitions of D, η_j and ζ_j.
ii) follows from i) since $T(\mu)y_j(\mu) = 0$ for $j = 1,\ldots,r$.
iii) is clear because of the biorthogonal relationships (1.5.14) and (1.5.13).
iv) immediately follows from i) and iii) if we also take the adjoints in the second equation of i). □

REMARK 1.8.2. The operator function D given by (1.5.15) differs from the reduced resolvent S_μ defined in Proposition 1.3.2 by a degenerate operator function. The operator $D(\mu)$ depends on the choice of the CSRFs. But for our purposes it is more appropriate than $S_\mu(\mu)$ since $T(\mu)D(\mu)T(\mu) = T(\mu)$ always holds. If μ is a semi-simple eigenvalue, then we can choose CSRFs of T and T^* at μ which consist of constant vector functions. In this case, D coincides with S_μ, and (1.3.6) is a special case of Proposition 1.8.1 ii).

PROPOSITION 1.8.3. *Let $T \in H(\Omega,\Phi(E,F))$, $\rho(T) \neq \emptyset$ and $\mu \in \sigma(T)$. Let $\{y_1,\ldots,y_r\}$ and $\{v_1,\ldots,v_r\}$ be biorthogonal CSRF of T and T^* at μ. Let D, η_j and ζ_j $(j = 1,\ldots,r)$ be as defined in Corollary 1.5.6. Then there are a neighbourhood $U \subset \Omega$ of μ and holomorphic operator functions $C_1 \in H(U,L(F))$, $C_2 \in H(U,L(E,F))$ and $D_1 \in H(U,L(E))$ such that $C_1(\lambda)$, $C_2(\lambda)$ and $D_1(\lambda)$ are invertible for all $\lambda \in U$ and such that*

$$T = C_1\big(T(\mu) + \sum_{j=1}^{r}(\cdot - \mu)^{m_j}\eta_j(\mu)\otimes\zeta_j(\mu)\big)D_1, \tag{1.8.1}$$

$$T = C_2\big(D(\mu)T(\mu) + \sum_{j=1}^{r}(\cdot - \mu)^{m_j}y_j(\mu)\otimes\zeta_j(\mu)\big)D_1 \tag{1.8.2}$$

hold in U.

Proof. Let $U' \subset \Omega$ be a neighbourhood of μ in which the operator function D defined by (1.5.15) is holomorphic. For $\lambda \in U'$ we define

$$\widetilde{C}_1(\lambda) := T(\lambda)D(\lambda)T(\mu)D(\mu) + \sum_{j=1}^{r}\eta_j(\lambda)\otimes v_j(\mu),$$

$$\widetilde{C}_2(\lambda) := T(\lambda)D(\lambda)T(\mu) + \sum_{j=1}^{r}\eta_j(\lambda)\otimes\zeta_j(\mu),$$

$$\widetilde{D}_1(\lambda) := D(\lambda)T(\mu) + \sum_{j=1}^{r}y_j(\lambda)\otimes\zeta_j(\mu).$$

Obviously, $\widetilde{C}_1 \in H(U',L(F))$, $\widetilde{C}_2 \in H(U',L(E,F))$ and $\widetilde{D}_1 \in H(U',L(E))$. Proposition 1.8.1 and $T^*(\mu)v_j(\mu) = 0$ yield

$$\widetilde{C}_1(\lambda)\Big(T(\mu) + \sum_{j=1}^{r}(\lambda-\mu)^{m_j}\eta_j(\mu)\otimes\zeta_j(\mu)\Big) = T(\lambda)D(\lambda)T(\mu) + \sum_{j=1}^{r}(\lambda-\mu)^{m_j}\eta_j(\lambda)\otimes\zeta_j(\mu) \tag{1.8.3}$$

for $\lambda \in U'$. Analogously, Proposition 1.8.1 and $T(\mu)y_j(\mu) = 0$ yield

$$\widetilde{C}_2(\lambda)\Big(D(\mu)T(\mu) + \sum_{j=1}^{r}(\lambda-\mu)^{m_j}y_j(\mu)\otimes\zeta_j(\mu)\Big) = T(\lambda)D(\lambda)T(\mu) + \sum_{j=1}^{r}(\lambda-\mu)^{m_j}\eta_j(\lambda)\otimes\zeta_j(\mu) \tag{1.8.4}$$

for $\lambda \in U'$. On the other hand,

$$T(\lambda)\widetilde{D}_1(\lambda) = T(\lambda)D(\lambda)T(\mu) + \sum_{j=1}^{r}(\lambda-\mu)^{m_j}\eta_j(\lambda)\otimes\zeta_j(\mu) \tag{1.8.5}$$

for $\lambda \in U'$. We shall prove that

$$\widetilde{C}_1(\mu),\ \widetilde{C}_2(\mu) \text{ and } \widetilde{D}_1(\mu) \text{ are invertible.} \tag{1.8.6}$$

Then, by Proposition 1.2.5, there is a neighbourhood $U \subset U'$ of μ such that $\widetilde{C}_1(\lambda)$, $\widetilde{C}_2(\lambda)$ and $\widetilde{D}_1(\lambda)$ are invertible for all $\lambda \in U$, and we define

$$C_1(\lambda) := \widetilde{C}_1(\lambda),\, C_2(\lambda) := \widetilde{C}_2(\lambda),\, D_1(\lambda) := \widetilde{D}_1(\lambda)^{-1}.$$

This proves the theorem because of (1.8.3), (1.8.4) and (1.8.5).

Now we are going to prove (1.8.6). Proposition 1.8.1 immediately yields

$$\widetilde{C}_1(\mu) = T(\mu)D(\mu) + \sum_{j=1}^{r}\eta_j(\mu)\otimes v_j(\mu) = \mathrm{id}_F\,,$$

$$\widetilde{D}_1(\mu) = D(\mu)T(\mu) + \sum_{j=1}^{r}y_j(\mu)\otimes\zeta_j(\mu) = \mathrm{id}_E\,.$$

Since

$$\widetilde{C}_2(\mu) = T(\mu) + \sum_{j=1}^{r}\eta_j(\mu)\otimes\zeta_j(\mu),$$

we infer with the aid of Proposition 1.8.1 and $T(\mu)y_j(\mu) = 0$ that

$$\widetilde{C}_2(\mu)\Big(D(\mu)T(\mu)D(\mu) + \sum_{j=1}^{r} y_j(\mu) \otimes v_j(\mu)\Big)$$
$$= T(\mu)D(\mu) + \sum_{j=1}^{r} \eta_j(\mu) \otimes v_j(\mu) = \mathrm{id}_F\,.$$

In the same way, Proposition 1.8.1 and $T^*(\mu)v_j(\mu) = 0$ yield

$$\Big(D(\mu)T(\mu)D(\mu) + \sum_{j=1}^{r} y_j(\mu) \otimes v_j(\mu)\Big)\widetilde{C}_2(\mu)$$
$$= D(\mu)T(\mu) + \sum_{j=1}^{r} y_j(\mu) \otimes \zeta_j(\mu) = id_E.$$

Hence $\widetilde{C}_2(\mu)$ is invertible. □

THEOREM 1.8.4. *Let $T \in H(\Omega, \Phi(E,F))$ such that $\rho(T) \neq \emptyset$. Let $\mu \in \sigma(T)$, $r :=$ nul $T(\mu)$, and let m_j $(j = 1,\ldots,r)$ be the partial multiplicities of T at μ. Then there are biorthogonal projections $P_i \in L(E)$ $(i = 0,\ldots,r)$, i.e.,*

$$P_iP_j = \delta_{ij}P_i \quad (i,j = 0,\ldots,r), \tag{1.8.7}$$

with $\dim R(P_i) = 1$ *for* $i = 1,\ldots,r$ *and*

$$\sum_{i=0}^{r} P_i = \mathrm{id}_E\,, \tag{1.8.8}$$

a neighbourhood $U \subset \Omega$ of μ, and operator functions $C \in H(U,L(E,F))$ and $D_1 \in H(U,L(E))$ such that $C(\lambda)$ and $D_1(\lambda)$ are invertible for all $\lambda \in U$ and

$$T = C\Big(P_0 + \sum_{j=1}^{r} (\cdot - \mu)^{m_j} P_j\Big) D_1 \tag{1.8.9}$$

holds in U. The right-hand side of formula (1.8.9) *is called a factorization of T at μ* (see [GS]).

Proof. With the notations of Proposition 1.8.3 we set $C := C_2$,

$$P_0 := D(\mu)T(\mu), \quad P_j := y_j(\mu) \otimes \zeta_j(\mu) \quad (j = 1,\ldots,r).$$

Then (1.8.9) is just the representation (1.8.2), and (1.8.8) follows from Proposition 1.8.1 i). Now we shall prove (1.8.7). $P_0^2 = P_0$ follows from Proposition 1.8.1 ii). For $j = 1,\ldots,r$ we have $P_0P_j = 0$ since y_j is a root function of T at μ, and $P_jP_0 = 0$ because of Proposition 1.8.1 iv). Finally, $P_iP_j = \delta_{ij}P_i$ is valid for $i,j = 1,\ldots,r$ because of Proposition 1.8.1 iii). □

Now let E and F be finite-dimensional spaces with $\dim E = \dim F$ and let $T \in H(\Omega, L(E,F))$. For fixed bases of E and F we consider the determinant

$$(\det T)(\lambda) = \det T(\lambda) \quad (\lambda \in \Omega).$$

Obviously, $\det T$ is a holomorphic function on Ω. With respect to different bases, the determinants only differ by a constant nonzero factor. Thus the zeros of $\det T$ and their multiplicities do not depend on the choice of the bases and we may speak of "the zeros" and "the multiplicities of the zeros" of "the determinant".

Since operators in finite-dimensional spaces are Fredholm operators, $\rho(T) \neq \emptyset$ implies that every $\mu \in \sigma(T)$ is a pole of T^{-1} and an eigenvalue of finite algebraic multiplicity.

PROPOSITION 1.8.5. *Let E and F be finite-dimensional spaces such that* $\dim E = \dim F$. *Let $T \in H(\Omega, L(E,F))$ and assume that $\rho(T) \neq \emptyset$. For $\mu \in \sigma(T)$ the algebraic multiplicity of T at μ is equal to the multiplicity of the zero of* $\det T$ *at μ.*

Proof. We shall apply Theorem 1.8.4. Let the P_j be as defined there. According to the representation (1.8.9), the multiplicity of the zero of the determinant of T at μ is equal to the multiplicity of the zero of the determinant of

$$S := P_0 + \sum_{j=1}^{r} (\cdot - \mu)^{m_j} P_j$$

at μ. From (1.8.7) and (1.8.8) we infer that there is a basis of E such that

$$S = \operatorname{diag}\big(1, \ldots, 1, (\cdot - \mu)^{m_1}, \ldots, (\cdot - \mu)^{m_r}\big)$$

with respect to this basis. Indeed, let $\{x_1, \ldots, x_s\}$ be a basis of $R(P_0)$ and choose $x_{s+j} \in R(P_j) \setminus \{0\}$ $(j = 1, \ldots, r)$. Then $\{x_1, \ldots, x_s, x_{s+1}, \ldots, x_{s+r}\}$ is a basis of E because of (1.8.7) and (1.8.8). Finally,

$$Sx_j = SP_0x_j = x_j \quad (j = 1, \ldots, s)$$

and

$$Sx_{s+j} = SP_jx_{s+j} = (\cdot - \mu)^{m_j} x_{s+j} \quad (j = 1, \ldots, r).$$

This completes the proof since $\det S = (\cdot - \mu)^m$ with respect to this basis, where $m = \sum_{j=1}^{r} m_j$ is the algebraic multiplicity of T at μ. □

1.9. The completion of biorthogonal systems of root functions

Let Ω be a domain in $\mathbb{C}$ and E and F be Banach spaces.

In Theorem 1.5.9 we have seen that two systems of root functions $\{y_1, \ldots, y_r\}$ and $\{v_1, \ldots, v_r\}$ of T and T^* at μ are canonical systems of root functions if $r = \operatorname{nul} T(\mu)$ and if they are biorthogonal.

Now assume that there are biorthogonal systems of root functions $\{y_1, \dots, y_{r'}\}$ and $\{v_1, \dots, v_{r'}\}$ of T and T^* at μ with $r' < r$. We ask if they can be completed to biorthogonal canonical systems of root functions, i. e., if there are biorthogonal systems of root functions $\{\tilde{y}_1, \dots, \tilde{y}_r\}$ and $\{\tilde{v}_1, \dots, \tilde{v}_r\}$ of T and T^* at μ such that $\{y_1, \dots, y_r\} \subset \{\tilde{y}_1, \dots, \tilde{y}_{r'}\}$ and $\{v_1, \dots, v_{r'}\} \subset \{\tilde{v}_1, \dots, \tilde{v}_r\}$. The next theorem shows that this is always possible.

THEOREM 1.9.1. *Let* $T \in H(\Omega, \Phi(E,F))$, $\rho(T) \neq \emptyset$ *and* $\mu \in \sigma(T)$. *Let* $0 < r' < r = \operatorname{nul} T(\mu)$, $y_1, \dots, y_{r'}$ *be root functions of* T *at* μ, $v_1, \dots, v_{r'}$ *be root functions of* T^* *at* μ, $k_j > 0$, $\nu(y_j) \geq k_j$, $\nu(v_j) \geq k_j$ $(j = 1, \dots, r')$. *Assume that*

$$\frac{\mathrm{d}^l}{\mathrm{d}\lambda^l} \langle \eta_i, v_j \rangle (\mu) = \delta_{ij} \delta_{0l} \quad (i, j = 1, \dots, r'; 0 \leq l \leq k_j - 1),$$

where

$$\eta_i = (\cdot - \mu)^{-k_i} T y_i \quad (i = 1, \dots, r').$$

Then there are biorthogonal canonical systems of root functions $\{\tilde{y}_1, \dots, \tilde{y}_r\}$ *and* $\{\tilde{v}_1, \dots, \tilde{v}_r\}$ *of* T *and* T^* *at* μ *such that* $\{y_1, \dots, y_r\} \subset \{\tilde{y}_1, \dots, \tilde{y}_{r'}\}$ *and* $\{v_1, \dots, v_{r'}\} \subset \{\tilde{v}_1, \dots, \tilde{v}_r\}$.

Proof. First we shall show that $\{y_1, \dots, y_{r'}\}$ can be completed to a canonical system of root functions of T at μ. Let

$$\zeta_j = (\cdot - \mu)^{-k_j} T^* v_j \quad (j = 1, \dots, r').$$

From the definition of η_i and ζ_j we immediately infer (see (1.5.18)) that the biorthogonal relationships can also be written in the form

$$\frac{\mathrm{d}^l}{\mathrm{d}\lambda^l} \langle y_i, \zeta_j \rangle (\mu) = \delta_{ij} \delta_{0l} \quad (i, j = 1, \dots, r'; 0 \leq l \leq k_i - 1).$$

We want to apply Proposition 1.4.8. The linear independence of the vectors $y_1(\mu), \dots, y_{r'}(\mu)$ is an immediate consequence of the above biorthogonal relationships. Let y be a root function of T at μ. We have to show that $\nu(y) \leq \max\{\nu(y_j) : j \in \Lambda\}$ if

$$y(\mu) = \sum_{i \in \Lambda} \alpha_i y_i(\mu),$$

where $\Lambda \subset \{1, \dots, r'\}$ and the α_i are complex numbers. Suppose that $\nu(y) > \max\{\nu(x_j) : j \in \Lambda\}$. Then $(\cdot - \mu)^{-k_j} T y$ has a zero at μ for each $j \in \Lambda$, which implies that

$$0 = \langle (\cdot - \mu)^{-k_j} T y, v_j \rangle (\mu) = \langle y(\mu), \zeta_j(\mu) \rangle = \alpha_j.$$

This is impossible since $y(\mu) \neq 0$. By Proposition 1.4.8 we know that there is a canonical system of root functions $\{\hat{y}_1, \dots, \hat{y}_r\}$ of T at μ such that $\{y_1, \dots, y_{r'}\} \subset \{\hat{y}_1, \dots, \hat{y}_r\}$. Now let $\{\hat{v}_1, \dots, \hat{v}_r\}$ be a canonical system of root functions of T^* at μ which is biorthogonal to $\{\hat{y}_1, \dots, \hat{y}_r\}$. Note that the number k_j is the multiplicity of y_j since $\langle \eta_j, v_j \rangle (\mu) = 1$ implies $\nu(y_j) \leq k_j$.

From Theorem 1.8.4 we know that there are a neighbourhood $U \subset \Omega$ of μ and operator functions $C \in H(U, L(E,F))$ and $D_1 \in H(U, L(E))$ such that $C(\lambda)$ and $D_1(\lambda)$ are invertible for all $\lambda \in U$ and such that

$$C^{-1}TD_1^{-1} = P_0 + \sum_{j=1}^{r} (\cdot - \mu)^{m_j} P_j =: \widehat{T},$$

where $P_j = \hat{y}_j(\mu) \otimes \hat{\zeta}_j(\mu)$ for $j = 1, \ldots, r$ and

$$D_1(\lambda)^{-1} = \widetilde{D}_1(\lambda) = D(\lambda)T(\mu) + \sum_{j=1}^{r} \hat{y}_j(\lambda) \otimes \hat{\zeta}_j(\mu) \quad (\lambda \in U),$$

see the proofs of Theorem 1.8.4 and Proposition 1.8.3 for the definition of P_j and $D_1(\lambda)$. For any two vector functions $y \in H(U,E)$ and $v \in H(U,F')$ we have

$$\langle Ty, v\rangle = \langle \widehat{T} D_1 y, C^* v\rangle.$$

Thus the systems of root functions $\{y_1', \ldots, y_s'\}$ of T at μ and $\{v_1', \ldots, v_s'\}$ of T^* at μ are biorthogonal if and only if the systems $\{D_1 y_1', \ldots, D_1 y_s'\}$ and $\{C^* v_1', \ldots, C^* v_s'\}$ are biorthogonal systems of root functions of $\widehat{T}$ and $\widehat{T}^*$ at μ. The definition of $\widetilde{D}_1(\lambda)$, $\hat{y}_i(\mu) \in N(T(\mu))$ and the biorthogonal relationships $\langle \hat{y}_i(\mu), \hat{\zeta}_j(\mu)\rangle = \delta_{ij}$ yield

$$\begin{aligned}\widetilde{D}_1(\lambda)\hat{y}_i(\mu) &= D(\lambda)T(\mu)\hat{y}_i(\mu) + \sum_{j=1}^{r} \langle \hat{y}_i(\mu), \hat{\zeta}_j(\mu)\rangle \hat{y}_j(\lambda) \\ &= \hat{y}_i(\lambda).\end{aligned}$$

Hence $\hat{y}_i(\mu) = D_1 \hat{y}_i$ for $i = 1, \ldots, r$.

Now we can reformulate the statement of the theorem as follows:

PROPOSITION 1.9.2. *For $\lambda \in \Omega$ let*

$$T(\lambda) = P_0 + \sum_{j=1}^{r} (\lambda - \mu)^{m_j} P_j,$$

where $m_1 \geq m_2 \geq \cdots \geq m_r > 0$, $P_0, P_1, \ldots, P_r$ are biorthogonal projections on E such that $P_i = x_i \otimes w_i \neq 0$ for $i = 1, \ldots, r$, where $x_i \in E$ and $w_i \in E'$, and

$$\mathrm{id}_E = \sum_{i=0}^{r} P_i.$$

Let $\{y_1, \ldots, y_{r'}\} \subset \{x_1, \ldots, x_r\}$ and let $v_1, \ldots, v_{r'}$ be root functions of T^* at μ with $\nu(v_j) \geq k_j$ $(j = 1, \ldots, r')$ such that

(1.9.1) $$\frac{\mathrm{d}^l}{\mathrm{d}\lambda^l} \langle y_i, v_j\rangle(\mu) = \delta_{ij}\delta_{0l} \quad (i, j = 1, \ldots, r'; 0 \leq l \leq k_j - 1),$$

where $k_j = m_{i_j}$ if $y_j = x_{i_j}$. Then there are biorthogonal canonical systems of root functions $\{\tilde{y}_1, \ldots, \tilde{y}_r\}$ and $\{\tilde{v}_1, \ldots, \tilde{v}_r\}$ of T and T^* at μ such that $\{y_1, \ldots, y_{r'}\} \subset \{\tilde{y}_1, \ldots, \tilde{y}_r\}$ and $\{v_1, \ldots, v_{r'}\} \subset \{\tilde{v}_1, \ldots, \tilde{v}_r\}$.

Proof. In the biorthogonal relationships (1.9.1) we wrote y_i instead of η_i. This is correct since $P_i x_i = x_i$ and hence

$$T(\lambda)x_i = T(\lambda)P_i x_i = (\lambda - \mu)^{m_i} x_i \quad (\lambda \in \Omega)$$

holds for $i = 1, \dots, r$ as $P_j P_i = \delta_{ij} P_i$ for $j = 0, \dots, r$. Also

$$T^*(\lambda)w_i = (\lambda - \mu)^{m_i} w_i \quad (\lambda \in \Omega)$$

holds for $i = 1, \dots, r$. Since

$$T^{-1} = P_0 + \sum_{j=1}^{r} (\cdot - \mu)^{-m_j} P_j,$$

$\{x_1, \dots, x_r\}$ and $\{w_1, \dots, w_r\}$ are biorthogonal CSRFs of T and T^* at μ by Theorem 1.5.9.

By assumption, there is a set of indices $\{i_1, \dots, i_{r'}\} \subset \{1, \dots, r\}$ such that $y_j = x_{i_j}$ for $j = 1, \dots, r'$. Now we choose $i_{r'+1}, \dots, i_r$ such that $\{i_{r'+1}, \dots, i_r\} = \{1, \dots, r\} \setminus \{i_1, \dots, i_{r'}\}$. We set

$$\tilde{v}_{i_j} := \sum_{k=0}^{k_j - 1} \frac{1}{k!} \frac{\mathrm{d}^k}{\mathrm{d}\lambda^k} v_j(\mu)(\cdot - \mu)^k \quad \text{for } j = 1, \dots, r'$$

and

$$\tilde{v}_{i_j} := w_{i_j} \quad \text{for } j = r' + 1, \dots, r.$$

From $k_j = m_{i_j}$ for $j = 1, \dots, r'$ and

$$\nu(\tilde{v}_{i_j}) = \nu(w_{i_j}) = m_{i_j} \quad \text{for } j = r' + 1, \dots, r$$

we infer $\nu(\tilde{v}_j) \geq m_j$ for $j = 1, \dots, r$. Hence $\{\tilde{v}_1, \dots, \tilde{v}_r\}$ is a canonical system of root functions of T^* at μ if the vectors $\tilde{v}_1(\mu), \dots, \tilde{v}_r(\mu)$ are linearly independent.

We state that for $j = 1, \dots, r'$ the identity

$$\tilde{v}_{i_j} = P_0^* \tilde{v}_{i_j} + w_{i_j} + \sum_{k=r'+1}^{r} \langle x_{i_k}, \tilde{v}_{i_j} \rangle w_{i_k} \tag{1.9.2}$$

holds. Indeed, we infer that

$$\begin{aligned} \tilde{v}_{i_j} &= P_0^* \tilde{v}_{i_j} + \sum_{i=1}^{r} \langle x_i, \tilde{v}_{i_j} \rangle w_i \\ &= P_0^* \tilde{v}_{i_j} + \sum_{k=1}^{r'} \langle y_k, \tilde{v}_{i_j} \rangle w_{i_k} + \sum_{k=r'+1}^{r} \langle x_{i_k}, \tilde{v}_{i_j} \rangle w_{i_k}, \end{aligned}$$

and (1.9.2) follows from the biorthogonal relationships (1.9.1), which yield that $\langle y_k, \tilde{v}_{i_j} \rangle = \delta_{kj}$ since $\langle y_k, \tilde{v}_{i_j} \rangle$ is a polynomial of degree less than k_j. Observing that $\tilde{v}_{i_j}(\mu) \in N(T^*(\mu))$ we obtain $P_0^* \tilde{v}_{i_j}(\mu) = 0$, which shows that $(\tilde{v}_{i_j}(\mu))_{j=1}^{r}$ is obtained from $(w_{i_j}(\mu))_{j=1}^{r}$ by a linear transformation whose coefficient matrix is a normed triangular matrix, and the linear independence of $\tilde{v}_1(\mu),\dots,v_r(\mu)$ follows from the linear independence of $w_1,\dots,w_r$.

According to Remark 1.5.8 we choose a canonical system of root functions $\{\tilde{y}_1,\dots,\tilde{y}_r\}$ of T at μ which is biorthogonal to $\{\tilde{v}_1,\dots,\tilde{v}_r\}$. In addition, the $\tilde{y}_j$ can be chosen to be polynomials of degree less than m_j for all $j=1,\dots,r$.

The proof of the proposition will be complete if we show that $y_j=\tilde{y}_{i_j}$ for $j=1,\dots,r'$. Let $j\in\{1,\dots,r'\}$. Since

$$P_0\tilde{y}_{i_j}=P_0T\tilde{y}_{i_j}$$

has a zero of order $\geq m_{i_j}$ at μ and is a polynomial of degree $<m_{i_j}$, $P_0\tilde{y}_{i_j}$ is zero. For $k=r'+1,\dots,r$ we have

$$\begin{aligned}P_{i_k}\tilde{y}_{i_j}&=\left(x_{i_k}\otimes\tilde{v}_{i_k}\right)\tilde{y}_{i_j}=\langle\tilde{y}_{i_j},\tilde{v}_{i_k}\rangle x_{i_k}\\&=\langle\tilde{y}_{i_j},(\cdot-\mu)^{-m_{i_k}}T^*\tilde{v}_{i_k}\rangle x_{i_k}.\end{aligned}$$

The biorthogonal relationships for $\{\tilde{y}_1,\dots,\tilde{y}_r\}$ and $\{\tilde{v}_1,\dots,\tilde{v}_r\}$ prove that $P_{i_k}\tilde{y}_{i_j}$ has a zero of order $\geq m_{i_j}$. Hence $P_{i_k}\tilde{y}_{i_j}=0$ for $k=r'+1,\dots,r$. Note that this implies $\langle\tilde{y}_{i_j},w_{i_k}\rangle=0$ for $k=r'+1,\dots,r$. Thus we have

$$\tilde{y}_{i_j}=\sum_{k=1}^{r'}P_{i_k}\tilde{y}_{i_j}=\sum_{k=1}^{r'}\langle\tilde{y}_{i_j},w_{i_k}\rangle x_{i_k}. \tag{1.9.3}$$

Now let $l\in\{1,\dots,r'\}$. From (1.9.2), $P_0\tilde{y}_{i_j}=0$, and $\langle\tilde{y}_{i_j},w_{i_k}\rangle=0$ for the numbers $k=r'+1,\dots,r$ it follows that

$$\begin{aligned}\langle\tilde{y}_{i_j},w_{i_l}\rangle&=(\cdot-\mu)^{-m_{i_l}}\langle\tilde{y}_{i_j},T^*w_{i_l}\rangle\\&=(\cdot-\mu)^{-m_{i_l}}\langle\tilde{y}_{i_j},T^*\tilde{v}_{i_l}\rangle-\sum_{k=r'+1}^{r}(\cdot-\mu)^{-m_{i_l}}\langle x_{i_k},\tilde{v}_{i_j}\rangle\langle\tilde{y}_{i_j},T^*w_{i_k}\rangle\\&=\langle\tilde{y}_{i_j},(\cdot-\mu)^{-m_{i_l}}T^*\tilde{v}_{i_l}\rangle.\end{aligned}$$

Since $\langle\tilde{y}_{i_j},w_{i_l}\rangle$ is a polynomial of order $<m_{i_j}$ we infer $\langle\tilde{y}_{i_j},w_{i_l}\rangle=\delta_{jl}$, and (1.9.3) gives $\tilde{y}_{i_j}=x_{ij}=y_j$. □

REMARK 1.9.3. If $y_{j_1}=\tilde{y}_{k_1}$ and $v_{j_2}=\tilde{v}_{k_2}$ in Theorem 1.9.1, then the biorthogonal relationships show that $j_1=j_2$ if and only if $k_1=k_2$.

1.10. The operator function $A+\lambda B$

Let E and F be Banach spaces. In this section we assume that

$$T(\lambda)=A+\lambda B,$$

where $\lambda\in\mathbb{C}$, A and B are in $L(E,F)$ and B is a compact operator. If $\rho(T)\neq\emptyset$, then $T\in H(\mathbb{C},\Phi(E,F))$, see e. g. [KA, Theorem IV.5.26].

Proposition 1.10.1. *Let $T(\lambda) = A + \lambda B$, where $\lambda \in \mathbb{C}$, A and B are in $L(E,F)$, and B is compact. Assume that $\rho(T) \neq \emptyset$ and let $\mu \in \sigma(T)$. Let the CSEAVs $\{y_l^{(j)} : 1 \le j \le r, 0 \le l \le m_j - 1\}$ of T at μ and $\{v_l^{(j)} : 1 \le j \le r, 0 \le l \le m_j - 1\}$ of T^* at μ be given. Then these CSEAVs are biorthogonal if and only if*

$$\langle By_{m_i-1}^{(i)}, v_m^{(j)} \rangle = \delta_{ij}\,\delta_{0m} \quad (1 \le i \le r, 1 \le j \le r, 0 \le m \le m_j - 1). \tag{1.10.1}$$

Proof. This immediately follows from (1.6.5) since on the left-hand side of (1.6.5) only the term for $k = 0$ and $q = 1$ is different from zero. □

Proposition 1.10.2. *Let $T(\lambda) = A + \lambda B$, where $\lambda \in \mathbb{C}$, A and B are in $L(E,F)$, and B is compact. Assume that $\rho(T) \neq \emptyset$ and let $\mu \in \sigma(T)$. Let $y_0, y_1, \ldots, y_k$ be a CEAV of T at μ. Then*

$$(A + \mu B)y_0 = 0, \tag{1.10.2}$$

$$(A + \mu B)y_{l+1} = -By_l \quad (l = 0, \ldots, k-1). \tag{1.10.3}$$

Proof. The definition of a CEAV yields that the function given by

$$\sum_{l=0}^{k} (\lambda - \mu)^l T(\lambda) y_l = \sum_{l=0}^{k} (\lambda - \mu)^l (A + \mu B) y_l + \sum_{l=0}^{k} (\lambda - \mu)^{l+1} B y_l$$

has a zero of order $\ge k+1$ at μ. This proves (1.10.2) and (1.10.3). □

Proposition 1.10.3. *Let $T(\lambda) = A + \lambda B$, where $\lambda \in \mathbb{C}$, A and B are in $L(E,F)$, and B is compact. Assume that $\rho(T) \neq \emptyset$ and let $\mu \in \sigma(T)$. Let the CSEAVs $\{y_l^{(j)} : 1 \le j \le r, 0 \le l \le m_j - 1\}$ of T at μ and $\{v_l^{(j)} : 1 \le j \le r, 0 \le l \le m_j - 1\}$ of T^* at μ be given. Then these CSEAVs are biorthogonal if and only if*

$$\langle By_l^{(i)}, v_{m_j-1-k}^{(j)} \rangle = \delta_{ij}\,\delta_{lk} \quad (1 \le i \le r, 0 \le l \le m_i - 1, 1 \le j \le r, 0 \le k \le m_j - 1). \tag{1.10.4}$$

Proof. We have to show that the relations (1.10.1) and (1.10.4) are equivalent. Obviously, the relations (1.10.4) for $l = m_i - 1$ and $k = m_j - 1 - m$ coincide with (1.10.1). For $l = 0, \ldots, m_i - 2$ and $m \le m_j - 1$ we have in view of (1.10.2) and (1.10.3) and the corresponding relations for a CEAV of T^* at μ that

$$\begin{aligned}\langle By_l^{(i)}, v_m^{(j)} \rangle &= -\langle (A + \mu B) y_{l+1}^{(i)}, v_m^{(j)} \rangle \\ &= -\langle y_{l+1}^{(i)}, (A^* + \mu B^*) v_m^{(j)} \rangle = \langle y_{l+1}^{(i)}, B^* v_{m-1}^{(j)} \rangle \\ &= \langle By_{l+1}^{(i)}, v_{m-1}^{(j)} \rangle,\end{aligned}$$

where $v_m^{(j)} := 0$ if $m < 0$. Now the relations (1.10.4) follow from (1.10.1) by a recursive application of the above identity. □

PROPOSITION 1.10.4. *Let $T(\lambda)=A+\lambda B$, where $\lambda \in \mathbb{C}$, A and B are in $L(E,F)$, and B is compact. Assume that $\rho(T)\neq\emptyset$ and let μ and $\tilde{\mu}$ be eigenvalues of T such that $\mu\neq\tilde{\mu}$. Let $\{y_l^{(j)}: 1\le j\le r, 0\le l\le m_j-1\}$ be a CSEAV of T at μ and $\{\tilde{v}_l^{(j)}: 1\le j\le \tilde{r}, 0\le l\le \tilde{m}_j-1\}$ be a CSEAV of T^* at $\tilde{\mu}$. Then*

$$\langle By_l^{(i)},\tilde{v}_k^{(j)}\rangle=0 \tag{1.10.5}$$
$$(1\le i\le r, 0\le l\le m_i-1, 1\le j\le \tilde{r}, 0\le k\le \tilde{m}_j-1).$$

Proof. In view of (1.10.2) and (1.10.3) and the corresponding formulas for a CEAV of T^* at μ we have for $l\le m_i-1$ and $k\le \tilde{m}_j-1$ that

$$\begin{aligned}(\mu-\tilde{\mu})\langle By_l^{(i)},\tilde{v}_k^{(j)}\rangle &= \langle\big[(A+\mu B)-(A+\tilde{\mu}B)\big]y_l^{(i)},\tilde{v}_k^{(j)}\rangle\\ &= -\langle By_{l-1}^{(i)},\tilde{v}_k^{(j)}\rangle+\langle y_l^{(i)},B^*\tilde{v}_{k-1}^{(j)}\rangle,\end{aligned}$$

where $y_m^{(i)}:=0$ and $\tilde{v}_m^{(j)}:=0$ for $m<0$. Now (1.10.5) follows by induction on $k+l$. □

A system of vectors $(y_\alpha)_{\alpha\in I}$ in a Banach space E is called a *minimal* if the closed linear hull of $(y_\alpha)_{\alpha\in I}$ is different from the closed linear hull of any proper subsystem of $(y_\alpha)_{\alpha\in I}$. In particular, for a series

$$\sum_{\alpha\in I} a_\alpha y_\alpha=0$$

with complex numbers a_α which converges in parenthesis we obtain $a_\alpha=0$ for all $\alpha\in I$ if $(y_\alpha)_{\alpha\in I}$ is minimal. This also shows that the vectors of a minimal system are linearly independent.

PROPOSITION 1.10.5. *Let $T(\lambda)=A+\lambda B$, where $\lambda\in\mathbb{C}$, A and B are in $L(E,F)$ and B is compact. Assume that $\rho(T)\neq\emptyset$. For each $\mu\in\sigma(T)$ choose a CSEAV $\{y_{\mu,l}^{(j)}: 1\le j\le r(\mu), 0\le l\le m_j(\mu)-1\}$ of T at μ. Then the system of vectors $\{y_{\mu,l}^{(j)}:\mu\in\sigma(T), 1\le j\le r(\mu), 0\le l\le m_j(\mu)-1\}$ is a minimal system in E.*

Proposition 1.10.5 is a particular case of the following proposition. It is included here because of its simpler formulation.

PROPOSITION 1.10.6. *Let $T(\lambda)=A+\lambda B$, where $\lambda\in\mathbb{C}$, A and B are in $L(E,F)$ and B is compact. Assume that $\rho(T)\neq\emptyset$. Let H be a Banach space which contains E such that the embedding $E\hookrightarrow H$ is continuous. Assume that B is the restriction of a continuous linear operator from H to F. For each $\mu\in\sigma(T)$ let $\{y_{\mu,l}^{(j)}: 1\le j\le r(\mu), 0\le l\le m_j(\mu)-1\}$ be a CSEAV of T at μ. Then the vectors $\{y_{\mu,l}^{(j)}:\mu\in\sigma(T), 1\le j\le r(\mu), 0\le l\le m_j(\mu)-1\}$ form a minimal system in H.*

Proof. Choose the CSEAV $\{v_{\mu,l}^{(j)}: 1\le j\le r(\mu), 0\le l\le m_j(\mu)-1\}$ of T^* at μ according to Theorem 1.6.5. Let $y_{\mu,l}^{(j)}$ be one of the eigenvectors or associated vectors of the given system. By Propositions 1.10.3 and 1.10.4 we know that $\langle y^{(j)}_{\mu,l}, B^* v^{(j)}_{\mu,m_j(\mu)-1-l}\rangle = 1$ and $\langle y^{(i)}_{\tilde\mu,k}, B^* v^{(j)}_{\mu,m_j(\mu)-1-l}\rangle = 0$ if $(\tilde\mu,k,i) \neq (\mu,l,j)$. We may assume without loss of generality that E is dense in H and therefore that $H' \subset E'$. The assumption on B says that $R(B^*) \subset H'$. Hence the bilinear form can be taken with respect to H and H'. This proves the minimality in H of the system of eigenvectors and associated vectors. □

The following simple example shows that λ-linearity of the operator function is crucial for minimality of a CSEAV.

EXAMPLE 1.10.7. We consider

$$T(\lambda) = \begin{pmatrix} 1 & 0 & 0 \\ 0 & \lambda & 0 \\ 0 & 0 & \lambda^2 \end{pmatrix}.$$

Then $T \in H(\mathbb{C}, L(\mathbb{C}^3))$, $\det T(\lambda) = \lambda^3$ and $\operatorname{nul} T(0) = 2$ are obvious. Set

$$y_{1,0} := \begin{pmatrix} 0 \\ 0 \\ 1 \end{pmatrix}, \quad y_{1,1} := \begin{pmatrix} 0 \\ 0 \\ 0 \end{pmatrix}, \quad y_{2,0} := \begin{pmatrix} 0 \\ 1 \\ 0 \end{pmatrix}.$$

From

$$T(\lambda) \begin{pmatrix} 0 \\ 0 \\ 1 \end{pmatrix} = \lambda^2 \begin{pmatrix} 0 \\ 0 \\ 1 \end{pmatrix}$$

and

$$T(\lambda) \begin{pmatrix} 0 \\ 1 \\ 0 \end{pmatrix} = \lambda \begin{pmatrix} 0 \\ 1 \\ 0 \end{pmatrix}$$

we infer that $\{y_{1,0}, y_{1,1}\}$ and $\{y_{2,0}\}$ are CEAVs of T at 0. Clearly, $\{y_{1,0}, y_{2,0}\}$ is a basis of $N(T(0))$, $m_1 \geq 2$, and $m_2 \geq 1$. By Proposition 1.8.5, $m_1 + m_2 = 3$, which proves that $m_1 = 2$ and $m_2 = 1$. Hence $\{y_{1,0}, y_{1,1}; y_{2,0}\}$ is a CSEAV of T at 0. Since the associated vector $y_{1,1}$ is zero, this CSEAV of T at 0 is not minimal.

PROPOSITION 1.10.8. *Let $T(\lambda) = A + \lambda B$, where $\lambda \in \mathbb{C}$, A and B are in $L(E,F)$ and B is compact. Assume that $\rho(T) \neq \emptyset$ and let $\mu \in \sigma(T)$. Then μ is a semi-simple eigenvalue of T if and only if for each eigenvector y of T at μ there is an eigenvector v of T^* at μ such that*

$$\langle By, v\rangle \neq 0.$$

Proof. The result immediately follows from the definition of a semi-simple eigenvalue. □

If $E = F$ is finite-dimensional and $B = -\operatorname{id}_E$, i.e. $T(\lambda) = A - \lambda \operatorname{id}_E$, then $\sigma(T)$ is also denoted by $\sigma(A)$, and eigenvectors and associated vectors of T are called eigenvectors and associated vectors of A.

THEOREM 1.10.9 (Jordan canonical form). *Let E be a finite-dimensional space, $\dim E = n$. Let $A \in L(E)$. Then $\sigma(A)$ consists of $p \le n$ eigenvalues $\mu_1, \ldots, \mu_p$. There is an invertible linear operator $C \in L(\mathbb{C}^n, E)$ such that*

$$C^{-1}AC = \bigoplus_{i=1}^{p} \bigoplus_{j=1}^{r_i} (\mu_i \delta_{kl} + \delta_{k+1,l})_{k,l=1}^{m_j(\mu_i)}, \tag{1.10.6}$$

where $r_i = \operatorname{nul}(A - \mu_i \operatorname{id}_E)$ $(i = 1, \ldots, p)$ and $m_1(\mu_i), \ldots, m_{r_i}(\mu_i)$ are the partial multiplicities of $A - \lambda \operatorname{id}_E$ at μ_i. This means that $C^{-1}AC$ is given by the block matrix

$$\begin{pmatrix} A_{11} & & & & & & \\ & A_{12} & & & & & \\ & & \ddots & & & 0 & \\ & & & A_{1r_1} & & & \\ & 0 & & & A_{21} & & \\ & & & & & \ddots & \\ & & & & & & A_{pr_p} \end{pmatrix},$$

where

$$A_{ij} = \begin{pmatrix} \mu_i & 1 & & & \\ & \cdot & \cdot & 0 & \\ & & \cdot & \cdot & \\ & & & \cdot & \cdot \\ & 0 & & \cdot & 1 \\ & & & & \mu_i \end{pmatrix}$$

is an $m_j(\mu_i) \times m_j(\mu_i)$ matrix for $i = 1, \ldots, p$ and $j = 1, \ldots, r_i$. A representation of C and C^{-1} is obtained as follows: For $i = 1, \ldots, p$ choose biorthogonal CSEAVs $\{y_{i,l}^{(j)} : 1 \le j \le r_i, 0 \le l \le m_j(\mu_i) - 1\}$ and $\{v_{i,l}^{(j)} : 1 \le j \le r_i, 0 \le l \le m_j(\mu_i) - 1\}$ of A and A^* at μ_i according to Corollary 1.6.6. Define

$$C := \left(y_{1,0}^{(1)} \cdots y_{1,m_1(\mu_1)-1}^{(1)} \cdots y_{1,0}^{(r_1)} \cdots y_{1,m_{r_1}(\mu_1)-1}^{(r_1)} \cdots y_{p,0}^{(r_p)} \cdots y_{p,m_{r_p}(\mu_p)-1}^{(r_p)}\right) \tag{1.10.7}$$

and

$$D := \left(v_{1,m_1(\mu_1)-1}^{(1)} \cdots v_{1,0}^{(1)} \cdots v_{1,m_{r_1}(\mu_1)-1}^{(r_1)} \cdots v_{1,0}^{(r_1)} \cdots v_{p,m_{r_p}(\mu_p)-1}^{(r_p)} \cdots v_{p,0}^{(r_p)}\right).$$

Then $C \in L(\mathbb{C}^n, E)$, $D^* \in L(E, \mathbb{C}^n)$, and (1.10.6) holds, where $C^{-1} = -D^*$.

Proof. The determinant of $A - \lambda \operatorname{id}_E$ is a polynomial of degree n. Since the eigenvalues of $A - \lambda \operatorname{id}_E$ are the zeros of $\det(A - \lambda \operatorname{id}_E)$, $\sigma(A)$ is finite and consists of at most n elements $\mu_1,\dots,\mu_p$. Choose biorthogonal CSEAVs as in the statement of the theorem. By Proposition 1.8.5,

$$m(\mu_i)=\sum_{j=1}^{r_i} m_j(\mu_i)$$

is the multiplicity of the zero of $\det(A-\lambda\,\mathrm{id}_E)$ at μ_i. Since the sum of the multiplicities of all zeros of $\det(A-\lambda\,\mathrm{id}_E)$ is equal to its degree, we obtain

$$\sum_{i=1}^{p}\sum_{j=1}^{r_i} m_j(\mu_i)=n.$$

Hence C given by (1.10.7) defines a linear operator $C:\mathbb{C}^n\to E$. From Proposition 1.10.5 we know that the vectors $y_{i,l}^{(j)}$ are linearly independent. Hence C is invertible. For each $k\in\{1,\dots,n\}$ there are unique numbers $i_k\in\{1,\dots,p\}$, $s_k\in\{1,\dots,r_{i_k}\}$ and $l_k\in\{0,\dots,m_{s_k}(\mu_{i_k})-1\}$ such that

$$k=\sum_{i=1}^{i_k-1} m(\mu_i)+\sum_{j=1}^{s_k-1} m_j(\mu_{i_k})+l_k+1.$$

Let e_k be the k-th unit vector in $\mathbb{C}^n$. Set $e'_{k-1}:=e_{k-1}$ if $l_k\neq 0$ and $e'_{k-1}:=0$ if $l_k=0$. From (1.10.2) and (1.10.3) we obtain

$$\begin{aligned}ACe_k&=Ay_{i_k,l_k}^{(s_k)}=\mu_{i_k}y_{i_k,l_k}^{(s_k)}+y_{i_k,l_k-1}^{(s_k)}\\&=\mu_{i_k}Ce_k+Ce'_{k-1}=C(\mu_{i_k}e_k+e'_{k-1}),\end{aligned}$$

which proves (1.10.6).

Finally $D\in L(\mathbb{C}^n,E')$, and therefore $D^*\in L(E,\mathbb{C}^n)$. Here we have used that we can identify E an E'' since E is finite-dimensional, see e. g. [KA, p. 15]. For $j,k\in\{1,\dots,n\}$ we have in view of Propositions 1.10.3 and 1.10.4 that

$$e_j^tD^*Ce_k=\langle e_j,D^*Ce_k\rangle=\langle De_j,Ce_k\rangle=-\delta_{jk}$$

for $j,k\in\{1,\dots,n\}$. This proves $C^{-1}=-D^*$. □

1.11. Abstract boundary eigenvalue operator functions

Let Ω be an open subset of $\mathbb{C}$. We consider the Banach spaces E, G, F_1, F_2, $F=F_1\times F_2$ and operator functions $T\in H(\Omega,L(E,F))$ and $Z\in H(\Omega,L(G,E))$. According to $F=F_1\times F_2$ we have $T_1\in H(\Omega,L(E,F_1))$ and $T_2\in H(\Omega,L(E,F_2))$ such that $T(\lambda)=\begin{pmatrix}T_1(\lambda)\\T_2(\lambda)\end{pmatrix}$ for $\lambda\in\Omega$. We assume that for all $\lambda\in\Omega$

$$\text{(1.11.1)}\qquad\begin{cases}\text{i) } T_1(\lambda)\text{ is right invertible,}\\ \text{ii) } Z(\lambda)\text{ is injective,}\\ \text{iii) } N(T_1(\lambda))=R(Z(\lambda)).\end{cases}$$

Condition (1.11.1) i) means that there is an operator $U(\lambda) \in L(F_1, E)$ such that $T_1(\lambda)U(\lambda) = \mathrm{id}_{F_1}$. We set

$$M(\lambda) := T_2(\lambda)Z(\lambda) \qquad (\lambda \in \Omega), \tag{1.11.2}$$

whence $M \in H(\Omega, L(G, F_2))$ by Corollary 1.2.4. We call T an *abstract boundary eigenvalue operator function*, Z a *"fundamental matrix" function* and M the *characteristic "matrix" function associated to* T (with respect to Z).

THEOREM 1.11.1. *There are operator functions*

$$C \in H(\Omega, L(F_2 \times F_1, F)), D \in H(\Omega, L(E, G \times F_1))$$

such that for $\lambda \in \Omega$ *the operators* $C(\lambda)$ *and* $D(\lambda)$ *are invertible and the factorization*

$$T(\lambda) = C(\lambda) \begin{pmatrix} M(\lambda) & 0 \\ 0 & id_{F_1} \end{pmatrix} D(\lambda)$$

holds.

Proof. By ŠUBIN [SU], see also BART [BA2, p.183], there is a holomorphic right inverse U of T_1, i. e., there is an operator function $U \in H(\Omega, L(F_1, E))$ such that

$$T_1(\lambda)U(\lambda) = \mathrm{id}_{F_1} \qquad (\lambda \in \Omega). \tag{1.11.3}$$

We shall show that the operator

$$(Z(\lambda), U(\lambda)) : G \times F_1 \to E \tag{1.11.4}$$

is invertible for all $\lambda \in \Omega$. First let $(x, y) \in G \times F_1$ and $Z(\lambda)x + U(\lambda)y = 0$. We apply $T_1(\lambda)$ to this equation and obtain $y = 0$ because of (1.11.1) iii) and (1.11.3). Hence $Z(\lambda)x = 0$, whence $x = 0$ since $Z(\lambda)$ is injective by (1.11.1) ii). Thus the operator (1.11.4) is injective. To prove its surjectivity let $x \in E$. From (1.11.3) we infer that

$$T_1(\lambda)\big(x - U(\lambda)T_1(\lambda)x\big) = 0.$$

The assumption (1.11.1) iii) yields that

$$x - U(\lambda)T_1(\lambda)x \in R(Z(\lambda))$$

whence $x \in R(Z(\lambda)) + R(U(\lambda))$. This proves the surjectivity. We set

$$D(\lambda) := (Z(\lambda), U(\lambda))^{-1} \qquad (\lambda \in \Omega). \tag{1.11.5}$$

From BANACH'S closed graph theorem and Proposition 1.2.5 it follows that D belongs to $H(\Omega, L(E, G \times F_1))$. We define

$$V(\lambda) := T_2(\lambda)U(\lambda) \qquad (\lambda \in \Omega) \tag{1.11.6}$$

and

$$C(\lambda) := \begin{pmatrix} 0 & \mathrm{id}_{F_1} \\ \mathrm{id}_{F_2} & V(\lambda) \end{pmatrix} \qquad (\lambda \in \Omega). \tag{1.11.7}$$

Obviously C belongs to $H(\Omega, L(F_2 \times F_1, F))$. For $\lambda \in \Omega$ the operator $C(\lambda)$ is invertible and

$$C^{-1}(\lambda) = \begin{pmatrix} -V(\lambda) & \mathrm{id}_{F_2} \\ \mathrm{id}_{F_1} & 0 \end{pmatrix}. \tag{1.11.8}$$

An easy calculation yields

$$\begin{aligned} C(\lambda)\begin{pmatrix} M(\lambda) & 0 \\ 0 & \mathrm{id}_{F_1} \end{pmatrix} &= \begin{pmatrix} 0 & \mathrm{id}_{F_1} \\ M(\lambda) & V(\lambda) \end{pmatrix} \\ &= \begin{pmatrix} T_1(\lambda) \\ T_2(\lambda) \end{pmatrix} (Z(\lambda), U(\lambda)) = T(\lambda) D^{-1}(\lambda) \end{aligned}$$

for $\lambda \in \Omega$. □

In the terminology of [GGK] and [KAS], Theorem 1.11.1 states that T is globally equivalent on Ω to the canonical F_1-extension of M.

In all applications of Theorem 1.11.1 it will be shown directly that the right inverse U is holomorphic. Therefore, the results of Šubin or Bart quoted in the proof are not essential for our purposes.

We assume now that there are Banach spaces X_1, X_2, Y_1, Y_2 and holomorphic operator functions

$$\begin{gathered} C \in H(\Omega, L(Y_1 \times Y_2, F)),\ D \in H(\Omega, L(E, X_1 \times X_2)), \\ M \in H(\Omega, L(X_1, Y_1)),\ J \in H(\Omega, L(X_2, Y_2)) \end{gathered}$$

such that

$$T(\lambda) = C(\lambda) \begin{pmatrix} M(\lambda) & 0 \\ 0 & J(\lambda) \end{pmatrix} D(\lambda) \qquad (\lambda \in \Omega). \tag{1.11.9}$$

We suppose that the operators $C(\lambda)$, $D(\lambda)$, and $J(\lambda)$ are invertible for all $\lambda \in \Omega$. There are operator functions

$$\begin{gathered} C_1 \in H(\Omega, L(F, Y_1)), C_2 \in H(\Omega, L(F, Y_2)), \\ D_1 \in H(\Omega, L(X_1, E)), D_2 \in H(\Omega, L(X_2, E)) \end{gathered}$$

such that

$$C^{-1}(\lambda) = \begin{pmatrix} C_1(\lambda) \\ C_2(\lambda) \end{pmatrix}, \quad D^{-1}(\lambda) = (D_1(\lambda), D_2(\lambda)) \qquad (\lambda \in \Omega).$$

It is obvious that $T \in H(\Omega, \Phi(E,F))$ if and only if $M \in H(\Omega, \Phi(X_1, Y_1))$ and that $\rho(T) = \rho(M)$. Also note that $\operatorname{nul} T(\mu) = \operatorname{nul} M(\mu)$ for all $\mu \in \sigma(T)$.

LEMMA 1.11.2. *Let Ω be a domain and assume that $T \in H(\Omega, \Phi(E,F))$ fulfils* (1.11.9). *Let $\rho(T) \neq \emptyset$ and $\mu \in \sigma(T)$. Let $\{c_1, \ldots, c_r\}$ and $\{d_1, \ldots, d_r\}$ be biorthogonal CSRFs of M and M^* at μ. Define*

$$y_j := D_1 c_j, \quad v_j := C_1^* d_j \qquad (j = 1, \ldots, r). \tag{1.11.10}$$

Then $\{y_1,\dots,y_r\}$ *and* $\{v_1,\dots,v_r\}$ *are biorthogonal CSRFs of* T *and* T^* *at* μ, $\nu(y_j)=\nu(v_j)=\nu(c_j)=\nu(d_j)=: m_j$ *for* $j=1,\dots,r$, *and the operator function*

$$T^{-1}-\sum_{j=1}^{r}(\cdot-\mu)^{-m_j}y_j\otimes v_j$$

is holomorphic at μ.

Proof. An easy calculation yields

$$\begin{aligned}T^{-1}&-\sum_{j=1}^{r}(\cdot-\mu)^{-m_j}y_j\otimes v_j\\&=(D_1,D_2)\begin{pmatrix}M^{-1}&0\\0&J^{-1}\end{pmatrix}\begin{pmatrix}C_1\\C_2\end{pmatrix}-\sum_{j=1}^{r}(\cdot-\mu)^{-m_j}(D_1c_j)\otimes(C_1^*d_j)\\&=D_1\left(M^{-1}-\sum_{j=1}^{r}(\cdot-\mu)^{-m_j}c_j\otimes d_j\right)C_1+D_2J^{-1}C_2.\end{aligned}$$

By Theorem 1.5.9 the operator function

$$M^{-1}-\sum_{j=1}^{r}(\cdot-\mu)^{-m_j}c_j\otimes d_j$$

is holomorphic at μ, whence

$$T^{-1}-\sum_{j=1}^{r}(\cdot-\mu)^{-m_j}y_j\otimes v_j$$

has the same property. The operators

$$D^{-1}(\mu)=(D_1(\mu),D_2(\mu))\text{ and }C^{-1^*}(\mu)=(C_1^*(\mu),C_2^*(\mu))$$

are invertible. Hence $D_1(\mu)$ and $C_1^*(\mu)$ are injective, which implies $y_j(\mu)\neq 0$ and $v_j(\mu)\neq 0$. From

$$Ty_j=TD_1c_j=TD^{-1}\begin{pmatrix}c_j\\0\end{pmatrix}=C\begin{pmatrix}M&0\\0&J\end{pmatrix}\begin{pmatrix}c_j\\0\end{pmatrix}=C\begin{pmatrix}Mc_j\\0\end{pmatrix}$$

we conclude that Ty_j has a zero of order m_j at μ. In the same way we obtain

$$T^*v_j=T^*C_1^*d_j=T^*C^{-1^*}\begin{pmatrix}d_j\\0\end{pmatrix}=D^*\begin{pmatrix}M^*&0\\0&J^*\end{pmatrix}\begin{pmatrix}d_j\\0\end{pmatrix}=D^*\begin{pmatrix}M^*d_j\\0\end{pmatrix},$$

whence T^*v_j has a zero of order m_j at μ. By Theorem 1.5.9 the proof is complete if we observe that $r=\operatorname{nul}M(\mu)=\operatorname{nul}T(\mu)$. □

THEOREM 1.11.3. *Let $T \in H(\Omega, \Phi(E,F))$ fulfil* (1.11.1) *and let Z, M, V be given as in* (1.11.1), (1.11.2) *and* (1.11.6). *Let $\rho(T) \neq \emptyset$ and $\mu \in \sigma(T)$. Let $\{c_1,\ldots,c_r\}$ and $\{d_1,\ldots,d_r\}$ be biorthogonal CSRFs of M and M^* at μ. Define*

$$y_j := Zc_j, \quad v_j := \begin{pmatrix} -V^*d_j \\ d_j \end{pmatrix} \qquad (j = 1,\ldots,r). \tag{1.11.11}$$

Then $\{y_1,\ldots,y_r\}$ and $\{v_1,\ldots,v_r\}$ are biorthogonal CSRFs of T and T^* at μ, $\nu(y_j) = \nu(v_j) = \nu(c_j) = \nu(d_j) =: m_j$ for $j = 1,\ldots,r$, and the operator function

$$T^{-1} - \sum_{j=1}^{r} (\cdot - \mu)^{-m_j} y_j \otimes v_j$$

is holomorphic at μ.

Proof. It is easy to verify the assumptions of Lemma 1.11.2. We set $X_1 := G$, $X_2 := F_1$, $Y_1 := F_2$ and $Y_2 := F_1$. By (1.11.5), $D^{-1} = (Z, U)$ whence $D_1 = Z$. From (1.11.8) we obtain

$$C^{-1*} = \begin{pmatrix} -V^* & \mathrm{id}_{F_1'} \\ \mathrm{id}_{F_2'} & 0 \end{pmatrix}$$

and hence

$$C_1^* = \begin{pmatrix} -V^* \\ \mathrm{id}_{F_2'} \end{pmatrix}. \qquad \square$$

1.12. Notes

Operator functions as considered in this chapter are also called operator bundles, operator pencils, or operator colligations. There are two different aspects of the spectral theory of holomorphic operator functions. Given that $T(\kappa) \in L(E)$, E a Banach space, depends holomorphically on the complex parameter κ, the question arises how the spectrum of the operator $T(\kappa)$ depends on the parameter κ. Notice that the spectral parameter λ occurs linearly in $T(\kappa) - \lambda\,\mathrm{id}_E$. Since we are not concerned with problems of this kind, we refer the reader to the monographs of KATO [KA, Chapter 7] and BAUMGÄRTEL [BG] for an extensive treatment of this question. We study $T(\lambda)$ with λ as the spectral parameter which in general is a nonlinear problem. The aspect considered here is the behavior of the resolvent of $T(\lambda)$ at its poles. Spectral properties of $T(\lambda)$ with λ as the spectral parameter, in particular the representation of the principal parts of the resolvent, were stated in Hilbert spaces for polynomial operator pencils $I - K(\lambda)$, $K(\lambda)$ compact, by KELDYSH [KE1] and proved in [KE2]. The discreteness of the spectrum in case $\rho(T) \neq \emptyset$ for holomorphic Fredholm operator valued functions was shown by ATKINSON [AT1] and SZ.-NAGY [SZN]. KELDYSH's theory was extended to holomorphic and meromorphic Fredholm operator valued functions by TROFIMOV [TRO], MARKUS and SIGAL [MAS], and GOHBERG and SIGAL [GS]. The proof in [GS] differs from ours in that a local factorization is used, whereas we obtain the local factorization in Section 1.8 from the representation of the principal part. We took this route since our main aim in this chapter is the representation of the principal parts of the resolvent. In the special case of an integral operator of Fredholm type whose kernel depends holomorphically on a parameter λ, the meromorphic dependence of the kernel of the inverse operator was shown by TAMARKIN [TA4]. Root functions were introduced by TROFIMOV [TRO]. Although root functions also occur in [GS], the first representation of the principal parts of the resolvent in terms of root functions has been proved in [MM1]. Further results in this direction appeared in [MM3, MM4, MM5]. A generalization to Fréchet spaces has been published in [MM2]. The structure of the resolvent in the λ-linear case is well-known, see e. g. KATO [KA, p. 181]. The representation of eigenvectors and associated vectors of an operator function by those of a related characteristic function were first published in [MM3]. The fact that in this case the given operator function is a global extension of its characteristic matrix function was shown by KAASHOEK [KAS].

Representations of the principal parts of meromorphic operator functions and in particular inverses of holomorphic operator functions play also a role in other publications. Apart from the literature cited above we just mention the contributions [BA1] by BART, [JW] by JEGGLE and WENDLAND and the recently published monograph [KM] by KOZLOV and MAZ'YA.

Chapter II

FIRST ORDER SYSTEMS OF ORDINARY DIFFERENTIAL EQUATIONS

This chapter contains basic elements of the theory of first order systems of ordinary linear differential equations. Since it is advantageous to treat boundary eigenvalue problems using weak derivatives in the corresponding differential equations, we give a short introduction to the theory of Sobolev spaces on intervals and prove properties which are needed in the sequel. A realization of the dual of such a Sobolev space is established (Theorem 2.2.5). For first order linear differential systems with coefficients depending holomorphically on a parameter the existence of a fundamental matrix function which depends holomorphically on this parameter is proved (Theorem 2.5.3). Further investigations concern the asymptotic behavior of fundamental matrix functions in case the coefficients of the first order differential system are asymptotically linear in the parameter (Theorem 2.8.2). These properties of the fundamental matrix functions are essential for asymptotic estimates of the GREEN'S function (the resolvent) of boundary eigenvalue problems.

2.1. Sobolev spaces on intervals

Throughout this chapter we assume that a and b are real numbers with $a < b$. Furthermore, let $1 \le p \le \infty$. There is a unique p' with $1 \le p' \le \infty$ and $1/p + 1/p' = 1$.

Let $I \subset \mathbb{R}$ be an interval. $C(I) = C^0(I)$ denotes the space of all continuous functions on I to $\mathbb{C}$. For a positive integer k, $C^k(I)$ denotes the space of k-times continuously differentiable functions on I. Let

$$C^\infty(I) := \bigcap_{k=1}^{\infty} C^k(I).$$

For $f \in C(I)$ the set

$$\operatorname{supp} f := \overline{\{x \in I : f(x) \neq 0\}}$$

is called the *support of* f, where the closure is taken with respect to I.

If I is compact, we set

$$|f|_{(k)} := \sum_{j=0}^{k} \max_{x \in I} |f^{(j)}(x)| \quad (f \in C^k(I)).$$

It is well-known that $(C^k(I), |\ |_{(k)})$ is a Banach space.

As usual, $L_p(I)$ denotes the space of measurable functions f on I (modulo functions which vanish almost everywhere) such that

$$|f|_p := \left(\int_I |f(x)|^p \,\mathrm{d}x\right)^{1/p} < \infty \quad (1 \le p < \infty),$$
$$|f|_\infty := \operatorname{ess\,sup}\{|f(x)| : x \in I\} < \infty \quad (p = \infty).$$

It is well-known that $L_p(I)$ is a Banach space with respect to the norm $|\ |_p$, see e.g. [HS, (13.11) and (20.14)].

For $f, g \in L_1(\mathbb{R})$ and almost all $x \in \mathbb{R}$, the function $y \mapsto f(x-y)g(y)$ belongs to $L_1(\mathbb{R})$; for these x the *convolution* $(f * g)(x)$ of f and g at x is defined by

$$(f * g)(x) := \int_{\mathbb{R}} f(x-y)g(y)\,\mathrm{d}y,$$

and $f * g$ belongs to $L_1(\mathbb{R})$, see e. g. [HS, (21.31)].

Let $L_1^{\mathrm{loc}}(I)$ be the set of measurable functions f on I (modulo functions which vanish almost everywhere) such that $f|_K \in L_1(K)$ for each compact subset K of I.

For simplicity of notation we identify an element of $L_1^{\mathrm{loc}}(I)$ with any of its representatives. Hence identities, inequalities etc. for $L_1^{\mathrm{loc}}(I)$-functions are to be understood almost everywhere.

If the interval is given explicitly, then we shall simplify the notation by omitting the outer parentheses in the definition of spaces over intervals, e. g. we write $C[a,b]$ instead of $C([a,b])$.

Now let I be open. A function $f \in C^\infty(I)$ is called a test function if its support is a compact subset of I. The space of all test functions on an open interval I is denoted by $C_0^\infty(I)$. We identify $C_0^\infty(I)$ with a subspace of $C_0^\infty(\mathbb{R})$ by setting $f = 0$ outside of I for each $f \in C_0^\infty(I)$. Thus

$$C_0^\infty(I) = \bigcup_{K \subset I,\,\text{compact}} C_0^\infty(K),$$

where

$$C_0^\infty(K) := \{f \in C_0^\infty(\mathbb{R}) : \operatorname{supp} f \subset K\}.$$

A linear functional u on $C_0^\infty(I)$ is called a *distribution on* I if for each compact set $K \subset I$ there are numbers $k \in \mathbb{N}$ and a $C \ge 0$ such that

$$|\langle \varphi, u \rangle| \le C \sum_{j=0}^{k} \sup_{x \in K} |\varphi^{(j)}(x)| \quad (\varphi \in C_0^\infty(K)),$$

where $\langle \varphi, u \rangle := u(\varphi)$. The space of distributions on I is denoted by $\mathscr{D}'(I)$.

For $u \in \mathscr{D}'(I)$ the *support of* u, denoted $\operatorname{supp} u$, is the set of points $x \in I$ such that for each neighbourhood $U \subset I$ of x there is a function $\varphi \in C_0^\infty(U)$ such that $\langle \varphi, u \rangle \ne 0$.

According to the definition

$$\langle \varphi, f \rangle := \int_I \varphi f \, \mathrm{d}x \quad (\varphi \in C_0^\infty(I),\, f \in L_1^{\mathrm{loc}}(I)), \tag{2.1.1}$$

we identify $L_1^{\mathrm{loc}}(I)$ with a subspace of $\mathscr{D}'(I)$, see e.g. [HÖ2, p. 37]. Then we also have $L_p(I) \subset \mathscr{D}'(I)$ $(1 \le p \le \infty)$ and $C(I) \subset \mathscr{D}'(I)$ with respect to the bilinear form (2.1.1). A distribution on I belonging to L_1^{loc} is called a *regular distribution.*

Let $u \in \mathscr{D}'(I)$. Then

$$\langle \varphi, u' \rangle := -\langle \varphi', u \rangle \quad (\varphi \in C_0^\infty(I)) \tag{2.1.2}$$

defines a distribution u' on I, called the *derivative in the sense of distributions* of u. For $k = 1, 2, \ldots$ we recursively define

$$u^{(k+1)} := \left(u^{(k)}\right)'.$$

Hence, for $k \in \mathbb{N}$, the k-th derivative $u^{(k)} \in \mathscr{D}'(I)$ of $u \in \mathscr{D}'(I)$ is well-defined. If $u \in C^k(I)$, then the k-th derivative $u^{(k)}$ in the sense of distributions coincides with the classical k-th derivative because of the formula for integration by parts.

Analogously, for $u \in \mathscr{D}'(I)$ and $\psi \in C^\infty(I)$,

$$\langle \varphi, \psi u \rangle := \langle \psi \varphi, u \rangle \quad (\varphi \in C_0^\infty(I)) \tag{2.1.3}$$

defines a unique distribution ψu on I.

DEFINITION 2.1.1. Let $I \subset \mathbb{R}$ be an open interval and $k \in \mathbb{N}$. The space

$$W_p^k(I) := \{ f \in L_p(I) : \forall j \in \{1, \ldots, k\}\ f^{(j)} \in L_p(I) \}$$

is called a *Sobolev space*. Here the derivatives $f^{(j)}$ are the derivatives in the sense of distributions. For $f \in W_p^k(I)$ we set

$$|f|_{p,k} := \sum_{j=0}^{k} |f^{(j)}|_p \,.$$

Note that $W_p^0(I) = L_p(I)$.

REMARK 2.1.2. Let $AC^{\mathrm{loc}}(I)$ be the set of functions f on I such that $f|_K$ is absolutely continuous for each compact subinterval K of I. Then, for $k > 0$,

$$W_p^k(I) = \{ f \in AC^{\mathrm{loc}}(I) : \forall j \in \{1, \ldots, k-1\}\ f^{(j)} \in AC^{\mathrm{loc}}(I),\, f^{(k)} \in L_p(I) \}.$$

Observing the fact that $f \in AC^{\mathrm{loc}}(I)$ if and only if f is the indefinite integral of a locally integrable function, Remark 2.1.2 will be an immediate consequence of Proposition 2.1.5 below.

PROPOSITION 2.1.3. *Let $I \subset \mathbb{R}$ be an open interval, $\gamma \in \overline{I}$ and $g \in L_p(I)$. Set*

$$G(x) := \int_\gamma^x g(t) \, \mathrm{d}t \quad (x \in \overline{I}). \tag{2.1.4}$$

Then G is continuous on $\overline{I}$ and $G' = g$ in $\mathscr{D}'(I)$.

More precisely, we should write $\left(G|_I\right)' = g$. But since a continuous function on $\bar{I}$ is uniquely determined by its values on I, we shall often identify G and $G|_I$.

Proof. Let $[\alpha,\beta]$ be any compact subinterval of $\bar{I}$ containing γ. Then $g|_{[\alpha,\beta]\cap I} \in L_p([\alpha,\beta]\cap I) \subset L_1([\alpha,\beta])$, which shows that G is well-defined. It is well-known that G is continuous on $[\alpha,\beta]$, see e.g. [HS, (18.1)]. This proves that G is continuous. Now let $\varphi \in C_0^\infty(I)$. Choose $\alpha,\beta \in \bar{I}$ such that $\alpha \le \gamma \le \beta$ and $\operatorname{supp}\varphi \subset [\alpha,\beta]$. With the aid of the theorem on integration by parts, see [HS, (18.19)], we obtain

$$\begin{aligned}\langle \varphi, G'\rangle &= -\langle \varphi', G\rangle \\ &= -\int_\alpha^\beta \varphi'(x)G(x)\,\mathrm{d}x = \int_\alpha^\beta \varphi(x)g(x)\,\mathrm{d}x \\ &= \langle \varphi, g\rangle,\end{aligned}$$

which proves $G' = g$. □

COROLLARY 2.1.4. *Let $k \in \mathbb{N}$ and $u \in \mathscr{D}'(a,b)$ such that $u' \in W_p^k(a,b)$. Then $u \in W_p^{k+1}(a,b)$.*

Proof. We must show that $u \in L_p(a,b)$. With $g := u' \in W_p^k(a,b)$ and G as in 2.1.4 we obtain $(u-G)' = 0$. Hence $u - G$ is a constant, see [HÖ2, Theorem 3.1.4]. By Proposition 2.1.3 this implies that u is continuous and hence belongs to $L_p(a,b)$. □

PROPOSITION 2.1.5. *Let $I \subset \mathbb{R}$ be an open interval and $k \in \mathbb{N}\setminus\{0\}$.*
i) *Let $f \in L_p(I)$ and $\gamma \in \bar{I}$. Then $f \in W_p^k(I)$ if and only if there are $g \in W_p^{k-1}(I)$ and $c \in \mathbb{C}$ such that*

$$f(x) = c + \int_\gamma^x g(t)\,\mathrm{d}t \quad (x \in I). \tag{2.1.5}$$

In this case, $g = f'$, f has a continuous extension to $\bar{I}$, which we also denote by f, and $c = f(\gamma)$.

ii)
$$W_p^k(I) \subset C^{k-1}(\bar{I}).$$

Proof. i): Let $f \in W_p^k(I)$. Since $k \ge 1$, we have $f' \in L_p(I)$ by definition of $W_p^k(I)$. With

$$G(x) := \int_\gamma^x f'(t)\,\mathrm{d}t \quad (x \in \bar{I}),$$

Proposition 2.1.3 yields $G' = f'$. Hence $G - f$ is a constant, see [HÖ2, Theorem 3.1.4], which proves (2.1.5) with $g = f'$. Since $g^{(j)} = f^{(j+1)} \in L_p(I)$ for $j = 1,\dots,k-1$, we have $g \in W_p^{k-1}(I)$. Conversely, if 2.1.5 holds with $g \in W_p^{k-1}(I) \subset L_p(I)$, then $f' = g$ by Proposition 2.1.3. Thus $f^{(j)} = g^{(j-1)} \in L_p(I)$ for $j = 1,\dots,k$. This proves $f \in W_p^k(I)$. Assume that (2.1.5) holds. Then f has a continuous

extension to $\bar{I}$ by Proposition 2.1.3, and (2.1.5) even holds for $x \in \bar{I}$. Evaluating this equation for $\gamma \in \bar{I}$ yields $f(\gamma) = c$.
ii): For $k = 1$ the statement has already been proved in part i). Now let $k > 1$ and suppose that ii) holds for $k-1$. Let $f \in W_p^k(I)$. Then $f' \in W_p^{k-1}(I)$. The representation (2.1.5) and ii) for $k-1$ complete the proof. □

PROPOSITION 2.1.6. *Let $I \subset \mathbb{R}$ be an open interval and $k \in \mathbb{N}$. Then $W_p^k(I)$ is a Banach space with respect to the norm $|\ |_{p,k}$.*

Proof. With respect to the mapping

$$f \mapsto (f, f', \ldots, f^{(k)}),$$

the space $W_p^k(I)$ is isomorphic to the subspace

$$R := \left\{ (f_j)_{j=0}^k : f_j \in L_p(I)\ (j = 0, \ldots, k),\ f_j' = f_{j+1}\ (j = 0, \ldots, k-1) \right\}$$

of $(L_p(I))^{k+1}$. Since $(L_p(I))^{k+1}$ is a product of Banach spaces and hence a Banach space, it is sufficient to prove that R is a closed subspace of $(L_p(I))^{k+1}$. For this let $((f_j^n)_{j=0}^k)_{n=0}^\infty$ be a sequence in R which converges to some $(f_j)_{j=0}^k$ in $(L_p(I))^{k+1}$. Let $\varphi \in C_0^\infty(I)$. Then, for $j = 0, \ldots, k$, HÖLDER'S inequality yields

$$\begin{aligned} |\langle \varphi, f_j^n - f_j \rangle| &= \left| \int_I \varphi(x)(f_j^n - f_j)(x)\, \mathrm{d}x \right| \\ &\le |\varphi|_{p'} |f_j^n - f_j|_p \to 0 \qquad (n \to \infty). \end{aligned}$$

Hence, for $j = 0, \ldots, k-1$,

$$\begin{aligned} \langle \varphi, f_j' \rangle &= -\langle \varphi', f_j \rangle = -\lim_{n\to\infty} \langle \varphi', f_j^n \rangle \\ &= \lim_{n\to\infty} \langle \varphi, f_{j+1}^n \rangle = \langle \varphi, f_{j+1} \rangle. \end{aligned}$$

This proves $f_j' = f_{j+1}$ for $j = 0, \ldots, k-1$, and we obtain $(f_j)_{j=0}^k \in R$. □

PROPOSITION 2.1.7. *For each $k \in \mathbb{N}$ and $1 \le p \le q \le \infty$ we have*

$$W_q^k(a,b) \subset W_p^k(a,b), \tag{2.1.6}$$

$$W_p^{k+1}(a,b) \subset C^k[a,b], \tag{2.1.7}$$

$$C^k[a,b] \subset W_p^k(a,b), \tag{2.1.8}$$

where the inclusions also hold topologically, i. e., the corresponding inclusion maps are continuous.

Proof. From [HS, (13.17)] we infer for $q < \infty$ that $L_q(a,b) \subset L_p(a,b)$, where the inclusion is continuous. A similar proof also holds for $q = \infty$. This immediately proves (2.1.6) and the continuity of the inclusion. The assertion (2.1.7) is a special case of Proposition 2.1.5 ii). The inclusion (2.1.8) is obvious since continuous functions on $[a,b]$ belong to $L_p(a,b)$.

Next we shall prove the continuity of the inclusion (2.1.7). For this we choose $f \in W_p^{k+1}(a,b)$ and $j \in \{0,\dots,k\}$. From Proposition 2.1.5 i) we infer

$$f^{(j)}(x) = f^{(j)}(a) + \int_a^x f^{(j+1)}(t)\,\mathrm{d}t \quad (x \in [a,b]). \tag{2.1.9}$$

Hence

$$|f^{(j)}(a)| \le |f^{(j)}(x)| + \int_a^b |f^{(j+1)}(t)|\,\mathrm{d}t \quad (x \in [a,b]).$$

We integrate and obtain

$$(b-a)|f^{(j)}(a)| \le \int_a^b |f^{(j)}(x)|\,\mathrm{d}x + (b-a)\int_a^b |f^{(j+1)}(x)|\,\mathrm{d}x.$$

This estimate, (2.1.9), and HÖLDER'S inequality yield

$$\begin{aligned}|f^{(j)}(x)| &\le (b-a)^{-1}\int_a^b |f^{(j)}(x)|\,\mathrm{d}x + 2\int_a^b |f^{(j+1)}(x)|\,\mathrm{d}x \\ &\le (b-a)^{-1/p}|f^{(j)}|_p + 2(b-a)^{1-1/p}|f^{(j+1)}|_p\,.\end{aligned}$$

Hence

$$|f|_{(k)} \le \left((b-a)^{-1/p} + 2(b-a)^{1-1/p}\right)|f|_{p,k+1}$$

holds for all $f \in W_p^{k+1}(a,b)$. This proves the continuity of the inclusion (2.1.7).

For $p = \infty$, the inclusion map in (2.1.8) is an isometry since $|\ |_{(k)}$ and $|\ |_{\infty,k}$ coincide on $C^k[a,b]$. Now let $p < \infty$, $f \in C^k[a,b]$ and $j \in \{0,\dots,k\}$. Then

$$|f^{(j)}|_p = \Big(\int_a^b |f^{(j)}(x)|^p\,\mathrm{d}x\Big)^{1/p} \le (b-a)^{1/p}|f^{(j)}|_{(0)} \quad (j = 0,\dots,k)$$

implies

$$|f|_{p,k} \le (b-a)^{1/p}|f|_{(k)}\,.$$

The continuity of the inclusion (2.1.8) is proved. □

PROPOSITION 2.1.8. *Let $k \in \mathbb{N}$. For $f \in W_p^k(a,b)$ we set*

$$(If)(x) := \int_a^x f(t)\,\mathrm{d}t.$$

Then $I \in L(W_p^k(a,b), W_p^{k+1}(a,b))$ with $|I| \le b-a+1$.

Proof. By Proposition 2.1.5 i) we have that I maps $W_p^k(a,b)$ into $W_p^{k+1}(a,b)$. Obviously, I is linear. For $p < \infty$, HÖLDER'S inequality yields

$$\begin{aligned}|If|_p &= \Big(\int_a^b \Big|\int_a^x f(t)\,\mathrm{d}t\Big|^p \mathrm{d}x\Big)^{\frac{1}{p}} \\ &\le \Big(\int_a^b (b-a)^{p-1}|f|_p^p\,\mathrm{d}x\Big)^{\frac{1}{p}} = (b-a)|f|_p\,.\end{aligned}$$

Obviously, $|If|_p \le (b-a)|f|_p$ also holds if $p=\infty$. Hence

$$|If|_{p,k+1} = |If|_p + |(If)'|_{p,k} \le (b-a)|f|_p + |f|_{p,k} \\ \le (b-a+1)|f|_{p,k}\,. \quad \square$$

2.2. The dual of $W_p^k(a,b)$ for $p<\infty$

If we consider distributions defined on different intervals, then we shall often use the notion $\langle\ ,\ \rangle_I$ for the bilinear form on $C_0^\infty(I)\times\mathscr{D}'(I)$ in order to distinguish between the diffent bilinear forms. Only for $I=(a,b)$ we shall never use a subscript.

PROPOSITION 2.2.1. *Let $k\in\mathbb{N}$. Let I and I' be open intervals with $I'\subset I$. A continuous linear map from $W_p^k(I)$ into $W_p^k(I')$ is given by $f\mapsto f|_{I'}$ for $f\in W_p^k(I)$. In particular, $(f|_{I'})^{(j)} = f^{(j)}|_{I'}$ holds for all $f\in W_p^k(I)$ and $j=0,\dots,k$. For $I=\mathbb{R}$ and $I'=(a,b)$ we write $\kappa_{p,k}f := f|_{(a,b)}$ for $f\in W_p^k(\mathbb{R})$. Thus*

$$\kappa_{p,k} : W_p^k(\mathbb{R}) \to W_p^k(a,b)$$

is a continuous linear map.

Proof. Let $f\in W_p^k(I)$ and $\varphi\in C_0^\infty(I')\subset C_0^\infty(I)$. For $j=0,\dots,k$ we have

$$\begin{aligned}\langle\varphi,(f|_{I'})^{(j)}\rangle_{I'} &= (-1)^j\langle\varphi^{(j)}, f|_{I'}\rangle_{I'} \\ &= (-1)^j\int_{I'}\varphi^{(j)}(x)f(x)\,\mathrm{d}x = (-1)^j\int_I\varphi^{(j)}(x)f(x)\,\mathrm{d}x \\ &= (-1)^j\langle\varphi^{(j)},f\rangle_I = \langle\varphi,f^{(j)}\rangle_I \\ &= \int_I\varphi(x)f^{(j)}(x)\,\mathrm{d}x = \int_{I'}\varphi(x)f^{(j)}(x)\,\mathrm{d}x \\ &= \langle\varphi,f^{(j)}|_{I'}\rangle_{I'}.\end{aligned}$$

Hence $(f|_{I'})^{(j)} = f^{(j)}|_{I'}\in L_p(I')$ for $j=0,\dots,k$, which proves $f|_{I'}\in W_p^k(I')$ and $|f|_{I'}|_{p,k}\le|f|_{p,k}\,$. $\square$

PROPOSITION 2.2.2. *Let $I\subset\mathbb{R}$ be an open interval, $\gamma\in I$ and $k\in\mathbb{N}\setminus\{0\}$. Let $f_1\in W_p^k(I\cap(-\infty,\gamma))$ and $f_2\in W_p^k(I\cap(\gamma,\infty))$. By Proposition* 2.1.5 ii) *the complex numbers $f_i^{(j)}(\gamma)$ are well-defined for $i=1,2$ and $j=0,\dots,k-1$. Set*

$$f(x) := \begin{cases} f_1(x) & \text{if } x\in I\cap(-\infty,\gamma], \\ f_2(x) & \text{if } x\in I\cap(\gamma,\infty). \end{cases} \tag{2.2.1}$$

Then $f\in W_p^k(I)$ if and only if $f_1^{(j)}(\gamma)=f_2^{(j)}(\gamma)$ for $j=0,\dots,k-1$. If this holds, then

$$f^{(j)}(x) = \begin{cases} f_1^{(j)}(x) & \text{if } x\in I\cap(-\infty,\gamma), \\ f_2^{(j)}(x) & \text{if } x\in I\cap(\gamma,\infty), \end{cases} \tag{2.2.2}$$

for $j=1,\dots,k$.

Proof. If $f \in W_p^k(I)$, then $f_1^{(j)}(\gamma) = f^{(j)}(\gamma) = f_2^{(j)}(\gamma)$ for $j = 0, \ldots, k-1$ since $f \in C^{k-1}(\bar{I})$ by Proposition 2.1.5 ii). And (2.2.2) immediately follows from Proposition 2.2.1 as $f^{(j)} \in L_p(I)$ for $j = 1, \ldots, k$.

Conversely, let $f_1^{(j)}(\gamma) = f_2^{(j)}(\gamma)$ for $j = 0, \ldots, k-1$. We shall prove $f \in W_p^l(I)$ for $l = 0, \ldots, k$ by induction. Obviously, $f \in L_p(I)$. Now let $l \in \{0, \ldots, k-1\}$ and suppose that $f \in W_p^l(I)$. Note that (2.2.2) holds for $j = l$ by the first part of the proof. Let $\varphi \in C_0^\infty(I)$ and choose $\alpha, \beta \in I$ such that $\alpha \leq \gamma \leq \beta$ and $\operatorname{supp} \varphi \subset [\alpha, \beta]$. With the aid of the theorem on integration by parts, see [HS, (18.19)], we infer that

$$\begin{aligned}
\langle \varphi, f^{(l+1)} \rangle &= -\langle \varphi', f^{(l)} \rangle \\
&= -\int_\alpha^\gamma \varphi'(x) f_1^{(l)}(x)\,\mathrm{d}x - \int_\gamma^\beta \varphi'(x) f_2^{(l)}(x)\,\mathrm{d}x \\
&= \int_\alpha^\gamma \varphi(x) f_1^{(l+1)}(x)\,\mathrm{d}x - \varphi(\gamma) f_1^{(l)}(\gamma) + \int_\gamma^\beta \varphi(x) f_2^{(l+1)}(x)\,\mathrm{d}x + \varphi(\gamma) f_2^{(l)}(\gamma) \\
&= \int_\alpha^\gamma \varphi(x) f_1^{(l+1)}(x)\,\mathrm{d}x + \int_\gamma^\beta \varphi(x) f_2^{(l+1)}(x)\,\mathrm{d}x \\
&= \langle \varphi, f_{(l+1)} \rangle,
\end{aligned}$$

where

$$f_{(l+1)}(x) := \begin{cases} f_1^{(l+1)}(x) & \text{if } x \in I \cap (-\infty, \gamma), \\ f_2^{(l+1)}(x) & \text{if } x \in I \cap (\gamma, \infty). \end{cases}$$

Hence $f^{(l+1)} = f_{(l+1)} \in L_p(I)$, and $f \in W_p^l(I)$ implies $f \in W_p^{l+1}(I)$. □

PROPOSITION 2.2.3. *Let $k \in \mathbb{N}$. Then there is a continuous linear map*

$$\tau_{p,k} : W_p^k(a,b) \to W_p^k(\mathbb{R})$$

such that $\kappa_{p,k} \tau_{p,k} = \mathrm{id}_{W_p^k(a,b)}$ and $\tau_{p,k} f$ has compact support for each $f \in W_p^k(a,b)$.

Proof. We choose a test function $\psi \in C_0^\infty(\mathbb{R})$ with $\psi = 1$ in a neighbourhood of $[a,b]$. For the existence of such a function see e.g. [HÖ2, Theorem 1.4.1]. For $f \in W_p^k(a,b)$ we set

$$f_c(x) := \sum_{j=0}^{k-1} \frac{1}{j!} f^{(j)}(c)(x-c)^j \quad (x \in \mathbb{R},\ c \in \{a,b\})$$

and

$$(\tau_{p,k} f)(x) := \begin{cases} \psi(x) f_a(x) & \text{if } x \leq a, \\ f(x) & \text{if } a < x < b, \\ \psi(x) f_b(x) & \text{if } x \geq b. \end{cases}$$

Obviously, $\tau_{p,k}f \in L_p(\mathbb{R}) \subset \mathscr{D}'(\mathbb{R})$ and $\operatorname{supp}(\tau_{p,k}f) \subset \operatorname{supp}\psi$ is compact. For $k=0$ the assertion is obvious. Now let $k \geq 1$. From $\psi f_c \in C_0^\infty(\mathbb{R})$ we immediately infer $\psi f_a|_{(-\infty,a)} \in W_p^k(-\infty,a)$ and $\psi f_b|_{(b,\infty)} \in W_p^k(b,\infty)$, see Proposition 2.2.1. Since $\psi = 1$ in a neighbourhood of $c=a$ or $c=b$, we have

$$(\psi f_c)^{(j)}(c) = f^{(j)}(c) \quad (j=0,\dots,k-1).$$

Applying Proposition 2.2.2 first to the interval $I=(-\infty,b)$ and $\gamma=a$ and then to the interval $I=\mathbb{R}$ and $\gamma=b$, we obtain $\tau_{p,k}f \in W_p^k(\mathbb{R})$. It is obvious that $\kappa_{p,k}\tau_{p,k} = \mathrm{id}_{W_p^k(a,b)}$. Choose $\alpha,\beta \in \mathbb{R}$ such that $\operatorname{supp}\psi \subset [\alpha,\beta]$. From

$$(\psi f_c)^{(l)} = \sum_{j=0}^{k-1} \frac{1}{j!} f^{(j)}(c)\left(\psi(\cdot - c)^j\right)^{(l)} \quad (l=0,\dots,k)$$

we infer that there are $C_1 \geq 0$, $C_2 \geq 0$ such that

$$|\psi f_a|_{(\alpha,a)}|_{p,k} \leq C_1 |f|_{(k-1)},$$
$$|\psi f_b|_{(b,\beta)}|_{p,k} \leq C_2 |f|_{(k-1)},$$

holds for all $f \in W_p^k(a,b)$. By (2.2.2) and the continuity of the inclusion (2.1.7),

$$\begin{aligned}|\tau_{p,k}f|_{p,k} &\leq |\psi f_a|_{(\alpha,a)}|_{p,k} + |f|_{p,k} + |\psi f_b|_{(b,\beta)}|_{p,k}\\ &\leq C|f|_{p,k}\end{aligned}$$

for some $C \geq 1$ and all $f \in W_p^k(a,b)$. □

The mapping $\tau_{p,k}$ depends on the choice of the test function ψ. Below, $\tau_{p,k}$ always means an arbitrary mapping fulfilling the assertion of Proposition 2.2.3.

For $1 \leq q \leq \infty$ and $f \in L_q(a,b)$ we set

$$f_e(x) := \begin{cases} f(x) & \text{if } x \in (a,b), \\ 0 & \text{if } x \in \mathbb{R}\setminus(a,b). \end{cases} \tag{2.2.3}$$

The function f_e is called the *canonical extension of f to $\mathbb{R}$*.

PROPOSITION 2.2.4. *Let $p<\infty$ and $k \in \mathbb{N}\setminus\{0\}$. Set*

$$W_{p'}^{-k}[a,b] := \left\{u \in \mathscr{D}'(\mathbb{R}) : \exists (u_j)_{j=0}^k \in (L_{p'}(a,b))^{k+1}\ u = \sum_{j=0}^k (u_j)_e^{(j)}\right\}.$$

Then, for $u = \sum_{j=0}^k (u_j)_e^{(j)} \in W_{p'}^{-k}[a,b]$ and $f \in W_p^k(a,b)$,

$$\langle f,u\rangle_{p,k} := \sum_{j=0}^k \int_a^b (-1)^j f^{(j)}(x) u_j(x)\,\mathrm{d}x \tag{2.2.4}$$

does not depend on the representation of u. For $\varphi \in C_0^\infty(\mathbb{R})$ we have

$$\langle \varphi|_{(a,b)}, u\rangle_{p,k} = \langle \varphi, u\rangle_{\mathbb{R}}. \tag{2.2.5}$$

Proof. Let $(u_j)_{j=0}^k \in (L_{p'}(a,b))^{k+1}$ and set $u := \sum\limits_{j=0}^{k} (u_j)_e^{(j)}$. For $\varphi \in C_0^\infty(\mathbb{R})$ we have

$$\langle \varphi, u\rangle_{\mathbb{R}} = \sum_{j=0}^{k} \langle \varphi, (u_j)_e^{(j)}\rangle_{\mathbb{R}} = \sum_{j=0}^{k} (-1)^j \langle \varphi^{(j)}, (u_j)_e\rangle_{\mathbb{R}}$$
$$= \sum_{j=0}^{k} \int_a^b (-1)^j \varphi^{(j)}(x) u_j(x)\,\mathrm{d}x$$

which proves (2.2.5). Let $f \in W_p^k(a,b)$. Since $\tau_{p,k} f \in W_p^k(\mathbb{R})$, there is a sequence $(\phi_n)_{n=1}^\infty$ in $C_0^\infty(\mathbb{R})$ such that

$$(\tau_{p,k} f * \phi_n)^{(j)} = (\tau_{p,k} f)^{(j)} * \phi_n \to (\tau_{p,k} f)^{(j)}$$

in $L_p(\mathbb{R})$ as $n \to \infty$ for $j = 0,\dots,k$, see [HÖ2, Theorem 1.3.2 and (4.2.5)]. Using $f = \kappa_{p,k}\tau_{p,k} f$ and $(\kappa_{p,k}\tau_{p,k} f)^{(j)} = (\tau_{p,k} f)^{(j)}|_{(a,b)}$ for $j = 0,\dots,k$, see Propositions 2.2.3 and 2.2.1, we infer that

$$\sum_{j=0}^{k} \int_a^b (-1)^j f^{(j)}(x) u_j(x)\,\mathrm{d}x = \lim_{n\to\infty} \sum_{j=0}^{k} \int_{\mathbb{R}} (-1)^j (\tau_{p,k} f * \phi_n)^{(j)}(x) (u_j)_e(x)\,\mathrm{d}x$$
$$= \lim_{n\to\infty} \langle \tau_{p,k} f * \phi_n, u\rangle_{\mathbb{R}},$$

which proves that $\langle f, u\rangle_{p,k}$ is well-defined. □

THEOREM 2.2.5. *Let $p < \infty$ and $k \in \mathbb{N}\setminus\{0\}$. For $u \in W_{p'}^{-k}[a,b]$ let*

$$|u|_{p',-k} := \inf\Big\{ \max_{j=0}^{k} |u_j|_{p'} : (u_j)_{j=0}^k \in (L_{p'}(a,b))^{k+1},\ u = \sum_{j=0}^{k} (u_j)_e^{(j)} \Big\}.$$

Then $(W_{p'}^{-k}[a,b], |\ |_{p',-k})$ is a Banach space and the dual of $(W_p^k(a,b), |\ |_{p,k})$ with respect to the bilinear form $\langle\ ,\ \rangle_{p,k}$ defined in (2.2.4).

Proof. For $u \in W_{p'}^{-k}[a,b]$ and $f \in W_p^k(a,b)$ we set

$$(Ju)(f) := \langle f, u\rangle_{p,k}.$$

Let $u = \sum\limits_{j=0}^{k} (u_j)_e^{(j)}$, where $u_j \in L_{p'}(a,b)$ $(j = 0,\dots,k)$. Then, by definition (2.2.4) of $\langle\ ,\ \rangle_{p,k}$ and HÖLDER'S inequality,

$$|(Ju)(f)| \le \max_{j=0}^{k} |u_j|_{p'} |f|_{p,k},$$

which proves that $Ju \in (W_p^k(a,b), |\ |_{p,k})'$ and $|Ju|'_{p,k} \le |u|_{p',-k}$. Here $|\ |'_{p,k}$ denotes the norm on $(W_p^k(a,b), |\ |_{p,k})'$. Hence

$$J \in L(W_{p'}^{-k}[a,b], (W_p^k(a,b))').$$

We have to prove that J is bijective and that J^{-1} is continuous. For the proof of the injectivity of J let $u \in W_{p'}^{-k}[a,b]$ such that $Ju = 0$. There is a representation $u = \sum_{j=0}^{k} (u_j)_e^{(j)}$, where $(u_j)_{j=0}^k \in (L_{p'}(a,b))^{k+1}$. For $\varphi \in C_0^\infty(\mathbb{R})$ we have by (2.2.5) that

$$\langle \varphi, u\rangle_{\mathbb{R}} = \langle \varphi|_{(a,b)}, u\rangle_{p,k} = (Ju)(\varphi|_{(a,b)}) = 0,$$

which proves $u = 0$.

For the proof of the surjectivity of J let $w \in (W_p^k(a,b))'$. Since $W_p^k(a,b)$ is isomorphic to the subspace R of $(L_p(a,b))^{k+1}$ defined in the proof of Proposition 2.1.6, there is a continuous linear functional v on R such that

$$w(f) = v((f^{(j)})_{j=0}^k)$$

for $f \in W_p^k(a,b)$. Since $L_{p'}(a,b)$ is the dual of $L_p(a,b)$, see [HS, (15.12) and (20.20)], the HAHN-BANACH theorem yields that there are $u_j \in L_{p'}(a,b)$ for $j = 0,\dots,k$ such that

$$w(f) = \sum_{j=0}^{k} (-1)^j \int_a^b f^{(j)}(x) u_j(x)\,\mathrm{d}x \quad (f \in W_p^k(a,b))$$

and

(2.2.6) $$|w|'_{p,k} = \max_{j=0}^{k} |u_j|_{p'}.$$

Set

$$u := \sum_{j=0}^{k} (u_j)_e^{(j)} \in W_{p'}^{-k}[a,b].$$

Then Proposition 2.2.4 yields

$$w(f) = \langle f, u\rangle_{p,k} = (Ju)(f) \quad (f \in W_p^k(a,b)),$$

which proves $w = Ju \in R(J)$. Finally we obtain from 2.2.6 that

$$|J^{-1}w|_{p',-k} = |u|_{p',-k} \le |w|'_{p,k},$$

whence J^{-1} is continuous. Since the dual of a Banach space is a Banach space, we also have proved that $(W_{p'}^{-k}[a,b], |\ |_{p',-k})$ is a Banach space. Note that the two norm estimates obtained in the proof show that J is an isometry. □

EXAMPLE 2.2.6 (Dirac distribution). Let $p < \infty$ and $c \in [a,b]$. Since $f \mapsto f(c)$ is a continuous linear functional on $C[a,b]$, it is also a continuous linear functional on $W_p^1(a,b)$ because of the continuity of the inclusion $W_p^1(a,b) \subset C[a,b]$, see (2.1.7). By Theorem 2.2.5, there is a distribution $\delta_c \in W_{p'}^{-1}[a,b]$ such that

$$f(c) = \langle f, \delta_c \rangle_{p,1} \quad (f \in W_p^1(a,b)).$$

δ_c is called the *Dirac distribution* or *Dirac measure at c*.

We can extend the definition of the space $W_{p'}^{-k}[a,b]$ to $k = 0$. In this case,

$$W_{p'}^0[a,b] = \{u_e : u \in L_{p'}(a,b)\}$$

is isomorphic to $L_{p'}(a,b)$.

PROPOSITION 2.2.7. *Let $k \in \mathbb{N}$, $l \in \mathbb{N} \setminus \{0\}$, and $u \in W_{p'}^{-k}[a,b]$. Then we have $u^{(l)} \in W_{p'}^{-k-l}[a,b]$.*

Proof. By assumption, there are $u_j \in L_{p'}(a,b)$ $(j = 0, \ldots, k)$ such that

$$u = \sum_{j=0}^{k} (u_j)_e^{(j)}.$$

Then

$$u^{(l)} = \sum_{j=l}^{k+l} (u_{j-l})_e^{(j)},$$

which proves $u^{(l)} \in W_{p'}^{-k-l}[a,b]$. □

Identifying $f \in L_{p'}(a,b)$ and f_e for $p < \infty$ we have $L_{p'}(a,b) \subset W_{p'}^{-k}(a,b)$ for all $k \in \mathbb{N}$. In this way we obtain

$$W_{p'}^l(a,b) \subset L_{p'}(a,b) \subset W_{p'}^{-k}[a,b] \tag{2.2.7}$$

for $k, l \in \mathbb{N} \setminus \{0\}$. Thus we may consider $y \in W_{p'}^l(a,b)$ as a distribution in $W_{p'}^{-k}[a,b]$ if we identify y and y_e.

But if we consider the derivatives of y as a distribution on (a,b) and $\mathbb{R}$, respectively, they are different in general, i. e., $y_e' \neq (y')_e$ (see Proposition 2.6.5). Thus we have to be careful in which space we take the derivative. For $f \in L_{p'}(a,b)$ and $j \in \mathbb{N}$, $f^{(j)}$ always means the j-th derivative in $\mathscr{D}'(a,b)$, and we write $f_e^{(j)}$ for the j-th derivative in $\mathscr{D}'(\mathbb{R})$.

PROPOSITION 2.2.8. *Let $p < \infty$ and $k > 0$. Then $L_1(a,b) \subset W_{p'}^{-k}[a,b]$.*

Proof. Let $f \in L_1(a,b)$. Choose $u_0 \in C_0^\infty(a,b)$ such that

$$\int_a^b u_0(t)\,\mathrm{d}t = \int_a^b f(t)\,\mathrm{d}t,$$

and define

$$u_1(x) := \int_a^x \big(f(t) - u_0(t)\big)\,\mathrm{d}t \quad (x \in (a,b)).$$

Then $u_0, u_1 \in W_1^1(a,b) \subset L_{p'}(a,b)$ with $u_1(a) = 0 = u_1(b)$, and it follows from Proposition 2.2.2 that $f_e = (u_0)_e + (u_1)'_e$. □

2.3. Multiplication in Sobolev spaces on the interval (a,b)

All of the results on scalar-valued functions in this section extend to vector- and matrix-valued functions since it is sufficient to consider their components.

PROPOSITION 2.3.1. *Let $1 \le p \le r \le \infty$ and $k \in \mathbb{N} \setminus \{0\}$. Then the multiplication operator*

$$\cdot : W_p^k(a,b) \times W_r^k(a,b) \to W_p^k(a,b)$$

is a continuous bilinear map and

$$(gf)' = g'f + gf' \tag{2.3.1}$$

for $g \in W_p^k(a,b)$ and $f \in W_r^k(a,b)$. Also, for $k = 0$ the multiplication operator

$$\cdot : L_p(a,b) \times L_\infty(a,b) \to L_p(a,b)$$

is a continuous bilinear map.

Proof. For $k = 0$ the result is clear since gf is measurable for all $g \in L_p(a,b)$ and $f \in L_\infty(a,b)$, see e.g. [HS, (11.17)], and $|gf|_p \le |g|_p |f|_\infty$ by definition of the L_p- and L_∞-norm.

Now let $k > 0$ and suppose that the multiplication operator from the product space $W_p^{k-1}(a,b) \times W_\infty^{k-1}(a,b)$ into $W_p^{k-1}(a,b)$ is continuous. Let $g \in W_p^k(a,b)$ and $f \in W_r^k(a,b)$. Since $f, g \in W_1^1(a,b)$, Proposition 2.1.5 i) and the theorem on integration by parts, see [HS, (18.19)], yield for $x \in [a,b]$ that

$$(gf)(x) = (gf)(a) + \int_a^x g(t) f'(t)\,\mathrm{d}t + \int_a^x g'(t) f(t)\,\mathrm{d}t.$$

By (2.1.6) we have $f' \in W_p^{k-1}(a,b)$, and (2.1.7) and (2.1.8) yield that g and f belong to $W_\infty^{k-1}(a,b)$. Hence, by assumption,

$$gf' + g'f \in W_p^{k-1}(a,b).$$

From Proposition 2.1.5 i) we infer $gf \in W_p^k(a,b)$ and $(gf)' = gf' + g'f$.

The continuity of the multiplication operator from $W_p^{k-1}(a,b) \times W_\infty^{k-1}(a,b)$ into $W_p^{k-1}(a,b)$ and the continuity of the inclusion maps in Proposition 2.1.7 yield that there are constants C_1, C_2, C_3, which do not depend on f and g, such that

$$\begin{aligned} |gf' + g'f|_{p,k-1} &\le C_1\big(|g|_{\infty,k-1}|f'|_{p,k-1} + |g'|_{p,k-1}|f|_{\infty,k-1}\big) \\ &\le C_2|g|_{p,k}|f'|_{r,k-1} + C_3|g'|_{p,k-1}|f|_{r,k}. \end{aligned}$$

Finally, the continuity of the inclusion $W_r^k(a,b) \hookrightarrow L_\infty(a,b)$ yields

$$\begin{aligned}|gf|_{p,k} &= |gf|_p + |(gf)'|_{p,k-1} \\ &\le C_4|g|_p|f|_{r,k} + (C_2+C_3)|g|_{p,k}|f|_{r,k} \\ &\le C_5|g|_{p,k}|f|_{r,k},\end{aligned}$$

where C_4 and C_5 do not depend on f and g. □

Since for $l > k$ the embedding $W_p^l(a,b) \hookrightarrow W_p^k(a,b)$ is continuous, Proposition 2.3.1 immediately yields

PROPOSITION 2.3.2. *Let $k \in \mathbb{N}$, $l \in \mathbb{N}\setminus\{0\}$ and $k \le l$. Then the multiplication operator*

$$\cdot : W_p^k(a,b) \times W_p^l(a,b) \to W_p^k(a,b) \tag{2.3.2}$$

is a continuous bilinear map.

PROPOSITION 2.3.3. *Let $k \in \mathbb{N}$, $l \in \mathbb{N}\setminus\{0\}$, $k \le l$ and $n \in \mathbb{N}\setminus\{0\}$. Then, with respect to the multiplication Ay ($A \in M_n(W_p^k(a,b))$, $y \in (W_p^l(a,b))^n$), $M_n(W_p^k(a,b))$ is isomorphic to a subspace of $L((W_p^l(a,b))^n,(W_p^k(a,b))^n)$.*

Proof. For $A \in M_n(W_p^k(a,b))$ and $y \in (W_p^l(a,b))^n$ let $T_Ay := Ay$. From the continuity of the bilinear map (2.3.2) it follows that $A \mapsto T_A$ is a continuous linear inclusion map from $M_n(W_p^k(a,b))$ to $L((W_p^l(a,b))^n,(W_p^k(a,b))^n)$. The definition of the matrix norm, see (1.1.2), gives

$$|A|_{p,k} \le n\sup\{|Ac|_{p,k} : c \in \mathbb{C}^n,\, |c| \le 1\}.$$

Identifying elements from $\mathbb{C}^n$ with constant functions we obtain for $c \in \mathbb{C}^n$

$$|Ac|_{p,k} = |T_Ac|_{p,k} \le |T_A|\,|c|_{p,l} = |T_A|\,(b-a)^{1/p}|c|,$$

which proves $|A|_{p,k} \le n(b-a)^{1/p}|T_A|$. □

Let $p < \infty$, $k \in \mathbb{N}$, $l \in \mathbb{N}\setminus\{0\}$, $k \le l$, and $g \in W_p^k(a,b)$. By Proposition 2.3.2, $g\cdot$ is a continuous operator from $W_p^l(a,b)$ to $W_p^k(a,b)$. Theorem 2.2.5 yields that its adjoint $(g\cdot)^*$ is an operator from $W_{p'}^{-k}[a,b]$ to $W_{p'}^{-l}[a,b]$. For $v \in L_{p'}(a,b)$ and $f \in W_p^l(a,b)$ we have

$$\begin{aligned}\langle f,(g\cdot)^*v_e\rangle_{p,l} &= \langle gf,v_e\rangle_{p,k} = \int_a^b g(x)f(x)v(x)\,\mathrm{d}x \\ &= \langle f,(gv)_e\rangle_{p,l},\end{aligned} \tag{2.3.3}$$

i.e., $(g\cdot)^*v_e = (gv)_e$ if $v \in L_{p'}(a,b)$. Here we have used that $gv \in L_1(a,b)$ and applied Proposition 2.2.8. We use the notation $gv_e = (gv)_e$ and extend (2.3.3) to

$$(g\cdot)^*u =: g\cdot u =: gu \tag{2.3.4}$$

for all $u \in W_{p'}^{-k}[a,b]$. Therefore we obtain

PROPOSITION 2.3.4. *Let $p < \infty$, $k \in \mathbb{N}$, $l \in \mathbb{N} \setminus \{0\}$, $k \leq l$, and $g \in W_p^k(a,b)$. Then the "multiplication" operator*

$$g\cdot : W_{p'}^{-k}[a,b] \to W_{p'}^{-l}[a,b]$$

is a continuous linear map.

2.4. Compact inclusion maps in Sobolev spaces on (a,b)

LEMMA 2.4.1. *Let $p > 1$. Then the inclusion map $W_p^k(a,b) \hookrightarrow C^{k-1}[a,b]$ is compact for each $k \in \mathbb{N} \setminus \{0\}$.*

Proof. Let

$$B_k := \{f \in W_p^k(a,b) : |f|_{p,k} \leq 1\}.$$

We have to show that B_k is a relatively compact subset of $C^{k-1}[a,b]$. Since $C^{k-1}[a,b]$ is isomorphic to the subspace

$$R = \{(f_j)_{j=0}^{k-1} : \forall j \in \{0\ldots,k-2\}\ f_j \in C^1[a,b],\ f_j' = f_{j+1}\}$$

of $(C[a,b])^k$ with respect to the mapping

$$f \mapsto (f, f', \ldots, f^{(k-1)}),$$

it is sufficient to show that for each $j \in \{0,\ldots,k-1\}$ the set $\{f^{(j)} : f \in B_k\}$ is relatively compact in $C[a,b]$. But in view of $\{f^{(j)} : f \in B_k\} \subset B_1$ for $j \in \{0,\ldots,k-1\}$ this holds if we prove the lemma for $k = 1$.

Thus let $k = 1$. By Proposition 2.1.7 there is $C \geq 0$ such that $|f|_{(0)} \leq C|f|_{p,1}$ for all $f \in W_p^1(a,b)$. Therefore $\{f(x) : f \in B_1\}$ is bounded for all $x \in [a,b]$. For $x, y \in [a,b]$ the representation in Proposition 2.1.5 i) and HÖLDER'S inequality yield

$$\begin{aligned} |f(y) - f(x)| &= \left| \int_x^y f'(t)\,\mathrm{d}t \right| \\ &\leq |y-x|^{1-1/p} |f|_{p,1} \\ &\leq |y-x|^{1-1/p} \end{aligned}$$

for all $f \in B$. Hence B_1 is an equicontinuous subset of $C([a,b])$. By ASCOLI'S theorem, see e.g. [DI1, 7.5.7], B_1 is a relatively compact subset of $C[a,b]$. □

THEOREM 2.4.2. *The inclusion map $W_p^k(a,b) \hookrightarrow W_p^{k-1}(a,b)$ is compact for each $k \in \mathbb{N} \setminus \{0\}$.*

Proof. If $p > 1$, the statement immediately follows from Lemma 2.4.1 and Proposition 2.1.7 since the composition of a compact linear operator and a continuous linear operator is a compact operator, see e. g. [KA, Theorem III.4.8]. Now let $p = 1$ and set

$$B := \{f \in W_1^k(a,b) : |f|_{1,k} \leq 1\}.$$

Since $W_1^{k-1}(a,b)$ is complete, it is sufficient to show that B is precompact in $W_1^{k-1}(a,b)$. For this let $\varepsilon > 0$ and $\tau_{1,k}$ be an extension operator according to Proposition 2.2.3. Set $\delta := \frac{\varepsilon}{3C}$, where C is the operator norm of $\tau_{1,k}$. Let $f \in B$. For $j = 0, \ldots, k-1$,

$$(\tau_{1,k} f)^{(j)}(x) = f^{(j)}(a) + \int_a^x (\tau_{1,k} f)^{(j+1)}(t)\,\mathrm{d}t \quad (x \in \mathbb{R})$$

by Proposition 2.1.5 i), and hence the following estimate holds for $y \in [-\delta, \delta]$:

$$\begin{aligned}\int_a^b |(\tau_{1,k} f)^{(j)}(x-y) - (\tau_{1,k} f)^{(j)}(x)|\,\mathrm{d}x &\leq \int_a^b \left| \int_x^{x-y} |(\tau_{1,k} f)^{(j+1)}(t)|\,\mathrm{d}t \right| \mathrm{d}x \\ &\leq \int_{a-\delta}^{b+\delta} |(\tau_{1,k} f)^{(j+1)}(t)| \left| \int_t^{t+y} \mathrm{d}x \right| \mathrm{d}t \\ &\leq \delta |(\tau_{1,k} f)^{(j+1)}|_1 .\end{aligned}$$

Set $f_\delta = (\tau_{1,k} f) * \phi_\delta$, where $\phi_\delta \in C_0^\infty[-\delta, \delta]$, $\phi_\delta \geq 0$ and $\int_{\mathbb{R}} \phi_\delta(x)\,\mathrm{d}x = 1$. The differentiation of convolutions yields $f_\delta^{(j)} = (\tau_{1,k} f)^{(j)} * \phi_\delta$. Hence

$$\begin{aligned}\left| f_\delta^{(j)}|_{(a,b)} - f^{(j)} \right|_1 &= \int_a^b |f_\delta^{(j)}(x) - f^{(j)}(x)|\,\mathrm{d}x \\ &= \int_a^b \left| \int_{-\delta}^{\delta} \left((\tau_{1,k} f)^{(j)}(x-y) - (\tau_{1,k} f)^{(j)}(x) \right) \phi_\delta(y)\,\mathrm{d}y \right| \mathrm{d}x \\ &\leq \int_{-\delta}^{\delta} \int_a^b |(\tau_{1,k} f)^{(j)}(x-y) - (\tau_{1,k} f)^{(j)}(x)|\,\mathrm{d}x\, \phi_\delta(y)\,\mathrm{d}y \\ &\leq \delta |(\tau_{1,k} f)^{(j+1)}|_1 ,\end{aligned}$$

which proves

(2.4.1) $$\left| f_\delta|_{(a,b)} - f \right|_{1,k-1} \leq \delta C = \frac{\varepsilon}{3}$$

for all $f \in B$. Let $M := \max_{|y| \leq \delta} \phi_\delta(y)$. Obviously, for $l = 0, \ldots, k$,

$$|f_\delta^{(l)}|_\infty \leq |(\tau_{1,k} f)^{(l)}|_1 M,$$

and thus

$$|f_\delta|_{\infty,k} \leq CM$$

for all $f \in B$. This proves that $B_\delta := \{ f_\delta|_{(a,b)} : f \in B \}$ is a bounded subset of $W_\infty^k(a,b)$. By Lemma 2.4.1 there are finitely many elements $f_1, \ldots, f_m$ in B such that for each $f \in B$ there is a $j \in \{1, \ldots, m\}$ such that

$$\left| f_\delta|_{(a,b)} - f_{j,\delta}|_{(a,b)} \right|_{(k-1)} \leq \frac{\varepsilon}{3(b-a)} .$$

Using (2.4.1) we infer that for each $f \in B$ there is a $j \in \{1,\ldots,m\}$ such that

$$\begin{aligned}|f-f_j|_{1,k-1} &\le \big|f-f_\delta|_{(a,b)}\big|_{1,k-1} + \big|f_\delta|_{(a,b)} - f_{j,\delta}|_{(a,b)}\big|_{1,k-1} + \big|f_j - f_{j,\delta}|_{(a,b)}\big|_{1,k-1}\\ &\le \varepsilon,\end{aligned}$$

which completes the proof. □

2.5. Fundamental matrices

In this section let Ω be a nonempty open subset of $\mathbb{C}$, $n \in \mathbb{N}\setminus\{0\}$ and $A \in H(\Omega, M_n(L_p(a,b)))$. The value of the matrix function A at $\lambda \in \Omega$ and $x \in (a,b)$ is denoted by $A(x,\lambda)$. We set

$$T^D(\lambda)y := y' - A(\cdot,\lambda)y \quad (y \in (W_p^1(a,b))^n,\ \lambda \in \Omega). \tag{2.5.1}$$

LEMMA 2.5.1. $T^D \in H(\Omega, L((W_p^1(a,b))^n, (L_p(a,b))^n))$.

Proof. Proposition 2.3.3 yields $A \in H(\Omega, L((W_p^1(a,b))^n, (L_p(a,b))^n))$. □

DEFINITION 2.5.2. Let $\lambda_0 \in \Omega$. A matrix $Y_0 \in M_n(W_p^1(a,b))$ is called a *fundamental matrix of* $T^D(\lambda_0)y = 0$ if for each $y \in N(T^D(\lambda_0))$ there is a $c \in \mathbb{C}^n$ such that $y = Y_0 c$.
A matrix function $Y : \Omega \to M_n(W_p^1(a,b))$ is called a *fundamental matrix function of* $T^D y = 0$ if $Y(\lambda)$ is a fundamental matrix of $T^D(\lambda)y = 0$ for each $\lambda \in \Omega$.

THEOREM 2.5.3. *Let* T^D *be given by* (2.5.1). *Then there is a fundamental matrix function* $Y \in H(\Omega, M_n(W_p^1(a,b)))$ *of* $T^D y = 0$ *with* $Y(a,\lambda) = \mathrm{id}_{\mathbb{C}^n}$ *for all* $\lambda \in \Omega$. *In addition,* $Y(\cdot,\lambda)$ *is invertible in* $M_n(W_p^1(a,b))$ *for all* $\lambda \in \Omega$.

Proof. For $\lambda \in \Omega$, $B \in M_n(L_p(a,b))$, $y \in (C[a,b])^n$, and $x \in [a,b]$ we set

$$((IB)y)(x) := \int_a^x B(t)y(t)\,dt\,.$$

Since $By \in (L_1(a,b))^n$, we obtain $(IB)y \in (C[a,b])^n$. HÖLDER'S inequality yields

$$|((IB)y)(x)| \le (b-a)^{1-\frac{1}{p}}|B|_p|y|_{(0)}$$

for each $x \in [a,b]$. This proves $I \in L(M_n(L_p(a,b)), L((C[a,b])^n))$. For $\lambda \in \Omega$, $y \in (C[a,b])^n$, and $x \in [a,b]$ we set

$$(K(\lambda)y)(x) := \int_a^x A(t,\lambda)y(t)\,dt\;.$$

Since $K(\lambda) = IA(\cdot,\lambda)$, Corollary 1.2.4 yields $K \in H(\Omega, L((C[a,b])^n))$. Let $\gamma \ge 0$ and set

$$\|y\|_\gamma := \max_{x\in[a,b]} |y(x)|e^{-\gamma(x-a)} \quad (y \in (C[a,b])^n).$$

For each $y \in (C[a,b])^n$ we have

$$\|y\|_\gamma \le |y|_{(0)} \le e^{\gamma(b-a)}\|y\|_\gamma,$$

which shows that the norms $|\ |_{(0)}$ and $\|\ \|_\gamma$ are equivalent. Hence $(C[a,b])^n$ is a Banach space with respect to the norm $\|\ \|_\gamma$.

Let $\lambda \in \Omega$. We shall prove that there is a $\gamma \ge 0$ such that $\|K(\lambda)\|_\gamma < 1$, where $\|K(\lambda)\|_\gamma$ is the operator norm of $K(\lambda)$ on $L((C[a,b])^n)$ induced by the norm $\|\ \|_\gamma$ on $(C[a,b])^n$. Since the function $t \mapsto |A(t,\lambda)|$ belongs to $L_p(a,b) \subset L_1(a,b)$, the function

$$x \mapsto \int_a^x |A(t,\lambda)|\,\mathrm{d}t$$

is uniformly continuous. Hence there is a $\delta \in (0, b-a)$ such that

$$\left|\int_\alpha^\beta |A(t,\lambda)|\,\mathrm{d}t\right| \le \frac{1}{4} \tag{2.5.2}$$

for all $\alpha, \beta \in [a,b]$ with $|\alpha - \beta| \le \delta$. For $y \in (C[a,b])^n$ and $x \in [a,b]$ we have

$$\begin{aligned}
|(K(\lambda)y)(x)|e^{-\gamma(x-a)} &= \left|\int_a^x A(t,\lambda)y(t)\,\mathrm{d}t\right| e^{-\gamma(x-a)} \\
&\le \int_a^x |A(t,\lambda)|e^{-\gamma(x-t)}|y(t)|e^{-\gamma(t-a)}\,\mathrm{d}t \\
&\le \int_a^x |A(t,\lambda)|e^{-\gamma(x-t)}\,\mathrm{d}t\,\|y\|_\gamma .
\end{aligned}$$

For $x \in [a, a+\delta]$, (2.5.2) implies

$$\int_a^x |A(t,\lambda)|e^{-\gamma(x-t)}\,\mathrm{d}t \le \frac{1}{4},$$

and for $x \in (a+\delta, b]$, (2.5.2) implies

$$\begin{aligned}
\int_a^x |A(t,\lambda)|e^{-\gamma(x-t)}\,\mathrm{d}t &\le \int_{x-\delta}^x |A(t,\lambda)|\,\mathrm{d}t + e^{-\gamma\delta}\int_a^{x-\delta} |A(t,\lambda)|\,\mathrm{d}t \\
&\le \frac{1}{4} + ne^{-\gamma\delta}|A(\cdot,\lambda)|_1 .
\end{aligned}$$

Choose $\gamma \ge 0$ such that $e^{-\gamma\delta}|A(\cdot,\lambda)|_1 \le \frac{1}{4n}$. Then the above estimates yield

$$\|K(\lambda)\|_\gamma \le \frac{1}{2},$$

and we obtain that $\mathrm{id}_{(C[a,b])^n} - K(\lambda)$ is invertible for all $\lambda \in \Omega$. From Proposition 1.2.5 we infer that $(\mathrm{id}_{(C[a,b])^n} - K)^{-1} \in H(\Omega, L((C[a,b])^n))$. For $k = 1, \ldots, n$ define

$$Y(\cdot,\lambda)e_k := (\mathrm{id}_{(C[a,b])^n} - K(\lambda))^{-1}e_k \quad (\lambda \in \Omega), \tag{2.5.3}$$

where e_k is the k-th unit vector in $\mathbb{C}^n$. Then $Y(\cdot,\lambda) \in M_n(C[a,b])$ and depends holomorphically on λ.

We apply $\mathrm{id}_{(C[a,b])^n} - K(\lambda)$ to $Y(\cdot,\lambda)$ and obtain

$$Y(x,\lambda) - \int_a^x A(t,\lambda)Y(t,\lambda)\,\mathrm{d}t = \mathrm{id}_{\mathbb{C}^n} \quad (x \in [a,b]). \tag{2.5.4}$$

As $A(\cdot,\lambda)Y(\cdot,\lambda) \in M_n(L_p(a,b))$, Proposition 2.1.5 i) gives $Y(\cdot,\lambda) \in M_n(W_p^1(a,b))$ and

$$Y'(\cdot,\lambda) = A(\cdot,\lambda)Y(\cdot,\lambda). \tag{2.5.5}$$

The multiplication from $M_n(L_p(a,b)) \times M_n(C([a,b]))$ to $M_n(L_p(a,b))$ is a continuous bilinear map. According to Proposition 1.2.3, $AY \in H(\Omega, M_n(L_p(a,b)))$. Proposition 1.2.3, Proposition 2.1.8 and (2.5.4) yield $Y \in H(\Omega, M_n(W_p^1(a,b)))$. From (2.5.4) we infer $Y(a,\lambda) = \mathrm{id}_{\mathbb{C}^n}$.

Let $\lambda \in \Omega$ and $y \in N(T^D(\lambda))$. Then, by Proposition 2.1.5 i),

$$y(x) = y(a) + \int_a^x A(t,\lambda)y(t)\,\mathrm{d}t \quad (x \in [a,b]),$$

and thus

$$((\mathrm{id}_{(C[a,b])^n} - K(\lambda))y)(x) = y(a) \quad (x \in [a,b]).$$

This proves

$$y = Y(\cdot,\lambda)y(a).$$

We still have to prove that $Y(\cdot,\lambda)$ is invertible in $M_n(W_p^1(a,b))$ for each $\lambda \in \Omega$. We consider the operator $\widetilde{T}^D(\lambda) \in L((W_p^1(a,b))^n, (L_p(a,b))^n)$ given by

$$\widetilde{T}^D(\lambda)y = y' + A^\mathsf{T}(\cdot,\lambda)y \quad (y \in (W_p^1(a,b))^n),$$

where A^T denotes the transposed matrix function of A. For fixed $\lambda \in \Omega$ we know that there exists a fundamental matrix $\widetilde{Y}_0$ of $\widetilde{T}^D(\lambda)y = 0$ with $\widetilde{Y}_0(a) = \mathrm{id}_{\mathbb{C}^n}$. Set $Y_0 := Y(\cdot,\lambda)$. Since $Y_0, \widetilde{Y}_0 \in M_n(W_p^1(a,b))$, Proposition 2.1.5 i) and Proposition 2.3.1 yield

$$\begin{aligned}(\widetilde{Y}_0^\mathsf{T} Y_0)(x) &= (\widetilde{Y}_0^\mathsf{T} Y_0)(a) + \int_a^x \widetilde{Y}_0^\mathsf{T}(t)Y_0'(t)\,\mathrm{d}t + \int_a^x \widetilde{Y}_0'^\mathsf{T}(t)Y_0(t)\,\mathrm{d}t \\ &= \mathrm{id}_{\mathbb{C}^n}\end{aligned} \tag{2.5.6}$$

since $Y_0' = A(\cdot,\lambda)Y_0$ and $\widetilde{Y}_0' = -A^\mathsf{T}(\cdot,\lambda)\widetilde{Y}_0$. This proves that $\widetilde{Y}_0^\mathsf{T}$ is the inverse of $Y(\cdot,\lambda)$. $\square$

PROPOSITION 2.5.4. *Let T^D be given by* (2.5.1), *$\lambda_0 \in \Omega$, and Y_1 be a fundamental matrix of $T^D(\lambda_0)y = 0$. Then $Y_0 \in M_n(W_p^1(a,b))$ is a fundamental matrix of $T^D(\lambda_0)y = 0$ if and only if there is an invertible matrix $C \in M_n(\mathbb{C})$ such that $Y_0 = Y_1 C$.*

Proof. Obviously, it is sufficient to consider the case $Y_1 = Y(\cdot,\lambda_0)$, where Y is the fundamental matrix function obtained in Theorem 2.5.3. Let Y_0 be a fundamental matrix of $T^D(\lambda_0)y = 0$. We define $C := Y_0(a)$. Let $d \in \mathbb{C}^n$. Since $T^D(\lambda_0)Y(\cdot,\lambda_0)d = 0$ by (2.5.5), there is a vector $c \in \mathbb{C}^n$ such that $Y_0c = Y(\cdot,\lambda_0)d$. We infer that $Cc = Y_0(a)c = d$, which implies that C is surjective and hence invertible and that $Y_0C^{-1} = Y(\cdot,\lambda_0)$.

Now let $Y_0 = Y(\cdot,\lambda_0)C$. For $y \in N(T^D(\lambda_0))$ there is a vector $d \in \mathbb{C}^n$ such that $y = Y(\cdot,\lambda_0)d$. Set $c := C^{-1}d$. Then $y = Y(\cdot,\lambda_0)Cc = Y_0c$. □

From Proposition 2.5.4 and (2.5.5) we immediately infer

COROLLARY 2.5.5. *Let T^D be given by* (2.5.1), *$\lambda_0 \in \Omega$, and Y_0 be a fundamental matrix of $T^D(\lambda_0)y = 0$. Then*

$$Y_0' - A(\cdot,\lambda_0)Y_0 = 0.$$

For each invertible matrix C in $M_n(\mathbb{C})$ there is exactly one fundamental matrix Y_C of $T^D(\lambda_0)y = 0$ with $Y_C(a) = C$. In particular, there is a unique fundamental matrix function Y of $T^Dy = 0$ with $Y(a,\lambda) = \mathrm{id}_{\mathbb{C}^n}$ for all $\lambda \in \Omega$.

PROPOSITION 2.5.6. *Let T^D be given by* (2.5.1) *and let Y be the fundamental matrix function of $T^Dy = 0$ with $Y(a,\lambda) = \mathrm{id}_{\mathbb{C}^n}$ for all $\lambda \in \Omega$. Set*

$$(Z(\lambda)c)(x) := Y(x,\lambda)c \quad (\lambda \in \Omega,\ c \in \mathbb{C}^n,\ x \in (a,b)).$$

Then $Z \in H(\Omega, L(\mathbb{C}^n, (W_p^1(a,b))^n))$, and $Z(\lambda)$ is injective for all $\lambda \in \Omega$.

Proof. By Theorem 2.5.3, $Y \in H(\Omega, M_n(W_p^1(a,b)))$. For $V \in M_n(W_p^1(a,b))$ and $c \in \mathbb{C}^n$, $|Vc|_{p,1} \le n|V|_{p,1}|c|$. Hence the "multiplication" from $M_n(W_p^1(a,b))$ to $L(\mathbb{C}^n, (W_p^1(a,b))^n)$ is continuous, which proves $Z \in H(\Omega, L(\mathbb{C}^n, (W_p^1(a,b))^n))$. Since $(Z(\lambda)c)(a) = c$, we obtain that $Z(\lambda)$ is injective. □

LEMMA 2.5.7. *Let T^D be given by* (2.5.1) *and let Y be the fundamental matrix function of $T^Dy = 0$ with $Y(a,\lambda) = \mathrm{id}_{\mathbb{C}^n}$ for all $\lambda \in \Omega$. For $\lambda \in \Omega$, $f \in (L_p(a,b))^n$ and $x \in (a,b)$ we set*

$$(U(\lambda)f)(x) := Y(x,\lambda)\int_a^x Y(t,\lambda)^{-1}f(t)\,\mathrm{d}t.$$

Then $U \in H(\Omega, L((L_p(a,b))^n, (W_p^1(a,b))^n))$ is a holomorphic right inverse of T^D.

Proof. From Theorem 2.5.3 we have that $Y \in H(\Omega, M_n(W_p^1(a,b)))$, and therefore $Y^{-1} \in H(\Omega, M_n(W_p^1(a,b)))$ by Theorem 2.5.3 and Propositions 2.3.3 and 1.2.5. From Propositions 2.3.3, 2.1.8 and Corollary 1.2.4 we infer that

$$f \mapsto \int_a^x Y(t,\lambda)^{-1}f(t)\,\mathrm{d}t$$

defines a continuous linear map from $(L_p(a,b))^n$ into $(W_p^1(a,b))^n$, which depends holomorphically on λ. An application of Proposition 2.3.3 and Corollary 1.2.4 shows $U \in H(\Omega, L((L_p(a,b))^n, (W_p^1(a,b))^n))$. Since $Y'(\cdot,\lambda) = A(\cdot,\lambda)Y(\cdot,\lambda)$, Propositions 2.3.1 and 2.1.5 i) yield

$$(U(\lambda)f)' = A(\cdot,\lambda)U(\lambda)f + Y(\cdot,\lambda)Y(\cdot,\lambda)^{-1}f \quad (f \in (L_p(a,b))^n),$$

whence

$$T^D(\lambda)U(\lambda)f = f \quad (f \in (L_p(a,b))^n). \quad \square$$

PROPOSITION 2.5.8. *Let $k \in \mathbb{N} \setminus \{0\}$ and $C \in M_n(W_p^k(a,b))$. Suppose that the matrix $C(x)$ is invertible for almost all $x \in (a,b)$ and that $C^{-1} \in M_n(L_{p'}(a,b))$. Then $C^{-1} \in M_n(W_p^k(a,b))$ and*

$$C^{-1\prime} = -C^{-1}C'C^{-1}. \tag{2.5.7}$$

Proof. Since $C^{-1}C' \in M_n(L_1(a,b))$, Corollary 2.5.5 yields that

$$y' + C^{-1}C'y = 0$$

has a fundamental matrix $Y \in M_n(W_1^1(a,b))$ with $Y(a) = C(a)^{-1}$. With the aid of Proposition 2.3.1 and Corollary 2.5.5 we infer

$$(CY)' = C'Y + CY' = 0. \tag{2.5.8}$$

Hence CY is constant, see [HÖ2, Theorem 3.1.4], and $C(a)Y(a) = \mathrm{id}_{\mathbb{C}^n}$ yields

$$C^{-1} = Y \in M_n(W_p^1(a,b)).$$

In addition, (2.5.8) implies (2.5.7). Since $k > 0$, (2.5.7) and Proposition 2.3.1 yield that $Y' \in M_n(L_p(a,b))$, whence $Y \in M_n(W_p^1(a,b))$. Now let $2 \le j \le k$ and suppose that $C^{-1} = Y \in M_n(W_p^{j-1}(a,b))$. From (2.5.7) and Proposition 2.3.1 it follows that $Y' \in M_n(W_p^{j-1}(a,b))$, which proves $Y \in M_n(W_p^j(a,b))$. By induction we obtain $Y \in M_n(W_p^k(a,b))$. $\square$

PROPOSITION 2.5.9. *Let $Y_0 \in M_n(W_p^1(a,b))$, $\lambda_0 \in \mathbb{C}$, and assume that*

$$Y_0' - A(\cdot,\lambda_0)Y_0 = 0$$

and that $Y_0(c)$ is invertible for some $c \in [a,b]$. Then Y_0 is a fundamental matrix of $T^D(\lambda_0)y = 0$.

Proof. Let Y be the fundamental matrix function of $T^D y = 0$ given in Theorem 2.5.3. Since $Y_1 := Y(\cdot,\lambda_0)$ is invertible in $M_n(W_p^1(a,b))$, Proposition 2.5.8 yields

$$\begin{aligned}(Y_1^{-1}Y_0)' &= -Y_1^{-1}Y_1'Y_1^{-1}Y_0 + Y_1^{-1}Y_0' \\ &= -Y_1^{-1}A(\cdot,\lambda_0)Y_1Y_1^{-1}Y_0 + Y_1^{-1}A(\cdot,\lambda_0)Y_0 = 0.\end{aligned}$$

Hence $Y_1^{-1}Y_0$ is constant, see [HÖ2, Theorem 3.1.4]. Since $Y_1^{-1}(c)Y_0(c)$ is invertible, an application of Proposition 2.5.4 completes the proof. $\square$

2.6. Regularity of solutions of differential equations

PROPOSITION 2.6.1. *Let* $k,n \in \mathbb{N}$, $p_i \in W_p^{k+i}(a,b)$ *for* $i = 0,\dots,n$, $p_n^{-1} \in L_{p'}(a,b)$, $\zeta \in L_p(a,b)$. *If* $k = 0$, *we additionally require that* $p_0\zeta \in L_1(a,b)$. *Assume that*

$$\sum_{i=0}^{n}(p_i\zeta)^{(i)} \in W_p^k(a,b). \tag{2.6.1}$$

Then $\zeta \in W_p^{k+n}(a,b)$.

Proof. In case $n = k = 0$ nothing has to be proved. For $n = 0$ and $k > 0$ the statement follows from Propositions 2.5.8 and 2.3.1. Let $n \geq 1$ and suppose that the statement holds for $n-1$. Since $p_i \in W_p^k(a,b) \subset L_\infty(a,b)$ if $k \geq 1$, we have for any k and $i = 0,\dots,n$ that $p_i\zeta \in L_1(a,b) \subset \mathscr{D}'(a,b)$. Thus $(p_i\zeta)^{(i)}$ is well-defined. Suppose that $\zeta \notin W_p^{k+n}(a,b)$. Then there is a $j \in \{0,\dots,k+n-1\}$ such that $\zeta \in W_p^j(a,b) \setminus W_p^{j+1}(a,b)$. Let $j' := \min\{j,k\}$. Set $q = p$ if $j' > 0$ and $q = 1$ if $j' = 0$. Then, in view of (2.6.1) and Proposition 2.3.1,

$$\Big(\sum_{i=0}^{n-1}(p_{i+1}\zeta)^{(i)}\Big)' = \sum_{i=0}^{n}(p_i\zeta)^{(i)} - p_0\zeta \in W_q^{j'}(a,b),$$

whence

$$\sum_{i=0}^{n-1}(p_{i+1}\zeta)^{(i)} \in W_q^{j'+1}(a,b)$$

by Corollary 2.1.4. Hence $\zeta \in W_q^{j'+1+n-1}(a,b) = W_q^{j'+n}(a,b)$ by induction hypothesis. In case $j' = 0$ we obtain $\zeta \in W_1^n(a,b) \subset L_\infty(a,b)$. Therefore $p_0\zeta \in L_p(a,b)$, and by repeating the above step we now can take $q = p$ also in case $j' = 0$. If $j' = j$, we obtain the contradiction $\zeta \in W_p^{j+n}(a,b) \subset W_p^{j+1}(a,b)$. If $j' = k$, we obtain the contradiction $\zeta \in W_p^{k+n}(a,b)$. □

PROPOSITION 2.6.2. *Let* $l \in \mathbb{N}$, $n \in \mathbb{N}\setminus\{0\}$, $p_i \in W_p^i(a,b)$ $(i = 0,\dots,n)$. *If* $l = 0$, *we additionally assume that* $p_i \in W^i_{\max\{p,p'\}}(a,b)$ $(i = 0,\dots,n)$. *Set*

$$q_i := \sum_{j=i}^{n}(-1)^{j-i}\binom{j}{i}p_j^{(j-i)} \quad (i = 0,\dots,n).$$

For $\zeta \in W_p^l(a,b)$ *we have*

$$\sum_{i=0}^{n} p_i\zeta^{(i)} = \sum_{i=0}^{n}(q_i\zeta)^{(i)}$$

in the sense of distributions, where $p_i\zeta^{(i)}$ *satisfies the equation*

$$\langle\varphi, p_i\zeta^{(i)}\rangle = (-1)^i\int_a^b(\varphi p_i)^{(i)}(x)\zeta(x)\,\mathrm{d}x \quad (\varphi \in C_0^\infty(a,b)) \tag{2.6.2}$$

if $p_i\zeta^{(i)}$ *defines a regular distribution; otherwise,* $p_i\zeta^{(i)}$ *is defined by* (2.6.2).

Proof. First we note that in case $p_i\zeta^{(i)}$ is a regular distribution the theorem on integration by parts, see [HS, (18.19)], yields (2.6.2). Now let $\varphi \in C_0^\infty(a,b)$ and $j \in \{0,\dots,n\}$. A recursive application of Proposition 2.3.1 to φp_i shows that the LEIBNIZ rule holds:

$$(\varphi p_i)^{(i)} = \sum_{j=0}^{i} \binom{i}{j} \varphi^{(j)} p_i^{(i-j)} \quad (i = 0,\dots,n).$$

Therefore

$$\begin{aligned}\langle \varphi, p_i\zeta^{(i)}\rangle &= (-1)^i \sum_{j=0}^{i} \binom{i}{j} \int_a^b \varphi^{(j)}(x) p_i^{(i-j)}(x)\zeta(x)\,\mathrm{d}x \\ &= (-1)^i \sum_{j=0}^{i} \binom{i}{j} \langle \varphi^{(j)}, p_i^{(i-j)}\zeta\rangle \\ &= \sum_{j=0}^{i} (-1)^{i-j} \binom{i}{j} \langle \varphi, (p_i^{(i-j)}\zeta)^{(j)}\rangle,\end{aligned}$$

which proves

$$p_i\zeta^{(i)} = \sum_{j=0}^{i} (-1)^{i-j} \binom{i}{j} (p_i^{(i-j)}\zeta)^{(j)}. \quad \square$$

REMARK 2.6.3. The distribution $p_i\zeta^{(i)}$ in Proposition 2.6.2 is regular in case $i \le l$. If $p > 1$ and $i > l$, then $(\zeta^{(i)})_e \in W_p^{l-i}[a,b]$ by (2.2.7), $p_i(\zeta^{(i)})_e \in W_p^{l-i}[a,b]$ in the sense of Proposition 2.3.4, and $p_i\zeta^{(i)}$ is the restriction of $p_i(\zeta^{(i)})_e$ to (a,b).

PROPOSITION 2.6.4. *Let $k,l \in \mathbb{N}$, $n \in \mathbb{N}\setminus\{0\}$, $p_i \in W_p^{k+i}(a,b)$ for $i = 0,\dots,n$, $p_n^{-1} \in L_{p'}(a,b)$, and $\zeta \in W_p^l(a,b)$. If $k = l = 0$, we additionally assume that $p_i \in W^i_{\max\{p,p'\}}(a,b)$ $(i = 0,\dots,n)$. Assume that*

$$\sum_{i=0}^{n} p_i\zeta^{(i)} \in W_p^k(a,b).$$

Then $\zeta \in W_p^{k+n}(a,b)$.

Proof. The assumptions of Proposition 2.6.2 are satisfied. Let $q_1,\dots,q_n$ be as defined there. For $\sum_{i=0}^{n} (q_i\zeta)^{(i)}$ the assumptions of Proposition 2.6.1 are satisfied. Therefore $\zeta \in W_p^{k+n}(a,b)$ follows from Proposition 2.6.1. $\square$

PROPOSITION 2.6.5. *Let $n \in \mathbb{N}$ and $u \in W_p^n(a,b)$. Then*

$$u_e^{(n)} = (u^{(n)})_e + \sum_{i=0}^{n-1} \left(u^{(i)}(a)\delta_a^{(n-1-i)} - u^{(i)}(b)\delta_b^{(n-1-i)}\right) \tag{2.6.3}$$

holds in $\mathscr{D}'(\mathbb{R})$, where u_e is the canonical extension of u.

Proof. The statement is trivial if $n=0$. Let $n=1$ and $\varphi \in C_0^\infty(\mathbb{R})$. Then, by the theorem on integration by parts, see [HS, (18.19)],

$$\begin{aligned}\langle \varphi, u_e' \rangle_{\mathbb{R}} &= -\int_{\mathbb{R}} \varphi'(x) u_e(x)\,\mathrm{d}x \\ &= \int_a^b \varphi(x)u'(x)\,\mathrm{d}x - \varphi(b)u(b) + \varphi(a)u(a) \\ &= \langle \varphi, (u')_e \rangle_{\mathbb{R}} - \langle \varphi, u(b)\delta_b \rangle_{\mathbb{R}} + \langle \varphi, u(a)\delta_a \rangle_{\mathbb{R}}.\end{aligned}$$

This proves (2.6.3) for $n=1$. Suppose that (2.6.3) holds for $n-1$, where $n \geq 2$. Then an application of (2.6.3) for $n-1$ and 1 yields

$$\begin{aligned}u_e^{(n)} &= \left(u_e^{(n-1)}\right)' \\ &= \left(u^{(n-1)}\right)_e' + \sum_{i=0}^{n-2} \left(u^{(i)}(a)\delta_a^{(n-2-i)} - u^{(i)}(b)\delta_b^{(n-2-i)}\right)' \\ &= \left(u^{(n)}\right)_e + u^{(n-1)}(a)\delta_a - u^{(n-1)}(b)\delta_b \\ &\quad + \sum_{i=0}^{n-2} \left(u^{(i)}(a)\delta_a^{(n-1-i)} - u^{(i)}(b)\delta_b^{(n-1-i)}\right). \quad \square\end{aligned}$$

2.7. Estimates of integrals with a complex parameter

We shall often deal with functions having a special asymptotic behaviour when the parameter λ tends to infinity. For this we introduce some notations.

Let U be an unbounded subset of $\mathbb{C}$, f be a function on U with values in $M_{k,n}(\mathbb{C})$ and g be a complex-valued function on U. We write

$$f(\lambda) = O(g(\lambda))$$

if there is a $C > 0$ such that $|f(\lambda)| \leq C|g(\lambda)|$ for $\lambda \in U$. The notation

$$f(\lambda) = o(g(\lambda))$$

means that $|f(\lambda)|\,|g(\lambda)|^{-1} \to 0$ as $\lambda \to \infty$ in U. Let $a \in M_{k,n}(\mathbb{C})$. We write

$$f(\lambda) = [a]$$

if $f(\lambda) - a = o(1)$.

Now let $f(\cdot,\lambda) \in M_{k,n}(L_p(a,b))$ for $\lambda \in U$ and, as above, g be a complex-valued function on U. We write

$$f(\cdot,\lambda) = \{O(g(\lambda))\}_p \quad \text{or} \quad f(\cdot,\lambda) = O(g(\lambda)) \text{ in } M_{k,n}(L_p(a,b))$$

if there is a $C > 0$ such that $|f(\cdot,\lambda)|_p \leq C|g(\lambda)|$ for $\lambda \in U$, and

$$f(\cdot,\lambda) = \{o(g(\lambda))\}_p \quad \text{or} \quad f(\cdot,\lambda) = o(g(\lambda)) \text{ in } M_{k,n}(L_p(a,b))$$

if $|f(\cdot,\lambda)|_p|g(\lambda)|^{-1} \to 0$ as $\lambda \to \infty$. For $h \in M_{k,n}(L_p(a,b))$, we write

$$f(\cdot,\lambda) = [h]_p$$

if $f(\cdot,\lambda) - h = \{o(1)\}_p$.

PROPOSITION 2.7.1. *Let $r \in L_p(a,b)$ such that $r(x) \geq 0$ for all $x \in (a,b)$ and $r^{-1} \in L_\infty(a,b)$. For $x \in [a,b]$ we set*

$$R(x) := \int_a^x r(\xi)\,\mathrm{d}\xi.$$

Let $1 \leq p_1 \leq \infty$ and $1 \leq p_2 \leq \infty$ such that $\frac{1}{p} + \frac{1}{p_1} - 1 = \frac{1}{p_2}$. Let $u \in L_p(a,b)$ and $v \in L_{p_1}(a,b)$. Then there is a function $h \in L_{p_2}(0,R(b))$ such that

$$|h|_{p_2} \leq |r^{-1}|_\infty^{1-1/p_2}|v|_{p_1}|u|_p$$

and

$$\int_a^b v(x)\exp\{\lambda R(x)\}\int_c^x \exp\{-\lambda R(\xi)\}u(\xi)\,\mathrm{d}\xi\,\mathrm{d}x = \int_0^{R(b)} \exp\{\varepsilon_c\lambda\tau\}h(\tau)\,\mathrm{d}\tau,$$

where $c = a$ or $c = b$, $\varepsilon_c = 1$ if $c = a$, $\varepsilon_c = -1$ if $c = b$.

Proof. Let $\rho : [0,R(b)] \to [a,b]$ be the inverse of the absolutely continuous increasing function $x \mapsto R(x)$. Then ρ is also a continuous increasing function. Since the function $\tau \to \frac{1}{r(\rho(\tau))}$ is the composite of a continuous and a bounded measurable function, it is a bounded measurable function. We define the function $f : [0,R(b)] \to \mathbb{R}$ by

$$f(t) := a + \int_0^t \frac{\mathrm{d}\tau}{r(\rho(\tau))}.$$

Since R is absolutely continuous with $R' = r$, the theorem on integration by substitution, see e.g. [HS, (20.5)] or [MS, 38.4], yields

$$\begin{aligned} f(R(x)) &= a + \int_0^{R(x)} \frac{\mathrm{d}\tau}{r(\rho(\tau))} \\ &= a + \int_a^x \frac{1}{r(t)}r(t)\,\mathrm{d}t = x \end{aligned}$$

for $x \in [a,b]$. Hence $f = \rho$, which proves $\rho \in W_\infty^1(0,R(b))$ and $\rho' = (r \circ \rho)^{-1}$, see Proposition 2.1.5 i).

Let $c = a$. We apply the substitutions $x = \rho(t)$, $\xi = \rho(\tau)$ and $\tau \mapsto t - \tau$ and obtain, again using the theorem on integration by substitution,

$$\begin{aligned}\int_a^b & v(x)\exp\{\lambda R(x)\}\int_a^x \exp\{-\lambda R(\xi)\}u(\xi)\,\mathrm{d}\xi\,\mathrm{d}x \\ &= \int_0^{R(b)} v(\rho(t))\exp\{\lambda t\}\int_a^{\rho(t)} \exp\{-\lambda R(\xi)\}u(\xi)\rho'(t)\,\mathrm{d}\xi\,\mathrm{d}t \\ &= \int_0^{R(b)} v(\rho(t))\int_0^t \exp\{\lambda(t-\tau)\}u(\rho(\tau))\rho'(\tau)\rho'(t)\,\mathrm{d}\tau\,\mathrm{d}t \\ &= \int_0^{R(b)} \exp\{\lambda\tau\}h(\tau)\,\mathrm{d}\tau,\end{aligned}$$

where

$$h(\tau) = \int_\tau^{R(b)} v(\rho(t))\rho'(t)u(\rho(t-\tau))\rho'(t-\tau)\,\mathrm{d}t.$$

The function h is the convolution of the two functions h_1 and h_2 given by

$$h_1(t) = \begin{cases}(v(\rho(t))\rho'(t) & \text{if } t \in (0, R(b)),\\ 0 & \text{if } t \in \mathbb{R}\setminus(0, R(b)),\end{cases}$$

and

$$h_2(t) = \begin{cases}(u(\rho(-t))\rho'(-t) & \text{if } t \in (-R(b), 0),\\ 0 & \text{if } t \in \mathbb{R}\setminus(-R(b), 0).\end{cases}$$

Hence we have

$$|h|_{p_2} \le |(v\circ\rho)\rho'|_{p_1}\,|(u\circ\rho)\rho'|_p,$$

see [HS, (21.31), (21.32), (21.33), and (21.56)]. For $1 \le q < \infty$ and $w \in L_q(a,b)$ the theorem on integration by substitution yields that

$$(2.7.1) \qquad |(w\circ\rho)\rho'|_q^q \le |\rho'|_\infty^{q-1}|(w\circ\rho)^q\rho'|_1 = |r^{-1}|_\infty^{q-1}|w|_q^q.$$

This completes the proof of the proposition in the case $c = a$.

If $c = b$, we have

$$\begin{aligned}\int_a^b & v(x)\exp\{\lambda R(x)\}\int_b^x \exp\{-\lambda R(\xi)\}u(\xi)\,\mathrm{d}\xi\,\mathrm{d}x \\ &= -\int_a^b u(\xi)\exp\{-\lambda R(\xi)\}\int_a^\xi \exp\{\lambda R(x)\}v(x)\,\mathrm{d}x\,\mathrm{d}\xi.\end{aligned}$$

Proceeding as above the proposition is proved. □

We need the following generalization of the RIEMANN-LEBESGUE lemma.

LEMMA 2.7.2. *Let $r \in L_1(a,b)$ such that $r(x) \ge 0$ for all $x \in (a,b)$ and $r^{-1} \in L_\infty(a,b)$. For $x \in [a,b]$ we set*

$$R(x) := \int_a^x r(\xi)\,\mathrm{d}\xi.$$

For $g \in L_p(a,b)$, $x,y \in [a,b]$ and $\lambda \in \mathbb{C}$ we define

$$F(g,x,y,\lambda) := \int_y^x \exp\{\lambda(R(x)-R(t))\}g(t)\,\mathrm{d}t$$

and

$$v(g,\lambda) := \max\{|F(g,x,y,\lambda)| : x,y \in [a,b],\ \Re(\lambda)(x-y) \leq 0\}.$$

We assert:
i) *Let $g_0, g(\lambda) \in L_p(a,b)$, where $g(\lambda) = g_0 + o(1)$ in $L_p(a,b)$. Then the estimate $v(g(\lambda),\lambda) = o(1)$ holds.*
ii) *There is a real number $C > 0$ such that*

$$v(g,\lambda) \leq C(1+|\Re(\lambda)|)^{1/p-1}|g|_p \quad (g \in L_p(a,b),\ \lambda \in \mathbb{C}).$$

iii) *Let $c(\lambda) = a$ if $\Re(\lambda) \leq 0$ or $c(\lambda) = b$ if $\Re(\lambda) \geq 0$. (In case $\Re(\lambda) = 0$ we can take either value.) Let $\tilde{p} \geq p$ and $1 \leq \hat{p} \leq \infty$ such that $\frac{1}{p} - \frac{1}{\tilde{p}} = \frac{1}{\hat{p}}$. Then there is a $C > 0$ such that*

$$|F(g,\cdot,c(\lambda),\lambda)|_{\tilde{p}} \leq C(1+|\Re(\lambda)|)^{1/\hat{p}-1}|g|_p \quad (g \in L_p(a,b),\ \lambda \in \mathbb{C}).$$

Proof. From the theorem on integration by substitution, see [HS, (20.5)], we infer

(2.7.2)
$$\begin{aligned}\int_y^\xi r(t)\exp\{\lambda(R(x)-R(t))\}\,\mathrm{d}t &= \int_{R(y)-R(x)}^{R(\xi)-R(x)} \exp\{-\lambda\tau\}\,\mathrm{d}\tau \\ &= \frac{1}{\lambda}\left(\exp\{\lambda(R(x)-R(y))\} - \exp\{\lambda(R(x)-R(\xi))\}\right)\end{aligned}$$

for ξ in the interval with the endpoints x and y.
i): Let $\varepsilon > 0$. Then there is a function $g_1 \in C^\infty[a,b]$ such that $|g_0 - g_1|_1 \leq \frac{\varepsilon}{4}$, see e.g. [HÖ2, Theorem 1.3.2]. There are a measurable set $M \subset (a,b)$ and a real number $K > 0$ such that

$$\int_M r(x)\,\mathrm{d}x \leq \frac{\varepsilon}{4\left(\left|\frac{g_1}{r}\right|_\infty + 1\right)},\ \int_M \mathrm{d}x \leq \frac{\varepsilon}{4(|g_1|_\infty + 1)},\ |r(1-\chi_M)|_\infty \leq K,$$

where χ_M is the characteristic function of M, see [HS, (12.34)] and its proof. According to [HÖ2, Theorem 1.3.2] we can choose a test function $\phi \in C_0^\infty(\mathbb{R})$ such that $\phi \geq 0$, $\int \phi(x)\,\mathrm{d}x = 1$ and $\left|(\frac{g_1}{r})_e * \phi - \frac{g_1}{r}\right|_1 \leq \frac{\varepsilon}{4K}$. Set $h := \left((\frac{g_1}{r})_e * \phi\right)\Big|_{[a,b]} \in C^\infty[a,b]$. From

$$\left|\int_{\mathbb{R}} \left(\frac{g_1}{r}\right)_e (x-y)\phi(y)\,\mathrm{d}y\right| \leq \left|\frac{g_1}{r}\right|_\infty \quad (x \in (a,b))$$

we infer $|h|_\infty \le \left|\frac{g_1}{r}\right|_\infty$. Then

$$\begin{aligned}|g_0 - hr|_1 &\le |g_0 - g_1|_1 + \left|\frac{g_1}{r} - h\right|_1 |r(1-\chi_M)|_\infty + |g_1\chi_M|_1 + |hr\chi_M|_1 \\ &\le \frac{\varepsilon}{4} + \frac{\varepsilon}{4} + \frac{\varepsilon}{4(|g_1|_\infty + 1)}|g_1|_\infty + |h|_\infty |r\chi_M|_1 \\ &\le \varepsilon .\end{aligned}$$

Note that

$$|\exp\{\lambda(R(x) - R(t))\}| = \exp\{\Re(\lambda)(R(x) - R(t))\} \le 1$$

for all $x, y \in [a,b]$ and t in the interval with endpoints x and y if $\Re(\lambda)(x-y) \le 0$. Hence

$$\left|\int_y^x \exp\{\lambda(R(x) - R(t))\}(g_0(t) - h(t)r(t))\,\mathrm{d}t\right| \le \varepsilon$$

for all $x, y \in [a,b]$ with $\Re(\lambda)(x-y) \le 0$. The formula for integration by parts, see [HS, (18.19)], and (2.7.2) yield

$$\begin{aligned}&\int_y^x \exp\{\lambda(R(x) - R(t))\}h(t)r(t)\,\mathrm{d}t \\ &= \frac{1}{\lambda}\left\{h(y)\exp\{\lambda(R(x) - R(y))\} - h(x) + \int_y^x \exp\{\lambda(R(x) - R(t))\}h'(t)\,\mathrm{d}t\right\} \\ &= \frac{1}{\lambda}C(x,y,\lambda),\end{aligned}$$

where $|C(x,y,\lambda)| \le 2|h|_\infty + |h_1'|$ for all $x, y \in [a,b]$ with $\Re(\lambda)(x-y) \le 0$. These two estimates prove $v(g_0, \lambda) \le 2\varepsilon$ if $|\lambda|$ is sufficiently large. By assumption, $v(g(\lambda) - g_0, \lambda) \le |g(\lambda) - g_0|_1 = o(1)$ as $\lambda \to \infty$. Then the assertion of part i) follows in view of $v(g(\lambda), \lambda) \le v(g_0, \lambda) + v(g(\lambda) - g_0, \lambda)$.

ii): The statement is obvious for $p = 1$. Now let $p > 1$ and $g \in L_p(a,b)$. With the aid of HÖLDER'S inequality we infer for $x, y \in [a,b]$ and $\lambda \in \mathbb{C}$ such that $\Re(\lambda)(x-y) \le 0$ and $|\Re(\lambda)| \ge 1$:

$$\begin{aligned}|F(g,x,y,\lambda)| &\le \left|\int_y^x \exp\{\Re(\lambda)(R(x) - R(t))\}|g(t)|\,\mathrm{d}t\right| \\ &= \left|\int_y^x |g(t)|\left(\frac{1}{r(t)}\right)^{1-1/p} r(t)^{1-1/p}\exp\{\Re(\lambda)(R(x) - R(t))\}\,\mathrm{d}t\right| \\ &\le \left|g(r^{-1})^{1-1/p}\right|_p \left|\int_y^x r(t)\exp\{p'\Re(\lambda)(R(x) - R(t))\}\,\mathrm{d}t\right|^{1-1/p} \\ &\le \left|r^{-1}\right|_\infty^{1-1/p} |g|_p |p'\Re(\lambda)|^{-1+1/p} \\ &\le 2^{1-1/p}|r^{-1}|_\infty^{1-1/p}(1 + |\Re(\lambda)|)^{-1+1/p}|g|_p .\end{aligned}$$

If $|\Re(\lambda)| \le 1$, we obtain

$$|F(g,x,y,\lambda)| \le (b-a)^{1-1/p}|g|_p \le (2(b-a))^{1-1/p}(1 + |\Re(\lambda)|)^{-1+1/p}|g|_p .$$

iii) For $p = \infty$ this follows from ii). Now let $p < \infty$. For $v \in L_{\tilde{p}'}(a,b)$ we obtain according to Proposition 2.7.1 that

$$\int_a^b v(x)F(g,x,c(\lambda),\lambda)\,\mathrm{d}x = \int_a^b v(x)\int_{c(\lambda)}^x \exp\{\lambda\,(R(x)-R(t))\}g(t)\,\mathrm{d}t\,\mathrm{d}x$$
$$= \int_0^{R(b)} \exp\{\varepsilon_{c(\lambda)}\lambda\tau\}h(\tau)\,\mathrm{d}\tau,$$

where $h \in L_{\hat{p}}(0,R(b))$ and $|h|_{\hat{p}} \le |r^{-1}|_\infty^{1-1/\hat{p}}|v|_{\tilde{p}'}|g|_p$ since $\frac{1}{p}+\frac{1}{\tilde{p}'}-1=\frac{1}{\hat{p}}$. In view of $\Re(\varepsilon_{c(\lambda)}\lambda) \le 0$, part ii) yields the estimate

$$\left|\int_0^{R(b)} \exp\{\varepsilon_{c(\lambda)}\lambda\tau\}h(\tau)\,\mathrm{d}\tau\right| = \left|\int_{R(b)}^0 \exp\{-\varepsilon_{c(\lambda)}\lambda(-\tau)\}h(\tau)\,\mathrm{d}\tau\right|$$
$$\le C'(1+|\Re(\lambda)|)^{1/\hat{p}-1}|h|_\infty,$$

where C' does not depend on h. Since $L_{p'}$ is the dual space of L_p, we infer

$$|F(g,\cdot,c(\lambda),\lambda)|_p \le C'|r^{-1}|_\infty(1+|\Re(\lambda)|)^{-1}|g|_p. \quad \square$$

REMARK 2.7.3. Let the notations be as in Lemma 2.7.2 and assume additionally that $r \in W_1^1(a,b)$. Then

$$v(g,\lambda) = O(\lambda^{-1})\left|\frac{g}{r}\right|_{1,1} \quad (g \in W_1^1(a,b)).$$

Proof. The estimate was essentially established in the proof of part ii) of Lemma 2.7.2 if we observe that we can take $h = \frac{g}{r}$ under the present assumptions. Here we also have to note that $\frac{g}{r}$ belongs to $W_1^1(a,b)$ by Propositions 2.5.8 and 2.3.1. $\square$

2.8. Asymptotic matrices

In this section we consider first order systems of differential equations

$$y' - \widetilde{A}(\cdot,\lambda)y = 0, \tag{2.8.1}$$

where, for some $k \in \mathbb{N}$ and $\gamma > 0$,

$$\widetilde{A}(\cdot,\lambda) = \sum_{j=-1}^{k} \lambda^{-j}A_{-j} + \lambda^{-k-1}A^k(\cdot,\lambda) \quad (|\lambda| \ge \gamma). \tag{2.8.2}$$

We shall construct a fundamental matrix of (2.8.1) which has an appropriate asymptotic behaviour for $\lambda \to \infty$. For this purpose we make the following

ASSUMPTION 2.8.1. We assume that

i) $A_1 \in M_n(W_p^k(a,b))$,

ii) $A_{-j} \in M_n(W_p^{k-j}(a,b)) \quad (j=0,\ldots,k)$,

iii) $A^k(\cdot,\lambda)$ belongs to $M_n(L_p(a,b))$ for $|\lambda| \geq \gamma$ and is bounded in $M_n(L_p(a,b))$ as $\lambda \to \infty$,

iv) A_1 has the diagonal form

$$A_1 = \begin{pmatrix} A_0^1 & & & & \\ & A_1^1 & & 0 & \\ & & \cdot & & \\ & 0 & & \cdot & \\ & & & & \cdot \\ & & & & & A_l^1 \end{pmatrix},$$

where $A_\nu^1 = r_\nu I_{n_\nu}$, $(\nu=0,\ldots,l)$, $\sum_{\nu=0}^{l} n_\nu = n$, I_{n_ν} is the $n_\nu \times n_\nu$ unit matrix.

v) For $\nu,\mu \in \{0,\ldots,l\}$ there are $\varphi_{\nu\mu} \in \mathbb{R}$ such that

$$r_\nu(x) - r_\mu(x) = |r_\nu(x) - r_\mu(x)|e^{i\varphi_{\nu\mu}} \tag{2.8.3}$$

holds for all $x \in [a,b]$. Finally we assume

$$(r_\nu - r_\mu)^{-1} \in L_\infty(a,b) \quad (\nu,\mu = 0,\ldots,l;\ \nu \neq \mu). \tag{2.8.4}$$

We set

$$R_\nu(x) := \int_a^x r_\nu(\xi)\,\mathrm{d}\xi \quad (\nu=0,\ldots,l; x \in [a,b]),$$

$$E_\nu(x,\lambda) := \exp(\lambda R_\nu(x))I_{n_\nu} \quad (\nu=0,\ldots,l; x \in [a,b]; \lambda \in \mathbb{C}),$$

$$E(x,\lambda) := \begin{pmatrix} E_0(x,\lambda) & & & & \\ & E_1(x,\lambda) & & 0 & \\ & & \cdot & & \\ & 0 & & \cdot & \\ & & & & \cdot \\ & & & & & E_l(x,\lambda) \end{pmatrix} \tag{2.8.5}$$

for $x \in [a,b]$ and $\lambda \in \mathbb{C}$.

For the matrices A_j and $P^{[r]}$, defined below, we form the block matrices

$$A_j =: (A_{j,\nu\mu})_{\nu,\mu=0}^{l} \quad \text{and} \quad P^{[r]} =: (P_{\nu\mu}^{[r]})_{\nu,\mu=0}^{l}$$

according to the block structure of A_1.

When applying the results of this section we shall not always assume that the dimensions n_ν are positive since some statements are easier to formulate if we allow n_ν to be zero. But $n_\nu = 0$ means that the corresponding entries do not occur. Therefore we may assume for the proofs in this section that all n_ν are positive.

THEOREM 2.8.2. *Let* Assumption 2.8.1 *be satisfied.*
A. *There are* $P^{[r]} \in M_n(W_p^{k+1-r}(a,b))$ $(r = 0,\dots,k)$ *such that the equations*

$$P^{[0]}A_1 - A_1 P^{[0]} = 0, \quad P^{[0]}(a) = I_n\,, \tag{2.8.6}$$

$$P^{[r]\prime} - \sum_{j=0}^{r} A_{-j} P^{[r-j]} + P^{[r+1]}A_1 - A_1 P^{[r+1]} = 0 \qquad (r = 0,\dots,k-1), \tag{2.8.7}$$

$$P_{\nu\nu}^{[k]\prime} - A_{0,\nu\nu}P_{\nu\nu}^{[k]} = \sum_{\substack{q=0\\ q\neq\nu}}^{l} A_{0,\nu q}P_{q\nu}^{[k]} + \sum_{j=1}^{k}\sum_{q=0}^{l} A_{-j,\nu q}P_{q\nu}^{[k-j]} \qquad (\nu = 0,\dots,l) \tag{2.8.8}$$

hold.
B. *For* $1 \le q \le \infty$ *we set*

$$\tau_q(\lambda) := \begin{cases} \max\limits_{\substack{\nu,\mu=0\\ \nu\neq\mu}}^{l} (1 + |\Re(\lambda e^{i\varphi_{\nu\mu}})|)^{-1+1/q} & \text{if } l > 0,\\ |\lambda|^{-1} & \text{if } l = 0. \end{cases} \tag{2.8.9}$$

For $r \in \{0,\dots,k\}$ *let the matrices* $P^{[r]}$ *belong to* $M_n(W_p^{k+1-r}(a,b))$ *and fulfil* (2.8.6), (2.8.7), (2.8.8). *We assert that for* $|\lambda| \ge \gamma$ *there is a matrix function* $B_k(\cdot,\lambda) \in M_n(W_p^1(a,b))$ *with the following properties:*
i) *For* $|\lambda| \ge \gamma$,

$$\widetilde{Y}(\cdot,\lambda) := \Big(\sum_{r=0}^{k} \lambda^{-r}P^{[r]} + \lambda^{-k}B_k(\cdot,\lambda)\Big)E(\cdot,\lambda) \tag{2.8.10}$$

is a fundamental matrix of the system (2.8.1).
ii) *For large* λ *we have the asymptotic estimates*

$$B_k(\cdot,\lambda) = \{o(1)\}_\infty\,, \tag{2.8.11}$$

$$B_k(\cdot,\lambda) = \{O(\tau_p(\lambda))\}_\infty\,, \tag{2.8.12}$$

$$B_k(\cdot,\lambda) = \{O(\tau_\infty(\lambda))\}_p \text{ if } k > 0, \tag{2.8.13}$$

$$\tfrac{1}{\lambda}B_k'(\cdot,\lambda) = \{o(1)\}_p\,, \tag{2.8.14}$$

$$\tfrac{1}{\lambda}B_k'(\cdot,\lambda) = \{O(\tau_p(\lambda))\}_p\,. \tag{2.8.15}$$

iii) Let $l > 0$. *If* $k = 0$ *and* $p \le \frac{3}{2}$, *then we additionally assume for* $\nu,\mu = 0,\dots,l$ *with* $\nu \neq \mu$ *that* $A_{0,\nu\mu} \in M_{n_\nu,n_\mu}(L_{p_{\nu\mu}}(a,b))$, *where* $1 \le p_{\nu\mu} \le \infty$ *are such that* $\frac{1}{p} + \frac{1}{p_{\nu q}} + \frac{1}{p_{q\mu}} < 2$ *for all* $\nu,\mu,q = 0,\dots,l$ *with* $\nu \neq q$ *and* $\mu \neq q$. *Then there is a number* $\varepsilon \in (0, 1-\frac{1}{p})$ *if* $p > 1$ *or* $\varepsilon = 0$ *if* $p = 1$ *such that*

$$B_0(\cdot,\lambda) = \left\{ O\left(\max_{\substack{\nu,\mu=0\\ \nu\neq\mu}}^{l} (1 + |\Re(\lambda e^{i\varphi_{\nu\mu}})|)^{-1/p-\varepsilon} \right) \right\}_p. \tag{2.8.16}$$

Proof. A. The equations (2.8.6), (2.8.7) and (2.8.8) are fulfilled if and only if the block submatrices of $P^{[r]}$ satisfy the relationships

$$(2.8.17)\quad P^{[0]}_{\nu\mu}=0\quad (\nu,\mu=0,\dots,l;\ \nu\neq\mu),\quad P^{[0]}_{\nu\nu}(a)=I_{n_\nu}\quad (\nu=0,\dots,l),$$

$$(2.8.18)\quad P^{[r]\prime}_{\nu\nu}-A_{0,\nu\nu}P^{[r]}_{\nu\nu}=\sum_{\substack{q=0\\ q\neq\nu}}^{l}A_{0,\nu q}P^{[r]}_{q\nu}+\sum_{j=1}^{r}\sum_{q=0}^{l}A_{-j,\nu q}P^{[r-j]}_{q\nu}$$
$$(\nu=0,\dots,l;\ r=0,\dots,k),$$

$$(2.8.19)\quad P^{[r+1]}_{\nu\mu}=(r_\nu-r_\mu)^{-1}\Big\{P^{[r]\prime}_{\nu\mu}-\sum_{j=0}^{r}\sum_{q=0}^{l}A_{-j,\nu q}P^{[r-j]}_{q\mu}\Big\}$$
$$(\nu,\mu=0,\dots,l;\ \nu\neq\mu;\ r=0,\dots,k-1).$$

Indeed, for $r=0,\dots,k$,

$$(2.8.20)\qquad P^{[r]}A_1-A_1P^{[r]}=\left((r_\mu-r_\nu)P^{[r]}_{\nu,\mu}\right)^{l}_{\nu,\mu=0}.$$

Hence (2.8.6) is fulfilled if and only if (2.8.17) holds, and (2.8.7) is fulfilled for $r\in\{0,\dots,k-1\}$ if and only if (2.8.18) and (2.8.19) hold for this r. Finally, (2.8.8) holds if and only if (2.8.18) holds for $r=k$.

We are going to solve the equations (2.8.17), (2.8.18), and (2.8.19). For $\nu\in\{0,\dots,l\}$ let $P^{[0]}_{\nu\nu}\in M_n(W^1_p(a,b))$ be the fundamental matrix of $y'-A_{0,\nu\nu}y=0$ with $P^{[0]}_{\nu\nu}(a)=I_{n_\nu}$, see Theorem 2.5.3. For $\nu\neq\mu$ we set $P^{[0]}_{\nu\mu}=0$. Then (2.8.17) and, for $r=0$, (2.8.18) are valid, see Corollary 2.5.5. Repeated application of Propositions 2.3.2 and 2.1.4 to (2.8.18) proves that $P^{[0]}\in M_n(W^{k+1}_p(a,b))$. Now we assume that $k>0$ and let $0\le m\le k-1$. Suppose that there exist $P^{[r]}_{\nu\mu}\in M_{n_\nu,n_\mu}(W^{k+1-r}_p(a,b))$ for $\nu,\mu=0,\dots,l$ and $r=0,\dots,m$ such that (2.8.17) is fulfilled, (2.8.18) holds for $r=0,\dots,m$, and (2.8.19) holds for $r=0,\dots,m-1$. For $\nu\neq\mu$ we define $P^{[m+1]}_{\nu\mu}$ by (2.8.19). For $\nu\neq\mu$, $r_\nu-r_\mu\in W^k_p(a,b)$ and $(r_\nu-r_\mu)^{-1}\in L_\infty(a,b)$. Hence $(r_\nu-r_\mu)^{-1}\in W^k_p(a,b)$ by Proposition 2.5.8. From Proposition 2.3.2 we infer $P^{[m+1]}_{\nu\mu}\in M_{n_\nu,n_\mu}(W^{k-m}_p(a,b))$. For $r=m+1$ the right hand side of (2.8.18) belongs to $M_{n_\nu}(W^{k-m-1}_p(a,b))\subset M_{n_\nu}(L_p(a,b))$ by Proposition 2.3.2. For the differential operator $y\mapsto y'-A_{0,\nu\nu}y$ we apply Lemma 2.5.7 to each column of this matrix function and obtain that there is a solution $P^{[m+1]}_{\nu\nu}\in M_n(W^1_p(a,b))$ of (2.8.18) for $r=m+1$. By a recursive application of Propositions 2.3.2 and 2.1.4 we infer $P^{[m+1]}_{\nu\nu}\in M_{n_\nu}(W^{k-m}_p[a,b])$. This completes the proof of part A.

B. Because of (2.8.17) and (2.8.18) for $r=0$, $P^{[0]}$ is the fundamental matrix of

$$y'-\operatorname{diag}(A_{0,11}\dots,A_{0,ll})y=0$$

with $P^{[0]}(a) = I_n$ by Proposition 2.5.9. Then Theorem 2.5.3 and the uniqueness of this fundamental matrix, see Corollary 2.5.5, yield that $P^{[0]}$ is invertible in $M_n(W_p^1(a,b))$. It follows that there is a $K > 0$ such that

$$|P^{[0]}|_\infty \leq \tfrac{K}{2}, \quad |P^{[0]^{-1}}|_\infty \leq \tfrac{K}{2}, \tag{2.8.21}$$

see Proposition 2.1.7. For $k \in \mathbb{N}$ we define

$$P_k(\cdot,\lambda) := \sum_{r=0}^{k} \lambda^{-r} P^{[r]} \quad (\lambda \in \mathbb{C}\setminus\{0\}) \tag{2.8.22}$$

and set $\widetilde{P}_k := P_k$ if $k > 0$. If $k = 0$ we choose a matrix $\widetilde{A}_0 = (\widetilde{A}_{0,\nu\mu})_{\nu,\mu=0}^{l}$ in $M_n(L_p(a,b))$ such that

$$\begin{cases} \widetilde{A}_{0,\nu\nu} = A_{0,\nu\nu} \quad (\nu = 0,\dots,l), \\ (r_\mu - r_\nu)^{-1}\widetilde{A}_{0,\nu\mu} \in M_{n_\nu,n_\mu}(C^\infty[a,b]) \quad (\nu,\mu = 0,\dots,l;\ \nu \neq \mu), \\ |A_0 - \widetilde{A}_0|_1 \leq (l+1)^{-2}K^{-6}. \end{cases} \tag{2.8.23}$$

This is possible since $C^\infty[a,b]$ is a dense subspace of $L_1(a,b)$; see the proof of Lemma 2.7.2 i) for a construction of such functions. In this case we set

$$P_{\nu\nu}^{[1]} := 0 \quad (\nu = 0,\dots,l),$$
$$P_{\nu\mu}^{[1]} := (r_\mu - r_\nu)^{-1}\widetilde{A}_{0,\nu\mu}P_{\mu\mu}^{[0]} \quad (\nu,\mu = 0,\dots,l;\ \nu \neq \mu),$$

which yields $P^{[1]} = (P_{\nu\mu}^{[1]})_{\nu,\mu=0}^{l} \in M_n(W_p^1(a,b))$ and the equation

$$P^{[0]'} - \widetilde{A}_0 P^{[0]} + P^{[1]}A_1 - A_1 P^{[1]} = 0. \tag{2.8.24}$$

We define

$$\widetilde{P}_0(\cdot,\lambda) := P^{[0]} + \lambda^{-1}P^{[1]} \quad (\lambda \in \mathbb{C}\setminus\{0\}).$$

Now let k be arbitrary. We set

$$S_k := P_k E, \quad \widetilde{S}_k := \widetilde{P}_k E,$$

where E is the matrix function defined in (2.8.5). Since $\widetilde{P}_k(\cdot,\lambda) \to P^{[0]}$ as $\lambda \to \infty$ in $M_n(W_p^1(a,b))$, there is a $\gamma_1 \geq \gamma$ such that $\widetilde{P}_k(\cdot,\lambda)$ is invertible in $M_n(W_p^1(a,b))$ for $|\lambda| \geq \gamma_1$ and

$$|\widetilde{P}_k(\cdot,\lambda)|_\infty \leq K, \quad |\widetilde{P}_k^{-1}(\cdot,\lambda)|_\infty \leq K \quad (|\lambda| \geq \gamma_1). \tag{2.8.25}$$

For $|\lambda| \geq \gamma_1$ we set

$$D_k(\cdot,\lambda) := \Big\{P^{[k]'} - \sum_{r=k}^{2k}\sum_{j=r-k}^{k} \lambda^{k-r}A_{-j}P^{[r-j]}\Big\}P_k(\cdot,\lambda)^{-1} - \lambda^{-1}A^k(\cdot,\lambda),$$

$\widetilde{D}_k := D_k$ if $k > 0$, and

$$\widetilde{D}_0(\cdot,\lambda) := \{(\widetilde{A}_0 - A_0)P^{[0]} + \lambda^{-1}(P^{[1]'} - A_0 P^{[1]})\}\widetilde{P}_0(\cdot,\lambda)^{-1} - \lambda^{-1}A^0(\cdot,\lambda).$$

Let $\kappa := \max\{1,k\}$. Then Proposition 2.3.1, $E'(\cdot,\lambda) = \lambda A_1 E(\cdot,\lambda)$ and the relationships (2.8.6), (2.8.7) and, if $k=0$, (2.8.24) yield

$$\begin{aligned}
&\widetilde{S}_k'(\cdot,\lambda) - \widetilde{A}(\cdot,\lambda)\widetilde{S}_k(\cdot,\lambda) \\
&\quad = \widetilde{P}_k'(\cdot,\lambda)E(\cdot,\lambda) + \widetilde{P}_k(\cdot,\lambda)E'(\cdot,\lambda) - \widetilde{A}(\cdot,\lambda)\widetilde{P}_k(\cdot,\lambda)E(\cdot,\lambda) \\
&\quad = \Big\{\sum_{r=0}^{\kappa} \lambda^{-r} P^{[r]\prime} + \lambda \sum_{r=0}^{\kappa} \lambda^{-r} P^{[r]} A_1 \\
&\qquad - \Big(\sum_{j=-1}^{k} \lambda^{-j} A_{-j} + \lambda^{-k-1} A^k(\cdot,\lambda)\Big) \sum_{r=0}^{\kappa} \lambda^{-r} P^{[r]}\Big\} E(\cdot,\lambda) \\
&\quad = \Big\{\sum_{r=0}^{\kappa} \lambda^{-r} P^{[r]\prime} + \sum_{r=0}^{\kappa-1} \lambda^{-r} P^{[r+1]} A_1 + \lambda P^{[0]} A_1 \\
&\qquad - \sum_{r=-1}^{k+\kappa} \sum_{j=\max\{-1,r-\kappa\}}^{\min\{k,r\}} \lambda^{-r} A_{-j} P^{[r-j]} - \lambda^{-k-1} A^k(\cdot,\lambda)\widetilde{P}_k(\cdot,\lambda)\Big\} E(\cdot,\lambda) \\
&\quad = \Big\{\lambda\Big(P^{[0]}A_1 - A_1 P^{[0]}\Big) \\
&\qquad + \sum_{r=0}^{\kappa-1} \lambda^{-r}\Big(P^{[r]\prime} - \sum_{j=0}^{r} A_{-j} P^{[r-j]} + P^{[r+1]} A_1 - A_1 P^{[r+1]}\Big) \\
&\qquad + \lambda^{-\kappa}\Big(P^{[\kappa]\prime} - \sum_{j=0}^{k} A_{-j} P^{[\kappa-j]}\Big) \\
&\qquad - \lambda^{-\kappa} \sum_{r=\kappa+1}^{k+\kappa} \sum_{j=r-\kappa}^{k} \lambda^{\kappa-r} A_{-j} P^{[r-j]} - \lambda^{-k-1} A^k(\cdot,\lambda)\widetilde{P}_k(\cdot,\lambda)\Big\} E(\cdot,\lambda) \\
&\quad = \lambda^{-k} \widetilde{D}_k(\cdot,\lambda)\widetilde{S}_k(\cdot,\lambda).
\end{aligned}$$

With $\kappa = k$ the same proof shows

$$S_k'(\cdot,\lambda) - \widetilde{A}(\cdot,\lambda)S_k(\cdot,\lambda) = \lambda^{-k} D_k(\cdot,\lambda) S_k(\cdot,\lambda).$$

For $\nu = 0,\dots,l$ let $I^{(\nu)}$ be the $n\times n$ matrix whose ν-th diagonal block is I_{n_ν} and whose other components are zero. For $\nu,\mu = 0,\dots,l$ and $\lambda \neq 0$ we set $x_{\nu\mu}(\lambda) = a$ if $\Re(\lambda e^{i\varphi_{\nu\mu}}) \le 0$ and $x_{\nu\mu}(\lambda) = b$ if $\Re(\lambda e^{i\varphi_{\nu\mu}}) > 0$.

Next we prove that the integral equation

$$\begin{aligned}
(2.8.26) \qquad C(x,\lambda) = I_n - \lambda^{-k} \sum_{\nu,\mu=0}^{l} \int_{x_{\nu\mu}(\lambda)}^{x} &\widetilde{S}_k(x,\lambda) I^{(\nu)} \widetilde{S}_k^{-1}(t,\lambda) \widetilde{D}_k(t,\lambda) \times \\
&\times C(t,\lambda) \widetilde{S}_k(t,\lambda) I^{(\mu)} \widetilde{S}_k^{-1}(x,\lambda)\, dt
\end{aligned}$$

has a unique solution in $M_n(C[a,b])$ for all sufficiently large λ. For this purpose we consider the continuous linear operator

$$T_\lambda : M_n(C[a,b]) \to M_n(C[a,b])$$

given by

$$\begin{aligned} T_\lambda(f)(x) :=& \lambda^{-k} \sum_{\nu,\mu=0}^{l} \int_{x_{\nu\mu}(\lambda)}^{x} \widetilde{S}_k(x,\lambda) I^{(\nu)} \widetilde{S}_k^{-1}(t,\lambda) \widetilde{D}_k(t,\lambda) \times \\ & \times f(t) \widetilde{S}_k(t,\lambda) I^{(\mu)} \widetilde{S}_k^{-1}(x,\lambda)\, dt, \end{aligned}$$

where $f \in M_n(C[a,b])$. Since E and $I^{(\nu)}$ are diagonal matrices,

$$E(x,\lambda) I^{(\nu)} E(t,\lambda)^{-1} = \exp\{\lambda(R_\nu(x) - R_\nu(t))\} I^{(\nu)}.$$

By assumption,

$$R_\nu(x) - R_\nu(t) + R_\mu(t) - R_\mu(x) = e^{i\varphi_{\nu\mu}} \int_t^x |r_\nu(\eta) - r_\mu(\eta)|\, d\eta,$$

whence the equation

$$\begin{aligned} T_\lambda(f)(x) = \lambda^{-k} \sum_{\nu,\mu=0}^{l} \int_{x_{\nu\mu}(\lambda)}^{x} & \exp\Big\{\lambda e^{i\varphi_{\nu\mu}} \int_t^x |r_\nu(\eta) - r_\mu(\eta)|\, d\eta\Big\} \times \\ & \times \widetilde{P}_k(x,\lambda) I^{(\nu)} \widetilde{P}_k^{-1}(t,\lambda) \widetilde{D}_k(t,\lambda) f(t) \widetilde{P}_k(t,\lambda) I^{(\mu)} \widetilde{P}_k^{-1}(x,\lambda)\, dt \end{aligned}$$

holds for $f \in M_n(C[a,b])$. By the choice of $x_{\nu\mu}(\lambda)$ we obtain

$$\begin{aligned} \Big|\exp\Big\{\lambda e^{i\varphi_{\nu\mu}} \int_t^x |r_\nu(\eta) - r_\mu(\eta)|\, d\eta\Big\}\Big| \\ = \exp\Big\{\Re(\lambda e^{i\varphi_{\nu\mu}}) \int_t^x |r_\nu(\eta) - r_\mu(\eta)|\, d\eta\Big\} \le 1 \end{aligned} \tag{2.8.27}$$

for t in the compact interval with the endpoints $x_{\nu\mu}(\lambda)$ and $x \in [a,b]$. Note that $|\widetilde{P}_k(\cdot,\lambda) I^{(\nu)}|_\infty \le |\widetilde{P}_k(\cdot,\lambda)|_\infty$. Then

$$|T_\lambda| \le |\lambda|^{-k} (l+1)^2 K^4 |\widetilde{D}_k(\cdot,\lambda)|_1 .$$

Since $A^k(\cdot,\lambda)$ is bounded in $M_n(L_p(a,b))$ as $\lambda \to \infty$, there is a constant $M > 0$ such that for all sufficiently large λ

$$|D_k(\cdot,\lambda)|_p \le M \quad \text{and} \quad |D_k(\cdot,\lambda)|_1 \le M, \tag{2.8.28}$$

whence

$$|T_\lambda| \le |\lambda|^{-k} (l+1)^2 K^4 M$$

follows for $k > 0$. For $k = 0$ we obtain with a suitable $M_0 > 0$ that

$$|\widetilde{D}_0(\cdot,\lambda)|_1 \le |\widetilde{A}_0 - A_0|_1 \tfrac{K^2}{2} + |\lambda|^{-1} M_0 \tag{2.8.28'}$$

for all sufficiently large λ. With the aid of the estimate in (2.8.23) we infer in this case that

$$|T_\lambda| \le \frac{1}{2} + |\lambda|^{-1}(l+1)^2 K^4 M_0.$$

We conclude in either case that there is a positive $\delta < 1$ and a number $\gamma_0 \ge \gamma$ such that $|T_\lambda| \le \delta$ for $|\lambda| \ge \gamma_0$. For $|\lambda| \ge \gamma_0$ the operator

$$\widehat{T}_\lambda := \mathrm{id}_{M_n(C[a,b])} + T_\lambda$$

is invertible by Proposition 1.1.4. As (2.8.26) holds if and only if $\widehat{T}_\lambda C(\cdot,\lambda) = I_n$, where $I_n := \mathrm{id}_{\mathbb{C}^n}$,

$$C(\cdot,\lambda) := \widehat{T}_\lambda^{-1} I_n$$

is the unique solution of (2.8.26). In addition, from

$$\widehat{T}_\lambda^{-1} - \mathrm{id}_{M_n(C[a,b])} = \sum_{j=1}^{\infty} (-1)^j T_\lambda^j,$$

see (1.1.6), we infer

$$|C(\cdot,\lambda) - I_n|_{(0)} \le (1-\delta)^{-1} |T_\lambda I_n|_{(0)} \quad (|\lambda| \ge \gamma_0). \tag{2.8.29}$$

From (2.8.26) and Propositions 2.3.2 and 2.1.8 we see that $C(\cdot,\lambda) \in M_n(W_p^1(a,b))$. Note that the components of $C(x,\lambda)$ are products of functions of the form $f_1(x)$ with $f_1 \in W_p^1(a,b)$ and $\int_{x_{\nu\mu}(\lambda)}^x f_2(t)\,\mathrm{d}t$ with $f_2 \in L_p(a,b)$. Therefore Propositions 2.3.1 and 2.1.3 and (2.5.7) yield that

$$\begin{aligned}
C'(x,\lambda) &= -\lambda^{-k}\widetilde{D}_k(x,\lambda)C(x,\lambda) \\
&\quad - \lambda^{-k} \sum_{\nu,\mu=0}^{l} \int_{x_{\nu\mu}(\lambda)}^{x} \widetilde{S}_k'(x,\lambda) I^{(\nu)} \widetilde{S}_k^{-1}(t,\lambda) \widetilde{D}_k(t,\lambda) C(t,\lambda) \widetilde{S}_k(t,\lambda) \times \\
&\quad \times I^{(\mu)} \widetilde{S}_k^{-1}(x,\lambda)\,\mathrm{d}t + \lambda^{-k} \sum_{\nu,\mu=0}^{l} \int_{x_{\nu\mu}(\lambda)}^{x} \widetilde{S}_k(x,\lambda) I^{(\nu)} \widetilde{S}_k^{-1}(t,\lambda) \times \\
&\quad \times \widetilde{D}_k(t,\lambda) C(t,\lambda) \widetilde{S}_k(t,\lambda) I^{(\mu)} \widetilde{S}_k^{-1}(x,\lambda) \widetilde{S}_k'(x,\lambda) \widetilde{S}_k^{-1}(x,\lambda)\,\mathrm{d}t \\
&= -\lambda^{-k}\widetilde{D}_k(x,\lambda)C(x,\lambda) + \widetilde{S}_k'(x,\lambda)\widetilde{S}_k^{-1}(x,\lambda)(C(x,\lambda) - I_n) \\
&\quad - (C(x,\lambda) - I_n)\widetilde{S}_k'(x,\lambda)\widetilde{S}_k^{-1}(x,\lambda) \\
&= -\lambda^{-k}\widetilde{D}_k(x,\lambda)C(x,\lambda) + \widetilde{S}_k'(x,\lambda)\widetilde{S}_k^{-1}(x,\lambda)C(x,\lambda) \\
&\quad - C(x,\lambda)\widetilde{S}_k'(x,\lambda)\widetilde{S}_k^{-1}(x,\lambda).
\end{aligned}$$

We define

$$\widetilde{Y} := C\widetilde{S}_k = C\widetilde{P}_k E \tag{2.8.30}$$

and infer from (2.3.1) that

$$\widetilde{Y}' = C'\widetilde{S}_k + C\widetilde{S}_k' = \widetilde{A}\widetilde{Y}$$

by inserting the right-hand sides of the matrix differential equations which we obtained above for $\widetilde{S}_k$ and C. If $k = 0$ we set

$$B_0(\cdot,\lambda) := \big(C(\cdot,\lambda) - I_n\big)\widetilde{P}_0(\cdot,\lambda) + \lambda^{-1}P^{[1]}, \tag{2.8.31}$$

and if $k \geq 1$ we set

$$B_k(\cdot,\lambda) := \lambda^k\big(C(\cdot,\lambda) - I_n\big)\widetilde{P}_k(\cdot,\lambda). \tag{2.8.31$'$}$$

These definitions and (2.8.30) immediately yield $B_k(\cdot,\lambda) \in M_n(W_p^1(a,b))$ and

$$\widetilde{Y}(\cdot,\lambda) = \Big\{P_k(\cdot,\lambda) + \lambda^{-k}B_k(\cdot,\lambda)\Big\}E(\cdot,\lambda). \tag{2.8.32}$$

This proves (2.8.10) for $|\lambda| \geq \gamma_0$. But for simplicity of notation we may assume that $\gamma_0 = \gamma$. Indeed, for any fundamental matrix there is a $B_k(\cdot,\lambda)$ such that (2.8.32) holds, and the values of $B_k(\cdot,\lambda)$ for $\gamma \leq |\lambda| \leq \gamma_0$ do not influence the asymptotic behaviour of B_k.

We have to prove the estimates for $B_k(\cdot,\lambda)$ as $\lambda \to \infty$. First we give the proof of (2.8.11), (2.8.12) and (2.8.16). In view of (2.8.31), (2.8.31$'$) and the uniform boundedness of $\widetilde{P}_k(\cdot,\lambda)$ in $M_n(L_\infty(a,b))$ for $|\lambda| \geq \gamma_0$, see (2.8.25), it is sufficient to estimate the matrix function $\lambda^k(C(\cdot,\lambda) - I_n)$. We set

$$Q^{[0]} := (\widetilde{A}_0 - A_0)$$

and, if $k \geq 1$,

$$Q^{[k]} := \Big\{P^{[k]'} - \sum_{j=0}^{k} A_{-j}P^{[k-j]}\Big\}P^{[0]^{-1}}.$$

For $f \in M_n(C[a,b])$ and $x \in [a,b]$ we define

$$\begin{aligned}
T_{\lambda,1}(f)(x) := \lambda^{-k} \sum_{\substack{\nu,\mu=0\\ \nu\neq\mu}}^{l} \int_{x_{\nu\mu}(\lambda)}^{x} &\exp\Big\{\lambda e^{i\varphi_{\nu\mu}} \int_t^x |r_\nu(\eta) - r_\mu(\eta)|\,\mathrm{d}\eta\Big\} \times\\
&\times P^{[0]}(x)I^{(\nu)}P^{[0]^{-1}}(t)Q^{[k]}(t)f(t)P^{[0]}(t)I^{(\mu)}P^{[0]^{-1}}(x)\,\mathrm{d}t,\\
T_{\lambda,2}(f)(x) := \lambda^{-k} \sum_{\nu=0}^{l} \int_{x_{\nu\nu}(\lambda)}^{x} &P^{[0]}(x)I^{(\nu)}P^{[0]^{-1}}(t)Q^{[k]}(t) \times\\
&\times f(t)P^{[0]}(t)I^{(\nu)}P^{[0]^{-1}}(x)\,\mathrm{d}t,
\end{aligned}$$

$$T_{\lambda,3}(f)(x) := \lambda^{-1-k} \sum_{\nu,\mu=0}^{l} \int_{x_{\nu\mu}(\lambda)}^{x} \exp\Big\{\lambda e^{i\varphi_{\nu\mu}} \int_t^x |r_\nu(\eta) - r_\mu(\eta)|\,\mathrm{d}\eta\Big\} \times$$
$$\times \Big\{\widetilde{P}_k(x,\lambda) I^{(\nu)} \widetilde{P}_k^{-1}(t,\lambda)\lambda\big(\widetilde{D}_k(t,\lambda) - Q^{[k]}(t)\big) f(t) \widetilde{P}_k(t,\lambda) I^{(\mu)} \widetilde{P}_k^{-1}(x,\lambda)$$
$$+ \lambda\big(\widetilde{P}_k(x,\lambda) - P^{[0]}(x)\big) I^{(\nu)} \widetilde{P}_k^{-1}(t,\lambda) Q^{[k]}(t) f(t) \widetilde{P}_k(t,\lambda) I^{(\mu)} \widetilde{P}_k^{-1}(x,\lambda)$$
$$+ P^{[0]}(x) I^{(\nu)} \lambda\big(\widetilde{P}_k^{-1}(t,\lambda) - P^{[0]^{-1}}(t)\big) Q^{[k]}(t) f(t) \widetilde{P}_k(t,\lambda) I^{(\mu)} \widetilde{P}_k^{-1}(x,\lambda)$$
$$+ P^{[0]}(x) I^{(\nu)} P^{[0]^{-1}}(t) Q^{[k]}(t) f(t) \lambda\big(\widetilde{P}_k(t,\lambda) - P^{[0]}(t)\big) I^{(\mu)} \widetilde{P}_k^{-1}(x,\lambda)$$
$$+ P^{[0]}(x) I^{(\nu)} P^{[0]^{-1}}(t) Q^{[k]}(t) f(t) P^{[0]}(t) I^{(\mu)} \lambda\big(\widetilde{P}_k^{-1}(x,\lambda) - P^{[0]^{-1}}(x)\big)\Big\}\,\mathrm{d}t.$$

Then

$$T_\lambda = T_{\lambda,1} + T_{\lambda,2} + T_{\lambda,3}.$$

The definition of $\widetilde{P}_k(\cdot,\lambda)$ yields that $\widetilde{P}_k(\cdot,\lambda) = P^{[0]} + \frac{1}{\lambda}O(1)$ in $M_n(W_p^1(a,b))$ as $\lambda \to \infty$. With the aid of Proposition 2.3.3, (1.1.3) and (1.1.4) we infer for $\lambda \to \infty$:

$$\begin{aligned}
&\widetilde{P}_k(\cdot,\lambda) = O(1) \text{ in } M_n(W_p^1(a,b)),\\
&\lambda\big(\widetilde{P}_k(\cdot,\lambda) - P^{[0]}\big) = O(1) \text{ in } M_n(W_p^1(a,b)),\\
&\widetilde{P}_k^{-1}(\cdot,\lambda) = O(1) \text{ in } M_n(W_p^1(a,b)),\\
&\lambda\big(\widetilde{P}_k^{-1}(\cdot,\lambda) - P^{[0]^{-1}}\big) = O(1) \text{ in } M_n(W_p^1(a,b)).
\end{aligned}$$

These estimates and the definitions of $\widetilde{D}_k$ and $Q^{[k]}$ yield

$$\lambda(\widetilde{D}_k(\cdot,\lambda) - Q^{[k]}) = O(1) \text{ in } M_n(L_p(a,b))$$

as $\lambda \to \infty$. It follows that $\lambda^{k+1}T_{\lambda,3}(f)(x)$ is a sum of terms of the form

$$Q_1(x,\lambda) \int_{x_{\nu\mu}(\lambda)}^{x} \exp\Big\{\lambda e^{i\varphi_{\nu\mu}} \int_t^x |r_\nu(\eta) - r_\mu(\eta)|\Big\}\,\mathrm{d}\eta\, Q_2(t,\lambda)\,\mathrm{d}t\, Q_3(x,\lambda),$$

where $Q_1(\cdot,\lambda), Q_3(\cdot,\lambda) = O(1)$ in $M_n(W_p^1(a,b))$ and $Q_2(\cdot,\lambda) = O(1)|f|_{(0)}$ in $M_n(L_p(a,b))$ as $\lambda \to \infty$. As $W_p^1(a,b)$ is contained continuously in $L_\infty(a,b)$, it follows that $|\lambda^k T_{\lambda,3}| = \{O(\frac{1}{\lambda})\}_\infty$. Since $|T_\lambda| \le \delta < 1$ for $|\lambda| \ge \gamma_0$, there are numbers $\gamma_1 \ge 1$ and $\delta_1, \delta_2 < 1$ such that $\delta_1 + \delta_2 < 1$, $|T_{\lambda,1} + T_{\lambda,2}| \le \delta_1$, and $|\lambda^k T_{\lambda,3}| \le \delta_2$ for $|\lambda| \ge \gamma_1$. We calculate

$$\begin{aligned}
\lambda^k(C(\cdot,\lambda) - I_n) &= \lambda^k \sum_{j=1}^{\infty} (-1)^j T_\lambda^j(I_n)\\
&= \lambda^k \sum_{j=1}^{\infty} (-1)^j (T_{\lambda,1} + T_{\lambda,2})^j(I_n) + T_{\lambda,4}(I_n),
\end{aligned}$$

where

$$
\begin{aligned}
|T_{\lambda,4}| &\le |\lambda|^k \sum_{j=1}^{\infty}\sum_{m=1}^{j}\binom{j}{m}|T_{\lambda,1}+T_{\lambda,2}|^{j-m}|T_{\lambda,3}|^m \\
&= |\lambda|^k \sum_{j=1}^{\infty}\{(|T_{\lambda,1}+T_{\lambda,2}|+|T_{\lambda,3}|)^j - |T_{\lambda,1}+T_{\lambda,2}|^j\} \\
&\le |\lambda^k T_{\lambda,3}| \sum_{j=1}^{\infty} j(|T_{\lambda,1}+T_{\lambda,2}|+|T_{\lambda,3}|)^{j-1} \\
&= O(|\lambda^k T_{\lambda,3}|) = O(\lambda^{-1}).
\end{aligned}
$$

This shows that $T_{\lambda,4}(I_n) = O(\lambda^{-1})$ in $M_n(C[a,b])$ and thus also in $M_n(L_p(a,b))$.

From (2.8.8) and (2.8.23) we conclude that the block diagonal of $Q^{[k]}$ is zero. This shows that $T_{\lambda,2}(I_n) = 0$ and $T_{\lambda,2}^2 = 0$. Therefore,

$$
\begin{aligned}
(2.8.33)\qquad f_\lambda &:= \lambda^k \sum_{j=1}^{\infty}(-1)^j(T_{\lambda,1}+T_{\lambda,2})^j(I_n) \\
&= -\lambda^k T_{\lambda,1}(I_n) + \lambda^k(T_{\lambda,1}+T_{\lambda,2})^2\sum_{j=0}^{\infty}(-1)^j(T_{\lambda,1}+T_{\lambda,2})^j(I_n) \\
&= -\lambda^k T_{\lambda,1}(I_n) + \lambda^k(T_{\lambda,1}^2 + T_{\lambda,1}T_{\lambda,2} + T_{\lambda,2}T_{\lambda,1})g_\lambda,
\end{aligned}
$$

where

$$
g_\lambda = \sum_{j=0}^{\infty}(-1)^j(T_{\lambda,1}+T_{\lambda,2})^j(I_n) = O(1) \text{ in } M_n(C[a,b])
$$

since $|T_{\lambda,1}+T_{\lambda,2}| \le \delta_1 < 1$.

It is clear that $T_{\lambda,1}$ and $T_{\lambda,2}$ are uniformly bounded with respect to λ. Therefore, in order to establish the estimates (2.8.11), (2.8.12), (2.8.13) and (2.8.16) it is sufficient to prove for $h \in M_n(C[a,b])$ that

(2.8.34) $\quad \lambda^k T_{\lambda,1}h$ satisfies $o(1)$ and $O(\tau_p(\lambda))|h|_{(0)}$ in $M_n(C[a,b])$,

(2.8.35) $\quad \lambda^k T_{\lambda,1}h$ and $\lambda^k T_{\lambda,2}T_{\lambda,1}h$ satisfy $O(\tau_\infty(\lambda))|h|_{(0)}$ in $M_n(L_p(a,b))$,

where (2.8.35) holds under the additional assumptions made in iii). To prove (2.8.11) we remark that it is sufficient to estimate $\lambda^k T_\lambda(I_n)$ in $M_n(C[a,b])$ because of (2.8.29). Hence it would be sufficient to consider $h = I_n$ in (2.8.34) for the estimate $o(1)$.

The estimates (2.8.34) and (2.8.35) are trivial in case $l = 0$ since $T_{\lambda,1} = 0$ in this case. Now let $l > 0$. To estimate $T_{\lambda,1}h$ we have to consider terms of the form

$$
\widetilde{F}(u,x,\lambda) = \int_{x_{\nu\mu}(\lambda)}^{x} \exp\Big\{\lambda e^{i\varphi_{\nu\mu}}\int_t^x |r_\nu(\eta) - r_\mu(\eta)|\,\mathrm{d}\eta\Big\}u(t)\,\mathrm{d}t,
$$

where $\nu \neq \mu$ and $u \in L_p(a,b)$ such that $|u|_p \le C|h|_{(0)}$ and $C > 0$ does not depend on h. For $x \in [a,b]$ we set

$$r(x) := |r_\nu(x) - r_\mu(x)|.$$

Let F be defined as in Lemma 2.7.2. Then

$$\widetilde{F}(u,x,\lambda) = F(u,x,x_{\nu\mu}(\lambda),\lambda e^{i\varphi_{\nu\mu}}).$$

As $\Re(\lambda e^{i\varphi_{\nu\mu}})(x - x_{\nu\mu}(\lambda)) \le 0$, Lemma 2.7.2 yields the estimates

$$\widetilde{F}(u,\cdot,\lambda) = o(1) \text{ as } \lambda \to \infty,$$

$$\widetilde{F}(u,\cdot,\lambda) = O((1+|\Re(\lambda e^{i\varphi_{\nu\mu}})|)^{-1+1/p})|h|_{(0)} \text{ in } L_\infty(a,b),$$

$$\widetilde{F}(u,\cdot,\lambda) = O((1+|\Re(\lambda e^{i\varphi_{\nu\mu}})|)^{-1})|h|_{(0)} \text{ in } L_p(a,b).$$

Since $|T_{\lambda,2}T_{\lambda,1}| = O(\lambda^{-2k})$, the estimate for $T_{\lambda,2}T_{\lambda,1}$ in (2.8.35) is trivial for $k > 0$. Therefore, let $k = 0$. The components of $T_{\lambda,2}T_{\lambda,1}h$ are sums of terms of the form

(2.8.36)

$$z(x,\lambda) = w(x)\int_{x_{\mu\mu}(\lambda)}^{x} v(\xi)\int_{x_{\nu\mu}(\lambda)}^{\xi} \exp\{\lambda e^{i\varphi_{\nu\mu}}\int_t^\xi |r_\nu(\eta) - r_\mu(\eta)|\,\mathrm{d}\eta\}u(t)\,\mathrm{d}t\,\mathrm{d}\xi,$$

where $\nu \neq \mu$, $w \in W_p^1(a,b)$ and $v \in L_{p_{\mu\nu}}(a,b)$ are functions depending only on the coefficient matrices in $T_{\lambda,1}$ and $T_{\lambda,2}$, and $u \in L_{p_{\nu q}}(a,b)$ for some $q \neq \nu$ satisfies $|u|_{p_{\nu q}} \le C|h|_{(0)}$, where $C > 0$ depends only on the coefficient matrices in $T_{\lambda,1}$ and $T_{\lambda,2}$. Here we take $p_{\mu\nu} = p_{\nu q} = p$ in case $p > \frac{3}{2}$. If $p > 1$ we choose $\varepsilon \in (0, 1 - \frac{1}{p})$ such that $\frac{1}{p} + \frac{1}{p_{\nu q}} + \frac{1}{p_{\mu\nu}} \le 2 - \varepsilon$ for all ν, μ, q under consideration. With r as above, $R(x) = \int_a^x r(t)\,\mathrm{d}t$, $x \in [a,b]$, and $g \in L_{p'}(a,b)$ we obtain

$$\int_a^b g(x)z(x,\lambda)\,\mathrm{d}x = \int_a^b v_1(\xi)\int_{x_{\nu\mu}(\lambda)}^{\xi} \exp\{\lambda e^{i\varphi_{\nu\mu}}(R(\xi) - R(t))\}u(t)\,\mathrm{d}t\,\mathrm{d}\xi,$$

where

$$v_1(\xi) = v(\xi)\int_\xi^{x'_{\mu\mu}(\lambda)} g(x)w(x)\,\mathrm{d}\xi$$

with $x'_{\mu\mu}(\lambda) = b$ if $x_{\mu\mu}(\lambda) = a$ and $x'_{\mu\mu}(\lambda) = a$ if $x_{\mu\mu}(\lambda) = b$. Since

$$\frac{1}{p_{\nu q}} - \frac{1}{p'_{\mu\nu}} = \frac{1}{p_{\nu q}} + \frac{1}{p_{\mu\nu}} - 1 \le 1 - \varepsilon - \frac{1}{p},$$

it follows from Lemma 2.7.2 iii) with suitable $p \ge p_{\nu q}$ and $\tilde{p} \le p'_{\mu\nu}$ that

$$\left|\int_a^b g(x)z(x,\lambda)\,\mathrm{d}x\right| \le C_1\left(1 + |\Re(\lambda e^{i\varphi_{\nu\mu}})|\right)^{-\frac{1}{p}-\varepsilon}|v_1|_{p_{\mu\nu}}|h|_{(0)}$$

$$\le C_2\left(1 + |\Re(\lambda e^{i\varphi_{\nu\mu}})|\right)^{-\frac{1}{p}-\varepsilon}|g|_{p'}|h|_{(0)},$$

which proves

$$|z(\cdot,\lambda)|_p \le C_2\big(1+|\Re(\lambda e^{i\varphi_{\nu\mu}})|\big)^{-\frac{1}{p}-\varepsilon}|h|_{(0)}.$$

Finally we shall prove the estimates (2.8.14) and (2.8.15). For $|\lambda| \ge \gamma_0$ we have

$$B_k(\cdot,\lambda) = \lambda^k[\widetilde{Y}(\cdot,\lambda) - S_k(\cdot,\lambda)]E^{-1}(\cdot,\lambda).$$

The differential equations fulfilled by $\widetilde{Y}(\cdot,\lambda)$, $E(\cdot,\lambda)$ and $S_k(\cdot,\lambda)$ yield

(2.8.37)

$$\begin{aligned} B_k'(\cdot,\lambda) &= \lambda^k\Big\{\widetilde{A}(\cdot,\lambda)[\widetilde{Y}(\cdot,\lambda) - S_k(\cdot,\lambda)] - \lambda^{-k}D_k(\cdot,\lambda)S_k(\cdot,\lambda)\Big\}E^{-1}(\cdot,\lambda) \\ &\quad - \lambda^{k+1}[\widetilde{Y}(\cdot,\lambda) - S_k(\cdot,\lambda)]E^{-1}(\cdot,\lambda)A_1 \\ &= \widetilde{A}(\cdot,\lambda)B_k(\cdot,\lambda) - D_k(\cdot,\lambda)P_k(\cdot,\lambda) - \lambda B_k(\cdot,\lambda)A_1. \end{aligned}$$

The estimates (2.8.14) and (2.8.15) for $\frac{1}{\lambda}B_k'(\cdot,\lambda)$ now follow from the estimates (2.8.11), (2.8.12) and (2.8.28). □

COROLLARY 2.8.3. *We assume that the conditions* i)–iii) *in* Assumption 2.8.1 *are sharpened such that the following properties hold for some* $\kappa \in \mathbb{N}$*:*

i′) $A_1 \in M_n(W_p^{k+\kappa}(a,b))$,

ii′) $A_{-j} \in M_n(W_p^{k+\kappa-j}(a,b)) \quad (j=0,\dots,k)$,

iii′) $A^k(\cdot,\lambda) \in M_n(W_p^\kappa(a,b))$ *if* $|\lambda| \ge \gamma$ *and* $A^k(\cdot,\lambda)$ *is bounded in* $M_n(W_p^\kappa(a,b))$ *as* $\lambda \to \infty$.

We assume that the matrix functions $P^{[r]}$ *belong to* $M_n(W_p^{k+1-r}(a,b))$ *for all* $r \in \{0,\dots,k\}$ *and fulfil* (2.8.6), (2.8.7) *and* (2.8.8). *For* $|\lambda| \ge \gamma$ *let the matrix function* $B_k(\cdot,\lambda)$ *be defined as in part B of Theorem* 2.8.2.

Then $P^{[r]} \in M_n(W_p^{k+\kappa+1-r}(a,b))$ *for* $r \in \{0,\dots,k\}$ *and* $B_k(\cdot,\lambda) \in M_n(W_p^{\kappa+1}(a,b))$ *for* $|\lambda| \ge \gamma$. *We have*

(2.8.38) $$\tfrac{1}{\lambda^\iota}B_k^{(\iota)}(\cdot,\lambda) = \{o(1)\}_p$$

and

(2.8.39) $$\tfrac{1}{\lambda^\iota}B_k^{(\iota)}(\cdot,\lambda) = \{O(\tau_p(\lambda))\}_p$$

for $|\lambda| \ge \gamma$ *and* $\iota \in \{0,\dots,\kappa+1\}$, *where* τ_p *is the function defined in* (2.8.9).

Proof. The case $\kappa = 0$ is part of Theorem 2.8.2. Therefore let $\kappa > 0$. With the aid of Proposition 2.3.2 and Corollary 2.1.4, the first assertion follows from (2.8.17), (2.8.18) and (2.8.19) by induction. In view of (2.8.37) we infer that the matrix function

$$\begin{aligned} B_k'(\cdot,\lambda) - \widetilde{A}(\cdot,\lambda)B_k(\cdot,\lambda) + \lambda B_k(\cdot,\lambda)A_1 &= -D_k(\cdot,\lambda)P_k(\cdot,\lambda) \\ &= -P^{[k]\prime} + \sum_{r=k}^{2k}\sum_{j=r-k}^{k}\lambda^{k-r}A_{-j}P^{[r-j]} + \lambda^{-1}A^k(\cdot,\lambda)\sum_{r=0}^{k}\lambda^{-r}P^{[r]} \end{aligned}$$

belongs to $M_n(W_p^\kappa(a,b))$ and is bounded in this space as $\lambda \to \infty$. An application of Proposition 2.3.2 and Corollary 2.1.4 proves $B_k(\cdot,\lambda) \in M_n(W_p^{\kappa+1}(a,b))$. The LEIBNIZ rule, which holds in view of (2.3.1), yields that

$$\tfrac{1}{\lambda^\iota} B_k^{(\iota)}(\cdot,\lambda) = \sum_{j=0}^{\iota-1} \binom{\iota-1}{j} \Big\{ \tfrac{1}{\lambda^{\iota-j}} \widetilde{A}^{(\iota-1-j)}(\cdot,\lambda) \tfrac{1}{\lambda^j} B_k^{(j)}(\cdot,\lambda) - \tfrac{1}{\lambda^j} B_k^{(j)}(\cdot,\lambda) \tfrac{1}{\lambda^{\iota-1-j}} A_1^{(\iota-1-j)} \Big\} - \tfrac{1}{\lambda^\iota} (D_k P_k)^{(\iota-1)}(\cdot,\lambda)$$

for $\iota = 1,\dots,\kappa+1$. The estimates (2.8.38) and (2.8.39) follow from these equations and (2.8.11), (2.8.12) by induction on ι. □

REMARK 2.8.4. Assume additionally that the coefficients $A_1, A_0, \dots, A_{-k}$ are indefinitely differentiable and that $A^k = 0$. Then the $P^{[j]}$ $(j = 0,\dots,k)$ and $B_k(\cdot,\lambda)$ belong to $M_n(C^\infty[a,b])$. Here k can be chosen arbitrarily large.

REMARK 2.8.5. Assume that $A_1 = r_1 I_n$ and write

$$\widetilde{A}(x,\lambda) =: \lambda A_1(x) + \widetilde{A}^0(x,\lambda).$$

Let the fundamental matrix $\widetilde{Y}(\cdot,\lambda)$ of (2.8.1) be as in Theorem (2.8.2) B. Then

$$P(\cdot,\lambda) := \widetilde{Y}(\cdot,\lambda) E(\cdot,\lambda)^{-1}$$

is a fundamental matrix of $y'(x) - \widetilde{A}^0(x,\lambda) y(x) = 0$.

Proof. We have

$$\widetilde{Y}'(x,\lambda) = \lambda A_1(x) \widetilde{Y}(x,\lambda) + \widetilde{A}^0(x,\lambda) \widetilde{Y}(x,\lambda)$$

and

$$\widetilde{Y}'(x,\lambda) = P'(x,\lambda) E(x,\lambda) + P(x,\lambda) \lambda A_1(x) E(x,\lambda)$$

as $E(\cdot,\lambda)$ is a fundamental matrix of $y'(x) - \lambda A_1(x) y(x) = 0$. Since $A_1(x)$ and $P(x,\lambda)$ commute, this proves that

$$P'(x,\lambda) = \widetilde{A}^0(x,\lambda) P(x,\lambda).$$

Finally, $P(a,\lambda) = \widetilde{Y}(a,\lambda)$ is invertible. Thus $P(\cdot,\lambda)$ is a fundamental matrix of $y' = \widetilde{A}_0(\cdot,\lambda) y$ by Proposition 2.5.9. □

REMARK 2.8.6. Assume that $A_1 = r_1 I_n$ and let $E(\cdot,\lambda)$ and $P^{[0]}$ be the fundamental matrices of

$$y'(x) - \lambda A_1(x) y(x) = 0 \quad \text{and} \quad y'(x) - A_0(x) y(x) = 0$$

with $E(a,\lambda) = I_n = P^{[0]}(a)$, respectively. Then $P^{[0]} E(\cdot,\lambda)$ is a fundamental matrix of

$$y'(x) - \big(\lambda A_1(x) + A_0(x)\big) y(x) = 0$$

as given in Theorem 2.8.2 B.

If the first order system of linear ordinary differential equations (2.8.1) is not of the form as considered in Remark 2.8.6, then, in general, one cannot find a $k \in \mathbb{N}$ such that $B_k(\cdot, \lambda) = 0$ for all sufficiently large λ.

If k can be chosen arbitrarily large (see e. g. Remark 2.8.4), then we may consider the formal series

$$\sum_{r=0}^{\infty} \lambda^{-r} P^{[r]}.$$

But it may happen that this series does not converge for any λ as is seen in the following example.

EXAMPLE 2.8.7. Let $\widetilde{A}(\cdot, \lambda) = \lambda A_1 + A_0$, where

$$A_1 = \begin{pmatrix} 1 & 0 \\ 0 & -1 \end{pmatrix}, \quad A_0 = \begin{pmatrix} 0 & \alpha \\ 0 & 0 \end{pmatrix},$$

and $\alpha \in C^\infty[a,b]$.

Then (2.8.17) and (2.8.18) yield $P_{22}^{[0]}(a) = 1$ and $P_{22}^{[r]\prime} = 0$ for $r \in \mathbb{N}$. Hence $P_{22}^{[0]}(x) = 1 =: c_0$ and $P_{22}^{[r]}(x) = c_r$ for $x \in [a,b]$ and $r \in \mathbb{N}$, where $c_1, c_2, \ldots$ are arbitrary complex numbers. From (2.8.17) and (2.8.19) we infer

$$P_{12}^{[0]} = 0 \quad \text{and} \quad P_{12}^{[r+1]} = \frac{1}{2}\{P_{12}^{[r]\prime} - \alpha c_r\} \text{ for } r \in \mathbb{N}.$$

This means that

$$P_{12}^{[0]} = 0 \quad \text{and} \quad P_{12}^{[r+1]} = -\sum_{j=0}^{r} 2^{-r-1+j} c_j \alpha^{(r-j)} \text{ for } r \in \mathbb{N}.$$

Now let us take $0 < a < 1$, $b = 1$ and $\alpha(x) = \frac{1}{x}$. Suppose that we can choose the c_r in such a way that

$$\sum_{r=0}^{\infty} \lambda^{-r} P^{[r]}$$

is pointwise convergent almost everywhere for sufficiently large λ. Then there is an $x \in [a,b]$ such that

$$\sum_{r=0}^{\infty} \lambda^{-r} P^{[r]}(x)$$

converges for sufficiently large λ, say $|\lambda| = d$. Hence

$$\sum_{r=0}^{\infty} \lambda^{-r} c_r = \sum_{r=0}^{\infty} \lambda^{-r} P_{22}^{[r]}(x)$$

converges for this λ. Then the sequence $(\lambda^{-r} c_r)_0^\infty$ is bounded for this λ, which means that there is a $C > 0$ such that

$$|c_r| \le C d^r \quad \text{for all } r \in \mathbb{N}. \tag{2.8.40}$$

In the same way, the convergence of

$$\sum_{r=0}^{\infty} \lambda^{-r} P_{12}^{[r]}(x)$$

yields

$$\left| \sum_{j=0}^{r} 2^{-r-1+j} c_j \alpha^{(r-j)}(x) \right| \le C d^{r+1} \quad \text{for } r \in \mathbb{N}, \tag{2.8.41}$$

where we may take the same constant C in (2.8.40) and (2.8.41). Since

$$\alpha^{(l)}(x) = (-1)^l l! x^{-l-1},$$

the estimates (2.8.40) and (2.8.41) imply

$$\begin{aligned} 2^{-r-1} c_0 r! x^{-r-1} &\le \sum_{j=1}^{r} 2^{-r-1+j} |c_j| (r-j)! x^{-r+j-1} + C d^{r+1} \\ &\le C \sum_{j=1}^{r} 2^{-r-1+j} (dx)^j (r-j)! x^{-r-1} + C d^{r+1}. \end{aligned}$$

Hence

$$1 \le C \sum_{j=1}^{r} 2^j (dx)^j \frac{(r-j)!}{r!} + C \frac{2^{r+1} (dx)^{r+1}}{r!}.$$

Since

$$\frac{(2dx)^{r+1}}{r!} \to 0 \text{ as } r \to \infty,$$

this implies that

$$\liminf_{r \to \infty} \sum_{j=1}^{r} (2dx)^j \frac{(r-j)!}{r!} > 0. \tag{2.8.42}$$

For $r = 1, 2, \ldots$ we set

$$b_r := \sum_{j=1}^{r} (2dx)^j \frac{(r-j)!}{r!}.$$

Then

$$\begin{aligned} b_r &= \frac{2dx}{r} + \sum_{j=2}^{r} (2dx)^j \frac{(r-j)!}{r!} \\ &= \frac{2dx}{r} \left(1 + \sum_{j=1}^{r-1} (2dx)^j \frac{(r-1-j)!}{(r-1)!} \right) \\ &= \frac{2dx}{r} (1 + b_{r-1}) \end{aligned}$$

holds for $r = 2, 3, \ldots$ For $r \ge 4dx$ we obtain

$$b_r \le \max\{1, b_{r-1}\}.$$

This proves that the sequence $(b_r)_{r=1}^{\infty}$ is bounded. Thus

$$b_r = \frac{2dx}{r}(1 + b_{r-1}) \to 0 \text{ as } r \to \infty$$

which contradicts (2.8.42). Hence the formal series

$$\sum_{r=0}^{\infty} \lambda^{-r} P^{[r]}$$

does not converge if the $P^{[r]}$ are chosen according to (2.8.6) and (2.8.7). □

If $k = 1$, then we have

$$P_1(\cdot,\lambda) = P^{[0]} + \lambda^{-1} P^{[1]} + \lambda^{-1} B_1(\cdot,\lambda),$$

where $B_1(\cdot,\lambda) = \{o(1)\}_\infty$. Hence there is a fundamental matrix function $\widetilde{Y}(\cdot,\lambda)$ of $y' - \widetilde{A}(\cdot,\lambda)y = 0$ such that

$$\widetilde{Y}(\cdot,\lambda)E(\cdot,\lambda)^{-1} = P^{[0]} + \{O(\lambda^{-1})\}_\infty.$$

The following example shows that this does not hold in general if $k = 0$.

EXAMPLE 2.8.8. Let $\widetilde{A}(\cdot,\lambda) = \lambda A_1 + A_0$, where

$$A_1 = \begin{pmatrix} 1 & 0 \\ 0 & -1 \end{pmatrix}, \quad A_0 = \begin{pmatrix} 0 & \alpha \\ 0 & 0 \end{pmatrix},$$

$a = 0$, $b = 1$, and $\alpha \in L_1(0,1)$. Let

$$\widetilde{Y}(\cdot,\lambda) = \begin{pmatrix} \tilde{y}_{11}(\cdot,\lambda) & \tilde{y}_{12}(\cdot,\lambda) \\ \tilde{y}_{21}(\cdot,\lambda) & \tilde{y}_{22}(\cdot,\lambda) \end{pmatrix}$$

be a fundamental matrix function of $y' - \widetilde{A}(\cdot,\lambda)y = 0$. It is easy to see that we have

$$y_{22}'(\cdot,\lambda) = -\lambda y_{22}(\cdot,\lambda) \tag{2.8.43}$$

and

$$y_{12}'(\cdot,\lambda) = \lambda y_{12}(\cdot,\lambda) + \alpha y_{22}(\cdot,\lambda). \tag{2.8.44}$$

The solution of equation (2.8.43) is

$$y_{22}(x,\lambda) = c_2(\lambda)e^{-\lambda x},$$

where c_2 is a complex-valued function. If we require that

$$\widetilde{Y}(\cdot,\lambda) = \left(P^{[0]} + \{O((1+|\Re(\lambda)|)^{-\eta})\}_\infty\right)E(\cdot,\lambda) \quad \text{with } \eta > 0, \tag{2.8.45}$$

where $P^{[0]}$ fulfils (2.8.6), then we obtain that $c_2(\lambda) = 1 + O((1+|\Re(\lambda)|)^{-\eta})$ as $|\lambda| \to \infty$. The solution of equation (2.8.44) is

$$y_{12}(x,\lambda) = e^{\lambda x}\left(c_1(\lambda) + c_2(\lambda)\int_0^x \alpha(t)e^{-2\lambda t}\,dt\right),$$

where c_1 is a complex-valued function. The representation in (2.8.45) leads to

$$e^{2\lambda x}\left(c_1(\lambda)+c_2(\lambda)\int_0^x \alpha(t)e^{-2\lambda t}\,\mathrm{d}t\right)=\{O((1+|\Re(\lambda)|)^{-\eta})\}_\infty(x) \tag{2.8.46}$$

as $|\lambda|\to\infty$. For $x=0$, (2.8.46) yields $c_1(\lambda)=O((1+|\Re(\lambda)|)^{-\eta})$. For our purposes it is sufficient to consider the case $\lambda<0$. Then (2.8.46) holds for $\lambda<0$ if and only if

$$\int_0^x \alpha(t)e^{2\lambda(x-t)}\,dt=\{O(|\lambda|^{-\eta})\}_\infty(x) \quad \text{as } \lambda\to-\infty.$$

Now let

$$\alpha(x)=x^{-1+\beta} \quad \text{with } \beta\in(0,1).$$

For $\lambda<-1$ and $x=-\lambda^{-1}$ we have

$$\begin{aligned}\int_0^x e^{2\lambda(x-t)}t^{-1+\beta}\,\mathrm{d}t &\geq x^{-1+\beta}\int_0^x e^{2\lambda(x-t)}\,\mathrm{d}t\\ &= x^{-1+\beta}\cdot\frac{1}{2|\lambda|}\cdot e^{2\lambda(x-t)}\Big|_0^x\\ &= \frac{1}{2}|\lambda|^{-\beta}(1-e^{-2}).\end{aligned}$$

Now assume that (2.8.45) holds whenever $\alpha\in L_p(0,1)$. Since α belongs to $L_p(0,1)$ for $\beta>1-1/p$, it follows that $\eta\leq 1-1/p$ in (2.8.45). Hence, for $k=0$, the estimate (2.8.12) is sharp in the sense that no $\eta>1-1/p$ exists such that

$$B_0(\cdot,\lambda)=\left\{O\left(\max_{\substack{\nu,\mu=0\\ \nu\neq\mu}}^{l}(1+|\Re(\lambda e^{i\varphi_{\nu\mu}})|)^{-\eta}\right)\right\}_\infty$$

holds for every first order system of differential equations (2.8.1) fulfilling the assumptions of this section. □

REMARK 2.8.9. The additional condition in Theorem 2.8.2 B. iii) is due to the fact that that the operator $T_{\lambda,2}$ defined in the proof of Theorem 2.8.2 does not have a bounded extension to $L(M_n(L_p(a,b)))$ if the coefficients of $Q^{[k]}$ do not belong to $L_{p'}(a,b)$.

The condition given in Theorem 2.8.2 B. iii) is satisfied if there is a number $q\in\{0,\dots,l\}$ such that the coefficients of $A_{0,\nu\mu}$ belong to $L_{p'}(a,b)$ for all pairs of numbers $(\nu,\mu)\in\Gamma$, where either $\Gamma=\{(\nu,\mu)\in\{0,\dots,l\}^2:\nu\neq\mu,\,\nu\neq q\}$ or $\Gamma=\{(\nu,\mu)\in\{0,\dots,l\}^2:\nu\neq\mu,\,\mu\neq q\}$.

Another condition under which (2.8.16) holds for $k=0$, $p\leq\frac{3}{2}$ and A_0 in $M_n(L_p(a,b))$ is that A_0 is a block triangular matrix. Indeed, in this case, (2.8.35) only needs to be shown for block triangular matrices h of the same form since g_λ has obviously this shape. But since $T_{\lambda,2}$ annihilates any block triangular matrix of the same shape as A_0, the crucial estimate in (2.8.35) is trivially fulfilled.

The asymptotic representation of the fundamental system obtained in Theorem 2.8.2 or Corollary 2.8.3 is stable with respect to change of variables:

REMARK 2.8.10. Let the assumptions be as in Theorem 2.8.2 or Corollary 2.8.3. Let $-\infty < c < d < \infty$ and $u : [c,d] \to [a,b]$ be surjective with $u \in W_\infty^{k+\kappa+1}(c,d)$ ($\kappa := 0$ for Theorem 2.8.2) and $\frac{1}{u'} \in L_\infty(c,d)$. Then

$$\widetilde{Y}(u(\cdot),\lambda) = \Big(\sum_{r=0}^{k} \lambda^{-r} P^{[r]} \circ u + \lambda^{-k} B_k(u(\cdot),\lambda)\Big) E(u(\cdot),\lambda)$$

has the same properties as $\widetilde{Y}(\cdot,\lambda)$; in particular, $P^{[0]} \circ u$ is invertible, and $B_k(u(\cdot),\lambda)$ satisfies the same estimates as $B_k(\cdot,\lambda)$.

Proof. In order to establish the estimates for $B_k(u(\cdot),\lambda)$ we note that, in view of the chain rule and LEIBNIZ' rule, it is sufficient to show that (2.8.38) and (2.8.39) also hold if we replace the variable x by $u(x)$. But this immediately follows from the formula on integration by substitution. □

2.9. Notes

The theory of Sobolev spaces over subsets of $\mathbb{R}^n$ is well-known, see e. g. ADAMS [AD]. However, here we only deal with intervals. In that case, we obtain stronger results and simpler proofs. Therefore, and in order to keep the book more self-contained, we have included Sections 2.1–2.4.

The definition of the fundamental matrix $Y(\cdot,\lambda)$ in Section 2.5 is not the standard one. But it is more convenient for our purposes since the conditions are kept at a minimum, and it is shown that the property $T^D(\cdot,\lambda)Y(\cdot,\lambda)c = 0$ for all $c \in \mathbb{C}^n$ and the invertibility of $Y(\cdot,\lambda)$ follow. The estimates in Section 2.7 are generalizations of the RIEMANN-LEBESGUE lemma and will be frequently used in the following chapters.

Asymptotic fundamental matrices and systems are the main ingredient to prove the convergence of expansions into eigenfunctions and associated functions. These asymptotic expansions for systems were obtained by WILDER [WI1], TAMARKIN [TA3], BIRKHOFF and LANGER [BIL], LANGER [LA9], WHYBURN [WHY1], and COLE [CO3], among others. In [LA9] the systems are considered in the complex domain, but essentially the same techniques as for intervals are applied.

In our presentation we tried to keep the regularity conditions on the coefficients as weak as possible. If all coefficients are infinitely differentiable, then the fundamental matrix can be chosen to be an asymptotic polynomial in $\frac{1}{\lambda}$ of arbitrary order, and the estimates and their proofs could be simplified. In most publications on asymptotics of solutions of differential equations much attention is given to a suitable choice of sectors, see e. g. [NA1, Chapter II]. By a suitable choice of the limit of integration $x_{\nu,\mu}(\lambda)$, see (2.8.26), we can avoid to consider these sectors. Our proof follows the approach of R. E. LANGER in [LA9] in the complex domain and of COLE [CO4] in the real domain.

The fundamental systems for n-th order differential equations will be discussed separately in the notes to Chapter VIII.

Chapter III

BOUNDARY EIGENVALUE PROBLEMS FOR FIRST ORDER SYSTEMS

In this chapter boundary eigenvalue problems for first order systems of ordinary linear differential equations are considered. The differential system as well as the boundary conditions are allowed to depend holomorphically on the eigenvalue parameter. The boundary conditions consist of terms at the endpoints and at interior points of the underlying interval and of an integral term. Such boundary eigenvalue problems are considered in suitable Sobolev spaces, so that both the differential operators and the boundary operators define bounded operators on Banach spaces. The assumptions on the boundary eigenvalue problems assure that these operators depend holomorphically on the eigenvalue parameter. In a canonical way we associate a holomorphic Fredholm operator valued function to such a boundary eigenvalue problem with the variable being the eigenvalue parameter. This operator function consists of two components, the first one is the differential operator function, the second one is the boundary operator function. Operator functions defined in this way are called boundary eigenvalue operator functions.

The theory of holomorphic Fredholm operator valued functions in Chapter I is applied to these boundary eigenvalue operator functions. As a first result we obtain that such an operator function is globally holomorphically equivalent to a canonical extension of the characteristic matrix function of the corresponding boundary eigenvalue problem (Theorem 3.1.2). The principal parts of the resolvent, i. e., the inverse of the boundary eigenvalue operator function, are expressed in terms of eigenfunctions and associated functions of this operator function and its adjoint (Theorem 3.1.4). The resolvent is defined on the direct sum of a space of vector functions and a finite-dimensional space of constants. On the space of vector functions, the resolvent is an integral operator whose kernel is the GREEN'S matrix; on the space of constants, it is a multiplication operator (Theorem 3.2.2).

The adjoint operator function of a boundary eigenvalue operator function defines the adjoint boundary eigenvalue problem (Theorem 3.3.1). The adjoint problem in this operator theoretical sense is obtained without further assumptions on the original boundary eigenvalue problem. The adjoint operator function maps the direct sum of a space of vector functions and a finite-dimensional space of constants into a space of distributions.

The realization of the original boundary eigenvalue problem in L_p-vector spaces leads to the adjoint boundary eigenvalue problem in parametrized form. This realization is achieved in the following way: Take the original boundary eigenvalue problem with homogeneous boundary conditions and associate to it the eigenvalue parameter family of closed linear operators whose domains consist of W_p^1-vector functions which fulfil the boundary conditions. These closed linear operators are not necessarily densely defined, and their domains may depend on the eigenvalue parameter. Consequently, the adjoints of these closed linear operators are closed linear relations but in general not operators. Additional assumptions are needed to assure that these adjoints form a family of operators, in which case they define the adjoint boundary eigenvalue problem in parametrized form. The relationships between the adjoint boundary eigenvalue problems in operator theoretical sense on one side and in parametrized form on the other side are discussed in detail (Theorems 3.4.3 and 3.4.5).

As a special case we consider two-point boundary eigenvalue problems. It is shown that the coefficients in the classical adjoint boundary conditions depend holomorphically on the eigenvalue parameter if the coefficients of the original boundary conditions have this property. We state that the classical adjoint boundary eigenvalue problem coincides with the adjoint problem in parametrized form. Root functions (eigenvectors and associated vectors) are defined for the above mentioned families of closed linear operators by taking root functions (eigenvectors and associated vectors) of the corresponding holomorphic boundary eigenvalue operator function. It is proved that the principal parts of the GREEN'S matrix can be represented in terms of eigenfunctions and associated functions of the family of closed linear operators for the realization of the boundary eigenvalue problem in L_p-vector spaces and the family of the adjoints of these operators (Theorem 3.5.11).

3.1. The boundary eigenvalue problem

Let Ω be a domain in $\mathbb{C}$, $-\infty < a < b < \infty$, $1 \le p \le \infty$, p' such that $1/p + 1/p' = 1$, and $n \in \mathbb{N} \setminus \{0\}$. Let $A \in H(\Omega, M_n(L_p(a,b)))$ and $T^R \in H(\Omega, L((W_p^1(a,b))^n, \mathbb{C}^n))$. We consider boundary eigenvalue problems of the form

$$\begin{cases} y' - A(\cdot,\lambda)y = 0, \\ T^R(\lambda)y = 0, \end{cases} \tag{3.1.1}$$

for $\lambda \in \Omega$. Here a solution $y \in (W_p^1(a,b))^n$ of the differential system in (3.1.1) is to be understood as a weak solution, i. e., a solution in the distributional sense. If, e. g., we take

$$T^R(\lambda)y = W^a(\lambda)y(a) + W^b(\lambda)y(b) \tag{3.1.2}$$

for $y \in (W_p^1(a,b))^n$ and $\lambda \in \Omega$, where $W^a, W^b \in H(\Omega, M_n(\mathbb{C}))$, then (3.1.1) is a *two-point boundary eigenvalue problem*. We define

$$(3.1.3) \qquad \begin{cases} T^D(\lambda)y := y' - A(\cdot,\lambda)y, \\ T(\lambda)y := \begin{pmatrix} T^D(\lambda)y \\ T^R(\lambda)y \end{pmatrix}, \end{cases} \qquad (y \in (W_p^1(a,b))^n, \lambda \in \Omega).$$

From Lemma 2.5.1 we know that $T^D \in H(\Omega, L((W_p^1(a,b))^n, (L_p(a,b))^n))$, whence $T \in H(\Omega, L((W_p^1(a,b))^n, (L_p(a,b))^n \times \mathbb{C}^n))$. Since (3.1.1) is a boundary eigenvalue problem, we call T given by (3.1.3) a *boundary eigenvalue operator function*. We choose the fundamental matrix function

$$(3.1.4) \qquad Y \in H(\Omega, M_n(W_p^1(a,b)))$$

with $Y(a,\lambda) = \mathrm{id}_{\mathbb{C}^n}$ for $\lambda \in \Omega$ according to Theorem 2.5.3. Define

$$(3.1.5) \qquad Z(\lambda)c := Y(\cdot,\lambda)c \quad (c \in \mathbb{C}^n,\ \lambda \in \Omega),$$

$$(3.1.6) \qquad (U(\lambda)f)(x) := Y(x,\lambda)\int_a^x Y(t,\lambda)^{-1} f(t)\,\mathrm{d}t$$

for $\lambda \in \Omega$, $f \in (L_p(a,b))^n$ and $x \in (a,b)$. Since the operator function Z belongs to $H(\Omega, L(\mathbb{C}^n, (W_p^1(a,b))^n))$ by Proposition 2.5.6, the *characteristic matrix function* M defined by

$$(3.1.7) \qquad M(\lambda) := T^R(\lambda)Z(\lambda) \quad (\lambda \in \Omega)$$

belongs to $H(\Omega, M_n(\mathbb{C}))$ by Corollary 1.2.4.

THEOREM 3.1.1. *T is an abstract boundary eigenvalue operator function in the sense of Section* 1.11.

Proof. We set $E := (W_p^1(a,b))^n$, $F_1 := (L_p(a,b))^n$, $G := F_2 := \mathbb{C}^n$, $T_1(\lambda) := T^D(\lambda)$ and $T_2(\lambda) := T^R(\lambda)$. We must prove that (1.11.1) holds. (1.11.1) i) and (1.11.1) ii) follow from Lemma 2.5.7 and Proposition 2.5.6, respectively. For the proof of (1.11.1) iii) let $\lambda \in \Omega$ and $y \in N(T^D(\lambda))$. Then, by Definition 2.5.2, there is a vector $c \in \mathbb{C}^n$ such that $y = Y(\cdot,\lambda)c = Z(\lambda)c$, which proves $y \in R(Z(\lambda))$. Conversely, let $y \in R(Z(\lambda))$. Then there is a vector $c \in \mathbb{C}^n$ such that $y = Z(\lambda)c = Y(\cdot,\lambda)c$. Corollary 2.5.5 proves $y \in N(T^D(\lambda))$. □

Since U depends holomorphically on λ by Lemma 2.5.7, we can apply Theorem 1.11.1 without using ŠUBIN'S result in the proof of that theorem:

THEOREM 3.1.2. *The operators*

$$\begin{pmatrix} 0 & \mathrm{id}_{(L_p(a,b))^n} \\ \mathrm{id}_{\mathbb{C}^n} & T^R(\lambda)U(\lambda) \end{pmatrix} \in L(\mathbb{C}^n \times (L_p(a,b))^n, (L_p(a,b))^n \times \mathbb{C}^n)$$

and

$$(Z(\lambda), U(\lambda)) \in L(\mathbb{C}^n \times (L_p(a,b))^n, (W_p^1(a,b))^n)$$

are invertible and depend holomorphically on $\lambda \in \Omega$. *The operator function* T *is holomorphically equivalent on* Ω *to the* $(L_p(a,b))^n$*-extension of* M*; more precisely, for* $\lambda \in \Omega$ *we have*

$$T(\lambda) = \begin{pmatrix} 0 & \mathrm{id}_{(L_p(a,b))^n} \\ \mathrm{id}_{\mathbb{C}^n} & T^R(\lambda)U(\lambda) \end{pmatrix} \begin{pmatrix} M(\lambda) & 0 \\ 0 & \mathrm{id}_{(L_p(a,b))^n} \end{pmatrix} (Z(\lambda), U(\lambda))^{-1}.$$

Proof. The statement is obvious from Theorem 1.11.1 in view of (1.11.5) and (1.11.7). □

Since M is an operator function from the finite-dimensional space $\mathbb{C}^n$ into itself and hence Fredholm operator valued with index zero, we immediately obtain

COROLLARY 3.1.3. *We have* $T \in H(\Omega, \Phi((W_p^1(a,b))^n, (L_p(a,b))^n \times \mathbb{C}^n))$ *and* $\operatorname{ind} T(\lambda) = 0$ *for all* $\lambda \in \Omega$.

THEOREM 3.1.4. *Let* M *be the characteristic matrix function given by* (3.1.7). *Assume that* $\rho(M) \neq \emptyset$. *Let* $\mu \in \sigma(M)$ *and* $r := \operatorname{nul} M(\mu)$. *Let* $\{c_1, \ldots, c_r\}$ *and* $\{d_1, \ldots, d_r\}$ *be biorthogonal CSRFs of* M *and* M^* *at* μ. *Define*

$$y_j := Zc_j, \quad v_j := \begin{pmatrix} -(T^RU)^*d_j \\ d_j \end{pmatrix} \qquad (j = 1, \ldots, r).$$

Then $\{y_1, \ldots, y_r\}$ *and* $\{v_1, \ldots, v_r\}$ *are biorthogonal CSRF of* T *and* T^* *at* μ, $\nu(y_j) = \nu(v_j) = \nu(c_j) = \nu(d_j) =: m_j$ *for* $j = 1, \ldots, r$, *and the operator function*

$$T^{-1} - \sum_{j=1}^{r} (\cdot - \mu)^{-m_j} y_j \otimes v_j$$

is holomorphic at μ.

Proof. This theorem is merely a restatement of Theorem 1.11.3 in the present context. □

PROPOSITION 3.1.5. *Let* $W \in H(\Omega, M_n(L_1(a,b)))$, $a_k \in [a,b]$ $(k \in \mathbb{N})$, $a_j \neq a_k$ $(k \neq j)$, $a_0 = a$, $a_1 = b$, $W^{(j)} \in H(\Omega, M_n(\mathbb{C}))$ $(j \in \mathbb{N})$ *such that*

$$\sum_{j=0}^{\infty} \sup_{\lambda \in K} |W^{(j)}(\lambda)| < \infty \tag{3.1.8}$$

for each compact subset K *of* Ω. *For* $\lambda \in \Omega$ *and* $y \in (W_p^1(a,b))^n$ *we set*

$$T^R(\lambda)y := \sum_{j=0}^{\infty} W^{(j)}(\lambda)y(a_j) + \int_a^b W(\xi,\lambda)y(\xi)\,\mathrm{d}\xi. \tag{3.1.9}$$

Then $T^R \in H(\Omega, L((W_p^1(a,b))^n, \mathbb{C}^n))$.

Proof. Integration as a map from $(L_1(a,b))^n$ to $\mathbb{C}^n$ is continuous. From Proposition 2.3.3 and Corollary 1.2.4 we infer that the assertion holds for the integral part. The assumption (3.1.8) implies that

$$y \mapsto \sum_{j=0}^{\infty} W^{(j)}(\lambda)y(a_j)$$

converges uniformly on compact subsets of Ω as a series of holomorphic operator functions in the Banach space $L((C([a,b]))^n, \mathbb{C}^n)$. Hence this series defines a holomorphic operator function in $L((C([a,b]))^n, \mathbb{C}^n)$, see [DI, (9.12.1)]. As $W_p^1(a,b)$ is contained continuously in $C[a,b]$, this completes the proof. □

REMARK 3.1.6. The statement of Proposition 3.1.5 even holds if we replace (3.1.8) by the weaker condition

$$\sup_{\lambda \in K} \sum_{j=0}^{\infty} |W^{(j)}(\lambda)| < \infty. \tag{3.1.10}$$

Proof. We apply VITALI'S theorem in order to obtain the holomorphy. □

3.2. The inhomogeneous boundary eigenvalue problem

Let T be the boundary eigenvalue operator function defined by (3.1.3), where T^R is given by (3.1.9). For $\lambda \in \rho(T)$, $f_1 \in (L_p(a,b))^n$ and $f_2 \in \mathbb{C}^n$ we set

$$R_1(\lambda)f_1 := T^{-1}(\lambda)(f_1, 0), \tag{3.2.1}$$

$$R_2(\lambda)f_2 := T^{-1}(\lambda)(0, f_2). \tag{3.2.2}$$

If $\lambda \in \rho(T)$ and $y \in (W_p^1(a,b))^n$, we obtain

$$y = T^{-1}(\lambda)T(\lambda)y = R_1(\lambda)T^D(\lambda)y + R_2(\lambda)T^R(\lambda)y. \tag{3.2.3}$$

We now give an explicit representation of $R_1(\lambda)$, $R_2(\lambda)$ and $T^{-1}(\lambda)$. For this purpose we set

$$F(x,\lambda) := \sum_{\substack{j=0 \\ a_j<x}}^{\infty} W^{(j)}(\lambda) + \int_a^x W(t,\lambda)\,\mathrm{d}t \quad (a \le x < b), \tag{3.2.4}$$

$$F(b,\lambda) := \sum_{j=0}^{\infty} W^{(j)}(\lambda) + \int_a^b W(t,\lambda)\,\mathrm{d}t. \tag{3.2.5}$$

The matrix function $F(\cdot,\lambda)$ is of bounded variation because of the assumptions (3.1.8) and $W(\lambda) \in M_n(L_1(a,b))$.

PROPOSITION 3.2.1. *Let* $\lambda \in \Omega$ *and* $y \in (W_p^1(a,b))^n$. *Then*

$$T^R(\lambda)y = \int_a^b \mathrm{d}_t F(t,\lambda)y(t) \quad (y \in (W_p^1(a,b))^n),$$

where the integral is the Riemann-Stieltjes integral of the vector function y *with respect to the integrator* $F(\cdot,\lambda)$.

Proof. The integral is well-defined since y is continuous, see [HS, (17.15), (17.16) and (8.7)]. There is a sequence $a = t_0^k < t_1^k < \cdots < t_{m_k}^k = b$ of subdivisions of $[a,b]$ such that

$$\sum_{i=1}^{m_k} \Bigg(\sum_{\substack{j=0 \\ t_{i-1}^k \le a_j < t_i^k}}^{\infty} W^{(j)}(\lambda)y(t_i^k) + \int_{t_{i-1}^k}^{t_i^k} W(t,\lambda)y(t_i^k)\,\mathrm{d}t \Bigg) + W^{(1)}(\lambda)y(b)$$

converges to $\int_a^b \mathrm{d}_t F(t,\lambda)y(t)$ as $k \to \infty$, where $\max\limits_{i=1}^{m_k}(t_i^k - t_{i-1}^k) \to 0$ as $k \to \infty$ can be assumed. On the other hand, the above sum converges to

$$\sum_{j=0}^{\infty} W^{(j)}(\lambda)y(a_j) + \int_a^b W(t,\lambda)y(t)\,\mathrm{d}t$$

by LEBESGUE'S dominated convergence theorem. Here we take the counting measure on $\mathbb{N}$ for the convergence of the sum and the LEBESGUE measure on $[a,b]$ for the convergence of the integral. □

Let $Y \in H(\Omega, M_n(W_p^1(a,b)))$ be the fundamental matrix function of the first order system $T^D y = 0$ with $Y(a,\lambda) = \mathrm{id}_{\mathbb{C}^n}$ for $\lambda \in \Omega$. Let M be the characteristic matrix function given by (3.1.7). For $\lambda \in \rho(T)$ the GREEN'S matrix of T is defined by

$$\text{(3.2.6)} \qquad G(x,\xi,\lambda) := \begin{cases} \displaystyle\int_{t=a}^{\xi} Y(x,\lambda)M^{-1}(\lambda)\,\mathrm{d}_t F(t,\lambda)Y(t,\lambda)Y^{-1}(\xi,\lambda) \\ \hfill (a \le \xi \le x \le b), \\ \displaystyle -\int_{t=\xi}^{b} Y(x,\lambda)M^{-1}(\lambda)\,\mathrm{d}_t F(t,\lambda)Y(t,\lambda)Y^{-1}(\xi,\lambda) \\ \hfill (a \le x < \xi \le b), \end{cases}$$

where the integrator is $F(\cdot,\lambda)$. From Proposition 2.5.4 we infer that the matrix functions $Y(x,\lambda)M^{-1}(\lambda)$ and $Y(t,\lambda)Y^{-1}(\xi,\lambda)$ do not depend on the choice of the fundamental matrix. Hence also the GREEN'S matrix does not depend on the choice of the fundamental matrix.

We set

$$\text{(3.2.7)} \qquad \widehat{G}(x,\lambda) := Y(x,\lambda)M^{-1}(\lambda) \quad (x \in [a,b],\ \lambda \in \rho(T))$$

and state

THEOREM 3.2.2. *For* $\lambda \in \rho(T)$, $f_1 \in (L_p(a,b))^n$, $f_2 \in \mathbb{C}^n$ *and* $x \in (a,b)$ *we have that* $G(x,\cdot,\lambda)$ *belongs to* $M_n(L_\infty(a,b))$ *and that*

$$(R_1(\lambda)f_1)(x) = \int_a^b G(x,\xi,\lambda)f_1(\xi)\,\mathrm{d}\xi, \tag{3.2.8}$$

$$(R_2(\lambda)f_2)(x) = \widehat{G}(x,\lambda)f_2, \tag{3.2.9}$$

$$(T^{-1}(\lambda)(f_1,f_2))(x) = \int_a^b G(x,\xi,\lambda)f_1(\xi)\,\mathrm{d}\xi + \widehat{G}(x,\lambda)f_2. \tag{3.2.10}$$

Proof. Since, for $\xi_1 < \xi_2$ in $[a,x]$ or $[x,b]$, respectively,

$$\Big|\int_{\xi_1}^{\xi_2} Y(x,\lambda)M^{-1}(\lambda)\,\mathrm{d}_tF(t,\lambda)Y(t,\lambda)Y^{-1}(\xi,\lambda)\Big|$$
$$\leq \Big[\max_{x,t,\xi\in[a,b]} |Y(x,\lambda)M^{-1}(\lambda)|\,|Y(t,\lambda)Y^{-1}(\xi,\lambda)|\Big]\int_{\xi_1}^{\xi_2} \mathrm{d}_t|F(t,\lambda)|,$$

$G(x,\cdot,\lambda)$ is of bounded variation for each $\lambda \in \Omega$ and $x \in [a,b]$. Since the real as well as the imaginary part of a function of bounded variation is the difference of two monotone functions, a function of bounded variation is measurable. Hence each component of $G(x,\cdot,\lambda)$ is measurable. This shows that $G(x,\cdot,\lambda)$ belongs to $M_n(L_\infty(a,b))$.

The proof of the theorem will be complete if we show (3.2.10). From

$$\begin{aligned}\int_a^b G(x,\xi,\lambda)f_1(\xi)\,\mathrm{d}\xi \\ &= Y(x,\lambda)\int_a^x\int_{t=a}^{\xi} M^{-1}(\lambda)\,\mathrm{d}_tF(t,\lambda)Y(t,\lambda)Y^{-1}(\xi,\lambda)f_1(\xi)\,\mathrm{d}\xi \\ &\quad - Y(x,\lambda)\int_x^b\int_{t=\xi}^{b} M^{-1}(\lambda)\,\mathrm{d}_tF(t,\lambda)Y(t,\lambda)Y^{-1}(\xi,\lambda)f_1(\xi)\,\mathrm{d}\xi\end{aligned} \tag{3.2.11}$$

and Propositions 2.3.1 and 2.1.8 it follows that $\int_a^b G(\cdot,\xi,\lambda)f_1(\xi)\,\mathrm{d}\xi$ belongs to $(W_p^1(a,b))^n$. On the right-hand side of (3.2.11) we add and subtract the term

$$Y(x,\lambda)\int_a^x\int_{t=\xi}^{b} M^{-1}(\lambda)\,\mathrm{d}_tF(t,\lambda)Y(t,\lambda)Y^{-1}(\xi,\lambda)f_1(\xi)\,\mathrm{d}\xi.$$

We obtain

$$\begin{aligned}\int_a^b G(x,\xi,\lambda)f_1(\xi)\,\mathrm{d}\xi &= Y(x,\lambda)\int_a^x Y^{-1}(\xi,\lambda)f_1(\xi)\,\mathrm{d}\xi \\ &\quad - Y(x,\lambda)\int_a^b\int_{t=\xi}^{b} M^{-1}(\lambda)\,\mathrm{d}_tF(t,\lambda)Y(t,\lambda)Y^{-1}(\xi,\lambda)f_1(\xi)\,\mathrm{d}\xi\end{aligned} \tag{3.2.12}$$

since, by Proposition 3.2.1,

$$M(\lambda) = \int_a^b \mathrm{d}_tF(t,\lambda)Y(t,\lambda).$$

Since the first term on the right hand side of (3.2.12) is $(U(\lambda)f_1)(x)$, where $U(\lambda)$ is a right inverse of $T^D(\lambda)$, and since $T^D(\cdot,\lambda)Y(\cdot,\lambda)=0$, it follows that

(3.2.13) $$T^D(\lambda)\Big(\int_a^b G(\cdot,\xi,\lambda)f_1(\xi)\,\mathrm{d}\xi+\widehat{G}(\cdot,\lambda)f_2\Big)=f_1.$$

Again from (3.2.12) we deduce in view of Proposition 3.2.1 and FUBINI'S theorem, applied to the measures $\mathrm{d}_xF(x,\lambda)$ and $\mathrm{d}\xi$, that

$$T^R(\lambda)\int_a^b G(\cdot,\xi,\lambda)f_1(\xi)\,\mathrm{d}\xi=\int_a^b \mathrm{d}_xF(x,\lambda)Y(x,\lambda)\int_a^x Y^{-1}(\xi,\lambda)f_1(\xi)\,\mathrm{d}\xi$$
$$-\int_a^b\int_{t=\xi}^b \mathrm{d}_tF(t,\lambda)Y(t,\lambda)Y^{-1}(\xi,\lambda)f_1(\xi)\,\mathrm{d}\xi=0,$$

whence

(3.2.14) $$T^R(\lambda)\Big(\int_a^b G(\cdot,\xi,\lambda)f_1(\xi)\,\mathrm{d}\xi+\widehat{G}(\cdot,\lambda)f_2\Big)=f_2.$$

This proves (3.2.10) since $T(\lambda)$ is invertible by assumption. □

3.3. The adjoint boundary eigenvalue problem

The *adjoint boundary eigenvalue problem in distributional sense* consists in finding nontrivial weak solutions $(u,d)\in(L_{p'}(a,b))^n\times\mathbb{C}^n$ of the differential equation

(3.3.1) $$u_e'+A^{\mathsf{T}}(\cdot,\lambda)u_e-T^{R^*}(\lambda)d=0$$

for $\lambda\in\Omega$, where u_e is the canonical extension of u. The following theorem justifies this definition of the adjoint boundary eigenvalue problem.

THEOREM 3.3.1. *Let the boundary eigenvalue operator function T be given by* (3.1.3) *and assume that $p<\infty$.*
Then $T^*\in H(\Omega,L((L_{p'}(a,b))^n\times\mathbb{C}^n,(W_{p'}^{-1}[a,b])^n))$ has the representation

(3.3.2) $$T^*(\lambda)(u,d)=-u_e'-A^{\mathsf{T}}(\cdot,\lambda)u_e+T^{R^*}(\lambda)d$$

($u\in(L_{p'}(a,b))^n$, $d\in\mathbb{C}^n$). If T^R has the form (3.1.9), *then*

(3.3.3) $$T^{R^*}(\lambda)=\sum_{j=0}^{\infty}W^{(j)\mathsf{T}}(\lambda)\delta_{a_j}+(W^{\mathsf{T}}(\cdot,\lambda))_e$$

and

(3.3.4) $$((T^RU)^*(\lambda)d)(t)$$
$$=Y^{-1}(t,\lambda)^{\mathsf{T}}\Big\{\sum_{j=1}^{\infty}Y(a_j,\lambda)^{\mathsf{T}}W^{(j)}(\lambda)^{\mathsf{T}}\chi_{(a,a_j)}(t)+\int_t^b Y(\xi,\lambda)^{\mathsf{T}}W(\xi,\lambda)^{\mathsf{T}}\,\mathrm{d}\xi\Big\}d.$$

Here $\chi_{(a,a_j)}$ is the characteristic function of the interval (a,a_j).

Proof. As for $n = 1$, we denote the canonical bilinear form on $(W_p^k(a,b))^n \times (W_{p'}^{-k}[a,b])^n$ by $\langle\ ,\ \rangle_{p,k}$. Since $L_{p'}(a,b)$ is the dual of $L_p(a,b)$ and $W_{p'}^{-1}[a,b]$ is the dual of $W_p^1(a,b)$ by Theorem 2.2.5, Proposition 1.2.6 yields that $T^* \in H(\Omega, L((L_{p'}(a,b))^n \times \mathbb{C}^n, (W_{p'}^{-1}[a,b])^n))$. Let $y \in (W_p^1(a,b))^n$ and $u \in (L_{p'}(a,b))^n$. Then we infer with the aid of the definition of $\langle\ ,\ \rangle_{p,1}$ in (2.2.4) and Proposition 2.3.4 that

$$\begin{aligned}\langle y, T^{D*}(\lambda)u\rangle_{p,1} &= \langle T^D(\lambda)y, u_e\rangle_{p,0}\\ &= \langle y', u_e\rangle_{p,0} - \langle A(\cdot,\lambda)y, u_e\rangle_{p,0}\\ &= -\langle y, u_e'\rangle_{p,1} - \langle y, A^\mathsf{T}(\cdot,\lambda)u_e\rangle_{p,1}.\end{aligned}$$

This proves

$$T^{D*}(\lambda)u = -u_e' - A^\mathsf{T}(\cdot,\lambda)u_e. \tag{3.3.5}$$

Since, for $d \in \mathbb{C}^n$,

$$\begin{aligned}\langle y, T^*(\lambda)(u,d)\rangle_{p,1} &= \langle T(\lambda)y, (u,d)\rangle\\ &= \langle T^D(\lambda)y, u_e\rangle_{p,0} + \langle T^R(\lambda)y, d\rangle\\ &= \langle y, T^{D*}(\lambda)u_e + T^{R*}(\lambda)d\rangle_{p,1},\end{aligned}$$

we obtain the representation (3.3.2).

Now let T^R be given by (3.1.9), $y \in (W_p^1(a,b))^n$ and $d \in \mathbb{C}^n$. The definition of the Dirac distribution, see Example 2.2.6, and Proposition 2.3.4 yield

$$\begin{aligned}\langle y, T^{R*}(\lambda)d\rangle_{p,1} &= \langle T^R(\lambda)y, d\rangle\\ &= \sum_{j=0}^{\infty} d^\mathsf{T} W^{(j)}(\lambda)y(a_j) + \int_a^b d^\mathsf{T} W(\xi,\lambda)y(\xi)\,\mathrm{d}\xi\\ &= \sum_{j=0}^{\infty} \langle y, W^{(j)\mathsf{T}}(\lambda)d\delta_{a_j}\rangle_{p,1} + \langle y, (W^\mathsf{T}(\cdot,\lambda))_e d\rangle_{p,1}\\ &= \left\langle y, \sum_{j=0}^{\infty} W^{(j)\mathsf{T}}(\lambda)d\delta_{a_j} + (W^\mathsf{T}(\cdot,\lambda))_e d\right\rangle_{p,1},\end{aligned}$$

which proves (3.3.3).

For $f \in (L_p(a,b))^n$ we obtain

$$\begin{aligned}T^R(\lambda)U(\lambda)f = &\sum_{j=0}^{\infty} W^{(j)}(\lambda)Y(a_j,\lambda)\int_a^{a_j} Y^{-1}(t,\lambda)f(t)\,\mathrm{d}t\\ &+ \int_a^b W(\xi,\lambda)Y(\xi,\lambda)\int_a^{\xi} Y^{-1}(t,\lambda)f(t)\,\mathrm{d}t\,\mathrm{d}\xi\end{aligned}$$

$$= \int_a^b \Big\{ \sum_{j=1}^{\infty} W^{(j)}(\lambda) Y(a_j,\lambda) Y^{-1}(t,\lambda) \chi_{(a,a_j)}(t)$$
$$+ \int_t^b W(\xi,\lambda) Y(\xi,\lambda)\,\mathrm{d}\xi\, Y^{-1}(t,\lambda) \Big\} f(t)\,\mathrm{d}t.$$

Now (3.3.4) immediately follows from

$$\langle f, (T^R U)^*(\lambda) d\rangle_{p,0} = d^{\mathsf{T}} T^R(\lambda) U(\lambda) f. \quad \square$$

3.4. The adjoint boundary eigenvalue problem in parametrized form

In this section let $p < \infty$. In order to define the *adjoint boundary eigenvalue problem in parametrized form* we consider the family of operators $T_0(\lambda)$ in $(L_p(a,b))^n$ defined by

(3.4.1) $$D(T_0(\lambda)) = \{ y \in (L_p(a,b))^n : y \in (W_p^1(a,b))^n, T^R(\lambda) y = 0 \}$$

and

(3.4.2) $$T_0(\lambda) y = T^D(\lambda) y \quad (y \in D(T_0(\lambda))).$$

This is the operator family which is classically considered together with the boundary eigenvalue problem (3.1.1). Note that the domain $D(T_0(\lambda))$ of $T_0(\lambda)$ may depend on λ since T^R depends on λ and that the domain of $T_0(\lambda)$ may be a nondense subspace of $(L_p(a,b))^n$. For example, the boundary eigenvalue operator $T^R(\lambda)$ given by $T^R(\lambda) y := \int_a^b y(t)\,\mathrm{d}t$ is a nonzero continuous linear operator from $(L_p(a,b))^n$ onto $\mathbb{C}^n$. Hence, in this case, $D(T_0(\lambda))$ is a nondense subspace of $(L_p(a,b))^n$.

Let $\rho'(T_0) := \{ \lambda \in \Omega : T_0(\lambda) \text{ is bijective} \}$ and, as usual for not neccessarily bounded operators, $\rho(T_0) := \{ \lambda \in \rho'(T_0) : T_0(\lambda)^{-1} \text{ is continuous} \}$.

THEOREM 3.4.1. i) *We have* $\rho(T_0) = \rho'(T_0) = \rho(T)$ *and* $T_0^{-1}(\lambda) f = T^{-1}(\lambda)(f,0)$ *for* $\lambda \in \rho(T)$ *and* $f \in (L_p(a,b))^n$.
ii) *Assume that* T^R *is of the form* (3.1.9) *and let* G *be the* GREEN'S *matrix given by* (3.2.6). *Then, for* $\lambda \in \rho(T_0)$ *and* $f \in (L_p(a,b))^n$,

(3.4.3) $$(T_0^{-1}(\lambda) f)(x) = \int_a^b G(x,\xi,\lambda) f(\xi)\,\mathrm{d}\xi.$$

Proof. i) Let $\lambda \in \rho'(T_0)$ and $y \in N(T(\lambda))$. Then $y \in (W_p^1(a,b))^n$ and $T^R(\lambda) y = 0$. Hence $y \in D(T_0(\lambda))$ and $T_0(\lambda) y = 0$. This proves $y = 0$ since $T_0(\lambda)$ is injective. We have proved that $T(\lambda)$ is injective. From Corollary 3.1.3 we know that $\operatorname{ind} T(\lambda) = 0$. This proves $\operatorname{def} T(\lambda) = \operatorname{nul} T(\lambda) = 0$, i. e., $T(\lambda)$ is bijective.

Let $\lambda \in \rho(T)$. For $y \in N(T_0(\lambda))$ we have $T^D(\lambda) y = 0$ and $T^R(\lambda) y = 0$, which proves $y = 0$ since $T(\lambda)$ is injective. For the proof of the surjectivity of $T_0(\lambda)$ let $f \in (L_p(a,b))^n$. Set $y := T^{-1}(\lambda)(f,0)$. Then $T^R(\lambda) y = 0$ shows $y \in D(T_0(\lambda))$,

and $T_0(\lambda)y = T^D(\lambda)y = f$ follows. The continuity of $T^{-1}(\lambda)$ immediately implies $T_0^{-1}(\lambda) \in L((L_p(a,b))^n)$, which also proves $\rho(T_0) = \rho'(T_0)$.
ii) is clear from i) and Theorem 3.2.2. $\square$

REMARK 3.4.2. For all $\lambda \in \Omega$, the operator $T_0(\lambda) : (L_p(a,b))^n \to (L_p(a,b))^n$ is closed.

Proof. If $\lambda \in \rho(T_0)$, this immediately follows from the continuity of $T_0^{-1}(\lambda)$, which was shown in Theorem 3.4.1 i). For arbitrary $\lambda \in \Omega$, let $y_k \in D(T_0(\lambda))$ ($k \in \mathbb{N}$) such that $y_k \to y$ in $(L_p(a,b))^n$ and $T_0(\lambda)y_k \to f$ in $(L_p(a,b))^n$ as $k \to \infty$. Since $U(\lambda)T^D(\lambda)y_k - y_k \in N(T^D(\lambda))$, there is a $c_k \in \mathbb{C}^n$ such that

$$U(\lambda)T_0(\lambda)y_k = y_k + Y(\cdot,\lambda)c_k \quad (k \in \mathbb{N}).$$

$U(\lambda)$ is continuous as an operator from $(L_p(a,b))^n$ into $(W_p^1(a,b))^n$, whence $U(\lambda)T_0(\lambda)y_k \to U(\lambda)f$ in $(W_p^1(a,b))^n$ and therefore also in $(L_p(a,b))^n$. Thus $(Y(\cdot,\lambda)c_k)_{k=0}^\infty$ converges in $(L_p(a,b))^n$. But since the set $\{Y(\cdot,\lambda)c : c \in \mathbb{C}^n\}$ is a finite-dimensional subspace of $(W_p^1(a,b))^n$, $(Y(\cdot,\lambda)c_k)_{k=0}^\infty$ also converges in $(W_p^1(a,b))^n$ since all norms on finite-dimensional spaces are equivalent, see [CON, Theorem III.3.1]. This shows that $(y_k)_{k=0}^\infty$ converges in $(W_p^1(a,b))^n$. Therefore $y \in (W_p^1(a,b))^n$, and the continuity of $T(\lambda)$ implies that $T^R(\lambda)y = 0$, i. e., $y \in D(T_0(\lambda))$, and $T_0(\lambda)y = T^D(\lambda)y = f$. $\square$

The adjoint $T_0^*(\lambda)$ is a linear relation in $(L_{p'}(a,b))^n$ defined by its graph

$$G(T_0^*(\lambda)) = (G(-T_0(\lambda)))^\perp,$$

i. e., for $u \in (L_{p'}(a,b))^n$ we have

$$u \in D(T_0^*(\lambda)) \Leftrightarrow \exists w \in (L_{p'}(a,b))^n \ \forall y \in D(T_0(\lambda)) \ \langle T_0(\lambda)y, u\rangle = \langle y, w\rangle$$

and

$$T_0^*(\lambda)u = \{w \in (L_{p'}(a,b))^n : \forall y \in D(T_0(\lambda)) \ \langle T_0(\lambda)y, u\rangle = \langle y, w\rangle\}.$$

Here $\langle\ ,\ \rangle$ is the canonical bilinear form on $(L_p(a,b))^n \times (L_{p'}(a,b))^n$.

Conditions to assure that the linear relations $T_0^*(\lambda)$ ($\lambda \in \Omega$) are operators will be given below. In that case,

$$T_0^*(\lambda)u = 0 \quad (u \in D(T_0^*(\lambda))$$

is called the *classical adjoint boundary eigenvalue problem* to the given problem $T_0(\lambda)y = 0$ ($y \in D(T_0(\lambda))$.

For $u \in \mathscr{D}'(\mathbb{R})$ we denote its restriction to (a,b) by u_r, i. e., $u_r \in \mathscr{D}'(a,b)$ is given by

$$\langle \varphi, u_r\rangle = \langle \varphi, u\rangle_{\mathbb{R}} \quad (\varphi \in C_0^\infty(a,b)).$$

Note that $(u_e)_r = u$ for $u \in L_{p'}(a,b)$.

For $y \in (W_p^1(a,b))^n$ and $u \in (W_{p'}^{-1}[a,b])^n$ with $u = v_e$ and $v \in (L_{p'}(\mathbb{R}))^n$ we have by (2.2.4) that

$$\langle y,u \rangle_{p,1} = \langle y,u \rangle_{p,0} = \langle y,u_r \rangle.$$

THEOREM 3.4.3. i) *Let $\lambda \in \Omega$ and $u \in (L_{p'}(a,b))^n$. Then $u \in D(T_0^*(\lambda))$ if and only if there is a vector $d \in \mathbb{C}^n$ such that $T^*(\lambda)(u,d) \in (L_{p'}(\mathbb{R}))^n$.*
ii) *Let $\lambda \in \Omega$ and $u \in D(T_0^*(\lambda))$. Then*

$$T_0^*(\lambda)u = \{(T^*(\lambda)(u,d))_r : d \in \mathbb{C}^n,\ T^*(\lambda)(u,d) \in (L_{p'}(\mathbb{R}))^n\}.$$

Proof. Let $u \in (L_{p'}(a,b))^n$ and $d \in \mathbb{C}^n$ such that $T^*(\lambda)(u,d) \in (L_{p'}(\mathbb{R}))^n$. For $y \in D(T_0(\lambda))$ we have $T^R(\lambda)y = 0$. Hence

$$\begin{aligned}\langle T_0(\lambda)y,u \rangle &= \langle T^D(\lambda)y,u \rangle + \langle T^R(\lambda)y,d \rangle \\ &= \langle T(\lambda)y,(u,d) \rangle \\ &= \langle y,T^*(\lambda)(u,d) \rangle_{p,1} \\ &= \langle y,(T^*(\lambda)(u,d))_r \rangle\ .\end{aligned}$$

This proves $u \in D(T_0^*(\lambda))$ and $(T^*(\lambda)(u,d))_r \in T_0^*(\lambda)u$.

Conversely, let $u \in D(T_0^*(\lambda))$. Then there is a $w \in (L_{p'}(a,b))^n$ such that

$$\langle T_0(\lambda)y,u \rangle = \langle y,w \rangle \tag{3.4.4}$$

for all $y \in D(T_0(\lambda))$. Hence, for $y \in D(T_0(\lambda))$,

$$\begin{aligned}0 &= \langle y,w \rangle - \langle T_0(\lambda)y,u \rangle \\ &= \langle y,w_e - T^{D^*}(\lambda)u \rangle_{p,1}\ .\end{aligned}$$

Since $T^{R^*}(\lambda)$ is defined on the finite-dimensional space $\mathbb{C}^n$, the range $R(T^{R^*}(\lambda))$ is finite-dimensional and thus closed. This implies

$$\begin{aligned}w_e - T^{D^*}(\lambda)u &\in (D(T_0(\lambda)))^{\perp((W_p^1(a,b))^n,(W_{p'}^{-1}[a,b])^n)} \\ &= (N(T^R(\lambda)))^{\perp((W_p^1(a,b))^n,(W_{p'}^{-1}[a,b])^n)} = R(T^{R^*}(\lambda)),\end{aligned}$$

see [KA, p. 234]. Thus there is a $d \in \mathbb{C}^n$ such that $w_e - T^{D^*}(\lambda)u = T^{R^*}(\lambda)d$. This proves $T^*(\lambda)(u,d) = w_e \in (L_{p'}(\mathbb{R}))^n$ and thus the distribution $(T^*(\lambda)(u,d))_r = w$ belongs to $(L_{p'}(a,b))^n$. Since $w \in T_0^*(\lambda)u$ satisfying (3.4.4) was arbitrary, we have

$$T_0^*(\lambda)u \subset \{(T^*(\lambda)(u,d))_r : d \in \mathbb{C}^n,\ T^*(\lambda)(u,d) \in (L_{p'}(\mathbb{R}))^n\}.$$

Together with the first part of the proof this shows that ii) holds. □

PROPOSITION 3.4.4. *Let $c_i \in [a,b]$ $(i \in \mathbb{N})$ with $c_i \neq c_j$ for $i \neq j$, $l \in \mathbb{N}$ and $\alpha_{i,\nu} \in \mathbb{C}$ $(i \in \mathbb{N}, \nu = 0,\dots,l)$ with $\sum_{i=0}^{\infty} |\alpha_{i,\nu}| < \infty$ for $\nu = 0,\dots,l$. Let*

$$\sum_{\nu=0}^{l}\sum_{i=0}^{\infty} \alpha_{i,\nu}\delta_{c_i}^{(\nu)} \in L_{p'}(\mathbb{R}).$$

Then $\alpha_{i,\nu} = 0$ for all $i \in \mathbb{N}$ and $\nu = 0,\dots,l$.

Proof. Since the operator norm of $\delta_{c_i}^{(\nu)}$ $(\nu = 0,\dots,l)$ as a continuous linear functional on $C^l([a,b])$ does not exceed 1, $\sum_{\nu=0}^{l}\sum_{i=0}^{\infty} \alpha_{i,\nu}\delta_{c_i}^{(\nu)} \in L(C^l([a,b]),\mathbb{C})$ is well-defined and also belongs to $W_q^{-l-1}[a,b]$ by Example 2.2.6 and Proposition 2.2.7. By assumption, $f := \sum_{\nu=0}^{l}\sum_{i=0}^{\infty} \alpha_{i,\nu}\delta_{c_i}^{(\nu)} \in L_{p'}(\mathbb{R})$. Choose a nonzero $\psi_0 \in C_0^\infty(\mathbb{R})$. Let $x_0 \in \mathbb{R}$ such that $|\psi_0^{(l)}(x_0)| = \max_{x\in\mathbb{R}} |\psi_0^{(l)}(x)| (> 0)$. There is a number $\rho > 0$ such that $\operatorname{supp}\psi_0 \subset [x_0 - \rho, x_0 + \rho]$. Let $\psi(x) := \left(\psi_0^{(l)}(x_0)\rho^l\right)^{-1}\psi_0(\rho x + x_0)$. Then $\psi \in C_0^\infty(\mathbb{R})$, $\psi^{(l)}(0) = 1$, $|\psi^{(l)}(x)| \le 1$ for $x \in \mathbb{R}$, and $\psi(x) = 0$ for $|x| \ge 1$. Inductively we obtain $|\psi^{(\nu)}(x)| \le 1$ for $\nu = l-1, l-2, \dots, 0$. Let $j \in \mathbb{N}$ and set $\varphi_k(x) := k^{-l}\psi(k(x - c_j))$ $(k = 1,2,\dots)$. Then $\varphi_k^{(l)}(c_j) = 1$, $|\varphi_k^{(\nu)}(x)| \le k^{\nu-l}$ for $x \in \mathbb{R}$, $\nu = 0,\dots,l$, $k = 1,2,\dots$, and $\varphi_k(x) = 0$ for $k = 1,2,\dots$ and $x \in \mathbb{R}$ with $|x - c_j| \ge \frac{1}{k}$. Thus LEBESGUE'S dominated convergence theorem yields

$$\int_a^b \varphi_k(x) f(x)\,\mathrm{d}x \to 0 \quad (k \to \infty)$$

and, with respect to the counting measure,

$$\sum_{\nu=0}^{l}\sum_{i=0}^{\infty} (-1)^\nu \alpha_{i,\nu}\varphi_k^{(\nu)}(c_i) \to (-1)^l \alpha_{j,l} \quad (k \to \infty),$$

which proves $\alpha_{j,l} = 0$ since

$$\begin{aligned}\int_a^b \varphi_k(x) f(x)\,\mathrm{d}x &= \langle \varphi_k, f\rangle_{\mathbb{R}} \\ &= \left\langle \varphi_k, \sum_{\nu=0}^{l}\sum_{i=0}^{\infty} \alpha_{i,\nu}\delta_{c_i}^{(\nu)} \right\rangle_{\mathbb{R}} = \sum_{\nu=0}^{l}\sum_{i=0}^{\infty} (-1)^\nu \alpha_{i,\nu}\varphi_k^{(\nu)}(c_i).\end{aligned}$$

A recursive application of this method yields the statement. □

THEOREM 3.4.5. *Let T^R be given by* (3.1.9) *and assume that $A(\cdot,\lambda)$ belongs to $M_n(L_\infty(a,b))$ and that $W(\cdot,\lambda)$ belongs to $M_n(L_{p'}(a,b))$ for all $\lambda \in \Omega$.*
i) *Let $\lambda \in \Omega$ and $u \in (L_{p'}(a,b))^n$. Then $u \in D(T_0^*(\lambda))$ if and only if there is a vector $d \in \mathbb{C}^n$ such that*

$$(3.4.5) \qquad -u_e' + \sum_{j=0}^{\infty} W^{(j)\mathsf{T}}(\lambda) d\delta_{a_j} \in (L_{p'}(\mathbb{R}))^n.$$

ii) *Let $\lambda \in \Omega$. Then $T_0^*(\lambda)$ is a linear operator if and only if for each $d \in \mathbb{C}^n$ the implication*

$$\forall j \in \mathbb{N}\ W^{(j)\mathsf{T}}(\lambda)d = 0 \Rightarrow W^{\mathsf{T}}(\cdot,\lambda)d = 0$$

holds.
iii) *Let $\lambda \in \Omega$. Then $T_0^*(\lambda)$ is a linear operator if $W(\cdot,\lambda) = 0$ or if for each $d \in \mathbb{C}^n \setminus \{0\}$ there is an integer $j \in \mathbb{N}$ such that $W^{(j)\mathsf{T}}(\lambda)d \neq 0$.*

Proof. i) is obvious because of Theorem 3.4.3 i), (3.3.2) and (3.3.3).
ii) We know from Theorem 3.4.3 ii) that

$$T_0^*(\lambda)(0) = \Big\{ \Big(\sum_{j=0}^{\infty} W^{(j)\mathsf{T}}(\lambda) d\delta_{a_j} + W^{\mathsf{T}}(\cdot,\lambda)d \Big)_r : d \in \mathbb{C}^n, \sum_{j=0}^{\infty} W^{(j)\mathsf{T}}(\lambda) d\delta_{a_j} \in (L_{p'}(\mathbb{R}))^n \Big\}.$$

Since $T_0^*(\lambda)$ is an operator if and only if $T_0^*(\lambda)(0) = \{0\}$, the result is obvious from Proposition 3.4.4.
iii) immediately follows from ii). □

COROLLARY 3.4.6. *Let T^R be given by* (3.1.9), *where the sum runs from 0 to k, $k \geq 1$. Assume that $A(\cdot,\lambda)$ belongs to $M_n(L_\infty(a,b))$ and that $W(\cdot,\lambda)$ belongs to $M_n(L_{p'}(a,b))$ for all $\lambda \in \Omega$. The set $[a,b] \setminus \{a_0,\ldots,a_k\}$ is the disjoint union of k open intervals $I_1,\ldots,I_k$. Let $\lambda \in \Omega$ and $u \in (L_{p'}(a,b))^n$. Then $u \in D(T_0^*(\lambda))$ if and only if there is a vector $d \in \mathbb{C}^n$ such that*

$$u|_{I_j} \in (W_{p'}^1(I_j))^n \quad (j = 1,\ldots,k),$$
$$u(a+) = W^{(0)\mathsf{T}}(\lambda)d, \quad u(b-) = -W^{(1)\mathsf{T}}(\lambda)d,$$
$$u(a_j+) - u(a_j-) = W^{(j)\mathsf{T}}(\lambda)d \quad (j = 2,\ldots,k),$$

where $u(c+)$ and $u(c-)$ are the right-hand limit and the left-hand limit, respectively. For such $d \in \mathbb{C}^n$, the function $T^*(\lambda)(u,d) \in T_0^*(\lambda)u$ is uniquely determined by

$$T^*(\lambda)(u,d)|_{I_j} = -(u|_{I_j})' - A^{\mathsf{T}}(\cdot,\lambda)u|_{I_j} + W^{\mathsf{T}}(\cdot,\lambda)|_{I_j} d$$

for $j = 1,\ldots,k$.

Proof. For $c \in \mathbb{R}$, the Heaviside function H_c is defined by $H_c(x) = 0$ if $x \le c$ and $H_c(x) = 1$ if $x > c$. We have $H_c \in L_1^{\mathrm{loc}}(\mathbb{R})$ and $H_c' = \delta_c$. Therefore, for $u \in (L_{p'}(a,b))^n$ and $d \in \mathbb{C}^n$,

$$\Big(-u_e + \sum_{j=0}^{k} W^{(j)\mathsf{T}}(\lambda) d H_{a_j}\Big)' = -u_e' + \sum_{j=0}^{k} W^{(j)\mathsf{T}}(\lambda) d\delta_{a_j},$$

Theorem 3.4.5 i), and Corollary 3.1.4 yield that $u \in D(T_0^*(\lambda))$ if and only if there is a vector $d \in \mathbb{C}^n$ such that

$$\Big(-u_e + \sum_{j=0}^{k} W^{(j)\mathsf{T}}(\lambda) d H_{a_j}\Big)\Big|_{(\alpha,\beta)} \in (W_{p'}^1(\alpha,\beta))^n,$$

where $\alpha < a < b < \beta$. To complete the proof we apply Proposition 2.2.2. □

Now let $\lambda \in \Omega$ and suppose that $T_0^*(\lambda)$ is an operator and that the assumptions of Corollary 3.4.6 are fulfilled. Then Corollary 3.4.6, Theorem 3.4.3 ii), and Theorem 3.3.1 show that λ is an eigenvalue and $u \in (L_{p'}(a,b))^n \setminus \{0\}$ an eigenfunction of the adjoint boundary eigenvalue problem in parametrized form if and only if there is a vector $d \in \mathbb{C}^n$ such that, for $j = 1,\dots,k$, $u|_{I_j} \in (W_{p'}^1(a,b))^n$,

$$(u|_{I_j})' + A^{\mathsf{T}}(\cdot,\lambda) u|_{I_j} - W^{\mathsf{T}}(\cdot,\lambda)|_{I_j} d = 0, \tag{3.4.6}$$

and the boundary conditions

$$\begin{cases} u(a+) = W^{(0)\mathsf{T}}(\lambda)d, \quad u(b-) = -W^{(1)\mathsf{T}}(\lambda)d, \\ u(a_j+) - u(a_j-) = W^{(j)\mathsf{T}}(\lambda)d \quad (j = 2,\dots,k); \end{cases} \tag{3.4.7}$$

are satisfied.

PROPOSITION 3.4.7. *Let M be the characteristic matrix function given by* (3.1.7) *and assume that $\rho(M) \neq \emptyset$. Let $\mu \in \sigma(M)$ and $r := \operatorname{nul} M(\mu)$. Let $\{c_1,\dots,c_r\}$ and $\{d_1,\dots,d_r\}$ be biorthogonal CSRF of M and M^* at μ. Define*

$$y_j := Zc_j, \quad u_j := -(T^R U)^* d_j \qquad (j = 1,\dots,r),$$

where Z and U are given by (3.1.5) *and* (3.1.6), *respectively. Let $m_j := \nu(c_j)$, the multiplicity of the root function c_j. Then the operator function*

$$T_0^{-1} - \sum_{j=1}^{r} (\cdot - \mu)^{-m_j} y_j \otimes u_j$$

is holomorphic at μ.

Proof. Let J be the canonical injection from $(L_p(a,b))^n$ into $(L_p(a,b))^n \times \mathbb{C}^n$. Since $T^{-1}J = T_0^{-1}$ by Theorem 3.4.1 and

$$(y_j \otimes v_j)J = y_j \otimes (J^* v_j) = y_j \otimes u_j$$

by Proposition 1.1.2, where v_j is defined as in Theorem 3.1.4, the result follows from Theorem 3.1.4 and Corollary 1.2.4. □

PROPOSITION 3.4.8. *Let* $\mu \in \sigma(T)$.
i) *Let* y_0 *be an eigenvector of* T *at* μ. *Then* $y_0 \in D(T_0(\mu))$ *and* $T_0(\mu)y_0 = 0$.
ii) *Assume that* $T_0^*(\mu)$ *is an operator and let* (u_0, d_0) *be an eigenvector of* T^* *at* μ. *Then* $u_0 \in D(T_0^*(\mu))$ *and* $T_0^*(\mu)u_0 = 0$.

Proof. i) From $T(\mu)y_0 = 0$ it follows that $T^R(\mu)y_0 = 0$, and thus $y_0 \in D(T_0(\mu))$.
ii) As $T^*(u_0, d_0) = 0 \in (L_{p'}(\mathbb{R}))^n$, the statement follows from Theorem 3.4.3. □

PROPOSITION 3.4.9. *Assume that* T^R *does not depend on* λ *and let* $\mu \in \sigma(T)$.
i) *Let* $(y_l)_{l=0}^h$ *be a CEAV of* T *at* μ. *Then* $y_l \in D(T_0(\mu))$ *for* $l = 0, \ldots, h$.
ii) *Assume in addition that* $A \in H(\Omega, M_n(L_\infty(a,b)))$ *and that* $T_0^*(\lambda)$ *is an operator for all* $\lambda \in \Omega$. *Let* $(u_l, d_l)_{l=0}^h$ *be a CEAV of* T^* *at* μ. *Then* $u_l \in D(T_0^*(\mu))$ *for* $l = 0, \ldots, h$.

Proof. i) By Definition 1.6.1, the function

$$T \sum_{l=0}^{h} (\cdot - \mu)^l y_l$$

has a zero of order $\geq h+1$ at μ. Then

$$\sum_{l=0}^{h} (\cdot - \mu)^l T^R(\mu) y_l$$

has a zero of order $\geq h+1$ at μ, which proves that $T^R(\mu)y_l = 0$, and hence $y_l \in D(T_0(\mu))$ for $l = 0, \ldots, h$.
ii) Let $k \in \{0, \ldots, h\}$. Since

$$T^* \sum_{l=0}^{h} (\cdot - \mu)^l (u_l, d_l)$$

has a zero of order $\geq h+1$ at μ,

$$\sum_{j=0}^{k} \frac{1}{j!} \Big(\frac{\partial^j}{\partial \lambda^j} T^* \Big)(\mu)(u_{k-j}, d_{k-j}) = 0.$$

With the aid of Theorem 3.3.1 we infer

$$\begin{aligned} T^*(\mu)(u_k, d_k) &= -\sum_{j=1}^{k} \frac{1}{j!} \Big(\frac{\partial^j}{\partial \lambda^j} T^* \Big)(\mu)(u_{k-j}, d_{k-j}) \\ &= \sum_{j=1}^{k} \frac{1}{j!} \Big(\frac{\partial^j}{\partial \lambda^j} A^{\mathsf{T}} \Big)(\cdot, \mu) \big(u_{k-j}\big)_e \in (L_{p'}(\mathbb{R}))^n. \end{aligned}$$

From Theorem 3.4.3 i) it follows that $u_k \in D(T_0^*(\mu))$. □

EXAMPLE 3.4.10. We consider $T(\lambda) \in L((W_p^1(0,1))^2, (L_p(0,1))^2 \times \mathbb{C}^2)$ given by

$$T^D(\lambda)y = y' - \lambda y \quad (y \in (W_p^1(0,1))^2),$$
$$T^R(\lambda)y = \begin{pmatrix} 0 & 1+2\lambda \\ 1 & 0 \end{pmatrix} y(0) + y(1).$$

Obviously,

$$Y(x,\lambda) = \begin{pmatrix} e^{\lambda x} & 0 \\ 0 & e^{\lambda x} \end{pmatrix}$$

is a fundamental matrix of $T^D(\lambda)y = 0$. Then

$$M(\lambda) = \begin{pmatrix} e^{\lambda} & 1+2\lambda \\ 1 & e^{\lambda} \end{pmatrix}.$$

From $\det M(\lambda) = e^{2\lambda} - 1 - 2\lambda$ we infer that $\rho(M) \neq \emptyset$ and that $\det M$ has a zero of order 2 at 0. The vector function

$$M(\lambda)\begin{pmatrix} 1+\lambda \\ -1 \end{pmatrix} = \begin{pmatrix} e^{\lambda} - 1 + \lambda e^{\lambda} - 2\lambda \\ 1 - e^{\lambda} + \lambda \end{pmatrix}$$

has a zero of order 2 at 0. Hence $\begin{pmatrix} 1+\lambda \\ -1 \end{pmatrix}$ is a root function of M at 0 of multiplicity 2, and by Proposition 1.8.5 it is also a CSRF of M at 0. In the same way,

$$\begin{pmatrix} e^{\lambda} & 1 \\ 1+2\lambda & e^{\lambda} \end{pmatrix}\begin{pmatrix} \frac{1}{2} - \frac{1}{3}\lambda \\ -\frac{1}{2} - \frac{1}{6}\lambda \end{pmatrix} = \begin{pmatrix} \frac{1}{2}e^{\lambda} - \frac{1}{3}\lambda e^{\lambda} - \frac{1}{2} - \frac{1}{6}\lambda \\ \frac{1}{2} + \frac{2}{3}\lambda - \frac{2}{3}\lambda^2 - \frac{1}{2}e^{\lambda} - \frac{1}{6}\lambda e^{\lambda} \end{pmatrix}$$

shows that $\begin{pmatrix} \frac{1}{2} - \frac{1}{3}\lambda \\ -\frac{1}{2} - \frac{1}{6}\lambda \end{pmatrix}$ is a CSRF of M^* at 0 of multiplicity 2. From

$$\begin{aligned} \frac{1}{\lambda^2}&\begin{pmatrix} \frac{1}{2} - \frac{1}{3}\lambda \\ -\frac{1}{2} - \frac{1}{6}\lambda \end{pmatrix}^{\mathsf{T}} M(\lambda)\begin{pmatrix} 1+\lambda \\ -1 \end{pmatrix} \\ &= \begin{pmatrix} \frac{1}{2} - \frac{1}{3}\lambda \\ -\frac{1}{2} - \frac{1}{6}\lambda \end{pmatrix}^{\mathsf{T}}\begin{pmatrix} \frac{3}{2} + \frac{2}{3}\lambda \\ -\frac{1}{2} - \frac{1}{6}\lambda \end{pmatrix} + \lambda^2 h_1(\lambda) \\ &= 1 + \lambda^2 h_2(\lambda), \end{aligned}$$

where h_1 and h_2 are holomorphic functions on $\mathbb{C}$, we see that the CSRFs are biorthogonal. According to Theorem 3.1.4 and (3.3.4),

$$\begin{pmatrix} (1+\lambda)e^{\lambda x} \\ -e^{\lambda x} \end{pmatrix} \quad \text{and} \quad \begin{pmatrix} \begin{pmatrix} (-\frac{1}{2} + \frac{1}{3}\lambda)e^{\lambda(1-x)} \\ (\frac{1}{2} + \frac{1}{6}\lambda)e^{\lambda(1-x)} \end{pmatrix} \\ \begin{pmatrix} \frac{1}{2} - \frac{1}{3}\lambda \\ -\frac{1}{2} - \frac{1}{6}\lambda \end{pmatrix} \end{pmatrix}$$

are biorthogonal CSRF of T and T^* at 0 of multiplicity 2. Hence

$$\left[\begin{pmatrix}1\\-1\end{pmatrix},\begin{pmatrix}1+x\\-x\end{pmatrix}\right] \quad \text{and} \quad \left[\begin{pmatrix}\begin{pmatrix}-\frac{1}{2}\\ \frac{1}{2}\end{pmatrix}\\ \begin{pmatrix}\frac{1}{2}\\ -\frac{1}{2}\end{pmatrix}\end{pmatrix},\begin{pmatrix}\begin{pmatrix}-\frac{1}{6}+\frac{1}{2}x\\ \frac{2}{3}-\frac{1}{2}x\end{pmatrix}\\ \begin{pmatrix}-\frac{1}{3}\\ -\frac{1}{6}\end{pmatrix}\end{pmatrix}\right]$$

are biorthogonal CSEAVs of T and T^* at 0. From

$$T^R(0)\begin{pmatrix}1+x\\-x\end{pmatrix}=\begin{pmatrix}0&1\\1&0\end{pmatrix}\begin{pmatrix}1\\0\end{pmatrix}+\begin{pmatrix}2\\-1\end{pmatrix}=\begin{pmatrix}2\\0\end{pmatrix}$$

we see that the associated vector $\begin{pmatrix}1+x\\-x\end{pmatrix}$ does not belong to the domain of $T_0(0)$. In view of Theorem 3.3.1 and Proposition 2.6.5 we have

$$\begin{aligned}
T^*(0)&\left(\begin{pmatrix}-\frac{1}{6}+\frac{1}{2}x\\ \frac{2}{3}-\frac{1}{2}x\end{pmatrix},\begin{pmatrix}d_1\\d_2\end{pmatrix}\right)\\
&=\begin{pmatrix}-\frac{1}{2}+\frac{1}{6}\delta_0+\frac{1}{3}\delta_1\\ \frac{1}{2}-\frac{2}{3}\delta_0+\frac{1}{6}\delta_1\end{pmatrix}+\begin{pmatrix}d_2\\d_1\end{pmatrix}\delta_0+\begin{pmatrix}d_1\\d_2\end{pmatrix}\delta_1\\
&=\begin{pmatrix}-\frac{1}{2}+(\frac{1}{6}+d_2)\delta_0+(\frac{1}{3}+d_1)\delta_1\\ \frac{1}{2}+(-\frac{2}{3}+d_1)\delta_0+(\frac{1}{6}+d_2)\delta_1\end{pmatrix}.
\end{aligned}$$

For any choice of $d_1,d_2\in\mathbb{C}$ not all the coefficients of δ_0 and δ_1 are zero. Hence, according to Theorem 3.4.3 i) and Proposition 3.4.4, we see that the first component $\begin{pmatrix}-\frac{1}{6}+\frac{1}{2}x\\ \frac{2}{3}-\frac{1}{2}x\end{pmatrix}$ of the associated vector of T^* at 0 does not belong to the domain of $T_0^*(0)$. □

The above result shows that the statements of Proposition 3.4.9 may be false if T^R depends on λ:

PROPOSITION 3.4.11. *There are two-point boundary eigenvalue problems* (3.1.1) *with associated operator function T given by* (3.1.3) *having the following properties:*

i) $\rho(T)\neq\emptyset$;

ii) *there is an eigenvalue $\mu\in\sigma(T)$ which is not semi-simple;*

iii) *for each CEAV (y_0,y_1) of T at μ, the associated vector y_1 does not belong to $D(T_0(\mu))$;*

iv) *for each CEAV (v_0,v_1) of T^* at μ, where $v_1=(u,d)\in(L_{p'}(\mathbb{R}))^n\times\mathbb{C}^n$, u does not belong to $D(T_0^*(\mu))$.*

Proof. The boundary eigenvalue problem considered in Example 3.4.10 fulfils i) and ii) with $\mu=0$.

iii) Let $(y_0^{(1)}, y_1^{(1)})$ be the CEAV of T at 0 defined in Example 3.4.10 and let $(y_0^{(2)}, y_1^{(2)})$ be an arbitrary CEAV of T at 0. Then $y_0^{(1)}$ and $y_0^{(2)}$ are eigenvectors. From $\operatorname{nul} T(0) = \operatorname{nul} M(0) = 1$ we infer $y_0^{(2)} = \alpha y_0^{(1)}$ for some $\alpha \in \mathbb{C} \setminus \{0\}$. Since

$$T(0)y_1^{(i)} + T'(0)y_0^{(i)} = 0 \qquad (i = 1,2),$$

we obtain that

$$T(0)y_1^{(2)} = \alpha T(0)y_1^{(1)} + \alpha T'(0)y_0^{(1)} - T'(0)y_0^{(2)} = \alpha T(0)y_1^{(1)}.$$

Hence

$$T^R(0)y_1^{(2)} = \alpha T^R(0)y_1^{(1)} = \begin{pmatrix} 2\alpha \\ 0 \end{pmatrix} \neq 0.$$

iv) Let $(v_0^{(1)}, v_1^{(1)})$ be the CEAV of T^* at 0 defined in Example 3.4.10 and let $(v_0^{(2)}, v_1^{(2)})$ be an arbitrary CEAV of T^* at 0. As in the proof of iii) we obtain

$$T^*(0)v_1^{(2)} = \alpha T^*(0)v_1^{(1)}$$

for a suitable $\alpha \in \mathbb{C} \setminus \{0\}$. We write

$$v_1^{(i)} = (u^{(i)}, d^{(i)}) \in (L_{p'}(\mathbb{R}))^n \times \mathbb{C}^n \qquad (i = 1,2).$$

Let $\tilde{d} \in \mathbb{C}^n$. Then

$$\begin{aligned} T^*(0)(u^{(2)}, \tilde{d}) &= T^*(0)v_1^{(2)} + T^*(0)(0, \tilde{d} - d^{(2)}) \\ &= \alpha T^*(0)\big(u^{(1)}, d^{(1)} + \alpha^{-1}(\tilde{d} - d^{(2)})\big) \notin (L_{p'}(\mathbb{R}))^n \end{aligned}$$

by the above example. Hence $u^{(2)} \notin D(T_0^*(0))$. $\square$

3.5. Two-point boundary eigenvalue problems in $(\mathbf{L_p(a,b)})^{\mathbf{n}}$

In this section let $p < \infty$ and

$$\begin{cases} T^D(\lambda)y = y' - A(\cdot,\lambda)y, \\ T^R(\lambda)y = W^a(\lambda)y(a) + W^b(\lambda)y(b), \end{cases} \tag{3.5.1}$$

for $\lambda \in \Omega$ and $y \in (W_p^1(a,b))^n$, where $A \in H(\Omega, M_n(L_\infty(a,b)))$ and $W^a, W^b \in H(\Omega, M_n(\mathbb{C}))$ with $\operatorname{rank}(W^a(\lambda), W^b(\lambda)) = n$ for all $\lambda \in \Omega$.

PROPOSITION 3.5.1. *There are $\widetilde{A}, \widetilde{B} \in H(\Omega, M_n(\mathbb{C}))$ such that the matrix*

$$\begin{pmatrix} W^a(\lambda)^\top & \widetilde{A}(\lambda) \\ W^b(\lambda)^\top & \widetilde{B}(\lambda) \end{pmatrix} \tag{3.5.2}$$

is invertible for all $\lambda \in \Omega$.

Proposition 3.5.1 is a special case of the following lemma, see also [BA2, Theorem 5.3].

LEMMA 3.5.2. *Let $k > l \geq 1$, $x_j \in H(\Omega, \mathbb{C}^k)$ $(j = 1,\dots,l)$ and assume that the vectors $x_1(\lambda),\dots,x_l(\lambda)$ are linearly independent for all $\lambda \in \Omega$. Then there are $x_j \in H(\Omega, \mathbb{C}^k)$ $(j = l+1,\dots,k)$ such that $x_1(\lambda),\dots,x_k(\lambda)$ are linearly independent for all $\lambda \in \Omega$. In case $x_1,\dots,x_l$ are polynomials in λ, $x_{l+1},\dots,x_k$ can be chosen to be polynomials.*

Proof. Set

$$X := (x_1,\dots,x_l).$$

Then $X \in H(\Omega, M_{k,l}(\mathbb{C}))$. With respect to the decomposition $\mathbb{C}^k = \mathbb{C}^l \oplus \mathbb{C}^{k-l}$ we write $X^\top =: (X_0, X_1)$. Since $\operatorname{rank} X(\lambda) = l$ for all $\lambda \in \Omega$, we may assume without loss of generality that $\operatorname{rank} X_0(\lambda_0) = l$ for some $\lambda_0 \in \Omega$. Thus the function $\det X_0$ is not identically zero. Let $B_0(\lambda)$ be the transpose of the matrix of the cofactors of $X_0(\lambda)$. Then B_0 is holomorphic and fulfils

$$X_0 B_0 = (\det X_0) I_l \, .$$

Let $c \in \{0\} \oplus (\mathbb{C}^{k-l} \setminus \{0\}) \subset \mathbb{C}^k$ and define $z \in H(\Omega, \mathbb{C}^k)$ by

$$z(\lambda) := \begin{pmatrix} B_0(\lambda) & -B_0(\lambda) X_1(\lambda) \\ 0 & (\det X_0(\lambda)) I_{k-l} \end{pmatrix} c \quad (\lambda \in \Omega).$$

Then $z(\lambda_0) \neq 0$ since the matrix

$$\begin{pmatrix} B_0(\lambda_0) & -B_0(\lambda_0) X_1(\lambda_0) \\ 0 & (\det X_0(\lambda_0)) I_{k-l} \end{pmatrix}$$

is invertible. This shows that the set of the zeros of z is a discrete subset of Ω. By WEIERSTRASS' theorem (see e.g. [BU, Theorem 7.32]) there is a holomorphic function $\gamma : \Omega \to \mathbb{C}$ such that the set of the zeros and their multiplicities coincide for γ and z. Hence

$$z_1 := \frac{z}{\gamma}$$

is a holomorphic function with $z_1(\lambda) \neq 0$ for all $\lambda \in \Omega$. In case $x_1,\dots,x_l$ are polynomials, also z is a polynomial. Then we can take a polynomial for γ, and z_1 is a polynomial. We have

$$\begin{aligned} X^\top z &= (X_0, X_1) \begin{pmatrix} B_0 & -B_0 X_1 \\ 0 & (\det X_0) I_{k-l} \end{pmatrix} c \\ &= ((\det X_0) I_l, 0) c = 0. \end{aligned}$$

Thus, for all $\lambda \in \Omega$,

$$z_1(\lambda) \in N(X^\top(\lambda)) = (R(X(\lambda)))^\perp.$$

In particular, $z_1(\lambda)^\top x_i(\lambda) = 0$ for $i = 1,\dots,l$ and $\lambda \in \Omega$. Since $z_1(\lambda) \neq 0$ for all $\lambda \in \Omega$, i. e., the components of z_1 do not have common zeros, and since every finitely generated ideal in $H(\Omega)$ is a principal ideal, see e g. [BU, Corollary 11.42], there is $x_{l+1} \in H(\Omega, \mathbb{C}^n)$ such that $z_1(\lambda)^\mathsf{T} x_{l+1} = 1$ for all $\lambda \in \Omega$. From the proof of this result we also see that x_{l+1} can be chosen to be a polynomial if $x_1, \dots, x_l$ are polynomials. It follows that $x_1(\lambda), \dots, x_{l+1}(\lambda)$ are linearly independent for all $\lambda \in \Omega$. The statement of the lemma follows by induction. □

Apart from T^D we consider $T^{D+} \in H(\mathbb{C}, L((W_{p'}^1(a,b))^n, (L_{p'}(a,b))^n))$ defined by

(3.5.3) $$T^{D+}(\lambda)y = -y' - A^\mathsf{T}(\cdot,\lambda)y \quad (\lambda \in \mathbb{C},\ y \in (W_{p'}^1(a,b))^n).$$

The differerential operator $T^{D+}(\lambda)$ is called the formally adjoint of the differential operator $T^D(\lambda)$.

PROPOSITION 3.5.3. *For all $u \in (W_{p'}^1(a,b))^n$ and $d \in \mathbb{C}^n$ we have*

(3.5.4) $$T^*(u,d) = (T^{D+}u)_e + \left(-u(a) + W^{a\mathsf{T}} d\right)\delta_a + \left(u(b) + W^{b\mathsf{T}} d\right)\delta_b.$$

Proof. With the aid of (3.3.5) and Proposition 2.6.5 we infer

$$\begin{aligned} T^{D*}u &= -u_e' - A^\mathsf{T} u_e \\ &= -(u')_e - u(a)\delta_a + u(b)\delta_b - A^\mathsf{T} u_e \\ &= (T^{D+}u)_e - u(a)\delta_a + u(b)\delta_b. \end{aligned}$$

Then (3.5.4) follows because of (3.3.2), (3.3.3) and the special boundary conditions (3.5.1). □

PROPOSITION 3.5.4. *Set*

$$H := \begin{pmatrix} -I_n & 0 \\ 0 & I_n \end{pmatrix}.$$

Then the LAGRANGE *identity*

$$\langle T^D(\lambda)y, u\rangle - \langle y, T^{D+}(\lambda)u\rangle = \begin{pmatrix} y(a) \\ y(b) \end{pmatrix}^\mathsf{T} H \begin{pmatrix} u(a) \\ u(b) \end{pmatrix}$$

holds for all $\lambda \in \Omega$, $y \in (W_p^1(a,b))^n$ and $u \in (W_{p'}^1(a,b))^n$.

Proof. Since $T^{D*}u = T^*(u,0)$, we immediately infer from Proposition 3.5.3 that

$$\begin{aligned} &\langle T^D(\lambda)y, u\rangle - \langle y, T^{D+}(\lambda)u\rangle \\ &= \langle y, T^{D*}(\lambda)u\rangle_{p,1} - \langle y, (T^{D+}(\lambda)u)_e\rangle_{p,0} \\ &= \langle y, -u(a)\delta_a + u(b)\delta_b\rangle_{p,1} \\ &= \begin{pmatrix} y(a) \\ y(b) \end{pmatrix}^\mathsf{T} H \begin{pmatrix} u(a) \\ u(b) \end{pmatrix}. \end{aligned}$$

We can also prove Proposition 3.5.4 without applying T^* if we use (2.3.1) and Proposition 2.1.5 i). □

With a matrix function of the form (3.5.2) which is invertible for all $\lambda \in \Omega$ we define

$$\begin{pmatrix} \widetilde{C}(\lambda) & \widetilde{D}(\lambda) \\ \widetilde{W}^a(\lambda) & \widetilde{W}^b(\lambda) \end{pmatrix} := \begin{pmatrix} W^a(\lambda)^\mathsf{T} & \widetilde{A}(\lambda) \\ W^b(\lambda)^\mathsf{T} & \widetilde{B}(\lambda) \end{pmatrix}^{-1} H, \tag{3.5.5}$$

where the matrix on the left-hand side is divided into $n \times n$ block matrices. In case $W^a(\lambda)$ and $W^b(\lambda)$ depend polynomially on $\lambda \in \mathbb{C}$, we infer from Lemma 3.5.2 that we can choose $\widetilde{A}(\lambda)$ and $\widetilde{B}(\lambda)$ to be polynomials in λ. Since the determinant of $\begin{pmatrix} W^a(\lambda)^\mathsf{T} & \widetilde{A}(\lambda) \\ W^b(\lambda)^\mathsf{T} & \widetilde{B}(\lambda) \end{pmatrix}$ is a polynomial which is invertible for all $\lambda \in \mathbb{C}$, it is constant. Therefore the matrix function (3.5.5) depends polynomially on λ. Hence we may assume that $\widetilde{W}^a(\lambda)$ and $\widetilde{W}^b(\lambda)$ depend polynomially on $\lambda \in \mathbb{C}$ if this holds for $W^a(\lambda)$ and $W^b(\lambda)$.

Analogous to the operator $T_0(\lambda)$ we define the operator $T_0^+(\lambda)$ in $(L_{p'}(a,b))^n$ by

$$\begin{aligned} &D(T_0^+(\lambda)) \\ &:= \{u \in (L_{p'}(a,b))^n : u \in (W_{p'}^1(a,b))^n, \widetilde{W}^a(\lambda)u(a) + \widetilde{W}^b(\lambda)u(b) = 0\} \end{aligned} \tag{3.5.6}$$

and

$$T_0^+(\lambda)u := T^{D+}(\lambda)u \quad (u \in D(T_0^+(\lambda))). \tag{3.5.7}$$

BIRKHOFF and LANGER [BL4, p. 64] considered adjoint boundary conditions in the case of invertible boundary matrices $W^a(\lambda)$ and $W^b(\lambda)$. In our notation, these adjoint boundary conditions are given by

$$W^{a\mathsf{T}}(\lambda)^{-1}u(a) + W^{b\mathsf{T}}(\lambda)^{-1}u(b) = 0.$$

It is easy to see that the boundary conditions are of the form (3.5.6), i. e., that $W^{a\mathsf{T}}(\lambda)^{-1} = \widetilde{W}^a(\lambda)$, $W^{b\mathsf{T}}(\lambda)^{-1} = \widetilde{W}^b(\lambda)$, if we set $\widetilde{A}(\lambda) := 0$, $\widetilde{B}(\lambda) := W^b(\lambda)^\mathsf{T}$. Then we also have $\widetilde{C}(\lambda) = -W^{a\mathsf{T}}(\lambda)^{-1}$, $\widetilde{D}(\lambda) = 0$.

By Theorem 3.4.5 iii), $T_0^*(\lambda)$ is a linear operator.

THEOREM 3.5.5. *For all $\lambda \in \Omega$ we have*
i) $T_0^+(\lambda) = T_0^*(\lambda)$,
ii) $(T_0^+(\lambda))^* = T_0(\lambda)$ *if* $p > 1$.

Proof. i) Let $y \in D(T_0(\lambda))$ and $u \in D(T_0^+(\lambda))$. From Proposition 3.5.4 we know that

$$\begin{aligned} \langle T_0(\lambda)y, u\rangle &= \langle T^D(\lambda)y, u\rangle \\ &= \langle y, T^{D+}(\lambda)u\rangle + \begin{pmatrix} y(a) \\ y(b) \end{pmatrix}^\mathsf{T} H \begin{pmatrix} u(a) \\ u(b) \end{pmatrix}. \end{aligned}$$

From (3.5.5) and the definitions of $D(T_0(\lambda))$ and $D(T_0^+(\lambda))$ we infer

$$\begin{aligned}
&\begin{pmatrix} y(a) \\ y(b) \end{pmatrix}^\mathsf{T} H \begin{pmatrix} u(a) \\ u(b) \end{pmatrix} \\
&= \begin{pmatrix} y(a) \\ y(b) \end{pmatrix}^\mathsf{T} \begin{pmatrix} W^a(\lambda)^\mathsf{T} & \widetilde{A}(\lambda) \\ W^b(\lambda)^\mathsf{T} & \widetilde{B}(\lambda) \end{pmatrix} \begin{pmatrix} \widetilde{C}(\lambda) & \widetilde{D}(\lambda) \\ \widetilde{W}^a(\lambda) & \widetilde{W}^b(\lambda) \end{pmatrix} \begin{pmatrix} u(a) \\ u(b) \end{pmatrix} \\
&= (0,*) \begin{pmatrix} * \\ 0 \end{pmatrix} = 0.
\end{aligned}$$

Thus

$$\langle T_0(\lambda)y,u\rangle = \langle y, T^{D+}(\lambda)u\rangle,$$

which proves $u \in D(T_0^*(\lambda))$ and $T_0^*(\lambda)u = T_0^+(\lambda)u$.

Conversely, let $u \in D(T_0^*(\lambda))$. We have to prove $u \in D(T_0^+(\lambda))$. From Corollary 3.4.6 we know that $u \in D(T_0^*(\lambda))$ implies that there is a vector $d \in \mathbb{C}^n$ with $u \in (W_{p'}^1(a,b))^n$, $u(a) = W^a(\lambda)^\mathsf{T} d$ and $u(b) = -W^b(\lambda)^\mathsf{T} d$. From (3.5.5) we infer

$$\begin{aligned}
&\widetilde{W}^a(\lambda)u(a) + \widetilde{W}^b(\lambda)u(b) \\
&= (0, I_n) \begin{pmatrix} \widetilde{C}(\lambda) & \widetilde{D}(\lambda) \\ \widetilde{W}^a(\lambda) & \widetilde{W}^b(\lambda) \end{pmatrix} \begin{pmatrix} u(a) \\ u(b) \end{pmatrix} \\
&= -(0, I_n) \begin{pmatrix} W^a(\lambda)^\mathsf{T} & \widetilde{A}(\lambda) \\ W^b(\lambda)^\mathsf{T} & \widetilde{B}(\lambda) \end{pmatrix}^{-1} \begin{pmatrix} -I_n & 0 \\ 0 & I_n \end{pmatrix} \begin{pmatrix} -W^a(\lambda)^\mathsf{T} d \\ W^b(\lambda)^\mathsf{T} d \end{pmatrix} \\
&= -(0, I_n) \begin{pmatrix} W^a(\lambda)^\mathsf{T} & \widetilde{A}(\lambda) \\ W^b(\lambda)^\mathsf{T} & \widetilde{B}(\lambda) \end{pmatrix}^{-1} \begin{pmatrix} W^a(\lambda)^\mathsf{T} & \widetilde{A}(\lambda) \\ W^b(\lambda)^\mathsf{T} & \widetilde{B}(\lambda) \end{pmatrix} \begin{pmatrix} I_n \\ 0 \end{pmatrix} d \\
&= -(0, I_n) \begin{pmatrix} I_n \\ 0 \end{pmatrix} d = 0,
\end{aligned}$$

which proves $u \in D(T_0^+(\lambda))$.

ii) Obviously, $-\left(-T^{D+}\right)^+(\lambda) = T^D(\lambda)$. From (3.5.5) and $H^\mathsf{T} = H$ we infer

$$\begin{pmatrix} \widetilde{A}(\lambda)^\mathsf{T} & \widetilde{B}(\lambda)^\mathsf{T} \\ W^a(\lambda) & W^b(\lambda) \end{pmatrix} = \begin{pmatrix} \widetilde{W}^a(\lambda)^\mathsf{T} & \widetilde{C}(\lambda)^\mathsf{T} \\ \widetilde{W}^b(\lambda)^\mathsf{T} & \widetilde{D}(\lambda)^\mathsf{T} \end{pmatrix}^{-1} H,$$

which proves $D(T_0(\lambda)) = D(-(-T_0^+)^+(\lambda))$. Hence $T_0(\lambda) = -(-T_0^+)^+(\lambda)$. Finally, apply part i) to $-T_0^+(\lambda)$ obtain $(T_0^+(\lambda))^* = -(-T_0^+)^+(\lambda) = T_0(\lambda)$. □

REMARK 3.5.6. Since $\widetilde{A}$ and $\widetilde{B}$ are not uniquely determined, also the boundary matrix functions $\widetilde{W}^a$ and $\widetilde{W}^b$ are not uniquely determined. But Theorem 3.5.5 i) shows that the definition of T_0^+ is unambiguous, i. e., the boundary conditions

$$\widetilde{W}^a(\lambda)u(a) + \widetilde{W}^b(\lambda)u(b) = 0$$

are uniquely determined by the boundary conditions

$$W^a(\lambda)y(a) + W^b(\lambda)y(b) = 0.$$

From Proposition 3.4.11 we know that there are two-point boundary value problems for which no associated vector of the corresponding operator functions T and T^* at some μ belongs to the domain of $D(T_0(\mu))$ and $D(T_0^*(\mu)) = D(T_0^+(\mu))$, respectively. Hence we have to go beyond the domain of $T_0(\mu)$ for the definition of a root function of T_0 at μ.

DEFINITION 3.5.7. Let $y \in H(\Omega, (W_p^1(a,b))^n)$ and $\mu \in \Omega$. The vector function y is called a root function of T_0 at μ if and only if $y(\mu) \neq 0$, $(T^D y)(\mu) = 0$ and $W^a(\mu)y(a,\mu) + W^b(\mu)y(b,\mu) = 0$. The minimum of the orders of the zeros of $T^D y$ and $W^a y(a,\cdot) + W^b y(b,\cdot)$ at μ is called the multiplicity of y.

From $T^R y = W^a y(a,\cdot) + W^b y(b,\cdot)$ we obtain

PROPOSITION 3.5.8. *Let $y \in H(\Omega, (W_p^1(a,b))^n)$, $\mu \in \Omega$ and $\nu \in \mathbb{N}$. Then y is a root function of T_0 of multiplicity ν at μ if and only if y is a root function of T of multiplicity ν at μ.*

Canonical systems of root functions of T_0 are defined in the same way as for T. Hence a system of root functions is a canonical system of root functions of T_0 at μ if and only if it is a canonical system of root functions of T at μ.

The situation is different for $T_0^+ = T_0^*$ and T^*.

PROPOSITION 3.5.9. *Let $(u,d) \in H(\Omega, (L_{p'}(a,b))^n \times \mathbb{C}^n)$ be a root function of T^* at μ of multiplicity ν. We may assume that u is a polynomial of order $\leq \nu - 1$. Then $u \in H(\Omega, (W_{p'}^1(a,b))^n)$, u is a root function of T_0^+ of multiplicity ν_1 at μ, and $d + \widetilde{C}u(a,\cdot) + \widetilde{D}u(b,\cdot)$ has a zero of order ν_2 at μ, where $\nu = \min\{\nu_1, \nu_2\}$.*

Proof. By assumption we have

$$u(\lambda) = \sum_{i=0}^{\nu-1} (\lambda - \mu)^i u_i \quad (\lambda \in \Omega),$$

where $u_i \in (L_{p'}(a,b))^n$ $(i = 0, \ldots, \nu - 1)$. In order to show $u_i \in (W_{p'}^1(a,b))^n$ set

$$d_i := \tfrac{1}{i!}\left(\tfrac{\mathrm{d}^i}{\mathrm{d}\lambda^i} d\right)(\mu) \quad (i = 0, \ldots, \nu - 1).$$

Since (u,d) is a root function of T^* of multiplicity ν at μ we have

$$T^*(\mu)(u_i, d_i) = -\sum_{j=1}^{i} \tfrac{1}{j!}\left(\tfrac{\mathrm{d}^j}{\mathrm{d}\lambda^j} T^*\right)(\mu)(u_{i-j}, d_{i-j}) \quad (i = 0, \ldots, \nu - 1). \tag{3.5.8}$$

From (3.3.2) and (3.3.3) we infer

$$\begin{aligned} u_i' &= -(T^*(\mu)(u_i, d_i))_r - A^\mathsf{T}(\cdot, \mu) u \\ &= -\sum_{j=0}^{i} \tfrac{1}{j!}\left(\tfrac{\partial^j}{\partial \lambda^j} A^\mathsf{T}\right)(\cdot, \mu) u_{i-j} \in (L_{p'}(a,b))^n. \end{aligned}$$

Hence $u_i \in (W_{p'}^1(a,b))^n$ by Corollary 2.1.4. From Proposition 3.5.3 we know that

$$T^*(u,d) = (T^{D^+} u)_e + \left(-u(a,\cdot) + W^{a\mathsf{T}} d\right)\delta_a + \left(u(b,\cdot) + W^{b\mathsf{T}} d\right)\delta_b .$$

Since (u,d) is a root function of T^* of multiplicity ν at μ, Proposition 3.4.4 yields that $-u(a,\cdot)+W^{a\mathsf{T}}d$ and $u(b,\cdot)+W^{b\mathsf{T}}d$ have a zero of order $\geq \nu$ at μ. Thus $T^{D^+}u$ also has a zero of order $\geq \nu$ at μ. In view of (3.5.5),

$$\begin{pmatrix} W^{a\mathsf{T}} & \widetilde{A} \\ W^{b\mathsf{T}} & \widetilde{B} \end{pmatrix} \begin{pmatrix} \widetilde{C} & \widetilde{D} \\ \widetilde{W}^a & \widetilde{W}^b \end{pmatrix} \begin{pmatrix} u(a,\cdot) \\ u(b,\cdot) \end{pmatrix} + \begin{pmatrix} W^{a\mathsf{T}} & \widetilde{A} \\ W^{b\mathsf{T}} & \widetilde{B} \end{pmatrix} \begin{pmatrix} d \\ 0 \end{pmatrix} \tag{3.5.9}$$

has a zero of order $\geq \nu$ at μ. Since $\begin{pmatrix} W^{a\mathsf{T}}(\lambda) & \widetilde{A}(\lambda) \\ W^{b\mathsf{T}}(\lambda) & \widetilde{B}(\lambda) \end{pmatrix}$ is invertible for all $\lambda \in \mathbb{C}$, we see that $\widetilde{W}^a u(a,\cdot)+\widetilde{W}^b u(b,\cdot)$ has a zero of order $\geq \nu$ at μ. Furthermore,

$$(\widetilde{C},\widetilde{D}) \begin{pmatrix} u(a,\cdot) \\ u(b,\cdot) \end{pmatrix} + d$$

has a zero of order $\geq \nu$ at μ. This together with $(u,d)(\mu) \neq 0$ proves $u(\cdot,\mu) \neq 0$. The above considerations immediately yield $\nu = \min\{\nu_1,\nu_2\}$. □

PROPOSITION 3.5.10. *Let $u \in H(\Omega,(W_p^1(a,b))^n)$ be a root function of T_0^+ of multiplicity ν at μ. Set $d := -\widetilde{C}u(a,\cdot)-\widetilde{D}u(b,\cdot)$. Then (u,d) is a root function of T^* of multiplicity ν at μ.*

Proof. By assumption, $\widetilde{W}^a u(a,\cdot)+\widetilde{W}^b u(b,\cdot)$ has a zero of order $\geq \nu$ at μ. Hence the matrix function (3.5.9) has a zero of order $\geq \nu$ at μ. With the aid of (3.5.5) we infer that $-u(a,\cdot)+W^{a\mathsf{T}}d$ and $u(b,\cdot)+W^{b\mathsf{T}}d$ have a zero of order $\geq \nu$ at μ. By assumption, this also holds for $T^{D^+}u$. For at least one of these three functions in $(W_{p'}^{-1}[a,b])^n$ the order of the zero at μ is exactly ν. In view of (3.5.4) and Proposition 3.4.4 the proof is complete. □

We define eigenvectors, associated vectors, and canonical systems of eigenvectors and associated vectors of T_0 and T_0^+ via corresponding root functions and canonical systems of root functions as we did for T and T^* in Section 1.6.

THEOREM 3.5.11. *Let $\mu \in \sigma(T_0)$ and let $\{y_h^{(i)} : 1 \leq i \leq r, 0 \leq h \leq m_i - 1\}$ be a canonical system of eigenvectors and associated vectors of T_0 at μ. Then there is a canonical system $\{u_h^{(i)} : 1 \leq i \leq r, 0 \leq h \leq m_i - 1\}$ of eigenvectors and associated vectors of T_0^+ at μ such that the principal part of the* GREEN'S *matrix $G(x,\xi,\cdot)$ at μ has the form*

$$\sum_{j=1}^{r} \sum_{l=0}^{m_j-1} (\cdot-\mu)^{l-m_j} \sum_{h=0}^{l} y_h^{(j)}(x) u_{l-h}^{(j)\mathsf{T}}(\xi). \tag{3.5.10}$$

If W^a and W^b do not depend on λ, then the biorthogonal relationships

$$\sum_{k=0}^{l} \sum_{q=1}^{m_i} \frac{1}{(k+q)!} \int_a^b u_{l-k}^{(j)\mathsf{T}}(x) \frac{\mathrm{d}^{k+q}A}{\mathrm{d}\lambda^{k+q}}(\mu) y_{m_i-q}^{(i)}(x)\,\mathrm{d}x = -\delta_{ij}\delta_{0l} \tag{3.5.11}$$

$$(1 \leq i \leq r,\ 1 \leq j \leq r,\ 0 \leq l \leq m_j - 1)$$

hold.

Proof. Since a CSEAV of T_0 at μ is a CSEAV of T at μ, by Theorem 1.6.5 there is a CSEAV $\{(u_h^{(i)}, d_h^{(i)}) : 1 \le i \le r, 0 \le h \le m_i - 1\}$ of T^* at μ such that

$$\sum_{j=1}^{r}\sum_{l=1}^{m_j}(\cdot - \mu)^{-l}\sum_{h=0}^{m_j-l} y_h^{(j)} \otimes (u_{m_j-l-h}^{(j)}, d_{m_j-l-h}^{(j)}) \tag{3.5.12}$$

is the principal part of T^{-1} at μ, and the biorthogonal relationships

$$\sum_{k=0}^{l}\sum_{q=1}^{m_i}\frac{1}{(k+q)!}\left\langle \frac{\mathrm{d}^{k+q}T}{\mathrm{d}\lambda^{k+q}}(\mu)y_{m_i-q}^{(i)}, (u_{l-k}^{(j)}, d_{l-k}^{(j)})\right\rangle = \delta_{ij}\delta_{0l} \tag{3.5.13}$$

$$(1 \le i \le r,\ 1 \le j \le r, 0 \le l \le m_j - 1)$$

hold.

By Proposition 3.5.9, for each $i \in \{1,\ldots,r\}$, $\{u_h^{(i)} : 0 \le h \le m_i - 1\}$ is a CEAV of T_0^+ at μ, and $d_0^{(i)} = -\widetilde{C}(\mu)u_0^{(i)}(a) - \widetilde{D}(\mu)u_0^{(i)}(b)$. Hence $u_0^{(1)},\ldots,u_0^{(r)}$ are linearly independent as $(u_0^{(1)}, d_0^{(1)}),\ldots,(u_0^{(r)}, d_0^{(r)})$ are linearly independent. Since the multiplicities of a CSEAV of T_0^+ at μ cannot exceed the multiplicities of a CSEAV of T^* at μ by Proposition 3.5.10, $\{u_h^{(i)} : 1 \le i \le r, 0 \le h \le m_i - 1\}$ is a CSEAV of T_0^+ at μ. Theorem 3.4.1 i) yields that the principal part of T_0^{-1} at μ is

$$\sum_{j=1}^{r}\sum_{l=1}^{m_j}(\lambda - \mu)^{-l}\sum_{h=0}^{m_j-l} y_h^{(j)} \otimes u_{m_j-l-h}^{(j)}.$$

Since the GREEN'S matrix is uniquely determined by (3.4.3), we obtain (3.5.10)—recall that we understand this identity to hold almost everywhere. Let

$$y_i := \sum_{l=0}^{m_i-1}(\cdot - \mu)^l y_l^{(i)}.$$

If W^a and W^b are constant, then $T^R y_i$ is a polynomial of degree $\le m_i - 1$ and has a zero of order $\ge m_i$ at μ. Hence $T^R y_i = 0$ for $i = 1,\ldots,r$. Thus (3.5.13) leads to (3.5.11). □

3.6. Notes

Historically, boundary eigenvalue problems have been investigated for scalar n-th order differential equations before they were considered for first order systems. However, since each boundary eigenvalue problem for an n-th order differential equation is equivalent to one for a first order system, we can use the results of this chapter also for n-th order differential equations. Of course, also higher order systems are equivalent to first order systems. In this monograph we do not consider systems of higher order differential equations. We just mention that in this case certain restrictions have to apply: for systems of differential equations of mixed order the associated operators may not be Fredholm, see [ALMS, Section 4]. But even if one only wants to consider n-th order scalar differential equations, it is often advisable also to deal with first order systems since it is in general easier to consider first order systems because solutions of first order systems of differential equations are easier to handle than solutions of n-th order differential equations.

It was first shown by Kaashoek [KAS] that the operator functions associated with two-point boundary eigenvalue problems are globally equivalent to an extension of their characteristic matrices, see Theorem I.2.

Since the boundary part is considered nonhomogeneously, the vector d occurs naturally in the adjoint boundary conditions, see Theorem 3.3.1. If one considers a homogeneous boundary part, the integral term in the boundary conditions of the original system together with this parametric vector occur in the adjoint system of differential equations, whereas the boundary conditions of the adjoint operator are interface conditions at those points where the original boundary conditions are taken, involving the parametric vector, see Corollary 3.4.6. This form of the adjoint problem was established by R. H. Cole in [CO3]. A. M. Krall [KR2] showed that Cole's definition coincides with the functional analytic adjoint operator. He called these adjoint problems differential-boundary problems since the boundary conditions are linked with the differential equation via the parameter. In a series of papers, [KR3]–[KR11], Krall investigated properties of these differential boundary operators. Krall's results were generalized by Möller in [MÖ1]–[MÖ4].

Chapter IV

BIRKHOFF REGULAR AND STONE REGULAR BOUNDARY EIGENVALUE PROBLEMS

This chapter deals with regular boundary eigenvalue problems for first order $n \times n$ systems of ordinary differential equations. These first order systems are asymptotically linear in the eigenvalue parameter λ. The leading matrix, i. e., the coefficient matrix of λ, is supposed to be a diagonal matrix whose nonzero diagonal elements as well as their nonzero differences are assumed to have constant arguments and to be bounded away from zero (almost everywhere). The boundary conditions are allowed to have infinitely many interior points and an integral term. The coefficients of the boundary conditions may depend on the eigenvalue parameter λ, with respect to which they are asymptotically constant as λ tends to infinity.

First Birkhoff regular boundary eigenvalue problems are dealt with (Definition 4.1.2). This regularity property is given in terms of the arguments of the nonzero diagonal elements of the leading matrix in the differential system, further by the zero-approximand of the λ-asymptotic fundamental matrix of the differential system, and by the limits of the coefficient matrices in the boundary conditions. If the leading matrix in the differential system is invertible, then Birkhoff regularity is easy to check since in this case it only depends on the arguments of the diagonal elements of the leading matrix in the differential system and on the limits of the coefficient matrices in the boundary conditions at the endpoints of the underlying interval. To illustrate the notion of Birkhoff regularity, some boundary eigenvalue problems for 2×2 differential systems are classified with respect to this property. One of the most important consequences of Birkhoff regularity of a boundary eigenvalue problem is the fact that the determinant of a suitable characteristic matrix function is bounded away from zero for λ in the union of circles Γ_ν $(\nu \in \mathbb{N})$ with centres at zero and radii ρ_ν which tend to infinity as ν tends to infinity (Theorem 4.3.9).

Stone regularity is also considered in this chapter. For a natural number s we call a boundary eigenvalue problem s-regular if there exists a sequence of circles Γ_ν $(\nu \in \mathbb{N})$ with centres at zero and radii ρ_ν which tend to infinity such that the determinant of an appropriate characteristic matrix multiplied by λ^s is bounded away from zero for λ in the union of the circles Γ_ν. Birkhoff regular boundary eigenvalue problems are 0-regular. A boundary eigenvalue problem is called Stone regular if it is s-regular for some natural number s.

In the next chapter eigenfunction expansions for regular boundary eigenvalue problems will be proved by the contour integral method, i. e., by integrating certain operator functions which essentially consist of the product of some negative power of λ and the resolvent of the boundary eigenvalue problem along the circles Γ_ν ($\nu \in \mathbb{N}$) which are given by the regularity assumptions. The exponent of this power of λ is determined by the order of regularity of the boundary eigenvalue problem. In the present fourth chapter we prepare the proofs of these expansion theorems by establishing several estimates of the operator functions which will be used in the contour integral method. These estimates yield the convergence of the contour integrals taken along the sequence of the regularity circles Γ_ν (Theorems 4.4.9 and 4.4.11). With the aid (of a special case) of the Hilbert transform these first results are sharpened for Birkhoff regular boundary eigenvalue problems. Under this regularity condition and some other technical assumptions it is shown that the sequence of certain contour integrals taken along the regularity circles Γ_ν ($\nu \in \mathbb{N}$) converge strongly in the L_p-norm to the identity operator on a subspace of $(L_p(a,b))^n$ (Theorem 4.6.9 for $1 < p < \infty$, Theorem 4.7.5 for $p = \infty$). In the case $1 < p < \infty$ this subspace is explicitly determined by the structure of the leading (diagonal) matrix in the differential system. For $p = \infty$ it is not that easy to describe: it consists of the class of continuous functions which are of bounded variation and fulfil certain boundary conditions.

4.1. Definitions and basic results

Let $-\infty < a < b < \infty$, $1 \leq p \leq \infty$ and $n \in \mathbb{N} \setminus \{0\}$. For sufficiently large complex numbers λ, say $|\lambda| \geq \gamma (> 0)$, we consider the boundary eigenvalue problem

$$y' - \big(\lambda A_1 + A_0 + \lambda^{-1} A^0(\cdot,\lambda)\big)y = 0, \tag{4.1.1}$$

$$\sum_{j=0}^{\infty} \widetilde{W}^{(j)}(\lambda) y(a_j) + \int_a^b \widetilde{W}(x,\lambda) y(x) \, dx = 0, \tag{4.1.2}$$

where y varies in $(W_p^1(a,b))^n$.

For the system of differential equations (4.1.1) we assume that the coefficient matrices A_0 and A_1 belong to $M_n(L_p(a,b))$, that $A^0(\cdot,\lambda)$ belongs to $M_n(L_p(a,b))$ for $|\lambda| \geq \gamma$ and depends holomorphically on λ there, and that

$$A^0(\cdot,\lambda) \text{ is bounded in } M_n(L_p(a,b)) \text{ as } \lambda \to \infty.$$

We suppose that A_1 is a diagonal matrix function, more precisely,

$$A_1 = \begin{pmatrix} A_0^1 & & & & \\ & A_1^1 & & 0 & \\ & & \cdot & & \\ & 0 & & \cdot & \\ & & & & \cdot & \\ & & & & & A_l^1 \end{pmatrix},$$

where l is a positive integer,

$$A_\nu^1 = r_\nu I_{n_\nu} \ (\nu = 0, \dots, l), \quad \sum_{\nu=0}^{l} n_\nu = n,$$

with $n_0 \in \mathbb{N}$ and $n_\nu \in \mathbb{N} \setminus \{0\}$ for $\nu = 1, \dots, l$. According to the block structure of A_1, we write $A_0 = (A_{0,\nu\mu})_{\nu,\mu=0}^{l}$.

For the diagonal elements of A_1 we assume:

I) $r_0 = 0$, and for $\nu, \mu = 0, \dots, l$ there are numbers $\varphi_{\nu\mu} \in [0, 2\pi)$ such that

$$(4.1.3) \qquad (r_\nu - r_\mu)^{-1} \in L_\infty(a,b) \quad \text{if } \nu \neq \mu,$$

$$(4.1.4) \qquad r_\nu(x) - r_\mu(x) = |r_\nu(x) - r_\mu(x)| e^{i\varphi_{\nu\mu}} \quad \text{a.e. in } (a,b).$$

Note that $\mu = 0$ gives $r_\nu^{-1} \in L_\infty(a,b)$ for $\nu = 1, \dots, l$ and

$$(4.1.5) \qquad r_\nu(x) = |r_\nu(x)| e^{i\varphi_\nu} \quad \text{a.e. in } (a,b) \quad (\nu = 1, \dots, l),$$

where $\varphi_\nu := \varphi_{\nu 0} = \varphi_{0\nu} \pm \pi$ for $\nu = 1, \dots, l$.

If $n_0 = 0$, then we need the conditions (4.1.3) and (4.1.4) only for $\nu, \mu \in \{1, \dots, l\}$. On the other hand, the conditions $r_\nu^{-1} \in L_\infty(a,b)$ for $\nu = 1, \dots, l$ and (4.1.5) are needed in any case. Hence it is no additional assumption if we take $\nu, \mu \in \{0, \dots, l\}$ in (4.1.3) and (4.1.4) also in the case $n_0 = 0$.

To give a more explicit representation of condition I), we consider the following conditions (we remind that identities and inequalities of functions are understood to hold almost everywhere):

II) $r_0 = 0$, and there are a number $\alpha \in \mathbb{C}$ and for $\nu = 0, \dots, l$ real-valued functions $\tilde{r}_\nu \in L_p(a,b)$ such that for all $\nu, \mu = 0, \dots, l$

$$(4.1.6) \qquad r_\nu = \alpha \tilde{r}_\nu,$$

$$(4.1.7) \qquad (r_\nu - r_\mu)^{-1} \in L_\infty(a,b) \text{ if } \nu \neq \mu,$$

$$(4.1.8) \qquad \tilde{r}_\nu - \tilde{r}_\mu \text{ is a positive or negative function if } \nu \neq \mu.$$

III) $r_0 = 0$, and there are a positive real-valued function $r \in L_p(a,b)$ such that $r^{-1} \in L_\infty(a,b)$ and $\alpha_\nu \in \mathbb{C}$ $(\nu = 0, \dots, l)$ such that $r_\nu = \alpha_\nu r$ and $\alpha_\nu \neq \alpha_\mu$ for $\nu, \mu = 0, \dots, l$ and $\nu \neq \mu$.

PROPOSITION 4.1.1. *Let* $r_1, \dots, r_l \in L_p(a,b)$. *Then* I) $\Leftrightarrow$ II) $\vee$ III).

Proof. I)⇒II)∨III): Observe that $\varphi_{\nu\mu}$, $\varphi_{\nu 0}$ and $\varphi_{\mu,0}$ for $\nu,\mu \in \{1,\ldots,l\}$ satisfy

(4.1.9) $|r_\nu(x) - r_\mu(x)|e^{i\varphi_{\nu\mu}} = r_\nu(x) - r_\mu(x) = \left(|r_\nu(x)| - e^{i(\varphi_{\mu 0} - \varphi_{\nu 0})}|r_\mu(x)|\right)e^{i\varphi_{\nu 0}}$

for $x \in (a,b)$. Condition (4.1.3) implies that $r_1 \neq 0$. If $\varphi_{\nu 0} - \varphi_{\mu 0} \in \pi\mathbb{Z}$ for all $\nu,\mu \in \{1,\ldots,l\}$, then we set $\alpha = e^{i\varphi_1}$ and $\tilde{r}_\nu := \alpha^{-1}r_\nu$. Then the functions $\tilde{r}_\nu$ are real-valued, (4.1.6) and (4.1.7) are obvious, and (4.1.8) follows from

$$\tilde{r}_\nu(x) - \tilde{r}_\mu(x) = |r_\nu(x) - r_\mu(x)|\alpha^{-1}e^{i\varphi_{\nu\mu}} \text{ a.e. in } (a,b)$$

since (4.1.9) implies that $\varphi_{\nu\mu} - \varphi_{\nu 0} \in \pi\mathbb{Z}$. Thus II) holds in this case. Now suppose that there are $\nu,\mu \in \{1,\ldots,l\}$ such that $\varphi_{\nu 0} - \varphi_{\mu 0} \notin \pi\mathbb{Z}$. For $x \in (a,b)$ we infer from (4.1.9) that

$$|r_\nu(x) - r_\mu(x)|e^{i(\varphi_{\nu\mu} - \varphi_{\nu 0})} + |r_\mu(x)|e^{i(\varphi_{\mu 0} - \varphi_{\nu 0})} = |r_\nu(x)|.$$

The imaginary part yields

$$|r_\nu(x) - r_\mu(x)|\sin(\varphi_{\nu\mu} - \varphi_{\nu 0}) + |r_\mu(x)|\sin(\varphi_{\mu 0} - \varphi_{\nu 0}) = 0.$$

Since $|r_\mu(x)| \neq 0$ a.e. and $\sin(\varphi_{\mu 0} - \varphi_{\nu 0}) \neq 0$ by assumption, we obtain that $\sin(\varphi_{\nu\mu} - \varphi_{\nu 0}) \neq 0$. Hence

$$r_\nu(x) = |r_\mu(x)|\left(e^{i\varphi_{\mu 0}} - \frac{\sin(\varphi_{\mu 0} - \varphi_{\nu 0})}{\sin(\varphi_{\nu\mu} - \varphi_{\nu 0})}e^{i\varphi_{\nu\mu}}\right).$$

We set $r(x) := |r_\mu(x)|$, $\alpha_\mu := e^{i\varphi_{\mu 0}}$, $\alpha_\nu := \left(e^{i\varphi_{\mu 0}} - \frac{\sin(\varphi_{\mu 0} - \varphi_{\nu 0})}{\sin(\varphi_{\nu\mu} - \varphi_{\nu 0})}e^{i\varphi_{\nu\mu}}\right)$, and $\alpha_0 = 0$. Now let $\kappa \in \{1,\ldots,l\} \setminus \{\nu,\mu\}$. Then $\varphi_{\kappa 0} - \varphi_{\mu 0} \notin \pi\mathbb{Z}$ or $\varphi_{\kappa 0} - \varphi_{\nu 0} \notin \pi\mathbb{Z}$. If $\varphi_{\kappa 0} - \varphi_{\mu 0} \notin \pi\mathbb{Z}$, then as above there is a complex number $\alpha_\kappa \in \mathbb{C}$ such that $r_\kappa(x) = \alpha_\kappa|r_\mu(x)|$. If $\varphi_{\kappa 0} - \varphi_{\nu 0} \notin \pi\mathbb{Z}$, then there is a complex number $\alpha'_\kappa \in \mathbb{C}$ such that $r_\kappa(x) = \alpha'_\kappa|r_\nu(x)|$. Set $\alpha_\kappa := \alpha'_\kappa|\alpha_\nu|$. This proves that III) holds since (4.1.3) implies that the α_κ are pairwise different.
II)⇒I) immediately follows from

$$r_\nu(x) - r_\mu(x) = \alpha(\tilde{r}_\nu(x) - \tilde{r}_\mu(x)).$$

III)⇒I) is obvious since $r^{-1} \in L_\infty(a,b)$. □

For the boundary conditions (4.1.2) we assume that $a_j \in [a,b]$ for $j \in \mathbb{N}$, that $a_j \neq a_k$ if $j \neq k$, and that $a_0 = a$, $a_1 = b$. We suppose that the matrix function $\widetilde{W}(\cdot,\lambda)$ belongs to $M_n(L_1(a,b))$ for $|\lambda| \geq \gamma$ and that there is $W_0 \in M_n(L_1(a,b))$ such that

(4.1.10) $$\widetilde{W}(\cdot,\lambda) - W_0 = O(\lambda^{-1}) \quad \text{in } M_n(L_1(a,b)) \text{ as } \lambda \to \infty.$$

Finally we assume that the $\widetilde{W}_j(\lambda)$ are $n \times n$ matrices, defined for $|\lambda| \geq \gamma$, and that there are $n \times n$ matrices $W_0^{(j)}$ such that the estimates

$$\sum_{j=0}^{\infty} |W_0^{(j)}| < \infty \tag{4.1.11}$$

and

$$\sum_{j=0}^{\infty} |\widetilde{W}^{(j)}(\lambda) - W_0^{(j)}| = O(\lambda^{-1}) \quad \text{as } \lambda \to \infty \tag{4.1.12}$$

hold. Since

$$\sum_{j=0}^{\infty} |\widetilde{W}^{(j)}(\lambda)| \leq \sum_{j=0}^{\infty} |W_0^{(j)}| + \sum_{j=0}^{\infty} |\widetilde{W}^{(j)}(\lambda) - W_0^{(j)}| < \infty, \tag{4.1.13}$$

the boundary conditions (4.1.2) are well-defined for $|\lambda| \geq \gamma$, see Section 3.1.

By Theorem 2.8.2 there is a fundamental matrix function

$$\widetilde{Y}(\cdot,\lambda) = \big(P^{[0]} + B_0(\cdot,\lambda)\big)E(\cdot,\lambda) \tag{4.1.14}$$

of the differential equation (4.1.1) having the following properties: The matrix function $E(\cdot,\lambda)$ belongs to $M_n(W_p^1(a,b))$ and

$$E(x,\lambda) = \operatorname{diag}(E_0(x,\lambda), E_1(x,\lambda), \ldots, E_l(x,\lambda)) \tag{4.1.15}$$

for $x \in [a,b]$ and $\lambda \in \mathbb{C}$, where

$$R_\nu(x) = \int_a^x r_\nu(\xi)\,\mathrm{d}\xi, \quad E_\nu(x,\lambda) = \exp(\lambda R_\nu(x)) I_{n_\nu}.$$

The matrix function $P^{[0]}$ belongs to $M_n(W_p^1(a,b))$ and has block diagonal form according to the block structure of A_1, i. e.,

$$P^{[0]} = \operatorname{diag}(P_{00}^{[0]}, P_{11}^{[0]}, \ldots, P_{ll}^{[0]}). \tag{4.1.16}$$

The diagonal elements $P_{\nu\nu}^{[0]}$ are uniquely given as solutions of the initial value problems

$$\begin{cases} P_{\nu\nu}^{[0]\prime} = A_{0,\nu\nu} P_{\nu\nu}^{[0]}, \\ P_{\nu\nu}^{[0]}(a) = I_{n_\nu}, \end{cases} \tag{4.1.17}$$

where the $n_\nu \times n_\nu$ matrix functions $A_{0,\nu\nu}$ are the block diagonal elements of A_0. The matrix function $B_0(\cdot,\lambda)$ belongs to $M_n(W_p^1(a,b))$ for $|\lambda| \geq \gamma$ and fulfils the estimates

$$\begin{cases} B_0(\cdot,\lambda) = \{o(1)\}_\infty, \\ B_0(\cdot,\lambda) = \{O(\tau_p(\lambda))\}_\infty, \end{cases} \quad \text{as } \lambda \to \infty, \tag{4.1.18}$$

where

$$\tau_p(\lambda) = \max_{\substack{\nu,\mu=0\\ \nu\neq\mu}}^{l} \left(1 + |\Re(\lambda e^{i\varphi_{\nu\mu}})|\right)^{-1+1/p}.$$

In case $1 < p \leq \frac{3}{2}$ we require that there are numbers $1 \leq p_{\nu\mu} \leq \infty$ for $\nu,\mu = 0,\dots,l$, $\nu \neq \mu$, such that

$$A_{0,\nu q} \in M_{n_\nu,n_q}(L_{p_{\nu q}}(a,b)) \text{ and } \frac{1}{p_{\nu q}} + \frac{1}{p_{q\mu}} < 2 - \frac{1}{p} \tag{4.1.19}$$

for all $\nu,\mu,q = 0,\dots,l$ with $\nu \neq q$ and $\mu \neq q$, where $A_{0,\nu q}$ is the (ν,q) block entry of A_0 according to the block structure of A_1. If $p > \frac{3}{2}$ or $1 < p \leq \frac{3}{2}$ and (4.1.19) holds, then there is a number $\varepsilon \in (0, 1-\frac{1}{p})$ such that

$$B_0(\cdot,\lambda) = \left\{ O\left(\max_{\substack{\nu,\mu=0\\ \nu\neq\mu}}^{l} \left(1 + |\Re(\lambda e^{i\varphi_{\nu\mu}})|\right)^{-1/p-\varepsilon} \right) \right\}_p \quad \text{as } \lambda \to \infty. \tag{4.1.20}$$

For the definition of Birkhoff regularity we need some further notations: For $\nu = 1,\dots,l$ let φ_ν be as defined in (4.1.5) and let $\lambda \in \mathbb{C}\setminus\{0\}$. We set

$$\delta_\nu(\lambda) := \begin{cases} 0 & \text{if } \Re(\lambda e^{i\varphi_\nu}) < 0, \\ 1 & \text{if } \Re(\lambda e^{i\varphi_\nu}) > 0, \\ 0 & \text{if } \Re(\lambda e^{i\varphi_\nu}) = 0 \text{ and } \Im(\lambda e^{i\varphi_\nu}) > 0, \\ 1 & \text{if } \Re(\lambda e^{i\varphi_\nu}) = 0 \text{ and } \Im(\lambda e^{i\varphi_\nu}) < 0. \end{cases} \tag{4.1.21}$$

For convenience let $\delta_0(\lambda) = \delta_1(\lambda)$. We define the block diagonal matrices

$$\begin{cases} \Delta(\lambda) := \operatorname{diag}(\delta_0(\lambda)I_{n_0},\dots,\delta_l(\lambda)I_{n_l}), \\ \Delta_0 := \operatorname{diag}(0\cdot I_{n_0}, I_{n_1},\dots,I_{n_l}), \end{cases} \tag{4.1.22}$$

which (by definition) reduce to

$$\Delta(\lambda) = \operatorname{diag}(\delta_1(\lambda)I_{n_1},\dots,\delta_l(\lambda)I_{n_l}), \quad \Delta_0 := I_n, \tag{4.1.23}$$

if $n_0 = 0$. Finally we set

$$\widetilde{M}_2 := \sum_{j=0}^{\infty} W_0^{(j)} P^{[0]}(a_j) + \int_a^b W_0(x) P^{[0]}(x)\,dx. \tag{4.1.24}$$

From the definition of the $\delta_\nu(\lambda)$ we immediately infer that

$$\Delta(-\lambda) = I_n - \Delta(\lambda).$$

This is one reason for the choice of δ_0. Another reason is that the values of Δ only depend on the values of $\delta_1,\dots,\delta_l$.

DEFINITION 4.1.2. The boundary eigenvalue problem (4.1.1), (4.1.2) is called *Birkhoff regular* if

$$W_0^{(0)}(I_n-\Delta(\lambda))\Delta_0+W_0^{(1)}\Delta(\lambda)\Delta_0+\widetilde{M}_2(I_n-\Delta_0) \tag{4.1.25}$$

is invertible for $\lambda\in\mathbb{C}\setminus\{0\}$.

We would like to mention that the first n_0 columns of the matrices in (4.1.25) do not depend on λ and are the corresponding columns of $\widetilde{M}_2$. For $j>n_0$ and $\lambda\in\mathbb{C}\setminus\{0\}$, the j-th column in the matrix (4.1.25) is the j-th column of $W_0^{(0)}$ if the j-th diagonal element of $\Delta(\lambda)$ is 0, and it is the j-th column of $W_0^{(1)}$ if the j-th diagonal element of $\Delta(\lambda)$ is 1.

Obviously, a necessary condition for Birkhoff regularity is that none of the first n_0 columns of $\widetilde{M}_2$ and none of the last $n-n_0$ columns of $W_0^{(0)}$ and $W_0^{(1)}$ is zero. For example, a sufficient condition for Birkhoff regularity is that $n_0=0$ and $W_0^{(1)}=\alpha W_0^{(0)}$ such that $\alpha\neq 0$ and $W_0^{(0)}$ is invertible. We shall call boundary conditions with $W_0^{(1)}=\alpha W_0^{(0)}$ *asymptotically periodic boundary conditions.*

For $\nu=1,\dots,l$ we define

$$\Lambda_\nu^1:=\begin{pmatrix} 0\cdot I_{n_0} & & & 0\\ & \delta_\nu^1 I_{n_1} & & \\ & & \ddots & \\ 0 & & & \delta_\nu^l I_{n_l}\end{pmatrix} \tag{4.1.26}$$

and

$$\Lambda_\nu^2:=\begin{pmatrix} 0\cdot I_{n_0} & & & 0\\ & (1-\delta_\nu^1) I_{n_1} & & \\ & & \ddots & \\ 0 & & & (1-\delta_\nu^l) I_{n_l}\end{pmatrix}, \tag{4.1.27}$$

where

$$\delta_\nu^\mu:=\begin{cases}1 & \text{if } \varphi_\mu\in[\varphi_\nu,\varphi_\nu+\pi)\,\mathrm{mod}(2\pi)\\ 0 & \text{if } \varphi_\mu\notin[\varphi_\nu,\varphi_\nu+\pi)\,\mathrm{mod}(2\pi)\end{cases} \tag{4.1.28}$$

for $\nu,\mu=1,\dots,l$. Here $\varphi_\mu\in Z\,\mathrm{mod}(2\pi)$ for a subset Z of $\mathbb{R}$ means that there is a number $a\in Z$ such that $\varphi_\mu-a\in 2\pi\mathbb{Z}$.

THEOREM 4.1.3. *The boundary eigenvalue problem* (4.1.1), (4.1.2) *is Birkhoff regular if and only if the matrices*

$$W_0^{(0)}\Lambda_\nu^1 + W_0^{(1)}\Lambda_\nu^2 + \widetilde{M}_2(I_n - \Delta_0) \tag{4.1.29}$$

and

$$W_0^{(0)}\Lambda_\nu^2 + W_0^{(1)}\Lambda_\nu^1 + \widetilde{M}_2(I_n - \Delta_0) \tag{4.1.30}$$

are invertible for all $\nu = 1,\dots,l$, *where* $W_0^{(0)}$ *and* $W_0^{(1)}$ *are uniquely determined by* $\widetilde{W}^{(j)}(\lambda) - W_0^{(j)} = O(\lambda)^{-1}$ *as* $\lambda \to \infty$ *for* $j = 0, 1$, *and* $\widetilde{M}_2$ *is defined in* (4.1.24).

First we are going to discuss the Birkhoff regularity conditions before we prove this theorem.

We shall call the matrices occuring in (4.1.25) or in (4.1.29) and (4.1.30), respectively, the *Birkhoff matrices* of the boundary eigenvalue problem (4.1.1), (4.1.2). If $A_1(x)$ is invertible a. e. in (a,b), i. e., $n_0 = 0$ and hence $\Delta_0 = I_n$, then the Birkhoff matrices only depend on A_1 and the boundary matrices

$$W_0^{(0)} = \lim_{\lambda\to\infty} W_0(\lambda), \quad W_0^{(1)} = \lim_{\lambda\to\infty} W_1(\lambda)$$

at the endpoints of the interval $[a,b]$. Thus, in this important special case, we do not need any information neither on the "boundary" matrices $W_j(\lambda)$ at the interior points ($j \notin \{0,1\}$) nor on the matrix $W(\cdot,\lambda)$ in the integral term in order to decide whether the boundary eigenvalue problem is Birkhoff regular or not.

The situation is different if $n_0 \neq 0$, i. e., $\Delta_0 \neq I_n$. In this case the limit matrices of all $W_j(\lambda)$ and $W(\cdot,\lambda)$ and the matrix $P^{[0]}$ have to be considered since they occur in $\widetilde{M}_2$. Fortunately,

$$I_n - \Delta_0 = \operatorname{diag}(I_{n_0}, 0 \cdot I_{n-n_0}),$$

so that we only need to know the $n_0 \times n_0$ block matrix $P_{00}^{[0]}$ which is defined by

$$\begin{cases} P_{00}^{[0]\prime} = A_{0,00} P_{00}^{[0]}, \\ P_{00}^{[0]}(a) = I_{n_0}, \end{cases}$$

see (4.1.17), and the first n_0 columns of the matrices $W_j^{(0)}$ and W_0.

REMARK 4.1.4. There are l_0 real numbers $0 \leq \chi_1 < \dots < \chi_{l_0} < 2\pi$ such that

$$\{\chi_\nu : \nu = 1,\dots,l_0\} = \{\varphi_\nu, \varphi_\nu + \pi : \nu = 1,\dots,l\} \bmod (2\pi).$$

Let $\chi_0 := \chi_{l_0} - 2\pi$ and set

$$\Sigma_k := \{\lambda \in \mathbb{C}\setminus\{0\} : -\frac{\pi}{2} - \chi_k \leq \arg\lambda < -\frac{\pi}{2} - \chi_{k-1}\} \quad (k = 1,\dots,l_0),$$

where, for convenience, $\arg\lambda$ is taken in the interval $[-\frac{5\pi}{2} - \chi_0, -\frac{\pi}{2} - \chi_0)$. Then $\mathbb{C}\setminus\{0\}$ is divided into the l_0 sectors Σ_k $(k = 1,\dots,l_0)$. The number l_0 is an even number with $2 \leq l_0 \leq 2l$.

Here $X = Z \bmod(2\pi)$ for two subsets X and Z of $\mathbb{R}$ means that for each $a \in X$ there is a number $b \in Z$ such that $a - b \in 2\pi\mathbb{Z}$ and for each $b \in Z$ there is a number $a \in X$ such that $a - b \in 2\pi\mathbb{Z}$.

PROPOSITION 4.1.5. *The matrix function* Δ *defined in* (4.1.22) *is constant on each* Σ_k $(k = 1, \ldots, l_0)$.

Proof. Let $k \in \{1, \ldots, l_0\}$ and $\nu \in \{1, \ldots, l\}$. Then

$$(\chi_{k-1}, \chi_k] \subset (\varphi_\nu, \varphi_\nu + \pi] \bmod(2\pi) \quad \text{or} \quad (\chi_{k-1}, \chi_k] \subset (\varphi_\nu - \pi, \varphi_\nu] \bmod(2\pi),$$

for otherwise

$$\varphi_\nu \in (\chi_{k-1}, \chi_k) \bmod(2\pi) \quad \text{or} \quad \varphi_\nu + \pi \in (\chi_{k-1}, \chi_k) \bmod(2\pi),$$

which contradicts the definition of χ_{k-1} and χ_k. If

$$(\chi_{k-1}, \chi_k] \subset (\varphi_\nu, \varphi_\nu + \pi] \bmod(2\pi),$$

then we obtain

$$\arg\lambda \in \Big[-\frac{3\pi}{2} - \varphi_\nu, -\frac{\pi}{2} - \varphi_\nu\Big) \bmod(2\pi)$$

for all $\lambda \in \Sigma_k$ and hence $\delta_\nu(\lambda) = 0$ for these λ. If

$$(\chi_{k-1}, \chi_k] \subset (\varphi_\nu - \pi, \varphi_\nu] \bmod(2\pi),$$

then we obtain

$$\arg\lambda \in \Big[-\frac{\pi}{2} - \varphi_\nu, \frac{\pi}{2} - \varphi_\nu\Big) \bmod(2\pi)$$

for all $\lambda \in \Sigma_k$ and hence $\delta_\nu(\lambda) = 1$ for these λ. □

PROPOSITION 4.1.6. *The matrix function* Δ *defined in* (4.1.22) *takes the values*

$$\Delta\big(-i\exp(-i\varphi_\nu)\big),\ I_n - \Delta\big(-i\exp(-i\varphi_\nu)\big) \quad (\nu = 1, \ldots, l).$$

Proof. Since $\exp(i(-\frac{\pi}{2} - \chi_k)) \in \Sigma_k$, Δ has the values

$$\Delta\big(\exp(i(-\frac{\pi}{2} - \chi_k))\big) = \Delta\big(-i\exp(-i\chi_k)\big) \quad (k = 1, \ldots, l_0)$$

by Remark 4.1.4 and Proposition 4.1.5. Since the values of all the χ_k are the values of all the φ_ν or $\varphi_\nu \pm \pi$, $\exp(-i(\varphi_\nu \pm \pi)) = -\exp(-i\varphi_\nu)$ and $\Delta(-\lambda) = I_n - \Delta(\lambda)$, the proposition is proved. □

Proof of Theorem 4.1.3. The definition of δ_μ immediately yields that the identity $\delta_\mu(-i\exp(-i\varphi_\nu)) = \delta_\mu\big(\exp(-i(\frac{\pi}{2} + \varphi_\nu))\big) = 1$ holds if and only if

$$-\frac{\pi}{2} - \varphi_\nu \in \Big[-\frac{\pi}{2} - \varphi_\mu, \frac{\pi}{2} - \varphi_\mu\Big) \bmod(2\pi),$$

i. e., we have

$$\delta_\mu(-i\exp(-i\varphi_\nu)) = 1 \Leftrightarrow \varphi_\mu \in [\varphi_\nu, \varphi_\nu + \pi) \bmod(2\pi) \tag{4.1.31}$$

for all $\mu, \nu = 1, \ldots, l$. From the definition of the δ_ν^μ we conclude that

$$\delta_\mu\bigl(-i\exp(-i\varphi_\nu)\bigr) = \delta_\nu^\mu \quad (\nu, \mu = 1, \ldots, l),$$

whence

$$\Lambda_\nu^1 = \Delta\bigl(-i\exp(-i\varphi_\nu)\bigr)\Delta_0, \quad \Lambda_\nu^2 = \bigl[I_n - \Delta\bigl(-i\exp(-i\varphi_\nu)\bigr]\Delta_0$$

for $\nu = 1, \ldots, l$, which proves Theorem 4.1.3 because of Proposition 4.1.6. □

We are going to investigate two special cases in more detail. First we consider the simple case that $n_0 = 0$ and $l = 1$, i. e., $A_1(x) = r_1(x)I_n$ a. e. in (a,b), where $r_1^{-1} \in L_\infty(a,b)$ and r_1 fulfils (4.1.5). Then Δ has only the values I_n and 0, and the problem (4.1.1), (4.1.2) is Birkhoff regular if and only if the matrices $W_0^{(0)}$ and $W_0^{(1)}$ are invertible. The second case is more important and treated in

PROPOSITION 4.1.7. *Let $l = n$ and $\Delta(\lambda) = \operatorname{diag}(\delta_1(\lambda), \ldots, \delta_n(\lambda))$ as given by* (4.1.23). *We suppose that*

$$\varphi_\nu = \tfrac{2\pi(\nu-1)}{n} \quad (\nu = 1, \ldots, n).$$

i) *If n is even, then the values of Δ are the diagonal matrices with $\frac{n}{2}$ consecutive ones and $\frac{n}{2}$ consecutive zeros in the diagonal in a cyclic arrangement.*
ii) *If n is odd, then the values of Δ are the diagonal matrices with $\frac{n+1}{2}$ consecutive ones and $\frac{n-1}{2}$ consecutive zeros in the diagonal and the diagonal matrices with $\frac{n-1}{2}$ consecutive ones and $\frac{n+1}{2}$ consecutive zeros in the diagonal, each in a cyclic arrangement.*

Proof. Since $l = n$, we have $n_0 = 0$. In this case, the values of Δ are the matrices Λ_ν^1 and Λ_ν^2 $(\nu = 1, \ldots, n)$. Obviously, for $\nu, \mu = 1, \ldots, l$ we have

$$\varphi_\mu \in [\varphi_\nu, \varphi_\nu + \pi) \operatorname{mod}(2\pi) \Leftrightarrow \mu \in \left[\nu, \nu + \frac{n}{2}\right) \operatorname{mod}(n).$$

If n is even, then $\varphi_{\nu+\frac{n}{2}} = (\varphi_\nu + \pi)\operatorname{mod}(2\pi)$ for $\nu = 1, \ldots, n$. Hence $\Lambda_\nu^2 = \Lambda_{\nu'}^1$, where $\nu' - \nu = \frac{n}{2} \operatorname{mod}(n)$, and we only have to consider the matrices Λ_ν^1. The definition of the diagonal elements δ_ν^μ immediately yields that the $\frac{n}{2}$ consecutive diagonal elements of Λ_ν^1 starting with the ν-th entry are 1 and that the others are 0, which proves the assertion in case n is even.

If n is odd, the same argument as above shows that the $\frac{n+1}{2}$ consecutive diagonal elements of Λ_ν^1 starting with the ν-th entry are 1 and that the others are 0. Since $\Lambda_\nu^2 = I_n - \Lambda_\nu^1$, the $\frac{n+1}{2}$ consecutive diagonal elements of Λ_ν^2 starting with the ν-th entry are 0 and the others are 1. □

We state the following obvious generalization of Proposition 4.1.7.

COROLLARY 4.1.8. *If $n_0 \neq 0$, $l = n - n_0$, and $\varphi_\nu = \frac{2\pi(\nu-1)}{l}$ $(\nu = 1, \ldots, l)$, then the assertions of Proposition* 4.1.7 *hold for the last l columns of Δ given by* (4.1.22), *where n has to be substituted by l.*

4.2. Examples of Birkhoff regular problems

In general, the boundary matrices are not asymptotically constant in the applications. Hence we multiply the boundary conditions, given in matrix form, by a matrix function C_2 from the left. Of course, if, for some given λ, $C_2(\lambda)$ is invertible, then the old boundary conditions at λ are equivalent to the new ones at λ. We shall call the boundary conditions obtained by multiplication with C_2 equivalent to the given ones if $C_2(\lambda)$ is invertible for all but finitely many numbers λ.

EXAMPLE 4.2.1. Here we consider the first order 2×2 system

$$y' - \lambda \begin{pmatrix} \alpha & 0 \\ 0 & \beta \end{pmatrix} y - \begin{pmatrix} a_{11} & a_{12} \\ a_{21} & a_{22} \end{pmatrix} y = 0,$$

where α and β are nonzero complex numbers and the functions a_{ij} belong to $L_p(0,1)$. To this system we first associate the boundary conditions

$$\begin{pmatrix} \lambda & 1 \\ \lambda & 2 \end{pmatrix} y(0) + \begin{pmatrix} \lambda & 0 \\ \lambda & 1 \end{pmatrix} y(1) = 0.$$

These conditions are not asymptotically constant. A simple method to make them asymptotically constant is the multiplication by λ^{-1}. But in this case the second columns of both boundary matrices are asymptotically zero, and the boundary eigenvalue problem would not be Birkhoff regular. If we only consider the λ-terms, we see that the first row and the second row are the same for both boundary matrices. Therefore we multiply the boundary conditions by the invertible matrix

$$\begin{pmatrix} 1 & 0 \\ -1 & 1 \end{pmatrix}$$

from the left and obtain

$$\begin{pmatrix} \lambda & 1 \\ 0 & 1 \end{pmatrix} y(0) + \begin{pmatrix} \lambda & 0 \\ 0 & 1 \end{pmatrix} y(1) = 0.$$

Now, for $\lambda \neq 0$, we multiply the boundary conditions by the matrix

$$\begin{pmatrix} \lambda^{-1} & 0 \\ 0 & 1 \end{pmatrix}$$

from the left, which yields

$$\begin{pmatrix} 1 & \lambda^{-1} \\ 0 & 1 \end{pmatrix} y(0) + \begin{pmatrix} 1 & 0 \\ 0 & 1 \end{pmatrix} y(1) = 0.$$

These boundary conditions are equivalent to the original boundary conditions and are asymptotically constant with

$$W_0^{(0)} = I_2 = W_0^{(1)}.$$

Hence, for any values of α and β, the boundary eigenvalue problem is Birkhoff regular; the Birkhoff matrices are equal to I_2.

Now we consider the boundary condition

$$\begin{pmatrix} 1 & 1 \\ 0 & 1 \end{pmatrix} y(0) + \begin{pmatrix} \gamma & 0 \\ 1 & 1 \end{pmatrix} y(1) = 0,$$

where γ is a complex number. In this case Δ takes the values

$$\begin{array}{ll} \begin{pmatrix} 1 & 0 \\ 0 & 1 \end{pmatrix} \text{ and } \begin{pmatrix} 0 & 0 \\ 0 & 0 \end{pmatrix} & \text{if } \arg\alpha = \arg\beta, \\ \begin{pmatrix} 1 & 0 \\ 0 & 0 \end{pmatrix} \text{ and } \begin{pmatrix} 0 & 0 \\ 0 & 1 \end{pmatrix} & \text{if } \arg\alpha = (\arg\beta + \pi) \operatorname{mod}(2\pi), \\ \text{and all these four matrices} & \text{if } \arg\alpha \neq \arg\beta \operatorname{mod}(\pi). \end{array}$$

As Birkhoff matrices we obtain

$$\begin{array}{ll} \begin{pmatrix} \gamma & 0 \\ 1 & 1 \end{pmatrix} \text{ and } \begin{pmatrix} 1 & 1 \\ 0 & 1 \end{pmatrix} & \text{if } \arg\alpha = \arg\beta, \\ \begin{pmatrix} \gamma & 1 \\ 1 & 1 \end{pmatrix} \text{ and } \begin{pmatrix} 1 & 0 \\ 0 & 1 \end{pmatrix} & \text{if } \arg\alpha = (\arg\beta + \pi) \operatorname{mod}(2\pi), \\ \text{and all these four matrices} & \text{if } \arg\alpha \neq \arg\beta \operatorname{mod}(\pi). \end{array}$$

Hence, in this case, the boundary eigenvalue problem is Birkhoff regular if and only if

$$\begin{cases} \gamma \neq 0 & \text{if } \arg\alpha = \arg\beta, \\ \gamma \neq 1 & \text{if } \arg\alpha = (\arg\beta + \pi) \operatorname{mod}(2\pi), \\ \gamma \neq 0, 1 & \text{if } \arg\alpha \neq \arg\beta \operatorname{mod}(\pi). \end{cases} \qquad \square$$

EXAMPLE 4.2.2. We consider the second order linear differential equation

$$\eta'' + (p_{10} + \lambda p_{11})\eta' + (p_{00} + \lambda p_{01} + \lambda^2 p_{02})\eta = 0 \tag{4.2.1}$$

for $\eta \in W_p^2(0,1)$. We assume that p_{11} and p_{02} are complex numbers and that the other coefficients belong to $L_p(a,b)$. To this second order equation we associate the first order 2×2 system

$$\tilde{y}' - \begin{pmatrix} 0 & 1 \\ -p_{00} - \lambda p_{01} - \lambda^2 p_{02} & -p_{10} - \lambda p_{11} \end{pmatrix} \tilde{y} = 0, \tag{4.2.2}$$

where $\tilde{y} \in (W_p^1(0,1))^2$. If $\eta \in W_p^2(0,1)$ is a solution of (4.2.1), then $\tilde{y} = \begin{pmatrix} \eta \\ \eta' \end{pmatrix}$ obviously belongs to $(W_p^1(0,1))^2$ and fulfils the equation (4.2.2). Vice versa, let $\tilde{y} = \begin{pmatrix} y_1 \\ y_2 \end{pmatrix} \in (W_p^1(0,1))^2$ be a solution of (4.2.2). The first component of this equation yields $y_1' = y_2$ whence $\eta := y_1 \in W_p^2(0,1)$. The second component of (4.2.2) shows that η solves (4.2.1).

The system (4.2.2) is not asymptotically linear in λ. In order to obtain an asymptotically linear system, we apply the transformation $\tilde{y} = C(\lambda)y$ with

$$C(\lambda) = \begin{pmatrix} 1 & 1 \\ \lambda r_1 & \lambda r_2 \end{pmatrix},$$

where r_1 and r_2 are the roots of the equation

$$\rho^2 + p_{11}\rho + p_{02} = 0.$$

We require that $C(\lambda)$ is invertible for $\lambda \neq 0$, i. e., that the roots r_1 and r_2 of the above quadratic equation are different. If one root is zero, we assume without loss of generality that then $r_1 = 0$. In order to obtain a system where the coefficient matrix of the derivative is the identity, we have to multiply the transformed system by $C(\lambda)^{-1}$ from the left which leads to

$$y' - C(\lambda)^{-1} \begin{pmatrix} 0 & 1 \\ -p_{00} - \lambda p_{01} - \lambda^2 p_{02} & -p_{10} - \lambda p_{11} \end{pmatrix} C(\lambda)y = 0.$$

An easy calculation yields

$$\begin{aligned} C(\lambda)^{-1} & \begin{pmatrix} 0 & 1 \\ -p_{00} - \lambda p_{01} - \lambda^2 p_{02} & -p_{10} - \lambda p_{11} \end{pmatrix} C(\lambda) \\ &= \lambda \begin{pmatrix} r_1 & 0 \\ 0 & r_2 \end{pmatrix} + (r_2 - r_1)^{-1} \begin{pmatrix} p_{01} + p_{10}r_1 & p_{01} + p_{10}r_2 \\ -p_{01} - p_{10}r_1 & -p_{01} - p_{10}r_2 \end{pmatrix} \\ &\quad + \lambda^{-1}(r_2 - r_1)^{-1} \begin{pmatrix} p_{00} & p_{00} \\ -p_{00} & -p_{00} \end{pmatrix}. \end{aligned}$$

Hence we obtain an asymptotically linear system

$$y' - \left(\lambda A_1 + A_0 + \lambda^{-1}A^0(\cdot,\lambda)\right)y = 0, \tag{4.2.3}$$

where

$$\begin{aligned} A_1 &= \begin{pmatrix} r_1 & 0 \\ 0 & r_2 \end{pmatrix}, \\ A_0 &= (r_2 - r_1)^{-1} \begin{pmatrix} p_{01} + p_{10}r_1 & p_{01} + p_{10}r_2 \\ -p_{01} - p_{10}r_1 & -p_{01} - p_{10}r_2 \end{pmatrix}, \\ A^0(\cdot,\lambda) &= (r_2 - r_1)^{-1} \begin{pmatrix} p_{00} & p_{00} \\ -p_{00} & -p_{00} \end{pmatrix}. \end{aligned}$$

We consider two different kinds of boundary conditions. First we take periodic boundary conditions

$$\eta(0) - \eta(1) = 0, \quad \eta'(0) - \eta'(1) = 0. \tag{4.2.4}$$

In terms of the first order system (4.2.2), these boundary conditions can be written as

$$\begin{pmatrix} 1 & 0 \\ 0 & 1 \end{pmatrix} \tilde{y}(0) + \begin{pmatrix} -1 & 0 \\ 0 & -1 \end{pmatrix} \tilde{y}(1) = 0.$$

The transformation of (4.2.2) to the system (4.2.3) changes these boundary conditions to

$$\begin{pmatrix} 1 & 1 \\ \lambda r_1 & \lambda r_2 \end{pmatrix} y(0) + \begin{pmatrix} -1 & -1 \\ -\lambda r_1 & -\lambda r_2 \end{pmatrix} y(1) = 0.$$

In order to make them asymptotically constant, we multiply this equation from the left by

$$\begin{pmatrix} 1 & 0 \\ 0 & \lambda^{-1} \end{pmatrix}$$

and obtain

$$\begin{pmatrix} 1 & 1 \\ r_1 & r_2 \end{pmatrix} y(0) + \begin{pmatrix} -1 & -1 \\ -r_1 & -r_2 \end{pmatrix} y(1) = 0. \tag{4.2.5}$$

If $r_1 \neq 0$ and $r_2 \neq 0$, i. e., $p_{02} \neq 0$, then (4.2.3), (4.2.5) is a boundary eigenvalue problem with $n_0 = 0$ and periodic boundary conditions, see page 135. Hence, in this case, the problem is Birkhoff regular.

Now let $r_1 = 0$, i. e., $p_{02} = 0$. Obviously $n_0 = 1$. Since the element in the upper left corner of A_0 is $r_2^{-1} p_{01}$, the element in the upper left corner of $P^{[0]}$ is the solution of $v' - r_2^{-1} p_{01} v = 0$ with $v(0) = 1$, whence

$$v(x) = \exp\left(r_2^{-1} \int_0^x p_{01}(t)\, dt \right) \quad (x \in [0,1]).$$

It is easy to see that

$$\begin{aligned} \widetilde{M}_2 &= \begin{pmatrix} 1 & 1 \\ 0 & r_2 \end{pmatrix} + \begin{pmatrix} -1 & -1 \\ 0 & -r_2 \end{pmatrix} \begin{pmatrix} v(1) & 0 \\ 0 & * \end{pmatrix} \\ &= \begin{pmatrix} 1 - v(1) & * \\ 0 & * \end{pmatrix}. \end{aligned}$$

Thus the Birkhoff matrices are

$$\begin{pmatrix} 1 - v(1) & 1 \\ 0 & r_2 \end{pmatrix} \quad \text{and} \quad \begin{pmatrix} 1 - v(1) & -1 \\ 0 & -r_2 \end{pmatrix}.$$

These matrices are invertible if and only if

$$\int_0^1 p_{01}(t)\, dt \notin 2\pi r_2 i \mathbb{Z}.$$

Thus the boundary eigenvalue problem for the first order system (4.2.3), (4.2.5) associated to the boundary eigenvalue problem (4.2.1), (4.2.4) with $p_{11}^2 \neq 4 p_{02}$ for a second order differential equation is Birkhoff regular if and only if

$$p_{02} \neq 0 \quad \text{or} \quad \int_0^1 p_{01}(t)\, dt \notin 2\pi p_{11} i \mathbb{Z},$$

where we have to note that $p_{11} \neq 0$ if $p_{02} = 0$.

Now we take separated boundary conditions:

$$\begin{aligned} \alpha_1\eta(0)+\alpha_2\eta'(0)&=0,\\ \beta_1\eta(1)+\beta_2\eta'(1)&=0, \end{aligned} \tag{4.2.6}$$

with $|\alpha_1|+|\alpha_2|>0$ and $|\beta_1|+|\beta_2|>0$. As above, this leads to the boundary conditions

$$\begin{pmatrix}\alpha_1 & \alpha_2\\ 0 & 0\end{pmatrix}\tilde{y}(0)+\begin{pmatrix}0 & 0\\ \beta_1 & \beta_2\end{pmatrix}\tilde{y}(1)=0$$

and

$$\begin{pmatrix}\alpha_1+\alpha_2\lambda r_1 & \alpha_1+\alpha_2\lambda r_2\\ 0 & 0\end{pmatrix}y(0)+\begin{pmatrix}0 & 0\\ \beta_1+\beta_2\lambda r_1 & \beta_1+\beta_2\lambda r_2\end{pmatrix}y(1)=0.$$

For any transformation, i. e., multiplication from the left by a λ-depending 2×2 matrix, which makes these boundary conditions asymptotically constant, these constant boundary matrices are not invertible. The discussion in Example 4.2.1 shows that the identity matrix and the zero matrix are values of Δ if $r_1r_2\neq 0$ and $\arg r_2\neq(\arg r_1+\pi)\operatorname{mod}(2\pi)$. Hence the boundary eigenvalue problem is not Birkhoff regular if $r_1r_2\neq 0$ and $\arg r_2\neq(\arg r_1+\pi)\operatorname{mod}(2\pi)$.

Now let us assume that $r_1r_2=0$ or $\arg r_2=(\arg r_1+\pi)\operatorname{mod}(2\pi)$. We consider the special case $\alpha_2=0$ and $\beta_2=0$. Then the boundary conditions do not depend on λ. In this case they are (asymptotically) constant. If $\alpha_2\neq 0$ or $\beta_2\neq 0$, we multiply the boundary conditions by $\widehat{C}(\lambda)$ from the left, where

$$\widehat{C}(\lambda)=\begin{cases}\begin{pmatrix}\lambda^{-1} & 0\\ 0 & 1\end{pmatrix} & \text{if } \alpha_2\neq 0 \text{ and } \beta_2=0,\\ \begin{pmatrix}1 & 0\\ 0 & \lambda^{-1}\end{pmatrix} & \text{if } \alpha_2=0 \text{ and } \beta_2\neq 0, \quad (\lambda\neq 0).\\ \begin{pmatrix}\lambda^{-1} & 0\\ 0 & \lambda^{-1}\end{pmatrix} & \text{if } \alpha_2\neq 0 \text{ and } \beta_2\neq 0,\end{cases}$$

In all these cases the asymptotic boundary matrices are

$$W_0^{(0)}=\begin{pmatrix}\gamma_1 & \gamma_2\\ 0 & 0\end{pmatrix} \quad\text{and}\quad W_0^{(1)}=\begin{pmatrix}0 & 0\\ \delta_1 & \delta_2\end{pmatrix}, \tag{4.2.7}$$

where, for $j=1,2$,

$$\gamma_j=\begin{cases}\alpha_2 r_j & \text{if } \alpha_2\neq 0,\\ \alpha_1 & \text{if } \alpha_2=0,\end{cases}$$

and

$$\delta_j=\begin{cases}\beta_2 r_j & \text{if } \beta_2\neq 0,\\ \beta_1 & \text{if } \beta_2=0.\end{cases}$$

If $r_1 = 0$, then

$$\begin{aligned}\widetilde{M}_2 &= \begin{pmatrix} \gamma_1 & \gamma_2 \\ 0 & 0 \end{pmatrix} + \begin{pmatrix} 0 & 0 \\ \delta_1 & \delta_2 \end{pmatrix} \begin{pmatrix} v(1) & 0 \\ 0 & * \end{pmatrix} \\ &= \begin{pmatrix} \gamma_1 & * \\ \delta_1 v(1) & * \end{pmatrix}.\end{aligned}$$

Thus the Birkhoff matrices are

$$\begin{pmatrix} \gamma_1 & \gamma_2 \\ \delta_1 v(1) & 0 \end{pmatrix} \quad \text{and} \quad \begin{pmatrix} \gamma_1 & 0 \\ \delta_1 v(1) & \delta_2 \end{pmatrix}.$$

It is always true that δ_2 and γ_2 are nonzero. Since $r_1 = 0$, the problem is not Birkhoff regular if $\alpha_2 \neq 0$ or $\beta_2 \neq 0$. But if $\alpha_2 = \beta_2 = 0$, then $\gamma_1 = \alpha_1 \neq 0$ and $\delta_1 = \beta_1 \neq 0$. Therefore, the problem is Birkhoff regular in this case.

If $r_1 \neq 0$ and $\arg r_2 = (\arg r_1 + \pi) \operatorname{mod}(2\pi)$, then the Birkhoff matrices are

$$\begin{pmatrix} \gamma_1 & 0 \\ 0 & \delta_2 \end{pmatrix} \quad \text{and} \quad \begin{pmatrix} 0 & \gamma_2 \\ \delta_1 & 0 \end{pmatrix}$$

by the discussion of the values of Δ in Example 4.2.1. Hence the problem is Birkhoff regular if $r_1 \neq 0$ and $\arg r_2 = (\arg r_1 + \pi) \operatorname{mod}(2\pi)$.

Now we shall express these conditions in terms of the coefficients of the differential equation. The condition

$$\arg r_2 = (\arg r_1 + \pi) \operatorname{mod}(2\pi) \quad \text{or} \quad r_1 = 0$$

is fulfilled if and only if 0 lies on the line segment connecting r_1 and r_2. Since the roots of

$$\rho^2 + p_{11}\rho + p_{02} = 0$$

are

$$\frac{1}{2}\left(-p_{11} \pm \sqrt{p_{11}^2 - 4p_{02}}\right),$$

this means that there is a $t \in [0,1]$ such that

$$\begin{aligned} 0 &= t\left\{\frac{1}{2}\left(-p_{11} + \sqrt{p_{11}^2 - 4p_{02}}\right)\right\} + (1-t)\left\{\frac{1}{2}\left(-p_{11} - \sqrt{p_{11}^2 - 4p_{02}}\right)\right\} \\ &= \frac{1}{2}\left\{-p_{11} + (2t-1)\sqrt{p_{11}^2 - 4p_{02}}\right\}. \end{aligned}$$

This holds if and only if

$$p_{11} = (2t-1)\sqrt{p_{11}^2 - 4p_{02}}.$$

Since $2t - 1$ varies in the interval $[-1,1]$, this condition is satisfied if and only if there is a $\tau \in [0,1]$ such that

$$p_{11}^2 = \tau(p_{11}^2 - 4p_{02}),$$

i. e., if there is a $\tau \in [0,1]$ such that

$$(1-\tau)p_{11}^2 = -4\tau p_{02}.$$

This holds if and only if 0 lies on the line segment with the endpoints p_{11}^2 and $4p_{02}$, i. e., if and only if $p_{11} = 0$, $p_{02} = 0$ or $2\arg p_{11} = (\arg p_{02} + \pi) \bmod (2\pi)$. Thus the boundary eigenvalue problem for the first order system (4.2.3) with the asymptotic boundary conditions (4.2.7) associated to the boundary eigenvalue problem (4.2.1), (4.2.6) with $p_{11}^2 \neq 4p_{02}$ for a second order differential equation is Birkhoff regular if and only if

$$p_{11} = 0, \quad p_{02} = 0 \quad \text{or} \quad 2\arg p_{11} = (\arg p_{02} + \pi) \bmod (2\pi),$$

where in the case $p_{02} = 0$ also $\alpha_2 = \beta_2 = 0$ has to be satisfied. For example, this is fulfilled if p_{02} is a negative real number and p_{11} is a real number. □

EXAMPLE 4.2.3. Suppose that the boundary eigenvalue problem (4.1.1), (4.1.2) satisfies the general assumptions stated in Section 4.1 and that $n_0 = 0$ and $l = 1$. Let

$$W_0^{(0)} = \begin{pmatrix} 1 & 1 & & & \\ & \cdot & \cdot & 0 & \\ & & \cdot & \cdot & \\ & 0 & & \cdot & 1 \\ & & & & 1 \end{pmatrix} \quad \text{and} \quad W_0^{(1)} = I_n.$$

Then the problem is Birkhoff regular.

We shall continue the discussion of this example on page 156.

EXAMPLE 4.2.4. Assume that $n_0 = n - l$, $\varphi_\nu = \frac{2\pi(\nu-1)}{l}$ $(\nu = 1,\dots,l)$ and that the boundary eigenvalue problem is *asymptotically separated*, i. e., $W_0^{(j)} = 0$ for $j \notin \{0,1\}$, $W_0 = 0$, and for each $j \in \{1,\dots,n\}$ either the j-th row of $W_0^{(0)}$ or the j-th row of $W_0^{(1)}$ is zero. We may assume without loss of generality that there is an $s \in \{0,\dots,n\}$ such that $e_j^\mathsf{T} W_0^{(0)} = 0$ for $j = 1,\dots,s$ and $e_j^\mathsf{T} W_0^{(1)} = 0$ for $j = s+1,\dots,n$. In this case, each of the Birkhoff matrices has the form

(4.2.8)

$\widetilde{M}_2$	$W_0^{(0)}$	$W_0^{(1)}$
$s \times n_0$	0	$s \times (n - n_0 - r)$
$(n-s) \times n_0$	$(n-s) \times r$	0

up to a permutation of the last l columns. Here the first line indicates from which matrices the columns are taken. The terms $s \times n_0$ etc. denote the size of the corresponding blocks, and $r = \frac{l}{2}$ if l is even, $r = \frac{l+1}{2}$ or $r = \frac{l-1}{2}$ if l is odd by Corollary 4.1.8. The last $n - n_0 - r$ columns of (4.2.8) are linearly dependent if its right upper block has more columns than rows. Thus a necessary condition for Birkhoff regularity is $s \geq n - n_0 - r$. The same consideration for the middle lower block of (4.2.8) yields $n - s \geq r$. Hence $n - n_0 - r \leq s \leq n - r$ is a necessary condition for Birkhoff regularity. If $l = n - n_0$ is even, then $\frac{n-n_0}{2} \le s \le \frac{n+n_0}{2}$ is a necessary condition for Birkhoff regularity. If $l = n - n_0$ is odd, then $\frac{n-n_0+1}{2} \le s \le \frac{n+n_0-1}{2}$ is a necessary condition for Birkhoff regularity.

Note that there are no Birkhoff regular asymptotically separated boundary eigenvalue problems if $n_0 = 0$ and n is odd.

In general, one cannot reduce the number of the matrices which are necessary to check for Birkhoff regularity as is seen in the following

THEOREM 4.2.5. *Let* $\lambda_k = \exp(i(-\frac{\pi}{2} - \chi_k))$ $(k = 1, \ldots, l_0)$, *where the* χ_k *are defined in Remark* 4.1.4. *For each* $k_0 \in \{1, \ldots, l_0\}$ *there are* $n \times n$ *matrices* A *and* B *such that*

$$\det\left[A\left(I_n - \Delta(\lambda_k)\right)\Delta_0 + B\Delta(\lambda_k)\Delta_0 + (AP^{[0]}(a) + BP^{[0]}(b))(I_n - \Delta_0)\right]$$

is zero if $k = k_0$ *and different from zero if* $k \in \{1, \ldots, l_0\} \setminus \{k_0\}$.

Proof. A pair (A, B) of $n \times n$ matrices fulfilling the property of the statement is said to be k_0-singular. If $n_0 \neq 0$, take $(n - n_0) \times (n - n_0)$ matrices A_0 and B_0 and set $A = P^{[0]}_{00}(a)^{-1} \oplus A_0$, $B = P^{[0]}_{00}(b)^{-1} \oplus B_0$. Let $\Delta'(\lambda) = \operatorname{diag}\left(\delta_1(\lambda)I_{n_1}, \ldots, \delta_l(\lambda)I_{n_l}\right)$. Then

$$\begin{aligned} &A\left(I_n - \Delta(\lambda_k)\right)\Delta_0 + B\Delta(\lambda_k)\Delta_0 + (AP^{[0]}(a) + BP^{[0]}(b))(I_n - \Delta_0) \\ &\quad = \begin{pmatrix} 2I_{n_0} & 0 \\ 0 & A_0\left(I_{n-n_0} - \Delta'(\lambda_k)\right) + B_0\Delta'(\lambda_k) \end{pmatrix}. \end{aligned}$$

Obviously, (A, B) is k_0-singular if and only if (A_0, B_0) is k_0-singular. Since the latter case corresponds to the case $n_0 = 0$, it is sufficient to consider the case $n_0 = 0$.

We have $\chi_{k_0} = \varphi_\nu$ or $\chi_{k_0} = (\varphi_\nu + \pi) \operatorname{mod}(2\pi)$ for some $\nu \in \{1, \ldots, l\}$. If $\chi_{k_0} \notin \{\varphi_1, \ldots, \varphi_l\}$, we choose $\nu \in \{1, \ldots, l\}$ such that $\chi_{k_0} = (\varphi_\nu + \pi) \operatorname{mod}(2\pi)$. There is a $k_1 \in \{1, \ldots, l_0\}$ such that $\chi_{k_1} = \varphi_\nu$. By the definition of the λ_k we obtain $\lambda_{k_1} = -\lambda_{k_0}$ since $\chi_{k_1} = (\chi_{k_0} + \pi) \operatorname{mod}(2\pi)$. Since

$$A\left(I_n - \Delta(\lambda)\right) + B\Delta(\lambda) = B\left(I_n - \Delta(-\lambda)\right) + A\Delta(-\lambda),$$

we infer that (A, B) is k_0-singular if (and only if) (B, A) is k_1-singular.

Hence it is sufficient to consider the case that $\chi_{k_0} = \varphi_\nu$ for some $\nu \in \{1, \ldots, l\}$. Without loss of generality we may assume $\nu = 1$. We consider matrices of the form

$$A = \begin{pmatrix} 1 & \alpha \\ 0 & I_{n-1} \end{pmatrix}, \quad B = \begin{pmatrix} \beta & \gamma \\ \delta & I_{n-1} \end{pmatrix}.$$

From the definition of δ_1 and from $\chi_{k_0} = \varphi_1$ we immediately infer that, for $k \in \{1, \ldots, l_0\}$, $\delta_1(\lambda_k) = 1$ holds if and only if

$$\chi_k \in (\chi_{k_0} - \pi, \chi_{k_0}] \operatorname{mod}(2\pi). \tag{4.2.9}$$

We note that this implies $\delta_1(\lambda_{k_0}) = 1$.

For those $k \in \{1,\dots,l\}$ for which $\delta_1(\lambda_k) = 0$, the first column of

$$A\big(I_n - \Delta(\lambda_k)\big) + B\Delta(\lambda_k) \tag{4.2.10}$$

is the first column of A, and the matrix (4.2.10) is a normed upper triangular matrix. Hence its determinant is nonzero.

We still have to show that there are suitable α, β, γ, δ such that the matrix (4.2.10) is invertible for those $k \in \{1,\dots,l\} \setminus \{k_0\}$ for which $\delta_1(\lambda_k) = 1$ and that it is not invertible for $k = k_0$. If $\chi_{k_0-1} = \chi_{k_0} - \pi$, then $\delta_1(\lambda_k) = 1$ if and only if $k = k_0$. In this case, the proposition is proved if we set $\alpha = 0$, $\beta = 0$, $\gamma = 0$, and $\delta = 0$.

Now we consider the case $\chi_{k_0-1} > \chi_{k_0} - \pi$. Then there is a $\mu \in \{1,\dots,l\}$ such that $\varphi_\mu = \chi_{k_0-1} \bmod(2\pi)$ or $\varphi_\mu = (\chi_{k_0-1} + \pi) \bmod(2\pi)$. Choose some j such that the diagonal element of Δ in the $j+1$-th column is the function δ_μ. From $\chi_{k_0-1} > \chi_{k_0} - \pi$ we infer that $\varphi_1 \neq \varphi_\mu$, i. e., we have $\mu \neq 1$. This implies $j > 0$. Now we set $\beta = 1$ and $\delta = (0,\dots,0,1,0,\dots,0)^\mathsf{T}$, where the 1 is at the j-th position. If $\varphi_\mu = \chi_{k_0-1} \bmod(2\pi)$ we set $\alpha = \delta^\mathsf{T}$ and $\gamma = 0$. If $\varphi_\mu = (\chi_{k_0-1} + \pi) \bmod(2\pi)$, we set $\gamma = \delta^\mathsf{T}$ and $\alpha = 0$.

First we consider the case $\varphi_\mu = \chi_{k_0-1} \bmod(2\pi)$. We have $\delta_\mu(\lambda_k) = 0$ if and only if

$$\chi_k \in (\chi_{k_0-1}, \chi_{k_0-1} + \pi] \bmod(2\pi). \tag{4.2.11}$$

Let $k \in \{1,\dots,l_0\}$ such that $\delta_1(\lambda_k) = 1$. From $\chi_{k_0} - \pi < \chi_{k_0-1} < \chi_{k_0}$ we infer that $(\chi_{k_0} - \pi, \chi_{k_0}]$ and $(\chi_{k_0-1}, \chi_{k_0-1} + \pi]$ are subsets of $(\chi_{k_0} - \pi, \chi_{k_0} + \pi]$. Since (4.2.9) and (4.2.11) can be written as

$$\begin{aligned} \chi_k - j_1 2\pi &\in (\chi_{k_0} - \pi, \chi_{k_0}], \\ \chi_k - j_2 2\pi &\in (\chi_{k_0-1}, \chi_{k_0-1} + \pi] \end{aligned}$$

for some $j_1, j_2 \in \mathbb{Z}$, this immediately implies $j_1 = j_2$. Hence $\delta_\mu(\lambda_k) = 0$ holds if and only if

$$\chi_k \in (\chi_{k_0-1}, \chi_{k_0}] \bmod(2\pi),$$

i. e., if and only if $k = k_0$.

For $k \in \{1,\dots,l_0\} \setminus \{k_0\}$ with $\delta_1(\lambda_k) = 1$ we thus have $\delta_\mu(\lambda_k) = 1$. In this case the first column and the $j+1$-th column of (4.2.10) are the corresponding columns of B. Hence (4.2.10) is a normed lower triangular matrix and thus invertible.

Finally, $\delta_1(\lambda_{k_0}) = 1$ and $\delta_\mu(\lambda_{k_0}) = 0$ imply that the first column of (4.2.10) is the first column of B and its $j+1$-th column is the $j+1$-th column of A. Since these columns coincide, the determinant of (4.2.10) is zero. This completes the proof in the case $\varphi_\mu = \chi_{k_0-1} \bmod(2\pi)$.

In the case $\varphi_\mu = (\chi_{k_0-1} + \pi) \operatorname{mod}(2\pi)$ we have to replace $\delta_\mu(\lambda_k) = 0$ by $\delta_\mu(\lambda_k) = 1$ in the above considerations. Hence $\delta_1(\lambda_k) = 1$ and $\delta_\mu(\lambda_k) = 1$ if and only if $k = k_0$. In this case the matrix (4.2.10) is a normed lower triangular matrix if $k \neq k_0$, and its determinant is zero if $k = k_0$. □

4.3. Estimates of the characteristic determinant

We introduce the fundamental matrix $\widehat{Y}(\cdot,\lambda)$ of (4.1.1) which is given by

$$\widehat{Y}(\cdot,\lambda) = \widetilde{Y}(\cdot,\lambda)\big(I_n - \Delta(\lambda) + E(b,\lambda)^{-1}\Delta(\lambda)\big) \quad (|\lambda| \geq \gamma), \tag{4.3.1}$$

where $\widetilde{Y}(\cdot,\lambda)$ is the fundamental matrix as considered in (4.1.14) and $\Delta(\lambda)$ is defined in (4.1.22). For $|\lambda| \geq \gamma$ we set

$$\widetilde{M}(\lambda) := \sum_{j=0}^{\infty} \widetilde{W}^{(j)}(\lambda)\widehat{Y}(a_j,\lambda) + \int_a^b \widetilde{W}(x,\lambda)\widehat{Y}(x,\lambda)\,\mathrm{d}x, \tag{4.3.2}$$

$$\widetilde{M}_0(\lambda) := W_0^{(0)}P^{[0]}(a)(I_n - \Delta(\lambda)) + W_0^{(1)}P^{[0]}(b)\Delta(\lambda), \tag{4.3.3}$$

$$\begin{aligned}\widetilde{M}_1(\lambda) := {} & \sum_{j=0}^{\infty} W_0^{(j)}P^{[0]}(a_j)E(a_j,\lambda)(I_n - \Delta(\lambda)) \\ & + \sum_{j=0}^{\infty} W_0^{(j)}P^{[0]}(a_j)E(a_j,\lambda)E(b,\lambda)^{-1}\Delta(\lambda),\end{aligned} \tag{4.3.4}$$

see (4.1.2) to (4.1.17) for the definition of the terms on the right-hand sides. We also need the matrix $\widetilde{M}_2$ defined in (4.1.24). Note that the matrices $\widetilde{M}_0(\lambda)$ and $\widetilde{M}_1(\lambda)$ are well-defined for all $\lambda \in \mathbb{C}\setminus\{0\}$.

PROPOSITION 4.3.1. *The boundary eigenvalue problem* (4.1.1), (4.1.2) *is Birkhoff regular if and only if*

$$M_0(\lambda) := \widetilde{M}_0(\lambda)\Delta_0 + \widetilde{M}_2(I_n - \Delta_0) \tag{4.3.5}$$

is invertible for all $\lambda \in \mathbb{C}\setminus\{0\}$.

Proof. For all $\lambda \in \mathbb{C}$,

$$P^{[0]}(a)(I_n - \Delta(\lambda))\Delta_0 + P^{[0]}(b)\Delta(\lambda)\Delta_0 + I_n - \Delta_0$$

is invertible and its inverse is

$$P^{[0]}(a)^{-1}(I_n - \Delta(\lambda))\Delta_0 + P^{[0]}(b)^{-1}\Delta(\lambda)\Delta_0 + I_n - \Delta_0.$$

For this we have to note that $P^{[0]}$ commutes with $\Delta(\lambda)$ and Δ_0 since $P^{[0]}$ is a block diagonal matrix. Now the assertion of the proposition follows from

$$\begin{aligned}M_0(\lambda) = {} & \big(W_0^{(0)}(I_n - \Delta(\lambda))\Delta_0 + W_0^{(1)}\Delta(\lambda)\Delta_0 + \widetilde{M}_2(I_n - \Delta_0)\big) \times \\ & \times \big(P^{[0]}(a)(I_n - \Delta(\lambda))\Delta_0 + P^{[0]}(b)\Delta(\lambda)\Delta_0 + I_n - \Delta_0\big)\end{aligned}$$

and the definition of Birkhoff regularity. □

From Proposition 4.1.5 and the definition of M_0 we immediately infer

REMARK 4.3.2. M_0 is constant on each Σ_k $(k = 1,\dots,l_0)$ and has at most l_0 different values, where l_0 and the sets Σ_k are defined in Remark 4.1.4.

PROPOSITION 4.3.3. *Let $c,d \in [a,b]$. We assert:*
i) $|E(c,\lambda)E(d,\lambda)^{-1}(I_n - \Delta(\lambda))| \le 1$ *if* $c \ge d$,
ii) $|E(c,\lambda)E(d,\lambda)^{-1}\Delta(\lambda))| \le 1$ *if* $c \le d$,
iii) $E(c,\lambda)E(d,\lambda)^{-1}(I_n - \Delta(\lambda))\Delta_0 = O\left(\max_{\nu=1}^{l}\left(1 + |\Re(\lambda e^{i\varphi_\nu})|\right)^{-1}\right)$ *if* $c > d$,
iv) $E(c,\lambda)E(d,\lambda)^{-1}\Delta(\lambda)\Delta_0 = O\left(\max_{\nu=1}^{l}\left(1 + |\Re(\lambda e^{i\varphi_\nu})|\right)^{-1}\right)$ *if* $c < d$.

Proof. Since all matrices under consideration are diagonal matrices, it is sufficient to prove the assertions for the diagonal elements. For $\nu = 0,\dots,l$ and $\lambda \in \mathbb{C}\setminus\{0\}$ it follows from

$$\left|\exp\{\lambda R_\nu(c)\}\exp\{-\lambda R_\nu(d)\}\right| = \exp\left\{\Re(\lambda e^{i\varphi_\nu})\left(|R_\nu(c)| - |R_\nu(d)|\right)\right\}$$

that

$$\begin{aligned}&\left|(1-\delta_\nu(\lambda))\exp\{\lambda R_\nu(c)\}\exp\{-\lambda R_\nu(d)\}\right| \\ &\qquad \le \exp\left\{|\Re(\lambda e^{i\varphi_\nu})|\left(|R_\nu(d)| - |R_\nu(c)|\right)\right\}\end{aligned}$$

and

$$\begin{aligned}&\left|\delta_\nu(\lambda)\exp\{\lambda R_\nu(c)\}\exp\{-\lambda R_\nu(d)\}\right| \\ &\qquad \le \exp\left\{|\Re(\lambda e^{i\varphi_\nu})|\left(|R_\nu(c)| - |R_\nu(d)|\right)\right\}.\end{aligned}$$

Now the assertions follow from the fact that $|R_\nu(\xi_1)| \ge |R_\nu(\xi_2)|$ for $\xi_1 \ge \xi_2$ and $\nu = 0,\dots,l$, that $|R_\nu(\xi_1)| > |R_\nu(\xi_2)|$ for $\xi_1 > \xi_2$ and $\nu = 1,\dots,l$ and that the set $\{(1+t)\exp(t\alpha) : t \ge 0\}$ is bounded if $\alpha < 0$. □

COROLLARY 4.3.4. i) *The matrix function $\widehat{Y}(\cdot,\lambda)$ is uniformly bounded in the space $M_n(C[a,b])$ for $|\lambda| \ge \gamma$.*
ii) *The matrix function $\widetilde{M}(\lambda)$ is bounded for $|\lambda| \ge \gamma$.*
iii) *The matrix function $\widetilde{M}_1$ is bounded on $\mathbb{C}\setminus\{0\}$.*

Proof. Using Proposition 4.3.3 i), ii) we obtain that i) follows from the estimate (4.1.18) of $B_0(\cdot,\lambda)$, that ii) follows from (4.1.10), (4.1.13), and part i), and that iii) follows from (4.1.11). □

PROPOSITION 4.3.5. *Let $f_1 \in (L_p(a,b))^n$. We assert that*

$$\int_\xi^b E(t,\lambda)E(\xi,\lambda)^{-1}\big(I_n - \Delta(\lambda)\big)\Delta_0 f_1(t)\,\mathrm{d}t,$$

$$\int_a^\xi E(\xi,\lambda)E(t,\lambda)^{-1}\big(I_n - \Delta(\lambda)\big)\Delta_0 f_1(t)\,\mathrm{d}t,$$

$$\int_a^\xi E(t,\lambda)E(\xi,\lambda)^{-1}\Delta(\lambda)\Delta_0 f_1(t)\,\mathrm{d}t$$

and

$$\int_\xi^b E(\xi,\lambda)E(t,\lambda)^{-1}\Delta(\lambda)\Delta_0 f_1(t)\,\mathrm{d}t$$

have the asymptotic behaviour

$$\{o(1)\}_\infty(\xi),$$

$$\left\{O\left(\max_{\nu=1}^{l}\left(1+|\Re(\lambda e^{i\varphi_\nu})|\right)^{-1+1/p}\right)\right\}_\infty(\xi)|f_1|_p,$$

$$\left\{O\left(\max_{\nu=1}^{l}\left(1+|\Re(\lambda e^{i\varphi_\nu})|\right)^{-1}\right)\right\}_p(\xi)|f_1|_p,$$

as $\lambda \to \infty$, where ξ varies in $[a,b]$ and the two latter estimates hold uniformly for all $f_1 \in (L_p(a,b))^n$.

Proof. Let $\nu \in \{1,\ldots,l\}$, g be the ν-th component of f_1, and F be defined as in Lemma 2.7.2 with $r(x) := |r_\nu(x)|$ $(x \in (a,b))$. Then Lemma 2.7.2 yields that

$$\begin{aligned}
&(1-\delta_\nu(\lambda))\int_\xi^b \exp\{\lambda\,(R_\nu(t)-R_\nu(\xi))\}g(t)\,\mathrm{d}t\\
&\qquad= -(1-\delta_\nu(\lambda))\int_b^\xi \exp\{-\lambda e^{i\varphi_\nu}\,(|R_\nu(\xi)|-|R_\nu(t)|)\}g(t)\,\mathrm{d}t\\
&\qquad= -(1-\delta_\nu(\lambda))F(g,\xi,b,-\lambda e^{i\varphi_\nu})
\end{aligned}$$

has the asserted asymptotic behaviour. In the same way we see that

$$\begin{aligned}
&\delta_\nu(\lambda)\int_a^\xi \exp\{\lambda\,(R_\nu(t)-R_\nu(\xi))\}g(t)\,\mathrm{d}t\\
&\qquad= \delta_\nu(\lambda)\int_a^\xi \exp\{-\lambda e^{i\varphi_\nu}\,(|R_\nu(\xi)|-|R_\nu(t)|)\}g(t)\,\mathrm{d}t\\
&\qquad= \delta_\nu(\lambda)F(g,\xi,a,-\lambda e^{i\varphi_\nu})
\end{aligned}$$

has the asserted asymptotic behaviour.

Since the nonzero components of the first and the third vector function in the assertion are of the form as considered above, the proposition holds for them. Since $\Delta(\lambda) = I_n - \Delta(-\lambda)$ and $E(\tau,\lambda) = E(\tau,-\lambda)^{-1}$, the two other vector functions in the assertion are of the same form. □

Obviously, the assertions of Proposition 4.3.5 also hold if we multiply the transposed vector function f_1^{T} from the left instead of f_1 from the right or if we take a matrix function instead of a vector function.

PROPOSITION 4.3.6. *Let* $\widetilde{M}$, $\widetilde{M}_0$, $\widetilde{M}_1$, $\widetilde{M}_2$, Δ_0 *be as defined in* (4.3.2)–(4.3.4), (4.1.24), (4.1.22). *We assert:*

i) $(\widetilde{M}(\lambda)-\widetilde{M}_0(\lambda))\Delta_0 = o(1)$ *as* $\min_{\nu=1}^{l}|\Re(\lambda e^{i\varphi_\nu})| \to \infty$.

ii) $(\widetilde{M}(\lambda)-\widetilde{M}_1(\lambda))\Delta_0 = o(1)$ *as* $\lambda \to \infty$.

iii) $(\widetilde{M}(\lambda)-\widetilde{M}_2)(I_n-\Delta_0) = o(1)$ *as* $\lambda \to \infty$.

iv) $(\widetilde{M}(\lambda)-\widetilde{M}_2)(I_n-\Delta_0) = O\left(\max_{\substack{\nu,\mu=0\\ \nu\neq\mu}}^{l}\left(1+|\Re(\lambda e^{i\varphi_{\nu\mu}})|\right)^{-1+1/p}\right)$.

v) *Let* $p>1$. *Suppose that* $W_0 \in M_n(L_q(a,b))$ *for some* $q>1$ *and that* a *and* b *are no accumulation points of the set* $\{a_j : j\in\mathbb{N},\ W_0^{(j)}\neq 0\}$. *Then*

$$(\widetilde{M}(\lambda)-\widetilde{M}_0(\lambda))\Delta_0 = O\left(\max_{\substack{\nu,\mu=0\\ \nu\neq\mu}}^{l}\left(1+|\Re(\lambda e^{i\varphi_{\nu\mu}})|\right)^{-\alpha}\right),$$

where $\alpha = \min\{1-1/p, 1-1/q\} > 0$.

Proof. Using the representation

$$\widetilde{Y}(\cdot,\lambda) = \left(P^{[0]}+B_0(\cdot,\lambda)\right)E(\cdot,\lambda)$$

of the fundamental matrix given by (4.1.14), we obtain

$$\begin{aligned}&\widetilde{W}^{(j)}(\lambda)\widetilde{Y}(a_j,\lambda)-W_0^{(j)}P^{[0]}(a_j)E(a_j,\lambda)\\ &= \left[\widetilde{W}^{(j)}(\lambda)B_0(a_j,\lambda)+(\widetilde{W}^{(j)}(\lambda)-W_0^{(j)})P^{[0]}(a_j)\right]E(a_j,\lambda).\end{aligned}$$

The estimates of B_0, the assumptions (4.1.11) and (4.1.12) on $\widetilde{W}_j$ and W_0^j, and Proposition 4.3.3 i), ii) for $c=a_j$ and $d=a$ or $d=b$, respectively, imply that

$$\sum_{j=0}^{\infty}\left[\widetilde{W}^{(j)}(\lambda)\widetilde{Y}(a_j,\lambda)-W_0^{(j)}P^{[0]}(a_j)E(a_j,\lambda)\right](I_n-\Delta(\lambda)) \tag{4.3.6}$$

and

$$\sum_{j=0}^{\infty}\left[\widetilde{W}^{(j)}(\lambda)\widetilde{Y}(a_j,\lambda)-W_0^{(j)}P^{[0]}(a_j)E(a_j,\lambda)\right]E(b,\lambda)^{-1}\Delta(\lambda) \tag{4.3.7}$$

are of the form $o(1)$ and $O\left(\max|_{\nu\neq\mu}^{\nu,\mu=0^l}\left(1+|\Re(\lambda e^{i\varphi_{\nu\mu}})|\right)^{-1+1/p}\right)$. In the same way,

$$\begin{aligned}&\widetilde{W}(x,\lambda)\widetilde{Y}(x,\lambda)-W_0(x)P^{[0]}(x)E(x,\lambda)\\ &= \left[\widetilde{W}(x,\lambda)B_0(x,\lambda)+\left(\widetilde{W}(x,\lambda)-W_0(x)\right)P^{[0]}(x)\right]E(x,\lambda)\end{aligned}$$

shows that

$$(4.3.8)\qquad \int_a^b \left[\widetilde{W}(x,\lambda)\widetilde{Y}(x,\lambda) - W_0(x)P^{[0]}(x)E(x,\lambda)\right] \mathrm{d}x\,(I_n - \Delta(\lambda))$$

and

$$(4.3.9)\qquad \int_a^b \left[\widetilde{W}(x,\lambda)\widetilde{Y}(x,\lambda) - W_0(x)P^{[0]}(x)E(x,\lambda)\right] \mathrm{d}x\,E(b,\lambda)^{-1}\Delta(\lambda)$$

are of the form $o(1)$ and $O\left(\max\limits_{\substack{\nu,\mu=0\\ \nu\neq\mu}}^{l}\left(1+|\Re(\lambda e^{i\varphi_{\nu\mu}})|\right)^{-1+1/p}\right)$. Since the equation $E(x,\lambda)(I_n - \Delta_0) = I_n - \Delta_0$ holds for all $x \in [a,b]$ and $\lambda \in \mathbb{C}$, the matrix function $\left(\widetilde{M}(\lambda) - \widetilde{M}_2\right)(I_n - \Delta_0)$ is the sum of the four terms (4.3.6)–(4.3.9) multiplied by $(I_n - \Delta_0)$ from the right. Thus the estimates of (4.3.6)–(4.3.9) prove iii) and iv).

Since $W_0 P^{[0]} \in M_n(L_1(a,b))$, the estimates in Proposition 4.3.5 yield

$$(4.3.10)\qquad \int_a^b W_0(x)P^{[0]}(x)E(x,\lambda)\,\mathrm{d}x\,(I_n - \Delta(\lambda))\Delta_0 = o(1) \text{ as } \lambda \to \infty$$

and

$$(4.3.11)\qquad \int_a^b W_0(x)P^{[0]}(x)E(x,\lambda)\,\mathrm{d}x\,E(b,\lambda)^{-1}\Delta(\lambda)\Delta_0 = o(1) \text{ as } \lambda \to \infty.$$

Since $\left(\widetilde{M}(\lambda) - \widetilde{M}_1(\lambda)\right)\Delta_0$ is the sum of the four terms (4.3.6)–(4.3.9) multiplied by Δ_0 from the right and the two terms (4.3.10) and (4.3.11), ii) follows from the estimates of (4.3.6)–(4.3.11).

Let $\varepsilon > 0$. Then there is a $j_0 \in \mathbb{N}$ such that

$$\left|\sum_{j=j_0+1}^{\infty} W_0^{(j)}P^{[0]}(a_j)E(\mathring{a}_j,\lambda)(I_n - \Delta(\lambda))\Delta_0\right| \leq \sum_{j=j_0+1}^{\infty} |W_0^{(j)}P^{[0]}(a_j)| \leq \frac{\varepsilon}{2}\,.$$

For sufficiently large α and all $\lambda \in \mathbb{C}$ with $\min\limits_{\nu=1}^{l}|\Re(\lambda e^{i\varphi_\nu})| \geq \alpha$ we obtain from Proposition 4.3.3 iii) with $c = a_j$ and $d = a$ that

$$\sum_{j=1}^{j_0} |W_0^{(j)}P^{[0]}(a_j)|\,|E(a_j,\lambda)(I_n - \Delta(\lambda))\Delta_0| \leq \frac{\varepsilon}{2}\,.$$

Hence

$$\sum_{j=1}^{\infty} W_0^{(j)}P^{[0]}(a_j)E(a_j,\lambda)(I_n - \Delta(\lambda))\Delta_0 = o(1) \text{ as } \min_{\nu=1}^{l}|\Re(\lambda e^{i\varphi_\nu})| \to \infty.$$

In the same way we obtain with the aid of Proposition 4.3.3 iv) with $c = a_j$ and $d = b$ that

$$\sum_{\substack{j=0\\ j\neq 1}}^{\infty} W_0^{(j)}P^{[0]}(a_j)E(a_j,\lambda)E(b,\lambda)^{-1}\Delta(\lambda)\Delta_0 = o(1) \text{ as } \min_{\nu=1}^{l}|\Re(\lambda e^{i\varphi_\nu})| \to \infty.$$

These two estimates and ii) prove i).

Suppose that the assumptions of v) hold. We apply Proposition 4.3.5 to the matrix functions on the left-hand sides of (4.3.10) and (4.3.11) and obtain

(4.3.12)
$$\int_a^b W_0(x)P^{[0]}(x)E(x,\lambda)\,\mathrm{d}x\,(I_n-\Delta(\lambda))\Delta_0 = O\left(\max_{\nu=1}^{l}\left(1+|\Re(\lambda e^{i\varphi_\nu})|\right)^{-1+1/q}\right)$$

and

(4.3.13)
$$\int_a^b W_0(x)P^{[0]}(x)E(t,\lambda)\,\mathrm{d}x\,E(b,\lambda)^{-1}\Delta(\lambda)\Delta_0 = O\left(\max_{\nu=1}^{l}\left(1+|\Re(\lambda e^{i\varphi_\nu})|\right)^{-1+1/q}\right).$$

By assumption,

$$\tilde{a} := \inf\{a_j : j\in\mathbb{N}\setminus\{0\},\ W_0^{(j)}\neq 0\} > a$$

and

$$\tilde{b} := \sup\{a_j : j\in\mathbb{N}\setminus\{1\},\ W_0^{(j)}\neq 0\} < b.$$

Hence

$$\begin{aligned}&\left|\sum_{j=1}^{\infty} W_0^{(j)}P^{[0]}(a_j)E(a_j,\lambda)(I_n-\Delta(\lambda))\Delta_0\right|\\ &\quad\le \left(\sum_{j=1}^{\infty}|W_0^{(j)}|\right)\left(\max_{x\in[a,b]}|P^{[0]}(x)|\right)|E(\tilde{a},\lambda)(I_n-\Delta(\lambda))\Delta_0|\end{aligned}$$

and

$$\begin{aligned}&\left|\sum_{\substack{j=0\\ j\neq 1}}^{\infty} W_0^{(j)}P^{[0]}(a_j)E(a_j,\lambda)E(b,\lambda)^{-1}\Delta(\lambda)\Delta_0\right|\\ &\quad\le \left(\sum_{\substack{j=0\\ j\neq 1}}^{\infty}|W_0^{(j)}|\right)\left(\max_{x\in[a,b]}|P^{[0]}(x)|\right)|E(\tilde{b},\lambda)E(b,\lambda)^{-1}\Delta(\lambda)\Delta_0|\end{aligned}$$

are of the form $O\left(\max_{\nu=1}^{l}\left(1+|\Re(\lambda e^{i\varphi_\nu})|\right)^{-1}\right)$ by Proposition 4.3.3. These estimates, the estimates (4.3.12) and (4.3.13) and the estimates of (4.3.6)–(4.3.9) prove part v). □

PROPOSITION 4.3.7. *Suppose the boundary eigenvalue problem* (4.1.1), (4.1.2) *is Birkhoff regular. Let* $\widetilde{M}$ *be given by* (4.3.2). *Then there are numbers* $\alpha > 0$ *and* $\delta > 0$ *such that* $|\det\widetilde{M}(\lambda)| \ge \delta$ *if* $\min_{\nu=1}^{l}|\Re(\lambda e^{i\varphi_\nu})| \ge \alpha$.

Proof. The matrix function M_0 is invertible and has only a finite number of different values for $\lambda \in \mathbb{C} \setminus \{0\}$ by Proposition 4.3.1 and Remark 4.3.2. Hence there is a positive number δ such that

$$|\det M_0(\lambda)| \geq 2\delta \quad \text{for all } \lambda \in \mathbb{C} \setminus \{0\}. \tag{4.3.14}$$

Proposition 4.3.6 i), iii) yields

$$\begin{aligned} \widetilde{M}(\lambda) &= \widetilde{M}_0(\lambda)\Delta_0 + \widetilde{M}_2(I_n - \Delta_0) + o(1) \\ &= M_0(\lambda) + o(1) \quad \text{as} \quad \min_{\nu=1}^{l} |\Re(\lambda e^{i\varphi_\nu})| \to \infty. \end{aligned}$$

The components of $M_0(\lambda)$ are bounded functions with respect to λ since M_0 has only a finite number of different values by Remark 4.3.2. Hence

$$\det \widetilde{M}(\lambda) \to \det M_0(\lambda) \quad \text{as} \quad \min_{\nu=1}^{l} |\Re(\lambda e^{i\varphi_\nu})| \to \infty,$$

i. e., there is a positive number α such that

$$|\det \widetilde{M}(\lambda) - \det M_0(\lambda)| \leq \delta \quad \text{if} \quad \min_{\nu=1}^{l} |\Re(\lambda e^{i\varphi_\nu})| \geq \alpha. \tag{4.3.15}$$

The assertion of the proposition follows from (4.3.14) and (4.3.15). □

PROPOSITION 4.3.8. *Let*

$$M_1(\lambda) := \widetilde{M}_1(\lambda)\Delta_0 + \widetilde{M}_2(I_n - \Delta_0). \tag{4.3.16}$$

Then

$$\det \widetilde{M}(\lambda) = \det M_1(\lambda) + o(1) \text{ as } \lambda \to \infty.$$

Proof. This is obvious from Proposition 4.3.6 ii), iii) since the components of $\widetilde{M}_1(\lambda)$ are bounded by Corollary 4.3.4. □

THEOREM 4.3.9. *Suppose that the boundary eigenvalue problem* (4.1.1), (4.1.2) *is Birkhoff regular. Let $\widetilde{M}$ be the characteristic matrix function given by* (4.3.2). *Then there are circles $\Gamma_\nu = \{\lambda \in \mathbb{C} : |\lambda| = \rho_\nu\}$ ($\nu \in \mathbb{N}$) with $\rho_\nu \nearrow \infty$ as $\nu \to \infty$ and a number $\delta > 0$ such that $|\det \widetilde{M}(\lambda)| \geq \delta$ for all $\lambda \in \bigcup_{\nu=0}^{\infty} \Gamma_\nu$.*

Proof. For $\lambda \in \mathbb{C}$ we set

$$\widehat{M}(\lambda) := \sum_{j=0}^{\infty} W_0^{(j)} P^{[0]}(a_j) E(a_j, \lambda) + \int_a^b W_0(x) P^{[0]}(x)(I_n - \Delta_0)\, dx.$$

We shall prove that $\det \widehat{M}$ is an exponential sum in the sense of Section A.2. We set

$$\hat{r}_q := r_\mu,\ \widehat{R}_q := R_\mu,\ \hat{\delta}_q := \delta_\mu \text{ and } \hat{\varphi}_q := \varphi_\mu \quad \text{for} \quad \sum_{\kappa=0}^{\mu-1} n_\kappa < q \leq \sum_{\kappa=0}^{\mu} n_\kappa,$$

$$
\begin{aligned}
(a_{mq}^{(0)})_{m,q=1}^{n} &:= W_0^{(0)} P^{[0]}(a_0)\Delta_0 \\
&\quad + \Big(\sum_{j=0}^{\infty} W_0^{(j)} P^{[0]}(a_j) + \int_a^b W_0(x) P^{[0]}(x)\,\mathrm{d}x\Big)(I_n - \Delta_0), \\
(a_{mq}^{(j)})_{m,q=1}^{n} &:= W_0^{(j)} P^{[0]}(a_j)\Delta_0 \quad (j \in \mathbb{N} \setminus \{0\}),
\end{aligned}
$$

and

$$
\alpha_{j_{n_0+1},\dots,j_n} := \sum_{\sigma\in S_n} \operatorname{sgn}(\sigma) \prod_{q=1}^{n_0} a_{\sigma(q),q}^{(0)} \prod_{q=n_0+1}^{n} a_{\sigma(q),q}^{(j_q)} \qquad (j_{n_0+1},\dots,j_n \in \mathbb{N}),
$$

where S_n is the set of permutations of the numbers $1,\dots,n$. We set

$$
\gamma_j := \max_{m,q=1}^{n} |a_{mq}^{(j)}| \qquad (j \in \mathbb{N}).
$$

From $\sum\limits_{j=0}^{\infty} |W_0^{(j)}| < \infty$ and $P^{[0]} \in M_n(W_p^1(a,b)) \subset M_n(C[a,b])$ we infer $\sum\limits_{j=0}^{\infty} \gamma_j < \infty$. Hence

$$
\begin{aligned}
\sum_{j_{n_0+1},\dots,j_n=0}^{\infty} |\alpha_{j_{n_0+1},\dots,j_n}| &\le n! \sum_{j_{n_0+1},\dots,j_n=0}^{\infty} \gamma_0^{n_0} \prod_{q=n_0+1}^{n} \gamma_{j_q} \\
&= n!\,\gamma_0^{n_0} \Big(\sum_{j=0}^{\infty} \gamma_j\Big)^{n-n_0} < \infty.
\end{aligned}
$$

Thus

$$
\begin{aligned}
\det \widehat{M}(\lambda) &= \sum_{\sigma\in S_n} \operatorname{sgn}(\sigma) \prod_{q=1}^{n} \Big(\sum_{j=0}^{\infty} a_{\sigma(q),q}^{(j)} \exp\{\lambda \widehat{R}_q(a_j)\}\Big) \\
&= \sum_{j_{n_0+1},\dots,j_n=0}^{\infty} \alpha_{j_{n_0+1},\dots,j_n} \prod_{q=n_0+1}^{n} \exp\{\lambda \widehat{R}_q(a_{j_q})\},
\end{aligned}
$$

where the sum is absolutely convergent. Hence $\det \widehat{M}(\lambda)$ is an exponential sum in the sense of Section A.2.

Let l_0, χ_k and Σ_k be as defined in Remark 4.1.4, $k = 1,\dots,l_0$. Let

$$
\theta_k := \{q \in \{n_0+1,\dots,n\} : \hat{\varphi}_q \in [\chi_k, \chi_k + \pi) \operatorname{mod}(2\pi)\}
$$

and set $\lambda_k := e^{-i(\frac{\pi}{2}+\chi_k)}$. Then $\lambda_k \in \Sigma_k$, $\hat{\delta}_q(\lambda_k) = 1$ if $q \in \theta_k$ and $\hat{\delta}_q(\lambda_k) = 0$ if $q \in \{n_0+1,\dots,n\} \setminus \theta_k$. Let

$$
\mathscr{E} := \Big\{ \sum_{k=n_0+1}^{n} \widehat{R}_k(a_{j_k}) : j_k \in \mathbb{N};\ k = n_0+1,\dots,n \Big\}. \tag{4.3.17}
$$

Since $\widehat{R}_k(a_{jk}) \in \overline{0,\widehat{R}_k(b)}$, the convex hull of $\mathscr{E}$ is a convex polygon by Theorem A.1.3. Set

$$b_k := \sum_{j\in\theta_k} \widehat{R}_j(b) \quad (k=1,\ldots,l_0).$$

Since the numbers $\chi_1,\ldots,\chi_{l_0}$ are the numbers $\varphi_1,\ldots,\varphi_{2m}$ of Section A.1, the set $\{b_k : k\in\{1,\ldots,l_0\}\}$ is the set $\widetilde{\mathscr{E}}$ of the vertices of $\mathscr{E}$. The representation of these points is unique by Theorem A.1.3 iii). Hence we obtain that the coefficient of $\exp\{\lambda b_k\}$ in $\det\widehat{M}$ is $\alpha_{\Delta(\lambda_k)} := \alpha_{j_{n_0+1},\ldots,j_n}$, where the numbers $j_{n_0+1},\ldots,j_n$ are given by $\Delta(\lambda_k) = \operatorname{diag}(*,\ldots,*,j_{n_0+1},\ldots,j_n)$ if we observe that $\widehat{R}_j(a_0)=0$ and $\widehat{R}_j(a_1)=\widehat{R}_j(b)$. Since (4.1.1), (4.1.2) is Birkhoff regular,

$$\alpha_{\Delta(\lambda_k)} = \det\big(W_0^{(0)}(I_n-\Delta(\lambda_k))\Delta_0 + W_0^{(1)}P^{[0]}(b)\Delta(\lambda_k)\Delta_0 + \widetilde{M}_2(I_n-\Delta_0)\big) \neq 0.$$

By Remark A.2.13 the exponential sum $\det\widehat{M}(\lambda)$ fulfils the assumptions of Theorem A.2.15, and we accordingly choose ε and $(\rho_n)_0^\infty$ from the statement of that theorem. Now let $\lambda\in\Sigma_k\cap\bigcup_{\nu=0}^{\infty}\Gamma_\nu$. From Proposition 4.1.5 we infer $\Delta(\lambda)=\Delta(\lambda_k)$. Hence

$$\det\big(I_n-\Delta(\lambda)+E(b,\lambda)^{-1}\Delta(\lambda)\big) = \exp\{-b_k\lambda\}.$$

Since

$$\begin{aligned} M_1(\lambda) &= \widetilde{M}_1(\lambda)\Delta_0 + \widetilde{M}_2(I_n-\Delta_0) \\ &= \widehat{M}(\lambda)\big(I_n-\Delta(\lambda)+E(b,\lambda)^{-1}\Delta(\lambda)\big), \end{aligned}$$

we obtain

$$\det M_1(\lambda) = \det\widehat{M}(\lambda)\exp\{-b_k\lambda\}.$$

Therefore Theorem A.2.15 yields

$$|\det M_1(\lambda)| \geq \varepsilon.$$

From Proposition 4.3.8 we coclude that $|\det\widetilde{M}(\lambda)-\det M_1(\lambda)| \leq \frac{\varepsilon}{2}$ holds for $|\lambda|\geq\rho_0$ if we choose ρ_0 sufficiently large. Then the assertion of the theorem follows with $\delta=\frac{\varepsilon}{2}$. □

We now continue the discussion of Example 4.2.3. For simplicity we assume $n=2$, $a=0$, $b=1$. We take the first order system

$$y' - \lambda y - \begin{pmatrix} 0 & \alpha \\ 0 & 0 \end{pmatrix} y = 0, \tag{4.3.18}$$

where α is a complex constant and λ varies in $\mathbb{C}$. An asymptotic fundamental matrix function in the sense of Theorem 2.8.2 B of this system is given by

$$Y(x,\lambda) = \begin{pmatrix} e^{\lambda x} & \alpha x e^{\lambda x} \\ 0 & e^{\lambda x} \end{pmatrix}, \tag{4.3.19}$$

where $k = 0$ and $B_0 = 0$. Indeed, this follows from Remark 2.8.6 since

$$\begin{pmatrix} e^{\lambda x} & 0 \\ 0 & e^{\lambda x} \end{pmatrix}$$

is the fundamental matrix of $y'(x) - \lambda y(x) = 0$ which is the identity at 0, and

$$\begin{pmatrix} 1 & \alpha x \\ 0 & 1 \end{pmatrix}$$

is the fundamental matrix of

$$y'(x) - \begin{pmatrix} 0 & \alpha \\ 0 & 0 \end{pmatrix} y(x) = 0$$

which is the identity at 0. The characteristic matrix with respect to the boundary conditions

$$\begin{pmatrix} 1 & 1 \\ 0 & \lambda \end{pmatrix} y(0) + \begin{pmatrix} 1 & 0 \\ 0 & \lambda + \beta \end{pmatrix} y(1) = 0 \tag{4.3.20}$$

is

$$\begin{aligned} M(\lambda) &= \begin{pmatrix} 1 & 1 \\ 0 & \lambda \end{pmatrix} + \begin{pmatrix} 1 & 0 \\ 0 & \lambda + \beta \end{pmatrix} \begin{pmatrix} e^{\lambda} & \alpha e^{\lambda} \\ 0 & e^{\lambda} \end{pmatrix} \\ &= \begin{pmatrix} e^{\lambda} + 1 & \alpha e^{\lambda} + 1 \\ 0 & (\lambda + \beta) e^{\lambda} + \lambda \end{pmatrix}. \end{aligned} \tag{4.3.21}$$

A multiplication of (4.3.20) by $\begin{pmatrix} 1 & 0 \\ 0 & \lambda^{-1} \end{pmatrix}$ from the left shows that this problem is of the form considered in Example 4.2.3 and hence Birkhoff regular.

First let $\beta = 0$. Then the eigenvalues of M are 0 and $(2k+1)\pi i$ $(k \in \mathbb{Z})$. For $\alpha = 1$ the dimension of the eigenspaces corresponding to the nonzero eigenvalues is 2. For $\alpha \neq 1$ the dimension of the eigenspaces corresponding to the nonzero eigenvalues is 1, and there is a chain of an eigenvector and an associated vector by Proposition 1.8.5.

For $\beta \neq 0$ we have two sequences of eigenvalues: $(2k+1)\pi i$ $(k \in \mathbb{Z})$, and the zeros of $(1 + \frac{\beta}{\lambda})e^{\lambda} + 1$.

Now we are going to deduce an asymptotic representation of the zeros of $(1 + \frac{\beta}{\lambda})e^{\lambda} + 1$ for large λ. Using the results of Section A.2, we see that they have the form

$$(2k+1)\pi i + o(1) \text{ as } |k| \to \infty \text{ for } k \in \mathbb{Z}.$$

But we shall give a more accurate representation and estimate. For this we note that

$$\left(1 + \frac{\beta}{\lambda}\right) e^{\lambda} + 1 = 0 \tag{4.3.22}$$

holds if and only if

$$e^{-\lambda+\pi i} = 1 + \frac{\beta}{\lambda}.$$

Taking the logarithm we see that (4.3.22) holds if and only if there is a $k \in \mathbb{Z}$ such that

$$-\lambda + \pi i = \log\left(1 + \frac{\beta}{\lambda}\right) - 2k\pi i,$$

where we choose the argument of the logarithm in the interval $[-\pi, \pi)$. For $k \in \mathbb{Z}$ we set

$$f_k(\lambda) = (2k+1)\pi i - \log\left(1 + \frac{\beta}{\lambda}\right) \quad (\lambda \in \mathbb{C} \setminus \{0, -\beta\}).$$

Thus we have to determine the fixed points of f_k for large λ. Since $\log\left(1 + \frac{\beta}{\lambda}\right)$ tends to zero as $\lambda \to \infty$, a large fixed point corresponds to a large value of $|k|$. On the other hand, for each $\lambda \in \mathbb{C} \setminus \{0\}$ we have

$$|f_k(\lambda)| \geq |\Im f_k(\lambda)| \geq 2(|k| - 1)\pi.$$

This shows that the fixed points of f_k are of the form

$$(2k+1)\pi i + o(1) \text{ as } |k| \to \infty.$$

Hence it is sufficient to consider the case

$$|k| \geq 1 + |\beta|\pi^{-1} \quad \text{and} \quad \lambda \in \overline{B}_1((2k+1)\pi i).$$

where $\overline{B}_1((2k+1)\pi i)$ is the closed disc with centre $(2k+1)\pi i$ and radius 1. In the following considerations of this example we shall always take k and λ according to these conditions. We have

$$|\lambda| \geq (2|k| - 1)\pi - 1 \geq 2(|k| - 1)\pi + 2 \geq 2|\beta| + 2. \tag{4.3.23}$$

The mean value theorem gives

$$\begin{aligned}\left|\log\left(1 + \frac{\beta}{\lambda}\right)\right| &\leq \frac{|\beta|}{|\lambda|} \sup_{|z| \leq \frac{|\beta|}{|\lambda|}} \frac{1}{|1+z|} \leq \frac{2|\beta|}{|\lambda|} \\ &\leq \frac{|\beta|}{(|k|-1)\pi + 1} \leq \frac{|\beta|}{|\beta| + 1}.\end{aligned}$$

From

$$f_k'(\lambda) = \frac{\beta}{\lambda^2(1 + \frac{\beta}{\lambda})}$$

we infer

$$\begin{aligned}\delta_k := &\sup_{z \in \overline{B}_1((2k+1)\pi i)} |f_k'(z)| \\ &\leq \frac{|\beta|}{2((|k|-1)\pi + 1)^2} \leq \frac{|\beta|}{2(|\beta| + 1)^2} \leq \frac{1}{8}.\end{aligned}$$

This estimate and

$$|f_k(\lambda) - (2k+1)\pi i| = \left|\log\left(1+\frac{\beta}{\lambda}\right)\right| \leq 1$$

show that $f_k : \overline{K}_1((2k+1)\pi i) \to \overline{K}_1((2k+1)\pi i)$ is a contractive mapping. By BANACH'S fixed point theorem, f_k has exactly one fixed point μ_k in the disk $\overline{B}_1((2k+1)\pi i)$. The a priori estimate yields

$$\begin{aligned}|\mu_k - f_k((2k+1)\pi i)| &\leq \frac{\delta_k}{1-\delta_k}|f_k((2k+1)\pi i) - (2k+1)\pi i| \\ &\leq \frac{|\beta|}{((|k|-1)\pi+1)^2}\left|\log\left(1+\frac{\beta}{(2k+1)\pi i}\right)\right| \\ &\leq \frac{|\beta|^2}{((|k|-1)\pi+1)^3}.\end{aligned}$$

Using the Taylor series expansion of $\log(1+z)$ we obtain

$$\begin{aligned}f_k((2k+1)\pi i) &= (2k+1)\pi i - \log\left(1+\frac{\beta}{(2k+1)\pi i}\right) \\ &= (2k+1)\pi i + \sum_{j=1}^{2}\frac{1}{j}\left(\frac{\beta i}{(2k+1)\pi}\right)^j + \frac{\hat{\gamma}_k}{((2k+1)\pi)^3},\end{aligned}$$

where $|\hat{\gamma}_k| \leq \dfrac{8|\beta|^3}{3}$ since

$$\sup_{|z|\leq\frac{1}{2}} \frac{1}{3!}\left|\frac{d^3}{dz^3}\log(1+z)\right| \leq \frac{8}{3}.$$

Thus we have

$$\mu_k = (2k+1)\pi i + \frac{\beta i}{(2k+1)\pi} - \frac{1}{2}\frac{\beta^2}{((2k+1)\pi)^2} + \frac{\gamma_k}{((|k|-1)\pi+1)^3},$$

where

$$|\gamma_k| \leq \frac{|\beta|^3}{3} + |\beta|^2$$

since $|(2k+1)|\pi \geq 2\big((|k|-1)\pi+1\big)$.

If we consider $\mu_k^1 := (2k+1)\pi i$ as a first approximation, we may take

$$\mu_k^2 := (2k+1)\pi i + \frac{\beta i}{(2k+1)\pi} - \frac{1}{2}\frac{\beta^2}{((2k+1)\pi)^2}$$

as a second approximation. Repeating the above method with μ_k^2 instead of μ_k^1, we get an approximation μ_k^3 such that $\mu_k^3 - (2k+1)\pi i$ is a polynomial in $(2k+1)^{-1}$ and $\mu_k^3 - \mu_k^2 = O(|k|^{-5})$. Proceeding in this way we see that $\mu_k - (2k+1)\pi i$ can be written as an asymptotic polynomial in $2k+1$ of arbitrary order. □

REMARK 4.3.10. The above example shows that there are Birkhoff regular eigenvalue problems with i) infinitely many eigenvalues for which the dimension of the eigenspace is larger that 1 or ii) infinitely many eigenvalues which are not semisimple or iii) the infimum of the distance of different eigenvalues is zero, i. e., there are sequences of eigenvalues $(\lambda_k)_1^\infty$ and $(\mu_k)_1^\infty$ such that $\lambda_k \to \infty$, $\mu_k \to \infty$, and $0 \neq \lambda_k - \mu_k \to 0$ as $k \to \infty$. Of course, it is possible to construct examples where i, ii), and iii) occur simultaneously.

4.4. Estimates of the Green's matrix

Besides Birkhoff regularity we shall also consider Stone regularity. In this section, no additional investigations are necessary if we prove the results for Stone regularity.

DEFINITION 4.4.1. Let $s \in \mathbb{N}$. The boundary eigenvalue problem (4.1.1), (4.1.2) is called *s-regular* if there are circles $\Gamma_\nu = \{\lambda \in \mathbb{C} : |\lambda| = \rho_\nu\}$ $(\nu \in \mathbb{N})$, where $(\rho_\nu)_{\nu\in\mathbb{N}}$ is a strictly increasing sequence of positive numbers with $\rho_\nu \nearrow \infty$ as $\nu \to \infty$, and a number $\delta > 0$ such that $|\lambda^s \det \widetilde{M}(\lambda)| \geq \delta$ for all $\lambda \in \bigcup_{\nu=0}^{\infty} \Gamma_\nu$, where $\widetilde{M}$ is the characteristic matrix function given by (4.3.2).
The boundary eigenvalue problem (4.1.1), (4.1.2) is called *Stone regular* if there is an integer $s \in \mathbb{N}$ such that the boundary eigenvalue problem (4.1.1), (4.1.2) is s-regular.

From Theorem 4.3.9 we know that Birkhoff regular problems are 0-regular. Throughout this section we suppose that the boundary eigenvalue problem (4.1.1), (4.1.2) is Stone regular.

Together with the boundary eigenvalue problem (4.1.1), (4.1.2) we consider the operator function

$$\text{(4.4.1)} \qquad \widetilde{T}(\lambda) := \begin{pmatrix} \widetilde{T}^D(\lambda) \\ \widetilde{T}^R(\lambda) \end{pmatrix} : (W_p^1(a,b))^n \to (L_p(a,b))^n \times \mathbb{C}^n \quad (|\lambda| \geq \gamma),$$

where γ is a fixed positive number and

$$\begin{aligned} \widetilde{T}^D(\lambda)y &= y' - \left(\lambda A_1 + A_0 + \lambda^{-1} A^0(\cdot,\lambda)\right)y, \\ \widetilde{T}^R(\lambda)y &= \sum_{j=0}^{\infty} \widetilde{W}^{(j)}(\lambda)y(a_j) + \int_a^b \widetilde{W}(x,\lambda)y(x)\,\mathrm{d}x, \end{aligned}$$

and y varies in $(W_p^1((a,b))^n$.

We have seen that we can take any fundamental matrix in the definition of the GREEN'S matrix function $G(\cdot,\cdot,\lambda)$ and in the definition of $\widehat{G}(\cdot,\lambda)$. Here we take the fundamental matrix $\widehat{Y}(\cdot,\lambda)$ as defined in (4.3.1). Theorem 3.2.2 and formula (3.2.12) yield

$$\widetilde{T}^{-1}(\lambda)(f_1,f_2)(x) = \widehat{Y}(x,\lambda)\int_a^x \widehat{Y}(\xi,\lambda)^{-1}f_1(\xi)\,\mathrm{d}\xi$$
$$+\widehat{Y}(x,\lambda)(\widetilde{T}^R(\lambda)(\widehat{Y}(\cdot,\lambda))^{-1}\left\{f_2-\int_a^b\int_{t=\xi}^b \mathrm{d}_t\widetilde{F}(t,\lambda)\widehat{Y}(t,\lambda)\widehat{Y}(\xi,\lambda)^{-1}f_1(\xi)\,\mathrm{d}\xi\right\}$$

for $\lambda \in \bigcup\limits_{\nu=0}^{\infty}\Gamma_\nu$, $f_1 \in (L_p(a,b))^n$, $f_2 \in \mathbb{C}^n$ and $x \in (a,b)$, where

$$\widetilde{F}(x,\lambda) := \sum_{\substack{j=0\\ a_j<x}}^{\infty} \widetilde{W}^{(j)}(\lambda) + \int_a^x \widetilde{W}(t,\lambda)\,\mathrm{d}t \tag{4.4.2}$$

for $a \le x < b$ and

$$\widetilde{F}(b,\lambda) := \sum_{j=0}^{\infty} \widetilde{W}^{(j)}(\lambda) + \int_a^b \widetilde{W}(t,\lambda)\,\mathrm{d}t. \tag{4.4.3}$$

We subtract and add the term (see Proposition 3.2.1)

$$\widehat{Y}(x,\lambda)\Delta(\lambda)\int_a^b \widehat{Y}(\xi,\lambda)^{-1}f_1(\xi)\,\mathrm{d}\xi$$
$$=\widehat{Y}(x,\lambda)\big(\widetilde{T}^R(\lambda)(\widehat{Y}(\cdot,\lambda))\big)^{-1}\int_a^b\int_a^b \mathrm{d}_t\widetilde{F}(t,\lambda)\widehat{Y}(t,\lambda)\Delta(\lambda)\widehat{Y}(\xi,\lambda)^{-1}f_1(\xi)\,\mathrm{d}\xi$$

and obtain with $f=(f_1,f_2)$ that

$$(\widetilde{T}^{-1}(\lambda)f)(x) = I_1(x,f_1,\lambda) + \widehat{Y}(x,\lambda)\widetilde{M}^{-1}(\lambda)I_3(f,\lambda), \tag{4.4.4}$$

where

$$I_1(x,f_1,\lambda) := \widehat{Y}(x,\lambda)\big(I_n-\Delta(\lambda)\big)\int_a^x \widehat{Y}(\xi,\lambda)^{-1}f_1(\xi)\,\mathrm{d}\xi \tag{4.4.5}$$
$$-\widehat{Y}(x,\lambda)\Delta(\lambda)\int_x^b \widehat{Y}(\xi,\lambda)^{-1}f_1(\xi)\,\mathrm{d}\xi\,,$$

$$I_3(f,\lambda) := f_2 - \int_a^b\int_{t=\xi}^b \mathrm{d}_t\widetilde{F}(t,\lambda)\widehat{Y}(t,\lambda)\big(I_n-\Delta(\lambda)\big)\widehat{Y}(\xi,\lambda)^{-1}f_1(\xi)\,\mathrm{d}\xi \tag{4.4.6}$$
$$+\int_a^b\int_{t=a}^{\xi} \mathrm{d}_t\widetilde{F}(t,\lambda)\widehat{Y}(t,\lambda)\Delta(\lambda)\widehat{Y}(\xi,\lambda)^{-1}f_1(\xi)\,\mathrm{d}\xi.$$

Note that we can take $\widetilde{Y}$ instead of $\widehat{Y}$ in (4.4.5) and (4.4.6).

The main task in this section will be to estimate I_1, $\widehat{Y}$, and I_3. For this we need some propositions. Define

$$I_2^0(x,\lambda) := P^{[0]}(x)\big((I_n-\Delta(\lambda))E(x,\lambda)+\Delta(\lambda)E(x,\lambda)E(b,\lambda)^{-1}\big). \tag{4.4.7}$$

PROPOSITION 4.4.2. i) *The matrix function* $\widehat{Y}(\cdot,\lambda) - I_2^0(\cdot,\lambda)$ *satisfies the estimates* $\{O(\tau_p(\lambda))\}_\infty$ *and* $\{o(1)\}_\infty$.
ii) *If* $p > 1$ *and, in case* $p \le \frac{3}{2}$, *the conditions* (4.1.19) *are satisfied, then there is a number* $\varepsilon \in (0, 1 - \frac{1}{p})$ *such that*

$$\widehat{Y}(\cdot,\lambda) - I_2^0(\cdot,\lambda) = \left\{ 0\left(\max_{\substack{\nu,\mu=0\\ \nu\neq\mu}}^{l} \left(1 + |\Re(\lambda e^{i\varphi_{\nu\mu}})|\right)^{-1/p-\varepsilon} \right) \right\}_p .$$

Proof. The asymptotic representation (4.1.14) of $\widetilde{Y}$ yields

$$\widehat{Y}(x,\lambda) = \left(P^{[0]}(x) + B_0(x,\lambda)\right)\left((I_n - \Delta(\lambda))E(x,\lambda) + \Delta(\lambda)E(x,\lambda)E(b,\lambda)^{-1}\right)$$

for $x \in (a,b)$ and $|\lambda| \ge \gamma$. The assertion of the proposition immediately follows from the estimates (4.1.18) and (4.1.20) of B_0 and from Proposition 4.3.3 i), ii).
□

PROPOSITION 4.4.3. *We have* $E(\cdot,\lambda)\widetilde{Y}(\cdot,\lambda)^{-1} = {P^{[0]}}^{-1} + \widetilde{B}_0(\cdot,\lambda)$, *where* $\widetilde{B}_0(\cdot,\lambda)$ *satisfies the estimates* $\{O(\tau_p(\lambda))\}_\infty$ *and* $\{o(1)\}_\infty$.

Proof. Since $P^{[0]}$ is invertible in $M_n(W_p^1(a,b))$ and hence in $M_n(L_\infty(a,b))$, and since $B_0(\cdot,\lambda) = \{o(1)\}_\infty$, the estimate

$$|B_0(x,\lambda){P^{[0]}}^{-1}(x)| \le \frac{1}{2}$$

holds for all $x \in (a,b)$ and all sufficiently large λ. Then we have the Neumann expansion

$$\begin{aligned} E(x,\lambda)\widetilde{Y}(\cdot,\lambda)^{-1} &= \left(P^{[0]}(x) + B_0(x,\lambda)\right)^{-1} \\ &= {P^{[0]}}^{-1}(x)\left(\sum_{j=0}^{\infty}(-1)^j\left(B_0(x,\lambda){P^{[0]}}^{-1}(x)\right)^j\right). \end{aligned}$$

With

$$\widetilde{B}_0(x,\lambda) := {P^{[0]}}^{-1}(x)\left(\sum_{j=1}^{\infty}(-1)^j\left(B_0(x,\lambda){P^{[0]}}^{-1}(x)\right)^j\right)$$

it follows that

$$(P^{[0]}(x) + B_0(x,\lambda))^{-1} = {P^{[0]}}^{-1}(x) + \widetilde{B}_0(x,\lambda)$$

and

$$\begin{aligned} |\widetilde{B}_0(x,\lambda)| &\le |{P^{[0]}}^{-1}(x)B_0(x,\lambda){P^{[0]}}^{-1}(x)| \sum_{j=0}^{\infty} |B_0(x,\lambda){P^{[0]}}^{-1}(x)|^j \\ &\le 2|{P^{[0]}}^{-1}(x)|^2|B_0(x,\lambda)|. \end{aligned}$$

Finally, $B_0(\cdot,\lambda) = \{O(\tau_p(\lambda))\}_\infty$, $B_0(\cdot,\lambda) = \{o(1)\}_\infty$, and the above estimate yields

$$\widetilde{B}_0(\cdot,\lambda) = \{O(\tau_p(\lambda))\}_\infty \text{ and } \widetilde{B}_0(\cdot,\lambda) = \{o(1)\}_\infty. \quad \square$$

We set

(4.4.8)

$$I_1^0(x,f_1,\lambda) := P^{[0]}(x)(I_n - \Delta(\lambda))E(x,\lambda)\int_a^x E(\xi,\lambda)^{-1}P^{[0]^{-1}}(\xi)f_1(\xi)\,\mathrm{d}\xi$$
$$- P^{[0]}(x)\Delta(\lambda)E(x,\lambda)\int_x^b E(\xi,\lambda)^{-1}P^{[0]^{-1}}(\xi)f_1(\xi)\,\mathrm{d}\xi\,.$$

PROPOSITION 4.4.4. *Let I_1 be as defined in* (4.4.5) *and let $f_1 \in (L_p(a,b))^n$. Then*
i) *$I_1(\cdot,f_1,\lambda) - I_1^0(\cdot,f_1,\lambda)$ is of the form $\{O(\tau_p(\lambda))\}_\infty |f_1|_p$ and $\{o(1)\}_\infty |f_1|_p$.*
ii) *Let $p > 1$. If $n_0 = 0$ or $p > \frac{3}{2}$ or $p \le \frac{3}{2}$ and the conditions in* (4.1.19) *are satisfied, then there is a number $\varepsilon > 0$ such that*

$$\Delta_0 I_1(\cdot,\Delta_0 f_1,\lambda) - \Delta_0 I_1^0(\cdot,\Delta_0 f_1,\lambda)$$
$$= \left\{ O\left(\max_{\substack{\nu,\mu=0\\ \nu\ne\mu}}^{l} \left(1 + |\Re(\lambda e^{i\varphi_{\nu\mu}})|\right)^{-1-\varepsilon} \right) \right\}_p |f_1|_p.$$

Proof. Let $x \in (a,b)$. The representation (4.1.14) of $\widetilde{Y}$ and Proposition 4.4.3 yield

$$\widetilde{Y}(x,\lambda)\left(I_n - \Delta(\lambda)\right)\int_a^x \widetilde{Y}(\xi,\lambda)^{-1}f_1(\xi)\,\mathrm{d}\xi \tag{4.4.9}$$
$$- P^{[0]}(x)\left(I_n - \Delta(\lambda)\right)E(x,\lambda)\int_a^x E(\xi,\lambda)^{-1}P^{[0]^{-1}}(\xi)f_1(\xi)\,\mathrm{d}\xi$$
$$= B_0(x,\lambda)\left(I_n - \Delta(\lambda)\right)E(x,\lambda)\int_a^x E(\xi,\lambda)^{-1}P^{[0]^{-1}}(\xi)f_1(\xi)\,\mathrm{d}\xi$$
$$+ P^{[0]}(x)\left(I_n - \Delta(\lambda)\right)E(x,\lambda)\int_a^x E(\xi,\lambda)^{-1}\widetilde{B}_0(\xi,\lambda)f_1(\xi)\,\mathrm{d}\xi$$
$$+ B_0(x,\lambda)\left(I_n - \Delta(\lambda)\right)E(x,\lambda)\int_a^x E(\xi,\lambda)^{-1}\widetilde{B}_0(\xi,\lambda)f_1(\xi)\,\mathrm{d}\xi\,.$$

Observing $B_0(\cdot,\lambda) = \{O(\tau_p(\lambda))\}_\infty$, $B_0(\cdot,\lambda) = \{o(1)\}_\infty$, $\widetilde{B}_0(\cdot,\lambda) = \{O(\tau_p(\lambda))\}_\infty$, $\widetilde{B}_0(\cdot,\lambda) = \{o(1)\}_\infty$, and the estimate in Proposition 4.3.3 i), this shows that the right-hand side of (4.4.9) is of the form $\{O(\tau_p(\lambda))\}_\infty(x)|f_1|_p$ and $\{o(1)\}_\infty(x)|f_1|_p$. Since analogous estimates hold for

$$\widetilde{Y}(x,\lambda)\Delta(\lambda)\int_x^b \widetilde{Y}(\xi,\lambda)^{-1}f_1(\xi)\,\mathrm{d}\xi \tag{4.4.10}$$
$$- P^{[0]}(x)\Delta(\lambda)E(x,\lambda)\int_x^b E(\xi,\lambda)^{-1}P^{[0]^{-1}}(\xi)f_1(\xi)\,\mathrm{d}\xi\,,$$

this proves part i).

For the proof of ii) we multiply (4.4.9) from the left by Δ_0 and replace f_1 by $\Delta_0 f_1$. Since the first term on the right-hand side is

$$\Delta_0 B_0(x,\lambda)\big(I_n - \Delta(\lambda)\big)\int_a^x E(x,\lambda)E(\xi,\lambda)^{-1}\Delta_0 P^{[0]^{-1}}(\xi)f_1(\xi)\,\mathrm{d}\xi,$$

Proposition 4.3.5 yields that it has the estimate

$$|B_0(x,\lambda)|\left\{O\left(\max_{\nu=1}^{l}\big(1+|\Re(\lambda e^{i\varphi_\nu})|\big)^{-1}\right)\right\}_p(x)|P^{[0]^{-1}}f_1|_p.$$

Similarly, the second term satisfies the estimate

$$|P^{[0]}(x)|\left\{O\left(\max_{\nu=1}^{l}\big(1+|\Re(\lambda e^{i\varphi_\nu})|\big)^{-1}\right)\right\}_p(x)|\widetilde{B}_0(\cdot,\lambda)f_1|_p.$$

From $B_0(\cdot,\lambda) = \{O(\tau_p(\lambda))\}_\infty$ and $\widetilde{B}_0(\cdot,\lambda) = \{O(\tau_p(\lambda))\}_\infty$ we obtain that the desired estimate is fulfilled for the first two terms. The last term can be estimated in the same way if $n_0 = 0$. Otherwise, we use (4.1.20) and Proposition 4.4.3. Analogous estimates of (4.4.10) complete the proof. □

PROPOSITION 4.4.5. *For $\xi \in (a,b)$ we have that*

$$\int_{t=\xi}^{b}\mathrm{d}_t\widetilde{F}(t,\lambda)\widetilde{Y}(t,\lambda)E(\xi,\lambda)^{-1}\big(I_n-\Delta(\lambda)\big)$$
$$-\sum_{\substack{j=0\\ a_j\ge\xi}}^{\infty} W_0^{(j)}P^{[0]}(a_j)E(a_j,\lambda)E(\xi,\lambda)^{-1}\big(I_n-\Delta(\lambda)\big)$$
$$-\int_\xi^b W_0(t)P^{[0]}(t)E(t,\lambda)E(\xi,\lambda)^{-1}\big(I_n-\Delta(\lambda)\big)\,\mathrm{d}t$$

and

$$\int_{t=a}^{\xi}\mathrm{d}_t\widetilde{F}(t,\lambda)\widetilde{Y}(t,\lambda)E(\xi,\lambda)^{-1}\Delta(\lambda)$$
$$-\sum_{\substack{j=0\\ a_j<\xi}}^{\infty} W_0^{(j)}P^{[0]}(a_j)E(a_j,\lambda)E(\xi,\lambda)^{-1}\Delta(\lambda)$$
$$-\int_a^\xi W_0(t)P^{[0]}(t)E(t,\lambda)E(\xi,\lambda)^{-1}\Delta(\lambda)\,\mathrm{d}t$$

fulfil the estimate $O(\tau_p(\lambda))$ and $o(1)$ uniformly.

Proof. Let $\alpha, \beta \in [a,b]$ with $\alpha < \beta$ and set $I := [\alpha, \beta)$ if $\beta \neq b$ and $I := [\alpha, \beta]$ if $\beta = b$. An obvious generalization of Proposition 3.2.1 yields

$$\begin{aligned}
(4.4.11) \qquad & \int_\alpha^\beta \mathrm{d}_t \widetilde{F}(t,\lambda)\widetilde{Y}(t,\lambda) \\
& \quad - \sum_{\substack{j=0 \\ a_j \in I}}^\infty W_0^{(j)} P^{[0]}(a_j)E(a_j,\lambda) - \int_\alpha^\beta W_0(t)P^{[0]}(t)E(t,\lambda)\,\mathrm{d}t \\
& = \sum_{\substack{j=0 \\ a_j \in I}}^\infty \left(\widetilde{W}^{(j)}(\lambda) - W_0^{(j)}\right)\left(P^{[0]}(a_j) + B_0(a_j,\lambda)\right)E(a_j,\lambda) \\
& \quad + \int_\alpha^\beta \left(\widetilde{W}(t,\lambda) - W_0(t)\right)\left(P^{[0]}(t) + B_0(t,\lambda)\right)E(t,\lambda)\,\mathrm{d}t \\
& \quad + \sum_{\substack{j=0 \\ a_j \in I}}^\infty W_0^{(j)} B_0(a_j,\lambda)E(a_j,\lambda) + \int_\alpha^\beta W_0(t)B_0(t,\lambda)E(t,\lambda)\,\mathrm{d}t\,.
\end{aligned}$$

For the proof of the estimate for the first term we set $\alpha = \xi$, $\beta = b$, and multiply (4.4.11) by $E(\xi,\lambda)^{-1}\left(I_n - \Delta(\lambda)\right)$ from the right. The assertion now follows from the estimates $\widetilde{W}(\cdot,\lambda) - W_0 = O(\lambda^{-1})$, $\sum_{j=0}^\infty \left|\widetilde{W}^{(j)} - W_0^{(j)}\right| = O(\lambda^{-1})$, $B_0(\cdot,\lambda) = \left\{O(\tau_p(\lambda))\right\}_\infty$, $B_0(\cdot,\lambda) = \left\{o(1)\right\}_\infty$, see (4.1.10), (4.1.12), (4.1.18) and Proposition 4.3.3 i) with $c = a_j$ or $c = t$ and $d = \xi$.

For the proof of the estimate for the second term we set $\alpha = a$, $\beta = \xi$, and multiply (4.4.11) with $E(\xi,\lambda)^{-1}\Delta(\lambda)$ from the right. The assertion now follows again from the estimates (4.1.10), (4.1.12), (4.1.18) and Proposition 4.3.3 ii) with $c = a_j$ or $c = t$ and $d = \xi$. □

LEMMA 4.4.6. *Suppose that the boundary eigenvalue problem* (4.1.1), (4.1.2) *is s-regular. Then the characteristic matrix function* $\widetilde{M}$ *given by* (4.3.2) *satisfies the estimate*

$$\widetilde{M}(\lambda)^{-1} = O(\lambda^s) \quad \left(\lambda \in \bigcup_{\nu=0}^\infty \Gamma_\nu\right),$$

where the circles Γ_ν *are as in Definition* 4.4.1.

Proof. The set $\left\{\widetilde{M}_1(\lambda)\Delta_0 + \widetilde{M}_2(I_n - \Delta_0) : \lambda \in \mathbb{C}\setminus\{0\}\right\}$ is bounded in $M_n(\mathbb{C})$ by Corollary 4.3.4 iii). Hence $\widetilde{M}(\lambda) = O(1)$ as $\lambda \to \infty$ by Proposition 4.3.6 ii), iii). Let $\widetilde{M}^{\mathrm{ad}}(\lambda)$ be the matrix of the cofactors of $\widetilde{M}(\lambda)$. Then the boundedness of $\widetilde{M}$ implies that $\widetilde{M}^{\mathrm{ad}}(\lambda) = O(1)$ as $\lambda \to \infty$. Since $\widetilde{M}(\lambda)^{-1} = (\det \widetilde{M}(\lambda))^{-1}\widetilde{M}^{\mathrm{ad}}(\lambda)$, the assertion of the proposition follows from the definition of s-regularity. □

For $f=(f_1,f_2)\in(L_p(a,b))^n\times\mathbb{C}^n$ we set

$$
\begin{aligned}
(4.4.12)\qquad I_3^0(f,\lambda) &:= f_2 \\
&- \sum_{j=0}^{\infty} W_0^{(j)} P^{[0]}(a_j)\int_a^{a_j} E(a_j,\lambda)E(\xi,\lambda)^{-1}(I_n-\Delta(\lambda))P^{[0]^{-1}}(\xi)f_1(\xi)\,d\xi \\
&- \int_a^b W_0(t)P^{[0]}(t)\int_a^t E(t,\lambda)E(\xi,\lambda)^{-1}(I_n-\Delta(\lambda))P^{[0]^{-1}}(\xi)f_1(\xi)\,d\xi\,dt \\
&+ \sum_{j=0}^{\infty} W_0^{(j)} P^{[0]}(a_j)\int_{a_j}^{b} E(a_j,\lambda)E(\xi,\lambda)^{-1}\Delta(\lambda)P^{[0]^{-1}}(\xi)f_1(\xi)\,d\xi \\
&+ \int_a^b W_0(t)P^{[0]}(t)\int_t^b E(t,\lambda)E(\xi,\lambda)^{-1}\Delta(\lambda)P^{[0]^{-1}}(\xi)f_1(\xi)\,d\xi\,dt.
\end{aligned}
$$

PROPOSITION 4.4.7. *The vector function $I_3(f,\lambda)-I_3^0(f,\lambda)$ satisfies the estimates $O(\tau_p(\lambda))|f|_{[p,n]}$ and $o(1)|f|_{[p,n]}$, where $|f|_{[p,n]}:=|f_1|_p+|f_2|_{\mathbb{C}^n}$, and I_3, I_3^0 are given by* (4.4.6), (4.4.12).

Proof. We multiply the matrix functions in the assertion of Proposition 4.4.5 by the vector function $E(\xi,\lambda)\widetilde{Y}(\xi,\lambda)^{-1}f_1(\xi)$ from the right and integrate from a to b with respect to ξ. From Propositions 4.4.5 and 4.4.3 we infer that these integrals are $O(\tau_p(\lambda))|f_1|_p$ and $o(1)|f_1|_p$, i. e.,

$$I_3(f,\lambda)-I_3^0(((I_n+P^{[0]}\widetilde{B}_0(\cdot,\lambda))f_1,f_2),\lambda)=O(\tau_p(\lambda))|f_1|_p$$

and

$$I_3(f,\lambda)-I_3^0(((I_n+P^{[0]}\widetilde{B}_0(\cdot,\lambda))f_1,f_2),\lambda)=o(1)|f_1|_p.$$

From Propositions 4.4.3 and 4.3.3 i), ii) and $\sum_{j=0}^{\infty}|W_0^{(j)}|<\infty$ we infer

$$I_3^0((P^{[0]}\widetilde{B}_0(\cdot,\lambda)f_1,0),\lambda)=O(\tau_p(\lambda))|f_1|_p$$

and

$$I_3^0((P^{[0]}\widetilde{B}_0(\cdot,\lambda)f_1,0),\lambda)=o(1)|f_1|_p.$$

This proves the proposition since I_3^0 is linear with respect to the first variable. □

In the following we shall use contour integrals of holomorphic vector valued functions. For this we briefly recall the definition and some properties of line integrals. A piecewise smooth path in $\mathbb{C}$ is a continuous and piecewise continuously differentiable mapping γ of a compact interval $[\alpha,\beta]$, $\alpha<\beta$, into $\mathbb{C}$. Let E be a Banach space and $h:\gamma([\alpha,\beta])\to E$ be continuous. Then the integral

$$\int_\gamma h(\lambda)\,d\lambda:=\int_\alpha^\beta h(\gamma(t))\gamma'(t)\,dt$$

is well-defined. Let $\gamma : [\alpha,\beta] \to \mathbb{C}$ and $\gamma_1 : [\alpha_1,\beta_1] \to \mathbb{C}$ be piecewise smooth paths. The path γ is called equivalent to γ_1 if there is a continuous strictly increasing function $\varphi : [\alpha_1,\beta_1] \to [\alpha,\beta]$ whith $\varphi(\alpha_1) = \alpha$ and $\varphi(\beta_1) = \beta$ such that $\gamma_1 = \gamma \circ \varphi$. A piecewise smooth curve is an equivalence class of piecewise smooth paths. For equivalent paths as above we have

$$\int_\alpha^\beta h(\gamma(t))\gamma'(t)\,\mathrm{d}t = \int_{\alpha_1}^{\beta_1} h(\gamma_1(t))\gamma_1'(t)\,\mathrm{d}t,$$

i.e., $\int_\gamma h(\lambda)\,d\lambda$ only depends on the curve. Hence we shall not distinguish between a smooth curve and its representative. A piecewise smooth contour is a piecewise smooth simply closed curve. We shall mostly take circles. They will be traversed anti-clockwise, i.e., they are given by the path

$$\gamma(t) = c + re^{it} \quad (t \in [0,2\pi]),$$

where c is the centre and r the radius of the circle. For a piecewise smooth curve $\gamma : [a,b] \to \mathbb{C}$ and a continuous function $h : \gamma([\alpha,\beta]) \to E$ we define

$$\int_\gamma h(\lambda)\,|\mathrm{d}\lambda| := \int_\alpha^\beta h(\gamma(t))|\gamma'(t)|\,\mathrm{d}t.$$

We have the following estimate, see e.g. [DIN, § 8, Proposition 4],

$$\left|\int_\gamma h(\lambda)\,\mathrm{d}\lambda\right| \le \int_\gamma |h(\lambda)|\,|\mathrm{d}\lambda|. \tag{4.4.13}$$

Note that for each $\lambda \in \rho(\widetilde{T})$ and $f = (f_1, f_2) \in (L_p(a,b))^n \times \mathbb{C}^n$ it follows from (4.4.4) that

$$\begin{aligned}\widetilde{T}^{-1}(\lambda)f &= I_1^0(\cdot,f_1,\lambda) + I_2^0(\cdot,\lambda)\widetilde{M}(\lambda)^{-1}I_3^0(f,\lambda)\\ &+I_1(\cdot,f_1,\lambda) - I_1^0(\cdot,f_1,\lambda) + \widehat{Y}(\cdot,\lambda)\widetilde{M}(\lambda)^{-1}\big(I_3(f,\lambda) - I_3^0(f,\lambda)\big)\\ &+\big(\widehat{Y}(\cdot,\lambda) - I_2^0(\cdot,\lambda)\big)\widetilde{M}(\lambda)^{-1}I_3^0(f,\lambda),\end{aligned} \tag{4.4.14}$$

where the terms on the right-hand side are defined in (4.3.1), (4.3.2), (4.4.5), (4.4.6), (4.4.7), (4.4.8), and (4.4.12).

PROPOSITION 4.4.8. *Let $s \in \mathbb{N}$ and assume that the boundary eigenvalue problem (4.1.1), (4.1.2) is s-regular. Then*

$$\widetilde{T}^{-1}(\lambda)f - I_1^0(\cdot,f_1,\lambda) - I_2^0(\cdot,\lambda)\widetilde{M}(\lambda)^{-1}I_3^0(f,\lambda) \tag{4.4.15}$$

for $f = (f_1, f_2) \in (L_p(a,b))^n \times \mathbb{C}^n$ fulfils the estimates $\{O(\lambda^s(\tau_p(\lambda))\}_\infty |f|_{[p,n]}$ and $\{o(\lambda^s)\}_\infty |f|_{[p,n]}$ on $\bigcup_\nu^\infty \Gamma_\nu$.

Proof. From Proposition 4.3.3 i), ii) it follows that $I_1^0(\cdot,f_1,\lambda) = \{O(1)\}_\infty |f_1|_p$ and $I_2^0(\cdot,\lambda) = \{O(1)\}_\infty$. In the same way, Proposition 4.3.3 i), ii) and $\sum_{j=0}^\infty |W_0^{(j)}| < \infty$

yields $I_3^0(f,\lambda) = O(1)|f|_{[p,n]}$. From (4.4.14), Proposition 4.4.4 i), Proposition 4.4.2, Lemma 4.4.6 and Proposition 4.4.7 the estimates of (4.4.15) follow. □

THEOREM 4.4.9. *Let $s \in \mathbb{N}$ and assume that the boundary eigenvalue problem (4.1.1), (4.1.2) is s-regular. Let $\lambda_0 \in \mathbb{C} \setminus \bigcup_{\nu=0}^{\infty} \Gamma_\nu$, where the circles Γ_ν are as in Definition 4.4.1 of s-regularity. Let $J_1 : (L_p(a,b))^n \to (L_p(a,b))^n \times \mathbb{C}^n$ and $J : (W_p^1(a,b))^n \to (L_p(a,b))^n$ be the canonical embeddings. We set*

$$S_{s,\nu} := \frac{1}{2\pi i} \oint_{\Gamma_\nu} (\lambda - \lambda_0)^{-s-1} \widetilde{T}^{-1}(\lambda)\, d\lambda \quad (\nu \in \mathbb{N}), \tag{4.4.16}$$

where $\widetilde{T}$ is given by (4.4.1).
i) *For all $\nu \in \mathbb{N}$ we have $S_{s,\nu} \in L((L_p(a,b))^n \times \mathbb{C}^n, (W_p^1(a,b))^n)$.*
ii) *For each compact set $G \subset \mathbb{C} \setminus \bigcup_{\nu=0}^{\infty} \Gamma_\nu$ we have*

$$\sup_{\mu \in G} \oint_{\Gamma_\nu} |\lambda - \mu|^{-s-2} |J\widetilde{T}^{-1}(\lambda)|\, |d\lambda| \to 0 \text{ as } \nu \to \infty.$$

iii) *We have $JS_{s+1,\nu} \to 0$ in $L((L_p(a,b))^n \times \mathbb{C}^n, (L_p(a,b))^n)$ as $\nu \to \infty$.*

Proof. i) is obvious since $\widetilde{T}^{-1}$ depends holomorphically and hence continuously on λ in $\bigcup_{\nu=0}^{\infty} \Gamma_\nu$.
ii) Proposition 4.4.8 and the estimates of $I_1^0(\cdot, f_1, \lambda)$, $I_2^0(\cdot, \lambda)$, and $I_3^0(f,\lambda)$ stated in the proof of Proposition 4.4.8 yield

$$|J\widetilde{T}^{-1}(\lambda)| = O(\lambda^s)$$

for $\lambda \in \bigcup_{\nu=0}^{\infty} \Gamma_\nu$. Since $\sup_{\mu \in G} |\lambda - \mu|^{-s-2} = O(|\lambda|^{-s-2})$ for these λ, we have

$$\begin{aligned}
&\sup_{\mu \in G} \oint_{\Gamma_\nu} |\lambda - \mu|^{-s-2} |J\widetilde{T}^{-1}(\lambda)|\, |d\lambda| \\
&= \oint_{\Gamma_\nu} O(|\lambda|^{-s-2}) O(\lambda^s)\, |d\lambda| \\
&= O(\rho_\nu^{-1}).
\end{aligned}$$

iii) immediately follows from ii). □

PROPOSITION 4.4.10. *We have*

$$I_3^0((\Delta_0 f_1, 0), \lambda) = O\left(\max_{j=1}^{l} (1 + |\Re(\lambda e^{i\varphi_j})|)^{-1+1/p} \right) |f_1|_p$$

for $f_1 \in (L_p(a,b))^n$, where I_3^0 is given by (4.4.12).

Proof. By Proposition 4.3.5 there is $C>0$ such that for all $f_1 \in (L_p(a,b))^n$ the $\mathbb{C}^n$-norm of the integrals with respect to ξ in $I_3^0((\Delta_0 f_1,0),\lambda)$ has the upper bound $C\max\limits_{j=1}^{l}\left(1+|\Re(\lambda e^{i\varphi_j})|\right)^{-1+1/p}|f_1|_p$. Now the assertion of the proposition immediately follows in view of (4.1.11). □

THEOREM 4.4.11. *Let $s\in\mathbb{N}$ and assume that the boundary eigenvalue problem (4.1.1), (4.1.2) is s-regular. Let $\lambda_0 \in \mathbb{C}\setminus \bigcup\limits_{\nu=0}^{\infty}\Gamma_\nu$, where the circles Γ_ν are as in Definition 4.4.1 of s-regularity. Let $\widetilde{J}_1 : (L_p(a,b))^n \to (L_p(a,b))^n\times\mathbb{C}^n$ be defined by $\widetilde{J}_1 f = (\Delta_0 f,0)$ $(f\in(L_p(a,b))^n)$ and let $\widetilde{J} : (W_p^1(a,b))^n \to (L_p(a,b))^n$ be defined by $\widetilde{J}y = \Delta_0 y$ $(y\in(W_p^1(a,b))^n)$, where Δ_0 is defined in (4.1.22). We set*

$$S_{s,\nu} := \frac{1}{2\pi i}\oint_{\Gamma_\nu}(\lambda-\lambda_0)^{-s-1}\widetilde{T}^{-1}(\lambda)\,\mathrm{d}\lambda \quad (\nu\in\mathbb{N}),$$

where $\widetilde{T}$ is given by (4.4.1).

i) *For $1<p<\infty$ and a compact set $G\subset\mathbb{C}\setminus\bigcup\limits_{\nu=0}^{\infty}\Gamma_\nu$ we have*

$$\sup_{\mu\in G}\oint_{\Gamma_\nu}|\lambda-\mu|^{-s-1}|\widetilde{J}\widetilde{T}^{-1}(\lambda)|\,|\mathrm{d}\lambda| \to 0 \text{ as } \nu\to\infty.$$

ii) *For $1<p<\infty$ we have $\widetilde{J}S_{s,\nu}\to 0$ in $L((L_p(a,b))^n\times\mathbb{C}^n,(L_p(a,b))^n)$ as $\nu\to\infty$.*

iii) *For $1\le p\le\infty$ and a compact set $G\subset\mathbb{C}\setminus\bigcup\limits_{\nu=0}^{\infty}\Gamma_\nu$ we have*

$$\sup_{\mu\in G}\oint_{\Gamma_\nu}|\lambda-\mu|^{-s-1}|\widetilde{J}\widetilde{T}^{-1}(\lambda)\widetilde{J}_1|\,|\mathrm{d}\lambda| \to 0 \text{ as } \nu\to\infty.$$

iv) *For $1\le p\le\infty$ we have $\widetilde{J}S_{s,\nu}\widetilde{J}_1\to 0$ in $L((L_p(a,b))^n)$ as $\nu\to\infty$.*

Proof. i) For $f=(f_1,f_2)\in(L_p(a,b))^n\times\mathbb{C}^n$ Proposition 4.4.8 yields

$$\begin{aligned}\widetilde{J}\widetilde{T}^{-1}(\lambda)f &= \Delta_0 I_1^0(\cdot,f_1,\lambda)+\Delta_0 I_2^0(\cdot,\lambda)\widetilde{M}(\lambda)^{-1}I_3^0(f,\lambda)\\ &\quad+\{O(\lambda^s\tau_p(\lambda))\}_\infty|f|_{[p,n]}.\end{aligned} \tag{4.4.17}$$

From Proposition 4.3.5 we obtain the estimate

$$\Delta_0 I_1^0(\cdot,f_1,\lambda) = \left\{O\left(\max_{j=1}^{l}\left(1+|\Re(\lambda e^{i\varphi_j})|\right)^{-1}\right)\right\}_p|f_1|_p. \tag{4.4.18}$$

Let $v\in(L_{p'}(a,b))^n$ where $1/p+1/p'=1$. By definition of I_2^0, The non-zero components of $\int_a^b v^{\mathsf{T}}(x)\Delta_0 I_2^0(x,\lambda)\,\mathrm{d}x$ are sums of integrals of the form

$$(1-\delta_j(\lambda))\int_a^b \tilde{v}(x)\exp\{\lambda e^{i\varphi_j}|R_j(x)|\}\,\mathrm{d}x$$

and

$$\delta_j(\lambda) \int_a^b \tilde{v}(x) \exp\{\lambda e^{i\varphi_j}(|R_j(x)| - |R_j(b)|)\} \,\mathrm{d}x,$$

where $j \in \{1,\dots,l\}$ and $\tilde{v}$ is a component of $v^\mathsf{T}\Delta_0 P^{[0]}$. By Lemma 2.7.2 ii), these integrals fulfil the estimates $O\left(\max_{j=1}^{l}\left(1+|\Re(\lambda e^{i\varphi_j})|\right)^{-1/p}\right)|v|_{p'}$. As $(L_{p'}(a,b))^n$ is the dual of $(L_p(a,b))^n$, this yields

$$\Delta_0 I_2^0(\cdot,\lambda) = \left\{O\left(\max_{j=1}^{l}\left(1+|\Re(\lambda e^{i\varphi_j})|\right)^{-1/p}\right)\right\}_p. \tag{4.4.19}$$

Note that the above proof also holds for $p=1$, and that (4.4.19) is also true for $p=\infty$. Finally, $I_3^0(f,\lambda) = O(1)|f|_{[p,n]}$ (see the proof of Theorem 4.4.9 ii)) and Lemma 4.4.6 give the estimate

$$|\widetilde{J}\widetilde{T}^{-1}(\lambda)| = O\left(\lambda^s \max_{\substack{\nu,\mu=0\\ \nu\neq\mu}}^{l}\left(1+|\Re(\lambda e^{i\varphi_{\nu\mu}})|\right)^{-\eta}\right),$$

where $\eta := \min\{1/p, 1-1/p\}$. Hence we have

$$\begin{aligned}
&\sup_{\mu\in G}\oint_{\Gamma_\nu} |\lambda-\mu|^{-s-1}|\widetilde{J}\widetilde{T}^{-1}(\lambda)|\,|\mathrm{d}\lambda| \\
&= \oint_{\Gamma_\nu} O(|\lambda|^{-s-1})\, O\left(\lambda^s \max_{\substack{\nu,\mu=0\\ \nu\neq\mu}}^{l}\left(1+|\Re(\lambda e^{i\varphi_{\nu\mu}})|\right)^{-\eta}\right)|\mathrm{d}\lambda| \\
&= \oint_{|\lambda|=1} O\left(\max_{\substack{\nu,\mu=0\\ \nu\neq\mu}}^{l}\left(1+\rho_\nu|\Re(\lambda e^{i\varphi_{\nu\mu}})|\right)^{-\eta}\right)|\mathrm{d}\lambda| \\
&= o(1)
\end{aligned}$$

as $\nu\to\infty$ by LEBESGUE'S dominated convergence theorem.
ii) immediately follows from i).
iii) For $f \in (L_p(a,b))^n$ Proposition 4.4.8 yields that

$$\begin{aligned}
\widetilde{J}\widetilde{T}^{-1}(\lambda)\widetilde{J}_1 f = \Delta_0 I_1^0(\cdot,\Delta_0 f,\lambda) + \Delta_0 I_2^0(\cdot,\lambda)\widetilde{M}^{-1}(\lambda) I_3^0((\Delta_0 f,0),\lambda)& \\
+ \{o(\lambda^s)\}_\infty |f|_p.&
\end{aligned}$$

Hence we obtain in view of (4.4.18), (4.4.19), Lemma 4.4.6 and Propositions 4.4.7 and 4.4.10 that

$$|\widetilde{J}\widetilde{T}^{-1}(\lambda)\widetilde{J}_1| = O\left(\lambda^s \max_{\substack{\nu,\mu=0\\ \nu\neq\mu}}^{l}\left(1+|\Re(\lambda e^{i\varphi_{\nu\mu}})|\right)^{-1}\right) + o(\lambda^s).$$

Proceeding as in part i), the assertion iii) is proved.
iv) immediately follows from iii). □

4.5. A special case of the Hilbert transform

In order to obtain better estimates for Birkhoff-regular operators we need two special cases of the Hilbert transform stated in Propositions 4.5.2 and 4.5.3 below. Since the proof of these propositions is easier than the proof of the existence and the boundedness of the Hilbert transform, we shall prove these propositions directly. The statement and the proof of Proposition 4.5.1 are extracted from the proof of [TI, Theorem 101].

PROPOSITION 4.5.1. *Let $1 < p < \infty$ and $-\infty < c < d < \infty$. For $f \in L_p(c,d)$ and $z \in \mathbb{C}$ with $\Im(z) > 0$ we define*

$$\phi_f(z) := \int_c^d \frac{f(t)}{t+z}\,\mathrm{d}t.$$

Then $\phi_f(\cdot + iy) \in L_p(-\infty,\infty)$ for all $y > 0$ and there is a constant $C > 0$ such that

$$|\phi_f(\cdot + iy)|_p \le C|f|_p$$

for all $f \in L_p(c,d)$ and all $y > 0$.

Proof. Since ϕ_f is holomorphic, $\phi_f(\cdot + iy)$ is measurable for all $f \in L_p(c,d)$ and all $y > 0$. Also $\phi_f(z) = O\left(\frac{1}{z}\right)$ as $|z| \to \infty$. Hence $\phi_f(\cdot + iy) \in L_p(-\infty,\infty)$ for all $f \in L_p(c,d)$ and all $y > 0$. Thus it is sufficient to prove that there is a constant $C_2 > 0$ such that $|\phi_f(\cdot + iy)|_p \le C_2|f|_p$ for each nonnegative $f \in L_p(c,d) \setminus \{0\}$.

In the following we need the function

$$z \mapsto z^\alpha = \exp\{\alpha \log z\} \ (z \in \mathbb{C} \setminus \{0\}),\ 0^\alpha := 0,$$

where $\alpha > 0$ and log is the principal value of the logarithm, i. e., $\log z = \log|z| + i\arg z$ for $z \in \mathbb{C} \setminus \{0\}$, where $\arg z \in [-\pi,\pi)$. This function is holomorphic on $\{z \in \mathbb{C} : \Im(z) < 0\}$ and continuous on $\{z \in \mathbb{C} : \Im(z) \le 0\}$.
i) First we consider the case $1 < p \le 2$. Let $2 \le p' < \infty$ such that $1/p + 1/p' = 1$. We set $C_1 := p2^{\frac{1}{2}(p-1)}$ and

$$C := \pi\left(\sup\left\{r > 0 : C_1(1 + r^{1-p}) - r\left|\cos\left(p\frac{\pi}{2}\right)\right| \ge 0\right\} + 1\right),$$

which is finite since $r^{1-p} \to 0$ as $r \to \infty$ and $\cos(p\frac{\pi}{2}) \ne 0$. Let $f \in L_p(c,d) \setminus \{0\}$ be real-valued and nonnegative. Then we write

$$\phi_f(z) =: u(x,y) - iv(x,y)$$

where $x, y \in \mathbb{R}$, $y > 0$, $z = x + iy$, and u, v are real-valued functions. For all $x \in \mathbb{R}$ and $y > 0$ we obtain

$$u(x,y) = \int_c^d \frac{f(t)(t+x)}{(t+x)^2 + y^2}\,\mathrm{d}t, \quad v(x,y) = y\int_c^d \frac{f(t)}{(t+x)^2 + y^2}\,\mathrm{d}t > 0.$$

This shows $\Im(\phi_f(z)) < 0$ for all $z \in \mathbb{C}$ with $\Im(z) > 0$. Hence

$$z \mapsto \phi_f(z)^p = \exp\{p \log \phi_f(z)\}$$

defines a holomorphic function on $\{z \in \mathbb{C} : \Im(z) > 0\}$. We consider the contour integral

$$\oint \phi_f(z)^p \, \mathrm{d}z = 0$$

along the straight line from $-R + iy$ to $R + iy$ and along the semicircle above it. Since $\phi_f(z) = O(\frac{1}{z})$, we obtain that

$$\int_{-R}^{R} \phi_f(x+iy)^p \, \mathrm{d}x = O(R^{-p}) \cdot \pi R = O(R^{1-p}) \quad \text{as } R \to \infty.$$

Hence

$$\int_{-\infty}^{\infty} \phi_f(x+iy)^p \, \mathrm{d}x = 0 \tag{4.5.1}$$

for all $y > 0$. From

$$\phi_f(x+iy)^p - u(x,y)^p = p \int_{u(x,y)}^{\phi_f(x+iy)} z^{p-1} \, \mathrm{d}z$$

we infer

$$\begin{aligned} \left|\phi_f(x+iy)^p - u(x,y)^p\right| &\le p v(x,y) \left(u(x,y)^2 + v(x,y)^2\right)^{\frac{1}{2}(p-1)} \\ &\le p v(x,y) \left(2 \max\{u(x,y)^2, v(x,y)^2\}\right)^{\frac{1}{2}(p-1)} \\ &\le C_1 \left(v(x,y)|u(x,y)|^{p-1} + v(x,y)^p\right). \end{aligned}$$

With the aid of (4.5.1) we conclude

$$\begin{aligned} \left|\int_{-\infty}^{\infty} u(x,y)^p \, \mathrm{d}x\right| &= \left|\int_{-\infty}^{\infty} \left(\phi_f(x+iy)^p - u(x,y)^p\right) \mathrm{d}x\right| \\ &\le C_1 \left(\int_{-\infty}^{\infty} v(x,y)|u(x,y)|^{p-1} \, \mathrm{d}x + \int_{-\infty}^{\infty} v(x,y)^p \, \mathrm{d}x\right). \end{aligned}$$

Since

$$e^{ip\frac{\pi}{2}} u(x,y)^p = \begin{cases} e^{ip\frac{\pi}{2}} |u(x,y)|^p & \text{if } u(x,y) > 0, \\ e^{-ip\frac{\pi}{2}} |u(x,y)|^p & \text{if } u(x,y) < 0, \end{cases}$$

we obtain

$$\begin{aligned} \left|\cos\left(p\frac{\pi}{2}\right)\right| \int_{-\infty}^{\infty} |u(x,y)|^p \, \mathrm{d}x &= \left|\Re\left(e^{ip\frac{\pi}{2}} \int_{-\infty}^{\infty} u(x,y)^p \, \mathrm{d}x\right)\right| \\ &\le \left|\int_{-\infty}^{\infty} u(x,y)^p \, \mathrm{d}x\right|. \end{aligned}$$

Then, by HÖLDER'S inequality,

$$\begin{aligned}\left|\cos\left(p\frac{\pi}{2}\right)\right| |u(\cdot,y)|_p^p &\le C_1\{|u(\cdot,y)^{\frac{p}{p'}}v(\cdot,y)|_1+|v(\cdot,y)|_p^p\}\\ &\le C_1\{|u(\cdot,y)^{\frac{p}{p'}}|_{p'}|v(\cdot,y)|_p+|v(\cdot,y)|_p^p\}\\ &= C_1\{|u(\cdot,y)|_p^{p-1}|v(\cdot,y)|_p+|v(\cdot,y)|_p^p\}.\end{aligned}$$

Dividing the above inequality by $|u(\cdot,y)|_p^{p-1}|v(\cdot,y)|_p$ if $u(\cdot,y)$ is not identically zero and setting $r:=|u(\cdot,y)|_p|v(\cdot,y)|_p^{-1}$ we obtain $|\cos(p\frac{\pi}{2})|r\le C_1(1+r^{1-p})$. Hence $r\le\left(\frac{C}{\pi}-1\right)$ by definition of C, i. e.,

$$|u(\cdot,y)|_p\le\left(\frac{C}{\pi}-1\right)|v(\cdot,y)|_p\,, \tag{4.5.2}$$

which trivially holds if u is identically zero. Applying HÖLDER'S inequality to

$$\frac{f}{((\cdot+x)^2+y^2)^{\frac{1}{p}}}\in L_p(c,d)\quad\text{and}\quad\frac{1}{((\cdot+x)^2+y^2)^{\frac{1}{p'}}}\in L_{p'}(c,d)$$

we obtain that

$$\begin{aligned}v(x,y)^p &\le y^p\int_c^d\frac{f(t)^p}{(t+x)^2+y^2}\,\mathrm{d}t\left\{\int_c^d\frac{\mathrm{d}t}{(t+x)^2+y^2}\right\}^{p-1}\\ &\le\pi^{p-1}y\int_c^d\frac{f(t)^p}{(t+x)^2+y^2}\,\mathrm{d}t\,.\end{aligned}$$

Hence

$$\begin{aligned}|v(\cdot,y)|_p^p &\le\pi^{p-1}y\int_c^d f(t)^p\int_{-\infty}^{\infty}\frac{\mathrm{d}x}{(t+x)^2+y^2}\,\mathrm{d}t\\ &=\pi^p|f|_p^p\,.\end{aligned}$$

Together with (4.5.2) we infer

$$|\phi_f(\cdot+iy)|_p\le C|f|_p\,.$$

ii) Now let $2<p<\infty$ and $1<p'<2$ such that $1/p+1/p'=1$. Applying part i) of the proof to p' we obtain that there is a $C>0$ such that

$$|\phi_g(\cdot+iy)|_{p'}\le C|g|_{p'}$$

for all $y>0$ and $g\in L_{p'}(\mathbb{R})$ with compact support. Here we have to note that the number C in part i) does not depend on c and d. Let $f\in L_p(c,d)$, $r>0$, $y>0$ and $g\in L_{p'}(-r,r)$. By FUBINI'S theorem and HÖLDER'S inequality we have

$$\begin{aligned}\left|\int_{-r}^{r}g(x)\phi_f(x+iy)\,\mathrm{d}x\right| &=\left|\int_c^d\int_{-r}^{r}\frac{g(x)}{t+x+iy}\,\mathrm{d}x\,f(t)\,\mathrm{d}t\right|\\ &=\left|\int_c^d f(t)\phi_g(t+iy)\,\mathrm{d}t\right|\le|f|_pC|g|_{p'}\,.\end{aligned}$$

Since $L_p(-r,r)$ is the dual of $L_{p'}(-r,r)$, we obtain $\phi_f(\cdot+iy)|_{(-r,r)} \in L_p(-r,r)$ and $|\phi_f(\cdot+iy)|_{(-r,r)}|_p \le C|f|_p$. B. LEVI'S theorem yields $\phi_f(\cdot+iy) \in L_p(\mathbb{R})$ and $|\phi_f(\cdot+iy)|_p \le C|f|_p$. □

PROPOSITION 4.5.2. *Let* $1 < p < \infty$, $c_1 > 0$ *and* $c_2 > 0$. *For* $f \in L_p(0,c_1)$ *we set*

$$\mathscr{H}(f)(x) := \int_0^{c_1} \frac{f(t)}{t+x}\,\mathrm{d}t \qquad (x \in (0,c_2)).$$

Then $\mathscr{H} \in L(L_p(0,c_1), L_p(0,c_2))$.

Proof. It is sufficient to consider the case $f \ge 0$. Let ϕ_f and u be as in Proposition 4.5.1. Since $u(\cdot,y)|_{(0,c_2)} \ge 0$ and $u(\cdot,y)|_{(0,c_2)} \nearrow \mathscr{H}(f)$ as $y \searrow 0$, B. LEVI'S theorem and Proposition 4.5.1 yield $\mathscr{H}(f) \in L_p(0,c_2)$ and

$$|\mathscr{H}(f)|_p = \lim_{y\searrow 0} \left| u(\cdot,y)|_{(0,c_2)} \right|_p \le C|f|_p. \quad \square$$

There are simpler proofs of Proposition 4.5.2, see e. g. [HLP, Theorem 316]. But since we need Proposition 4.5.1 to prove Proposition 4.5.3, we have also used it for the proof of Proposition 4.5.2.

PROPOSITION 4.5.3. *Let* $1 < p < \infty$ *and* $r \in L_p(a,b)$ *such that* $r \ge 0$ *and* $r^{-1} \in L_\infty(a,b)$. *Set*

$$R(x) := \int_a^x r(t)\,\mathrm{d}t \qquad (x \in [a,b])$$

and, for $\alpha > 0$, $f \in L_p(a,b)$ *and* $x \in (a,b)$,

$$\psi_{f,\alpha}(x) := \int_{-\alpha}^{\alpha} \int_a^b \exp\{i\tau(R(x) - R(\xi))\} f(\xi)\,\mathrm{d}\xi\,\mathrm{d}\tau.$$

Then $\psi_{f,\alpha} \in L_p(a,b)$, *and there is a constant* $C > 0$ *such that*

$$|\psi_{f,\alpha}|_p \le C|f|_p$$

for all $\alpha > 0$ *and* $f \in L_p(a,b)$.

Proof. Since $\psi_{f,\alpha}$ depends continuously on x it is clear that $\psi_{f,\alpha} \in L_p(a,b)$. Let $g \in L_{p'}(a,b)$ where $1/p + 1/p' = 1$. Applying the transformations ρ which is the inverse of $x \mapsto R(x)$, and $t = -R(\xi)$, we obtain with the aid of the theorem on integration by substitution, see [HS, (20.5)], that

$$\text{(4.5.3)} \qquad \int_a^b g(x)\psi_{f,\alpha}(x)\,\mathrm{d}x$$
$$= \int_0^{R(b)} g(\rho(x))\rho'(x) \int_{-\alpha}^{\alpha} \int_{-R(b)}^{0} e^{i\tau(x+t)} f(\rho(-t))\rho'(-t)\,\mathrm{d}t\,\mathrm{d}\tau\,\mathrm{d}x.$$

For $h \in L_p(-R(b),0)$ and $z \in \mathbb{C}$ we define

$$\tilde{\Psi}_{h,\alpha}(z) := \int_{-\alpha}^{\alpha} \int_{-R(b)}^{0} e^{i\tau(t+z)} h(t)\, \mathrm{d}t\, \mathrm{d}\tau.$$

For $\Im(z) > 0$, integration with respect to τ yields

$$\tilde{\Psi}_{h,\alpha}(z) = \int_{-R(b)}^{0} \frac{h(t)}{i(t+z)} \{e^{i\alpha(t+z)} - e^{-i\alpha(t+z)}\}\, \mathrm{d}t.$$

We set $h_\alpha(t) := h(t)e^{i\alpha t}$. Then, with the notation ϕ_f from Proposition 4.5.1,

$$\tilde{\Psi}_{h,\alpha}(z) = -ie^{i\alpha z}\phi_{h_\alpha}(z) + ie^{-i\alpha z}\phi_{h_{-\alpha}}(z),$$

and Proposition 4.5.1 implies that there is a constant $C' > 0$ such that

$$|\tilde{\Psi}_{h,\alpha}(\cdot + iy)|_{(0,R(b))}|_p \leq (e^{-\alpha y} + e^{\alpha y})C'|h|_p.$$

Since $\tilde{\Psi}_{h,\alpha}(\cdot + iy)|_{(0,R(b))}$ converges uniformly to $\tilde{\Psi}_{h,\alpha}|_{(0,R(b))}$ as $y \to 0$, we obtain

$$|\tilde{\Psi}_{h,\alpha}|_{(0,R(b))}|_p \leq 2C'|h|_p.$$

Hence (4.5.3) yields in view of HÖLDER'S inequality and (2.7.1) that

$$\begin{aligned}\left|\int_a^b g(x)\psi_{f,\alpha}(x)\, \mathrm{d}x\right| &\leq 2C'|(g\circ\rho)\rho'|_{p'}|(f\circ\rho)\rho'|_p \\ &\leq 2C'|r^{-1}|_\infty|g|_{p'}|f|_p.\end{aligned}$$

Now RIESZ' theorem yields the assertion of the proposition with $C := 2C'|r^{-1}|_\infty$. □

Proposition 4.5.3 does not hold for $p = \infty$. But since $p = \infty$ is the case of uniform convergence, it is desirable to include this case. For this we have to restrict the class of functions to functions of bounded variation, see e. g. [HS, p. 266] for a definition of these functions. Here we use the fact that f is of bounded variation if and only if there are bounded nonnegative nondecreasing functions f_1, f_2, f_3, f_4 such that

$$f = f_1 - f_2 + if_3 - if_4. \tag{4.5.4}$$

We denote the set of all functions of bounded variation on $[a,b]$ by $BV[a,b]$. It is a Banach space with respect to the norm

$$|f|_{BV} = \inf\{|f_1|_\infty + |f_2|_\infty + |f_3|_\infty + |f_4|_\infty\}, \tag{4.5.5}$$

where the infimum is taken over all decompositions (4.5.4) having the above properties. The norm on $BV[a,b]$ is usually defined as

$$|f|'_{BV} = |f(a)| + \sup\left\{\sum_{\nu=1}^{j} |f(x_\nu) - f(x_{\nu-1})| : a = x_0 < x_1 < \cdots < x_j = b\right\}.$$

The two norms are equivalent since

$$|f|'_{BV} \le |f|_{BV} \le 2|f|'_{BV}.$$

PROPOSITION 4.5.4. *Let $r \in L_\infty(a,b)$ such that $r \ge 0$ and $r^{-1} \in L_\infty(a,b)$. Set*

$$R(x) := \int_a^x r(t)\,\mathrm{d}t \qquad (x \in [a,b])$$

and, for $\alpha > 0$, $f \in L_\infty(a,b)$ and $x \in (a,b)$,

$$\psi_{f,\alpha}(x) := \int_{-\alpha}^{\alpha}\int_a^b \exp\{i\tau(R(x)-R(\xi))\}f(\xi)\,\mathrm{d}\xi\,\mathrm{d}\tau.$$

Then $\psi_{f,\alpha} \in BV[a,b]$, and there are constants $C_\alpha > 0$ $(\alpha > 0)$ such that

$$|\psi_{f,\alpha}|_{BV} \le C_\alpha |f|_\infty$$

for all $\alpha > 0$ and $f \in L_\infty(a,b)$, and a constant $C > 0$ such that

$$|\psi_{f,\alpha}|_\infty \le C|f|_{BV}$$

for all $\alpha > 0$ and $f \in BV[a,b]$.

Proof. The function $\psi_{f,\alpha}$ belongs to $W^1_\infty(a,b)$ and

$$\psi'_{f,\alpha}(x) = \int_{-\alpha}^{\alpha}\int_a^b i\tau r(x)\exp\{i\tau(R(x)-R(\xi))\}f(\xi)\,\mathrm{d}\xi\,\mathrm{d}\tau.$$

In view of Proposition 2.1.5 i) this implies

$$\frac{1}{2}|\psi_{f,\alpha}|_{BV} \le |\psi_{f,\alpha}|'_{BV} = |f(a)| + \int_a^b |\psi'_{f,\alpha}(x)|\,\mathrm{d}x \le \alpha^2|r|_\infty|f|_1.$$

Therefore the first estimate holds. We have proved it for functions in $L_1(a,b)$, but we shall only apply it to functions in $L_\infty(a,b)$.

Let ρ be the inverse of $x \mapsto R(x)$ and $\eta = R(x)$, $x \in [a,b]$. Then integrating with respect to τ and the theorem on integration by substitution, see [HS, (20.5)], yield

$$\begin{aligned}\psi_{f,\alpha}(x) &= 2\int_a^b \frac{\sin\{\alpha(\eta - R(\xi))\}}{\eta - R(\xi)}f(\xi)\,\mathrm{d}\xi \\ &= 2\int_0^{R(b)} \frac{\sin(\alpha(\eta - t))}{\eta - t}f(\rho(t))\rho'(t)\,\mathrm{d}t.\end{aligned}$$

We set $h(t) = f(\rho(t))\rho'(t)$ and obtain that h is of bounded variation with $|h|_{BV} \le |f|_{BV}|r^{-1}|_\infty$. Let $h = h_1 - h_2 + ih_3 - ih_4$ be the decomposition (4.5.4) for h, and set $\hat{f}_j(\xi) := h_i(R(\xi))r(\xi)$ for $j = 1,\ldots,4$. With the substitutions $\alpha(\eta - t) \mapsto t$ and $\alpha(\eta - t) \mapsto -t$ we obtain

$$\begin{aligned}\psi_{\hat{f}_j,\alpha}(x) =& 2\int_0^{\alpha\eta} \frac{\sin t}{t} h_j\left(\dot{\eta} - \frac{t}{\alpha}\right) \mathrm{d}t + 2\int_0^{\alpha(R(b)-\eta)} \frac{\sin t}{t} |h_j|_\infty \,\mathrm{d}t \\ &- 2\int_0^{\alpha(R(b)-\eta)} \frac{\sin t}{t}\left(|h_j|_\infty - h_j\left(\eta + \frac{t}{\alpha}\right)\right) \mathrm{d}t.\end{aligned}$$

Each of the above integrals is an integral of the product of $\frac{\sin t}{t}$ with a nonnegative nonincreasing function. Hence an upper bound of the absolute value of each of these integrals is the corresponding integral over $[0, \pi]$. This yields

$$|\psi_{\hat{f}_j,\alpha}(x)| \le 6\pi |h_j|_{BV} \le 6\pi |h|_{BV} \le 6\pi |r^{-1}|_\infty |f|_{BV},$$

and the desired estimate holds with $C = 24\pi |r^{-1}|_\infty$. □

LEMMA 4.5.5. *Let $c > 0$ and H_c be the class of all continuous real-valued functions h on an interval $(0, c)$ with the following properties:*
$(0, c)$ can be divided into k_h subintervals such that h is monotonic and does not change sign on any of these subintervals;

$$\sup_{t\in(0,c)} |th(t)| =: C_h < \infty.$$

Then for each $0 < \beta \le 1$ there is a constant $C > 0$ such that

$$\left|\int_0^c \sin(rt)h(t)f(t)\,\mathrm{d}t\right| \le CC_h k_h |f|_{BV}, \tag{4.5.6}$$

$$\left|\int_0^c \left(\cos(\beta rt) - \chi_{[0,\frac{\pi}{2r}]}(t)\right)h(t)f(t)\,\mathrm{d}t\right| \le CC_h k_h |f|_{BV}, \tag{4.5.7}$$

$$\left|\int_0^c \left(e^{-rt} - \chi_{[0,\frac{\pi}{2r}]}(t)\right)h(t)f(t)\,\mathrm{d}t\right| \le CC_h |f|_\infty, \tag{4.5.8}$$

$$\left|\int_0^c \left(e^{-\alpha rt} - 1\right)\chi_{[0,\frac{\pi}{2r}]}(t)h(t)f(t)\,\mathrm{d}t\right| \le CC_h |f|_\infty, \tag{4.5.9}$$

holds for all $c > 0$, $h \in H_c$, $f \in BV[0, c]$, $r > 0$, and $\alpha \in [0, 1]$. Here χ denotes the characteristic function.

Proof. We shall prove the four estimates with different constants C. Then we take the maximum of these four constants as a common estimate. For the proof of (4.5.6) it is sufficient to consider nonnegative nondecreasing $f \in BV[0, c]$ and a subinterval $[c_1, c_2] \subset [0, c]$ on which h is monotonic and does not change sign; we may even assume that h is nonnegative. Then

$$\int_{c_1}^{c_2} \sin(rt)h(t)f(t)\,\mathrm{d}t = \int_{c_1}^{c_2} \sin(rt)h(t)|f|_\infty\,\mathrm{d}t - \int_{c_1}^{c_2} \sin(rt)h(t)\left(|f|_\infty - f(t)\right)\mathrm{d}t.$$

Taking the left-hand side if h is nondecreasing and the right-hand side if h is nonincreasing, we see that in each of these integrals the integrand is the product of

$\sin(rt)$ with a nonnegative monotonic function. An upper bound for the modulus of such an integral is obtained if we take the modulus of the integral over a subinterval $[q\frac{\pi}{r},(q+1)\frac{\pi}{r}]\cap[c_1,c_2]$ for a suitable nonnegative integer q. From the assumptions on h we infer

$$\left|\int_{c_1}^{c_2}\sin(rt)h(t)f(t)\,\mathrm{d}t\right|\le\int_{q\frac{\pi}{r}}^{(q+1)\frac{\pi}{r}}r\left|\frac{\sin(rt)}{rt}\right|\mathrm{d}t\,2C_h|f|_\infty\le 2\pi C_h|f|_\infty.$$

This proves (4.5.6) with $C=2\pi$.

To prove (4.5.7) we calculate for $c'=\min\{c,\frac{\pi}{2r}\}$

$$\left|\int_0^{c'}(1-\cos(rt))h(t)f(t)\,\mathrm{d}t\right|$$
$$\le\int_0^{\frac{\pi}{2r}}\frac{\sin^2(rt)}{t(1+\cos(rt))}\,\mathrm{d}t\,C_h|f|_\infty\le\int_0^{\frac{\pi}{2r}}r^2t\,\mathrm{d}t\,C_h|f|_\infty\le\frac{\pi^2}{8}C_h|f|_\infty.$$

In case $c>\frac{\pi}{2r}$ we obtain from (4.5.6) that

$$\left|\int_{\frac{\pi}{2r}}^{c}\cos(rt)h(t)f(t)\,\mathrm{d}t\right|=\left|\int_0^{c-\frac{\pi}{2r}}\sin(rt)h\left(t+\frac{\pi}{2r}\right)f\left(t+\frac{\pi}{2r}\right)\mathrm{d}t\right|$$
$$\le 2\pi C_h k_h|f|_{BV}$$

since

$$\left|h\left(t+\frac{\pi}{2r}\right)t\right|\le\left|h\left(t+\frac{\pi}{2r}\right)\left(t+\frac{\pi}{2r}\right)\right|\le C_h.$$

This proves (4.5.7) for $\beta=1$. If $0<\beta<1$, then we replace r by βr. The difference to the integral in (4.5.7) can be estimated by

$$\int_{\frac{\pi}{2r}}^{\frac{\pi}{2\beta r}}\frac{\mathrm{d}t}{t}C_h|f|_\infty=|\log\beta|C_h|f|_\infty.$$

This completes the proof of (4.5.7).

Again with $c'=\min\{c,\frac{\pi}{2r}\}$ and for arbitrary $f\in BV[a,b]$ (we can even take $f\in L_\infty(a,b)$), the estimate

$$\left|\int_0^{c'}(1-e^{-\alpha rt})h(t)f(t)\,\mathrm{d}t\right|\le\int_0^{\frac{\pi}{2r}}\frac{1-e^{-\alpha rt}}{t}\,\mathrm{d}t\,C_h|f|_\infty\le\frac{\alpha\pi}{2}C_h|f|_\infty$$

and, in case $c>\frac{\pi}{2r}$, the estimate

$$\left|\int_{\frac{\pi}{2r}}^{c}e^{-rt}h(t)f(t)\,\mathrm{d}t\right|\le\int_{\frac{\pi}{2}}^{rc}\frac{e^{-s}}{s}\,\mathrm{d}s\,C_h|f|_\infty\le\frac{2}{\pi}e^{-\frac{\pi}{2}}C_h|f|_\infty$$

prove the inequalities (4.5.8) and (4.5.9). □

PROPOSITION 4.5.6. *Let $\gamma_1, \gamma_2 \in \mathbb{R}$ and Σ be a sector in the complex plane which is bounded by the rays* $\arg\lambda = \gamma_1$ *and* $\arg\lambda = \gamma_2$. *Let $\eta_1, \eta_2 \in \mathbb{R}$ be such that $\gamma_\nu + \eta_\mu \in [\frac{\pi}{2}, \frac{3}{2}\pi] \operatorname{mod}(2\pi)$ for $\nu, \mu \in \{1,2\}$. If $\eta_1 \neq \eta_2$ we require $\gamma_2 - \gamma_1 \notin \pi\mathbb{Z}$. Let $c_1, c_2 > 0$. For $f \in BV[0,c_1]$, $r > 0$, and $x \in [0,c_2]$ we define*

$$\psi_{f,r}(x) := \int_{\substack{\lambda\in\Sigma\\ |\lambda|=r}} \exp\{\lambda e^{i\eta_1}x\} \int_0^{c_1} \exp\{\lambda e^{i\eta_2}\tau\} f(\tau)\,\mathrm{d}\tau\,\mathrm{d}\lambda.$$

Then there is a constant $C > 0$ such that

$$|\psi_{f,r}|_\infty \leq C|f|_{BV}$$

for all $r > 0$ and $f \in BV[0,c_1]$.

Proof. Integrating with respect to λ yields

$$|\psi_{f,r}(x)| = \left| \int_0^{c_1} \frac{\exp\{e^{i\gamma_2} r(e^{i\eta_1}x + e^{i\eta_2}\tau)\} - \exp\{e^{i\gamma_1} r(e^{i\eta_1}x + e^{i\eta_2}\tau)\}}{e^{i\eta_1}x + e^{i\eta_2}\tau} f(\tau)\,\mathrm{d}\tau \right|.$$

Subtracting and adding $\chi_{[0,\frac{\pi}{2r}]}(\tau + x)$ in the numerator, it is sufficient to prove the estimate for

$$\int_0^{c_1} \frac{\exp\{e^{i\gamma} r(e^{i\eta_1}x + e^{i\eta_2}\tau)\} - \chi_{[0,\frac{\pi}{2r}]}(\tau + x)}{e^{i\eta_1}x + e^{i\eta_2}\tau} f(\tau)\,\mathrm{d}\tau, \tag{4.5.10}$$

where $\gamma = \gamma_1$ or $\gamma = \gamma_2$. Since $\gamma_2 - \gamma_1 \notin \pi\mathbb{Z}$ in case $\eta_1 \neq \eta_2$ and $\gamma_\nu + \eta_\mu \in [\frac{\pi}{2}, \frac{3}{2}\pi] \operatorname{mod}(2\pi)$, we infer $\eta_2 - \eta_1 \notin \pi + 2\pi\mathbb{Z}$. Replacing $\gamma + \eta_\nu$ with η_ν we thus have to estimate (4.5.10) for $\gamma = 0$, $\eta_1, \eta_2 \in [\frac{\pi}{2}, \frac{3}{2}\pi]$ and $\eta_2 - \eta_1 \notin \pi + 2\pi\mathbb{Z}$. The denominator becomes $e^{-i\gamma}(e^{i\eta_1}x + e^{i\eta_2}\tau)$. Since the constant $e^{-i\gamma}$ has modulus 1, we can omit it in the sequel. We further simplify the numerator by writing it as

$$\begin{aligned} &\exp\{r(e^{i\eta_1}x + e^{i\eta_2}\tau)\} - \exp\{r(e^{i\eta_1}x + e^{i\eta_1}\tau)\} \\ &+ \exp\{r(e^{i\eta_1}x + e^{i\eta_1}\tau)\} - \chi_{[0,\frac{\pi}{2r}]}(\tau + x). \end{aligned}$$

In the first two terms, we factor out the τ-independent term $\exp\{re^{i\eta_1}x\}$, whose modulus does not exceed 1. We write the remaining part of this term as

$$\exp\{re^{i\eta_2}\tau\} - \chi_{[0,\frac{\pi}{2r}]}(\tau) - \left[\exp\{re^{i\eta_1}\tau\} - \chi_{[0,\frac{\pi}{2r}]}(\tau)\right].$$

Hence we have to estimate the integrals

$$\int_0^{c_1} \frac{\exp\{re^{i\eta_j}\tau\} - \chi_{[0,\frac{\pi}{2r}]}(\tau)}{e^{i\eta_1}x + e^{i\eta_2}\tau} f(\tau)\,\mathrm{d}\tau \quad (j = 1,2)$$

and

$$\int_0^{c_1} \frac{\exp\{r(e^{i\eta_1}x + e^{i\eta_1}\tau)\} - \chi_{[0,\frac{\pi}{2r}]}(\tau + x)}{e^{i\eta_1}x + e^{i\eta_2}\tau} f(\tau)\,\mathrm{d}\tau.$$

With the transformation $\tau + x \mapsto \tau$, the latter integral can be written as

$$\int_0^{c_1+x} \frac{\exp\{re^{i\eta_1}\tau\} - \chi_{[0,\frac{\pi}{2r}]}(\tau)}{(e^{i\eta_1} - e^{i\eta_2})x + e^{i\eta_2}\tau} f(\tau - x)\, d\tau,$$

where $f(\xi) := 0$ if $\xi < 0$. Since the norm in $BV[0,c_1]$ of $\tau \mapsto f(\tau)$ and the norm in $BV[0,c_1+x]$ of $\tau \mapsto f(\tau - x)$ coincide, it is therefore sufficient to prove that

$$\left| \int_0^{c} \frac{\exp\{re^{i\eta_j}\tau\} - \chi_{[0,\frac{\pi}{2r}]}(\tau)}{wx+\tau} f(\tau)\, d\tau \right| \le C|f|_{BV} \quad (j = 1,2),$$

where w is a complex number which is not a negative real and the constant C is independent of x and c. Again, we split up the numerator. We have $e^{i\eta_j} = -\alpha + i\beta$, where α and β are real numbers with $\alpha \ge 0$. If $\beta \ne 0$, then we write

$$\begin{aligned}\exp\{re^{i\eta_j}\tau\} - \chi_{[0,\frac{\pi}{2r}]}(\tau) &= \exp(-\alpha r\tau)\big(\cos(\beta r\tau) + i\sin(\beta r\tau) - \chi_{[0,\frac{\pi}{2r}]}(\tau)\big)\\ &\quad + \big(\exp(-\alpha r\tau) - 1\big)\chi_{[0,\frac{\pi}{2r}]}(\tau).\end{aligned}$$

If $\beta = 0$, then we simply have

$$\exp(-r\tau) - \chi_{[0,\frac{\pi}{2r}]}(\tau).$$

Hence the result follows from Lemma 4.5.5 if we show that the real and imaginary parts h_1, h_2 of

$$h(t) = \frac{\exp(-st)}{wx+t} \tag{4.5.11}$$

are functions of class H_c, where an upper bound for k_{h_j} and C_{h_j} $(j = 1,2)$ can be found which does not depend on $x \ge 0$, $c > 0$, and $s \ge 0$. This is obvious if $w = 0$. For $w \ne 0$ we have

$$|h(t)t| \le \frac{1}{|w\frac{x}{t} + 1|} \le \begin{cases} 1 & \text{if } \Re(w) > 0,\\ \frac{|w|}{|\Im(w)|} & \text{if } \Re(w) \le 0.\end{cases}$$

From

$$h(t) = \frac{\exp(-st)(\overline{w}x+t)}{|wx+t|^2},$$

$$h'(t) = -\exp(-st)\frac{[s(wx+t)+1](\overline{w}x+t)^2}{|wx+t|^4},$$

we infer that h_1 and h_2 have at most one zero and three turning points. Thus the numbers k_{h_1} and k_{h_2} are at most 5. □

4.6. Improved estimates of the Green's matrix

PROPOSITION 4.6.1. *Let $f \in (L_p(a,b))^n$. Then each component of the matrix function $I_3^0((\Delta_0 f,0),\lambda)$ defined in* (4.4.12) *is a sum of terms of the form*

$$\delta_\nu(\lambda)\int_0^{|R_\nu(b)|}\exp\{-\lambda e^{ivarphi_\nu}\tau\}u(\tau)\,\mathrm{d}\tau \tag{4.6.1}$$

and

$$(1-\delta_\nu(\lambda))\int_0^{|R_\nu(b)|}\exp\{\lambda e^{i\varphi_\nu}\tau\}u(\tau)\,\mathrm{d}\tau, \tag{4.6.2}$$

where $\nu \in \{1,\dots,l\}$, $u \in L_p(a,b)$, and $|u|_p \le C|f|_p$ for some $C>0$ which does not depend on f.

Proof. Let ρ_ν be the inverse function of $x \mapsto |R_\nu(x)|$. In the proof of Proposition 2.7.1 we have shown that $\rho_\nu \in W_\infty^1(0,|R_\nu(b)|)$. The theorem on integration by substitution, see [HS, (20.5)], shows for $g \in L_p(a,b)$ that

$$\int_{a_j}^b \exp\{\lambda R_\nu(a_j)\}\exp\{-\lambda R_\nu(\xi)\}g(\xi)\,\mathrm{d}\xi = \int_0^{|R_\nu(b)|}\exp\{-\lambda e^{i\varphi_\nu}\tau\}\tilde{g}(\tau)\,\mathrm{d}\tau,$$

where

$$\tilde{g}(\tau) = \begin{cases} g\big(\rho_\nu(\tau+|R_\nu(a_j)|)\big)\rho_\nu'(\tau+|R_\nu(a_j)|) & \text{if } 0 \le \tau \le |R_\nu(b)|-|R_\nu(a_j)|,\\ 0 & \text{if } |R_\nu(b)|-|R_\nu(a_j)| < \tau \le |R_\nu(b)|, \end{cases}$$

and $\tilde{g} \in L_p(0,|R_\nu(b)|)$ with $|\tilde{g}|_p \le C_1|g|_p$ for some $C_1>0$ which does not depend on g, see (2.7.1). This representation and a similar representation for the integration over $[a,a_j]$ prove that the components of the sums in $I_3^0((\Delta_0 f,0),\lambda)$ are sums of terms of the form (4.6.1) and (4.6.2) which fulfil the estimate $|u|_p \le C_2|f|_p$ for some C_2 independent of u in view of assumption (4.1.11). From Proposition 2.7.1 we infer that this also holds for the double integrals in $I_3^0((\Delta_0 f,0),\lambda)$. □

Let χ_k $(k=1,\dots,l_0)$ be as defined in Remark 4.1.4. Set

$$\delta^k(\lambda) := \begin{cases} 0 & \text{if } \Re(\lambda e^{i\chi_k}) < 0,\\ 1 & \text{if } Re(\lambda e^{i\chi_k}) > 0,\\ 0 & \text{if } \Re(\lambda e^{i\chi_k}) = 0 \text{ and } \Im(\lambda e^{i\chi_k}) > 0,\\ 1 & \text{if } \Re(\lambda e^{i\chi_k}) = 0 \text{ and } \Im(\lambda e^{i\chi_k}) < 0. \end{cases}$$

The following result is obvious:

REMARK 4.6.2. For each $\nu \in \{1,\dots,l\}$, $\lambda \in \mathbb{C}\setminus\{0\}$ and $\tau \in \mathbb{R}$ we have

$$\delta_\nu(\lambda)\exp\{-\lambda e^{i\varphi_\nu}\tau\} = \delta^k(\lambda)\exp\{-\lambda e^{i\chi_k}\tau\}$$

if $\chi_k = \varphi_\nu$, and

$$(1-\delta_\nu(\lambda))\exp\{\lambda e^{i\varphi_\nu}\tau\} = \delta^k(\lambda)\exp\{-\lambda e^{i\chi_k}\tau\}$$

if $\chi_k = (\varphi_\nu+\pi)\,\mathrm{mod}(2\pi)$.

PROPOSITION 4.6.3. *Let $1 < p < \infty$ and suppose that the boundary eigenvalue problem* (4.1.1), (4.1.2) *is Birkhoff regular. Then we obtain*

$$\left| \oint_{|\lambda|=r} \Delta_0 I_2^0(\cdot,\lambda) M_0^{-1}(\lambda) I_3^0((\Delta_0 f,0),\lambda)\,\mathrm{d}\lambda \right|_p = O(1)|f|_p \ \text{as } r \to \infty,$$

where f varies in $(L_p(a,b))^n$ and Δ_0, M_0, I_2^0, I_3^0 are defined in (4.1.22), (4.3.5), (4.4.7), (4.4.12).

Proof. By Remark 4.3.2, M_0 is constant in each Σ_m $(m = 1,\dots,l_0)$. From Proposition 4.6.1 and Remark 4.6.2 we infer that in every sector Σ_m $(m = 1,\dots,l_0)$ each component of $\Delta_0 I_2^0(x,\lambda) M_0^{-1}(\lambda) I_3^0((\Delta_0 f,0),\lambda)$ is a linear combination of terms of the form

$$\begin{aligned} &w(x)\big((1-\delta_\nu(\lambda))\exp\{\lambda R_\nu(x)\} + \delta_\nu(\lambda)\exp\{\lambda(R_\nu(x)-R_\nu(b))\}\big) \times \\ &\quad\times\ \delta^k(\lambda)\int_0^{c_1} \exp\{-\lambda e^{i\chi_k}\tau\}u(\tau)\,\mathrm{d}\tau \quad (\nu = 1,\dots,n;\ k = 1,\dots,l_0), \end{aligned}$$

where $w \in L_\infty(a,b)$ is a component of the matrix function $P^{[0]}$, c_1 is one of the numbers $|R_1(b)|,\dots,|R_l(b)|$, $u \in L_p(0,c_1)$, and $|u|_p \le C|f|_p$ for some $C > 0$ not depending on f. Let ρ_j be the inverse of $x \mapsto |R_j(x)|$. We apply the transformations $x \mapsto \rho_j(\xi)$ and $x \mapsto \rho_j(|R_j(b)| - \xi)$, respectively, and obtain as in the proof of Proposition 4.6.1 with the aid of the theorem on integration by substitution, see [HS, (20.5)], that Proposition 4.6.3 is proved if we show that

(4.6.3)
$$\left| \int_{\substack{\lambda\in\Sigma_m\\ |\lambda|=r}} \delta^j(\lambda)\delta^k(\lambda)\exp\{-\lambda e^{i\chi_j}\cdot\} \int_0^{c_1} \exp\{-\lambda e^{i\chi_k}\tau\}u(\tau)\,\mathrm{d}\tau\,\mathrm{d}\lambda \right|_p = O(1)|u|_p$$

for $j,k \in \{1,\dots,l_0\}$ and for $u \in L_p(0,c_1)$, where $c_1, c_2 > 0$ and the norm on the left hand side is taken in $L_p(0,c_2)$. Since δ^j, δ^k are constant on Σ_m (see Proposition 4.1.5), we only have to consider those $j,k \in \{1,\dots,l_0\}$ for which $\delta^j(\lambda) = \delta^k(\lambda) = 1$ for all $\lambda \in \Sigma_m$.

First let $j \ne k$. Since $\Re(\lambda e^{i\chi_j}) \ge 0$ and $\Re(\lambda e^{i\chi_k}) \ge 0$ as $\delta^j(\lambda) = \delta^k(\lambda) = 1$, Lemma 2.7.2 ii) yields that there is a $C > 0$ such that, for $v \in L_{p'}(0,c_2)$,

$$\begin{aligned} &\left| \int_0^{c_2} v(\xi) \int_{\substack{\lambda\in\Sigma_m\\ |\lambda|=r}} \exp\{-\lambda e^{i\chi_j}\xi\} \int_0^{c_1} \exp\{-\lambda e^{i\chi_k}\tau\}u(\tau)\,\mathrm{d}\tau\,\mathrm{d}\lambda\,\mathrm{d}\xi \right| \\ &\le C \int_{\substack{\lambda\in\Sigma_m\\ |\lambda|=r}} \big(1+|\Re(\lambda e^{i\chi_j})|\big)^{-1/p}\big(1+|\Re(\lambda e^{i\chi_k})|\big)^{1/p-1} |\mathrm{d}\lambda|\,|v|_{p'}|u|_p \\ &= C \int_{-\frac{\pi}{2}-\chi_m}^{-\frac{\pi}{2}-\chi_{m-1}} \left(\frac{1}{r} + |\cos(\varphi+\chi_j)|\right)^{-1/p} \left(\frac{1}{r} + |\cos(\varphi+\chi_k)|\right)^{1/p-1} \mathrm{d}\varphi\,|v|_{p'}|u|_p. \end{aligned}$$

The integrand is bounded by the function

$$g(\varphi) = |\cos(\varphi+\chi_j)|^{-1/p}|\cos(\varphi+\chi_k)|^{1/p-1}.$$

We have $\chi_j - \chi_k \notin \{-\pi, \pi\}$ since $\delta^j(\lambda)\delta^k(\lambda) = 1$ for λ in the non-empty set Σ_m. We infer that the function g is piecewise bounded by functions of the form $C\cos(\cdot + \alpha)^{-\beta}$ with $\beta \in \{1/p, 1-1/p\}$. Since

$$x \mapsto \left|\cos(x + \tfrac{\pi}{2})\right|^{-\beta} = \left(\frac{\cos(x+\frac{\pi}{2})}{|x|}\right)^{-\beta} |x|^{-\beta}$$

is the product of a bounded function and an integrable function in a neighbourhood of 0, g is an integrable function, which proves (4.6.3) for $j \neq k$.

Now let $j = k$. We calculate

$$\begin{aligned}
&\int_{\substack{\lambda\in\Sigma_m\\|\lambda|=r}} \exp\{-\lambda e^{i\chi_j}\xi\} \int_0^{c_1} \exp\{-\lambda e^{i\chi_j}\tau\} u(\tau)\, \mathrm{d}\tau\, \mathrm{d}\lambda \\
&\quad = \int_0^{c_1} \int_{\substack{\lambda\in\Sigma_m\\|\lambda|=r}} \exp\{-\lambda e^{i\chi_j}(\xi+\tau)\}\, \mathrm{d}\lambda\, u(\tau)\, \mathrm{d}\tau \\
&\quad = \int_0^{c_1} (e^{i\chi_j}(\xi+\tau))^{-1} \times \\
&\qquad \times \left(\exp\left\{re^{i(\frac{\pi}{2}-\chi_m+\chi_j)}(\xi+\tau)\right\} - \exp\left\{re^{i(\frac{\pi}{2}-\chi_{m-1}+\chi_j)}(\xi+\tau)\right\}\right) u(\tau)\, \mathrm{d}\tau \\
&\quad = e^{-i\chi_j} \exp\left\{re^{i(\frac{\pi}{2}-\chi_m+\chi_j)}\xi\right\} \int_0^{c_1} \frac{1}{\xi+\tau} \exp\left\{re^{i(\frac{\pi}{2}-\chi_m+\chi_j)}\tau\right\} u(\tau)\, \mathrm{d}\tau \\
&\quad\quad - e^{-i\chi_j} \exp\left\{re^{i(\frac{\pi}{2}-\chi_{m-1}+\chi_j)}\xi\right\} \int_0^{c_1} \frac{1}{\xi+\tau} \exp\left\{re^{i(\frac{\pi}{2}-\chi_{m-1}+\chi_j)}\tau\right\} u(\tau)\, \mathrm{d}\tau.
\end{aligned}$$

Since $\Re(\lambda e^{i\chi_j})$ is nonnegative for all $\lambda \in \Sigma_m$, we obtain that $\Re\left(e^{i(\frac{\pi}{2}-\chi_m+\chi_j)}\right)$ and $\Re\left(e^{i(\frac{\pi}{2}-\chi_{m-1}+\chi_j)}\right)$ are nonpositive. Thus (4.6.3) is proved because of Proposition 4.5.2. □

PROPOSITION 4.6.4. *Let $\beta > 0$. Then*

$$\int_{|\lambda|=r} (1 + |\Re(\lambda)|)^{-1-\beta}\, |\mathrm{d}\lambda| = O(1) \text{ as } r \to \infty.$$

Proof. We calculate

$$\begin{aligned}
\int_{|\lambda|=r} (1 + |\Re(\lambda)|)^{-1-\beta}\, |\mathrm{d}\lambda| &= \int_0^{2\pi} r(1 + r|\cos\varphi|)^{-1-\beta}\, \mathrm{d}\varphi \\
&\le 4\int_0^{\frac{\pi}{4}} r(r\cos\varphi)^{-1-\beta}\, \mathrm{d}\varphi + 8\int_{\frac{\pi}{4}}^{\frac{\pi}{2}} r\sin\varphi(1 + r\cos\varphi)^{-1-\beta}\, \mathrm{d}\varphi \\
&\le \pi r\left(\frac{r}{2}\right)^{-1-\beta} + \frac{8}{\beta}(1 + r\cos\varphi)^{-\beta}\Big|_{\frac{\pi}{4}}^{\frac{\pi}{2}} \\
&\le 2^{\beta+1}\pi r^{-\beta} + \frac{8}{\beta}. \quad \square
\end{aligned}$$

PROPOSITION 4.6.5. *Let $1 < p < \infty$. Then we obtain for $f \in (L_p(a,b))^n$*

$$\left| \oint_{|\lambda|=r} I_1^0(\cdot, \Delta_0 f, \lambda)\, \mathrm{d}\lambda \right|_p = O(1)|f|_p \text{ as } r \to \infty,$$

where Δ_0 and I_1^0 are defined in (4.1.22) *and* (4.4.8).

Proof. We set

$$\gamma_{r,\nu}^+ := \{\lambda \in \mathbb{C} : |\lambda| = r, \Re(\lambda e^{i\varphi_\nu}) > 0\},$$
$$\gamma_{r,\nu}^- := \{\lambda \in \mathbb{C} : |\lambda| = r, \Re(\lambda e^{i\varphi_\nu}) < 0\}.$$

Let $I^{(\nu)}$ be the $n \times n$-matrix whose ν-th diagonal block is I_{n_ν} and whose other components are zero. We have

$$\Delta_0 E(x,\lambda) = \sum_{\nu=1}^{l} I^{(\nu)} E(x,\lambda) = \sum_{\nu=1}^{l} I^{(\nu)} \exp\{\lambda R_\nu(x)\}.$$

Since Δ_0 and ${P^{[0]}}^{-1}$ commute, we obtain

$$\begin{aligned}
&\oint_{|\lambda|=r} I_1^0(x, \Delta_0 f, \lambda)\, \mathrm{d}\lambda \\
&= \oint_{|\lambda|=r} P^{[0]}(x)(I_n - \Delta(\lambda))E(x,\lambda) \int_a^x E(\xi,\lambda)^{-1} {P^{[0]}}^{-1}(\xi)\Delta_0 f(\xi)\, \mathrm{d}\xi\, \mathrm{d}\lambda \\
&- \oint_{|\lambda|=r} P^{[0]}(x)\Delta(\lambda)E(x,\lambda) \int_x^b E(\xi,\lambda)^{-1} {P^{[0]}}^{-1}(\xi)\Delta_0 f(\xi)\, \mathrm{d}\xi\, \mathrm{d}\lambda \\
&= \sum_{\nu=1}^{l} \int_{\gamma_{r,\nu}^-} P^{[0]}(x) I^{(\nu)} \int_a^x \exp\{\lambda(R_\nu(x) - R_\nu(\xi))\} {P^{[0]}}^{-1}(\xi) f(\xi)\, \mathrm{d}\xi\, \mathrm{d}\lambda \\
&- \sum_{\nu=1}^{l} \int_{\gamma_{r,\nu}^+} P^{[0]}(x) I^{(\nu)} \int_x^b \exp\{\lambda(R_\nu(x) - R_\nu(\xi))\} {P^{[0]}}^{-1}(\xi) f(\xi)\, \mathrm{d}\xi\, \mathrm{d}\lambda \\
&= -\sum_{\nu=1}^{l} e^{i(\frac{\pi}{2} - \varphi_\nu)} P^{[0]}(x) I^{(\nu)} \int_{-r}^{r} \int_a^b \exp\{i\tau(|R_\nu(x)| - |R_\nu(\xi)|)\} {P^{[0]}}^{-1}(\xi) f(\xi)\, \mathrm{d}\xi\, \mathrm{d}\tau
\end{aligned}$$

where we have used that the contour integrals along the semicircle $\gamma_{r,\nu}^+$ and the line segment $\overline{re^{i(\frac{\pi}{2}-\varphi_\nu)}, -re^{i(\frac{\pi}{2}-\varphi_\nu)}}$ and the contour integrals along the semicircle $\gamma_{r,\nu}^-$ and the line segment $\overline{-re^{i(\frac{\pi}{2}-\varphi_\nu)}, re^{i(\frac{\pi}{2}-\varphi_\nu)}}$ are 0 by CAUCHY'S theorem. Now the assertion of the proposition immediately follows from Proposition 4.5.3. □

PROPOSITION 4.6.6. *Let $1 < p < \infty$ and suppose that there is a number $\tilde{p} > p$ such that $A_{0,0\nu} \in M_{n_0,n_\nu}(L_{\tilde{p}}(a,b))$ for $\nu = 1,\dots,l$. Here $A_{0,0\nu}$ is the block of A_0 with index $(0,\nu)$ according to the block structure of A_1. Then there is a number $\eta < 1/p$ such that*

$$I_3((\Delta_0 f,0),\lambda) - I_3^0((\Delta_0 f,0),\lambda) = O\left(\max_{\substack{\nu,\mu=0\\ \nu\neq\mu}}^{l}\left(1+|\Re(\lambda e^{i\psi_{\nu\mu}})|\right)^{-1+\eta}\right)|f|_p$$

for $f \in (L_p(a,b))^n$, where Δ_0, I_3, I_3^0 are defined in (4.1.22), (4.4.6), (4.4.12).

Proof. We have

$$\begin{aligned}
&I_3((\Delta_0 f,0),\lambda) - I_3^0((\Delta_0 f,0),\lambda)\\
&= I_3((\Delta_0 f,0),\lambda) - I_3^0(((I_n + P^{[0]}\widetilde{B}_0(\cdot,\lambda))\Delta_0 f,0),\lambda)\\
&\quad + I_3^0((P^{[0]}\widetilde{B}_0(\cdot,\lambda)\Delta_0 f,0),\lambda).
\end{aligned}$$

We multiply the matrix functions in Proposition 4.4.5 by $E(\xi,\lambda)\widetilde{Y}(\xi,\lambda)^{-1}\Delta_0 f(\xi)$ from the right and integrate from a to b. In view of (4.4.11) we obtain with the aid of Proposition 4.4.3, (4.1.10), (4.1.12) and Proposition 4.3.3 i), ii) that

$$\begin{aligned}
&I_3((\Delta_0 f,0),\lambda) - I_3^0(((I_n + P^{[0]}\widetilde{B}_0(\cdot,\lambda))\Delta_0 f,0),\lambda)\\
&= -\sum_{j=0}^{\infty} W_0^{(j)} B_0(a_j,\lambda)\int_a^{a_j} E(a_j,\lambda)(I_n-\Delta(\lambda))\widetilde{Y}(\xi,\lambda)^{-1}\Delta_0 f(\xi)\,d\xi\\
&\quad - \int_a^b W_0(t)B_0(t,\lambda)\int_a^t E(t,\lambda)(I_n-\Delta(\lambda))\widetilde{Y}(\xi,\lambda)^{-1}\Delta_0 f(\xi)\,d\xi\,dt\\
&\quad + \sum_{j=0}^{\infty} W_0^{(j)} B_0(a_j,\lambda)\int_{a_j}^{b} E(a_j,\lambda)\Delta(\lambda)\widetilde{Y}(\xi,\lambda)^{-1}\Delta_0 f(\xi)\,d\xi\\
&\quad + \int_a^b W_0(t)B_0(t,\lambda)\int_t^b E(t,\lambda)\Delta(\lambda)\widetilde{Y}(\xi,\lambda)^{-1}\Delta_0 f(\xi)\,d\xi\,dt\\
&\quad + O(\lambda^{-1})|f|_p.
\end{aligned}$$

Since $B_0(\cdot,\lambda)$ and $\widetilde{B}_0(\cdot,\lambda)$ are of the form $\{O(\tau_p(\lambda))\}_\infty$, see (4.1.18) and Proposition 4.4.3, we infer that

$$\begin{aligned}
&I_3((\Delta_0 f,0),\lambda) - I_3^0(((I_n + P^{[0]}\widetilde{B}_0(\cdot,\lambda))\Delta_0 f,0),\lambda) =\\
&\quad -\sum_{j=0}^{\infty} W_0^{(j)} B_0(a_j,\lambda)\int_a^{a_j} E(a_j,\lambda)E(\xi,\lambda)^{-1}(I_n-\Delta(\lambda))P^{[0]^{-1}}(\xi)\Delta_0 f(\xi)\,d\xi\\
&\quad - \int_a^b W_0(t)B_0(t,\lambda)\int_a^t E(t,\lambda)E(\xi,\lambda)^{-1}(I_n-\Delta(\lambda))P^{[0]^{-1}}(\xi)\Delta_0 f(\xi)\,d\xi\,dt
\end{aligned}$$

$$+\sum_{j=0}^{\infty} W_0^{(j)} B_0(a_j,\lambda)\int_{a_j}^{b} E(a_j,\lambda)E(\xi,\lambda)^{-1}\Delta(\lambda)P^{[0]^{-1}}(\xi)\Delta_0 f(\xi)\,\mathrm{d}\xi$$
$$+\int_a^b W_0(t)B_0(t,\lambda)\int_t^b E(t,\lambda)E(\xi,\lambda)^{-1}\Delta(\lambda)P^{[0]^{-1}}(\xi)\Delta_0 f(\xi)\,\mathrm{d}\xi\,\mathrm{d}t$$
$$+\left\{O(\lambda^{-1})+O(\tau_p(\lambda))^2\right\}|f|_p.$$

With the aid of Proposition 4.3.5 it follows that

$$I_3((\Delta_0 f,0),\lambda)-I_3^0(((I_n+P^{[0]}\widetilde{B}_0(\cdot,\lambda))\Delta_0 f,0),\lambda)$$
$$=\left\{O(\lambda^{-1})+O(\tau_p(\lambda))^2\right.$$
$$\left.+\{B_0(\cdot,\lambda)\}_\infty O\left(\max_{\substack{\nu,\mu=0\\ \nu\neq\mu}}^{l}\left(1+|\Re(\lambda e^{i\varphi_{\nu\mu}})|\right)^{-1+1/p}\right)\right\}|f|_p.$$

Hence

$$I_3((\Delta_0 f,0),\lambda)-I_3^0(((I_n+P^{[0]}\widetilde{B}_0(\cdot,\lambda))\Delta_0 f,0),\lambda)$$
$$=O\left(\max_{\substack{\nu,\mu=0\\ \nu\neq\mu}}^{l}\left(1+|\Re(\lambda e^{i\varphi_{\nu\mu}})|\right)^{-1+\eta'}\right)|f|_p$$

where $\eta'=\max\{0,-1+2/p\}$. From the definition of $\widetilde{B}_0$ (see the proof of Proposition 4.4.3) we infer

$$P^{[0]}\widetilde{B}_0(\cdot,\lambda)=-B_0(\cdot,\lambda)P^{[0]^{-1}}+\sum_{j=2}^{\infty}(-1)^j\left(B_0(\cdot,\lambda)P^{[0]^{-1}}\right)^j$$
$$=-B_0(\cdot,\lambda)P^{[0]^{-1}}+\{O(\tau_p(\lambda))^2\}_\infty.$$

Thus we obtain the estimate

$$I_3((\Delta_0 f,0),\lambda)-I_3^0((\Delta_0 f,0),\lambda)$$
$$=-I_3^0((B_0(\cdot,\lambda)P^{[0]^{-1}}\Delta_0 f,0),\lambda)+O\left(\max_{\substack{\nu,\mu=0\\ \nu\neq\mu}}^{l}\left(1+|\Re(\lambda e^{i\varphi_{\nu\mu}})|\right)^{-1+\eta'}\right)|f|_p.$$

The proof of the estimate of $T_{\lambda,1}h$ and $T_{\lambda,2}T_{\lambda,1}h$ in (2.8.35) yields that

$$(I_n-\Delta_0)B_0(\cdot,\lambda)=\{O(\tau_{\tilde{p}}(\lambda))\}_\infty$$

since $(I_n-\Delta_0)Q^{[0]}\in M_n(L_{\tilde{p}}(a,b))$ by the assumption on the $A_{0,0\nu}$. From

$$B_0(\cdot,\lambda)\Delta_0=\Delta_0 B_0(\cdot,\lambda)\Delta_0+(I_n-\Delta_0)B_0(\cdot,\lambda)\Delta_0$$

we infer

$$I_3((\Delta_0 f,0),\lambda) - I_3^0((\Delta_0 f,0),\lambda)$$
$$= -I_3^0((\Delta_0 B_0(\cdot,\lambda)P^{[0]^{-1}}\Delta_0 f,0),\lambda) + O\left(\max_{\substack{\nu,\mu=0\\ \nu\neq\mu}}^{l}\left(1+|\Re(\lambda e^{i\varphi_{\nu\mu}})|\right)^{-1+\eta}\right)|f|_p$$

where $\eta := \max\{\eta', 1/\tilde{p}\} < 1/p$. Finally, Proposition 4.4.10 yields

$$I_3^0((\Delta_0 B_0(\cdot,\lambda)P^{[0]^{-1}}\Delta_0 f,0),\lambda) = O(\tau_p(\lambda))^2|f|_p. \quad \square$$

LEMMA 4.6.7. *Let $1 < p < \infty$. Suppose that $A_1 \in M_n(L_\infty(a,b))$ and that there is a number $\tilde{p} > p$ such that $A_{0,0\nu} \in M_{n_0,n_\nu}(L_{\tilde{p}}(a,b))$ for $\nu = 1,\ldots,l$. Here $A_{0,0\nu}$ is the block of A_0 with index $(0,\nu)$ according to the block structure of A_1. If $p \leq \frac{3}{2}$ we require that the condition* (4.1.19) *holds. Let W_0 be defined by* (4.1.10) *and suppose that $W_0 \in M_n(L_q(a,b))$ for some number $q > 1$ and that a and b are no accumulation points of the set of points $\{a_j : j \in \mathbb{N}, W_0^{(j)} \neq 0\}$. Suppose that the boundary eigenvalue problem* (4.1.1), (4.1.2) *is Birkhoff regular and choose the curves Γ_ν $(\nu \in \mathbb{N})$ according to Theorem* 4.3.9. *We define*

$$P_\nu f := -\frac{1}{2\pi i}\oint_{\Gamma_\nu} \widetilde{J}\widetilde{T}^{-1}(\lambda)(A_1 f,0)\,d\lambda \qquad (f \in (L_p(a,b))^n),$$

where $\widetilde{J} : (W_p^1(a,b))^n \to (L_p(a,b))^n$ is defined by $\widetilde{J}y = \Delta_0 y$ and Δ_0 is given by (4.1.22). *Then $\{P_\nu : \nu \in \mathbb{N}\}$ is bounded in $L((L_p(a,b))^n)$.*

Proof. By $A_1 = \Delta_0 A_1$, (4.4.14), Proposition 4.4.4 ii), (4.4.19), Proposition 4.4.2 ii), Lemma 4.4.6, Proposition 4.6.6, and Proposition 4.4.10 it follows that

$$\Delta_0\widetilde{T}^{-1}(\lambda)(A_1 f,0) = \Delta_0 I_1^0(\cdot,A_1 f,\lambda) + \Delta_0 I_2^0(\cdot,\lambda)\widetilde{M}^{-1}(\lambda)I_3^0((A_1 f,0),\lambda)$$
$$+\left\{O\left(\max_{\substack{\nu,\mu=0\\ \nu\neq\mu}}^{l}\left(1+|\Re(\lambda e^{i\varphi_{\nu\mu}})|\right)^{-1-\beta'}\right)\right\}_p |f|_p$$

for $f \in (L_p(a,b))^n$ and $\lambda \in \bigcup_{\nu=0}^{\infty}\Gamma_\nu$, where $\beta' := \min\{\varepsilon, 1-1/p, 1/p-\eta\} > 0$, ε is as in Proposition 4.4.2 ii), and η is as in Proposition 4.6.6. From Proposition 4.3.6 iv), v) and the definition of M_0 in (4.3.5) we obtain

$$\widetilde{M}(\lambda) - M_0(\lambda) = O\left(\max_{\substack{\nu,\mu=0\\ \nu\neq\mu}}^{l}\left(1+|\Re(\lambda e^{i\psi_{\nu\mu}})|\right)^{-\beta}\right),$$

where $\beta := \min\{\beta', \alpha\} > 0$ and α is as in Proposition 4.3.6 v). Hence Lemma 4.4.6 with $s = 0$, Proposition 4.3.1 and Remark 4.3.2 yield

$$\widetilde{M}^{-1}(\lambda) - M_0^{-1}(\lambda) = \widetilde{M}^{-1}(\lambda)\big(M_0(\lambda) - \widetilde{M}(\lambda)\big)M_0^{-1}(\lambda) \tag{4.6.4}$$
$$= O\left(\max_{\substack{\nu\mu=0\\ \nu\neq\mu}}^{l}\big(1+|\Re(\lambda e^{i\varphi_{\nu\mu}})|\big)^{-\beta}\right)$$

for $\lambda \in \bigcup_{\nu=0}^{\infty}\Gamma_\nu$. This estimate, (4.4.19) and Proposition 4.4.10 yield

$$\Delta_0\widetilde{T}^{-1}(\lambda)(A_1f,0) = \Delta_0 I_1^0(\cdot,A_1f,\lambda) + \Delta_0 I_2^0(\cdot,\lambda)M_0^{-1}(\lambda)I_3^0((A_1f,0),\lambda)$$
$$+\left\{O\left(\max_{\substack{\nu,\mu=0\\ \nu\neq\mu}}^{l}\big(1+|\Re(\lambda e^{i\varphi_{\nu\mu}})|\big)^{-1-\beta}\right)\right\}_p |f|_p$$

for $f \in (L_p(a,b))^n$ and $\lambda \in \bigcup_{\nu=0}^{\infty}\Gamma_\nu$. The statement of the theorem now follows from Proposition 4.6.5, Proposition 4.6.3, Proposition 4.6.4, and (4.4.13) with respect to the Banach space $M_n(L_p(a,b))$. □

LEMMA 4.6.8. *Let E and F be Banach spaces such that F is contained continuously in E. Denote the corresponding embedding by J_F. For $\nu \in \mathbb{N}$ let $P_\nu \in L(E)$. Suppose that $\{P_\nu J_F : \nu \in \mathbb{N}\}$ is bounded in $L(F,E)$. Let $H \subset F$ such that $P_\nu w \to w$ in E as $\nu \to \infty$ for all $w \in H$. Then $P_\nu z \to z$ in E as $\nu \to \infty$ holds for all $z \in \overline{H}^F$.*

Proof. We may assume without loss of generality that $|J_F| \le 1$. Let $z \in \overline{H}^F$ and $\varepsilon > 0$. Set $M := \sup_{\nu\in\mathbb{N}} |P_\nu J_F|$. Choose $w \in H$ such that $|z-w|_F \le \frac{\varepsilon}{M+2}$ and $\nu_0 \in \mathbb{N}$ such that $|P_\nu w - w|_E \le \frac{\varepsilon}{M+2}$ for $\nu \ge \nu_0$. We obtain

$$|P_\nu z - z|_E \le |P_\nu J_F|\,|z-w|_F + |P_\nu w - w|_E + |z-w|_E \le \varepsilon. \quad \square$$

THEOREM 4.6.9. *Let $1 < p < \infty$. Suppose that $A_1 \in M_n(L_\infty(a,b))$ and that there is a number $\tilde{p} > p$ such that $A_{0,0\nu} \in M_{n_0,n_\nu}(L_{\tilde{p}}(a,b))$ for $\nu = 1,\dots,l$. Here $A_{0,0\nu}$ is the block of A_0 with index $(0,\nu)$ according to the block structure of A_1. If $p \le \frac{3}{2}$ we require that the condition (4.1.19) holds. Let W_0 be defined by (4.1.10) and suppose that $W_0 \in M_n(L_q(a,b))$ for some number $q > 1$ and that a and b are no accumulation points of the set of points $\{a_j : j \in \mathbb{N}, W_0^{(j)} \neq 0\}$. Suppose that the boundary eigenvalue problem (4.1.1), (4.1.2) is Birkhoff regular and choose the curves Γ_ν ($\nu \in \mathbb{N}$) according to Theorem 4.3.9. We define*

$$P_\nu f := -\frac{1}{2\pi i}\oint_{\Gamma_\nu} \widetilde{J}\widetilde{T}^{-1}(\lambda)(A_1f,0)\,d\lambda \qquad (f \in (L_p(a,b))^n),$$

where $\widetilde{J} : (W_p^1(a,b))^n \to (L_p(a,b))^n$ is defined by $\widetilde{J}y = \Delta_0 y$ and Δ_0 is given by (4.1.22). Then $\lim_{\nu\to\infty} P_\nu f = f$ holds for all $f \in (L_p(a,b))^n$ with $f = \Delta_0 f$.

Proof. We are going to show that the assumptions of the foregoing lemma are fulfilled. Here we take $E = F = (L_p(a,b))^n$. The boundedness of $\{P_\nu J_F : \nu \in \mathbb{N}\}$ was shown in Lemma 4.6.7. Let $\Omega := \{\lambda \in \mathbb{C} : |\lambda| > \gamma\}$. For $\lambda, \mu \in \Omega$ we define

$$B(\lambda,\mu)y := \begin{pmatrix} -(\lambda^{-1}A^0(\cdot,\lambda) - \mu^{-1}A^0(\cdot,\mu))y \\ (\widetilde{T}^R(\lambda) - \widetilde{T}^R(\mu))y \end{pmatrix} \qquad (y \in (W_p^1(a,b))^n).$$

Then $B(\lambda,\mu) \in L((W_p^1(a,b))^n, (L_p(a,b))^n \times \mathbb{C}^n)$, and the operators $B(\lambda,\mu)$ are uniformly bounded in $L((W_p^1(a,b))^n, (L_p(a,b))^n \times \mathbb{C}^n)$ for $\lambda, \mu \in \Omega$. Set $J_0 f := (A_1 f, 0)$ $(f \in (L_p(a,b))^n$. Then

$$J_0\widetilde{J} = -(\lambda-\mu)^{-1}(\widetilde{T}(\lambda) - \widetilde{T}(\mu) - B(\lambda,\mu)) \qquad (\lambda,\mu \in \Omega,\, \lambda \neq \mu)$$

since $A_1 = A_1\Delta_0$. For $\lambda_0 \in \Omega \setminus \bigcup_{\nu=0}^{\infty} \Gamma_\nu$ we have with $S_{0,\nu}$ as defined in (4.4.15) that

$$\begin{aligned}\widetilde{J}S_{0,\nu}\widetilde{T}(\lambda_0) &= \frac{1}{2\pi i}\oint_{\Gamma_\nu} \frac{1}{\lambda-\lambda_0}\widetilde{J}\widetilde{T}^{-1}(\lambda)\widetilde{T}(\lambda)\,\mathrm{d}\lambda \\ &+ \frac{1}{2\pi i}\oint_{\Gamma_\nu} \widetilde{J}\widetilde{T}^{-1}(\lambda)J_0\widetilde{J}\,\mathrm{d}\lambda - \frac{1}{2\pi i}\oint_{\Gamma_\nu}\frac{1}{\lambda-\lambda_0}\widetilde{J}\widetilde{T}^{-1}(\lambda)B(\lambda,\lambda_0)\,\mathrm{d}\lambda.\end{aligned}$$

With the aid of Theorem 4.4.11 i) we infer

$$\begin{aligned}&\left|\oint_{\Gamma_\nu}\frac{1}{\lambda-\lambda_0}\widetilde{J}\widetilde{T}^{-1}(\lambda)B(\lambda,\mu)\,\mathrm{d}\lambda\right| \\ &\qquad \leq \oint_{\Gamma_\nu} |\lambda-\lambda_0|^{-1}|\widetilde{J}\widetilde{T}^{-1}(\lambda)|\,|\mathrm{d}\lambda| \sup_{\lambda\in\Gamma_\nu}|B(\lambda,\lambda_0)| \to 0\end{aligned}$$

as $\nu \to \infty$. Thus, in view of Theorem 4.4.11 ii),

$$\widetilde{J} = \lim_{\nu\to\infty} P_\nu \widetilde{J}.$$

Hence we can take $H = \Delta_0(W_p^1(a,b))^n$. Since $C^1([a,b]) \subset W_p^1(a,b)$ is a dense subspace of $L_p(a,b)$, see [HÖ2, p. 17], we obtain $\overline{H} = \Delta_0(L_p(a,b))^n$. □

4.7. Uniform estimates of the Green's matrix

In this section we consider $p = \infty$, i. e., the norm of uniform convergence. In particular, all general assumptions involving p are understood to hold for $p = \infty$ unless otherwise stated.

PROPOSITION 4.7.1. *Suppose the boundary eigenvalue problem* (4.1.1), (4.1.2) *is Birkhoff-regular. Then we obtain*

$$\left|\oint_{|\lambda|=r} \Delta_0 I_2^0(\cdot,\lambda)M_0^{-1}(\lambda)I_3^0((\Delta_0 f,0),\lambda)\,\mathrm{d}\lambda\right|_\infty = O(1)|f|_{BV} \text{ as } r\to\infty,$$

where f varies in $(BV[a,b])^n$ and Δ_0, M_0, I_2^0, I_3^0 are given by (4.1.22), (4.3.5), (4.4.7), (4.4.12).

Proof. First we note that Proposition 4.6.1 also holds if we replace $L_p(a,b)$ and its norm by $BV[a,b]$ and its norm. Also note that $BV[a,b]$ is a Banach algebra and that $P^{[0]}$ and ${P^{[0]}}^{-1}$ belong to $M_n(W^1_\infty(a,b)) \subset M_n(BV[a,b])$. For the application of Proposition 2.7.1 in the proof of Proposition 4.6.1 we have to observe that the convolution from $L_1(a,b) \times BV[a,b]$ is a continuous bilinear map into $BV[a,b]$. Proceeding as in the proof of Proposition 4.6.3 we see that it is sufficient to show that there is a constant C such that

$$\left| \int_{\substack{\lambda\in\Sigma_m \\ |\lambda|=r}} \delta^j(\lambda)\delta^k(\lambda)\exp\{-\lambda e^{i\chi_j}x\} \int_0^{c_1} \exp\{-\lambda e^{i\chi_k}\tau\}u(\tau)\,\mathrm{d}\tau\,\mathrm{d}\lambda \right| \le C|u|_{BV}$$

holds for $x \in [0,c_2]$, $u \in BV[0,c_1]$ and $r > 0$, where the notations and conditions are as in the proof of Proposition 4.6.3. Now we are going to show that the assumptions of Proposition 4.5.6 are satisfied. We set $\gamma_1 = -\frac{\pi}{2} - \chi_m$, $\gamma_2 = -\frac{\pi}{2} - \chi_{m-1}$, $\eta_1 = \chi_j + \pi$, $\eta_2 = \chi_k + \pi$. Since, for sufficiently small $\varepsilon > 0$,

$$\delta^j(e^{i\gamma_1}) = \delta^j(e^{i(\gamma_2-\varepsilon)}) = \delta^k(e^{i\gamma_1}) = \delta^k(e^{i(\gamma_2-\varepsilon)}) = 1,$$

we infer $\gamma_\nu + \chi_\mu \in [-\frac{\pi}{2}, \frac{\pi}{2}] \operatorname{mod}(2\pi)$ for $\nu = 1,2$ and $\mu = j,k$, and it follows that $\gamma_\nu + \eta_\mu \in [\frac{\pi}{2}, \frac{3}{2}\pi] \operatorname{mod}(2\pi)$ for $\nu,\mu \in \{1,2\}$. We have $\gamma_2 - \gamma_1 \notin \pi\mathbb{Z}$ unless $l_0 = 2$. If $j \ne k$ and $l_0 = 2$, then $\delta^j(\lambda)\delta^k(\lambda) = 0$ for all $\lambda \in \mathbb{C}\setminus\{0\}$. Hence only $j = k$ has to be considered in case $\gamma_1 - \gamma_2 \in \pi\mathbb{Z}$. Therefore Proposition 4.5.6 is applicable, and the desired estimate holds. □

PROPOSITION 4.7.2. *For $f \in (BV[a,b])^n$ we have*

$$\left| \oint_{|\lambda|=r} I_1^0(\cdot,\Delta_0 f,\lambda)\,\mathrm{d}\lambda \right|_\infty = O(1)|f|_{BV} \quad \text{as } r \to \infty,$$

where Δ_0, I_1^0 are defined in (4.1.22), (4.4.8).

Proof. This is the same proof as for Proposition 4.6.5. Only in the last step we have to apply Proposition 4.5.4 instead of Proposition 4.5.3. □

PROPOSITION 4.7.3. *Suppose that $r_1,\ldots,r_l \in W_1^1(a,b)$ and that $A_{0,0\nu}$ belongs to $M_{n_0,n_\nu}(W_1^1(a,b))$ for $\nu = 1,\ldots,l$, where $A_{0,0\nu}$ is the block of A_0 with index $(0,\nu)$ according to the block structure of A_1. Then*

$$\begin{aligned} &I_3((\Delta_0 f,0),\lambda) - I_3^0((\Delta_0 f,0),\lambda) \\ &\quad = \left\{ O(\lambda^{-1}) + O\left(\max_{\substack{\nu,\mu=0 \\ \nu\ne\mu}}^{l} \left(1 + |\Re(\lambda e^{i\varphi_{\nu\mu}})|\right)^{-2} \right) \right\} |f|_p \end{aligned}$$

for $f \in (L_p(a,b))^n$, where Δ_0, I_3, I_3^0 are defined in (4.1.22), (4.4.6), (4.4.12).

Proof. As in the proof of Proposition 4.6.6 we obtain

$$\begin{aligned} I_3((\Delta_0 f,0),\lambda) - I_3^0((\Delta_0 f,0),\lambda) &= I_3^0(((I_n-\Delta_0)B_0(\cdot,\lambda)P^{[0]^{-1}}\Delta_0 f,0),\lambda) \\ &= O(\lambda^{-1}) + O(\tau_\infty(\lambda))^2. \end{aligned}$$

Then the required estimate will follow from

$$|(I_n-\Delta_0)B_0(\cdot,\lambda)|_\infty = O(\lambda^{-1}) + O(\tau_\infty(\lambda))^2. \tag{4.7.1}$$

To obtain this estimate we return to the proof of the estimates of $B_0(\cdot,\lambda)$ in Section (2.8). It is sufficient to show that, with f_λ defined in (2.8.33),

$$(I_n-\Delta_0)f_\lambda = -(I_n-\Delta_0)T_{\lambda,1}(I_n) + (I_n-\Delta_0)(T_{\lambda,1}^2 + T_{\lambda,1}T_{\lambda,2} + T_{\lambda,2}T_{\lambda,1})g_\lambda$$

satisfies the estimate $\{O(\lambda^{-1}) + O(\tau_\infty(\lambda))^2)\}_\infty$, where $g_\lambda = \{O(1)\}_\infty$.

We have the estimates $(I_n-\Delta_0)T_{\lambda,1}(I_n) = \{O(\lambda^{-1})\}_\infty$ in view of Remark 2.7.3 and $T_{\lambda,1}^2 g_\lambda = \{O(\tau_\infty(\lambda))^2\}_\infty$. Also $T_{\lambda,2}$ is bounded from $M_n(L_\infty(a,b))$ to $M_n(W_\infty^1(a,b))$ and independent of λ, which implies $T_{\lambda,1}T_{\lambda,2}g_\lambda = \{O(\lambda^{-1})\}_\infty$ in view of Remark 2.7.3. Therefore we still have to consider $T_{\lambda,2}T_{\lambda,1}g_\lambda$, i. e., we have to consider the function $z(x,\lambda)$ as in (2.8.36):

$$z(x,\lambda) = w(x)\int_{x_{\mu\mu}(\lambda)}^{x} v(\xi)\int_{x_{\nu\mu}(\lambda)}^{\xi} \exp\left\{\lambda e^{i\varphi_{\nu\mu}}\int_t^\xi |r_\nu(\eta)-r_\mu(\eta)|\,\mathrm{d}\eta\right\} u(t)\,\mathrm{d}t\,\mathrm{d}\xi.$$

Writing the outer integral as an integral from $x_{\mu\mu}(\lambda)$ to $x_{\nu\mu}(\lambda)$ and an integral from $x_{\nu\mu}(\lambda)$ to x, we see that we may assume that $x_{\mu\mu}(\lambda) = x_{\nu\mu}(\lambda)$ in order to find an L_∞-estimate of $z(x,\lambda)$. We cannot apply Remark 2.7.3 directly. Therefore we interchange the order of integration in $z(x,\lambda)$ and obtain

$$z(x,\lambda) = w(x)\int_{x_{\nu\mu}(\lambda)}^{x} u(t)z_1(x,t,\lambda)\,\mathrm{d}t,$$

where

$$z_1(x,t,\lambda) = \int_t^x v(\xi)\exp\left\{\lambda e^{i\varphi_{\nu\mu}}\int_t^\xi |r_\nu(\eta)-r_\mu(\eta)|\,\mathrm{d}\eta\right\}\mathrm{d}\xi,$$

$|u|_\infty \le C|g_\lambda|_\infty = O(1)$, and $v\in W_1^1(a,b)$ contains the components of $(I_n-\Delta_0)A_0$. Remark 2.7.3 shows that $z_1(x,t,\lambda) = O(\lambda^{-1})$ uniformly for all $x\in[a,b]$ and t between $x_{\nu\mu}(\lambda)$ and x, and the estimate (4.7.1) is proved. □

LEMMA 4.7.4. *Suppose that* $A_1 \in M_n(W_1^1(a,b))$ *and that* $A_{0,0\nu} \in M_{n_0,n_\nu}(W_1^1(a,b))$ *for* $\nu = 1,\ldots,l$, *where* $A_{0,0\nu}$ *is the block of* A_0 *with index* $(0,\nu)$ *according to the block structure of* A_1. *Suppose that the matrix function* W_0 *defined in* (4.1.10) *belongs to* $M_n(L_q(a,b))$ *for some* $q > 1$ *and that* a *and* b *are no accumulation points of the set* $\{a_j : j \in \mathbb{N}, W_0^{(j)} \neq 0\}$. *Suppose that the boundary eigenvalue problem* (4.1.1), (4.1.2) *is Birkhoff regular and choose the curves* Γ_ν $(\nu \in \mathbb{N})$ *according to Theorem* 4.3.9. *We define*

$$P_\nu f := -\frac{1}{2\pi i}\oint_{\Gamma_\nu} \widetilde{J}\widetilde{T}^{-1}(\lambda)(A_1 f, 0)\,\mathrm{d}\lambda \qquad (f \in (L_\infty(a,b))^n),$$

where $\widetilde{J}\colon (W_\infty^1(a,b))^n \to (L_\infty(a,b))^n$ *is defined by* $\widetilde{J}y = \Delta_0 y$ *and* Δ_0 *is given by* (4.1.22). *Then* $\{P_\nu|_{(BV[a,b])^n} : \nu \in \mathbb{N}\}$ *is bounded in* $L((BV[a,b])^n, (L_\infty(a,b))^n)$.

Proof. We use estimates similar to those in the proof of Lemma 4.6.7. Using Proposition 4.7.3 instead of Proposition 4.6.6 we obtain

$$\Delta_0\widetilde{T}^{-1}(\lambda)(A_1 f,0) = \Delta_0 I_1^0(\cdot,A_1 f,\lambda) + \Delta_0 I_2^0(\cdot,\lambda)\widetilde{M}^{-1}(\lambda)I_3^0((A_1 f,0),\lambda) + \left\{O(\lambda^{-1}) + O\left(\max_{\substack{\nu,\mu=0\\ \nu\neq\mu}}^{l}\left(1+|\Re(\lambda e^{i\varphi_{\nu\mu}})|\right)^{-2}\right)\right\}_\infty |f|_\infty$$

and then

$$\Delta_0\widetilde{T}^{-1}(\lambda)(A_1 f,0) = \Delta_0 I_1^0(\cdot,A_1 f,\lambda) + \Delta_0 I_2^0(\cdot,\lambda)M_0^{-1}(\lambda)I_3^0((A_1 f,0),\lambda) + \left\{O(\lambda^{-1}) + O\left(\max_{\substack{\nu,\mu=0\\ \nu\neq\mu}}^{l}\left(1+|\Re(\lambda e^{i\varphi_{\nu\mu}})|\right)^{-1-\beta}\right)\right\}_\infty |f|_\infty$$

for $f \in (L_\infty(a,b))^n$, $\lambda \in \bigcup_{\nu=0}^{\infty}\Gamma_\nu$ and some $\beta > 0$. The statement of the theorem now follows from Propositions 4.7.2, 4.7.1, and 4.6.4. □

Apart from the operator function $\widetilde{T}$ we shall also consider

$$\widetilde{T}_0 = \begin{pmatrix} \widetilde{T}_0^D \\ \widetilde{T}_0^R \end{pmatrix},$$

where

$$\widetilde{T}_0^D(\lambda)y := y' - (\lambda A_1 + A_0)y$$

and

$$\widetilde{T}_0^R y := \sum_{j=0}^{\infty} W_0^{(j)} y(a_j) + \int_a^b W_0(x)y(x)\,\mathrm{d}x$$

for $y \in (W_\infty^1(a,b))^n$. Note that $\widetilde{T}_0^R y$ can also be defined for $y \in (C[a,b])^n$.

THEOREM 4.7.5. *Suppose that* $A_1 \in M_n(W_1^1(a,b))$ *and* $A_{0,0\nu} \in M_{n_0,n_\nu}(W_1^1(a,b))$ *for* $\nu = 1,\dots,l$, *where* $A_{0,0\nu}$ *is the block of* A_0 *with index* $(0,\nu)$ *according to the block structure of* A_1. *Suppose that the matrix function* W_0 *defined in* (4.1.10) *belongs to* $M_n(L_q(a,b))$ *for some* $q > 1$ *and that* a *and* b *are no accumulation points of the set* $\{a_j : j \in \mathbb{N}, W_0^{(j)} \neq 0\}$. *Suppose that the boundary eigenvalue problem* (4.1.1), (4.1.2) *is Birkhoff regular and choose the curves* Γ_ν $(\nu \in \mathbb{N})$ *according to Theorem* 4.3.9. *We define*

$$P_\nu f := -\frac{1}{2\pi i}\oint_{\Gamma_\nu} \widetilde{J}\widetilde{T}^{-1}(\lambda)(A_1 f, 0)\,\mathrm{d}\lambda \qquad (f \in (L_\infty(a,b))^n),$$

where $\widetilde{J} : (W_\infty^1(a,b))^n \to (L_\infty(a,b))^n$ *is defined by* $\widetilde{J}y = \Delta_0 y$ *and* Δ_0 *is given by* (4.1.22). *Then* $\lim_{\nu\to\infty} P_\nu f = f$ *holds for all* $f \in (C[a,b] \cap BV[a,b])^n$ *satisfying* $f = \Delta_0 f$ *and*

$$(4.7.2) \qquad B(\lambda)f := \Delta_0 M_0(\lambda)^{-1} I_3^0(((\Delta_0 - I_n)A_0 f, \widetilde{T}_0^R f), \lambda) = 0$$

for some $\lambda \in \mathbb{C}\setminus\{0\}$, *where* M_0, I_3^0 *are given by* (4.3.5), (4.4.12). *The condition* $B(\lambda)f = 0$ *is independent of* λ, *and there are exactly* $n - n_0$ *linearly independent linear functionals given by* $B(\lambda)$.

Proof. We shall show that the assumptions of Lemma 4.6.8 are fulfilled if we take $E = (L_\infty(a,b))^n$, $F = (BV[a,b])^n$, and

$$H = \{y \in \Delta_0(W_\infty^1(a,b))^n : B(\lambda)y = 0\}.$$

The uniform boundedness of the operators $P_\nu J_F$ was shown in Lemma 4.7.4. Next we prove that $P_\nu y \to y$ as $\nu \to \infty$ in $(L_\infty(a,b))^n$ for all $y \in H$. In the following we shall use that the condition $B(\lambda)y = 0$ is independent of $\lambda \in \mathbb{C}\setminus\{0\}$. The proof of the independence will be postponed to the end of the proof of Theorem 4.7.5. Let $\Omega := \{\lambda \in \mathbb{C} : |\lambda| > \gamma\}$. For $\lambda, \mu \in \Omega$ we define

$$B(\lambda,\mu)y := \begin{pmatrix} -(\lambda^{-1}A^0(\cdot,\lambda) - \mu^{-1}A^0(\cdot,\mu))y \\ (\widetilde{T}^R(\lambda) - \widetilde{T}^R(\mu))y \end{pmatrix} \qquad (y \in (W_\infty^1(a,b))^n).$$

Then $B(\lambda,\mu) \in L((W_\infty^1(a,b))^n, (L_\infty(a,b))^n \times \mathbb{C}^n)$, and the operators $B(\lambda,\mu)$ are uniformly bounded in $L((W_\infty^1(a,b))^n, (L_\infty(a,b))^n \times \mathbb{C}^n)$ for $\lambda, \mu \in \Omega$. Furthermore, $|B(\lambda,\mu)| = O(\lambda^{-1}) + O(\mu^{-1})$ as $\lambda, \mu \to \infty$. This is clear for the first component, and it follows from (4.1.10) and (4.1.12) for the second component. The representation (4.4.17) also holds in case $p = \infty$. We immediately infer that $|\widetilde{J}\widetilde{T}^{-1}(\lambda)| = O(1)$ on $\bigcup_{\nu=0}^{\infty} \Gamma_\nu$, whence

$$|\widetilde{J}S_{0,\nu}(\lambda_0)| = O\Big(\frac{\rho_\nu}{\rho_\nu - |\lambda_0|}\Big).$$

Here we have written $S_{0,\nu}(\lambda_0)$ instead of $S_{0,\nu}$ since we shall vary λ_0 and have to observe the dependence of $S_{0,\nu}$ on λ_0. Since $|\widetilde{T}(\lambda_0)-\widetilde{T}_0(\lambda_0)|=O(\lambda_0^{-1})$, we infer

$$|\widetilde{J}S_{0,\nu}(\lambda_0)[\widetilde{T}(\lambda_0)-\widetilde{T}_0(\lambda_0)]|=O\Big(\frac{\rho_\nu}{|\lambda_0|(\rho_\nu-|\lambda_0|)}\Big).$$

For $\nu\in\mathbb{N}$ and $\lambda\in\Omega\setminus\Gamma_\nu$ we set

$$P_{\nu,\lambda_0}:=\frac{1}{2\pi i}\oint_{\Gamma_\nu}\frac{1}{\lambda-\lambda_0}\widetilde{J}\widetilde{T}^{-1}(\lambda)B(\lambda,\lambda_0)\,d\lambda.$$

Then

$$|P_{\nu,\lambda_0}|=O\Big(\frac{\rho_\nu}{\rho_\nu-|\lambda_0|}\Big)\big(O(\rho_\nu^{-1})+O(|\lambda_0|^{-1})\big).$$

As in the proof of Theorem 4.6.9 we obtain that

$$\widetilde{J}-P_\nu\widetilde{J}=\widetilde{J}S_{0,\nu}(\lambda_0)\widetilde{T}(\lambda_0)+P_{\nu,\lambda_0}$$

holds for every $\nu\in\mathbb{N}$ and every $\lambda_0\in\Omega$ such that $|\lambda_0|<\rho_\nu$. Also, by Theorem 4.4.11 iv) we have

$$\left|\widetilde{J}S_{0,\nu}(\lambda_0)\begin{pmatrix}\Delta_0\widetilde{T}_0^D(\lambda_0)\\0\end{pmatrix}\right|=o(1)\text{ as }\nu\to\infty$$

for all $\lambda_0\in\Omega\setminus\bigcup_{\nu=0}^{\infty}\Gamma_\nu$. This shows that

$$\limsup_{\nu\to\infty}\left|\widetilde{J}-P_\nu\widetilde{J}-\widetilde{J}S_{0,\nu}(\lambda_0)\begin{pmatrix}(I_n-\Delta_0)\widetilde{T}_0^D(\lambda_0)\\\widetilde{T}_0^R\end{pmatrix}\right|=O(|\lambda_0|^{-1})$$

for all $\lambda_0\in\Omega\setminus\bigcup_{\nu=0}^{\infty}\Gamma_\nu$. Again from the estimates in the proof of Theorem 4.4.11 i) we infer that only the terms coming from

$$\Delta_0I_2^0(\cdot,\lambda)M_0(\lambda)^{-1}I_3^0(((I_n-\Delta_0)(y'-A_0y),\widetilde{T}_0^Ry),\lambda)$$

in $\widetilde{J}S_{0,\nu}(\lambda_0)\begin{pmatrix}(I_n-\Delta_0)\widetilde{T}_0^D(\lambda_0)\\\widetilde{T}_0^R\end{pmatrix}y$ do not necessarily tend to zero as $\nu\to\infty$.

Hence $P_\nu y\to y$ in $(L_\infty(a,b))^n$ as $\nu\to\infty$ for all $y\in H$. Therefore, we can apply Lemma 4.6.8, and the convergence result in Theorem 4.7.5 follows if we show that

$$\overline{H}=\{y\in\Delta_0(C[a,b]\cap BV[a,b])^n: B(\lambda)y=0\},$$

where the closure is taken in $(BV[a,b])^n$. Since H consists of continuous functions and since the BV-norm is stronger than the L_∞-norm, we have that $\overline{H}\subset(C[a,b])^n$. Also, $B(\lambda):(C[a,b])^n\to\mathbb{C}^n$ is continuous. Hence $B(\lambda)y=0$ for all $y\in\overline{H}$.

We still have to show that every function $z\in\Delta_0(C[a,b]\cap BV[a,b])^n$ with $B(\lambda)z=0$ belongs to $\overline{H}$. To this end we first note that $C^1[a,b]$ is a dense subset of $C[a,b]\cap BV[a,b]$ with respect to the norm of bounded variation. It is sufficient to show this for nonnegative nondecreasing functions f in $C[a,b]\cap BV[a,b]$. Set $f(x)=f(a)$ for $x<a$ and $f(x)=f(b)$ for $x>b$. Take any sequence of nonnegative functions $(\phi_j)_{j=0}^\infty$ in $C_0^\infty(\mathbb{R})$ with $\int_{\mathbb{R}}\phi_j(x)\,\mathrm{d}x=1$ and $\operatorname{supp}\phi_j\subset[-\varepsilon_j,\varepsilon_j]$ with $\varepsilon_j\to 0$ as $j\to\infty$. Then the convolution $f*\phi_j$ belongs to $C^\infty(\mathbb{R})$ and tends to f uniformly, see e. g. [HÖ2, Theorem 1.3.2] and its proof. Also, since f is nonnegative and nondecreasing, all the $f*\phi_j$ have the same property. Hence $f*\phi_j$ also converges in $BV[a,b]$ to f.

Let $\{y_1,\dots,y_{j_0}\}\subset\Delta_0(W_\infty^1(a,b))^n$ such that $\{B(\lambda)y_1,\dots,B(\lambda)y_{j_0}\}$ is a basis of $\{B(\lambda)y : y\in\Delta_0(W_\infty^1(a,b))^n\}$. Now let $z\in\Delta_0(C[a,b]\cap BV[a,b])^n$ such that $B(\lambda)z=0$. Since $C^1[a,b]\subset W_\infty^1(a,b)$, the denseness result from the previous paragraph shows that there is a sequence $(z_k)_0^\infty$ in $\Delta_0(W_\infty^1(a,b))^n$ with $z_k\to z$ in $\Delta_0(C[a,b]\cap BV[a,b])^n$. There are $\alpha_{j,k}\in\mathbb{C}$ such that

$$B(\lambda)z_k=\sum_{j=1}^{j_0}\alpha_{j,k}B(\lambda)y_j\quad(k\in\mathbb{N}).$$

From $B(\lambda)z=0$ we infer $B(\lambda)z_k\to 0$ as $k\to\infty$. Hence $\alpha_{j,k}\to 0$ for $j=1,\dots,j_0$ as $k\to\infty$ because of the linear independence of $B(\lambda)y_1,\dots,B(\lambda)y_{j_0}$. This shows that

$$z_k-\sum_{j=1}^{j_0}\alpha_{j,k}y_j$$

belongs to H and tends to z as $k\to\infty$, and $z\in\overline{H}$ is proved.

Now we shall prove that the condition $B(\lambda)f=0$ does not depend on λ. Let $I_3^{0,1}$ and $I_3^{0,2}$ be defined as I_3^0 as in (4.4.12) with $\Delta(\lambda)$ replaced by arbitrary diagonal matrices $\Delta^1(\lambda)$ and $\Delta^2(\lambda)$, respectively. Then $E(t,\lambda)(I_n-\Delta_0)=I_n-\Delta_0$ yields

$$\begin{aligned}I_3^{0,1}&(((\Delta_0-I_n)A_0f,\widetilde{T}_0^Rf),\lambda)-I_3^{0,2}(((\Delta_0-I_n)A_0f,\widetilde{T}_0^Rf),\lambda)\\&=\widetilde{M}_2(\Delta_0-I_n)(\Delta^1(\lambda)-\Delta^2(\lambda))\int_a^b P^{[0]^{-1}}(\xi)A_0(\xi)f(\xi)\,\mathrm{d}\xi\\&=M_0(\lambda)(\Delta_0-I_n)(\Delta^1(\lambda)-\Delta^2(\lambda))\int_a^b P^{[0]^{-1}}(\xi)A_0(\xi)f(\xi)\,\mathrm{d}\xi.\end{aligned}$$

This shows that $B^1(\lambda)=B^2(\lambda)$, where $B^1(\lambda)$ and $B^2(\lambda)$ are defined as $B(\lambda)$ with I_3^0 replaced by $I_3^{0,1}$ and $I_3^{0,2}$, respectively. Taking $\Delta^1(\lambda)=\Delta(\lambda)$ and $\Delta^2(\lambda)=0$ we infer that

$$B(\lambda)f=\Delta_0M_0(\lambda)^{-1}I_3^{0,2}(((\Delta_0-I_n)A_0f,\widetilde{T}_0^Rf),\lambda),$$

where it is easy to see that $I_3^{0,2}(((\Delta_0-I_n)A_0f,\widetilde{T}_0^Rf),\lambda)$ does not depend on λ. Let $\lambda,\lambda_0\in\mathbb{C}\setminus\{0\}$. Then

$$M_0(\lambda_0)^{-1}M_0(\lambda)=M_0(\lambda_0)^{-1}\widetilde{M}_0(\lambda)\Delta_0+M_0(\lambda_0)^{-1}\widetilde{M}_2(I_n-\Delta_0)$$

implies that

$$M_0(\lambda_0)^{-1}M_0(\lambda)(I_n-\Delta_0)=M_0(\lambda_0)^{-1}\widetilde{M}_2(I_n-\Delta_0).$$

The right-hand side shows that this matrix is independent of λ, whereas the left-hand side for $\lambda=\lambda_0$ shows that it is $I_n-\Delta_0$. Hence, with λ_0 fixed, it follows that

$$M_0(\lambda)^{-1}M_0(\lambda)=\begin{pmatrix} I_{n_0} & D_1(\lambda)\\ 0 & D_2(\lambda)\end{pmatrix}$$

with suitable matrix functions $D_1(\lambda)$ and $D_2(\lambda)$. Now

$$\begin{pmatrix} I_{n_0} & 0\\ 0 & D_2(\lambda)\end{pmatrix}\Delta_0 M_0(\lambda)^{-1}=\Delta_0\begin{pmatrix} I_{n_0} & D_1(\lambda)\\ 0 & D_2(\lambda)\end{pmatrix}M_0(\lambda)^{-1}=\Delta_0 M_0(\lambda_0)^{-1}$$

shows that $B(\lambda)f=0$ is independent of λ since $D_2(\lambda)$ is invertible.

This also shows that the number of conditions given by $B(\lambda)f=0$ is at most $n-n_0$ since Δ_0 has rank $n-n_0$. We still have to prove that it is at least $n-n_0$. To this end we observe that, for any $c_0,c_1\in\mathbb{C}^n$ and any $\varepsilon>0$, we can find $f_\varepsilon\in(C[a,b]\cap BV[a,b])^n$ such that $f_\varepsilon(a)=c_0$, $f_\varepsilon(b)=c_1$, $|f_\varepsilon|_1\le\varepsilon$ and

$$\widetilde{T}_0^R f_\varepsilon=W_0^{(0)}f_\varepsilon(a)+W_0^{(1)}f_\varepsilon(b)=W_0^{(0)}c_0+W_0^{(1)}c_1$$

(note that we assume that a and b are no accumulation points of the a_j for which $W_0^{(j)}\neq 0$). For example, we can take for f_ε the function whose graph consists of the line segments connecting the points (a,c_0), $(a+\delta_\varepsilon,0)$, $(b-\delta_\varepsilon,0)$, (b,c_1), where δ_ε is a sufficiently small positive number. Then

$$\lim_{\varepsilon\to 0}B(\lambda)f_\varepsilon=\Delta_0 M_0(\lambda)^{-1}\left(W_0^{(0)}c_0+W_0^{(1)}c_1\right)$$

yields that it is sufficient to show that

$$\Delta_0 M_0(\lambda)^{-1}(W_0^{(0)},W_0^{(1)})$$

has rank $n-n_0$. But this is obvious since

$$\begin{aligned}\Delta_0&=\Delta_0 M_0(\lambda)^{-1}M_0(\lambda)\Delta_0=\Delta_0 M_0(\lambda)^{-1}\widetilde{M}_0(\lambda)\Delta_0\\ &=\Delta_0 M_0(\lambda)^{-1}(W_0^{(0)},W_0^{(1)})\begin{pmatrix} P^{[0]}(a)(I_n-\Delta(\lambda))\Delta_0\\ P^{[0]}(b)\Delta(\lambda)\Delta_0\end{pmatrix}.\qquad\square\end{aligned}$$

REMARK 4.7.6. i) If $n_0=0$, then $\Delta_0=I_n$, and the condition $B(\lambda)f=0$ reduces to $\widetilde{T}_0^R f=0$.

ii) Let Λ be a diagonal matrix whose first n_0 diagonal elements are zero and whose other diagonal elements are zero or one. Then it is immediately clear from the proof of Theorem 4.7.5 that $\lim_{\nu\to\infty}\Lambda P_\nu f=f$ holds for $f\in\Lambda(C(a,b)\cap BV[a,b])^n$ if $\Lambda M_0(\lambda)^{-1}I_3^0(((\Delta_0-I_n)A_0 f,\widetilde{T}_0^R f),\lambda)=0$ for $\lambda\in\mathbb{C}\setminus\{0\}$. From Theorem 4.7.5 we know that there are exactly $n-n_0$ conditions given by $B(\lambda)f=0$. Therefore, for such Λ with rank $\Lambda = m$, the number of conditions lies between m and $n - n_0$. In general, the conditions depend on λ and have to be evaluated for each sector Σ_k (see Remark 4.1.4).

REMARK 4.7.7. In case $\widetilde{T}_0^R f = 0$ is a two-point boundary condition we have

$$B_0(\lambda)f := I_3^0(((\Delta_0 - I_n)A_0 f, \widetilde{T}_0^R f), \lambda) = W_0^{(0)} f(a) + W_0^{(1)} f(b)$$
$$- \left[W_0^{(0)}\Delta(\lambda) - W_0^{(1)} P^{[0]}(b)\left(I_n - \Delta(\lambda)\right)\right](I_n - \Delta_0) \int_a^b P^{[0]^{-1}}(\xi) A_0(\xi) f(\xi)\, \mathrm{d}\xi.$$

As was seen in the proof of Theorem 4.7.5 we can take $\Delta(\lambda)(I_n - \Delta_0) = 0$ or $\Delta(\lambda)(I_n - \Delta_0) = I_n - \Delta_0$.

REMARK 4.7.8. If $y \in (W_\infty^1(a,b))^n$ and $\lim_{\nu\to\infty} P_\nu y = y$ in $(L_\infty(a,b))^n$, then it follows that $y \in \Delta_0(W_\infty^1(a,b))^n$ and $B(\lambda)y = 0$, where P_ν and $B(\lambda)$ are as in Theorem 4.7.5 and Δ_0 is defined in (4.1.22).

Proof. Since $R(P_\nu) \subset \Delta_0(W_\infty^1(a,b))^n$, the condition $y \in \Delta_0(W_\infty^1(a,b))^n$ is necessary. And from the proof of Theorem 4.7.5 we immediately infer that $\lim_{\nu\to\infty} P_\nu y = y$ holds for $y \in \Delta_0(W_\infty^1(a,b))^n$ if and only if

$$\lim_{\lambda_0\to\infty} \limsup_{\nu\to\infty} \left| \oint_{|\lambda|=1} \frac{1}{\lambda - \frac{\lambda_0}{\rho_\nu}} \Delta_0 I_2^0(x, \rho_\nu\lambda) M_0(\lambda)^{-1} \times \right.$$
$$\left. \times\ I_3^0(((\Delta_0 - I_n)A_0 y, \widetilde{T}_0^R y), \lambda)\, \mathrm{d}\lambda \right| = 0$$

uniformly for all $x \in [a,b]$. It is easy to see that the limit superior does not depend on λ_0. For simplicity, we thus may take $\lambda_0 = 0$. We also may multiply $I_2^0(x, \rho_\nu\lambda)$ by $P^{[0]^{-1}}(x)$ from the left. Hence we have that $\lim_{\nu\to\infty} P_\nu y = y$ holds for $y \in \Delta_0(W_\infty^1(a,b))^n$ if and only if

$$(\widetilde{I}^\nu y)(x) :=$$
$$\oint_{|\lambda|=1} \frac{1}{\lambda} \Delta_0\left((I_n - \Delta(\lambda))E(x, \rho_\nu\lambda) + \Delta(\lambda)E(x, \rho_\nu\lambda)E(b, \rho_\nu\lambda)^{-1}\right) M_0(\lambda)^{-1} \times$$
$$\times\ I_3^0(((\Delta_0 - I_n)A_0 y, \widetilde{T}_0^R y), \lambda)\, \mathrm{d}\lambda$$

tends to 0 as $\nu \to \infty$ uniformly for all $x \in [a,b]$. The components of this integral are of the form

$$\int_{\frac{\pi}{2}}^{\frac{3}{2}\pi} \exp\{e^{i\varphi}\rho_\nu |R_j(x)|\} m_1(\varphi)\, \mathrm{d}\varphi \tag{4.7.3}$$
$$+ \int_{-\frac{\pi}{2}}^{\frac{\pi}{2}} \exp\{e^{i\varphi}\rho_\nu(|R_j(x)| - |R_j(b)|)\} m_2(\varphi)\, \mathrm{d}\varphi,$$

where m_1 and m_2 are step functions consisting of those parts of the components of $B(\lambda)y$ which only depend on $\arg\lambda$, and $j \in \{1,\dots,l\}$. We are going to show that $m_1 = 0$ and $m_2 = 0$. Assume that the first integral in (4.7.3) is different from zero for some $\nu_0 \in \mathbb{N}$ and $x_{\nu_0} \in [a,b]$. Since $|R_j|$ is strictly increasing and continuous with $R_j(a) = 0$, there is a (unique) $x_\nu \in [a,b)$ such that $\rho_\nu|R_j(x_\nu)| = \rho_{\nu_0}|R_j(x_{\nu_0})|$ for $\nu > \nu_0$. From $\rho_\nu|R_j(b)| \to \infty$ as $\nu \to \infty$ we infer that, with these x_ν, the second integral tends to zero as $\nu \to \infty$ by LEBESGUE'S dominated convergence theorem. This shows that $\widetilde{I}^\nu y(x_\nu)$ would not tend to zero as $\nu \to \infty$. A similar argument with the second integral shows that both integrals in (4.7.3) must be zero for all $\nu \in \mathbb{N}$ and all $x \in [a,b]$.

It follows that

$$\int_{\frac{\pi}{2}}^{\frac{3}{2}\pi} \exp\{e^{i\varphi}\rho\} m_j(\varphi)\,\mathrm{d}\varphi = 0$$

for $j = 1,2$ and all $\rho \geq 0$. Differentiating with respect to ρ gives

$$\int_{\frac{\pi}{2}}^{\frac{3}{2}\pi} e^{ik\varphi} \exp\{e^{i\varphi}\rho\} m_j(\varphi)\,\mathrm{d}\varphi = 0 \tag{4.7.4}$$

for all nonnegative integers k. Integrating with respect to ρ gives

$$\int_{\frac{\pi}{2}}^{\frac{3}{2}\pi} e^{-i\varphi}\left[\exp\{e^{i\varphi}\rho_1\} - \exp\{e^{i\varphi}\rho\}\right] m_j(\varphi)\,\mathrm{d}\varphi = 0,$$

where $0 \leq \rho \leq \rho_1$. For $\rho_1 \to \infty$,

$$\int_{\frac{\pi}{2}}^{\frac{3}{2}\pi} e^{-i\varphi} \exp\{e^{i\varphi}\rho_1\} m_j(\varphi)\,\mathrm{d}\varphi \to 0$$

by LEBESGUE'S dominated convergence theorem. Hence

$$\int_{\frac{\pi}{2}}^{\frac{3}{2}\pi} e^{-i\varphi} \exp\{e^{i\varphi}\rho\} m_j(\varphi)\,\mathrm{d}\varphi = 0$$

for all $\rho \geq 0$. Repeating this integration we see that (4.7.4) holds for all integers k. Since $\{e^{2ik\varphi} : k \in \mathbb{Z}\}$ is a basis in $L_2\left(\frac{\pi}{2}, \frac{3}{2}\pi\right)$, we infer that $\exp\{e^{i\varphi}\rho\} m_j(\varphi) = 0$ and hence $m_1 = 0$ and $m_2 = 0$. This shows that $B(\lambda)y = 0$. □

EXAMPLE 4.7.9. We consider the boundary eigenvalue problem

$$y' = \lambda A_1 y + A_0 y, \quad W^{(0)}y(0) + W^{(1)}y(1) = 0,$$

$$A_1 = \begin{pmatrix} 0 & 0 & 0 \\ 0 & 1 & 0 \\ 0 & 0 & -1 \end{pmatrix}, \quad A_0 = \begin{pmatrix} 0 & \alpha & \beta \\ 0 & 0 & 0 \\ 0 & 0 & 0 \end{pmatrix},$$

$$W^{(0)} = \begin{pmatrix} 0 & 0 & -1 \\ -1 & 1 & 0 \\ 0 & 0 & 1 \end{pmatrix}, \quad W^{(1)} = \begin{pmatrix} 1 & 0 & -1 \\ 0 & 0 & 1 \\ 0 & 1 & 1 \end{pmatrix},$$

where α and β are complex numbers. We shall determine the conditions given by $B(\lambda)y=0$. In this example, $P^{[0]}(x)$ is the identity matrix for all x since $l=2$, $n_0=n_1=n_2=1$, and the diagonal elements of A_0 are zero. Also, A_0 is constant, and we obtain in view of Remark 4.7.7 that

$$B_0(\lambda)y = W^{(0)}y(0)+W^{(1)}y(1) \\ -\left[W^{(0)}\Delta(\lambda)-W^{(1)}(I_n-\Delta(\lambda))\right](I_n-\Delta_0)A_0\int_a^b y(\xi)\,\mathrm{d}\xi.$$

We have

$$\Delta(1)=\begin{pmatrix}1&0&0\\0&1&0\\0&0&0\end{pmatrix},\quad \Delta(-1)=\begin{pmatrix}0&0&0\\0&0&0\\0&0&1\end{pmatrix}.$$

M_0 has the two values

$$M_0(1)=\begin{pmatrix}1&0&-1\\-1&0&0\\0&1&1\end{pmatrix},\quad M_0(-1)=\begin{pmatrix}1&0&-1\\-1&1&1\\0&0&1\end{pmatrix}.$$

Since $M_0(1)$ and $M_0(-1)$ are invertible, the problem is Birkhoff regular. From

$$M_0(1)^{-1}=\begin{pmatrix}0&-1&0\\1&1&1\\-1&-1&0\end{pmatrix},\quad M_0(-1)^{-1}=\begin{pmatrix}1&0&1\\1&1&0\\0&0&1\end{pmatrix}$$

and

$$B_0(1)y=\begin{pmatrix}0&0&-1\\-1&1&0\\0&0&1\end{pmatrix}y(0)+\begin{pmatrix}1&0&-1\\0&0&1\\0&1&1\end{pmatrix}y(1) \\ +\begin{pmatrix}0&0&0\\0&\alpha&\beta\\0&0&0\end{pmatrix}\int_0^1 y(\xi)\,\mathrm{d}\xi,$$

$$B_0(-1)y=\begin{pmatrix}0&0&-1\\-1&1&0\\0&0&1\end{pmatrix}y(0)+\begin{pmatrix}1&0&-1\\0&0&1\\0&1&1\end{pmatrix}y(1) \\ +\begin{pmatrix}0&\alpha&\beta\\0&0&0\\0&0&0\end{pmatrix}\int_0^1 y(\xi)\,\mathrm{d}\xi$$

it follows that

$$B(1)y = \begin{pmatrix} 0 & 0 & 0 \\ -1 & 1 & 0 \\ 1 & -1 & 1 \end{pmatrix} y(0) + \begin{pmatrix} 0 & 0 & 0 \\ 1 & 1 & 1 \\ -1 & 0 & 0 \end{pmatrix} y(1)$$
$$+ \begin{pmatrix} 0 & 0 & 0 \\ 0 & \alpha & \beta \\ 0 & -\alpha & -\beta \end{pmatrix} \int_0^1 y(\xi)\, \mathrm{d}\xi,$$

$$B(-1)y = \begin{pmatrix} 0 & 0 & 0 \\ -1 & 1 & -1 \\ 0 & 0 & 1 \end{pmatrix} y(0) + \begin{pmatrix} 0 & 0 & 0 \\ 1 & 0 & 0 \\ 0 & 1 & 1 \end{pmatrix} y(1)$$
$$+ \begin{pmatrix} 0 & 0 & 0 \\ 0 & \alpha & \beta \\ 0 & 0 & 0 \end{pmatrix} \int_0^1 y(\xi)\, \mathrm{d}\xi.$$

We have to consider arbitrary $y = (0, y_2, y_3) \in \Delta_0(C[0,1] \cap BV[0,1])^3$. We see that the conditions given by $B(1)y = 0$ and $B(-1)y = 0$ are the same. Of course, we already know this from Theorem 4.7.5. Explicitly, the conditions given by $B(\lambda)y = 0$ are

$$y_2(0) - y_3(0) + \alpha \int_0^1 y_2(\xi)\, \mathrm{d}\xi + \beta \int_0^1 y_3(\xi)\, \mathrm{d}\xi = 0,$$
$$y_3(0) + y_2(1) + y_3(1) = 0.$$

In case $\alpha = \beta = 0$ it is easy to see that any $y \in \Delta_0(C[0,1] \cap BV[0,1])^3$ satisfying the given boundary conditions also satisfies $B(\lambda)y = 0$. But there is no obvious way how to deduce the conditions $B(\lambda)y = 0$ from the boundary conditions.

Now let us assume that we only want the convergence for the third component. In view of Remark 4.7.6 ii) we have to consider only the third component of $B(-1)$ and $B(1)$. This gives the two conditions

$$y_3(0) + y_3(1) = 0,$$
$$y_3(0) - \beta \int_0^1 y_3(\xi)\, \mathrm{d}\xi = 0,$$

that is, here we have to consider both $B(1)$ and $B(-1)$ to find a minimal set of conditions. Also, if we only consider the second component, we obtain two conditions. In this case, there is no reduction of the number of conditions obtained for general $(0, y_2, y_3)$. But there are other boundary conditions for which the number of conditions for y_2 or y_3 alone reduces to one.

4.8. Notes

The boundary conditions in this section are kept as general as possible and may include (infinitely many) interior point boundary conditions as well as integral terms. In this sense, they coincide with those in COLE [CO3]. Although this makes many proofs more involved (and certain restrictions may have to be imposed), the interior point boundary terms and the integral term do not enter into the condition for Birkhoff regularity if A_1 is invertible. The regularity condition in COLE [CO3, p. 541] yields Stone regularity as defined in Section 4.4, see Lemma 5.7.8. Our estimates of the resovent follow the approach of LANGER [LA9] and COLE [CO2], [CO3]. The main idea for the estimates of the inverse of the characteristic matrix is to write all terms as a product of bounded matrices, where the factorization may be different for different values of the variables and the parameters. This splitting is been taken care of in the matrices $\Delta(\lambda)$, so that one can handle different cases with one formula. This, and staying with matrices, if possible, makes the proofs more manageable.

Eigenfunction expansions are often stated as being equiconvergent with Fourier series. Although we do not make any statements of that form, Proposition 4.6.5 may be considered as a result on the convergence of the Fourier expansion. In this light, Proposition 4.6.5 may be seen as the main estimate, whereas the other estimates would be perturbation results stating equiconvergence with expansions for a simplified problem.

Chapter V

EXPANSION THEOREMS FOR REGULAR BOUNDARY EIGENVALUE PROBLEMS FOR FIRST ORDER SYSTEMS

In this chapter eigenfunction expansions for regular boundary eigenvalue problems for first order $n \times n$ systems of ordinary differential equations are proved. The boundary conditions are allowed to contain countably many interior points and also an integral term. The first order differential system depends linearly on the eigenvalue parameter λ and the coefficients in the boundary conditions are $n \times n$ matrix polynomials in λ. The notions Birkhoff regularity and Stone regularity are introduced for such boundary eigenvalue problems in terms of the corresponding notions defined in the preceding chapter for boundary eigenvalue problems with asymptotically constant boundary conditions (Definitions 5.2.1 and 5.5.1). For this the λ-polynomial boundary conditions of this chapter have to be transformed to asymptotically constant boundary conditions as considered in the fourth chapter. Indeed, it depends on this transformation whether a boundary eigenvalue problem of the present type is Birkhoff regular or Stone regular, and sometimes an appropriate choice is not obvious. A method to check Birkhoff regularity is deduced, first for the more important and simpler case of two-point boundary eigenvalue problems (Theorem 5.2.2), and afterwards in the general case (Theorem 5.2.3). Also a procedure is described by which Stone regularity can be checked. The actual conditions which have to be satisfied are rather sophisticated, and the calculations which have to be performed are very laborious.

Under the assumption that the endpoints of the underlying interval are no accumulation points of the interior points occurring in the boundary conditions, it is shown for Birkhoff regular boundary eigenvalue problems in the case $1 < p < \infty$ that certain components of the vector functions in $(L_p(a,b))^n$ are expandable into series of the corresponding eigenfunctions and associated functions (Theorem 5.3.2). These series are L_p-convergent. If the leading matrix A_1 in the differential system is invertible, each vector function in $(L_p(a,b))^n$ is expandable. For $p = \infty$, which means uniform convergence of the eigenfunction expansions, a more restrictive result holds (Theorem 5.3.3). As in the case of Fourier series, only continuous vector functions which are of bounded variation and fulfil certain boundary conditions can be expanded.

For Stone regular boundary eigenvalue problems the situation is more complicated. In this case the resolvent behaves like some nonnegative power of λ on the regularity circles Γ_ν ($\nu \in \mathbb{N}$). It is shown that for $1 < p \le \infty$ certain components of the eigenfunction expansions of such problems converge to these components of a given vector function if this function is smooth enough, i. e., belongs to some Sobolev space of sufficiently high order, and fulfils certain boundary conditions (Theorems 5.6.7, 5.6.9, 5.6.10 and 5.6.11). These boundary conditions are defined by an iterative procedure.

Finally, the notion of strong s-regularity is introduced. This concept yields some improvements of the above mentioned expansion theorems.

5.1. First order systems which are linear in the eigenvalue parameter

Let $-\infty < a < b < \infty$, $1 \le p \le \infty$ and $n \in \mathbb{N}\setminus\{0\}$. Here we consider the boundary eigenvalue problem

$$y' - (\lambda A_1 + A_0)y = 0, \tag{5.1.1}$$

$$\sum_{j=0}^{\infty} W^{(j)}(\lambda)y(a_j) + \int_a^b W(x,\lambda)y(x)\,\mathrm{d}x = 0, \tag{5.1.2}$$

where y varies in $(W_p^1(a,b))^n$.

For the system of differential equations (5.1.1) we assume that the coefficient matrices A_0 and A_1 belong to $M_n(L_p(a,b))$. As in Section 4.1 we suppose that A_1 is a diagonal matrix,

$$A_1 = \begin{pmatrix} A_0^1 & & & & \\ & A_1^1 & & 0 & \\ & & \cdot & & \\ & 0 & & \cdot & \\ & & & & \cdot \\ & & & & & A_l^1 \end{pmatrix},$$

where l is a positive integer,

$$A_\nu^1 = r_\nu I_{n_\nu} \ (\nu = 0,\ldots,l), \quad \sum_{\nu=0}^{l} n_\nu = n,$$

where $n_0 \in \mathbb{N}$ and $n_\nu \in \mathbb{N}\setminus\{0\}$ for $\nu = 1,\ldots,l$. According to the block structure of A_1, we write $A_0 = (A_{0,\nu\mu})_{\nu,\mu=1}^{l}$. For the diagonal elements of A_1 we assume: $r_0 = 0$, and for $\nu,\mu = 0,\ldots,l$, there are numbers $\varphi_{\nu\mu} \in [0,2\pi)$ such that

$$\begin{aligned} &(r_\nu - r_\mu)^{-1} \in L_\infty(a,b) \text{ if } \nu \ne \mu, \\ &r_\nu(x) - r_\mu(x) = |r_\nu(x) - r_\mu(x)|e^{i\varphi_{\nu\mu}} \text{ a. e. in } (a,b). \end{aligned}$$

Note that $\mu = 0$ gives $r_\nu^{-1} \in L_\infty(a,b)$ for $\nu = 1,\dots,l$ and

$$r_\nu(x) = |r_\nu(x)| e^{i\varphi_\nu} \text{ a.e. in } (a,b) \quad (\nu = 1,\dots,l),$$

where $\varphi_\nu := \varphi_{\nu 0} = \varphi_{0\nu} \pm \pi$ for $\nu = 1,\dots,l$.

For the boundary conditions (5.1.2) we assume that $a_j \in [a,b]$ for $j \in \mathbb{N}$, that $a_j \neq a_k$ if $j \neq k$, and that $a_0 = a$, $a_1 = b$. We suppose that the matrix function $W(\cdot,\lambda)$ is a polynomial with coefficients in $M_n(L_1(a,b))$ and that the $W^{(j)}(\lambda)$ are polynomials in $M_n(\mathbb{C})$ with a common upper bound for their degrees.

The first order system (5.1.1) fulfils the assumptions of Section 4.1, but the boundary conditions (5.1.2) are not asymptotically constant if W or one of the $W^{(j)}$ depends on λ. In order to obtain the conditions (4.1.10)–(4.1.12) we require that there is an $n \times n$ matrix polynomial $C_2(\lambda)$ whose determinant is not identically zero such that the following properties hold:
There is a matrix function $W_0 \in M_n(L_1(a,b))$ such that

$$C_2^{-1}(\lambda)W(\cdot,\lambda) - W_0 = O(\lambda^{-1}) \quad \text{in } M_n(L_1(a,b)) \text{ as } \lambda \to \infty, \tag{5.1.3}$$

and there are $n \times n$ matrices $W_0^{(j)}$, $j \in \mathbb{N}$, such that the estimates

$$\sum_{j=0}^{\infty} |W_0^{(j)}| < \infty \tag{5.1.4}$$

and

$$\sum_{j=0}^{\infty} \left| C_2^{-1}(\lambda)W^{(j)}(\lambda) - W_0^{(j)} \right| = O(\lambda^{-1}) \quad \text{as } \lambda \to \infty \tag{5.1.5}$$

hold.

We have to check that the boundary conditions (5.1.2) are well-defined. Since the determinant of $C_2(\lambda)$ is a polynomial and since $C_2(\lambda)$ is invertible if its determinant is nonzero, $C_2(\lambda)$ is invertible for all sufficiently large λ. Hence condition (5.1.3) makes sense. From

$$\begin{aligned}\sum_{j=0}^{\infty} |W^{(j)}(\lambda)| &\le \sum_{j=0}^{\infty} \left| C_2^{-1}(\lambda)W^{(j)}(\lambda) \right| |C_2(\lambda)| \\ &= O(|C_2(\lambda)|)\end{aligned}$$

as $\lambda \to \infty$ (see Section 4.1) we infer that

$$\sum_{j=0}^{\infty} W^{(j)}(\lambda) y(a_j) \qquad (y \in (W_p^1(a,b))^n) \tag{5.1.6}$$

is locally uniformly bounded and convergent for sufficiently large λ, say $|\lambda| \ge \gamma$. Since $C_2(\lambda)$ is invertible for all sufficiently large λ, we may assume that $C_2(\lambda)$ is invertible for $|\lambda| \geq \gamma$. For $|\lambda| < \gamma$ and $k \in \mathbb{N}$ we obtain by the maximum modulus principle that

$$\Big|\sum_{j=0}^{k} W^{(j)}(\lambda)y(a_j)\Big| \leq \sup_{|\mu|=\gamma} \sum_{j=0}^{k} |W^{(j)}(\lambda)|\,|y(a_j)|.$$

Hence (5.1.6) is locally uniformly bounded on $\mathbb{C}$. By VITALI'S theorem, (5.1.6) converges for all $\lambda \in \mathbb{C}$ and represents a holomorphic vector function. Thus

$$T^R(\lambda)y = \sum_{j=0}^{\infty} W^{(j)}(\lambda)y(a_j) + \int_a^b W(x,\lambda)y(x)\,dx \quad (y \in W_1^p(a,b))^n \tag{5.1.7}$$

defines a holomorphic operator function T^R on $L((W_p^1(a,b))^n, \mathbb{C}^n)$. Here we have to note that the strong holomorphy, i. e., the holomorphy of $T^R y$ for each function $y \in (W_p^1(a,b))^n$, implies the (norm) holomorphy of T^R, see e. g. [KA1, Theorems III.1.37 and III.3.12]. For $|\lambda| \geq \gamma$ and $y \in W_1^p(a,b))^n$ we set

$$\widetilde{T}^R(\lambda)y = \sum_{j=0}^{\infty} C_2^{-1}(\lambda)W_j(\lambda)y(a_j) + \int_a^b C_2^{-1}(\lambda)W(x,\lambda)y(x)\,dx. \tag{5.1.8}$$

Then $\widetilde{T}^R(\lambda)$ belongs to $L(W_p^1(a,b))^n, \mathbb{C}^n)$ and depends holomorphically on λ for $\lambda \in \Omega := \{\lambda \in \mathbb{C} : |\lambda| > \gamma\}$.

5.2. Birkhoff regular first order systems

DEFINITION 5.2.1. The boundary eigenvalue problem (5.1.1), (5.1.2) is called *Birkhoff regular* if there is an $n \times n$ matrix polynomial $C_2(\lambda)$ fulfilling the assumptions (5.1.3)–(5.1.5) such that the differential system (5.1.1) with the boundary conditions $\widetilde{T}^R(\lambda)y = 0$ is Birkhoff regular in the sense of Definition 4.1.2.

Now we present a method how to check Birkhoff regularity. If ν is the maximum of the orders of the polynomials W and $W^{(j)}$, then we can take $C_2(\lambda) = \lambda^\nu I_n$. In this case

$$C_2^{-1}(\lambda)W(\lambda) - W_0 = O(\lambda^{-1})$$

and

$$C_2^{-1}(\lambda)W^{(j)}(\lambda) - W_0^{(j)} = O(\lambda^{-1}) \quad (j \in \mathbb{N})$$

holds for suitable W_0 and $W_0^{(j)}$. But in order to obtain Birkhoff regularity one often has to take a more sophisticated matrix polynomial C_2, see e.g. the examples in Section 4.2.

First we consider two-point boundary eigenvalue problems with $n_0 = 0$. Thus $W = 0$ and $W^{(j)} = 0$ for $j \geq 2$. Then we write

$$\widehat{W}(\lambda) := (W^{(0)}(\lambda), W^{(1)}(\lambda)) := \begin{pmatrix} w_1(\lambda) \\ \vdots \\ w_n(\lambda) \end{pmatrix}.$$

If one component $w_j(\lambda)$ is identically zero, then the matrix $\widehat{W}(\lambda)$ has rank $n-1$, and so $C_2^{-1}(\lambda)\widehat{W}(\lambda)$ has rank $n-1$ for any $C_2(\lambda)$ such that $C_2^{-1}(\lambda)\widehat{W}(\lambda)$ is asymptotically constant. But then the determinants of all $n \times n$ submatrices of $C_2^{-1}(\lambda)\widehat{W}(\lambda)$ are identically zero. Thus the determinants of all $n \times n$ submatrices of

$$(W_0^{(0)}, W_0^{(1)}) = \lim_{\lambda \to \infty} C_2^{-1}(\lambda)\widehat{W}(\lambda)$$

are zero, which proves that none of the corresponding Birkhoff matrices—which are certain $n \times n$ submatrices of $\left(W_0^{(0)}, W_0^{(1)}\right)$—is invertible.

Hence a necessary condition for Birkhoff regularity is that none of the $w_j(\lambda)$ is identically zero.

Let ν_j be the degree of the vector polynomial w_j and let w_j^0 be its coefficient of λ^{ν_j}. If the w_j^0 $(j = 1, \ldots, n)$ are linearly dependent, then there are $\alpha_1, \ldots, \alpha_n \in \mathbb{C}$ such that

$$\sum_{j=1}^{n} \alpha_j w_j^0 = 0 \quad \text{and} \quad (\alpha_1, \ldots, \alpha_n) \neq 0.$$

We choose a number $j_0 \in \{1, \ldots, n\}$ with

$$\nu_{j_0} = \max\{\nu_j : j \in \{1, \ldots, n\},\ \alpha_j \neq 0\}.$$

We may assume that $\alpha_{j_0} = 1$. We define $c(\lambda) = (c_1(\lambda), \ldots, c_n(\lambda))$ by

$$c_j(\lambda) := \begin{cases} 0 & \text{if } \alpha_j = 0 \text{ or } j = j_0, \\ \lambda^{\nu_{j_0} - \nu_j}\alpha_j & \text{if } \alpha_j \neq 0 \text{ and } j \neq j_0. \end{cases}$$

Then

$$C(\lambda) := I_n - e_{j_0} c(\lambda)$$

is invertible with

$$C^{-1}(\lambda) = I_n + e_{j_0} c(\lambda)$$

since $c(\lambda)e_{j_0} = 0$. Now we consider $C^{-1}(\lambda)\widehat{W}(\lambda)$. This is again a matrix polynomial. For $j \in \{1, \ldots, n\} \setminus \{j_0\}$, the j-th row of $C^{-1}(\lambda)\widehat{W}(\lambda)$ is the j-th row of $\widehat{W}(\lambda)$, and the j_0-th row of $C^{-1}(\lambda)\widehat{W}(\lambda)$ is

$$\sum_{j=0}^{n} \lambda^{\nu_{j_0} - \nu_j} \alpha_j w_j(\lambda).$$

From the definition of the ν_j we know that this row is a polynomial of degree less or equal ν_{j_0}, and its coefficient of $\lambda^{\nu_{j_0}}$ is

$$\sum_{j=1}^{n} \alpha_j w_j^0 = 0.$$

Hence the j_0-th row of $C^{-1}(\lambda)\widehat{W}(\lambda)$ is a polynomial of degree less than ν_{j_0}. If the j_0-th row of $C^{-1}(\lambda)\widehat{W}(\lambda)$ is identically zero, then we already know that the boundary eigenvalue problem cannot be Birkhoff regular. If it is nonzero, then we repeat the above procedure. The sum of the degrees of the vector polynomials formed by the rows decreases strictly. After a finite number of steps we thus obtain an $n \times 2n$ matrix function $\widetilde{C}^{-1}\widehat{W}$ with the following properties:

i) $\widetilde{C}$ is a polynomial and $\widetilde{C}(\lambda)$ is invertible for all $\lambda \in \mathbb{C}$,

ii) either one of the rows of $\widetilde{C}^{-1}(\lambda)\widehat{W}(\lambda)$ is identically zero or for each $j \in \{1,\dots,n\}$ the j-th row of $\widetilde{C}^{-1}(\lambda)\widehat{W}(\lambda)$ is a polynomial, say of degree ν_j, and if its coefficient of λ^{ν_j} is denoted by w_j^0, then we have that $w_1^0,\dots,w_n^0$ are linearly independent.

Boundary conditions fulfilling these properties are called *normalized.* Since this procedure is done by multiplying the boundary matrices with an invertible matrix polynomial from the left, it would be no restriction to require that the boundary conditions are normalized.

Finally we multiply the normalized boundary conditions by

$$\operatorname{diag}(\lambda^{-\nu_1},\dots,\lambda^{-\nu_n})$$

from the left. Then we obtain the representation

$$C_2^{-1}(\lambda)\widehat{W}(\lambda) = (W_0^{(0)}, W_0^{(1)}) + O(\lambda^{-1}),$$

where $C_2(\lambda)$ is the product of $\operatorname{diag}(\lambda^{\nu_1},\dots,\lambda^{\nu_n})$ and a finite number of matrices of the form $I_n - e_{j_0}c(\lambda)$, i.e., $C_2(\lambda)$ is a matrix polynomial and invertible for $\lambda \neq 0$.

THEOREM 5.2.2. *We consider the two-point boundary eigenvalue problem*

$$(5.2.1) \qquad y' - (\lambda A_1 + A_0)y = 0, \quad W^{(0)}(\lambda)y(a) + W^{(1)}(\lambda)y(b) = 0$$

and assume that A_1 is invertible, i. e., $n_0 = 0$. Then the two-point boundary eigenvalue problem (5.2.1) *is Birkhoff regular if and only if the following two properties hold:*

i) *There is a matrix polynomial C_2 whose determinant is not identically zero such that*

$$C_2^{-1}(\lambda)\big(W^{(0)}(\lambda), W^{(1)}(\lambda)\big) = (W_0^{(0)}, W_0^{(1)}) + O(\lambda^{-1}) \quad \text{as } \lambda \to \infty,$$

where $(W_0^{(0)}, W_0^{(1)})$ is an $n \times 2n$ matrix of rank n.

ii) *For any matrix polynomial C_2 fulfilling* i) *the boundary eigenvalue problem* (5.2.1) *is Birkhoff regular in the sense of Definition* 4.1.2.

Proof. The sufficiency of the conditions is obvious by the definition of Birkhoff regularity. We have already seen above that i) is necessary. Now let the problem be Birkhoff regular and C_2 be a matrix polynomial fulfilling i), i. e.,

$$C_2^{-1}(\lambda)\widehat{W}(\lambda) = \widehat{W}_2 + O(\lambda^{-1}),$$

where $\widehat{W}(\lambda) := \big(W^{(0)}(\lambda), W^{(1)}(\lambda)\big)$ and $\operatorname{rank}\widehat{W}_2 = n$. By definition of Birkhoff regularity, there is a matrix polynomial C_1 such that

$$C_1^{-1}(\lambda)\widehat{W}(\lambda) = \widehat{W}_1 + O(\lambda^{-1})$$

with $\operatorname{rank}\widehat{W}_1 = n$ such that

$$\widehat{W}_1 \begin{pmatrix} I_n - \Delta(\lambda) \\ \Delta(\lambda) \end{pmatrix} \text{ is invertible for all } \lambda \in \mathbb{C}\setminus\{0\}.$$

The matrix $C_2^{\mathrm{ad}}(\lambda)$ is the matrix of the cofactors of $C_2(\lambda)$, i. e., its entries are the determinants of $(n-1)\times(n-1)$ submatrices of $C_2(\lambda)$. Hence C_2^{ad} is a matrix polynomial. Since the determinant of $C_2(\lambda)$ is a polynomial which is not identically zero, we have

$$\det C_2(\lambda) = \alpha\lambda^r\big(1 + O(\lambda^{-1})\big) \quad \text{as } \lambda \to \infty$$

for some integer r and some $\alpha \neq 0$. Hence

$$(\det C_2(\lambda))^{-1} = \alpha^{-1}\lambda^{-r}\big(1 + O(\lambda^{-1})\big) \quad \text{as } \lambda \to \infty,$$

which proves that

$$C_2^{-1}(\lambda)C_1(\lambda) = (\det C_2(\lambda))^{-1}C_2^{\mathrm{ad}}(\lambda)C_1(\lambda) = \lambda^s C + O(\lambda^{s-1}) \text{ as } \lambda \to \infty$$

for some integer s and a nonzero $n\times n$-matrix C. Then

$$\begin{aligned} &C_2^{-1}(\lambda)C_1(\lambda)C_1^{-1}(\lambda)\widehat{W}(\lambda) \\ &\quad = \big(\lambda^s C + O(\lambda^{s-1})\big)\big(\widehat{W}_1 + O(\lambda^{-1})\big) \\ &\quad = \lambda^s C\widehat{W}_1 + O(\lambda^{s-1}). \end{aligned}$$

Since $\widehat{W}_1$ has rank n and C is nonzero, the matrix $C\widehat{W}_1$ is nonzero. Hence $s = 0$ and $C\widehat{W}_1 = \widehat{W}_2$. Since $\operatorname{rank}\widehat{W}_2 = n$, we infer that C is invertible. But this proves that

$$\widehat{W}_2 \begin{pmatrix} I_n - \Delta(\lambda) \\ \Delta(\lambda) \end{pmatrix} = C\widehat{W}_1 \begin{pmatrix} I_n - \Delta(\lambda) \\ \Delta(\lambda) \end{pmatrix}$$

is invertible for all $\lambda \in \mathbb{C}\setminus\{0\}$. □

If $s \in \mathbb{N}$ and f is a complex-valued function defined on an unbounded subset U of $\mathbb{C}$, then we call f an *asymptotic polynomial of order* s (with respect to $\frac{1}{\lambda}$) if there are $f_j \in \mathbb{C}$ ($j = 0, \ldots, s$) such that

$$f(\lambda) = \sum_{j=0}^{s} \lambda^{-j} f_j + \lambda^{-s} o(1)$$

as $\lambda \to \infty$. f is called an *asymptotic polynomial* if it is an asymptotic polynomial of some order s. Asymptotic polynomials for vector-valued functions are defined analogously.

THEOREM 5.2.3. *Let* $n_0 = 0$. *The boundary eigenvalue problem* (5.1.1), (5.1.2) *is Birkhoff regular if and only if the following three properties hold:*
i) *There is a matrix polynomial* C_2 *whose determinant is not identically zero such that*

$$C_2^{-1}(\lambda)\big(W^{(0)}(\lambda), W^{(1)}(\lambda)\big) = \big(W_0^{(0)}, W_0^{(1)}\big) + O(\lambda^{-1}) \quad \textit{as } \lambda \to \infty$$

and $(W_0^{(0)}, W_0^{(1)})$ *is an* $n \times 2n$ *matrix of rank* n.
ii) *For any matrix polynomial* C_2 *fulfilling* i) *the two-point boundary eigenvalue problem* (5.1.1), $W_0^{(0)} y(a) + W_0^{(1)} y(b) = 0$ *is Birkhoff regular in the sense of Definition* 4.1.2.
iii) *For any matrix polynomial* C_2 *fulfilling* i) *the estimates*

$$C_2^{-1}(\lambda) W(\cdot, \lambda) = O(1) \quad \textit{in } M_n(L_1(a,b))$$

and, for $j \in \mathbb{N}$,

$$C_2^{-1}(\lambda) W^{(j)}(\lambda) = O(1) \quad \textit{in } M_n(\mathbb{C})$$

hold, and the properties (5.1.4) *and* (5.1.5) *are fulfilled, where the matrices* $W_0^{(j)}$ *are uniquely determined by* C_2. *If* $W^{(j)} \neq 0$ *only for finitely many* j, *then* (5.1.4) *and* (5.1.5) *are automatically satisfied.*

Proof. First we suppose that there is a matrix polynomial C_2 fulfilling i)–iii). Here we have to note that

$$C_2^{-1}(\lambda) = \lambda^s \widetilde{C}_2(\lambda),$$

where s is a suitable integer and $\widetilde{C}_2(\lambda)$ is an asymptotic polynomial of arbitrary order. Indeed,

$$\det C_2(\lambda) = \lambda^r \Big(\alpha_0 + \sum_{j=1}^{r} \alpha_j \lambda^{r-j}\Big)$$

with $r \in \mathbb{N}$ and $\alpha_0 \neq 0$ shows that $(\det C_2(\lambda))^{-1}$ is an asymptotic polynomial of arbitrary order. As in the proof of Theorem 5.2.2 we infer that

$$C_2^{-1}(\lambda) = \big(\det C_2(\lambda)\big)^{-1} C_2^{\mathrm{ad}}(\lambda)$$

has the representation stated above. Thus $C_2^{-1}(\lambda)W(\cdot,\lambda)$ and the $C_2^{-1}(\lambda)W^{(j)}(\lambda)$ are asymptotic polynomials which proves the representations (5.1.3) and

$$C_2^{-1}(\lambda)W^{(j)}(\lambda) = W_0^{(j)} + O(\lambda^{-1}) \quad (j \in \mathbb{N}).$$

Conversely, let the problem (5.1.1), (5.1.2) be Birkhoff regular. Then also the two-point boundary eigenvalue problems (5.1.1), $W^{(0)}(\lambda)y(a) + W^{(1)}(\lambda)y(b) = 0$ is Birkhoff regular. Hence condition ii) is necessary by Theorem 5.2.2. Let C_2 be any matrix polynomial fulfilling i). We have to prove that iii) is fulfilled. Choose some matrix polynomial C_1 such that the boundary eigenvalue problem (5.1.1), $C_1^{-1}(\lambda)T^R(\lambda)y = 0$ is Birkhoff regular in the sense of Definition 4.1.2. Since the corresponding two-point boundary eigenvalue problem is also Birkhoff regular, we know from the proof of Theorem 5.2.2 that

$$C_2^{-1}(\lambda)C_1(\lambda) = C + O(\lambda^{-1}),$$

where C is an invertible $n \times n$ matrix. But this immediately proves that iii) does not only hold with respect to C_1 but also with respect to C_2. □

REMARK 5.2.4. In case $n_0 \neq 0$ we have to substitute

$$\left(W^{(0)}(\lambda)\Delta_0 + T^R(\lambda)P^{[0]}(I_n - \Delta_0), W^{(1)}(\lambda)\Delta_0 + T^R(\lambda)P^{[0]}(I_n - \Delta_0)\right)$$

for $\left(W^{(0)}(\lambda), W^{(1)}(\lambda)\right)$ in condition i) of Theorem 5.2.2 or 5.2.3, respectively. Also, the boundary condition in ii) has to be replaced by

$$\sum_{j=0}^{\infty} W_0^{(j)} y(a_j) + \int_a^b W_0(t)y(t)\,\mathrm{d}t = 0.$$

We leave the details to the reader.

5.3. Expansion theorems for Birkhoff regular problems

We suppose that the boundary eigenvalue problem (5.1.1), (5.1.2) is Birkhoff regular and associate to it the operator function

$$T = \begin{pmatrix} T^D \\ T^R \end{pmatrix},$$

where

$$T^D(\lambda)y = y' - A_0 y - \lambda A_1 y, \tag{5.3.1}$$

$$T^R(\lambda)y = \sum_{j=0}^{\infty} W^{(j)}(\lambda)y(a_j) + \int_a^b W(x,\lambda)y(x)\,\mathrm{d}x \tag{5.3.2}$$

for $\lambda \in \mathbb{C}$ and $y \in (W_p^1(a,b))^n$. We choose a matrix polynomial C_2 according to Definition 5.2.1 of Birkhoff regularity and set

$$\widetilde{T} = \begin{pmatrix} \mathrm{id}_{(L_p(a,b))^n} & 0 \\ 0 & C_2^{-1} \end{pmatrix} T. \tag{5.3.3}$$

Then the boundary eigenvalue problem $\widetilde{T}(\lambda)y = 0$ is Birkhoff regular in the sense of Definition 4.1.2. By Theorem 4.3.9 and the discussion following Definition 4.4.1 we obtain that there are circles Γ_ν centred at 0 with radii ρ_ν ($\nu \in \mathbb{N}$) such that $\rho_\nu \nearrow \infty$ as $\nu \to \infty$ and $\widetilde{T}(\lambda)$ is invertible for all $\lambda \in \Gamma_\nu$ and $\nu \in \mathbb{N}$. Since we may assume that $C_2(\lambda)$ is invertible for these λ, $T(\lambda)$ is invertible for all $\lambda \in \Gamma_\nu$ and $\nu \in \mathbb{N}$. Hence, for $f \in (L_p(a,b))^n$,

$$Q_\nu f := -\frac{1}{2\pi i} \oint_{\Gamma_\nu} \widetilde{J} T^{-1}(\lambda)(A_1 f, 0)\, \mathrm{d}\lambda \qquad (\nu \in \mathbb{N}) \tag{5.3.4}$$

is well-defined, where $\widetilde{J}\colon (W_p^1(a,b))^n \to (L_p(a,b))^n$ is given by $\widetilde{J}y = \Delta_0 y$. Because of (5.3.3) we obtain

$$\widetilde{T}^{-1}(\lambda)(f_1, f_2) = T^{-1}(\lambda)(f_1, C_2(\lambda) f_2) \tag{5.3.5}$$

for $\lambda \in \rho(\widetilde{T})$ and $(f_1, f_2) \in (L_p(a,b))^n \times \mathbb{C}^n$. We immediately infer that

$$Q_\nu = P_\nu \qquad (\nu \in \mathbb{N}),$$

where the P_ν are defined in Lemma 4.6.7, i. e.,

$$P_\nu f = -\frac{1}{2\pi i} \oint_{\Gamma_\nu} \widetilde{J}\widetilde{T}^{-1}(\lambda)(A_1 f, 0)\, \mathrm{d}\lambda \qquad (\nu \in \mathbb{N}).$$

As an immediate consequence of Theorem 4.6.9 we obtain

THEOREM 5.3.1. *Let $1 < p < \infty$. Suppose that $A_1 \in M_n(L_\infty(a,b))$ and that there is a number $\tilde{p} > p$ such that $A_{0,0\nu} \in M_{n_0,n_\nu}(L_{\tilde{p}}(a,b))$ for $\nu = 1, \ldots, l$. Here $A_{0,0\nu}$ is the block of A_0 with index $(0,\nu)$ according to the block structure of A_1. If $p \le \frac{3}{2}$ we require that the condition* (4.1.22) *holds. Let W_0 be defined by* (5.1.3) *and suppose that W_0 belongs to $M_n(L_q(a,b))$ for some $q > 1$. Assume that a and b are no accumulation points of the set $\{a_j : j \in \mathbb{N}, W_0^{(j)} \neq 0\}$. Suppose that the boundary eigenvalue problem* (5.1.1), (5.1.2) *is Birkhoff regular and let Q_ν be defined by* (5.3.4). *Then $\lim\limits_{\nu\to\infty} Q_\nu f = f$ in $(L_p(a,b))^n$ holds for all $f \in (L_p(a,b))^n$ with $f = \Delta_0 f$, where Δ_0 is defined in* (4.1.22).

Since T is a Fredholm operator function, we can represent the principal parts of T^{-1} in terms of eigen- and associated vectors of T and T^*. Since an eigenvector or associated vector v of T^* belongs to $(L_{p'}(a,b))^n \times \mathbb{C}^n$, we can write $v = (u,d)$ with $u \in (L_{p'}(a,b))^n$ and $d \in \mathbb{C}^n$. For an eigenvalue λ_κ of T we define $r(\lambda_\kappa) := \dim N(T(\lambda_\kappa))$ and let $m_{\kappa,j}$ ($j = 1, \ldots, r_\kappa$) denote the partial multiplicities of T at λ_κ. Theorem 5.3.1 and Corollary 1.6.6 lead to

THEOREM 5.3.2. *Let $1 < p < \infty$. Suppose that $A_1 \in M_n(L_\infty(a,b))$ and that there is a number $\tilde{p} > p$ such that $A_{0,0\nu} \in M_{n_0,n_\nu}(L_{\tilde{p}}(a,b))$ for $\nu = 1,\dots,l$. Here $A_{0,0\nu}$ is the block of A_0 with index $(0,\nu)$ according to the block structure of A_1. If $p \le \frac{3}{2}$ we require that the condition (4.1.22) holds. Let W_0 be defined by (5.1.3) and suppose that W_0 belongs to $M_n(L_q(a,b))$ for some $q > 1$. Assume that a and b are no accumulation points of the set $\{a_j : j \in \mathbb{N}, W_0^{(j)} \ne 0\}$. Suppose that the boundary eigenvalue problem (5.1.1), (5.1.2) is Birkhoff regular and choose the curves Γ_ν ($\nu \in \mathbb{N}$) according to Theorem 4.3.9. Let $\lambda_0, \lambda_1, \dots$ be the eigenvalues of T and let*

$$\{y_{\kappa,l}^{(j)} : j = 1,\dots,r(\lambda_\kappa);\, l = 0,\dots,m_{\kappa,j} - 1\}$$

and

$$\{(u_{\kappa,l}^{(j)}, d_{\kappa,l}^{(j)}) : j = 1,\dots,r(\lambda_\kappa);\, l = 0,\dots,m_{\kappa,j} - 1\}$$

be biorthogonal CSEAVs of T and T^* at λ_κ. Then

$$f = -\lim_{\nu\to\infty} \sum_{\substack{\kappa\in\mathbb{N} \\ |\lambda_\kappa|<\rho_\nu}} \left(\sum_{j=1}^{r(\lambda_\kappa)} \sum_{h=0}^{m_{\kappa,j}-1} y_{\kappa,h}^{(j)} \int_a^b u_{\kappa,m_{\kappa,j}-1-h}^{(j)}(x)^\mathsf{T} A_1(x) f(x)\,\mathrm{d}x \right)$$

holds in $(L_p(a,b))^n$ for all $f \in (L_p(a,b))^n$ with $f = \Delta_0 f$, where Δ_0 is defined in (4.1.22).

Now we consider the case $p = \infty$. In the above theorem, we used the eigenfunctions and the associated functions of the adjoint operator function to find the coefficients of the expansion. But for $p = \infty$ we want to refrain from this since the dual of $L_\infty(a,b)$ has a much more complicated representation than the dual of $L_p(a,b)$ for $p < \infty$.

The choice of a suitable p for the boundary eigenvalue operator depends on the functions we want to expand and the regularity of the coefficients in the differential equation. If T is defined for p_0, then it is also defined for all $p < p_0$. But the characteristic matrix of T does not depend on p. Hence the partial multiplicities are independent of p, and any CSEAV of T corresponding to p_0 is also a CSEAV of T corresponding to $p < p_0$. In addition, a CSEAV of T^* for p is also a CSEAV of T^* for p_0 if $p_0 < \infty$. In particular, the principal part of T^{-1} at a pole is independent of p as a linear combination of tensor products of the eigenvectors and associated vectors of T and T^*.

Since $T^{-1}(\lambda)$ for $p = \infty$ is the restriction of $T^{-1}(\lambda)$ for any $p < \infty$, say $p = 2$, we can consider $p = 2$ to find the residues of T^{-1} in the case $p = \infty$. Thus we do not need the representation of the dual of $L_\infty(a,b)$ or even the dual of $W_\infty^1(a,b)$ in the following theorem.

The analog of Theorem 5.3.1 is obtained for $p = \infty$ if we use Theorem 4.7.5. And with the above considerations we obtain

THEOREM 5.3.3. *Suppose that* $A_1 \in M_n(W_1^1(a,b))$ *and* $A_{0,0\nu} \in M_{n_0,n_\nu}(W_1^1(a,b))$ *for* $\nu = 1,\dots,l$. *Here* $A_{0,0\nu}$ *is the block of* A_0 *with index* $(0,\nu)$ *according to the block structure of* A_1. *Let* W_0 *be defined by* (5.1.3) *and suppose that* W_0 *belongs to* $M_n(L_q(a,b))$ *for some* $q > 1$. *Assume that* a *and* b *are no accumulation points of the set* $\{a_j : j \in \mathbb{N}, W_0^{(j)} \neq 0\}$. *Suppose that the boundary eigenvalue problem* (5.1.1), (5.1.2) *is Birkhoff regular and choose the curves* Γ_ν $(\nu \in \mathbb{N})$ *according to Theorem* 4.3.9. *Let* $\lambda_0, \lambda_1, \dots$ *be the eigenvalues of* T *and let*

$$\{y_{\kappa,l}^{(j)} : j = 1,\dots,r(\lambda_\kappa);\, l = 0,\dots,m_{\kappa,j}-1\}$$

and

$$\{(u_{\kappa,l}^{(j)}, d_{\kappa,l}^{(j)}) : j = 1,\dots,r(\lambda_\kappa);\, l = 0,\dots,m_{\kappa,j}-1\}$$

be biorthogonal CSEAVs of T *and* T^* *at* λ_κ, *respectively. Then*

$$f = -\lim_{\nu\to\infty} \sum_{\substack{\kappa\in\mathbb{N}\\ |\lambda_\kappa|<\rho_\nu}} \left(\sum_{j=1}^{r(\lambda_\kappa)} \sum_{h=0}^{m_{\kappa,j}-1} y_{\kappa,h}^{(j)} \int_a^b u_{\kappa,m_{\kappa,j}-1-h}^{(j)}(x)^{\mathsf{T}} A_1(x) f(x)\,\mathrm{d}x \right)$$

holds in $(C(a,b))^n$ *for all* $f \in (C[a,b] \cap BV[a,b])^n$ *with* $f = \Delta_0 f$ *and*

$$B(\lambda)f = \Delta_0 M_0(\lambda)^{-1} I_3^0(((\Delta_0 - I_n)A_0 f, \widetilde{T}_0^R f), \lambda) = 0$$

for some $\lambda \in \mathbb{C}\setminus\{0\}$, *where* Δ_0 *and* I_3^0 *are defined in* (4.1.22) *and* (4.4.12),

$$\widetilde{T}_0^R f = \sum_{j=0}^{\infty} W_0^{(j)} f(a_j) + \int_a^b W_0(t) f(t)\,\mathrm{d}t,$$

and W_0, $W_0^{(j)}$ $(j \in \mathbb{N})$ *are defined in* (5.1.3)–(5.1.5).

5.4. Examples for expansions in eigenfunctions and associated functions

We continue the discussion of the boundary eigenvalue problem (4.3.18), (4.3.20) given by

$$y' - \lambda y - \begin{pmatrix} 0 & \alpha \\ 0 & 0 \end{pmatrix} y = 0,$$

$$\begin{pmatrix} 1 & 1 \\ 0 & \lambda \end{pmatrix} y(0) + \begin{pmatrix} 1 & 0 \\ 0 & \lambda+\beta \end{pmatrix} y(1) = 0.$$

We have shown that this boundary eigenvalue problem is Birkhoff regular. For the expansion of arbitrary functions with respect to eigenfunctions and associated functions of this boundary eigenvalue problem we have to determine biorthogonal CSEAVs of the corresponding boundary eigenvalue operator function T and its adjoint T^*. But since we know the characteristic matrix of the boundary eigenvalue problem explicitly, we shall first determine biorthogonal CSEAVs of M and M^*.

We have seen that the algebraic and the geometric multiplicities of the eigenvalues depend on the values of α and β. Therefore, we shall consider three cases. First let us state some general properties. A fundamental matrix function of the differential system (4.3.18) is

$$Y(x,\lambda)=\begin{pmatrix} e^{\lambda x} & \alpha x e^{\lambda x} \\ 0 & e^{\lambda x} \end{pmatrix}$$

(see (4.3.19)). The characteristic matrix is given by (4.3.21):

$$M(\lambda)=\begin{pmatrix} e^{\lambda}+1 & \alpha e^{\lambda}+1 \\ 0 & (\lambda+\beta)e^{\lambda}+\lambda \end{pmatrix}.$$

According to Theorem 3.1.4 we need the operator function $(T^R U)^*$ in order to determine the eigenfunctions and associated functions of T^* from the eigenvectors and associated vectors of M^*. In (3.3.4) we have calculated

$$\big((T^R U)^*(\lambda)d\big)(x)=Y^{-1}(x,\lambda)^{\mathsf{T}}Y(1,\lambda)^{\mathsf{T}}W^{(1)}(\lambda)^{\mathsf{T}}d.$$

Since $W^{(1)}(\lambda)$ is the coefficient matrix of $y(1)$ in (4.3.20), we obtain

$$\big((T^R U)^*(\lambda)d\big)(x)=\begin{pmatrix} e^{\lambda(1-x)} & 0 \\ \alpha(1-x)e^{\lambda(1-x)} & (\lambda+\beta)e^{\lambda(1-x)} \end{pmatrix}d.$$

The asymptotic boundary conditions were obtained with $C_2(\lambda)=\begin{pmatrix} 1 & 0 \\ 0 & \lambda \end{pmatrix}$. Thus Remark 4.7.6 i) yields that the boundary condition $B(\lambda)f=0$ in Theorem 5.3.3 is equivalent to $\begin{pmatrix} 1 & 1 \\ 0 & 1 \end{pmatrix} f(0)+f(1)=0$.

If $\beta=0$, the eigenvalues of M are 0, $\lambda_k:=(2k-1)\pi i$ and $\lambda_{-k}:=(-2k+1)\pi i$ for $k=1,2,\dots$.

CASE I. $\beta=0$ and $\alpha=1$. For $k\in\mathbb{Z}\setminus\{0\}$ we set

$$V_k(\lambda):=\frac{1}{\lambda-\lambda_k}M(\lambda)$$

and obtain

$$V_k(\lambda_k)=\begin{pmatrix} -1 & -1 \\ 0 & -\lambda_k \end{pmatrix}$$

since

$$\lim_{\lambda\to\lambda_k}\frac{e^{\lambda}+1}{\lambda-\lambda_k}=-1.$$

We set

$$c_k^{(1)}:=\begin{pmatrix} 1 \\ 0 \end{pmatrix},\quad c_k^{(2)}:=\begin{pmatrix} 0 \\ 1 \end{pmatrix},$$

$$d_k^{(1)}:=V_k(\lambda_k)^{-1\mathsf{T}}\begin{pmatrix} 1 \\ 0 \end{pmatrix}=\begin{pmatrix} -1 \\ \frac{1}{\lambda_k} \end{pmatrix},\quad d_k^{(2)}:=V_k(\lambda_k)^{-1\mathsf{T}}\begin{pmatrix} 0 \\ 1 \end{pmatrix}=\begin{pmatrix} 0 \\ -\frac{1}{\lambda_k} \end{pmatrix}.$$

Since $M(\lambda_k) = 0$, any nonzero element of $\mathbb{C}^2$ is an eigenvector of M or M^*, respectively. From

$$\begin{pmatrix} d_k^{(1)\mathsf{T}} \\ d_k^{(2)\mathsf{T}} \end{pmatrix} V_k(\lambda_k) \begin{pmatrix} c_k^{(1)} & c_k^{(2)} \end{pmatrix} = I_2$$

we immediately infer that $\{c_k^{(1)}, c_k^{(2)}\}$ and $\{d_k^{(1)}, d_k^{(2)}\}$ are biorthogonal CSEAVs of M and M^* at λ_k.

Now we have to find the corresponding eigenfunctions of T and T^*. We set

$$y_k^{(j)} := Y(x,\lambda_k)c_k^{(j)},\ u_k^{(j)} := -(T^R U)^*(\lambda_k)d_k^{(j)},\ v_k^{(j)} := \begin{pmatrix} u_k^{(j)} \\ d_k^{(j)} \end{pmatrix} \qquad (j = 1,2).$$

From Theorem 3.1.4 we know that $\{y_k^{(1)}, y_k^{(2)}\}$ and $\{v_k^{(1)}, v_k^{(2)}\}$ are biorthogonal CSEAVs of T and T^* at λ_k. An easy calculation yields

$$y_k^{(1)}(x) = \begin{pmatrix} e^{\lambda_k x} \\ 0 \end{pmatrix}, \quad y_k^{(2)}(x) = \begin{pmatrix} x e^{\lambda_k x} \\ e^{\lambda_k x} \end{pmatrix},$$

$$u_k^{(1)}(x) = \begin{pmatrix} -e^{-\lambda_k x} \\ x e^{-\lambda_k x} \end{pmatrix}, \quad u_k^{(2)}(x) = \begin{pmatrix} 0 \\ -e^{-\lambda_k x} \end{pmatrix}.$$

We still have to consider the eigenvalue 0. For this α can be arbitrary. Since 0 is a simple eigenvalue and $M(0) = \begin{pmatrix} 2 & \alpha+1 \\ 0 & 0 \end{pmatrix}$, $d := \begin{pmatrix} 0 \\ 1 \end{pmatrix}$ is a CSEAV of M^* at μ. But as

$$u := (T^R U)^*(0)d = 0,$$

the eigenvalue 0 does not contribute to the expansion. In view of Theorem 5.3.2 we thus obtain that the eigenfunction expansion

$$f = -\lim_{\nu\to\infty} \left\{ \sum_{k=1}^{\nu} \sum_{\substack{j=1,2 \\ l=\pm k}} y_l^{(j)} \int_0^1 u_l^{(j)}(x)^{\mathsf{T}} f(x)\,dx \right\}$$

holds for all $f \in (L_p(0,1))^2$, $1 < p < \infty$, and the series converges in $(L_p(0,1))^2$. In case $p = \infty$ we infer from Theorem 5.3.3 that the above series converges in $(L_\infty(0,1))^2$ for all $f \in \left(C[a,b] \cap BV[a,b]\right)^2$ satisfying $\begin{pmatrix} 1 & 1 \\ 0 & 1 \end{pmatrix} f(0) + f(1) = 0$.

For example, if we take $f(x) = \begin{pmatrix} 1 \\ 0 \end{pmatrix}$, then the first component yields

$$1 = \frac{4}{\pi} \sum_{k=1}^{\infty} \frac{\sin((2k-1)\pi x)}{2k-1}$$

in $L_p(0,1)$ for $1 < p < \infty$. But the right hand side is the Fourier series on $(-1,1)$ of the function which is 1 on $(0,1)$ and -1 on $(-1,0)$. Therefore, the above expansion does not hold uniformly on $[0,1]$.

CASE II. $\beta = 0$ and $\alpha \neq 1$. As we have already seen in the discussion of Case I, the eigenvalue 0 does not contribute to the eigenfunction expansion. For $k \in \mathbb{Z} \setminus \{0\}$ we set

$$c_k(\lambda) := \begin{pmatrix} 1 \\ \dfrac{\lambda - \lambda_k}{1-\alpha} \end{pmatrix}.$$

Since

$$e^{\lambda} + 1 = -(\lambda - \lambda_k) - \frac{1}{2}(\lambda - \lambda_k)^2 + O((\lambda - \lambda_k)^3)$$

as $\lambda \to \lambda_k$, we obtain

$$\begin{aligned} M(\lambda)c_k(\lambda) &= \begin{pmatrix} e^{\lambda} + 1 + \dfrac{\alpha e^{\lambda} + 1}{1-\alpha}(\lambda - \lambda_k) \\ \dfrac{\lambda}{1-\alpha}(e^{\lambda} + 1)(\lambda - \lambda_k) \end{pmatrix} \\ &= \begin{pmatrix} +\left(\dfrac{1}{2} - \dfrac{1}{1-\alpha}\right)(\lambda - \lambda_k)^2 + O((\lambda - \lambda_k)^3) \\ -\dfrac{\lambda_k}{1-\alpha}(\lambda - \lambda_k)^2 - \dfrac{1 + \frac{1}{2}\lambda_k}{1-\alpha}(\lambda - \lambda_k)^3 + O((\lambda - \lambda_k)^4) \end{pmatrix} \\ &= O((\lambda - \lambda_k)^2). \end{aligned}$$

This proves that c_k is a root function of M at λ_k of order 2. It is easy to see that

$$d_k := \begin{pmatrix} -(\lambda - \lambda_k) \\ \dfrac{\alpha - 1}{\lambda_k} + \left(\dfrac{1}{\lambda_k} + \dfrac{1-\alpha}{\lambda_k^2}\right)(\lambda - \lambda_k) \end{pmatrix}$$

is a root function of M^* at λ_k of order 2. Furthermore we obtain that

$$d_k(\lambda)^{\mathsf{T}}(\lambda - \lambda_k)^{-2} M(\lambda)c_k(\lambda) = 1 + O((\lambda - \lambda_k))^2.$$

Since the dimension of the null space of $M(\lambda_k)$ is 1, this proves that c_k and d_k are biorthogonal CSRFs of M and M^* at μ.

Now we calculate

$$\begin{aligned} Y(x,\lambda)c_k(\lambda) &= \begin{pmatrix} e^{\lambda x} & \alpha x e^{\lambda x} \\ 0 & e^{\lambda x} \end{pmatrix} \begin{pmatrix} 1 \\ \dfrac{\lambda - \lambda_k}{1-\alpha} \end{pmatrix} \\ &= \begin{pmatrix} e^{\lambda_k x} + \dfrac{x}{1-\alpha} e^{\lambda_k x}(\lambda - \lambda_k) \\ \dfrac{1}{1-\alpha} e^{\lambda_k x}(\lambda - \lambda_k) \end{pmatrix} + O((\lambda - \lambda_k)^2) \end{aligned}$$

and

$$
\begin{aligned}
&-\left((T^R U)^*(\lambda)d_k(\lambda)\right)(x)\\
&= \begin{pmatrix} e^{\lambda(1-x)} & 0 \\ \alpha(1-x)e^{\lambda(1-x)} & \lambda e^{\lambda(1-x)} \end{pmatrix} \begin{pmatrix} \lambda-\lambda_k \\ \dfrac{1-\alpha}{\lambda_k} - \left(\dfrac{1}{\lambda_k}+\dfrac{1-\alpha}{\lambda_k^2}\right)(\lambda-\lambda_k) \end{pmatrix}\\
&= \begin{pmatrix} e^{\lambda_k(1-x)}(\lambda-\lambda_k) \\ (1-\alpha)e^{\lambda_k(1-x)} - xe^{\lambda_k(1-x)}(\lambda-\lambda_k) \end{pmatrix} + O((\lambda-\lambda_k)^2).
\end{aligned}
$$

For $k \in \mathbb{Z}\setminus\{0\}$ we set

$$
y_{k,0}(x) := \begin{pmatrix} e^{\lambda_k x} \\ 0 \end{pmatrix}, \quad y_{k,1}(x) := \begin{pmatrix} \dfrac{x}{1-\alpha} e^{\lambda_k x} \\ \dfrac{1}{1-\alpha} e^{\lambda_k x} \end{pmatrix},
$$

$$
d_{k,0} := \begin{pmatrix} 0 \\ \dfrac{\alpha-1}{\lambda_k} \end{pmatrix}, \quad d_{k,1}(x) := \begin{pmatrix} -1 \\ \dfrac{1}{\lambda_k}+\dfrac{1-\alpha}{\lambda_k^2} \end{pmatrix},
$$

$$
u_{k,0}(x) := \begin{pmatrix} 0 \\ -(1-\alpha)e^{-\lambda_k x} \end{pmatrix}, \quad u_{k,1}(x) := \begin{pmatrix} -e^{-\lambda_k x} \\ xe^{-\lambda_k x} \end{pmatrix}.
$$

Then $\{y_{k,0}, y_{k,1}\}$ and $\left\{\begin{pmatrix} u_{k,0} \\ d_{k,0} \end{pmatrix}, \begin{pmatrix} u_{k,1} \\ d_{k,1} \end{pmatrix}\right\}$ are biorthogonal canonical systems of eigenfunctions and associated functions at λ_k. With the aid of Theorem 5.3.2 we obtain that

$$
f = -\lim_{\nu\to\infty}\left\{ \sum_{k=1}^{\nu} \sum_{\substack{j=1,2\\ l=\pm k}} y_{l,j} \int_0^1 u_{l,1-j}(x)^{\mathsf{T}} f(x)\,\mathrm{d}x \right\}
$$

holds for all $f \in (L_p(0,1))^2$, $1<p<\infty$, and the series converges in $(L_p(0,1))^2$. Furthermore, we know from Theorem 5.3.3 that the series converges in $(L_\infty(0,1))^2$ for all $f \in (C[a,b]\cap BV[a,b])^2$ with $\begin{pmatrix} 1 & 1 \\ 0 & 1 \end{pmatrix} f(0) + f(1) = 0$.

CASE III. $\beta \neq 0$. We already know that the eigenvalues are λ_k ($k \in \mathbb{Z}\setminus\{0\}$) as defined above and additionally the zeros of $(\lambda+\beta)e^\lambda+\lambda$. We assume that the zeros of $(\lambda+\beta)e^\lambda+\lambda$ are simple, i. e., that this function and its derivative $(\lambda+\beta+1)e^\lambda+1$ do not have common zeros. This condition holds if and only if $(\lambda+\beta)e^\lambda+\lambda$ and $e^\lambda+1-\lambda$ or, equivalently, $e^\lambda+1-\lambda$ and $(\lambda+\beta)(\lambda-1)+\lambda$ have no common zeros. Thus we exclude those $\beta = \frac{\mu}{1-\mu}-\mu$ where μ is any root of $e^\lambda+1-\lambda=0$. Then the zeros of $(\lambda+\beta)e^\lambda+\lambda$ and of $e^\lambda+1$ are different and simple. Thus the boundary eigenvalue problem has only simple eigenvalues. We have already given an estimate for the large zeros of $(\lambda+\beta)e^{\lambda}+\lambda$ (we slightly change the notation):

$$\begin{cases} \mu_k = (2k-1)\pi i + o(1) & \text{as } k \to +\infty, \\ \mu_k = (2k+1)\pi i + o(1) & \text{as } k \to -\infty. \end{cases} \tag{5.4.1}$$

Hence, for sufficiently large natural numbers k_0, the zeros of $(\lambda+\beta)e^{\lambda}+\lambda$ with $|\Re(\lambda)| > 2k_0\pi$ or $|\Im(\lambda)| > 2k_0\pi$ are exactly

$$\mu_{k_0+1}, \mu_{-k_0-1}, \mu_{k_0+2}, \mu_{-k_0-2}, \dots$$

We shall show that $(\lambda+\beta)e^{\lambda}+\lambda$ has exactly $2k_0+1$ zeros with $|\Re(\lambda)| \le 2k_0\pi$ and $|\Im(\lambda)| \le 2k_0\pi$ if k_0 is sufficiently large. Thus we can denote the zeros of $(\lambda+\beta)e^{\lambda}+\lambda$ by μ_k ($k \in \mathbb{Z}$) such that the asymptotic behaviour (5.4.1) holds. We know that $\lambda e^{\lambda}+\lambda$ has exactly $2k_0+1$ simple zeros with $|\Re(\lambda)| \le 2k_0\pi$ and $|\Im(\lambda)| \le 2k_0\pi$, namely $(-2k_0+1)\pi i, (-2k_0+3)\pi i, \dots, (2k_0-3)\pi i, (2k_0-1)\pi i$, and 0. Let $k_0 \in \mathbb{N}$ such that $|\beta| < k_0\pi$. For $\Im(\lambda) = \pm 2k_0\pi$ we have $e^{\lambda} = e^{\Re(\lambda)} > 0$, and hence

$$|\beta e^{\lambda}| < |\lambda|\,|e^{\lambda}+1|.$$

For $\Re(\lambda) = 2k_0\pi$ we have because of $|e^{\lambda}| > 2$ that

$$\begin{aligned} |\beta e^{\lambda}| &< \frac{1}{2}|\lambda|\,|e^{\lambda}| < |\lambda|\left(|e^{\lambda}|-1\right) \\ &\le |\lambda|\,|e^{\lambda}+1|. \end{aligned}$$

For $\Re(\lambda) = -2k_0\pi$ we have because of $|e^{\lambda}| < \frac{1}{2}$ that

$$\begin{aligned} |\beta e^{\lambda}| &< \frac{1}{2}|\lambda| < |\lambda|(1-|e^{\lambda}|) \\ &\le |\lambda|\,|e^{\lambda}+1|. \end{aligned}$$

Hence ROUCHÉ'S theorem yields that $(\lambda+\beta)e^{\lambda}+\lambda$ and $\lambda e^{\lambda}+\lambda$ have the same number of zeros inside the rectangle $|\Re(\lambda)| \le 2k_0\pi$ and $|\Im(\lambda)| \le 2k_0\pi$.

For $k \in \mathbb{Z}\setminus\{0\}$, $c_k := \begin{pmatrix} 1 \\ 0 \end{pmatrix}$ is an eigenvector of M at λ_k, and $d_k := \begin{pmatrix} -1 \\ \dfrac{\alpha-1}{\beta} \end{pmatrix}$ is an eigenvector of M^* at λ_k. From

$$(\lambda-\lambda_k)^{-1}M(\lambda)c_k|_{\lambda=\lambda_k} = \begin{pmatrix} -1 \\ 0 \end{pmatrix}$$

we immediately infer that c_k and d_k are a biorthogonal CSEAVs of M and M^* at λ_k. The corresponding CSEAVs y_k and $\begin{pmatrix} u_k \\ d_k \end{pmatrix}$ of the boundary eigenvalue problem are given by

$$y_k(x) := Y(x,\lambda_k)c_k = \begin{pmatrix} e^{\lambda_k x} \\ 0 \end{pmatrix}$$

and

$$u_k(x) := -\big((T^R U)^*(\lambda_k)d_k\big)(x) = \begin{pmatrix} -e^{-\lambda_k x} \\ -\Big(\alpha(1-x) + (1-\alpha)\dfrac{\lambda_k+\beta}{\beta}\Big)e^{-\lambda_k x} \end{pmatrix}.$$

For $k \in \mathbb{Z}$,

$$\tilde{c}_k := \begin{pmatrix} -\dfrac{\alpha e^{\mu_k}+1}{e^{\mu_k}+1} \\ 1 \end{pmatrix}$$

is an eigenvector of M at μ_k, and

$$\tilde{d}_k := \begin{pmatrix} 0 \\ \dfrac{1}{e^{\mu_k}+1-\mu_k} \end{pmatrix}$$

is an eigenvector of M^* at μ_k. An easy calculation yields

$$\begin{aligned} &\tilde{d}_k^{\mathsf{T}}(\lambda-\mu_k)^{-1}M(\lambda)\tilde{c}_k|_{\lambda=\mu_k} \\ &= \frac{1}{e^{\mu_k}+1-\mu_k}(\lambda-\mu_k)^{-1}\big((\lambda+\beta)e^{\lambda}+\lambda\big)\big|_{\lambda=\mu_k} \\ &= \frac{e^{\mu_k}+(\mu_k+\beta)e^{\mu_k}+1}{e^{\mu_k}+1-\mu_k} \\ &= 1, \end{aligned}$$

which proves that $\tilde{c}_k$ and $\tilde{d}_k$ are biorthogonal CSEAVs of M and M^* at μ_k. We set

$$\tilde{y}_k(x) := Y(x,\mu_k)\tilde{c}_k = \begin{pmatrix} \dfrac{\alpha x - \alpha(1-x)e^{\mu_k} - 1}{e^{\mu_k}+1}e^{\mu_k x} \\ e^{\mu_k x} \end{pmatrix}$$

$$\tilde{u}_k(x) := -\big((T^R U)^*(\mu_k)\tilde{d}_k\big)(x) = \begin{pmatrix} 0 \\ \dfrac{\mu_k}{e^{\mu_k}+1-\mu_k}e^{-\mu_k x} \end{pmatrix}.$$

With the aid of Theorem 5.3.2 we obtain that

$$f = -\lim_{\nu\to\infty}\left\{\sum_{k=1}^{\nu}\sum_{l=\pm k} y_l \int_0^1 u_l(x)^{\mathsf{T}} f(x)\,\mathrm{d}x + \sum_{k=-\nu}^{\nu} \tilde{y}_l \int_0^1 \tilde{u}_l(x)^{\mathsf{T}} f(x)\,\mathrm{d}x\right\}$$

holds for all $f \in (L_p(0,1))^2$, $1 < p < \infty$, and the series converges in $(L_p(0,1))^2$. Furthermore, we know from Theorem 5.3.3 that the series converges in $(L_\infty(0,1))^2$ for all $f \in (C[a,b] \cap BV[a,b])^2$ with $\begin{pmatrix} 1 & 1 \\ 0 & 1 \end{pmatrix} f(0) + f(1) = 0$.

In case $\beta = \frac{\mu}{1-\mu} - \mu$ for some root μ of $e^\lambda + 1 - \lambda = 0$, μ is a non-simple eigenvalue. Hence we have at most two eigenvalues for which a CSEAV contains an associated vector. We shall not pursue this exceptional case further.

5.5. Stone regular boundary eigenvalue problems

In Section 4.4 we have defined Stone regularity for the boundary eigenvalue problem (4.1.1), (4.1.2). In an analogous manner to Definition 5.2.1 we define Stone regularity for the boundary eigenvalue problem (5.1.1), (5.1.2):

DEFINITION 5.5.1. Let $s \in \mathbb{N}$. The boundary eigenvalue problem (5.1.1), (5.1.2) is called *s-regular* if there is an $n \times n$ matrix polynomial $C_2(\lambda)$ satisfying the assumptions made at the beginning of Section 5.1 such that the boundary eigenvalue problem (5.1.1), $\widetilde{T}^R(\lambda)y = 0$ is s-regular in the sense of Definition 4.4.1, where $\widetilde{T}^R(\lambda)$ is defined in (5.1.8). The boundary eigenvalue problem (5.1.1), (5.1.2) is called *Stone regular* if it is s-regular for some $s \in \mathbb{N}$.

We now deduce a method how to check Stone regularity. As in Section 5.1 we can construct a matrix polynomial C_2 such that

$$\text{(5.5.1)} \qquad \begin{cases} C_2^{-1}(\lambda)W^{(j)}(\lambda) = W_0^{(j)} + O(\lambda^{-1}), \\ C_2^{-1}(\lambda)W(\cdot,\lambda) = W_0 + O(\lambda^{-1}) \text{ in } M_n(L_1(a,b)) \end{cases}$$

as $\lambda \to \infty$, and such that the following alternative holds:
i) either one row of

$$C_2^{-1}(\lambda)\big(W(\cdot,\lambda), W^{(1)}(\lambda), W^{(2)}(\lambda), \dots\big)$$

is identically zero, or
ii) the rows of

$$\big(W_0, W_0^{(1)}, W_0^{(2)}, \dots\big)$$

are linearly independent.

In the first case, the determinant of the characteristic matrix is identically zero, and hence the problem cannot be Stone regular.

Thus we shall assume that the condition ii) holds. We suppose that the assumptions of Theorem 2.8.2 are fulfilled for some $k > 0$. First we shall see that the asymptotic behaviour of the boundary matrices (5.5.1) can be described more precisely. In the proof of Theorem 5.2.3 we have seen that $C_2^{-1}(\lambda) = \lambda^s \widetilde{C}_2(\lambda)$, where $\widetilde{C}_2(\lambda)$ is an asymptotic polynomial of arbitrary order. Since (5.5.1) holds, there are $n \times n$ matrices $W_\nu^{(j)}$ $(\nu = 1,\dots,k;\ j \in \mathbb{N})$ and $n \times n$ matrix functions $W_\nu \in M_n(L_1(a,b))$ such that

$$(5.5.2)\qquad \begin{cases} C_2^{-1}(\lambda)W^{(j)}(\lambda) - \sum_{\nu=0}^{k} \lambda^{-\nu} W_\nu^{(j)} = O(\lambda^{-k-1}) \quad (j \in \mathbb{N}) \\ C_2^{-1}(\lambda)W(\cdot,\lambda) - \sum_{\nu=0}^{k} \lambda^{-\nu} W_\nu = O(\lambda^{-k-1}) \quad \text{in } M_n(L_1(a,b)) \end{cases}$$

as $\lambda \to \infty$. If only finitely many matrix functions $W^{(j)}$ are different from zero, then we obtain

$$\sum_{j=0}^{\infty} \left| C_2^{-1}(\lambda)W^{(j)}(\lambda) - \sum_{\nu=0}^{k} \lambda^{-\nu} W_\nu^{(j)} \right| = O(\lambda^{-k-1})$$

as $\lambda \to \infty$. This immediately implies that

$$\sum_{j=0}^{\infty} \widehat{W}^{(j)}(\lambda)y(a_j) - \sum_{\nu=0}^{k} \lambda^{-\nu} W_\nu^{(j)} y(a_j) = O(\lambda^{-k-1})|y|_{(0)}$$

holds, where y varies in $(C([a,b]))^n$ and

$$\widehat{W}^{(j)}(\lambda) := C_2^{-1}(\lambda)W^{(j)}(\lambda) \quad (j \in \mathbb{N})$$

for sufficiently large λ. Now we shall show that this estimate also holds if infinitely many matrix functions $W^{(j)}$ are different from zero and if the estimates (5.1.4) and (5.1.5) hold with respect to C_2. The matrix functions $\widehat{W}^{(j)}$ $(j \in \mathbb{N})$ are holomorphic in a neighbourhood of ∞, and

$$\widehat{W}^{(j)}(\infty) = W_0^{(j)} \quad (j \in \mathbb{N})$$

by (5.1.5). The estimates (5.1.4) and (5.1.5) give

$$\begin{aligned} \sum_{j=0}^{\infty} |\widehat{W}^{(j)}(\lambda)y(a_j)| &\le \Big(\sum_{j=0}^{\infty} |C_2^{-1}(\lambda)W^{(j)}(\lambda) - W_0^{(j)}| + \sum_{j=0}^{\infty} |W_0^{(j)}| \Big) |y|_{(0)} \\ &= O(1)|y|_{(0)} \end{aligned}$$

for $y \in (C[a,b])^n$ as $\lambda \to \infty$. This proves that

$$\sum_{j=0}^{\infty} \widehat{W}^{(j)} y(a_j)$$

is uniformly bounded and convergent in a neighbourhood of ∞. By VITALI'S theorem there is a neighbourhood of ∞ where $\sum_{j=0}^{\infty} \widehat{W}^{(j)} y(a_j)$ converges uniformly and defines a holomorphic function at ∞. Because of the uniform convergence we have for sufficiently large r and $\nu \in \mathbb{N}$ that

$$\frac{1}{2\pi i}\oint_{|\lambda|=r} \lambda^{\nu-1}\sum_{j=0}^{\infty}\widehat{W}^{(j)}(\lambda)y(a_j)\,\mathrm{d}\lambda = \sum_{j=0}^{\infty}\frac{1}{2\pi i}\oint_{|\lambda|=r}\lambda^{\nu-1}\widehat{W}^{(j)}(\lambda)y(a_j)\,\mathrm{d}\lambda$$
$$= \sum_{j=0}^{\infty} W_\nu^{(j)} y(a_j).$$

In view of the above estimate this shows that

$$y \mapsto \sum_{j=0}^{\infty} W_\nu^{(j)} y(a_j) \qquad (y \in (C[a,b])^n)$$

converges unconditionally in $L((C[a,b])^n, \mathbb{C}^n)$ and that

$$\left|\sum_{j=0}^{\infty} W_\nu^{(j)} y(a_j)\right|_{(0)} = O(r^\nu)|y|_{(0)}.$$

Taking constant functions, e. g. the unit vectors in $\mathbb{C}^n$, we obtain that $\sum_{j=0}^{\infty} W_\nu^{(j)}$ converges unconditionally and hence also absolutely, i. e.,

$$\sum_{j=0}^{\infty} |W_\nu^{(j)}| < \infty \qquad (\nu = 0,\dots,k). \tag{5.5.3}$$

The Taylor expansion of $\sum_{j=0}^{\infty} \widehat{W}^{(j)}(\lambda)y(a_j)$ at ∞ yields that

$$\sum_{j=0}^{\infty}\widehat{W}^{(j)}(\lambda)y(a_j) - \sum_{\nu=0}^{k}\lambda^{-\nu}\sum_{j=0}^{\infty}W_\nu^{(j)}y(a_j) \tag{5.5.4}$$
$$= \sum_{\nu=k+1}^{\infty}\lambda^{-\nu}\sum_{j=0}^{\infty}W_\nu^{(j)}y(a_j) = O(\lambda^{-k-1})|y|_{(0)}$$

as $\lambda \to \infty$ for $y \in (C([a,b]))^n$.

PROPOSITION 5.5.2. *Let $k \in \mathbb{N}$ be such that the assumptions of Theorem* 2.8.2 *are satisfied. The matrix functions $P^{[r]}$ $(r = 0,\dots,k)$ and $E(\cdot,\lambda)$ are as in Theorem* 2.8.2. *The matrices $W_\nu^{(j)}$ and the matrix functions W_ν are as given by* (5.5.2). *In addition, we assume that $W_\nu \in M_n(W_1^{k-\nu}(a,b))$ for $\nu = 0,\dots,k$. For $r = 0,\dots,k$ we set*

$$\left(u_{m,q,r}^{[0]}\right)_{m,q=1}^{n} := \sum_{\nu=0}^{r} W_\nu P^{[r-\nu]} \in M_n(W_1^{k-r}(a,b)).$$

With $\hat{r}_q$ $(q = n_0+1,\dots,n)$ given by $A_1 =: \mathrm{diag}(0,\dots,0,\hat{r}_{n_0+1},\dots,\hat{r}_n)$ we define

$$u_{m,q,r}^{[j+1]} := -\left(\frac{u_{m,q,r}^{[j]}}{\hat{r}_q}\right)'$$

for $m \in \{1,\dots,n\}$, $q \in \{n_0+1,\dots,n\}$, $r \in \{0,\dots,k-1\}$, *and* $j = 0,\dots k-r-1$. *For* $m = 1,\dots,n$ *and* $r = 0,\dots,k$ *we set*

$$\hat{w}^{(0)}_{m,q,r} := \begin{cases} -\sum\limits_{j=0}^{r-1} \dfrac{u^{[j]}_{m,q,r-j-1}(a)}{\hat{r}_q(a)} & \text{for } q = n_0+1,\dots,n,\\ 0 & \text{for } q = 1,\dots,n_0,\end{cases}$$

$$\hat{w}^{(1)}_{m,q,r} := \begin{cases} \sum\limits_{j=0}^{r-1} \dfrac{u^{[j]}_{m,q,r-j-1}(b)}{\hat{r}_q(b)} & \text{for } q = n_0+1,\dots,n,\\ 0 & \text{for } q = 1,\dots,n_0.\end{cases}$$

For $j = 0, 1$ *and* $r \in \{1,\dots,k\}$ *we set*

$$\widehat{W}^{(j)}_r := (\hat{w}^{(j)}_{m,q,r})^n_{m,q=1}.$$

For $j > 1$ *and* $r \in \{0,\dots,k\}$ *let* $\widehat{W}^{(j)}_r := 0$. *We set*

$$\widehat{M}_{1,k}(\lambda) := \sum_{j=0}^{\infty}\sum_{r=0}^{k} \lambda^{-r}\widetilde{W}^{(j)}_r E(a_j,\lambda) + \sum_{r=0}^{k}\lambda^{-r}\sum_{\nu=0}^{r}\int_a^b W_\nu(t)P^{[r-\nu]}(t)\,\mathrm{d}t\,(I_n-\Delta_0), \tag{5.5.5}$$

where

$$\widetilde{W}^{(j)}_r = \sum_{\nu=0}^{r} W^{(j)}_\nu P^{[r-\nu]}(a_j) + \widehat{W}^{(j)}_r \tag{5.5.6}$$

and Δ_0 *is defined in* (4.1.22). *Let*

$$\widetilde{M}(\lambda) = \widetilde{T}^R(\lambda)\widetilde{Y}(\lambda)\big(I_n - \Delta(\lambda) + E(b,\lambda)^{-1}\Delta(\lambda)\big) \tag{5.5.7}$$

denote the characteristic matrix of the boundary eigenvalue problem (5.1.1), $\widetilde{T}^R(\lambda)y = 0$, *where* $\widetilde{T}^R(\lambda)$ *is defined in* (5.1.8) *and the fundamental matrix function*

$$\widetilde{Y}(\cdot,\lambda) = \Big(\sum_{r=0}^{k}\lambda^{-r}P^{[r]} + \lambda^{-k}B_k(\cdot,\lambda)\Big)E(\cdot,\lambda)$$

is as in Theorem 2.8.2. *Then*

$$\widetilde{M}(\lambda) - \widehat{M}_{1,k}(\lambda)\big(I_n - \Delta(\lambda) + E(b,\lambda)^{-1}\Delta(\lambda)\big) = o(\lambda^{-k}), \tag{5.5.8}$$

where $\Delta(\lambda)$ *is defined in* (4.1.22).

Proof. For sufficiently large λ, the matrix function

$$\sum_{j=0}^{\infty}\widehat{W}^{(j)}(\lambda)\widetilde{Y}(a_j,\lambda) - \sum_{j=0}^{\infty}\sum_{r=0}^{k}\lambda^{-r}\sum_{\nu=0}^{r}W^{(j)}_\nu P^{[r-\nu]}(a_j)E(a_j,\lambda)$$

is well-defined by (5.5.3). For $c \in \mathbb{C}^n$ we set

$$R^{(j)}(\lambda)c := \widehat{W}^{(j)}(\lambda)c - \sum_{\nu=0}^{k} \lambda^{-\nu} W_\nu^{(j)} c.$$

Then we obtain

$$\begin{aligned}
&\widehat{W}^{(j)}(\lambda)\widetilde{Y}(a_j,\lambda) - \sum_{r=0}^{k} \lambda^{-r} \sum_{\nu=0}^{r} W_\nu^{(j)} P^{[r-\nu]}(a_j)E(a_j,\lambda) \\
&= \Big(\sum_{\nu=0}^{k} \lambda^{-\nu} W_\nu^{(j)} + R^{(j)}(\lambda)\Big)\Big(\sum_{r=0}^{k} \lambda^{-r} P^{[r]}(a_j) + \lambda^{-k} B_k(a_j,\lambda)\Big)E(a_j,\lambda) \\
&\quad - \sum_{r=0}^{k} \lambda^{-r} \sum_{\nu=0}^{r} W_\nu^{(j)} P^{[r-\nu]}(a_j)E(a_j,\lambda) \\
&= R^{(j)}(\lambda)\sum_{r=0}^{k} \lambda^{-r} P^{[r]}(a_j)E(a_j,\lambda) + \lambda^{-k}\widehat{W}^{(j)}(\lambda)B_k(a_j,\lambda)E(a_j,\lambda) \\
&\quad + \sum_{r=k+1}^{2k} \lambda^{-r} \sum_{\nu=r-k}^{k} W_\nu^{(j)} P^{[r-\nu]}(a_j)E(a_j,\lambda).
\end{aligned}$$

Let $D_j(\lambda)$ be either $E(a_j,\lambda)(I_n - \Delta(\lambda))$ or $E(a_j,\lambda)E(b,\lambda)^{-1}\Delta(\lambda)$. The estimate (5.5.4) and Proposition 4.3.3 i), ii) for $c = a_j$ and $d = a$ or $d = b$, respectively, yield that

$$\begin{aligned}
\Big|\sum_{j=0}^{\infty} R^{(j)}(\lambda)\sum_{r=0}^{k} \lambda^{-r} P^{[r]}(a_j)D_j(\lambda)\Big| &= O(\lambda^{-k-1})\Big|\sum_{r=0}^{k} \lambda^{-r} P^{[r]} D_j(\lambda)\Big|_{(0)} \\
&= O(\lambda^{-k-1}).
\end{aligned}$$

With the aid of the estimates (2.8.11) and (2.8.12) of B_k we also infer that

$$\sum_{j=0}^{\infty} \lambda^{-k}\widehat{W}^{(j)}(\lambda)B_k(a_j,\lambda)D_j(\lambda) = O(1)\lambda^{-k}|B_k(\cdot,\lambda)D_j(\lambda)|_{(0)}$$

is of the form $o(\lambda^{-k})$ and $O(\lambda^{-k}\tau_p(\lambda))$ as $\lambda \to \infty$. And the estimates (5.5.3) yield that

$$\Big|\sum_{j=0}^{\infty} \sum_{r=k+1}^{2k} \lambda^{-r} \sum_{\nu=r-k}^{k} W_\nu^{(j)} P^{[r-\nu]}(a_j)D_j(\lambda)\Big| = O(\lambda^{-k-1}).$$

Altogether we obtain that

$$\sum_{j=0}^{\infty}\Big[\widehat{W}^{(j)}(\lambda)\widetilde{Y}(a_j,\lambda) - \sum_{r=0}^{k} \lambda^{-r} \sum_{\nu=0}^{r} W_\nu^{(j)} P^{[r-\nu]}(a_j)E(a_j,\lambda)\Big](I_n - \Delta(\lambda))$$

and

$$\sum_{j=0}^{\infty}\Big[\widehat{W}^{(j)}(\lambda)\widetilde{Y}(a_j,\lambda)-\sum_{r=0}^{k}\lambda^{-r}\sum_{\nu=0}^{r}W_\nu^{(j)}P^{[r-\nu]}(a_j)E(a_j,\lambda)\Big]E(b,\lambda)^{-1}\Delta(\lambda)$$

are of the form $o(\lambda^{-k})$ and $O(\lambda^{-k}\tau_p(\lambda))$ as $\lambda\to\infty$. A similar proof shows that the same estimates hold for

$$\Big[\int_a^b C_2(\lambda)^{-1}W(t,\lambda)\widetilde{Y}(t,\lambda)\,\mathrm{d}t-\sum_{r=0}^{k}\lambda^{-r}\sum_{\nu=0}^{r}\int_a^b W_\nu(t)P^{[r-\nu]}(t)E(t,\lambda)\,\mathrm{d}t\Big]\times$$
$$\times\,(I_n-\Delta(\lambda))$$

and

$$\Big[\int_a^b C_2(\lambda)^{-1}W(t,\lambda)\widetilde{Y}(t,\lambda)\,\mathrm{d}t-\sum_{r=0}^{k}\lambda^{-r}\sum_{\nu=0}^{r}\int_a^b W_\nu(t)P^{[r-\nu]}(t)E(t,\lambda)\,\mathrm{d}t\Big]\times$$
$$\times\,E(b,\lambda)^{-1}\Delta(\lambda).$$

We thus obtain

$$\begin{aligned}(5.5.9)\qquad &\widetilde{M}(\lambda)-\sum_{j=0}^{\infty}\sum_{r=0}^{k}\lambda^{-r}\sum_{\nu=0}^{r}W_\nu^{(j)}P^{[r-\nu]}(a_j)E(a_j,\lambda)(I_n-\Delta(\lambda))\\ &-\sum_{j=0}^{\infty}\sum_{r=0}^{k}\lambda^{-r}\sum_{\nu=0}^{r}W_\nu^{(j)}P^{[r-\nu]}(a_j)E(a_j,\lambda)E(b,\lambda)^{-1}\Delta(\lambda)\\ &-\sum_{r=0}^{k}\lambda^{-r}\sum_{\nu=0}^{r}\int_a^b W_\nu(t)P^{[r-\nu]}(t)E(t,\lambda)\,\mathrm{d}t\,(I_n-\Delta(\lambda))\\ &-\sum_{r=0}^{k}\lambda^{-r}\sum_{\nu=0}^{r}\int_a^b W_\nu(t)P^{[r-\nu]}(t)E(t,\lambda)\,\mathrm{d}t\,E(b,\lambda)^{-1}\Delta(\lambda)\\ &=o(\lambda^{-k}).\end{aligned}$$

Let $r\in\{0,\dots,k\}$. Theorem 2.8.2 yields $P^{[r-\nu]}\in M_n(W_p^{k+1-r+\nu}(a,b))$ for $\nu=0,\dots,r$. By Propositions 2.3.1 and 2.1.7 we have

$$\big(u_{m,q,r}^{[0]}\big)_{m,q=1}^{n}\in M_n(W_1^{k-r}(a,b)).$$

The functions $u_{m,q,r}^{[j+1]}\in W_1^{k-r-j}(a,b)$ are well-defined because of Propositions 2.3.1 and 2.5.8 since $\hat{r}_q\in W_p^k(a,b)$. An integration by parts, see [HS, (18.19)], yields

$$\begin{aligned}\int_a^b u_{m,q,r}^{[j]}(t)\exp\{\lambda\widehat{R}_q(t)\}\,\mathrm{d}t&=\frac{1}{\lambda}\frac{u_{m,q,r}^{[j]}(t)}{\hat{r}_q(t)}\exp\{\lambda\widehat{R}_q(t)\}\Big|_a^b\\ &+\frac{1}{\lambda}\int_a^b u_{m,q,r}^{[j+1]}(t)\exp\{\lambda\widehat{R}_q(t)\}\,\mathrm{d}t\end{aligned}$$

for $j = 0, \dots, k-r-1$, where $\widehat{R}_q(t) = \int_a^t \hat{r}_q(\tau)\,\mathrm{d}\tau$. Hence we obtain

$$\begin{aligned}
&\int_a^b u_{m,q,r}^{[0]}(t)\exp\{\lambda \widehat{R}_q(t)\}\,\mathrm{d}t \\
&\quad = \sum_{j=0}^{k-r-1} \lambda^{-j-1}\Big(\frac{u_{m,q,r}^{[j]}(b)}{\hat{r}_q(b)}\exp\{\lambda\widehat{R}_q(b)\} - \frac{u_{m,q,r}^{[j]}(a)}{\hat{r}_q(a)}\Big) \\
&\quad + \lambda^{-k+r}\int_a^b u_{m,q,r}^{[k-r]}(t)\exp\{\lambda\widehat{R}_q(t)\}\,\mathrm{d}t
\end{aligned}$$

by a recursive application of the foregoing equation. In the proof of Proposition 4.3.5 we have seen that, for $\nu = 1, \dots, l$ and q such that $\hat{r}_q = r_\nu$,

$$(1-\delta_\nu(\lambda))\int_a^b u_{m,q,r}^{[k-r]}\exp\{\lambda\widehat{R}_q(t)\}\,\mathrm{d}t$$

and

$$\delta_\nu(\lambda)\int_a^b u_{m,q,r}^{[k-r]}\exp\{\lambda(\widehat{R}_q(t)-\hat{R}_q(b))\}\,\mathrm{d}t$$

are $o(1)$ as $\lambda \to \infty$. Altogether the representation (5.5.8) is proved. □

PROPOSITION 5.5.3. *Let the notations and assumptions be as in Proposition 5.5.2. We set*

$$\begin{aligned}
(a_{mq}^{(0)}(\lambda))_{m,q=1}^n &:= \sum_{r=0}^k \lambda^{-r}\widetilde{W}_r^{(0)}\Delta_0 \\
&+ \Big(\sum_{j=0}^\infty\sum_{r=0}^k \lambda^{-r}\widetilde{W}_r^{(j)} + \sum_{r=0}^k \lambda^{-r}\sum_{\nu=0}^r\int_a^b W_\nu(t)P^{[r-\nu]}(t)\,\mathrm{d}t\Big)(I_n - \dot{\Delta}_0), \\
(a_{mq}^{(j)}(\lambda))_{m,q=1}^n &:= \sum_{r=0}^k \lambda^{-r}\widetilde{W}_r^{(j)}\Delta_0 \quad (j \in \mathbb{N}\setminus\{0\}),
\end{aligned}$$

and

$$\alpha_{j_{n_0+1},\dots,j_n}(\lambda) := \sum_{\sigma\in S_n}\mathrm{sgn}(\sigma)\prod_{q=1}^{n_0} a_{\sigma(q),q}^{(0)}(\lambda)\prod_{q=n_0+1}^{n} a_{\sigma(q),q}^{(j_q)}(\lambda), \tag{5.5.10}$$

where $j_{n_0+1}, \dots, j_n \in \mathbb{N}$ and S_n is the set of permutations of the numbers $1, \dots, n$. Let $\widehat{M}_{1,k}(\lambda)$ be as defined in (5.5.5). Then

$$\begin{aligned}
\det\widehat{M}_{1,k}(\lambda) &= \sum_{\sigma\in S_n}\mathrm{sgn}(\sigma)\prod_{q=1}^n\Big(\sum_{j=0}^\infty a_{\sigma(q),q}^{(j)}(\lambda)\exp\{\lambda\widehat{R}_q(a_j)\}\Big) \\
&= \sum_{j_{n_0+1},\dots,j_n=0}^\infty \alpha_{j_{n_0+1},\dots,j_n}(\lambda)\prod_{q=n_0+1}^n \exp\{\lambda\widehat{R}_q(a_{j_q})\},
\end{aligned}$$

where the sum is absolutely convergent.

Proof. The representation is as in the proof of Theorem 4.3.9, where this was done for $k = 0$. Since all the estimates also hold here in view of (5.5.3), the convergence also holds for $k > 0$. □

The following criterion states a sufficient condition for Stone regularity which is essentially due to COLE [CO2], [CO4].

LEMMA 5.5.4. *Let $k \in \mathbb{N}$ be such that the assumptions of Theorem* 2.8.2 *are satisfied. The matrix functions $P^{[r]}$ $(r = 0, \dots, k)$ and $E(\cdot, \lambda)$ are given according to Theorem* 2.8.2. *The matrices $W_\nu^{(j)}$ and the matrix functions W_ν are as given by* (5.5.2). *In addition, we assume that W_ν belongs to $M_n(W_1^{k-\nu}(a,b))$ for $\nu = 0, \dots, k$. For $\nu = 1, \dots, l$ and $\mu = 1, 2$ we set*

$$\begin{aligned} W_{\nu\mu}(\lambda) &:= \sum_{r=0}^{k} \lambda^{-r} (\widetilde{W}_r^{(0)} \Lambda_\nu^\mu + \widetilde{W}_r^{(1)} \Lambda_\nu^{3-\mu}) \\ &\quad + \sum_{r=0}^{k} \lambda^{-r} \sum_{j=0}^{\infty} \widetilde{W}_r^{(j)} (I_n - \Delta_0) \\ &\quad + \sum_{r=0}^{k} \lambda^{-r} \sum_{\nu=0}^{r} \int_a^b W_\nu(t) P^{[r-\nu]}(t) \, \mathrm{d}t \, (I_n - \Delta_0), \\ \tilde{b}_{\nu\mu}(\lambda) &:= \det W_{\nu\mu}(\lambda), \end{aligned}$$

where Λ_ν^μ is given by (4.1.26) *or* (4.1.27), *respectively, and $\widetilde{W}_r^{(j)}$ is defined in* (5.5.6). *The functions $\tilde{b}_{\nu\mu}$ are asymptotic polynomials. Suppose that*

$$\tilde{b}_{\nu\mu}(\lambda) = \lambda^{-s_{\nu\mu}} [b_{\nu\mu}]$$

and

$$b_{\nu\mu} \neq 0$$

for $\nu = 1, \dots, l$ and $\mu = 1, 2$. Suppose that the exponential sum $\det \widehat{M}_{1,k}$ *is weakly regular in the sense of Definition* A.2.12, *where $\widehat{M}_{1,k}$ is defined in* (5.5.5). *If*

$$s := \max\{s_{\nu\mu} : \nu = 1, \dots, l;\ \mu = 1, 2\} \leq k,$$

then the boundary eigenvalue problem (5.1.1), (5.1.2) *is s-regular.*

Proof. Let $A_1 =: \operatorname{diag}(0, \dots, 0, \hat{r}_{n_0+1}, \dots, \hat{r}_n)$ and

$$\widehat{R}_q(t) := \int_a^t \hat{r}_q(\tau) \, \mathrm{d}\tau \quad (t \in [a,b],\ q \in \{n_0 + 1, \dots, n\}).$$

By Theorem A.1.3 the convex hull of

$$\mathscr{E} = \Big\{ \sum_{q=n_0+1}^{n} \widehat{R}_q(a_{j_q}) : j_{n_0+1}, \dots, j_n \in \mathbb{N} \Big\}$$

is a convex polygon, and the set of vertices $\widetilde{\mathscr{E}}$ of this convex polygon has the representation

$$\widetilde{\mathscr{E}} = \Big\{ \sum_{q=n_0+1}^{n} \widehat{R}_q(a_{\hat{\delta}_q(\lambda)}) : \lambda \in \mathbb{C}\setminus\{0\}\Big\},$$

see (A.1.3), where $(*,\dots,*,\hat{\delta}_{n_0+1}(\lambda),\dots,\hat{\delta}_n(\lambda)) := \Delta(\lambda)$ and $\Delta(\lambda)$ is defined in (4.1.22). Let $\widehat{\mathscr{E}}$ be as given by (A.2.36). Since, for each $c \in \mathscr{E}$, the coefficient function $\tilde{b}_c(\lambda)$ of $\exp(\lambda c)$ in $\det \widehat{M}_{1,k}(\lambda)$ is a polynomial in λ^{-1}, we have either $\tilde{b}_c(\lambda) = \lambda^{-\nu_c}[b_c]$ with $b_c \neq 0$ or $\tilde{b}_c(\lambda) = 0 = \lambda^{-s}[0]$. In the latter case c does not belong to $\widehat{\mathscr{E}}$. Therefore the assumptions of Theorem A.2.15 are satisfied for the exponential sum $\det \widehat{M}_{1,k}(\lambda)$. Here we have to note that the estimate (A.2.4) holds because of (5.5.3) and that the $\tilde{b}_{\nu\mu}$ are the coefficients of $\exp(\lambda c)$ for $c \in \widetilde{\mathscr{E}}$. Thus we obtain for

$$M_{1,k}(\lambda) = \widehat{M}_{1,k}(\lambda)\big(I_n - \Delta(\lambda) + E(b,\lambda)^{-1}\Delta(\lambda)\big)$$

that there are a number $\varepsilon > 0$ and circles Γ_ν ($\nu \in \mathbb{N}$) such that

$$|\lambda^s \det M_{1,k}(\lambda)| \geq |\lambda^{-\nu(\lambda)} \det M_{1,k}(\lambda)| \geq \varepsilon$$

for $\lambda \in \bigcup_{\nu=0}^{\infty} \Gamma_\nu$, where $\nu(\lambda) = -\nu_c$ for a suitable $c \in \widetilde{E}$. Here we have assumed without loss of generality that the radius ρ_0 is greater or equal 1. Since the estimate $M_{1,k}(\lambda) - \widetilde{M}(\lambda) = o(\lambda^{-k})$ holds by Proposition 5.5.2, $\widetilde{M}(\lambda)$ is bounded with respect to λ by Corollary 4.3.4 ii), and $s \leq k$, we may assume that

$$|\det M_{1,k}(\lambda) - \det \widetilde{M}(\lambda)| \leq |\lambda|^{-k}\frac{\varepsilon}{2} \leq |\lambda|^{-s}\frac{\varepsilon}{2}$$

for $\lambda \in \bigcup_{\nu=0}^{\infty} \Gamma_\nu$. Then

$$|\lambda^s \det \widetilde{M}(\lambda)| \geq \frac{\varepsilon}{2}$$

for $\lambda \in \bigcup_{\nu=0}^{\infty} \Gamma_\nu$. □

If there are infinitely many j for which $W^{(j)} \neq 0$, then one has to consider infinitely many polynomials in λ^{-1} in order to decide whether the exponential sum is weakly regular. Since the actual formulation of these conditions would be lengthy, we shall not give it here. Instead, we shall consider the special case that only finitely many $W^{(j)}$ are different from zero. In this case, the exponential sum is weakly regular, see Definition A.2.12 and Remark A.2.9. Therefore Lemma 5.5.4 yields

THEOREM 5.5.5. *Let $k \in \mathbb{N}$ be such that the assumptions of Theorem 2.8.2 are satisfied. The matrix functions $P^{[r]}$ $(r = 0,\dots,k)$ and $E(\cdot,\lambda)$ are given according to Theorem 2.8.2. The matrices $W_\nu^{(j)}$ and the matrix functions W_ν are as given by (5.5.2). In addition, we assume that $W_\nu \in M_n(W_1^{k-\nu}(a,b))$ for $\nu = 0,\dots,k$. Suppose that there are only finitely many j such that $W^{(j)} \neq 0$. For $\nu = 1,\dots,l$ and $\mu = 1,2$ we set*

$$\begin{aligned}
W_{\nu\mu}(\lambda) := & \sum_{r=0}^{k} \lambda^{-r}\left(\widetilde{W}_r^{(0)}\Lambda_\nu^\mu + \widetilde{W}_r^{(1)}\Lambda_\nu^{3-\mu}\right) \\
& + \sum_{r=0}^{k} \lambda^{-r} \sum_{j=0}^{\infty} \widetilde{W}_r^{(j)}(I_n - \Delta_0) \\
& + \sum_{r=0}^{k} \lambda^{-r} \sum_{\nu=0}^{r} \int_a^b W_\nu(t) P^{[r-\nu]}(t)\,\mathrm{d}t\,(I_n - \Delta_0),
\end{aligned}$$

$$\tilde{b}_{\nu\mu}(\lambda) := \det W_{\nu\mu}(\lambda),$$

where Λ_ν^μ is given by (4.1.26) or (4.1.27), respectively, and $\widetilde{W}_r^{(j)}$ is defined in (5.5.6). The functions $\tilde{b}_{\nu\mu}$ are asymptotic polynomials. Suppose that

$$\tilde{b}_{\nu\mu}(\lambda) = \lambda^{-s_{\nu\mu}}[b_{\nu\mu}]$$

and

$$b_{\nu\mu} \neq 0$$

for $\nu = 1,\dots,l$ and $\mu = 1,2$. If

$$s := \max\{s_{\nu\mu} : \nu = 1,\dots,l;\ \mu = 1,2\} \leq k,$$

then the boundary eigenvalue problem (5.1.1), (5.1.2) is s-regular.

COROLLARY 5.5.6. *Let $k \in \mathbb{N}$ be such that the assumptions of Theorem 2.8.2 are satisfied. Suppose that A_1 is invertible, that there is no integral term in the boundary conditions (5.1.2), and that only finitely many $W^{(j)}$ are different from zero. The matrices $W_\nu^{(j)}$ are as given by (5.5.2). For $\nu = 1,\dots,l$ and $\mu = 1,2$ set*

$$\widetilde{W}_{\nu\mu}(\lambda) := \sum_{r=0}^{k} \lambda^{-r}\left(\widetilde{\widetilde{W}}{}_r^{(0)}\Lambda_\nu^\mu + \widetilde{\widetilde{W}}{}_r^{(1)}\Lambda_\nu^{3-\mu}\right)$$

$$\tilde{b}_{\nu\mu}(\lambda) := \det \widetilde{W}_{\nu\mu}(\lambda),$$

where Λ_ν^μ is given by (4.1.26) or (4.1.27), respectively, and

$$\widetilde{\widetilde{W}}{}_r^{(j)} = \sum_{\nu=0}^{r} W_\nu^{(j)} P^{[r-\nu]}(a_j) P^{[0]^{-1}}(a_j).$$

The functions $\tilde{b}_{\nu\mu}$ are asymptotic polynomials. Suppose that

$$\tilde{b}_{\nu\mu}(\lambda) = \lambda^{-s_{\nu\mu}}[b_{\nu\mu}]$$

and

$$b_{\nu\mu} \neq 0$$

for $\nu = 1, \ldots, l$ *and* $\mu = 1, 2$. *If*

$$s := \max\{s_{\nu\mu} : \nu = 1, \ldots, l;\ \mu = 1, 2\} \leq k,$$

then the boundary eigenvalue problem (5.1.1), (5.1.2) *is s-regular.*

Proof. Since A_1 is invertible, we have $\Lambda_\nu^\mu + \Lambda_\nu^{3-\mu} = I_n$. It is also true that Λ_ν^μ and $P^{[0]}$ commute. Therefore $P^{[0]}(a_0)\Lambda_\nu^\mu + P^{[0]}(a_1)\Lambda_\nu^{3-\mu}$ is invertible, and the asymptotic polynomials $\tilde{b}_{\nu\mu}$ in Theorem 5.5.5 and Corollary 5.5.6 differ by a nonzero constant factor. In view of Theorem 5.5.5 the proof is complete. □

In the formulation of Corollary 5.5.6 we have replaced $P^{[r]}$ by $P^{[r]}P^{[0]^{-1}}$. As we have seen, this is advantageous for Birkhoff regular boundary eigenvalue problems since $P^{[0]}P^{[0]^{-1}} = I_n$. In general, if we would take $P^{[0]^{-1}}P^{[r]}$, then solving the differential equation (2.8.18) would reduce to a simple integration. The following proposition shows that this is also true if we assume that the $A_{0,\nu\nu}$ are scalar matrix functions. This trivially holds if $l = n$.

PROPOSITION 5.5.7. *Let the assumptions be as in Theorem* 2.8.2. *Suppose that the matrices* $A_{0,\nu\nu}$ $(\nu = 0, \ldots, l)$, *which are the block diagonals of* A_0 *with index* (ν, ν) *according to the block structure of* A_1, *are scalar matrix functions. Set* $\widehat{P}^{[r]} := P^{[r]}P^{[0]^{-1}}$. *Then the conditions* (2.8.18) *and* (2.8.19) *are equivalent to*

$$\widehat{P}_{\nu\nu}^{[r]\prime} = \sum_{\substack{q=0\\ q\neq\nu}}^{l} A_{0,\nu q}\widehat{P}_{q\nu}^{[r]} + \sum_{j=1}^{r}\sum_{q=0}^{l} A_{-j,\nu q}\widehat{P}_{q\nu}^{[r-j]} \qquad (\nu = 0, \ldots, l;\ r = 1, \ldots, k), \tag{5.5.11}$$

$$\widehat{P}_{\nu\mu}^{[r+1]} = (r_\nu - r_\mu)^{-1}\Big\{\widehat{P}_{\nu\mu}^{[r]\prime} + \widehat{P}_{\nu\mu}^{[r]}A_{0,\mu\mu} - \sum_{j=0}^{r}\sum_{q=0}^{l} A_{-j,\nu q}\widehat{P}_{q\mu}^{[r-j]}\Big\} \qquad (\nu, \mu = 0, \ldots, l;\ \nu \neq \mu;\ r = 0, \ldots, k-1). \tag{5.5.12}$$

Proof. We have

$$\begin{aligned}
\widehat{P}_{\nu\nu}^{[r]\prime} &= P_{\nu\nu}^{[r]\prime}P_{\nu\nu}^{[0]^{-1}} - P_{\nu\nu}^{[r]}P_{\nu\nu}^{[0]^{-1}}P_{\nu\nu}^{[0]\prime}P_{\nu\nu}^{[0]^{-1}}\\
&= P_{\nu\nu}^{[r]\prime}P_{\nu\nu}^{[0]^{-1}} - P_{\nu\nu}^{[r]}P_{\nu\nu}^{[0]^{-1}}A_{0,\nu\nu}\\
&= \big(P_{\nu\nu}^{[r]\prime} - A_{0,\nu\nu}P_{\nu\nu}^{[r]}\big)P_{\nu\nu}^{[0]^{-1}}
\end{aligned}$$

and

$$\begin{aligned}
P_{\nu\mu}^{[r]\prime} &= \widehat{P}_{\nu\mu}^{[r]\prime}P_{\mu\mu}^{[0]} + \widehat{P}_{\nu\mu}^{[r]}P_{\mu\mu}^{[0]\prime}\\
&= \big(\widehat{P}_{\nu\mu}^{[r]\prime} + \widehat{P}_{\nu\mu}^{[r]}A_{0,\mu\mu}\big)P_{\mu\mu}^{[0]}. \quad \square
\end{aligned}$$

5.6. Expansion theorems for Stone regular problems

Here we suppose that the assumptions of Theorem 2.8.2 hold for some $k \geq 0$. According to the decomposition $\mathbb{C}^n = \mathbb{C}^{n_0} \times \mathbb{C}^{n-n_0}$ we write

$$A_0 =: \begin{pmatrix} A_{11}^{[0]} & A_{12}^{[0]} \\ A_{21}^{[0]} & A_{22}^{[0]} \end{pmatrix}$$

and

$$A_1 =: \begin{pmatrix} 0 & 0 \\ 0 & \Omega \end{pmatrix}.$$

PROPOSITION 5.6.1. *Let $f \in (W_p^{k+1}(a,b))^n$. We set*

$$f =: f^{[0]} =: \begin{pmatrix} f_1^{[0]} \\ f_2^{[0]} \end{pmatrix} \in (W_p^{k+1}(a,b))^{n_0} \times (W_p^{k+1}(a,b))^{n-n_0}$$

and suppose that

$$f_1^{[0]\prime} - A_{11}^{[0]} f_1^{[0]} - A_{12}^{[0]} f_2^{[0]} = 0.$$

Then there are $f^{[j]} \in (W_p^{k+1-j}(a,b))^n$ for $j \in \{1,\dots,k+1\}$ such that

$$f^{[j]\prime} - A_0 f^{[j]} - A_1 f^{[j+1]} = 0 \quad (j = 0,\dots,k). \tag{5.6.1}$$

Proof. We are going to prove that there are functions $f_1^{[j]} \in (W_p^{k+1-j}(a,b))^{n_0}$ and $f_2^{[j]} \in (W_p^{k+1-j}(a,b))^{n-n_0}$ $(j = 1,\dots,k+1)$ such that

$$f_1^{[j]\prime} - A_{11}^{[0]} f_1^{[j]} - A_{12}^{[0]} f_2^{[j]} = 0, \tag{5.6.2}$$

$$f_2^{[j+1]} = \Omega^{-1}\big(f_2^{[j]\prime} - A_{21}^{[0]} f_1^{[j]} - A_{22}^{[0]} f_2^{[j]}\big) \tag{5.6.3}$$

for $j = 0,\dots,k$. By assumption, $f_1^{[0]}$ and $f_2^{[0]}$ satisfy (5.6.2).

Let $0 \leq l \leq k$ and assume that there are functions $f_1^{[j]} \in (W_p^{k+1-j}(a,b))^{n_0}$ and $f_2^{[j]} \in (W_p^{k+1-j}(a,b))^{n-n_0}$ for $j = 1,\dots,l$ such that (5.6.2) holds for $j = 0,\dots,l$ and such that (5.6.3) holds for $j = 0,\dots,l-1$. Note that the components of $A_{11}^{[0]}$, $A_{12}^{[0]}$, $A_{21}^{[0]}$ and $A_{22}^{[0]}$, i.e., the components of A_0, belong to $W_p^k(a,b)$ by the assumptions of Theorem 2.8.2. Hence

$$f_2^{[l]\prime} - A_{21}^{[0]} f_1^{[l]} - A_{22}^{[0]} f_2^{[l]} \in (W_p^{k+1-(l+1)}(a,b))^{n-n_0}$$

by Proposition 2.3.2. Since the components of Ω^{-1} belong to $W_p^k(a,b)$ if $k > 0$ and to $L_\infty(a,b)$ if $k = 0$, Propositions 2.3.2 and 2.3.1, respectively, yield that (5.6.3) defines a unique $f_2^{[l+1]} \in \big(W_p^{k+1-(l+1)}(a,b)\big)^{n-n_0}$. If $l < k$, we have, again in view of Proposition 2.3.2, that

$$A_{12}^{[0]} f_2^{[l+1]} \in (W_p^{k-l}(a,b))^{n_0} \subset (L_p(a,b))^{n_0}.$$

Lemma 2.5.7 yields that there is a solution $f_1^{[l+1]} \in (W_p^1(a,b))^{n_0}$ of (5.6.2) with $j = l+1$. From

$$f_1^{[l+1]'} = A_{11}^{[0]} f_1^{[l+1]} + A_{12}^{[0]} f_2^{[l+1]}$$

and Proposition 2.3.2 we inductively infer that $f_1^{[l+1]'} \in (W_p^m(a,b))^{n_0}$ and hence $f_1^{[l+1]} \in (W_p^{m+1}(a,b))^{n_0}$ for $m = 1,\dots,k-l$. For $m = k-l$ we obtain $f_1^{[l+1]} \in \left(W_p^{k+1-l}(a,b)\right)^{n_0} \subset \left(W_p^{k+1-(l+1)}(a,b)\right)^{n_0}$. Hence there are $f_1^{[j]} \in \left(W_p^{k+1-j}(a,b)\right)^{n_0}$ $(j = 0,\dots,k)$ and $f_2^{[j]} \in \left(W_p^{k+1-j}(a,b)\right)^{n-n_0}$ $(j = 0,\dots,k+1)$ such that (5.6.2) and (5.6.3) hold. Finally we set $f_1^{[k+1]} := 0$ and $f^{[j]} := \begin{pmatrix} f_1^{[j]} \\ f_2^{[j]} \end{pmatrix}$ for $j = 1,\dots,k+1$. Then (5.6.2) and (5.6.3) immediately prove (5.6.1). □

REMARK 5.6.2. If $n_0 = 0$, then the condition

$$f_1^{[0]'} - A_{11}^{[0]} f_1^{[0]} - A_{12}^{[0]} f_2^{[0]} = 0$$

is trivially fulfilled since it is a condition in a 0-dimensional vector space. Hence we can always find $f^{[j]}$ for $j = 1,\dots,k+1$, and they are recursively given by

$$f^{[j+1]} = A_1^{-1} f^{[j]'} - A_1^{-1} A_0 f^{[j]}.$$

REMARK 5.6.3. Let $f_2^{[0]} \in (W_p^{k+1}(a,b))^{n-n_0}$ and $A_0, A_1 \in (W_p^k(a,b))^n$. Then there is a function $f_1^{[0]} \in (W_p^{k+1}(a,b))^{n_0}$ such that

$$f_1^{[0]'} - A_{11}^{[0]} f_1^{[0]} - A_{12}^{[0]} f_2^{[0]} = 0. \tag{5.6.4}$$

Proof. By Lemma 2.5.7 there is a solution $f_1^{[0]} \in (W_p^1(a,b))^{n_0}$ of (5.6.4). From (5.6.4) we inductively obtain $f_1^{[0]} \in (W^{k+1}(a,b))^{n_0}$. □

From Proposition 5.6.1 and Remark 5.6.3 we immediately infer

REMARK 5.6.4. Let $f_2^{[0]} \in (W_p^{k+1}(a,b))^{n-n_0}$ and $A_0, A_1 \in (W_p^k(a,b))^n$. Then there are $f^{[j]} \in (W_p^{k+1-j}(a,b))^n$ $(j = 0,\dots,k+1)$ such that (5.6.1) holds and such that $\Delta_0 f^{[0]} = \begin{pmatrix} 0 \\ f_2^{[0]} \end{pmatrix}$, where $\Delta_0 = \begin{pmatrix} 0 & 0 \\ 0 & I_{n-n_0} \end{pmatrix}$.

In Section 5.1 we have seen that

$$T^R(\lambda) = O(|C_2(\lambda)|) \quad \text{in } L((W_p^1(a,b))^n, \mathbb{C}^n) \text{ as } \lambda \to \infty.$$

Since $|C_2(\lambda)|$ has a polynomial growth and since T^R is holomorphic on $\mathbb{C}$, it is a polynomial by LIOUVILLE'S theorem. Hence there is a $q \in \mathbb{N}$ such that

$$T^R(\lambda) = \sum_{r=0}^{q} \lambda^r T_r^R, \tag{5.6.5}$$

where $T_r^R \in L((W_p^1(a,b))^n, \mathbb{C}^n)$ for $r = 0, \ldots, q$. For $\lambda \in \rho(T)$ we set $R_1(\lambda)f_1 = T^{-1}(\lambda)(f_1, 0)$ $(f_1 \in (L_p(a,b))^n)$ and $R_2(\lambda)f_2 := T^{-1}(\lambda)(0, f_2)$ $(f_2 \in \mathbb{C}^n)$, see (3.2.1) and (3.2.2).

PROPOSITION 5.6.5. *Let $f \in (W_p^{k+1}(a,b))^n$. We set*

$$f =: f^{[0]} =: \begin{pmatrix} f_1^{[0]} \\ f_2^{[0]} \end{pmatrix} \in (W_p^{k+1}(a,b))^{n_0} \times (W_p^{k+1}(a,b))^{n-n_0}$$

and assume that

$$f_1^{[0]\prime} - A_{11}^{[0]} f_1^{[0]} - A_{12}^{[0]} f_2^{[0]} = 0.$$

For $j \in \{1, \ldots, k+1\}$ choose functions $f^{[j]} \in \left(W_p^{k+1-j}(a,b)\right)^n$ according to Proposition 5.6.1 *such that the equations* (5.6.1) *hold. Let $s_1 \in \mathbb{N}$ such that $s_1 \le k$. Then*

$$R_1(\lambda)A_1 f = -\sum_{j=0}^{s_1} \lambda^{-j-1} f^{[j]} + \lambda^{-s_1-1} R_1(\lambda) A_1 f^{[s_1+1]} + \sum_{j=0}^{s_1} \lambda^{-j-1} R_2(\lambda) T^R(\lambda) f^{[j]}. \tag{5.6.6}$$

Suppose that the "boundary conditions"

$$\sum_{r=0}^{\min\{q, s_1+1-j\}} T_r^R f^{[j+r-1]} = 0 \quad (j = 1, \ldots, \min\{s_1+1, s_2\}) \tag{5.6.7}$$

are fulfilled for some $s_2 \in \mathbb{N}$. Then

$$\sum_{j=0}^{s_1} \lambda^{-j-1} T^R(\lambda) f^{[j]} = \sum_{j=0}^{q-1} \lambda^j \sum_{r=j+1}^{\min\{q, s_1+1+j\}} T_r^R f^{[r-j-1]} + \sum_{j=s_2+1}^{s_1+1} \lambda^{-j} \sum_{r=0}^{\min\{q, s_1+1-j\}} T_r^R f^{[j+r-1]} \tag{5.6.8}$$

and

$$\begin{aligned} \frac{1}{\lambda} f = &- R_1(\lambda)A_1 f - \sum_{j=1}^{s_1} \lambda^{-j-1} f^{[j]} + \lambda^{-s_1-1} R_1(\lambda) A_1 f^{[s_1+1]} \\ &+ R_2(\lambda) \sum_{j=0}^{q-1} \lambda^j \sum_{r=j+1}^{\min\{q, s_1+1+j\}} T_r^R f^{[r-j-1]} \\ &+ R_2(\lambda) \sum_{j=s_2+1}^{s_1+1} \lambda^{-j} \sum_{r=0}^{\min\{q, s_1+1-j\}} T_r^R f^{[j+r-1]}. \end{aligned} \tag{5.6.9}$$

Proof. The relationships (3.2.3) and (5.6.1) lead to

$$\begin{aligned} R_1(\lambda)A_1 f^{[j]} &= -\frac{1}{\lambda}R_1(\lambda)\big(T^D(\lambda)f^{[j]} - f^{[j]\prime} + A_0 f^{[j]}\big) \\ &= -\frac{1}{\lambda}f^{[j]} + \frac{1}{\lambda}R_2(\lambda)T^R(\lambda)f^{[j]} + \frac{1}{\lambda}R_1(\lambda)A_1 f^{[j+1]} \end{aligned}$$

for $j = 0,\dots,k$. Then a recursive substitution for $R_1(\lambda)A_1 f^{[j]}$ for $j = 1,\dots,s_1$ yields (5.6.6). We calculate

$$\begin{aligned} \sum_{j=0}^{s_1} \lambda^{-j-1}T^R(\lambda)f^{[j]} &= \sum_{j=0}^{s_1}\sum_{r=0}^{q} \lambda^{r-j-1}T_r^R f^{[j]} \\ &= \sum_{r=0}^{q}\sum_{j-r+1=-r+1}^{s_1+1-r} \lambda^{r-j-1}T_r^R f^{[j]} \\ &= \sum_{r=0}^{q}\sum_{j=-r+1}^{s_1+1-r} \lambda^{-j}T_r^R f^{[j+r-1]} \\ &= \sum_{j=-q+1}^{s_1+1} \lambda^{-j} \sum_{r=\max\{0,-j+1\}}^{\min\{q,s_1+1-j\}} T_r^R f^{[j+r-1]} \\ &= \sum_{j=-q+1}^{0} \lambda^{-j} \sum_{r=-j+1}^{\min\{q,s_1+1-j\}} T_r^R f^{[j+r-1]} \\ &\quad + \sum_{j=1}^{s_1+1} \lambda^{-j} \sum_{r=0}^{\min\{q,s_1+1-j\}} T_r^R f^{[j+r-1]}. \end{aligned}$$

Then (5.6.8) immediately follows from (5.6.7), and (5.6.9) is a consequence of (5.6.6) and (5.6.8). □

The following definition is a refinement of the definition of s-regularity, see e. g. Theorem 5.6.9 below.

DEFINITION 5.6.6. Let T be given by (5.3.1) and (5.3.2) and let $s_1, s_2 \in \mathbb{N}$. Let Λ be a diagonal matrix whose diagonal elements are 0 or 1. T is said to be (s_1, s_2, Λ)-regular if there are circles $\Gamma_\nu = \{\lambda \in \mathbb{C} : |\lambda| = \rho_\nu\}$ $(\nu \in \mathbb{N})$ with $\rho_\nu \nearrow \infty$ as $\nu \to \infty$ such that

$$\oint_{\Gamma_\nu} |\lambda|^{-s_1-1}|J_\Lambda R_1(\lambda)A_1 f_1||\mathrm{d}\lambda| \to 0 \quad \text{as} \quad \nu \to \infty$$

and

$$\oint_{\Gamma_\nu} |\lambda|^{-s_2-1}|J_\Lambda R_2(\lambda)f_2||\mathrm{d}\lambda| \to 0 \quad \text{as} \quad \nu \to \infty,$$

where $f_1 \in (L_p(a,b))^n$ satisfies the condition that $A_1 f_1 \in (L_p(a,b))^n$, $f_2 \in \mathbb{C}^n$, the map $J : (W_p^1(a,b))^n \to (L_p(a,b))^n$ is the canonical embedding, $J_\Lambda = \Lambda J$, and R_1, R_2 are given by (3.2.1) and (3.2.2).

THEOREM 5.6.7. *Let T be given by* (5.3.1) *and* (5.3.2) *and let $s_1, s_2 \in \mathbb{N}$. Let Λ be a diagonal matrix whose diagonal elements are* 0 *or* 1. *Suppose that T is (s_1, s_2, Λ)-regular and that $A_0, A_1 \in M_n(W_p^{s_1}(a,b))$. Let $f \in (W_p^{s_1+1}(a,b))^n$. We set*

$$f =: f^{[0]} =: \begin{pmatrix} f_1^{[0]} \\ f_2^{[0]} \end{pmatrix} \in (W_p^{s_1+1}(a,b))^{n_0} \times (W_p^{s_1+1}(a,b))^{n-n_0}$$

and assume that

$$f_1^{[0]\prime} - A_{11}^{[0]} f_1^{[0]} - A_{12}^{[0]} f_2^{[0]} = 0.$$

For $j \in \{1, \dots, s_1+1\}$ we choose functions $f^{[j]} \in (W_p^{s_1+1-j}(a,b))^n$ according to Proposition 5.6.1 *such that* (5.6.1) *holds. Suppose that the "boundary conditions"*

$$\sum_{r=0}^{\min\{q, s_1+1-j\}} T_r^R f^{[j+r-1]} = 0 \quad (j = 1, \dots, \min\{s_1+1, s_2\}) \tag{5.6.10}$$

are fulfilled, where

$$T^R(\lambda) = \sum_{r=0}^{q} \lambda^r T_r^R.$$

Choose the curves Γ_ν ($\nu \in \mathbb{N}$) according to the definition of (s_1, s_2, Λ)-regularity. Then the expansion

$$\begin{aligned} \Lambda f = \lim_{\nu\to\infty} \sum_{\mu\in\sigma(T)\cap\operatorname{int}\Gamma_\nu} \Big\{ & -\Lambda(\operatorname{res}_\mu R_1) A_1 f \\ & + \operatorname{res}_{\lambda=\mu} \Big(\Lambda R_2(\lambda) \sum_{j=0}^{q-1} \lambda^j \sum_{r=j+1}^{\min\{q, s_1+1+j\}} T_r^R f^{[r-j-1]} \Big) \Big\} \end{aligned} \tag{5.6.11}$$

holds in $(L_p(a,b))^n$.

Proof. The assumptions of Proposition 5.6.5 are fulfilled with $k = s_1$. We multiply (5.6.9) by Λ from the left and integrate along the curves Γ_ν. From Definition 5.6.6 we obtain

$$\oint_{\Gamma_\nu} \lambda^{-s_1-1} \Lambda R_1(\lambda) A_1 f^{[s_1+1]} \, d\lambda \to 0$$

and for $j = s_2+1, s_2+2, \dots, s_1+1$, $r = 0, \dots, \min\{q, s_1+1-j\}$

$$\oint_{\Gamma_\nu} \lambda^{-j} \Lambda R_2(\lambda) T_r^R f^{[j+r-1]} \, d\lambda \to 0$$

in $(L_p(a,b))^n$ as $\nu \to \infty$. These estimates and the residue theorem complete the proof. □

If $s_2 = 0$ in Theorem 5.6.7, then the condition (5.6.10) is void. In this case we do not need that the functions to be expanded fulfil some boundary conditions. This can always be achieved by introducing an auxiliar eigenvalue:

COROLLARY 5.6.8. *Let T be given by (5.3.1) and (5.3.2) and let $s_1, s_2 \in \mathbb{N}$ with $s_2 > 0$. Let Λ be a diagonal matrix whose diagonal elements are 0 or 1. Suppose that T is (s_1, s_2, Λ)-regular and that $A_0, A_1 \in M_n(W_p^{s_1}(a,b))$. For the function $f \in (W_p^{s_1+1}(a,b))^n$ we set*

$$f =: f^{[0]} =: \begin{pmatrix} f_1^{[0]} \\ f_2^{[0]} \end{pmatrix} \in (W_p^{s_1+1}(a,b))^{n_0} \times (W_p^{s_1+1}(a,b))^{n-n_0}$$

and assume that

$$f_1^{[0]\prime} - A_{11}^{[0]} f_1^{[0]} - A_{12}^{[0]} f_2^{[0]} = 0.$$

For $j \in \{1, \ldots, s_1+1\}$ choose the functions $f^{[j]} \in (W_p^{s_1+1-j}(a,b))^n$ according to Proposition 5.6.1 such that (5.6.1) holds. Choose the curves Γ_ν ($\nu \in \mathbb{N}$) according to the definition of (s_1, s_2, Λ)-regularity. Let $\lambda_0 \in \mathbb{C} \setminus \sigma(T)$. Then the expansion

$$\begin{aligned}\Lambda f = \lim_{\nu\to\infty} \sum_{\mu\in\sigma(T)\cap \operatorname{int}\Gamma_\nu} \Big\{ &-\Lambda(\operatorname{res}_\mu R_1) A_1 f \\ &+ \operatorname{res}_{\lambda=\mu} \Big((\lambda-\lambda_0)^{-s_2} \Lambda R_2(\lambda) \sum_{j=0}^{q-1} \lambda^j \sum_{r=j+1}^{\min\{q, s_1+1+j\}} T_r^R f^{[r-j-1]} \Big) \Big\} \\ &+ \operatorname{res}_{\lambda=\lambda_0} \Big((\lambda-\lambda_0)^{-s_2} \Lambda R_2(\lambda) \sum_{j=0}^{q-1} \lambda^j \sum_{r=j+1}^{\min\{q, s_1+1+j\}} T_r^R f^{[r-j-1]} \Big)\end{aligned}$$

holds in $(L_p(a,b))^n$. If $\lambda_0 \in \sigma(T)$, then we obtain the expansion

$$\begin{aligned}\Lambda f = \lim_{\nu\to\infty} \sum_{\mu\in\sigma(T)\cap \operatorname{int}\Gamma_\nu} \Big\{ &-\Lambda(\operatorname{res}_\mu R_1) A_1 f \\ &+ \operatorname{res}_{\lambda=\mu} \Big((\lambda-\lambda_0)^{-s_0} \Lambda R_2(\lambda) \sum_{j=0}^{q-1} \lambda^j \sum_{r=j+1}^{\min\{q, s_1+1+j\}} T_r^R f^{[r-j-1]} \Big) \Big\}.\end{aligned}$$

Proof. We replace $T^R(\lambda)$ by $(\lambda-\lambda_0)^{s_2} T^R(\lambda)$, i.e., we consider the boundary eigenvalue operator function

$$T_1(\lambda) = \begin{pmatrix} T^D(\lambda) \\ (\lambda-\lambda_0)^{s_2} T^R(\lambda) \end{pmatrix}.$$

Obviously,

$$T_1^{-1}(\lambda) = (R_1(\lambda), (\lambda-\lambda_0)^{-s_2} R_2(\lambda)).$$

Thus T_1 is $(s_1, 0, \Lambda)$-regular and the corollary immediately follows from Theorem 5.6.7. □

THEOREM 5.6.9. *Let T be given by* (5.3.1) *and* (5.3.2) *and let $s \in \mathbb{N}$. Suppose that T is s-regular in the sense of Definiton* 5.5.1 *and that $A_0, A_1 \in M_n(W_p^{s+1}(a,b))$. Choose a matrix polynomial C_2 according to Definition* 5.5.1 *and let $\kappa \in \mathbb{Z}$ be such that*

$$C_2^{-1}(\lambda) = O(|\lambda|^\kappa) \quad \text{as } \lambda \to \infty.$$

Let $f \in (W_p^{s+2}(a,b))^n)$. We set

$$f =: f^{[0]} =: \begin{pmatrix} f_1^{[0]} \\ f_2^{[0]} \end{pmatrix} \in (W_p^{s+2}(a,b))^{n_0} \times (W_p^{s+2}(a,b))^{n-n_0}$$

and assume that

$$f_1^{[0]\prime} - A_{11}^{[0]} f_1^{[0]} - A_{12}^{[0]} f_2^{[0]} = 0.$$

For $j \in \{1,\dots,s+2\}$ choose functions $f^{[j]} \in (W_p^{s+2-j}(a,b))^n$ according to Proposition 5.6.1 *such that* (5.6.1) *holds. Suppose that the "boundary conditions"*

$$\sum_{r=0}^{\min\{q,s+2-j\}} T_r^R f^{[j+r-1]} = 0 \quad (j = 1,\dots,\min\{s+2, s+\kappa+1\}) \tag{5.6.12}$$

are fulfilled, where

$$T^R(\lambda) = \sum_{r=0}^{q} \lambda^r T_r^R.$$

Choose the curves Γ_ν ($\nu \in \mathbb{N}$) according to the definition of s-regularity. Then the expansion

$$\begin{aligned} f = \lim_{\nu\to\infty} \sum_{\mu \in \sigma(T) \cap \int \Gamma_\nu} \Big\{ &-(\operatorname{res}_\mu R_1) A_1 f \\ &+ \operatorname{res}_{\lambda=\mu} \Big(R_2(\lambda) \sum_{j=0}^{q-1} \lambda^j \sum_{r=j+1}^{\min\{q,s+2+j\}} T_r^R f^{[r-j-1]} \Big) \Big\} \end{aligned} \tag{5.6.13}$$

holds in $(L_p(a,b))^n$.

Proof. We have already seen in Section 5.1 that we can choose a number κ such that $C_2^{-1}(\lambda) = O(|\lambda|^\kappa)$ as $\lambda \to \infty$. From (5.3.3) we know that

$$\begin{aligned} R_1(\lambda) A_1 f_1 &= \widetilde{T}^{-1}(\lambda)(A_1 f_1, 0) \quad (f_1 \in L_p(a,b))^n), \\ R_2(\lambda) f_2 &= \widetilde{T}^{-1}(\lambda)(0, C_2^{-1}(\lambda) f_2) \quad (f_2 \in \mathbb{C}^n). \end{aligned}$$

From Theorem 4.4.9 ii) we obtain

$$\oint_{\Gamma_\nu} |\lambda|^{-s-2} |J R_1(\lambda) A_1 f_1| \, |\mathrm{d}\lambda| \to 0 \quad \text{as } \nu \to \infty$$

and, for some $C > 0$,

$$\oint_{\Gamma_\nu} |\lambda|^{-(s+\kappa+1)-1} |J R_2(\lambda) f_2| \, |\mathrm{d}\lambda| \le C \oint_{\Gamma_\nu} |\lambda|^{-s-2} |J \widetilde{T}^{-1}(\lambda)| \, |f_2| \, |\mathrm{d}\lambda| \to 0$$

as $\nu \to \infty$. Hence T is $(s+1, s_2, I)$-regular, where $s_2 := \max\{0, s+\kappa+1\}$. Now the statement of this theorem immediately follows from Theorem 5.6.7. □

THEOREM 5.6.10. *Let T be given by* (5.3.1) *and* (5.3.2) *and let $s \in \mathbb{N}$. Suppose that T is s-regular in the sense of Definiton* 5.5.1 *and that $A_0, A_1 \in M_n(W_p^s(a,b))$. Choose a matrix polynomial C_2 according to Definition* 5.5.1 *and let $\kappa \in \mathbb{Z}$ be such that*

$$C_2^{-1}(\lambda) = O(|\lambda|^\kappa) \quad \textit{as } \lambda \to \infty.$$

Let $f \in (W_p^{s+1}(a,b))^n)$. We set

$$f =: f^{[0]} =: \begin{pmatrix} f_1^{[0]} \\ f_2^{[0]} \end{pmatrix} \in (W_p^{s+1}(a,b))^{n_0} \times (W_p^{s+1}(a,b))^{n-n_0}$$

and assume that

$$f_1^{[0]\prime} - A_{11}^{[0]} f_1^{[0]} - A_{12}^{[0]} f_2^{[0]} = 0.$$

For $j \in \{1, \ldots, s+1\}$ choose functions $f^{[j]} \in (W_p^{s+1-j}(a,b))^n$ according to Proposition 5.6.1 *such that* (5.6.1) *holds. Suppose that the "boundary conditions"*

$$\sum_{r=0}^{\min\{q,s+1-j\}} T_r^R f^{[j+r-1]} = 0 \quad (j = 1, \ldots, \min\{s+1, s+\kappa+1\}) \tag{5.6.14}$$

are fulfilled, where

$$T^R(\lambda) = \sum_{r=0}^{q} \lambda^r T_r^R.$$

Choose the curves Γ_ν ($\nu \in \mathbb{N}$) according to the definition of s-regularity. Then the expansion

$$\begin{aligned} \Delta_0 f = \lim_{\nu \to \infty} \sum_{\mu \in \sigma(T) \cap \int \Gamma_\nu} \Big\{ & -\operatorname{res}_\mu (\Delta_0 R_1) A_1 f \\ & + \operatorname{res}_{\lambda=\mu} \Big(\Delta_0 R_2(\lambda) \sum_{j=0}^{q-1} \lambda^j \sum_{r=j+1}^{\min\{q,s+1+j\}} T_r^R f^{[r-j-1]} \Big) \Big\} \end{aligned} \tag{5.6.15}$$

holds in $(L_p(a,b))^n$, where $\Delta_0 = \begin{pmatrix} 0 & 0 \\ 0 & I_{n-n_0} \end{pmatrix} \in M_n(\mathbb{C}^{n_0} \times \mathbb{C}^{n-n_0})$.

Proof. Arguing as in the proof of Theorems 5.6.9 and 5.6.7 we still have to prove that

$$\oint_{\Gamma_\nu} |\lambda|^{-s-1} |\widetilde{J} R_1(\lambda) A_1 f_1| |d\lambda| \to 0 \quad \text{as } \nu \to \infty.$$

But this is true in view of Theorem 4.4.11 iii). Here we have to note that $A_1 = \Delta_0 A_1$ implies $\widetilde{T}^{-1}(\lambda)(A_1 f_1, 0) = \widetilde{T}^{-1}(\lambda) \widetilde{J}_1 A_1 f_1$. □

THEOREM 5.6.11. *Let $1 < p < \infty$, let T be given by* (5.3.1) *and* (5.3.2) *and let $s \in \mathbb{N}$. Suppose that T is s-regular in the sense of Definition* 5.5.1 *and that A_0, A_1 belong to $M_n(W_p^s(a,b))$. Choose a matrix polynomial C_2 according to Definition* 5.5.1 *and let $\kappa \in \mathbb{Z}$ be such that*

$$C_2^{-1}(\lambda) = O(|\lambda|^\kappa) \quad \textit{as } \lambda \to \infty.$$

Let $f \in (W_p^{s+1}(a,b))^n)$ with $\Delta_0 f = f$. Then $f =: \begin{pmatrix} 0 \\ f^{[2]} \end{pmatrix}$, and we choose some $f_1 \in W_p^{s+1}(a,b))^{n_0}$ such that

$$f_1^{[0]\prime} - A_{11}^{[0]} f_1^{[0]} - A_{12}^{[0]} f_2^{[0]} = 0.$$

Set $f^{[0]} := \begin{pmatrix} f_1^{[0]} \\ f_2^{[0]} \end{pmatrix}$. For $j \in \{1,\dots,s+1\}$ choose $f^{[j]} \in (W_p^{s+1-j}(a,b))^n$ according to Proposition 5.6.1 *such that* (5.6.1) *holds. Suppose that the "boundary conditions"*

$$\sum_{r=0}^{\min\{q,s+1-j\}} T_r^R f^{[j+r-1]} = 0 \quad (j = 1,\dots,\min\{s+1, s+\kappa\}) \tag{5.6.16}$$

are fulfilled, where

$$T^R(\lambda) = \sum_{r=0}^{q} \lambda^r T_r^R.$$

Choose the curves Γ_ν ($\nu \in \mathbb{N}$) according to the definition of s-regularity. Then the expansion

$$\begin{aligned} \Delta_0 f = \lim_{\nu\to\infty} \sum_{\mu \in \sigma(T) \cap \int \Gamma_\nu} \Big\{ &-(\operatorname{res}_\mu \Delta_0 R_1) A_1 f \\ &+ \operatorname{res}_{\lambda=\mu} \Big(\Delta_0 R_2(\lambda) \sum_{j=0}^{q-1} \lambda^j \sum_{r=j+1}^{\min\{q,s+1+j\}} T_r^R f^{[r-j-1]} \Big) \Big\} \end{aligned} \tag{5.6.17}$$

holds in $(L_p(a,b))^n$, where $\Delta_0 = \begin{pmatrix} 0 & 0 \\ 0 & I_{n-n_0} \end{pmatrix} \in M_n(\mathbb{C}^{n_0} \times \mathbb{C}^{n-n_0})$.

Proof. In Remark 5.6.4 we have seen that we can define $f^{[0]},\dots,f^{[s+1]}$ fulfilling the statements of Proposition 5.6.1. Since $f = \Delta_0 f^{[0]}$ and $A_1 f = A_1 f^{[0]}$, a revision of the proofs of Theorems 5.6.7, 5.6.9, and 5.6.10 shows that it is sufficient to prove that

$$\oint_{\Gamma_\nu} |\lambda|^{-s-1} |\Delta_0 C_2^{-1}(\lambda) R_2(\lambda) f_2| \, |\mathrm{d}\lambda| \to 0 \quad \text{as } \nu \to \infty,$$

where the norm is taken in $L(\mathbb{C}^n, (L_p(a,b))^n)$. But this estimate holds by Theorem 4.4.11 i). □

REMARK 5.6.12. In case $s = 0$, i. e., if the problem is Birkhoff regular, Theorems 5.6.10 and 5.6.11 give expansions for those cases in which the expansion theorems 5.3.2 and 5.3.3 are not applicable; e. g. if a or b is an accumulation point of those a_j for which $W_0^{(j)} \neq 0$.

Finally, we note that the residues which have to be calculated in the expansion theorems can be expressed by the eigenvectors and associated vectors of T and T^*. For simplicity of notation we shall work with the corresponding root functions. Let μ be any eigenvalue of T and $\{y_1,\dots,y_r\}$, $\{v_1,\dots,v_r\}$ be biothogonal CSRFs of T and T^* at μ. We know that these CSRFs can be expressed by biorthogonal CSRFs $\{c_1,\dots,c_r\}$ and $\{d_1,\dots,d_r\}$ of the characteristic matrix function M of T and its adjoint M^*, see Theorem 3.1.4. We have

$$y_j = Yc_j, \quad u_j := -(T^R U)^* d_j, \quad \text{and} \quad v_j = \begin{pmatrix} u_j \\ d_j \end{pmatrix}.$$

In Section 5.3 we have seen that the residues of $R_1 A_1$ can be expressed in terms of y_j and u_j, see Theorem 5.3.2. The principal part of T^{-1} at μ is

$$\sum_{j=1}^{r} (-\mu)^{-m_j} y_j \otimes v_j,$$

where the m_j are the partial multiplicities. Since

$$R_2 = T^{-1}\widetilde{J}_2,$$

where $\widetilde{J}_2 : \mathbb{C}^n \to (L_p(a,b))^n \times \mathbb{C}^n$ is the canonical embedding, and since the map $\widetilde{J}_2^* : (L_{p'}(a,b))^n \times \mathbb{C}^n \to \mathbb{C}^n$ is the canonical projection, we obtain because of Proposition 1.1.2 that the principal part of R_2 at μ is given by

$$\sum_{j=1}^{r} (-\mu)^{-m_j} y_j \otimes d_j.$$

5.7. Improved expansion theorems for Stone regular problems

DEFINITION 5.7.1. Let $\widetilde{M}(\lambda)$ be the characteristic matrix defined in (5.5.7) associated with the boundary eigenvalue problem (5.1.1), $\widetilde{T}^R(\lambda)y = 0$, where $\widetilde{T}^R$ is defined in (5.1.8). The boundary eigenvalue problem (5.1.1), (5.1.2) is called *strongly s-regular* if there are finitely many numbers $\alpha_1 < \alpha_2 < \cdots < \alpha_{t+1} = \alpha_1 + 2\pi$, a positive number β, and curves $\Gamma_\nu = \{\lambda \in \mathbb{C} : |\lambda| = \rho_\nu\}$ $(\nu \in \mathbb{N})$ with $\rho_\nu \nearrow \infty$ as $\nu \to \infty$, such that

$$\widetilde{M}(\lambda)^{-1} = \lambda^s \left(M_{0,s}(\lambda) + O\left(\max_{j=1}^{t} \left(1 + |\Re(\lambda e^{i\alpha_j})|\right)^{-\beta} \right) \right)$$

holds on $\bigcup_{\nu \in \mathbb{N}} \Gamma_\nu$, where the matrix function $M_{0,s}$ is constant on each of the sectors $\{\lambda \in \mathbb{C} \setminus \{0\} : \alpha_j < \arg \lambda < \alpha_{j+1}\}$ $(j = 1,\dots,t)$.

REMARK 5.7.2. i) Each s-regular boundary eigenvalue problem is strongly $s+1$-regular.
ii) Each Birkhoff regular problem satisfying the hypotheses of Theorem 5.3.2 is strongly 0-regular, see (4.6.4).

THEOREM 5.7.3. *Let $1 < p < \infty$ and let s be a positive integer. Suppose that A_0 and A_1 belong to $M_n(W_p^s(a,b))$. Suppose that the boundary eigenvalue problem* (5.1.1), (5.1.2) *is strongly s-regular. Let $\lambda_0, \lambda_1, \ldots$ be the eigenvalues of T and let*

$$\{y_{\kappa,l}^{(j)} : j = 1,\ldots,r(\lambda_\kappa); l = 0,\ldots,m_{\kappa,j}-1\}$$

and

$$\{(u_{\kappa,l}^{(j)}, d_{\kappa,l}^{(j)}) : j = 1,\ldots,r(\lambda_\kappa); l = 0,\ldots,m_{\kappa,j}-1\}$$

be biorthogonal CSEAVs of T and T^* at λ_κ, respectively. Then

$$f = -\lim_{\nu\to\infty} \sum_{\substack{\kappa\in\mathbb{N}\\ |\lambda_\kappa|<\rho_\nu}} \left(\sum_{j=1}^{r(\lambda_\kappa)} \sum_{h=0}^{m_{\kappa,j}-1} y_{\kappa,h}^{(j)} \int_a^b u_{\kappa,m_{\kappa,j}-1-h}^{(j)}(x)^\mathsf{T} A_1(x) f(x)\,\mathrm{d}x \right)$$

holds for all $f \in (W_p^s(a,b))^n$ with $f = \Delta_0 f$ and

$$\sum_{j=0}^{s-1} \lambda^{-j-1} C_2(\lambda)^{-1} T^R(\lambda) \tilde{f}^{[j]} = O(\lambda^{-s-1}),$$

where the series converges in parenthesis in $(L_p(a,b))^n$, $\tilde{f} \in (W_p^s(a,b))^n$ is chosen such that $\Delta_0 \tilde{f} = f$ and the construction in Remark 5.6.4 *holds, and* $\Delta_0 = \begin{pmatrix} 0 & 0 \\ 0 & I_{n-n_0} \end{pmatrix} \in M_n(\mathbb{C}^n)$.

Proof. Consider the space

$$F := \left\{ f \in (W_p^s(a,b))^n : \sum_{j=0}^{s-1} \lambda^{-j-1} C_2(\lambda)^{-1} T^R(\lambda) \tilde{f}^{[j]} = O(\lambda^{-s-1}) \right\}.$$

Here we have to note that $\tilde{f}^{[0]}, \ldots, \tilde{f}^{[s-1]}$ belong to $(W_p^1(a,b))^n$ if f belongs to $(W_p^s(a,b))^n$.

If $n_0 > 0$, then $\tilde{f}^{[0]}, \ldots, \tilde{f}^{[s-1]}$ are not uniquely defined. To make them unique we require that, in addition to (5.6.2), $(I_n - \Delta_0)\tilde{f}^{[0]}(a), \ldots, (I_n - \Delta_0)\tilde{f}^{[s-1]}(a)$ are certain continuous linear functionals depending on $f \in F$. For any particular choice made for $(I_n - \Delta_0)\tilde{f}^{[0]}(a), \ldots, (I_n - \Delta_0)\tilde{f}^{[s-1]}(a)$ for any nonzero f, we can always satisfy this requirement by the HAHN-BANACH theorem. Of course, we could simply take $(I_n - \Delta_0)\tilde{f}_1^{[0]}(a) = 0, \ldots, (I_n - \Delta_0)\tilde{f}_1^{[s-1]}(a) = 0$. But since it is desirable to have F as large as possible, a different choice might be better.

In view of the definition of $\tilde{f}^{[j]}$ and Lemma 2.5.7 we obtain that the maps $f \mapsto \tilde{f}^{[j]}$ from $(W_p^s(a,b))^n$ into $(W_p^{s-j}(a,b))^n$ are continuous for $j = 0, \dots, s$. For $f \in F$ we obtain

$$\begin{aligned}
&\left| \int_{\Gamma_\nu} \sum_{j=0}^{s-1} \lambda^{-j-1} \widetilde{J} R_2(\lambda) T^R(\lambda) \tilde{f}^{[j]} \, d\lambda \right| \\
&\le \int_{\Gamma_\nu} |\lambda|^{-s-1} |\widetilde{J} R_2(\lambda) C_2(\lambda)| \left| \sum_{j=0}^{s-1} \lambda^{s-j} C_2^{-1}(\lambda) T^R(\lambda) \tilde{f}^{[j]} \right| |d\lambda| \\
&= \int_{\Gamma_\nu} O(|\lambda|^{-s-1}) |\widetilde{J} R_2(\lambda) C_2(\lambda)| |d\lambda| |f|_{p,s} \\
&= O(1) |f|_{p,s}
\end{aligned}$$

by Theorem 4.4.11 i) since the problem is s-regular.

Since F is a closed subspace of $(W_p^s(a,b))^n$, it is a Banach space with respect to the norm induced by $(W_p^s(a,b))^n$. Let $J_F : F \to (L_p(a,b))^n$ be the canonical embedding from F into $E = (L_p(a,b))^n$. Let $P_\nu (= Q_\nu)$ be defined as in Section 5.3. From (5.6.6) with $s_1 = s - 1$ we infer

$$-P_\nu J_F f = -f + \frac{1}{2\pi i} \oint_{\Gamma_\nu} \lambda^{-s} \Delta_0 R_1(\lambda) A_1 \tilde{f}^{[s]} \, d\lambda + O(1) |f|_{p,s}.$$

We have

$$\lambda^{-s} \widetilde{M}(\lambda)^{-1} = M_{0,s}(\lambda) + O\big(\max_{j=1}^{t} (1 + |\Re(\lambda e^{i\alpha_j})|)^{-\beta}\big),$$

where $M_{0,s}(\lambda)$ is constant on sectors. And that is exactly what we used in the proof of Proposition 4.6.3. The actual location of the sectors was inessential; the sector Σ_m can be replaced by any smaller one. An obvious modification of Lemma 4.6.7 leads to

$$\left| \frac{1}{2\pi i} \oint_{\Gamma_\nu} \lambda^{-s} \Delta_0 R_1(\lambda) A_1 \tilde{f}^{[s]} \, d\lambda \right|_p = O(1) |\tilde{f}^{[s]}|_p = O(1) |f|_{p,s},$$

and the boundedness of $\{P_\nu J_F : \nu \in \mathbb{N}\}$ follows.

Now let $f \in (W_p^{s+1}(a,b))^n \cap F$. Then we have by (5.6.6) that

$$P_\nu J_F f = f - \frac{1}{2\pi i} \oint_{\Gamma_\nu} \lambda^{-s} \Delta_0 R_1(\lambda) A_1 \tilde{f}^{[s]} \, d\lambda + \widetilde{J} S_{s,\nu}(0,c) + O(\lambda^{-1}),$$

where $S_{s,\nu}$ is defined in Theorem 4.4.11 and $\widetilde{T}$ in the definition of $S_{s,\nu}$ is given by (5.3.3), and where $c \in \mathbb{C}^n$ (and depends on $\tilde{f}_1^{[0]}, \dots, \tilde{f}_1^{[s-1]}$). Here $S_{s,\nu}$ is taken with respect to $\lambda_0 = 0$. From Theorem 4.4.11 ii) we know that $\widetilde{JS}_{s,\nu}(0,c) \to 0$ as $\nu \to \infty$. As in the proof of Theorem 4.6.9 we infer, now for a suitable $\lambda_0 \neq 0$,

$$\widetilde{JS}_{s,\nu}\widetilde{T}(\lambda_0) = \frac{1}{2\pi i}\oint_{\Gamma_\nu} (\lambda - \lambda_0)^{-s-1}\widetilde{J}\widetilde{T}^{-1}(\lambda)\widetilde{T}(\lambda)\,\mathrm{d}\lambda + \frac{1}{2\pi i}\oint_{\Gamma_\nu} (\lambda - \lambda_0)^{-s}\widetilde{J}\widetilde{T}^{-1}(\lambda)J_0\widetilde{J}\,\mathrm{d}\lambda - \frac{1}{2\pi i}\oint_{\Gamma_\nu} (\lambda - \lambda_0)^{-s-1}\widetilde{J}\widetilde{T}^{-1}(\lambda)B(\lambda,\lambda_0)\,\mathrm{d}\lambda. \tag{5.7.1}$$

Since $s > 0$ we conclude with the same estimates as in Theorem 4.6.9 that

$$\frac{1}{2\pi i}\oint_{\Gamma_\nu} (\lambda - \lambda_0)^{-s}\widetilde{J}\widetilde{T}^{-1}(\lambda)J_0\widetilde{J}\,\mathrm{d}\lambda \to 0$$

as $\nu \to 0$. Applying this to $f^{[s]}$ we see that $P_\nu J_F f \to f$ as $\nu \to \infty$.

The proof of the theorem will be complete by Lemma 4.6.8 if we show that $(W_p^{s+1}(a,b))^n \cap F$ is dense in F. Since $W_p^{s+1}(a,b))^n$ is dense in $(W_p^s(a,b))^n$ and F is a finite-codimensional subspace of $(W_p^s(a,b))^n$, the result will be a consequence of the following lemma.

LEMMA 5.7.4. *Let E be a Banach space, F a closed and finite-codimensional subspace of E, and H a dense subspace of E. Then $H \cap F$ is dense in F.*

Proof. Let M be a (finite-dimensional) complementary space of F in E and let P be the projection of E onto M along F. Let $Q : P(H) \to H$ be a linear operator such that PQ is the identity on $P(H)$. Since $P(H) \subset M$ is finite-dimensional, Q is continuous. Also P is continuous by the closed graph theorem since F and M are closed. Now let $x \in F$ and choose $(x_n)_0^\infty \subset H$ such that $x_n \to x$ as $n \to \infty$. By definition of Q we have $y_n := x_n - QPx_n \in H$ for $n \in \mathbb{N}$. Also $Py_n = 0$ since PQ is the identity. Thus $y_n \in H \cap F$. Now $QPx_n \to QPx = 0$ since Q and P are continuous and $Px = 0$. This shows that $y_n \to x$ as $n \to \infty$. □

REMARK 5.7.5. The condition

$$\sum_{j=0}^{s-1} \lambda^{-j-1} C_2(\lambda)^{-1} T^R(\lambda) \tilde{f}^{[j]} = O(\lambda^{-s-1})$$

in Theorem 5.7.3 can be given in a more explicit form. For this let us write the asymptotic boundary conditions in the form

$$C_2(\lambda)^{-1}T^R(\lambda) = \widetilde{T}^R(\lambda) =: \sum_{r=0}^{s-1} \lambda^{-r}\widetilde{T}_r^R + O(\lambda^{-s}).$$

Then

$$\sum_{j=0}^{s-1} \lambda^{-j-1} C_2(\lambda)^{-1} T^R(\lambda) \tilde{f}^{[j]} = O(\lambda^{-s-1})$$

is equivalent to

$$\sum_{r=0}^{j} \widetilde{T}_r^R \tilde{f}^{[j-r]} = 0 \quad (j = 0, \ldots, s-1).$$

THEOREM 5.7.6. *Let s be a positive integer. Suppose that A_0 and A_1 belong to $M_n(W_\infty^s(a,b))$. Suppose that the boundary eigenvalue problem* (5.1.1), (5.1.2) *is strongly s-regular. Let $\lambda_0, \lambda_1, \ldots$ be the eigenvalues of T and let*

$$\{y_{\kappa,l}^{(j)} : j = 1, \ldots, r(\lambda_\kappa); l = 0, \ldots, m_{\kappa,j} - 1\}$$

and

$$\{(u_{\kappa,l}^{(j)}, d_{\kappa,l}^{(j)}) : j = 1, \ldots, r(\lambda_\kappa); l = 0, \ldots, m_{\kappa,j} - 1\}$$

be biorthogonal CSEAVs of T and T^* at λ_κ, respectively. Then

$$f = -\lim_{\nu \to \infty} \sum_{\substack{\kappa \in \mathbb{N} \\ |\lambda_\kappa| < \rho_\nu}} \left(\sum_{j=1}^{r(\lambda_\kappa)} \sum_{h=0}^{m_{\kappa,j}-1} y_{\kappa,h}^{(j)} \int_a^b u_{\kappa,m_{\kappa,j}-1-h}^{(j)}(x)^{\mathsf{T}} A_1(x) f(x)\,\mathrm{d}x \right)$$

holds for all $f \in (C^s[a,b])^n$ with $f^{(s)} \in (BV[a,b])^n$, $f = \Delta_0 f$, and

$$\sum_{r=0}^{j} \widetilde{T}_r^R \tilde{f}^{[j-r]} = 0 \quad (j = 0, \ldots, s), \tag{5.7.2}$$

where the series converges in parenthesis in $(C[a,b])^n$, $\tilde{f} \in (W_p^s(a,b))^n$ is chosen such that $\Delta_0 \tilde{f} = f$ and the construction in Remark 5.6.4 *holds.*

Proof. Let E be the set of all $f \in (C^s[a,b])^n$ such that $f^{(s)} \in (BV[a,b])^n$. Then E is a Banach space with respect to the norm $|f|_\infty + |f^{(s)}|_{(BV[a,b])^n}$, and F, the subset of E consisting of functions satisfying (5.7.2), is a closed finite-codimensional subspace of E. Let J_F be the canonical embedding from F into $(L_\infty(a,b))^n$. Combining the proofs of Theorem 5.7.3 and Lemma 4.7.4 we obtain that $\{P_\nu J_F : \nu \in \mathbb{N}\}$ is bounded in $L(F, (L_\infty(a,b))^n)$. Here we have to note that $f^{(s)} \in (BV[a,b])^n$ implies $\tilde{f}^{[s]} \in (BV[a,b])^n$. Since also $\tilde{f}^{[s+1]}$ is defined for $f \in (W_\infty^{s+1}(a,b))^n$, we can take the iteration one step further for these f and obtain with the aid of (5.6.6) that

$$P_\nu J_F f = f - \frac{1}{2\pi i} \oint_{\Gamma_\nu} \lambda^{-s-1} \Delta_0 R_1(\lambda) A_1 \tilde{f}^{[s+1]}\,\mathrm{d}\lambda + o(1).$$

Here we have used that (5.7.2), (5.3.3), and Theorem 4.4.9 ii) imply that

$$\left| \oint_{\Gamma_\nu} \sum_{j=0}^{s} \lambda^{-j-1} R_2(\lambda) T^R(\lambda) \tilde{f}^{[j]}\,\mathrm{d}\lambda \right|_\infty \to 0 \text{ as } \nu \to \infty.$$

Taking now the representation (5.7.1) with s replaced by $s+1$, an application of Theorem 4.4.9 iii) shows that $P_\nu J_F f \to f$ as $\nu \to \infty$ for all $f \in F \cap (W_\infty^{s+2}(a,b))^n$. As in the proof of Theorem 4.7.5 we can show that $(C_0^\infty(a,b))^n$ is dense in E. Therefore, an application of Lemmas 4.6.8 and 5.7.4 completes the proof. □

REMARK 5.7.7. i) If we suppose that A_0 and A_1 belong to $M_n(W_1^1(a,b))$, then also the case $s = 0$ is covered by Theorem 5.7.6.
ii) If we consider Example 4.7.9 in light of Theorem 5.7.6, then we obtain that we can expand functions $(0, f_2, f_3)^\mathsf{T}$ for which there is a function f_1 such that $f_1' = \alpha f_2 + \beta f_3$ and, for $\tilde{f} = (f_1, f_2, f_3)^\mathsf{T}$, $W^{(0)}\tilde{f}(0) + W^{(1)}\tilde{f}(1) = 0$. If we choose

$$f_1(x) = f_2(0) + f_3(1) + \alpha \int_0^x f_2(\xi)\,\mathrm{d}\xi + \beta \int_0^x f_3(\xi)\,\mathrm{d}\xi,$$

then we obtain exactly the conditions which were deduced in Example 4.7.9.

The next result states a sufficient condition of an s-regular problem to be strongly s-regular.

LEMMA 5.7.8. *Let s be a positive integer. Suppose that $A_0, A_1 \in M_n(W_p^s(a,b))$, where $p > 1$, and that the W_ν given in* (5.5.2) *belong to $M_n(W_{\tilde{q}}^{s-\nu}(a,b))$ for some $\tilde{q} > 1$ and $\nu = 0, \ldots, s$. Suppose that a and b are no accumulation points of $\{a_j : W^{(j)}(a_j) \neq 0\}$. We consider the boundary eigenvalue problem* (5.1.1), (5.1.2) *and the determinant*

$$\det \widehat{M}_{1,s} = \sum_{c \in \mathscr{E}} \tilde{b}_c(\lambda) \exp(c\lambda),$$

where $\tilde{b}_c(\lambda) = \lambda^{-\nu_c}[b_c]$ and $\widehat{M}_{1,s}$ is defined in (5.5.5). *Suppose that $\nu_c = -s$ for all $c \in \mathscr{E}$ and $b_c \neq 0$ for $c \in \widetilde{\mathscr{E}}$, the set of vertices of $\mathscr{E}$. Then the boundary eigenvalue problem is strongly s-regular.*

Proof. Without loss of generality we may assume $p = \tilde{q}$. Let

$$\widehat{M}_{0,s}(\lambda) = \widehat{M}_{1,s}(\lambda)\big(I_n - \Delta(\lambda) + E(b,\lambda)^{-1}\Delta(\lambda)\big).$$

In (5.5.8) we have proved that

$$\widetilde{M}(\lambda) - \widehat{M}_{0,s}(\lambda) = \lambda^{-s} o(1).$$

In the proof of Proposition 5.5.2 we have seen that we also obtain

$$\widetilde{M}(\lambda) - \widehat{M}_{0,s}(\lambda) = \lambda^{-s} O(\tau_p(\lambda)).$$

The assumptions imply that the problem is s-regular, and thus $|\lambda^s \det \widetilde{M}(\lambda)| \geq \delta$ and $|\lambda^s \det \widehat{M}_{0,s}(\lambda)| \geq \delta$ on $\bigcup_{\nu\in\mathbb{N}} \Gamma_\nu$ for some $\delta > 0$. Hence $\widetilde{M}(\lambda)^{-1} = O(\lambda^s)$ and $\widehat{M}_{0,s}(\lambda)^{-1} = O(\lambda^s)$ on $\bigcup_{\nu\in\mathbb{N}} \Gamma_\nu$. This implies

$$\begin{aligned}\widetilde{M}(\lambda)^{-1} - \widehat{M}_{0,s}(\lambda)^{-1} &= \widetilde{M}(\lambda)^{-1}\big(\widehat{M}_{0,s}(\lambda) - \widetilde{M}(\lambda)\big)\widehat{M}_{0,s}(\lambda)^{-1} \\ &= \lambda^s O(\tau_p(\lambda)) \quad \text{on} \quad \bigcup_{\nu\in\mathbb{N}} \Gamma_\nu .\end{aligned}$$

Hence it is sufficient to show that there is a matrix function $M_{0,s}(\lambda)$ which is constant on sectors such that

$$\widehat{M}_{0,s}(\lambda)^{-1} = \lambda^s\big(M_{0,s}(\lambda) + O(\tau_p(\lambda))\big).$$

Arguing as in the proof of Proposition 4.3.6 iv),v) we infer that

$$\widehat{M}_{0,s}(\lambda)^{-1} = \big(\det \widehat{M}_{0,s}(\lambda)\big)^{-1}\widehat{M}_{0,s}^{\mathrm{ad}}(\lambda),$$

where

$$\widehat{M}_{0,s}^{\mathrm{ad}}(\lambda) = \widehat{M}_{2,s}(\lambda) + O(\tau_p(\lambda))$$

and the matrix $\widehat{M}_{2,s}(\lambda)$ is constant on sectors. In the same way it follows that $\tilde{b}_c(\lambda) = \lambda^{-\nu_c}(b_c + O(\tau_p(\lambda)))$, where $0 \le \nu_c \le s$. An application of Theorem A.3.1 completes the proof. Here we have to note that the additional assumption on $\mathscr{E}$ in Theorem A.3.1 are satisfied in view of Proposition A.1.6 and its proof and Corollary A.1.4. □

5.8. Notes

We recall that historically n-th order scalar differential equations were considered before first order sytems of differential equations. Also, n-th order equations have attracted more attention than first order systems, mainly because of its greater relevance in applications. See the notes of Chapter VII and Chapter VIII for more details.

The first expansion theorems for systems have been obtained with respect to uniform convergence, see the paper [BIL] of BIRKHOFF and LANGER. In the sequel, important generalizations have been published by R. E. LANGER [LA5], [LA6], [LA9] and R. H. COLE [CO2], [CO3], [CO4]. In most of these publications, in case of two-point boundary conditions, the eigenfunction expansions are stated as being pointwise convergent or locally uniformly convergent in the interior of the interval (a,b). However, we are mostly interested in either L_p convergence or uniform convergence on the whole interval. Then the behavior of the functions at the boundary becomes important; indeed, we see that we have to impose some auxiliary boundary conditons for the expandable functions if we consider uniform convergence or Stone regular problems. The boundary conditons which have to be satisfied for Stone regular problems were found explicitly in [MM5]. These boundary conditions are fulfilled if the functions which are to be expanded and their derivatives up to a certain order vanish at the endpoints of the interval (a,b).

Further results, in particular concerning completeness, minimality and basisness of the eigenfunctions and associated functions for systems of of differential equations of type (5.1.1) with λ-polynomial boundary conditions (5.1.2) have very recently been published by C. TRETTER in [TR8]. The proofs of these results are based on the spectral theory for linear operator pencils $A - \lambda B$, on a new linearization method for a class of λ-nonlinear boundary eigenvalue problems developed by C. TRETTER in [TR9] and [TR7] and on sharp asymptotic estimates of the Green's matrix function as published in [MM5].

Chapter VI

n-TH ORDER DIFFERENTIAL EQUATIONS

In this chapter boundary eigenvalue problems for n-th order ordinary linear differential equations are considered. The differential equation as well as the boundary conditions are allowed to depend holomorphically on the eigenvalue parameter. The boundary conditions consist of terms at the endpoints and at interior points of the underlying interval and of an integral term. Such boundary eigenvalue problems are considered in suitable Sobolev spaces, so that both the differential operators and the boundary operators define bounded operators on Banach spaces. The assumptions on the boundary eigenvalue problems assure that these operators depend holomorphically on the eigenvalue parameter. In a canonical way a holomorphic Fredholm operator valued function is associated to such a boundary eigenvalue problem with independent variable being the eigenvalue parameter. This operator function consists of two components, the first one is the differential operator function, the second one is the boundary operator function. Operator functions defined in this way are briefly called boundary eigenvalue operator functions.

The results of this chapter are the analogs of those proved for boundary eigenvalue problems for first order differential systems in the third chapter. Some of the present statements are derived by the usual transformation of boundary eigenvalue problems for n-th order differential equations to such problems for first order $n \times n$ differential systems. A notable feature of this transformation is the fact that the characteristic matrix functions of the original boundary eigenvalue problem for the n-th order differential equation and the associated problem for the first order system coincide. Other results of this chapter are proved directly without reference to first order systems.

We obtain that a boundary eigenvalue operator function associated to an n-th order differential equation is globally holomorphically equivalent to a canonical extension of the characteristic matrix function of the corresponding boundary eigenvalue problem (Theorem 6.3.2). The principal parts of the resolvent, i. e., the inverse of the given boundary eigenvalue operator function, is expressed in terms of the eigenfunctions and associated functions of this operator function and its adjoint (Theorem 6.3.4). As in the third chapter, inhomogeneous boundary conditions are treated in a natural way. The resolvent is defined on the direct sum of an L_p-function space and a finite-dimensional space of constants. On the function space, the resolvent is an integral operator whose kernel is the Green's function; on the space of constants, it is a multiplication operator (Theorem 6.4.1).

The adjoint operator function of a boundary eigenvalue operator function defines the adjoint boundary eigenvalue problem (Theorem 6.5.1). For the adjoint problem in this operator theoretical sense no additional assumptions on the original boundary eigenvalue problem are needed. The adjoint operator function maps the direct sum of an L_p-function space and a finite-dimensional space of constants into a space of distributions.

The realization of the original boundary eigenvalue problem within an L_p-function space is achieved in the following way: Take the original boundary eigenvalue problem with homogeneous boundary conditions and associate to it a family of closed linear operators whose domains consist of W_p^1-functions which fulfil the boundary conditions. These closed linear operators are not necessarily densely defined and their domains may depend on the eigenvalue parameter. The adjoints of these closed linear operators are in general not operators but closed linear relations. Under additional assumptions these adjoints form a family of operators, in which case they yield the adjoint boundary eigenvalue problem in the parametrized form. The relationships between the adjoint boundary eigenvalue problem in operator theoretical sense and the corresponding problem in parametrized form is thoroughly discussed (Theorems 6.6.4 and 6.6.5).

Finally, the special case of two-point boundary eigenvalue problems is considered. We state that the classical adjoint boundary eigenvalue problem coincides with the adjoint problem in the parametrized form. Root functions (eigenvectors and associated vectors) are defined for the above mentioned families of closed linear operators by taking root functions (eigenvectors and associated vectors) of the corresponding holomorphic boundary eigenvalue operator function. It is shown that the principal parts of the GREEN'S function can be represented in terms of eigenfunctions and associated functions of the family of closed linear operators and the family of the adjoints of these operators (Theorem 6.7.8).

6.1. Differential equations and systems

In this chapter let Ω be a nonempty open subset of $\mathbb{C}$, $-\infty < a < b < \infty$, $1 \le p \le \infty$, $1 \le p' \le \infty$ such that $1/p + 1/p' = 1$, and $n \in \mathbb{N}$, $n \ge 2$. By e_j we denote the j-th unit vector in $\mathbb{C}^n$. We consider the scalar n-th order differential equation

$$\eta^{(n)} + \sum_{i=0}^{n-1} p_i(\cdot,\lambda)\eta^{(i)} = 0 \quad (\eta \in W_p^n(a,b), \lambda \in \Omega), \tag{6.1.1}$$

where $p_i \in H(\Omega, L_p(a,b))$ $(i = 0, \ldots, n-1)$. Together with this differential equation we consider the differential operator

$$L^D(\lambda)\eta := \eta^{(n)} + \sum_{i=0}^{n-1} p_i(\cdot,\lambda)\eta^{(i)} \quad (\eta \in W_p^n(a,b), \lambda \in \Omega). \tag{6.1.2}$$

LEMMA 6.1.1. $L^D \in H(\Omega, L(W_p^n(a,b), L_p(a,b)))$.

Proof. From Proposition 2.3.3 we infer that $p_i \in H(\Omega, L(W_p^{n-i}(a,b), L_p(a,b)))$ for $i = 0, \ldots, n-1$. $\square$

We associate a first order system to the n-th order differential equation. This system is defined by the operator

$$T^D(\lambda)y := y' - A(\cdot,\lambda)y \quad (y \in (W_p^1(a,b))^n, \lambda \in \mathbb{C}), \tag{6.1.3}$$

where

$$A := (\delta_{i,j-1} - \delta_{i,n}p_{j-1})_{i,j=1}^n = \begin{pmatrix} 0 & 1 & & & \\ & \cdot & \cdot & 0 & \\ & & \cdot & \cdot & \\ & 0 & & 0 & 1 \\ -p_0 & \cdot & \cdot & \cdot & -p_{n-1} \end{pmatrix}. \tag{6.1.4}$$

PROPOSITION 6.1.2. *Let $\eta \in W_p^n(a,b)$, $\lambda \in \Omega$, and set*

$$y := \begin{pmatrix} \eta \\ \eta' \\ \vdots \\ \eta^{(n-1)} \end{pmatrix}.$$

Then $y \in (W_p^1(a,b))^n$ and

$$T^D(\lambda)y = \begin{pmatrix} 0 \\ \vdots \\ 0 \\ L^D(\lambda)\eta \end{pmatrix}.$$

Proof. The assertions $y \in (W_p^1(a,b))^n$ and

$$e_i^{\mathsf{T}} A(\cdot,\lambda) = e_{i+1}^{\mathsf{T}} \qquad (i = 1, \ldots, n-1) \tag{6.1.5}$$

are obvious. For $i = 1, \ldots, n-1$, (6.1.3) and (6.1.5) yield

$$e_i^{\mathsf{T}} T^D(\lambda)y = e_i^{\mathsf{T}} y' - e_i^{\mathsf{T}} A(\cdot,\lambda)y = \eta^{(i)} - e_{i+1}^{\mathsf{T}} y = 0.$$

Finally we obtain

$$\begin{aligned} e_n^\mathsf{T} T^D(\lambda) y &= e_n^\mathsf{T} y' + \sum_{i=0}^{n-1} p_i(\cdot,\lambda) e_{i+1}^\mathsf{T} y \\ &= \eta^{(n)} + \sum_{i=0}^{n-1} p_i(\cdot,\lambda)\eta^{(i)} = L^D(\lambda)\eta. \quad \square \end{aligned}$$

PROPOSITION 6.1.3. *Let* $y \in (W_p^1(a,b))^n$, $\lambda \in \Omega$, *and assume that* $e_i^\mathsf{T} T^D(\lambda) y = 0$ *for* $i = 1,\dots,n-1$. *Then* $\eta := e_1^\mathsf{T} y \in W_p^n(a,b)$,

$$y = \begin{pmatrix} \eta \\ \eta' \\ \vdots \\ \eta^{(n-1)} \end{pmatrix}, \tag{6.1.6}$$

and

$$L^D(\lambda)\eta = e_n^\mathsf{T} T^D(\lambda) y. \tag{6.1.7}$$

Proof. Let $i \in \{1,\dots,n-1\}$. By assumption and from (6.1.5) we obtain

$$e_i^\mathsf{T} y' = e_i^\mathsf{T} y' - e_i^\mathsf{T} T^D(\lambda) y = e_i^\mathsf{T} A(\cdot,\lambda) y = e_{i+1}^\mathsf{T} y. \tag{6.1.8}$$

This proves $\eta \in W_p^i(a,b)$ and $e_i^\mathsf{T} y = \eta^{(i-1)}$ for $i = 1,\dots,n$. Indeed, this is true for $i = 1$. Assume that $\eta \in W_p^i(a,b)$ and $e_i^\mathsf{T} y = \eta^{(i-1)}$ holds for some $i < n$. Then (6.1.8) yields

$$\eta^{(i)} = e_i^\mathsf{T} y' = e_{i+1}^\mathsf{T} y \in W_p^1(a,b)$$

which proves $\eta \in W_p^{i+1}(a,b)$, see Corollary 2.1.4. Thus $\eta \in W_p^n(a,b)$, and the equation (6.1.6) holds. Because of (6.1.6), the equation (6.1.7) immediately follows from Proposition 6.1.2. $\square$

DEFINITION 6.1.4. Let $\lambda_0 \in \Omega$ and $\eta_1,\dots,\eta_n \in W_p^n(a,b)$. Then $\{\eta_1,\dots,\eta_n\}$ is called a *fundamental system* of $L^D(\lambda_0)\eta = 0$ if for each $\eta \in N(L^D(\lambda_0))$ there are $c_j \in \mathbb{C}$ $(j = 1,\dots,n)$ such that

$$\eta = \sum_{j=1}^{n} c_j \eta_j.$$

A function $(\eta_1,\dots,\eta_n) : \Omega \to M_{1,n}(W_p^n(a,b))$ is called a *fundamental system function* of $L^D y = 0$ if $\{\eta_1(\lambda),\dots,\eta_n(\lambda)\}$ is a fundamental system of $L^D(\lambda)y = 0$ for each $\lambda \in \Omega$.

LEMMA 6.1.5. *Let $\lambda_0 \in \Omega$ and $Y_0 \in M_n(W_p^1(a,b))$ be a fundamental matrix of $T^D(\lambda_0)y = 0$. Then $\{e_1^\top Y_0 e_1, \dots, e_1^\top Y_0 e_n\}$ is a fundamental system of $L^D(\lambda_0)\eta = 0$, and*

$$(e_1^\top Y_0 e_j)^{(i-1)} = e_i^\top Y_0 e_j \tag{6.1.9}$$

holds for $i = 1, \dots, n$ and $j = 1, \dots, n$.

Proof. For each $j \in \{1, \dots, n\}$, $Y_0 e_j$ fulfils the assumptions of Proposition 6.1.3. Thus $e_1^\top Y_0 e_1, \dots, e_1^\top Y_0 e_n \in W_p^n(a,b)$, and (6.1.9) holds. Now let $\eta \in N(L^D(\lambda_0))$ and set $y := (\eta, \eta', \dots, \eta^{(n-1)})^\top$. Then $y \in N(T^D(\lambda_0))$ by Proposition 6.1.2. Definition 2.5.2 yields a vector $c = (c_1, \dots, c_n)^\top \in \mathbb{C}^n$ such that $y = Y_0 c$. It follows that

$$\eta = e_1^\top y = e_1^\top Y_0 c = \sum_{j=1}^n c_j e_1^\top Y_0 e_j. \quad \square$$

LEMMA 6.1.6. *Let $\lambda_0 \in \Omega$ and $\eta_1, \dots, \eta_n \in W_p^n(a,b)$ such that $\{\eta_1, \dots, \eta_n\}$ is a fundamental system of $L^D(\lambda_0) = 0$. Then $\left(\eta_j^{(i-1)}\right)_{i,j=1}^n \in M_n(W_p^1(a,b))$ is a fundamental matrix of $T^D(\lambda_0)y = 0$.*

Proof. Let $y \in N(T^D(\lambda_0))$. Proposition 6.1.3 yields that $\eta := e_1^\top y \in W_p^n(a,b)$, $y = (\eta, \eta', \dots, \eta^{(n-1)})^\top$ and $L^D(\lambda_0)\eta = e_n^\top T^D(\lambda_0)y = 0$. Hence there is a vector $c = (c_1, \dots, c_n)^\top \in \mathbb{C}^n$ such that

$$\eta = \sum_{j=1}^n c_j \eta_j.$$

This proves

$$y = \begin{pmatrix} \eta \\ \eta' \\ \vdots \\ \eta^{(n-1)} \end{pmatrix} = (\eta_j^{(i-1)})_{i,j=1}^n c. \quad \square$$

PROPOSITION 6.1.7. *Let $\lambda_0 \in \Omega$ and $\eta_1, \dots, \eta_n \in W_p^n(a,b)$. Then the following conditions are equivalent:*
i) *$\eta_1, \dots, \eta_n$ are linearly independent, $L^D(\lambda_0)\eta_j = 0$ for each $j \in \{1, \dots, n\}$, and for each $\eta \in N(L^D(\lambda_0))$ there are $c_j \in \mathbb{C}$ $(j = 1, \dots, n)$ such that*

$$\eta = \sum_{j=1}^n c_j \eta_j;$$

ii) *$\{\eta_1, \dots, \eta_n\}$ is a fundamental system of $L^D(\lambda_0)\eta = 0$;*
iii) *$(\eta_j^{(i-1)})_{i,j=1}^n$ is a fundamental matrix of $T^D(\lambda_0)y = 0$.*

Proof. i) $\Rightarrow$ ii) is clear by definition of a fundamental system and ii) $\Rightarrow$ iii) follows from Lemma 6.1.6.

Assume that iii) holds. For $j = 1,\dots,n$ we set $y_j := (\eta_j, \eta_j', \dots, \eta_j^{(n-1)})^\mathsf{T}$. From Corollary 2.5.5 we infer $T^D(\lambda_0)y_j = 0$ and hence, by Proposition 6.1.2, $L^D(\lambda_0)\eta_j = 0$ for $j = 1,\dots,n$. Since a fundamental matrix is invertible by Theorem 2.5.3 and Proposition 2.5.4, $y_1,\dots,y_n$ are linearly independent. This implies that $\eta_1,\dots,\eta_n$ are linearly independent. An application of Lemma 6.1.5 completes the proof. $\square$

THEOREM 6.1.8. *There is a fundamental system function $(\eta_1,\dots,\eta_n)$ of $L^D\eta = 0$ such that $\eta_j^{(i-1)}(a,\lambda) = \delta_{i,j}$ for $\lambda \in \Omega$ and $i,j = 1,\dots,n$. Furthermore, the fundamental system function is uniquely determined and depends holomorphically on $\lambda \in \Omega$. More precisely, we have $\eta_j \in H(\Omega, W_p^n(a,b))$ for $j = 1,\dots,n$.*

Proof. By Theorem 2.5.3 there is a fundamental matrix function Y of $T^D y = 0$ such that $Y(a,\lambda) = I_n$ for all $\lambda \in \Omega$. For $j = 1,\dots,n$ we set $\eta_j := e_1^\mathsf{T} Y e_j$. By Lemma 6.1.5 we obtain that $(\eta_1,\dots,\eta_n)$ is a fundamental system function. In addition, (6.1.9) and $Y(a,\lambda) = I_n$ yield $\eta_j^{(i-1)}(a,\lambda) = \delta_{ij}$ for $\lambda \in \Omega$ and $i,j = 1,\dots,n$.

Now let $(\eta_1,\dots,\eta_n)$ be any fundamental system function of $L^D\eta = 0$ with $\eta_j^{(i-1)}(a,\lambda) = \delta_{ij}$ for $\lambda \in \Omega$ and $i,j = 1,\dots,n$. By Corollary 2.5.5 there is a unique fundamental matrix function Y of $T^D y = 0$ with $Y(a,\lambda) = I_n$ for $\lambda \in \Omega$. Since $(\eta_j^{(i-1)})_{i,j=1}^n$ is a fundamental matrix function with these properties by Lemma 6.1.6, we obtain that $(\eta_1,\dots,\eta_n)$ is uniquely determined.

Since Y depends holomorphically on λ by Theorem 2.5.3, it follows that $\eta_j^{(i-1)} \in H(\Omega, W_p^1(a,b))$ for $i,j = 1,\dots,n$. For $h \in \mathbb{N}$, the indefinite integral defines a continuous linear map from $W_p^h(a,b)$ to $W_p^{h+1}(a,b)$ by Proposition 2.1.8. From $\eta_j^{(i-1)}(a) = \delta_{ij}$ and Proposition 2.1.5 i) we know that

$$\eta_j^{(i-1)}(x,\lambda) = \delta_{ij} + \int_a^x \eta_j^{(i)}(t,\lambda)\,\mathrm{d}t \quad (x \in (a,b))$$

for $j = 1,\dots,n$ and $i = 1,\dots,n-1$. Hence we obtain in view of Corollary 1.2.4 that

$$\eta_j^{(n-2)} \in H(\Omega, W_p^2(a,b)),\dots,\eta_j \in H(\Omega, W_p^n(a,b)). \quad \square$$

6.2. Boundary conditions

Let $L^R \in H(\Omega, L(W_p^n(a,b),\mathbb{C}^n))$. Suppose that $p < \infty$. We fix some $\lambda_0 \in \Omega$ and $l \in \{1,\dots,n\}$. By Theorem 2.2.5 there are $u_j \in L_{p'}(a,b)$ $(j = 0,\dots,n)$ such that $e_l^\mathsf{T} L^R(\lambda_0) = \sum_{j=0}^{n} (u_j)_e^{(j)}$ and

$$e_l^\mathsf{T} L^R(\lambda_0)\eta = \langle \eta, e_l^\mathsf{T} L^R(\lambda_0)\rangle_{p,n} = \sum_{j=0}^{n} \int_a^b (-1)^j \eta^{(j)}(x) u_j(x)\,\mathrm{d}x$$

for each $\eta \in W_p^n(a,b)$. Hence

$$e_l^\mathsf{T} L^R(\lambda_0)\eta = \langle \eta, (u_0)_e + (u_1)_e'\rangle_{p,1} + \sum_{j=2}^{n} \langle \eta^{(j-1)}, (-1)^{(j-1)}(u_j)_e'\rangle_{p,1}$$

for each $\eta \in W_p^n(a,b)$. This proves that for each $\lambda \in \Omega$ there is an operator $T^R(\lambda) \in L((W_p^1(a,b))^n, \mathbb{C}^n)$ such that

$$L^R(\lambda)\eta = T^R(\lambda)\begin{pmatrix} \eta \\ \eta' \\ \vdots \\ \eta^{(n-1)} \end{pmatrix} \tag{6.2.1}$$

holds for all $\eta \in W_p^n(a,b)$.

In applications, the boundary conditions are mostly given in a form such that it is easy to give a representation (6.2.1) with $T^R \in H(\Omega, L((W_p^1(a,b))^n, \mathbb{C}^n)$. For example, let

$$L^R(\lambda)\eta := \left(\sum_{j=1}^{n} (\alpha_{ij}(\lambda)\eta^{(j-1)}(a) + \beta_{ij}(\lambda)\eta^{(j-1)}(b))\right)_{i=1}^{n} \quad (\eta \in W_p^n(a,b)),$$

where the α_{ij} and β_{ij} are complex valued functions and where L^R depends holomorphically on λ. Choosing functions η for which exactly one of the values $\eta^{(j-1)}(a)$, $\eta^{(j-1)}(b)$ $(j = 1,\dots,n)$ is different from zero we see that the α_{ij} and β_{ij} are holomorphic functions. Then $T^R(\lambda)$, defined by

$$T^R(\lambda)y := \left(\alpha_{ij}(\lambda)\right)_{i,j=1}^{n} y(a) + \left(\beta_{ij}(\lambda)\right)_{i,j=1}^{n} y(b) \quad (y \in W_p^n(a,b)),$$

depends holomorphically on λ.

Now we shall show that, if $p < \infty$, we can always choose $T^R(\lambda)$ in such a way that it also depends holomorphically on λ. This immediately follows from

PROPOSITION 6.2.1. *Let $l \in \mathbb{N}$, $1 \le p < \infty$ and $v \in H(\Omega, W_{p'}^{-l}[a,b])$. Then there are $u_0, \dots, u_l \in H(\Omega, L_{p'}(a,b))$ such that*

$$v(\lambda) = \sum_{i=0}^{l} u_i(\lambda)_e^{(i)} \qquad (\lambda \in \Omega).$$

Proof. We shall prove the proposition by induction on l. In case $l = 0$ nothing has to be proved. Assume that the assertion holds for $l-1$ and let $v \in H(\Omega, W_{p'}^{-l}[a,b])$. For each $\lambda \in \Omega$ there are $v_0(\lambda), \dots, v_l(\lambda) \in L_{p'}(a,b)$ such that

$$v(\lambda) = \sum_{i=0}^{l} v_i(\lambda)_e^{(i)} \qquad (\lambda \in \Omega).$$

For $\lambda \in \Omega$ and $x \in \mathbb{R}$ we set

$$w(\lambda)(x) := \int_a^x \left[v_0(\lambda)_e(t) - \frac{1}{b-a} \langle 1, v(\lambda) \rangle_{p,l} \chi_{(a,b)}(t) \right] \mathrm{d}t,$$

where $\chi_{(a,b)}$ is the characteristic function of (a,b). Since

$$w(\lambda)(b) = \int_a^b v_0(\lambda)(t)\,\mathrm{d}t - \langle 1, v(\lambda) \rangle_{p,l} = 0$$

by definition of the bilinear form $\langle\ ,\ \rangle_{p,n}$ (see (2.2.4)), we have $w(\lambda)(x) = 0$ for $x \in \mathbb{R} \setminus (a,b)$. From Proposition 2.1.5 i) it follows that $w(\lambda) \in W_p^1(\mathbb{R}) \subset L_p(\mathbb{R})$ and

$$w(\lambda)' = v_0(\lambda)_e - \frac{1}{b-a} \langle 1, v(\lambda) \rangle_{p,l} \chi_{(a,b)}.$$

Then

$$\tilde{v}(\lambda) := w(\lambda) + \sum_{i=1}^{l} v_i(\lambda)_e^{(i-1)} \in W_{p'}^{-l+1}[a,b].$$

We note that

$$\tilde{v}(\lambda)' = w(\lambda)' + \sum_{i=1}^{l} v_i(\lambda)_e^{(i)} = v(\lambda) - \frac{1}{b-a} \langle 1, v(\lambda) \rangle_{p,l} \chi_{(a,b)} \tag{6.2.2}$$

depends holomorphically on λ in the Banach space $W_{p'}^{-l}[a,b]$. Let $\varphi \in W_p^{l-1}(a,b)$ and set

$$\psi(x) := \int_a^x \varphi(t)\,\mathrm{d}t \qquad (x \in (a,b)).$$

Then $\psi \in W_p^l(a,b)$ and

$$\langle \varphi, \tilde{v}(\lambda) \rangle_{p,l-1} = \langle \psi', \tilde{v}(\lambda) \rangle_{p,l-1} = \langle -\psi, \tilde{v}(\lambda)' \rangle_{p,l}.$$

Hence $\tilde{v}$ is (weakly) holomorphic in $W_{p'}^{-l+1}[a,b]$. The induction hypothesis yields that there are $u_1,\dots,u_l \in H(\Omega, L_{p'}(a,b))$ such that

$$\tilde{v}(\lambda) = \sum_{i=0}^{l-1} u_{i+1}(\lambda)_e^{(i)} \qquad (\lambda \in \Omega).$$

We set

$$u_0(\lambda) := \frac{1}{b-a}\langle 1, v(\lambda)\rangle_{p,l}\chi_{(a,b)}.$$

Then $u_0 \in H(\Omega, L_{p'}(a,b))$. From (6.2.2) we obtain

$$v(\lambda) = u_0(\lambda) + \tilde{v}(\lambda)' = \sum_{i=0}^{l} u_i(\lambda)_e^{(i)} \quad (\lambda \in \Omega). \quad \square$$

We would like to note that the "canonical way" to associate T^R to L^R does not always yield a holomorphic operator function T^R. For example, the identity

$$\int_a^b \eta'(t)\,\mathrm{d}t = \eta(b) - \eta(a)$$

yields that, for any complex valued function α on $\mathbb{C}$,

$$L^R(\lambda)\eta := \begin{pmatrix} \eta(a) \\ \eta(b) + \alpha(\lambda)\big(\eta(a) - \eta(b)\big) + \alpha(\lambda)\int_a^b \eta'(t)\,\mathrm{d}t \end{pmatrix}$$

defines a holomorphic operator function

$$L^R \in H(\mathbb{C}, L(W_2^2(a,b), \mathbb{C}^2)).$$

But the operator function T^R, given by

$$T^R(\lambda)y := \begin{pmatrix} y_1(a) \\ y_1(b) + \alpha(\lambda)\big(y_1(a) - y_1(b)\big) + \alpha(\lambda)\int_a^b y_2(t)\,\mathrm{d}t \end{pmatrix},$$

where $y = \begin{pmatrix} y_1 \\ y_2 \end{pmatrix} \in (W_2^1(a,b))^2$, does not depend holomorphically on λ if α does not depend holomorphically on λ.

We shall assume also in the case $p = \infty$ that $L^R \in H(\Omega, L(W_p^n(a,b), \mathbb{C}^n))$ is given in such a way that (6.2.1) holds for some $T^R \in H(\Omega, L((W_p^1(a,b))^n, \mathbb{C}^n))$.

6.3. The boundary eigenvalue operator function

Let L^D and L^R be as defined in Sections 6.1 and 6.2. We call

(6.3.1) $$L = (L^D, L^R) \in H(\Omega, L(W_p^n(a,b), L_p(a,b) \times \mathbb{C}^n))$$

a *boundary eigenvalue operator function*.

Let $\{\eta_1,\dots,\eta_n\}$ be the fundamental system function of $L^D\eta = 0$ given by Theorem 6.1.8 and set $Y := \left(\eta_j^{(i-1)}\right)_{i,j=1}^n$. Define

$$Z_L(\lambda)c := \big(\eta_1(\cdot,\lambda),\dots,\eta_n(\cdot,\lambda)\big)c = e_1^\intercal Y(\cdot,\lambda)c \quad (c\in\mathbb{C}^n, \lambda\in\Omega) \tag{6.3.2}$$

and

$$(U_L(\lambda)f)(x) := e_1^\intercal Y(x,\lambda)\int_a^x Y(t,\lambda)^{-1}e_n f(t)\,\mathrm{d}t \quad (f\in L_p(a,b)). \tag{6.3.3}$$

From Lemma 6.1.6 we know that Y is a fundamental matrix function of $T^D y = 0$. Let $\lambda\in\Omega$ and $U(\lambda)$ be the right inverse of $T^D(\lambda)$ given by (3.1.6). Then

$$U_L(\lambda) = e_1^\intercal U(\lambda)e_n. \tag{6.3.4}$$

A characteristic matrix function of L is defined by

$$M(\lambda) = L^R(\lambda)Z_L(\lambda). \tag{6.3.5}$$

Note that M is also a characteristic matrix function of the associated first order boundary eigenvalue operator function $T = (T^D, T^R)$ given by (6.1.3) and (6.2.1).

THEOREM 6.3.1. *L is an abstract boundary eigenvalue operator function in the sense of Section* 1.11.

Proof. We set $E := W_p^n(a,b)$, $F_1 := L_p(a,b)$, $G := F_2 := \mathbb{C}^n$, $T_1 := L^D$, $T_2 := L^R$. We have to prove that (1.11.1) holds. For this let $\lambda\in\Omega$. For each $f\in L_p(a,b)$ we have

$$T^D(\lambda)U(\lambda)e_n f = e_n f.$$

Thus we can apply Proposition 6.1.3 and obtain $e_1^\intercal U(\lambda)e_n f\in W_p^n(a,b)$ and

$$\left(e_1^\intercal U(\lambda)e_n f\right)^{(j-1)} = e_j^\intercal U(\lambda)e_n f \quad (j=1,\dots,n). \tag{6.3.6}$$

Then (6.1.7) and (6.3.4) yield

$$L^D(\lambda)U_L(\lambda)f = e_n^\intercal T^D(\lambda)U(\lambda)e_n f = f$$

for each $f\in L_p(a,b)$, i. e., $U_L(\lambda)$ is a right inverse of $L^D(\lambda)$.

For each $\lambda\in\Omega$, $Z_L(\lambda)$ is injective since $\eta_1,\dots,\eta_n$ are linearly independent by Proposition 6.1.7.

For the proof of (1.11.1) iii) let $\lambda\in\Omega$ and $\eta\in N(L^D(\lambda))$. Then, by Definition 6.1.4, there is a vector $c\in\mathbb{C}^n$ such that $\eta = (\eta_1(\cdot,\lambda),\dots,\eta_n(\cdot,\lambda))c = Z_L(\lambda)c$, which proves $\eta\in R(Z_L(\lambda))$. Conversely, let $\eta\in R(Z_L(\lambda))$. Then there is a vector $c\in\mathbb{C}^n$ such that $\eta = Z_L(\lambda)c = (\eta_1(\cdot,\lambda),\dots,\eta_n(\cdot,\lambda))c$. Proposition 6.1.7 proves $\eta\in N(L^D(\lambda))$.

We shall show that U_L is even a holomorphic right inverse. As in the proof of Proposition 2.1.6 we define

$$R := \left\{(f_j)_{j=0}^k : f_j\in L_p(a,b)\ (j=0,\dots,k),\ f_j' = f_{j+1}\ (j=0,\dots,k-1)\right\}.$$

Because of the isomorphism proved in that proposition it is sufficient to show that for $j=0,\dots,n$

$$(\lambda,f)\mapsto (U_L(\lambda)f)^{(j)} \qquad (\lambda\in\Omega,\ f\in L_p(a,b))$$

defines a holomorphic map in $L(L_p(a,b),L_p(a,b))$. For $j=0,\dots,n-1$ this follows from (6.3.6) since $U\in H(\Omega,L((L_p(a,b))^n,(W_p^1(a,b))^n))$. Since this also yields that

$$(\lambda,f)\mapsto (U_L(\lambda)f)^{(n-1)} \qquad (\lambda\in\Omega,\ f\in L_p(a,b))$$

defines a holomorphic map in $L(L_p(a,b),W_p^1(a,b))$, we finally obtain that the assertion also holds for $j=n$. □

As in Section 3.1 we apply Theorem 1.11.1 and obtain

THEOREM 6.3.2. *The boundary eigenvalue operator function L given by* (6.3.1) *is holomorphically equivalent on* Ω *to the* $L_p(a,b)$*-extension of M; more precisely, for* $\lambda\in\Omega$ *we have*

$$L(\lambda)=\begin{pmatrix}0 & \mathrm{id}_{L_p(a,b)}\\ \mathrm{id}_{\mathbb{C}^n} & L^R(\lambda)U_L(\lambda)\end{pmatrix}\begin{pmatrix}M(\lambda) & 0\\ 0 & \mathrm{id}_{L_p(a,b)}\end{pmatrix}(Z_L(\lambda),U_L(\lambda))^{-1},$$

and the operators

$$\begin{pmatrix}0 & \mathrm{id}_{L_p(a,b)}\\ \mathrm{id}_{\mathbb{C}^n} & L^R(\lambda)U_L(\lambda)\end{pmatrix}\in L(\mathbb{C}^n\times L_p(a,b),L_p(a,b)\times\mathbb{C}^n)$$

and

$$(Z_L(\lambda),U_L(\lambda))\in L(\mathbb{C}^n\times L_p(a,b),W_p^n(a,b))$$

are invertible and depend holomorphically on λ.

COROLLARY 6.3.3. *The boundary eigenvalue operator function L is Fredholm operator valued and* $\rho(L)=\rho(M)=\rho(T)$.

Proof. The first assertion and $\rho(L)=\rho(M)$ immediately follow from Theorem 6.3.2 since $M(\lambda)$ is an operator in finite dimensional spaces. As M is also a characteristic matrix function of T, we have $\rho(M)=\rho(T)$ by Theorem 3.1.2. □

In the same way as Theorem 3.1.4 we obtain

THEOREM 6.3.4. *Let M be the characteristic matrix function given by* (6.3.5). *Suppose that* $\rho(M)\neq\emptyset$. *Let* $\mu\in\sigma(M)$ *and* $r:=\operatorname{nul}M(\mu)$. *Let* $\{c_1,\dots,c_r\}$ *and* $\{d_1,\dots,d_r\}$ *be biorthogonal CSRFs of M and* M^* *at* μ. *Define*

$$\eta_j:=Z_Lc_j,\quad v_j:=\begin{pmatrix}-(L^RU_L)^*d_j\\ d_j\end{pmatrix} \qquad (j=1,\dots,r).$$

Then $\{\eta_1,\ldots,\eta_r\}$ *and* $\{v_1,\ldots,v_r\}$ *are biorthogonal CSRF of* L *and* L^* *at* μ, $\nu(\eta_j)=\nu(v_j)=\nu(c_j)=\nu(d_j)=: m_j$ $(j=1,\ldots,r)$, *and the operator function*

$$L^{-1}-\sum_{j=1}^{r}(\cdot-\mu)^{-m_j}\eta_j\otimes v_j$$

is holomorphic at μ.

PROPOSITION 6.3.5. *Let* $W\in H(\Omega,M_n(L_1(a,b)))$, $a_k\in[a,b]$ $(k\in\mathbb{N})$, $a_j\neq a_k$ $(k\neq j)$, $a_0=a$, $a_1=b$, $W^{(j)}\in H(\Omega,M_n(\mathbb{C}))$ $(j\in\mathbb{N})$ *such that*

(6.3.7) $$\sum_{j=0}^{\infty}\sup_{\lambda\in K}|W^{(j)}(\lambda)|<\infty$$

for each compact subset K *of* Ω. *For* $\lambda\in\Omega$ *and* $\eta\in W_p^n(a,b)$ *we set*

(6.3.8) $$L^R(\lambda)\eta:=\sum_{j=0}^{\infty}W^{(j)}(\lambda)\begin{pmatrix}\eta(a_j)\\ \vdots\\ \eta^{(n-1)}(a_j)\end{pmatrix}+\int_a^b W(\xi,\lambda)\begin{pmatrix}\eta(\xi)\\ \vdots\\ \eta^{(n-1)}(\xi)\end{pmatrix}\mathrm{d}\xi.$$

Then $L^R\in H(\Omega,L(W_p^n(a,b),\mathbb{C}^n))$.

Proof. This follows from Proposition 3.1.5. □

6.4. The inverse of the boundary eigenvalue operator function

Let L be the boundary eigenvalue operator function defined by (6.3.1), where L^R is given by (6.3.8). For $\lambda\in\rho(L)$, $f_1\in L_p(a,b)$ and $f_2\in\mathbb{C}^n$ we set

(6.4.1) $$K_1(\lambda)f_1:=L^{-1}(\lambda)(f_1,0),$$

(6.4.2) $$K_2(\lambda)f_2:=L^{-1}(\lambda)(0,f_2).$$

As in Section 3.2 we set

(6.4.3) $$F(x,\lambda):=\sum_{\substack{j=0\\ a_j<x}}^{\infty}W^{(j)}(\lambda)+\int_a^x W(t,\lambda)\,\mathrm{d}t\quad(a\le x<b),$$

(6.4.4) $$F(b,\lambda):=\sum_{j=0}^{\infty}W^{(j)}(\lambda)+\int_a^b W(t,\lambda)\,\mathrm{d}t.$$

Let $(\eta_1,\ldots,\eta_n)\in H(\mathbb{C},M_{1,n}(W_p^n(a,b)))$ be the fundamental system function of $L^D\eta=0$ with $\eta_j^{(i-1)}(a,\lambda)=\delta_{ij}$ for $\lambda\in\mathbb{C}$, and set $Y:=\left(\eta_j^{(i-1)}\right)_{i,j=1}^{n}$. Let M be the characteristic matrix function defined by (6.3.5). For $\lambda \in \rho(L)$, the GREEN'S function of L is defined by

$$G(x,\xi,\lambda) := \begin{cases} \displaystyle\int_{t=a}^{\xi} e_1^{\mathsf{T}} Y(x,\lambda) M^{-1}(\lambda)\,\mathrm{d}_t F(t,\lambda) Y(t,\lambda) Y^{-1}(\xi,\lambda) e_n \\ \qquad\qquad (a \le \xi \le x \le b), \\ \displaystyle -\int_{t=\xi}^{b} e_1^{\mathsf{T}} Y(x,\lambda) M^{-1}(\lambda)\,\mathrm{d}_t F(t,\lambda) Y(t,\lambda) Y^{-1}(\xi,\lambda) e_n \\ \qquad\qquad (a \le x < \xi \le b), \end{cases} \tag{6.4.5}$$

where the integrator is $F(\cdot,\lambda)$. We set

$$\widehat{G}(x,\lambda) := e_1^{\mathsf{T}} Y(x,\lambda) M^{-1}(\lambda) \quad (x \in [a,b],\ \lambda \in \rho(T)). \tag{6.4.6}$$

THEOREM 6.4.1. *Let L be the boundary eigenvalue operator function as defined in* (6.3.1)*, and let K_1, K_2 be given by* (6.4.1), (6.4.2). *For $\lambda \in \rho(L)$, $f_1 \in L_p(a,b)$, $f_2 \in \mathbb{C}^n$ and $x \in (a,b)$ we have*

$$(K_1(\lambda)f_1)(x) = \int_a^b G(x,\xi,\lambda) f_1(\xi)\,\mathrm{d}\xi, \tag{6.4.7}$$

$$(K_2(\lambda)f_2)(x) = \widehat{G}(x,\lambda) f_2, \tag{6.4.8}$$

$$(L^{-1}(\lambda)(f_1,f_2))(x) = \int_a^b G(x,\xi,\lambda) f_1(\xi)\,\mathrm{d}\xi + \widehat{G}(x,\lambda) f_2. \tag{6.4.9}$$

Proof. We only have to prove (6.4.9). By Corollary 6.3.3 we have $\lambda \in \rho(T)$. From

$$T^D(\lambda) T^{-1}(\lambda)(e_n f_1, f_2) = e_n f_1$$

we infer that $y := T^{-1}(\lambda)(e_n f_1, f_2)$ fulfils the assumptions of Proposition 6.1.3. Hence (6.1.7) yields

$$L^D(\lambda) e_1^{\mathsf{T}} T^{-1}(\lambda)(e_n f_1, f_2) = e_n^{\mathsf{T}} T^D(\lambda) T^{-1}(\lambda)(e_n f_1, f_2) = f_1, \tag{6.4.10}$$

and (6.1.6) and (6.2.1) prove that

$$L^R(\lambda) e_1^{\mathsf{T}} T^{-1}(\lambda)(e_n f_1, f_2) = T^R(\lambda) T^{-1}(\lambda)(e_n f_1, f_2) = f_2. \tag{6.4.11}$$

Since $L(\lambda)$ is invertible by assumption, we have

$$L^{-1}(\lambda)(f_1,f_2) = e_1^{\mathsf{T}} T^{-1}(\lambda)(e_n f_1, f_2). \tag{6.4.12}$$

This proves (6.4.9) in view of (3.2.10). □

6.5. The adjoint of the boundary eigenvalue problem

The adjoint boundary eigenvalue problem in distributional sense consists in finding nontrivial weak solutions $(u,d) \in L_{p'}(a,b) \times \mathbb{C}^n$ of the differential equation

$$(6.5.1) \qquad (-1)^n u_e^{(n)} + \sum_{i=0}^{n-1} (-1)^i \big(p_i(\cdot,\lambda)u\big)_e^{(i)} + L^{R*}(\lambda)d = 0$$

for $\lambda \in \Omega$, where u_e is the canonical extension of u. The following theorem justifies this definition of the adjoint boundary eigenvalue problem.

THEOREM 6.5.1. *Let the boundary eigenvalue operator function L be given by (6.3.1). If $p < \infty$ then $L^* \in H(\Omega, L(L_{p'}(a,b) \times \mathbb{C}^n, W_{p'}^{-n}[a,b]))$ is given by*

$$(6.5.2) \qquad L^*(\lambda)(u,d) = (-1)^n u_e^{(n)} + \sum_{i=0}^{n-1} (-1)^i \big(p_i(\cdot,\lambda)u\big)_e^{(i)} + L^{R*}(\lambda)d$$

$(u \in L_{p'}(a,b),\ d \in \mathbb{C}^n)$, where u_e is the canonical extension of u. If L^R has the form (6.3.8), then

$$(6.5.3) \quad L^{R*}(\lambda) = \sum_{i=1}^{n} (-1)^{i-1} \Big(\sum_{j=0}^{\infty} \big(W^{(j)}(\lambda)e_i\big)^{\mathsf{T}} \delta_{a_j}^{(i-1)} + \big((W(\cdot,\lambda))_e^{(i-1)} e_i\big)^{\mathsf{T}} \Big)$$

and, for $d \in \mathbb{C}^n$,

$$(6.5.4) \quad ((L^R U_L)^*(\lambda)d)(t) = \big(Y^{-1}(t,\lambda)e_n\big)^{\mathsf{T}} \times$$
$$\times \left\{ \sum_{j=1}^{\infty} Y(a_j,\lambda)^{\mathsf{T}} W^{(j)}(\lambda)^{\mathsf{T}} \chi_{(a,a_j)}(t) + \int_t^b Y(\xi,\lambda)^{\mathsf{T}} W(\xi,\lambda)^{\mathsf{T}}\, \mathrm{d}\xi \right\} d.$$

Here $\chi_{(a,a_j)}$ is the characteristic function of the interval (a,a_j).

Proof. Let $\eta \in W_p^n(a,b)$ and $u \in L_{p'}(a,b)$. Then we infer with the aid of Proposition 2.3.4 for $k = 0$, $l = 1$ and Theorem 2.2.5 that

$$\begin{aligned}\langle \eta, L^{D*}(\lambda)u \rangle_{p,n} &= \langle L^D(\lambda)\eta, u_e \rangle_{p,0} \\ &= \langle \eta^{(n)}, u_e \rangle_{p,0} + \sum_{i=0}^{n-1} \langle \eta^{(i)}, p_i(\cdot,\lambda)u_e \rangle_{p,1} \\ &= \langle \eta, (-1)^n u_e^{(n)} \rangle_{p,n} + \sum_{i=0}^{n-1} \langle \eta, (-1)^i \big(p_i(\cdot,\lambda)u\big)_e^{(i)} \rangle_{p,n}.\end{aligned}$$

This proves

$$L^{D*}(\lambda)u = (-1)^n u_e^{(n)} + \sum_{i=0}^{n-1} (-1)^i \big(p_i(\cdot,\lambda)u\big)_e^{(i)}.$$

Since, for $d \in \mathbb{C}^n$,

$$\begin{aligned}\langle \eta, L^*(\lambda)(u,d)\rangle &= \langle L(\lambda)\eta,(u,d)\rangle \\ &= \langle L^D(\lambda)\eta,u\rangle + \langle L^R(\lambda)\eta,d\rangle \\ &= \left\langle \eta, L^{D^*}(\lambda)u + L^{R^*}(\lambda)d\right\rangle,\end{aligned}$$

we obtain the representation (6.5.2). For a function $\eta \in W_p^n(a,b)$ we set $y := (\eta,\eta',\dots,\eta^{(n-1)})^\intercal$ and obtain

$$\begin{aligned}\langle \eta, L^{R^*}(\lambda)d\rangle_{p,n} &= \langle L^R(\lambda)\eta,d\rangle_{\mathbb{C}^n} \\ &= \langle T^R(\lambda)y,d\rangle_{\mathbb{C}^n} \\ &= \left\langle y, T^{R^*}(\lambda)d\right\rangle_{p,1} \\ &= \sum_{i=1}^n \left\langle \eta^{(i-1)}, e_i^\intercal T^{R^*}(\lambda)d\right\rangle_{p,1} \\ &= \sum_{i=1}^n \left\langle \eta, (-1)^{i-1}\left(e_i^\intercal T^{R^*}(\lambda)d\right)^{(i-1)}\right\rangle_{p,n}.\end{aligned}$$

This equation and (3.3.3) yield (6.5.3).

Since

$$L^R(\lambda)U_L(\lambda) = T^R(\lambda)U(\lambda)e_n$$

by (6.2.1), (6.3.4), and (6.3.6), we have

$$(L^R U_L)^*(\lambda) = e_n^\intercal (T^R U)^*(\lambda).$$

With the aid of (3.3.4) we obtain the representation (6.5.4). □

6.6. The adjoint boundary eigenvalue problem in parametrized form

In this section let $p < \infty$. Let L be given by (6.3.1) and define $L_0(\lambda)$ in $L_p(a,b)$ by

(6.6.1) $$D(L_0(\lambda)) = \{\eta \in W_p^n(a,b) : L^R(\lambda)\eta = 0\} \subset L_p(a,b)$$

and

(6.6.2) $$L_0(\lambda)\eta = L^D(\lambda)\eta \quad (\eta \in D(L_0(\lambda))).$$

As for first order systems considered in Section 3.4, the domain $D(L_0(\lambda))$ of $L_0(\lambda)$ may depend on λ and may be a non-dense subspace of $L_p(a,b)$.

Let $\rho(L_0) := \{\lambda \in \Omega : L_0(\lambda) \text{ is bijective}\}$, $\sigma(L_0) := \Omega \setminus \rho(L_0)$.

THEOREM 6.6.1. *Let L be the boundary eigenvalue operator function given by* (6.3.1) *and let L_0 be its restriction in $L_p(a,b)$ with homogeneous boundary conditions as given by* (6.6.1) *and* (6.6.2).

i) $\rho(L_0) = \rho(L)$ *and* $L_0^{-1}(\lambda)f = L^{-1}(\lambda)(f,0)$ *for* $\lambda \in \rho(L)$ *and* $f \in L_p(a,b)$. *If* $\lambda \in \rho(L_0)$, *then* $L_0^{-1}(\lambda)$ *is continuous.*

ii) *Suppose that* L^R *is of the form* (6.3.8) *and let* G *be the* GREEN'S *function given by* (6.4.5). *Then, for* $\lambda \in \rho(L_0)$ *and* $f \in L_p(a,b)$,

$$(L_0^{-1}(\lambda)f)(x) = \int_a^b G(x,\xi,\lambda)f(\xi)\,\mathrm{d}\xi.$$

Proof. Up to some changes in notations and references, the proof coincides with the proof of Theorem 3.4.1. □

As in Section 3.4 we can prove

PROPOSITION 6.6.2. *For all* $\lambda \in \Omega$ *the operator* $L_0(\lambda) : L_p(a,b) \to L_p(a,b)$ *is closed.*

In the same way as Proposition 3.4.7 we prove

PROPOSITION 6.6.3. *Let* M *be the characteristic matrix function given by* (6.3.5) *and suppose that* $\rho(M) \neq \emptyset$. *Let* $\mu \in \sigma(M)$ *and* $r := \operatorname{nul} M(\mu)$. *Let* $\{c_1,\ldots,c_r\}$ *and* $\{d_1,\ldots,d_r\}$ *be biorthogonal CSRF of* M *and* M^* *at* μ. *Define*

$$\eta_j := Z_L c_j, \quad u_j := -(L^R U_L)^* d_j \qquad (j = 1,\ldots,r),$$

where Z_L *and* U_L *are given by* (6.3.2) *and* (6.3.3), *respectively. Let* $m_j := \nu(\eta_j)$, *the multiplicity of the root function* c_j. *Then the operator function*

$$L_0^{-1} - \sum_{j=1}^{r} (\cdot - \mu)^{-m_j} \eta_j \otimes u_j$$

is holomorphic at μ.

The adjoint $L_0^*(\lambda)$ is a linear relation in $L_{p'}(a,b)$ defined by its graph

$$G(L_0^*(\lambda)) = (G(-L_0(\lambda)))^{\perp},$$

i. e.,

$$u \in D(L_0^*(\lambda)) \Leftrightarrow \exists w \in L_{p'}(a,b) \forall y \in D(L_0(\lambda))\ \langle L_0(\lambda)y, u\rangle = \langle y, w\rangle$$

and

$$L_0^*(\lambda)u = \{w \in L_{p'}(a,b) : \forall y \in D(L_0(\lambda))\ \langle L_0(\lambda)y, u\rangle = \langle y, w\rangle\}.$$

Here $\langle\ ,\ \rangle$ is the canonical bilinear form on $L_p(a,b) \times L_{p'}(a,b)$.

In the same way as Theorem 3.4.3 we prove

THEOREM 6.6.4. i) *Let* $\lambda \in \Omega$ *and* $u \in L_{p'}(a,b)$. *Then* $u \in D(L_0^*(\lambda))$ *if and only if there is a vector* $d \in \mathbb{C}^n$ *such that* $L^*(\lambda)(u,d) \in L_{p'}(\mathbb{R})$.
ii) *Let* $\lambda \in \Omega$ *and* $u \in D(L_0^*(\lambda))$. *Then*

$$L_0^*(\lambda)u = \{(L^*(\lambda)(u,d))_r : d \in \mathbb{C}^n,\ L^*(\lambda)(u,d) \in L_{p'}(\mathbb{R})\}.$$

THEOREM 6.6.5. *Let L^R be given by (6.3.8) and suppose that $W(\cdot,\lambda)e_i$ belongs to $(W_{p'}^{i-1}(a,b))^n$ for $i=1,\dots,n$.*
i) *Let $\lambda\in\Omega$ and $u\in L_{p'}(a,b)$. Then $u\in D(L_0^*(\lambda))$ if and only if there is a vector $d\in\mathbb{C}^n$ such that*

$$(-1)^n u_e^{(n)}+\sum_{i=0}^{n-1}(-1)^i(p_i(\cdot,\lambda)u)_e^{(i)}+\sum_{i=1}^{n}(-1)^{i-1}\sum_{j=0}^{\infty}(W^{(j)}(\lambda)e_i)^{\mathsf T}d\delta_{a_j}^{(i-1)}$$
$$+\sum_{i=1}^{n-1}\sum_{m=i+1}^{n}(-1)^{m-1}\Big((W(\cdot,\lambda)^{(m-i-1)}e_m)^{\mathsf T}(a)d\delta_a^{(i-1)}$$
$$-(W(\cdot,\lambda)^{(m-i-1)}e_m)^{\mathsf T}(b)d\delta_b^{(i-1)}\Big)$$

belongs to $L_{p'}(\mathbb{R})$.
ii) *Let $\lambda\in\Omega$. Then $L_0^*(\lambda)$ is a linear operator if $W(\cdot,\lambda)=0$ or if for each $d\in\mathbb{C}^n\setminus\{0\}$ one of the following five properties holds:*
There is an integer $j\in\mathbb{N}$, $j\geq 2$, such that $W^{(j)}(\lambda)^{\mathsf T}d\neq 0$;
$e_n^{\mathsf T}W^{(0)}(\lambda)^{\mathsf T}d\neq 0$;
$e_n^{\mathsf T}W^{(1)}(\lambda)^{\mathsf T}d\neq 0$;
There is a number $i\in\{1,\dots,n-1\}$ such that

$$(-1)^{i-1}(W^{(0)}(\lambda)e_i)^{\mathsf T}d+\sum_{m=i+1}^{n}(-1)^{m-1}\big((W(\cdot,\lambda)^{(m-1-i)}e_m\big)^{\mathsf T}(a)d\neq 0;$$

There is a number $i\in\{1,\dots,n-1\}$ such that

$$(-1)^{i-1}(W^{(1)}(\lambda)e_i)^{\mathsf T}d+\sum_{m=i+1}^{n}(-1)^{m-1}\big((W(\cdot,\lambda)^{(m-1-i)}e_m\big)^{\mathsf T}(b)d\neq 0.$$

Proof. i) is obvious because of Theorem 6.6.4 i), (6.5.2), (6.5.3), and Proposition 2.6.5.
ii) For $d\in\mathbb{C}^n$ we have $L^*(\lambda)(0,d)=L^{R^*}(\lambda)d$. In case $W(\cdot,\lambda)=0$, the property $L^{R^*}(\lambda)d\in L_{p'}(a,b)$ immediately implies $L^{R^*}(\lambda)d=0$ by Proposition 3.4.4. And if one of the other five conditions is satisfied, then it follows by part i) and Proposition 3.4.4 that $L^{R^*}(\lambda)d\notin L_{p'}(a,b)$ if $d\neq 0$. In both cases, Theorem 6.6.4 ii) yields $L^*(\lambda)(0)=\{0\}$, i.e., $L^*(\lambda)$ is an operator. □

COROLLARY 6.6.6. *Let $p_i(\cdot,\lambda) \in W^i_{\max\{p,p'\}}(a,b)$ for $i = 1,\dots,n-1$. Let L^R be given by* (6.3.8), *where the sum runs from* 0 *to k, $k \geq 1$, and $W(\cdot,\lambda)e_i$ belongs to $(W^{i-1}_{p'}(a,b))^n$ for $i = 1,\dots,n$. The set $[a,b] \setminus \{a_0,\dots,a_k\}$ is the disjoint union of k open intervals $I_1,\dots,I_k$. We set $p_n := 1$ and define*

$$h_{lm} := \begin{cases} \sum\limits_{i=l+m+1}^{n} (-1)^{i-1-l}\binom{i-1-l}{m} p_i^{(i-1-l-m)} & \text{if } 0 \leq l \leq n-1; 0 \leq m \leq n-1-l \\ 0 & \text{if } 1 \leq l \leq n-1; n-l \leq m \leq n-1, \end{cases}$$

$$H_x(\lambda) := \left(h_{lm}(x,\lambda)\right)_{l,m=0}^{n-1}.$$

Then $\zeta \in D(L_0^*(\lambda))$ if and only if there is a vector $d \in \mathbb{C}^n$ such that

$$\begin{aligned}
&\zeta|_{I_j} \in W^n_{p'}(I_j) \quad (j = 1,\dots,k), \\
&H_a(\lambda)(\zeta^{(i)}(a+))_{i=0}^{n-1} = W^{(0)}(\lambda)^\intercal d \\
&\qquad - \sum_{l=0}^{n-2} \sum_{m=l+2}^{n} (-1)^{m-l} e_{l+1} \left(W(\cdot,\lambda)^{(m-l)} e_m\right)^\intercal (a) d, \\
&H_b(\lambda)(\zeta^{(i)}(b-))_{i=0}^{n-1} = -W^{(1)}(\lambda)^\intercal d \\
&\qquad - \sum_{l=0}^{n-2} \sum_{m=l+2}^{n} (-1)^{m-l} e_{l+1} \left(W(\cdot,\lambda)^{(m-l)} e_m\right)^\intercal (b) d, \\
&H_{a_j}(\lambda)\left(\zeta^{(i)}(a_j+) - \zeta^{(i)}(a_j-)\right)_{i=0}^{n-1} = W^{(j)}(\lambda)^\intercal d \quad (j = 2,\dots,k).
\end{aligned}$$

Proof. From Theorem 6.6.5 we infer that

$$(\zeta|_{I_j})^{(n)} + \sum_{i=0}^{n-1} (-1)^{n-i} \left(p_i(\cdot,\lambda)\zeta|_{I_j}\right)^{(i)} \in L_{p'}(I_j)$$

for $\zeta \in D(L_0^*(\lambda))$ and $j = 1,\dots,k$. Proposition 2.6.1 shows that $\zeta|_{I_j} \in W^n_{p'}(I_j)$ for $j = 1,\dots,k$. Therefore, in the following, we may suppose that $\zeta \in L_{p'}(a,b)$ satisfies this property. Note that

$$(p_i(\cdot,\lambda)\zeta)_e = \sum_{j=1}^{k} \left(p_i(\cdot,\lambda)\zeta|_{I_j}\right)_e.$$

Therefore, by Theorem 6.6.5 and Proposition 2.6.5, $u \in D(L_0^*(\lambda))$ if and only if there is a vector $d \in \mathbb{C}^n$ such that

$$\sum_{j=1}^{k}\sum_{i=1}^{n}(-1)^i\sum_{l=0}^{i-1}\left\{(p_i(\cdot,\lambda)\zeta)^{(i-1-l)}(\alpha_j+)\delta_{\alpha_j}^{(l)} - (p_i(\cdot,\lambda)\zeta)^{(i-1-l)}(\beta_j-)\delta_{\beta_j}^{(l)}\right\}$$
$$-\sum_{i=1}^{n}(-1)^i\sum_{j=0}^{k}(W^{(j)}(\lambda)e_i)^{\mathsf{T}}d\delta_{a_j}^{(i-1)}$$
$$+\sum_{i=1}^{n-1}\sum_{m=i+1}^{n}(-1)^{m-1}\Big(\big(W(\cdot,\lambda)^{(m-i-1)}e_m\big)^{\mathsf{T}}(a)d\delta_a^{(i-1)}$$
$$-\big(W(\cdot,\lambda)^{(m-i-1)}e_m\big)^{\mathsf{T}}(b)d\delta_b^{(i-1)}\Big) \in L_{p'}(\mathbb{R}),$$

where $I_j =: (\alpha_j, \beta_j)$. In view of Proposition 3.4.4 this holds if and only if all the coefficients of $\delta_{a_j}^{(l)}$ are zero for $l = 0, \ldots, n-1$ and $j = 1, \ldots, k$. Let us consider $j = 0$; the cases $j = 1$ and $j = 2, \ldots, k$ are similar. So, for $j = 0$, we have to satisfy

(6.6.3)
$$\sum_{i=l+1}^{n}(-1)^{i-1-l}(p_i(\cdot,\lambda)\zeta)^{(i-1-l)}(a+) - (W^{(0)}(\lambda)e_{l+1})^{\mathsf{T}}d$$
$$+\sum_{m=l+2}^{n}(-1)^{m-l}\big(W(\cdot,\lambda)^{(m-l-2)}e_m\big)^{\mathsf{T}}(a)d = 0$$

for $l = 0, \ldots, n-1$. Applying LEIBNIZ' rule to $(p_i(\cdot,\lambda)\zeta)^{(i-1-l)}$ we obtain

$$\sum_{i=l+1}^{n}(-1)^{i-1-l}(p_i(\cdot,\lambda)\zeta)^{(i-1-l)} = \sum_{m=0}^{n-1}h_{lm}(\cdot,\lambda)\zeta^{(m)}$$

for $l = 0, \ldots, n-1$, which shows that (6.6.3) is equivalent to

$$H_a(\lambda)(\zeta^{(i)}(a+))_{i=0}^{n-1}$$
$$= W^{(0)}(\lambda)^{\mathsf{T}}d - \sum_{l=0}^{n-2}\sum_{m=l+2}^{n}(-1)^{m-l}e_{l+1}\big(W(\cdot,\lambda)^{(m-l-2)}e_m\big)^{\mathsf{T}}(a)d. \quad \square$$

Now let $\lambda \in \Omega$ and suppose that the assumptions of Corollary 6.6.6 are fulfilled. Then λ is an eigenvalue and the nonzero function $\zeta \in L_{p'}(a,b)$ is an eigenfunction of the adjoint boundary eigenvalue problem in parametrized form if and only if there is a vector $d \in \mathbb{C}^n$ such that, for $j = 1, \ldots, k$, $\zeta|_{I_j} \in W_{p'}^n(a,b)$,

(6.6.4)
$$(-1)^n(\zeta|_{I_j})^{(n)} + \sum_{i=0}^{n-1}(-1)^i(p_i(\lambda)\zeta|_{I_j})^{(i)} + \sum_{i=1}^{n}(-1)^{i-1}d^{\mathsf{T}}(W(\cdot,\lambda)|_{I_j}e_i)^{(i-1)} = 0,$$

and the boundary conditions

$$H_a(\lambda)(\zeta^{(i)}(a+))_{i=0}^{n-1} = W^{(0)}(\lambda)^{\mathsf{T}}d - \sum_{l=0}^{n-2}\sum_{m=l+2}^{n}(-1)^{m-l}e_{l+1}\left(W(\cdot,\lambda)^{(m-l)}e_m\right)^{\mathsf{T}}(a)d,$$

$$H_b(\lambda)(\zeta^{(i)}(b-))_{i=0}^{n-1} = -W^{(1)}(\lambda)^{\mathsf{T}}d + \sum_{l=0}^{n-2}\sum_{m=l+2}^{n}(-1)^{m-l}e_{l+1}\left(W(\cdot,\lambda)^{(m-l)}e_m\right)^{\mathsf{T}}(b)d, \tag{6.6.5}$$

$$H_{a_j}(\lambda)\left(\zeta^{(i)}(a_j+) - \zeta^{(i)}(a_j-)\right)_{i=0}^{n-1} = W^{(j)}(\lambda)^{\mathsf{T}}d \quad (j=2,\dots,k)$$

are satisfied.

Analogously to Proposition 3.4.8 we obtain

PROPOSITION 6.6.7. *Suppose that* $\rho(L) \neq \emptyset$ *and let* $\mu \in \sigma(L)$.
i) *Let* η_0 *be an eigenvector of* L *at* μ. *Then* $\eta_0 \in D(L_0(\mu))$ *and* $L_0(\mu)\eta_0 = 0$.
ii) *Assume that* $L_0^*(\mu)$ *is an operator. Let* (u_0, d_0) *be an eigenvector of* L^* *at* μ. *Then* $u_0 \in D(L_0^*(\mu))$ *and* $L_0^*(\mu)u_0 = 0$.

PROPOSITION 6.6.8. *Suppose that* L^R *does not depend on* λ.
i) *Let* $\mu \in \sigma(L)$ *and* $(\eta_k)_{k=0}^h$ *be a CEAV of* L *at* μ. *Then* $\eta_k \in D(L_0(\mu))$ *for* $k = 0,\dots,h$.
ii) *Suppose in addition that* $p_0 \in H(\Omega, L_\infty(a,b))$, *that the* p_i *do not depend on* λ *for* $i = 1,\dots,n-1$, *and that* $L_0^*(\lambda)$ *is an operator for all* $\lambda \in \Omega$. *Let* $\mu \in \sigma(L)$ *and* $(u_k, d_k)_{k=0}^h$ *be a CEAV of* L^* *at* μ. *Then* $u_k \in D(L_0^*(\mu))$ *for* $k = 0,\dots,h$.

Proof. The proof of part i) is similar to the proof of part i) in Proposition 3.4.9 and therefore omitted.
ii) For $k = 0,\dots,h$ we obtain as in the proof of Proposition 3.4.9 and with the aid of (6.5.2) that

$$L^*(\mu)(u_k,d_k) = -\sum_{j=1}^{k}\frac{1}{j!}\left(\frac{\partial^j}{\partial\lambda^j}L^*\right)(\mu)(u_{k-j},d_{k-j})$$

$$= -\sum_{j=1}^{k}\frac{1}{j!}\frac{\partial^j p_0}{\partial\lambda^j}(\cdot,\mu)u_{k-j} \in L_{p'}(a,b). \quad \square$$

The next example shows that the statement of Proposition 6.6.8 ii) is not necessarily true if one of the functions $p_1,\dots,p_{n-1}$ depends on λ.

EXAMPLE 6.6.9. We consider $L(\lambda) \in L(W_p^2(0,1), L_p(0,1)\times\mathbb{C}^2)$ given by

$$L^D(\lambda)\eta = \eta'' - 2\lambda\eta' + \lambda^2\eta,$$

$$L^R(\lambda)\eta = \begin{pmatrix} -1 & -1 \\ 0 & 0 \end{pmatrix}\begin{pmatrix} \eta(0) \\ \eta'(0) \end{pmatrix} + \begin{pmatrix} \eta(1) \\ \eta'(1) \end{pmatrix}.$$

Obviously, $\{e^{\lambda x} - \lambda x e^{\lambda x}, x e^{\lambda x}\}$ is a fundamental system of $L^D(\lambda)\eta = 0$. The corresponding fundamental matrix is

$$Y(x,\lambda) = \begin{pmatrix} e^{\lambda x} - \lambda x e^{\lambda x} & x e^{\lambda x} \\ -\lambda^2 x e^{\lambda x} & e^{\lambda x} + \lambda x e^{\lambda x} \end{pmatrix}$$

and fulfils $Y(0,\lambda) = I_2$. The characteristic determinant is given by

$$\begin{aligned} M(\lambda) &= \begin{pmatrix} -1 & -1 \\ 0 & 0 \end{pmatrix} + \begin{pmatrix} e^{\lambda} - \lambda e^{\lambda} & e^{\lambda} \\ -\lambda^2 e^{\lambda} & e^{\lambda} + \lambda e^{\lambda} \end{pmatrix} \\ &= \begin{pmatrix} e^{\lambda} - 1 - \lambda e^{\lambda} & e^{\lambda} - 1 \\ -\lambda^2 e^{\lambda} & e^{\lambda} + \lambda e^{\lambda} \end{pmatrix}. \end{aligned}$$

From $\det M(\lambda) = e^{2\lambda} - e^{\lambda} - \lambda e^{\lambda} - \lambda^2 e^{\lambda}$ we infer $\rho(M) \neq \emptyset$ and that $\det M$ has a zero of order 2 at 0. The vector function given by

$$M(\lambda)\begin{pmatrix} 1 \\ 0 \end{pmatrix} = \begin{pmatrix} e^{\lambda} - 1 - \lambda e^{\lambda} \\ -\lambda^2 e^{\lambda} \end{pmatrix}$$

has a zero of order 2 at 0. Hence $\begin{pmatrix} 1 \\ 0 \end{pmatrix}$ is a root function of M at 0 of multiplicity 2, and by Proposition 1.8.5 it is also a CSRF of M at 0. In the same way,

$$\begin{aligned} M^*(\lambda)\begin{pmatrix} -2 - \frac{8}{3}\lambda \\ 2\lambda \end{pmatrix} &= \begin{pmatrix} e^{\lambda} - 1 - \lambda e^{\lambda} & -\lambda^2 e^{\lambda} \\ e^{\lambda} - 1 & e^{\lambda} + \lambda e^{\lambda} \end{pmatrix}\begin{pmatrix} -2 - \frac{8}{3}\lambda \\ 2\lambda \end{pmatrix} \\ &= \begin{pmatrix} (e^{\lambda} - 1 - \lambda e^{\lambda})(-2 - \frac{8}{3}\lambda) - 2\lambda^3 e^{\lambda} \\ 2(1 - e^{\lambda} + \lambda e^{\lambda} + \lambda^2 e^{\lambda}) - \frac{8}{3}\lambda(e^{\lambda} - 1) \end{pmatrix} \end{aligned}$$

shows that $\begin{pmatrix} -2 - \frac{8}{3}\lambda \\ 2\lambda \end{pmatrix}$ is a CSRF of M^* at 0 of multiplicity 2. From

$$\begin{aligned} &\frac{1}{\lambda^2}\begin{pmatrix} -2 - \frac{8}{3}\lambda \\ 2\lambda \end{pmatrix}^{\mathsf{T}} M(\lambda)\begin{pmatrix} 1 \\ 0 \end{pmatrix} \\ &\qquad = \begin{pmatrix} -2 - \frac{8}{3}\lambda \\ 2\lambda \end{pmatrix}^{\mathsf{T}}\begin{pmatrix} -\frac{1}{2} - \frac{1}{3}\lambda \\ -1 \end{pmatrix}\lambda^2 h_1(\lambda) \\ &\qquad = 1 + \lambda^2 h_2(\lambda), \end{aligned}$$

where h_1 and h_2 are holomorphic functions on $\mathbb{C}$, we see that the CSRFs are biorthogonal. It is easy to see that

$$Y^{-1}(x,\lambda) = \begin{pmatrix} e^{-\lambda x} + \lambda x e^{-\lambda x} & -x e^{-\lambda x} \\ \lambda^2 x e^{-\lambda x} & e^{-\lambda x} - \lambda x e^{-\lambda x} \end{pmatrix}.$$

Now 6.5.4 yields

$$\begin{aligned}((L^RU_L)^*(\lambda)d)(x) &= (Y^{-1}(x,\lambda)e_2)^\mathsf{T} Y(1,\lambda)^\mathsf{T} W^{(1)^\mathsf{T}} d\\
&= \begin{pmatrix} -xe^{-\lambda x} \\ e^{-\lambda x} - \lambda x e^{-\lambda x}\end{pmatrix}^\mathsf{T} \begin{pmatrix} e^\lambda - \lambda e^\lambda & -\lambda^2 e^\lambda \\ e^\lambda & e^\lambda + \lambda e^\lambda \end{pmatrix} d\\
&= \begin{pmatrix} \{-x(1-\lambda) + (1-\lambda x)\} e^{\lambda(1-x)} \\ \{\lambda^2 x + (1-\lambda x)(1+\lambda)\} e^{\lambda(1-x)}\end{pmatrix} d\\
&= \begin{pmatrix} (1-x)e^{\lambda(1-x)} \\ (1+\lambda-\lambda x)e^{\lambda(1-x)}\end{pmatrix} d.\end{aligned}$$

Hence

$$\left((L^RU_L)^*(\lambda)\begin{pmatrix} -2-\frac{8}{3}\lambda \\ 2\lambda\end{pmatrix}\right)(x) = \left(2x-2-\frac{2}{3}\lambda+\frac{8}{3}\lambda x + 2\lambda^2 - 2\lambda^2 x\right)e^{\lambda(1-x)}.$$

According to Theorem 6.3.4,

$$\begin{pmatrix} 2-2x \\ \begin{pmatrix} -2 \\ 0\end{pmatrix}\end{pmatrix}, \begin{pmatrix} \frac{8}{3} - \frac{20}{3}x + 2x^2 \\ \begin{pmatrix} -\frac{8}{3} \\ 2\end{pmatrix}\end{pmatrix}$$

is a CEAV of L^* at 0. In view of Theorem 6.5.1 we have

$$\begin{aligned}&L^*(\lambda)(u,d)\\
&= u_e'' + 2\lambda u_e' + \lambda^2 u + \sum_{i=1}^{2}(-1)^{i-1}\left((W^{(0)}e_i)^\mathsf{T}\delta_0^{(i-1)} + (W^{(1)}e_i)^\mathsf{T}\delta_1^{(i-1)}\right)d\\
&= u_e'' + 2\lambda u_e' + \lambda^2 u + \left\{\begin{pmatrix} -1 \\ 0\end{pmatrix}^\mathsf{T}\delta_0 + \begin{pmatrix} 1 \\ 0\end{pmatrix}^\mathsf{T}\delta_0' + \begin{pmatrix} 1 \\ 0\end{pmatrix}^\mathsf{T}\delta_1 - \begin{pmatrix} 0 \\ 1\end{pmatrix}^\mathsf{T}\delta_1'\right\}d.\end{aligned}$$

From Proposition 2.6.5 we infer

$$u_e'' = (u'')_e + u(0)\delta_0' - u(1)\delta_1' + u'(0)\delta_0 - u'(1)\delta_1.$$

Thus we obtain for $d = (d_1,d_2)^\mathsf{T} \in \mathbb{C}^2$

$$\begin{aligned}&L^*(0)\left(\frac{8}{3} - \frac{20}{3}x + 2x^2, d\right)\\
&\quad = 4 + \frac{8}{3}\delta_0' + 2\delta_1' - \frac{20}{3}\delta_0 + \frac{8}{3}\delta_1 - d_1\delta_0 + d_1\delta_0' + d_1\delta_1 - d_2\delta_1'.\end{aligned}$$

Since for any choice of d_1, d_2 this distribution does not belong to $L_{p'}(a,b)$, it follows from Theorem 6.6.4 ii) that the first component $\frac{8}{3} - \frac{20}{3}x + 2x^2$ of the associated vector of L^* at 0 does not belong to the domain of $L_0^*(0)$. □

6.7. Two-point boundary eigenvalue problems in $\mathbf{L_p(a,b)}$

In this section let $p < \infty$ and

$$L^D(\lambda)\eta = \eta^{(n)} + \sum_{i=0}^{n-1} p_i(\cdot,\lambda)\eta^{(i)},$$

$$L^R(\lambda)\eta = W^a(\lambda)\begin{pmatrix}\eta(a)\\ \vdots \\ \eta^{(n-1)}(a)\end{pmatrix} + W^b(\lambda)\begin{pmatrix}\eta(b)\\ \vdots \\ \eta^{(n-1)}(b)\end{pmatrix}$$

for $\lambda \in \Omega$ and $\eta \in W_p^n(a,b)$, where $p_i \in H(\Omega, W^i_{\max\{p,p'\}}(a,b))$ $(1 \le i \le n-1)$ and $W^a, W^b \in H(\Omega, M_n(\mathbb{C}))$. We suppose that $\operatorname{rank}(W^a(\lambda), W^b(\lambda)) = n$ for all $\lambda \in \Omega$.

Apart from L^D we consider $L^{D+} \in H(\Omega, L(W^n_{p'}(a,b), L_{p'}(a,b)))$ defined by

$$L^{D+}(\lambda)\eta = (-1)^n\eta^{(n)} + \sum_{i=0}^{n-1}(-1)^i(p_i(\cdot,\lambda)\eta)^{(i)}$$

for $\lambda \in \Omega$ and $\eta \in W^n_{p'}(a,b)$.

PROPOSITION 6.7.1. *Let $l \in \mathbb{N}$, $v \in W_p^l(a,b)$ and $w \in W_{p'}^l(a,b)$. Then*

$$\begin{aligned}&\langle v^{(l)}, w\rangle - (-1)^l\langle v, w^{(l)}\rangle \\ &= \sum_{i=0}^{l-1}(-1)^i\big(v^{(l-1-i)}(b)w^{(i)}(b) - v^{(l-1-i)}(a)w^{(i)}(a)\big).\end{aligned}$$

Proof. For $l = 0$ nothing has to be proved. Now let $l = 1$. In view of (2.3.1) and Proposition 2.1.5 i) we obtain

$$\begin{aligned}\langle v', w\rangle + \langle v, w'\rangle &= \int_a^b (v'w + vw')\,\mathrm{d}x = \int_a^b (vw)'\,\mathrm{d}x \\ &= v(b)w(b) - v(a)w(a).\end{aligned}$$

Assume that the statement holds for some $l \ge 1$. Then

$$\begin{aligned}\langle v^{(l+1)}, w\rangle - (-1)^{l+1}\langle v, w^{(l+1)}\rangle &= \langle v^{(l+1)}, w\rangle - (-1)^l\langle v', w^{(l)}\rangle \\ &\quad + (-1)^l\Big(\langle v', w^{(l)}\rangle + \langle v, w^{(l+1)}\rangle\Big) \\ &= \sum_{i=0}^{l-1}(-1)^i\big(v^{(l-i)}(b)w^{(i)}(b) - v^{(l-i)}(a)w^{(i)}(a)\big) \\ &\quad + (-1)^l\big(v(b)w^{(l)}(b) - v(a)w^{(l)}(a)\big). \quad \square\end{aligned}$$

PROPOSITION 6.7.2. *We set* $p_n := 1$ *and*

$$H(\lambda) := \begin{pmatrix} -H_a(\lambda) & 0 \\ 0 & H_b(\lambda) \end{pmatrix},$$

where $H_x(\lambda)$ *has been defined in Corollary* 6.6.6. *Then* $H \in H(\Omega, M_{2n}(\mathbb{C}))$, $H(\lambda)$ *is invertible for all* $\lambda \in \Omega$, *and*

$$(6.7.1) \qquad \langle L^D(\lambda)\eta, \zeta\rangle - \langle \eta, L^{D^+}(\lambda)\zeta\rangle = \begin{pmatrix} \eta(a) \\ \vdots \\ \eta^{(n-1)}(a) \\ \eta(b) \\ \vdots \\ \eta^{(n-1)}(b) \end{pmatrix}^{\mathsf{T}} H(\lambda) \begin{pmatrix} \zeta(a) \\ \vdots \\ \zeta^{(n-1)}(a) \\ \zeta(b) \\ \vdots \\ \zeta^{(n-1)}(b) \end{pmatrix}$$

holds for all $\lambda \in \mathbb{C}$, $\eta \in W_p^n(a,b)$ *and* $\zeta \in W_{p'}^n(a,b)$.

Equation (6.7.1) is called the LAGRANGE *identity*. We call $H(\lambda)$ the LAGRANGE *matrix* of $L^D(\lambda)$.

Proof. Since $p_i \in H(\Omega, W_1^i(a,b))$ for $i = 1, \dots, n$, we have that h_{lm} belongs to $H(\Omega, W_1^1(a,b))$ and hence $h_{lm} \in H(\Omega, C[a,b])$ by Proposition 2.1.7. This proves that $h_{lm}(x,\cdot)$ is a holomorphic function for each $x \in [a,b]$. Therefore H depends holomorphically on λ. Since $h_{lm} = 0$ if $l+m \ge n$, $H_x(\lambda)$ is an upper left triangular matrix, and $h_{l,n-1-l} = (-1)^{n-1-l} p_n = (-1)^{n-1-l}$ shows that the corresponding diagonal elements are nonzero. Hence $H_x(\lambda)$ is invertible for all $\lambda \in \Omega$ and all $x \in [a,b]$. Thus also $H(\lambda)$ is invertible for all $\lambda \in \Omega$. For the proof of (6.7.1) we calculate with the aid of Proposition 6.7.1

$$\begin{aligned}
&\langle L^D(\lambda)\eta, \zeta\rangle - \langle \eta, L^{D^+}(\lambda)\zeta\rangle \\
&= \sum_{i=1}^{n} \Big(\langle \eta^{(i)}, p_i(\cdot,\lambda)\zeta\rangle - (-1)^i \langle \eta, (p_i(\cdot,\lambda)\zeta)^{(i)}\rangle \Big) \\
&= \sum_{i=1}^{n}\sum_{l=0}^{i-1} (-1)^l \Big(\eta^{(i-1-l)}(b)(p_i(\cdot,\lambda)\zeta)^{(l)}(b) - \eta^{(i-1-l)}(a)(p_i(\cdot,\lambda)\zeta)^{(l)}(a) \Big).
\end{aligned}$$

Because of (2.3.1) we can apply LEIBNIZ' rule to $p_i\zeta$ and obtain that

$$\begin{aligned}
&\sum_{i=1}^{n}\sum_{l=0}^{i-1} (-1)^l \eta^{(i-1-l)}(x)(p_i(\cdot,\lambda)\zeta)^{(l)}(x) \\
&= \sum_{l=0}^{n-1} \eta^{(l)}(x) \sum_{i=l+1}^{n} (-1)^{i-1-l} (p_i(\cdot,\lambda)\zeta)^{(i-1-l)}(x) \\
&= \sum_{l=0}^{n-1} \eta^{(l)}(x) \sum_{m=0}^{n-1} h_{lm}(x,\lambda)\zeta^{(m)}(x)
\end{aligned}$$

$$= \begin{pmatrix} \eta(x) \\ \vdots \\ \eta^{(n-1)}(x) \end{pmatrix}^{\mathsf{T}} H_x(\lambda) \begin{pmatrix} \zeta(x) \\ \vdots \\ \zeta^{(n-1)}(x) \end{pmatrix}. \qquad \square$$

For $\eta \in W_p^n(a,b)$ and $\zeta \in W_{p'}^n(a,b)$ we briefly write $y := (\eta,\dots,\eta^{(n-1)})^{\mathsf{T}}$ and $u := (\zeta,\dots,\zeta^{(n-1)})^{\mathsf{T}}$.

We note that the LAGRANGE identity can be written as

(6.7.2)

$$L^{D*}\zeta = \left(L^{D+}\zeta\right)_e + \sum_{i=1}^{n}(-1)^{i-1}\left(\delta_a^{(i-1)}e_i^{\mathsf{T}}(-H_a)u(a,\cdot) + \delta_b^{(i-1)}e_i^{\mathsf{T}}H_b u(b,\cdot)\right)$$

for $\zeta \in W_{p'}^n(a,b)$. Furthermore, (6.5.3) immediately yields that

$$L^{R*}d = \sum_{i=1}^{n}(-1)^{i-1}\left(e_i^{\mathsf{T}}W^{a\mathsf{T}}d\delta_a^{(i-1)} + e_i^{\mathsf{T}}W^{b\mathsf{T}}d\delta_b^{(i-1)}\right) \tag{6.7.3}$$

holds for $d \in \mathbb{C}^n$.

By Proposition 3.5.1 there is an invertible matrix $Q(\lambda) \in M_{2n}(\mathbb{C})$ which depends holomorphically on λ such that

$$Q(\lambda) = \begin{pmatrix} W^a(\lambda)^{\mathsf{T}} & \widetilde{A}(\lambda) \\ W^b(\lambda)^{\mathsf{T}} & \widetilde{B}(\lambda) \end{pmatrix} \tag{6.7.4}$$

with suitable $\widetilde{A},\widetilde{B} \in H(\Omega, M_n(\mathbb{C}))$.

We define

$$\begin{pmatrix} \widetilde{C}(\lambda) & \widetilde{D}(\lambda) \\ \widetilde{W}^a(\lambda) & \widetilde{W}^b(\lambda) \end{pmatrix} := \begin{pmatrix} W^a(\lambda)^{\mathsf{T}} & \widetilde{A}(\lambda) \\ W^b(\lambda)^{\mathsf{T}} & \widetilde{B}(\lambda) \end{pmatrix}^{-1} H(\lambda), \tag{6.7.5}$$

where the matrix on the left hand side is divided into $n \times n$-block-matrices.

The operator $L_0^+(\lambda)$ in $L_{p'}(a,b)$ is defined by

(6.7.6) $D(L_0^+(\lambda))$

$$:= \left\{ \zeta \in W_{p'}^n(a,b) : \widetilde{W}^a(\lambda)\begin{pmatrix} \zeta(a) \\ \vdots \\ \zeta^{(n-1)}(a) \end{pmatrix} + \widetilde{W}^b(\lambda)\begin{pmatrix} \zeta(b) \\ \vdots \\ \zeta^{(n-1)}(b) \end{pmatrix} = 0 \right\}$$

$\subset L_{p'}(a,b)$

and

$$L_0^+(\lambda)\zeta := L^{D+}(\lambda)\zeta \quad (\zeta \in D(L_0^+(\lambda))). \tag{6.7.7}$$

By Theorem 6.6.5 ii), $L_0^*(\lambda)$ is a linear operator.

THEOREM 6.7.3. *We consider the families of operators* $L_0(\lambda)$ *and* $L_0^+(\lambda)$ *defined by* (6.6.1), (6.6.2) *and* (6.7.6), (6.7.7), *respectively. For all* $\lambda \in \Omega$ *we have*
i) $L_0^+(\lambda) = L_0^*(\lambda)$,
ii) $(L_0^+(\lambda))^* = L_0(\lambda)$ *if* $p > 1$.

Proof. i) Let $\eta \in D(L_0(\lambda))$ and $\zeta \in D(L_0^+(\lambda))$. From Proposition 6.7.2 we infer

$$\begin{aligned}\langle L_0(\lambda)\eta,\zeta\rangle &= \langle L^D(\lambda)\eta,\zeta\rangle \\ &= \langle \eta, L^{D^+}(\lambda)\zeta\rangle + \begin{pmatrix} y(a) \\ y(b)\end{pmatrix}^{\mathsf{T}} H(\lambda)\begin{pmatrix} u(a) \\ u(b)\end{pmatrix}.\end{aligned}$$

From (6.7.5) and the definitions of $D(L_0(\lambda))$ and $D(L_0^+(\lambda))$ we infer

$$\begin{aligned}&\begin{pmatrix} y(a) \\ y(b)\end{pmatrix}^{\mathsf{T}} H(\lambda)\begin{pmatrix} u(a) \\ u(b)\end{pmatrix} \\ &= \begin{pmatrix} y(a) \\ y(b)\end{pmatrix}^{\mathsf{T}} \begin{pmatrix} W^a(\lambda)^{\mathsf{T}} & \widetilde{A}(\lambda) \\ W^b(\lambda)^{\mathsf{T}} & \widetilde{B}(\lambda)\end{pmatrix}\begin{pmatrix} \widetilde{C}(\lambda) & \widetilde{D}(\lambda) \\ \widetilde{W}^a(\lambda) & \widetilde{W}^b(\lambda)\end{pmatrix}\begin{pmatrix} u(a) \\ u(b)\end{pmatrix} \\ &= (0,*)\begin{pmatrix} * \\ 0\end{pmatrix} = 0.\end{aligned}$$

Thus

$$\langle L_0(\lambda)\eta,\zeta\rangle = \langle \eta, L^{D^+}(\lambda)\zeta\rangle,$$

which proves $\zeta \in D(L_0^*(\lambda))$ and $L_0^*(\lambda)\zeta = L_0^+(\lambda)\zeta$.

Conversely, let $\zeta \in D(L_0^*(\lambda))$. We have to prove that $\zeta \in D(L_0^+(\lambda))$. Corollary 6.6.6 shows that $\zeta \in W_{p'}^n(a,b)$ and that there is a vector $d \in \mathbb{C}^n$ such that

$$H_a(\lambda)u(a) = W^a(\lambda)^{\mathsf{T}} d \quad \text{and} \quad H_b(\lambda)u(b) = -W^b(\lambda)^{\mathsf{T}} d. \tag{6.7.8}$$

From (6.7.5) we infer

$$\begin{aligned}&\widetilde{W}^a(\lambda)u(a) + \widetilde{W}^b(\lambda)u(b) \\ &= (0,I_n)\begin{pmatrix} \widetilde{C}(\lambda) & \widetilde{D}(\lambda) \\ \widetilde{W}^a(\lambda) & \widetilde{W}^b(\lambda)\end{pmatrix}\begin{pmatrix} u(a) \\ u(b)\end{pmatrix} \\ &= (0,I_n)\begin{pmatrix} W^a(\lambda)^{\mathsf{T}} & \widetilde{A}(\lambda) \\ W^b(\lambda)^{\mathsf{T}} & \widetilde{B}(\lambda)\end{pmatrix}^{-1}\begin{pmatrix} -H_a(\lambda) & 0 \\ 0 & H_b(\lambda)\end{pmatrix}\begin{pmatrix} u(a) \\ u(b)\end{pmatrix} \\ &= -(0,I_n)\begin{pmatrix} W^a(\lambda)^{\mathsf{T}} & \widetilde{A}(\lambda) \\ W^b(\lambda)^{\mathsf{T}} & \widetilde{B}(\lambda)\end{pmatrix}^{-1}\begin{pmatrix} W^a(\lambda)^{\mathsf{T}} & \widetilde{A}(\lambda) \\ W^b(\lambda)^{\mathsf{T}} & \widetilde{B}(\lambda)\end{pmatrix}\begin{pmatrix} I_n \\ 0\end{pmatrix} d \\ &= -(0,I_n)\begin{pmatrix} I_n \\ 0\end{pmatrix} d = 0,\end{aligned}$$

which proves that $\zeta \in D(L_0^+(\lambda))$.
ii) Since $L_p(a,b)$ is reflexive and $L_0(\lambda)$ is closed by Proposition 6.6.2, we have $L_0(\lambda)^{**} = L_0(\lambda)$, see [KA, Theorem III.5.29]. Therefore ii) follows from i). □

DEFINITION 6.7.4. Let $\eta \in H(\Omega, W_p^n(a,b))$ and $\mu \in \Omega$. η is called a *root function* of L_0 at μ if and only if $\eta(\mu) \neq 0$, $(L^D\eta)(\mu) = 0$ and $W^a(\mu)(\eta^{(i)}(a,\mu))_{i=0}^{n-1} + W^b(\mu)(\eta^{(i)}(b,\mu))_{i=0}^{n-1} = 0$. The minimum of the orders of the zero of $L^D\eta$ and $W^a(\eta^{(i)}(a,\cdot))_{i=0}^{n-1} + W^b(\eta^{(i)}(b,\cdot))_{i=0}^{n-1}$ at μ is called the *multiplicity* of η.

From $L^R\eta = W^a(\eta^{(i)}(a,\cdot))_{i=0}^{n-1} + W^b(\eta^{(i)}(b,\cdot))_{i=0}^{n-1}$ we obtain

PROPOSITION 6.7.5. *Let $\eta \in H(\Omega, W_p^n(a,b))$, $\mu \in \Omega$ and $\nu \in \mathbb{N}$. Then η is a root function of L_0 of multiplicity ν at μ if and only if η is a root function of L of multiplicity ν at μ.*

Canonical systems of root functions of L_0 are defined in the same way as for L. Hence a system of root functions is a canonical system of root functions of L_0 at μ if and only if it is a canonical system of root functions of L at μ.

The situation is different for $L_0^+ = L_0^*$ and L^*.

PROPOSITION 6.7.6. *Let $(\zeta, d) \in H(\Omega, L_{p'}(a,b) \times \mathbb{C}^n)$ be a root function of L^* of multiplicity ν at μ. We may assume that ζ is a polynomial of order $\leq \nu - 1$. Then $\zeta \in H(\Omega, W_{p'}^n(a,b))$, ζ is a root function of L_0^+ of multiplicity $\geq \nu$ at μ, and $d + \widetilde{C}u(a,\cdot) + \widetilde{D}u(b,\cdot)$ has a zero of order $\geq \nu$ at μ.*

Proof. By assumption

$$\zeta(\lambda) = \sum_{i=0}^{\nu-1} (\lambda - \mu)^i \zeta_i \quad (\lambda \in \mathbb{C}),$$

where $\zeta_i \in L_{p'}(a,b)$ $(i = 0, \ldots, \nu - 1)$. First we shall show that ζ_i belongs to $W_{p'}^n(a,b)$. For this, define

$$d_i := \tfrac{1}{i!}\left(\tfrac{\mathrm{d}^i}{\mathrm{d}\lambda^i} d\right)(\mu) \quad (i = 0, \ldots, \nu - 1).$$

Since (ζ, d) is a root function of L^* of multiplicity ν at μ, we have

$$L^*(\mu)(\zeta_i, d_i) = -\sum_{j=1}^{i} \tfrac{1}{j!}\left(\tfrac{\mathrm{d}^j}{\mathrm{d}\lambda^j} L^*\right)(\mu)(\zeta_{i-j}, d_{i-j}) \quad (i = 0, \ldots, \nu - 1).$$

Since the restriction of L^{R^*} to $\mathscr{D}'(a,b)$ is zero, we obtain

$$(6.7.9) \qquad \left(L^{D^*}(\mu)\zeta_i\right)_r = -\sum_{j=1}^{i} \tfrac{1}{j!}\left(\left(\tfrac{\mathrm{d}^j}{\mathrm{d}\lambda^j} L^{D^*}\right)(\mu)\zeta_{i-j}\right)_r.$$

For $i = 0$, (6.7.9) yields $\left(L^{D^*}(\mu)\zeta_0\right)_r = 0$ and hence $\zeta_0 \in W_{p'}^n(a,b)$ by (6.5.2) and Proposition 2.6.1. Now, for $i = 1$, the right hand side of (6.7.9) belongs to $L_{p'}(a,b)$, and Proposition 2.6.1 yields $\zeta_1 \in W_{p'}^n(a,b)$. Repeating this procedure we obtain $\zeta_i \in W^n_{p'}(a,b)$ for $i = 0,1,\dots,\nu-1$. As in the proof of Corollary 6.6.6 we obtain that

$$H_a u(a,\cdot) - W^{a\mathsf{T}} d \quad \text{and} \quad H_b u(b,\cdot) + W^{b\mathsf{T}} d \tag{6.7.10}$$

have a zero of order $\geq \nu$ at μ. From the representations (6.7.2) and (6.7.3) we infer that $L^{D+}\zeta$ has a zero of order $\geq \nu$ at μ. Hence ζ is a root function of L_0^+ of order $\geq \nu$ at μ. From (6.7.10) and the definition of H we obtain that

$$\begin{pmatrix} W^{a\mathsf{T}} & \widetilde{A} \\ W^{b\mathsf{T}} & \widetilde{B} \end{pmatrix} \begin{pmatrix} \widetilde{C} & \widetilde{D} \\ \widetilde{W}^a & \widetilde{W}^b \end{pmatrix} \begin{pmatrix} u(a,\cdot) \\ u(b,\cdot) \end{pmatrix} + \begin{pmatrix} W^{a\mathsf{T}} & \widetilde{A} \\ W^{b\mathsf{T}} & \widetilde{B} \end{pmatrix} \begin{pmatrix} d \\ 0 \end{pmatrix} \tag{6.7.11}$$

has a zero of order $\geq \nu$ at μ. Hence it follows in view of the invertibility of $\begin{pmatrix} W^{a\mathsf{T}} & \widetilde{A} \\ W^{b\mathsf{T}} & \widetilde{B} \end{pmatrix}$ that $\widetilde{C}u(a,\cdot) + \widetilde{D}u(b,\cdot) + d$ has a zero of order $\geq \nu$ at μ. □

PROPOSITION 6.7.7. *Let $\zeta \in H(\Omega, W^n_p(a,b))$ be a root function of L_0^+ of multiplicity ν at μ. Set $d := -\widetilde{C}u(a,\cdot) - \widetilde{D}u(b,\cdot)$. Then (ζ, d) is a root function of L^* of multiplicity $\geq \nu$ at μ.*

Proof. By assumption, $\widetilde{W}^a u(a,\cdot) + \widetilde{W}^b u(b,\cdot)$ has a zero of order $\geq \nu$ at μ. Hence the matrix function (6.7.11) has a zero of order $\geq \nu$ at μ. Now the assertion is clear because of (6.7.2), (6.7.3), (6.7.5) and (6.5.2). □

A canonical system of eigenfunctions and associated functions of the family of operators $L_0(\lambda)$ is defined by taking a canonical system of eigenfunctions and associated functions of the holomorphic boundary eigenvalue operator function L.

THEOREM 6.7.8. *We consider the families of operators $L_0(\lambda)$ and $L_0^+(\lambda)$ defined by* (6.6.1), (6.6.2) *and* (6.7.6), (6.7.7), *respectively. Assume that $\mu \in \sigma(L_0)$ and let $\{\eta_{i,h} : 1 \leq i \leq r, 0 \leq h \leq m_i - 1\}$ be a canonical system of eigenvectors and associated vectors of L_0 at μ. Then there is a canonical system of eigenvectors and associated vectors $\{\zeta_{i,h} : 1 \leq i \leq r, 0 \leq h \leq m_i - 1\}$ of L_0^+ at μ such that the principal part of the* GREEN'S *function $G(x,\xi,\cdot)$ at μ has the form*

$$\sum_{i=1}^{r} \sum_{j=0}^{m_i-1} (\cdot - \mu)^{j-m_i} \sum_{k=0}^{j} \eta_{i,k}(x) \zeta_{i,j-k}(\xi). \tag{6.7.12}$$

If W^a and W^b do not depend on λ, then the biorthogonal relationships

$$\sum_{k=0}^{m} \frac{1}{k!} \int_a^b \left(\frac{\partial^k}{\partial \lambda^k} \theta_{ih}\right)(x,\mu) \zeta_{j,m-k}(x)\,\mathrm{d}x = \delta_{ij}\,\delta_{m_i-h,m} \tag{6.7.13}$$

$(1 \leq h \leq m_i; 0 \leq m \leq m_j - 1; i,j = 1,\dots,r)$ hold, where

$$\theta_{ih} := L^D \sum_{l=0}^{m_i-1} (\cdot - \mu)^{l-h} \eta_{i,m}.$$

Proof. We set

$$\eta_i(\lambda) := \sum_{h=0}^{m_i-1} (\lambda-\mu)^h \eta_{i,h} \quad (i=1,\ldots,r).$$

$\{\eta_1,\ldots,\eta_r\}$ is a CSRF of L at μ by Propositions 1.6.2 and 6.7.5. By Theorem 1.5.4 there are polynomials $(\zeta_i,d_i):\mathbb{C}\to L_{p'}(a,b)\times\mathbb{C}^n$ of degree $< m_i$ such that $\{(\zeta_1,d_1),\ldots,(\zeta_r,d_r)\}$ is a CSRF of L^* at μ,

$$L^{-1} - \sum_{i=1}^{r} (\cdot-\mu)^{-m_i} \eta_i \otimes (\zeta_i,d_i) \tag{6.7.14}$$

is holomorphic at μ and the biorthogonal relationships

$$\frac{1}{m!}\frac{\mathrm{d}^m}{\mathrm{d}\lambda^m}\langle\tilde{\theta}_{ih},(\zeta_j,d_j)\rangle(\mu) = \delta_{ij}\delta_{m_i-h,m} \tag{6.7.15}$$

hold for $1\le h\le m_i$, $0\le m\le m_j-1$, $i,j=1,\ldots,r$, where we use the notation $\tilde{\theta}_{ih} := (\cdot-\mu)^{-h}L\eta_i$. By Proposition 6.7.6, $\zeta_1,\ldots,\zeta_r$ are root functions of L_0^+ at μ and $d_i(\mu) = -\widetilde{C}(\mu)u_i(a,\mu) - \widetilde{D}(\mu)u_i(b,\mu)$. Hence $\zeta_1(\mu),\ldots,\zeta_r(\mu)$ are linearly independent as $(\zeta_1,d_1)(\mu),\ldots,(\zeta_r,d_r)(\mu)$ are linearly independent. Since the multiplicities of a CSRF of L_0^+ at μ cannot exceed the multiplicities of a CSRF of L^* at μ by Proposition 6.7.7, $\{\zeta_1,\ldots,\zeta_r\}$ is a CSRF of L_0^+ at μ. We set

$$\zeta_i(\lambda) =: \sum_{h=0}^{m_i-1} (\lambda-\mu)^h \zeta_{i,h} \quad (i=1,\ldots,r)$$

and infer that $\{\zeta_{i,h} : 1\le i\le r, 0\le h\le m_i-1\}$ is a canonical system of eigenvectors and associated vectors of L_0^+ at μ. By Theorem 6.6.1 i) and (6.7.14) the principal part of L_0^{-1} at μ is equal to the principal part of

$$\sum_{i=1}^{r} (\cdot-\mu)^{-m_i}\eta_i\otimes\zeta_i$$

at μ, and Theorem 6.6.1 ii) yields that the principal part of $G(x,\xi,\cdot)$ at μ is

$$\begin{aligned}
&\sum_{i=1}^{r}\sum_{j=0}^{m_i-1} (\cdot-\mu)^{j-m_i}\frac{1}{j!}\frac{\mathrm{d}^j}{\mathrm{d}\lambda^j}\big(\eta_i(x,\cdot)\zeta_i(\xi,\cdot)\big)(\mu)\\
&=\sum_{i=1}^{r}\sum_{j=0}^{m_i-1} (\cdot-\mu)^{j-m_i}\sum_{k=0}^{j}\eta_{i,k}(x)\zeta_{i,j-k}(\xi).
\end{aligned}$$

If W^a and W^b are constant, then $L^R\eta_i$ is a polynomial of degree $\le m_i-1$ and has a zero of order $\ge m_i$ at μ. Hence $L^R\eta_i=0$ for $i=1,\ldots,r$. Thus $\tilde{\theta}_{ih}=(\theta_{ih},0)$, and (6.7.15) leads to (6.7.13). □

6.8. Notes

Mostly, the operators associated with boundary eigenvalue problems are considered as operators from $L_p(a,b)$ to $L_p(a,b)$. In that case, it it sufficient to consider somewhat weaker conditions on the coefficients, see [NA2, Chapter V]. However, in general it is much more advantageous to have bounded operators, and we are therefore going to use the operator $L : W_p^n(a,b) \to L_p(a,b) \times \mathbf{C^n}$ in subsequent chapters. An important advantage of this approach is the fact that the adjoint operator $L^*(\lambda)$ is defined on the whole space $L_{p'}(a,b) \times \mathbb{C}^n$ and is a bounded operator with values in some space of distributions. As a consequence, the associated adjoint boundary eigenvalue problem is defined without any restrictions. This implies that the eigenvectors and associated vectors of the adjoint problem are always defined. The associated vectors may not belong to the domain of the classical adjoint problem. This explains why in the classical approach problems containing associated vectors are mostly disallowed.

Adjoint boundary conditions for two-point boundary eigenvalue problems were introduced by G. D. Birkhoff in [BI2]. G. FROBENIUS has shown in [FRO] that the adjoint differential expression A' of a differential expression of the form $A = \sum_{j=0}^{n} A_j D^j$ is uniquely determined by the identity $vA(u) - uA'(v) = DA(u,v)$, where $A(u,v) = \sum_{j,k}^{n} (-1)^k \left(D^j u\right) D^k \left(A_{j+k+1} v\right)$.

Chapter VII

REGULAR BOUNDARY EIGENVALUE PROBLEMS FOR n-TH ORDER EQUATIONS

This chapter deals with eigenfunction expansions for regular boundary eigenvalue problems for n-th order ordinary differential equations. The coefficients in the differential equation as well as in the boundary conditions depend polynomially on the eigenvalue parameter λ. As in the fifth chapter, the boundary conditions are allowed to contain countably many interior points and also an integral term. For such boundary eigenvalue problems the notions Birkhoff regularity and Stone regularity are defined in terms of the corresponding notions introduced in the fourth chapter for boundary eigenvalue problems for first order $n \times n$ differential systems with asymptotically linear parameter dependence (Definitions 7.3.1 and 7.6.1). To this end, in a first step the n-th order differential equation depending polynomially on the eigenvalue parameter λ is transformed to a first order $n \times n$ differential system which is asymptotically linear in λ and has a leading matrix in diagonal form satisfying the specific assumptions in the fourth chapter (Theorem 7.2.4). In a second step, the boundary conditions, which depend polynomially on λ, are transformed to conditions which are asymptotically constant as λ tends to infinity. An efficient method to check Birkhoff regularity of boundary eigenvalue problems for n-th order differential equations is presented (Theorems 7.3.2 and 7.3.3).

Under the assumption that the endpoints of the underlying interval are no accumulation points of the interior points of the boundary conditions it is shown for Birkhoff regular boundary eigenvalue problems in the case $1 < p < \infty$ that functions in $L_p(a,b)$ are expandable into series of corresponding eigenfunctions and associated functions. These series are L_p-convergent (Theorem 7.4.3). With respect to uniform convergence, i. e., if $p = \infty$, continuous functions which are of bounded variation and fulfil certain boundary conditions can be expanded (Theorem 7.4.4). Eigenfunction expansions are also established for Stone regular boundary eigenvalue problems. It is shown that for $1 < p \leq \infty$ these expansions converge to the given function if this function is smooth enough, i. e., belongs to some Sobolev space of sufficiently high order, and fulfils certain boundary conditions (Theorems 7.6.5 and 7.6.6). These boundary conditions are defined by some iterative procedure in terms of the coefficients of the given differential equation and of the boundary conditions. The convergence proofs of these eigenfunction expansions are based on the contour integral method and make use of convergence results from the fourth chapter concerning certain sequences of contour integrals of the resolvent of the transformed boundary eigenvalue problem.

7.1. General assumptions

Let $-\infty < a < b < \infty$, $1 \le p \le \infty$, $n \in \mathbb{N}$, $n \ge 2$, and let $a_j \in [a,b]$ $(j \in \mathbb{N})$ such that $a_j \ne a_k$ $(j \ne k)$, $a_0 = a$, and $a_1 = b$. Let

(7.1.1)
$$p_i(\cdot,\lambda) = \sum_{j=0}^{n-i} \lambda^j \pi_{n-i,j} \quad (i = 0,\dots,n-1),$$

where $\pi_{n-i,j} \in L_p(a,b)$ $(i = 0,\dots,n-1,\ j = 0,\dots,n-i)$. We assume $\pi_{n-i,n-i} \ne 0$ for some $i \in \{0,\dots,n-1\}$. Let w_{ki} $(i,k = 1,\dots,n)$ be polynomials in λ with coefficients in $L_1(a,b)$ and $w_{ki}^{(j)}$ $(j \in \mathbb{N};\ i,k = 1,\dots,n)$ be polynomials in λ with complex coefficents. Suppose that

$$\sum_{j=0}^{\infty} \sup_{|\lambda| \le r} |w_{ki}^{(j)}(\lambda)| < \infty$$

for all $r > 0$ and $i,k = 1,\dots,n$.

For $\lambda \in \mathbb{C}$ and $\eta \in W_p^n(a,b)$ we consider the boundary eigenvalue problem

(7.1.2)
$$\eta^{(n)} + \sum_{i=0}^{n-1} p_i(\cdot,\lambda)\eta^{(i)} = 0,$$

(7.1.3)
$$\sum_{i=1}^{n}\sum_{j=0}^{\infty} w_{ki}^{(j)}(\lambda)\eta^{(i-1)}(a_j) + \sum_{i=1}^{n}\int_a^b w_{ki}(\xi,\lambda)\eta^{(i-1)}(\xi)\,\mathrm{d}\xi = 0 \quad (k = 1,\dots,n).$$

The function π defined by

(7.1.4)
$$\pi(\cdot,\rho) := \rho^n + \sum_{i=0}^{n-1} \rho^i \pi_{n-i,n-i} \quad (\rho \in \mathbb{C})$$

is called the *characteristic function* of the differential equation (7.1.2). Together with the boundary eigenvalue problem (7.1.2), (7.1.3) we consider the operators

(7.1.5)
$$L^D(\lambda)\eta := \eta^{(n)} + \sum_{i=0}^{n-1} p_i(\cdot,\lambda)\eta^{(i)},$$

(7.1.6)
$$L^R(\lambda)\eta := \Big(\sum_{i=1}^{n}\sum_{j=0}^{\infty} w_{ki}^{(j)}(\lambda)\eta^{(i-1)}(a_j) + \sum_{i=1}^{n}\int_a^b w_{ki}(\xi,\lambda)\eta^{(i-1)}(\xi)\,\mathrm{d}\xi\Big)_{k=1}^{n},$$

where $\lambda \in \mathbb{C}$ and $\eta \in W_p^n(a,b)$, and

$$L := (L^D, L^R). \tag{7.1.7}$$

From Lemma 6.1.1 and Proposition 6.3.5 we infer

PROPOSITION 7.1.1. $L \in H(\mathbb{C}, L(W_p^n(a,b), L_p(a,b) \times \mathbb{C}^n))$.

Together with the n-th order differential equation (7.1.2) we consider the associated first order system $T^D(\lambda)y = 0$, where T^D is given by (6.1.3), (6.1.4). We shall assume that there are a matrix function $C(\cdot,\lambda) \in M_n(W_p^1(a,b))$ depending polynomially on λ and a positive real number γ such that

$$C(\cdot,\lambda) \text{ is invertible in } M_n(W_p^1(a,b)) \text{ if } |\lambda| \geq \gamma \tag{7.1.8}$$

and such that the equation

$$C^{-1}(\cdot,\lambda)T^D(\lambda)C(\cdot,\lambda)y = y' - \widetilde{A}(\cdot,\lambda)y =: \widetilde{T}^D(\lambda)y \tag{7.1.9}$$

holds for $|\lambda| \geq \gamma$ and $y \in (W_p^1(a,b))^n$, where

$$\widetilde{A}(\cdot,\lambda) = \lambda A_1 + A_0 + \lambda^{-1}A^0(\cdot,\lambda) \quad (|\lambda| \geq \gamma) \tag{7.1.10}$$

fulfils the assumptions made in Section 4.1.

We shall also consider the following sharpened form of (7.1.10): For some $\kappa \in \mathbb{N}$ let

$$\widetilde{A}(\cdot,\lambda) = \sum_{j=-1}^{\kappa} \lambda^{-j}A_{-j} + \lambda^{-\kappa-1}A^{\kappa}(\cdot,\lambda) \quad (|\lambda| \geq \gamma), \tag{7.1.11}$$

where

$$A_1 \in M_n(W_p^{\kappa}(a,b)), \tag{7.1.12}$$

$$A_{-j} \in M_n(W_p^{\kappa-j}(a,b)) \quad (j = 0,\dots,k), \tag{7.1.13}$$

$$A^{\kappa}(\cdot,\lambda) \in M_n(L_p(a,b)) \text{ for } |\lambda| \geq \gamma \text{ and is bounded in } M_n(L_p(a,b)) \text{ as } \lambda \to \infty. \tag{7.1.14}$$

Condition (7.1.8) is fulfilled if for each $x \in [a,b]$ and $|\lambda| \geq \gamma$ the matrix $C(x,\lambda)$ is invertible: Since $C(\cdot,\lambda)$ is continuous in $[a,b]$, $C^{-1}(\cdot,\lambda) \in M_n(L_\infty(a,b))$ if $|\lambda| \geq \gamma$. Proposition 2.5.8 yields $C^{-1}(\cdot,\lambda) \in M_n(W_p^1(a,b))$ for $|\lambda| \geq \gamma$.

PROPOSITION 7.1.2. *Suppose that C fulfils (7.1.8) and let $r \in \mathbb{N} \setminus \{0\}$ such that C is a matrix polynomial with coefficients in $M_n(W_p^r(a,b))$. Then we have*

$$\det C^{-1}(\cdot,\lambda) = \lambda^{\tilde{q}}\left(c_0 + \tfrac{1}{\lambda}c_1(\cdot,\lambda)\right)$$

for $|\lambda| \geq \gamma$, where $\tilde{q} \in \mathbb{Z}$, c_0 is an invertible element in $W_p^r(a,b)$, and c_1 is an asymptotic polynomial in $W_p^r(a,b)$ of arbitrary order. There is a number $\hat{q} \in \mathbb{Z}$ such that

$$C^{-1}(\cdot,\lambda) = \lambda^{\hat{q}}C_0(\cdot,\lambda),$$

where C_0 is an asymptotic polynomial in $M_n(W_p^r(a,b))$ of arbitrary order.

Proof. Since C is a polynomial with coefficients in $M_n(W_p^r(a,b))$, there is a non-negative integer s such that

$$\det C(x,\lambda) = \sum_{j=0}^{s} \gamma_j(x)\lambda^j$$

for $x \in [a,b]$ and $\lambda \in \mathbb{C}$, where $\gamma_j \in W_p^r(a,b)$ for $j \in \{0,\dots,s\}$ and $\gamma_s \neq 0$. If $\gamma_s(x) \neq 0$ for all $x \in [a,b]$, then there exists a $\delta > 0$ such that $|\gamma_s(x)| \geq \delta$ for all $x \in [a,b]$ since γ_s is continuous. Then the NEUMANN series expansion

$$\lambda^s(\det C(x,\lambda))^{-1} = \frac{1}{\gamma_s(x)} \sum_{k=0}^{\infty} (-1)^k \left(\sum_{j=0}^{s-1} \frac{\gamma_j(x)}{\gamma_s(x)} \lambda^{j-s} \right)^k$$

holds if λ is sufficiently large. From Proposition 2.5.8 we know that γ_s^{-1} belongs to $W_p^r(a,b)$. Hence $\lambda^s(\det C(\cdot,\lambda))^{-1}$ is an asymptotic polynomial in $W_p^r(a,b)$ of arbitrary order which tends to γ_s^{-1} as $\lambda \to \infty$. This proves the first assertion; the second assertion follows by CRAMER'S rule.

Assume that γ_s has a zero in $[a,b]$. Since γ_s is continuous and not identically zero in $[a,b]$, there is a sequence $(x_\nu)_0^\infty$ in (a,b) converging to some $z \in [a,b]$ such that $\gamma_s(x_\nu) \neq 0$ $(\nu \in \mathbb{N})$ and $\gamma_s(z) = 0$. Since $\det C(z,\lambda) \neq 0$ for sufficiently large λ, there is a number $j_0 \in \{0,\dots,s-1\}$ such that $\gamma_{j_0}(z) \neq 0$ and $\gamma_j(z) = 0$ for $j \in \{j_0+1,\dots,s\}$. We define

$$\rho_0 := \begin{cases} \max\left\{ \dfrac{|\gamma_j(x)|}{|\gamma_{j_0}(z)|} : x \in [a,b],\ j \in \{0,\dots,j_0-1\} \right\} & \text{if } j_0 > 0, \\ 0 \qquad \text{if } j_0 = 0, \end{cases}$$

and $\rho := \max\{4j_0\rho_0, \gamma, 1\}$. Since $\gamma_j(x_\nu) \to 0$ as $\nu \to \infty$ for each $j \in \{j_0+1,\dots,s\}$ and $\gamma_{j_0}(x_\nu) \to \gamma_{j_0}(z) \neq 0$ as $\nu \to \infty$, there is an integer $\nu \in \mathbb{N}$ such that

$$|\gamma_{j_0}(x_\nu)| \geq \frac{1}{2}|\gamma_{j_0}(z)| \qquad \text{and} \qquad \sum_{j=j_0+1}^{s} |\gamma_j(x_\nu)|\rho^{j-j_0} < \frac{|\gamma_{j_0}(x_\nu)|}{4}.$$

From the definitions of ρ_0 and ρ we infer that

$$\left| \sum_{\substack{j=0 \\ j \neq j_0}}^{s} \gamma_j(x_\nu)\lambda^j \right| < |\gamma_{j_0}(x_\nu)\lambda^{j_0}|$$

for $|\lambda| = \rho$. Hence ROUCHÉ'S theorem yields that $\det C(x_\nu,\lambda)$ has exactly j_0 zeros in the open disk with centre 0 and radius ρ. Since $j_0 < s$ and $\gamma_s(x_\nu) \neq 0$, there is at least one λ with $|\lambda| > \rho$ and $\det C(x_\nu,\lambda) = 0$. This contradicts the assumption (7.1.8), which implies that $C(x_\nu,\lambda)$ is invertible for $|\lambda| > \rho \geq \gamma$. □

7.2. Asymptotic linearizations

The most crucial assumption of those in Section 7.1 is that A_1 is a diagonal matrix fulfilling the conditions (4.1.3), (4.1.4) and (4.1.5). In the sequel we are looking for necessary and sufficient conditions in terms of the problem (7.1.2) to fulfil this assumption. First we look for an "asymptotic linearization" of A with respect to λ. The easiest way to do so is by multiplying A with a diagonal matrix

$$C_0(\lambda) = \operatorname{diag}(\lambda^{\nu_1}, \dots, \lambda^{\nu_n}) \quad (\nu_1, \dots, \nu_n \in \mathbb{Z}) \tag{7.2.1}$$

from the right and with its inverse from the left. We have

$$C_0^{-1}(\lambda)A(\cdot,\lambda)C_0(\lambda) = \left(\delta_{i,j-1}\lambda^{\nu_j-\nu_i} - \delta_{i,n}p_{j-1}(\cdot,\lambda)\lambda^{\nu_j-\nu_n}\right)_{i,j=1}^n. \tag{7.2.2}$$

Let $\deg p_j$ be the degree of the polynomial p_j with respect to λ. Then we must have

$$\nu_{i+1} - \nu_i \le 1 \quad \text{and} \quad \deg p_j + \nu_{j+1} - \nu_n \le 1 \tag{7.2.3}$$

for $i = 1, \dots, n-1$ and $j = 0, \dots, n-1$. For $j = 0, \dots, n-1$ we infer that

$$\deg p_j \le 1 + \nu_n - \nu_{j+1} = 1 + \sum_{i=j+1}^{n-1} (\nu_{i+1} - \nu_i) \le n - j. \tag{7.2.4}$$

This is the reason for assumption (7.1.1).

PROPOSITION 7.2.1. *Let $n \in \mathbb{N}\setminus\{0\}$, $b_1, \dots, b_n \in \mathbb{C}$,*

$$B_k := (\delta_{i,j-1} - \delta_{i,n}b_{n-j+1})_{i,j=n+1-k}^n \quad (k = 1, \dots, n)$$

and

$$a_k(\rho) := \det(\rho - B_k) \quad (k = 1, \dots, n).$$

We assert:

i) *For $\rho \in \mathbb{C}$ and $k = 2, \dots, n$ we have*

$$a_k(\rho) = \rho a_{k-1}(\rho) + b_k.$$

ii) *For $\rho \in \mathbb{C}$ and $k = 1, \dots, n$ we have*

$$a_k(\rho) = \rho^k + \sum_{j=1}^k \rho^{k-j} b_j.$$

Proof. i) Let $k \in \{2, \dots, n\}$. Expanding $a_k(\rho)$ with respect to the first column we obtain

$$\begin{aligned} a_k(\rho) =& \rho \det(\rho\delta_{i,j} - \delta_{i,j-1} + \delta_{i,n}b_{n-j+1})_{i,j=n+2-k}^n \\ &+ (-1)^{k-1}b_k \det(\rho\delta_{i-1,j} - \delta_{i-1,j-1})_{i,j=n+2-k}^n \\ =& \rho a_{k-1}(\rho) + b_k. \end{aligned}$$

ii) For $k = 1$ we have

$$a_1(\rho) = \rho + b_1.$$

Assume that the assertion holds for some $k \in \{1, \ldots, n-1\}$. With the aid of i) we obtain

$$\begin{aligned} a_{k+1}(\rho) &= \rho a_k(\rho) + b_{k+1} \\ &= \rho^{k+1} + \sum_{j=1}^{k} \rho^{k-j+1} b_j + b_{k+1} \\ &= \rho^{k+1} + \sum_{j=1}^{k+1} \rho^{(k+1)-j} b_j. \quad \square \end{aligned}$$

PROPOSITION 7.2.2. i) *Assume that* $A(\cdot, \lambda)$ *has an asymptotic linearization, i. e., that there is an invertible matrix function* $C_0(\lambda)$, $|\lambda| \geq \gamma > 0$, *such that*

$$C_0^{-1}(\lambda) A(\cdot, \lambda) C_0(\lambda) = \lambda \widehat{A} + O(1). \tag{7.2.5}$$

Then

$$\det(\rho - \widehat{A}) = \pi(\cdot, \rho), \tag{7.2.6}$$

where π *is the characteristic function given by* (7.1.4).
ii) *Additionally, let* C_1 *be invertible in* $M_n(W_p^1(a,b))$ *and set*

$$C(x, \lambda) = C_0(\lambda) C_1(x) \quad (x \in [a,b],\ |\lambda| \geq \gamma).$$

Then the matrix function $\widetilde{A}$ *given by* (7.1.9) *satisfies*

$$\widetilde{A}(\cdot, \lambda) = \lambda A_1 + \widetilde{A}_0(\cdot, \lambda), \tag{7.2.7}$$

where

$$A_1 = C_1^{-1} \widehat{A} C_1 \in M_n(L_p(a,b)), \tag{7.2.8}$$

$\widetilde{A}_0 = O(1)$ *in* $M_n(L_p(a,b))$ *as* $\lambda \to \infty$, *and*

$$\pi(\cdot, \rho) = \det(\rho - A_1). \tag{7.2.9}$$

iii) *Additionally to* i) *and* ii) *we suppose that* C_0 *and* A_1 *are diagonal matrices,*

$$A_1 = \operatorname{diag}(0, \ldots, 0, r_1, \ldots, r_l), \tag{7.2.10}$$

where $r_1, \ldots, r_l$ *are not identically zero. Then*

$$\pi(x, \rho) = \rho^{n_0} \prod_{i=1}^{l} (\rho - r_j(x)) \quad (x \in (a,b)), \tag{7.2.11}$$

where $n_0 = n - l$. *Furthermore,* $l \geq 1$, *and* $r_j(x) \neq r_m(x)$ *for all* $j, m \in \{1, \ldots, l\}$ *and* $x \in (a,b)$ *such that* $r_j(x) \neq 0$, $r_m(x) \neq 0$, *and* $j \neq m$.

Proof. i) By Proposition 7.2.1 we have

$$\begin{aligned}\lambda^{-n}\det(\rho\lambda - A(\cdot,\lambda)) &= \rho^n + \sum_{i=0}^{n-1}\rho^i\lambda^{i-n}p_i(\cdot,\lambda)\\ &= \rho^n + \sum_{i=0}^{n}\rho^i\left(\pi_{n-i,n-i} + \frac{1}{\lambda}O(1)\right).\end{aligned}$$

On the other hand, (7.2.5) yields

$$\begin{aligned}\lambda^{-n}\det(\rho\lambda - A(\cdot,\lambda)) &= \det\left(\rho - \widehat{A} + O\left(\frac{1}{\lambda}\right)\right)\\ &= \det(\rho - \widehat{A}) + \frac{1}{\lambda}\sum_{i=0}^{n-1}\rho^i O(1).\end{aligned}$$

For $\lambda\to\infty$ we obtain $\det(\rho - \widehat{A}) = \pi(\cdot,\rho)$.
ii) In view of LEIBNIZ' rule (2.3.1) we have for $y\in(W_p^1(a,b))^n$ that

$$\begin{aligned}\widetilde{T}^D(\lambda)y &= C_1^{-1}C_0(\lambda)^{-1}T^D(\lambda)\big(C_0(\lambda)C_1y\big)\\ &= C_1^{-1}(C_1y)' - C_1^{-1}\big(\lambda\widehat{A} + O(1)\big)C_1y\\ &= y' - \lambda C_1^{-1}\widehat{A}C_1y + O(1)y,\end{aligned} \tag{7.2.12}$$

whence the representation of $\widetilde{A}$ holds, and (7.2.9) follows from (7.2.6).
iii) Obviously, by a suitable choice of C_1, we can write A_1 in the form (7.2.10), where $r_1,\dots,r_l$ are not identically zero. Then (7.2.11) follows from (7.2.9). Since we suppose that $\pi_{n-i,n-i}\neq 0$ for some $i\in\{1,\dots,n\}$, not all the r_j can be zero, i. e., we must have $l\geq 1$. In the matrix A given by (6.1.4), the $(n-1)\times(n-1)$ submatrix in the upper left corner is a triangular matrix with zeros in the diagonal. By (7.2.5) this also holds for $\widehat{A}$ since we suppose that $C_0(\lambda)$ is a diagonal matrix. Therefore the $(n-1)\times(n-1)$ submatrix of $\rho-\widehat{A}$ in the upper left corner is invertible for $\rho\neq 0$ since it is a triangular matrix with diagonal elements ρ. Hence the rank of $\rho-\widehat{A}$ is at least $n-1$ if $\rho\neq 0$. In view of (7.2.8) we infer that $r_j(x)-A_1$ has rank $n-1$ whenever $r_j(x)\neq 0$. □

PROPOSITION 7.2.3. *Let $n_0\in\{0,\dots,n-1\}$, $l := n-n_0$, and suppose that*

$$\pi(\cdot,\rho) = \rho^{n_0}\pi_l(\cdot,\rho),$$

where

$$\pi_l(\cdot,\rho) = \rho^l + \sum_{j=1}^{l}\rho^{l-j}\pi_{j,j}. \tag{7.2.13}$$

Suppose that for all $x \in [a,b]$ *the roots of* $\pi_l(x,\rho) = 0$ *are simple and nonzero and that there is* $\kappa \in \mathbb{N} \setminus \{0\}$ *such that* $\pi_{1,1},\dots,\pi_{l,l} \in W_p^\kappa(a,b)$. *Then there are* $r_1,\dots,r_l \in W_p^\kappa(a,b)$ *such that*

$$\pi_l(x,\rho) = \prod_{j=1}^{l} (\rho - r_j(x)) \tag{7.2.14}$$

holds for all $x \in [a,b]$ *and* $\rho \in \mathbb{C}$. *In addition, we have that* $r_j^{-1} \in W_p^\kappa(a,b)$ *for* $j = 1,\dots,l$.

Proof. For $x \in [a,b]$ let $\alpha_{x,1},\dots,\alpha_{x,l}$ be the roots of $\pi_l(x,\rho) = 0$. We set

$$\varepsilon_x := \tfrac{1}{2}\min\{|\alpha_{x,j} - \alpha_{x,k}|, |\alpha_{x,j}| : j,k = 1,\dots,l; j \neq k\} > 0$$

and

$$\gamma_x := \bigcup_{j=1}^{l} \{\rho \in \mathbb{C} : |\rho - \alpha_{x,j}| = \varepsilon_x\}.$$

Then $\pi_l(x,\rho) \neq 0$ for all $\rho \in \gamma_x$. Since γ_x is compact, there is a $\delta_x > 0$ such that $\pi_l(\xi,\rho) \neq 0$ for all $\xi \in [x - \delta_x, x + \delta_x] \cap [a,b]$ and $\rho \in \gamma_x$.

Let $x_0 \in [a,b]$. For $x \in [x_0 - \delta_{x_0}, x_0 + \delta_{x_0}] \cap [a,b]$ and $j = 1,\dots,l$ we define

$$\beta_{x_0,j}(x) := \frac{1}{2\pi i} \oint_{|\rho - \alpha_{x_0,j}| = \varepsilon_{x_0}} \rho \, \frac{\frac{\partial \pi_l}{\partial \rho}(x,\rho)}{\pi_l(x,\rho)} \, \mathrm{d}\rho. \tag{7.2.15}$$

The functions $\beta_{x_0,j}$ are continuous since the integrand depends continuously on x. Choosing δ_{x_0} sufficiently small we may assume that $|\beta_{x_0,j}(x) - \beta_{x_0,j}(x_0)| < \varepsilon_{x_0}$ for $j = 1,\dots,l$ and $x \in [x_0 - \delta_{x_0}, x_0 + \delta_{x_0}] \cap [a,b]$. From $\pi_l(x,\rho) = \prod\limits_{k=1}^{l} (\rho - \alpha_{x,k})$ we infer

$$\sum_{k=1}^{l} \frac{1}{\rho - \alpha_{x,k}} = \frac{\frac{\partial \pi_l}{\partial \rho}(x,\rho)}{\pi_l(x,\rho)}.$$

The residue theorem yields

$$\beta_{x_0,j}(x) = \sum_{\substack{k=1 \\ |\alpha_{x,k} - \alpha_{x_0,j}| < \varepsilon_{x_0}}}^{l} \alpha_{x,k} \tag{7.2.16}$$

for $j = 1,\dots,l$ and $x \in [x_0 - \delta_{x_0}, x_0 + \delta_{x_0}] \cap [a,b]$. For $x = x_0$ and $j = 1,\dots,l$ we obtain $\beta_{x_0,j}(x_0) = \alpha_{x_0,j}$. From

$$|\beta_{x_0,j}(x)| \geq |\beta_{x_0,j}(x_0)| - |\beta_{x_0,j}(x_0) - \beta_{x_0,j}(x)| > 2\varepsilon_{x_0} - \varepsilon_{x_0} > 0$$

for $j = 1,\dots,l$ and $x \in [x_0 - \delta_{x_0}, x_0 + \delta_{x_0}] \cap [a,b]$ and (7.2.16) we infer for each $j \in \{1,\dots,l\}$ and $x \in [x_0 - \delta_{x_0}, x_0 + \delta_{x_0}] \cap [a,b]$ that there is at least one $k \in \{1,\dots,l\}$ such that $|\alpha_{x,k} - \alpha_{x_0,j}| < \varepsilon_{x_0}$. Since the disks $\{\rho \in \mathbb{C} : |\rho - \alpha_{x_0,j}| < \varepsilon_{x_0}\}$ are pairwise disjoint, we obtain $\{\beta_{x_0,1}(x),\ldots,\beta_{x_0,l}(x)\} = \{\alpha_{x,1},\ldots,\alpha_{x,l}\}$. Hence, for all $x \in [x_0-\delta_{x_0}, x_0+\delta_{x_0}] \cap [a,b]$, we obtain $\pi_l(x,\rho) = \prod_{j=1}^{l}(\rho - \beta_{x_0,j}(x))$.

Let $J_1, J_2 \subset [a,b]$ be intervals, $\alpha_1,\ldots,\alpha_l \in C(J_1)$ and $\beta_1,\ldots,\beta_l \in C(J_2)$. Assume that there is a $c \in J_1 \cap J_2$ such that, for all $j = 1,\ldots,l$, $\alpha_j(c) = \beta_j(c) = \alpha_{c,j}$, $\pi_l(x,\alpha_j(x)) = 0$ $(x \in J_1)$, $\pi_l(x,\beta_j(x)) = 0$ $(x \in J_2)$, and $\alpha_{c,j} \neq \alpha_{c,k}$ if $k = 1,\ldots,l$ and $k \neq j$. We set

$$\begin{aligned} I := &\{x \in J_1 \cap J_2 : \exists\, j \in \{1,\ldots,l\}\ \alpha_j(x) \neq \beta_j(x)\} \\ &\cup \{x \in J_1 \cap J_2 : \pi_l(x,\rho) \neq \prod_{j=1}^{l}(\rho - \alpha_j(x))\}. \end{aligned}$$

We shall prove that $I = \emptyset$. Assume that $I \cap [a,c] \neq \emptyset$. The case $I \cap [c,b] \neq \emptyset$ can be treated analogously. Because of the continuity of π_l and the α_j and β_j, the set I and thus also the set $I \cap [a,c)$ is an open subset of $J_1 \cap J_2$. Furthermore $c \notin I$. Hence $x_0 := \sup(I \cap [a,c)) \notin I$. Since the roots of $\pi_l(x_0,\rho) = 0$ are pairwise different, we obtain $\beta_j(x_0) = \alpha_j(x_0) \neq \alpha_k(x_0) = \beta_k(x_0)$ for $j \neq k$. Because of the continuity of the α_j and β_j we obtain for all $x \in I \cap [a,x_0)$, sufficiently close to x_0, that $\beta_j(x) \neq \alpha_k(x)$, $\alpha_j(x) \neq \alpha_k(x)$, $\beta_j(x) \neq \beta_k(x)$ for $j \neq k$. Since the $\alpha_j(x)$ and $\beta_j(x)$ are roots of $\pi_l(x,\rho)$, we obtain that the sets $\{\alpha_1(x),\ldots,\alpha_l(x)\}$ and $\{\beta_1(x),\ldots,\beta_l(x)\}$ are contained in $\{\alpha_{x,1},\ldots,\alpha_{x,l}\}$. Since each of these three sets consists of l elements, these sets are equal. Hence $x \notin I$. This is a contradiction, and $I = \emptyset$ is proved.

Thus there are a maximal subinterval $J \subset [a,b]$ such that $a \in J$ and unique continuous functions $r_1,\ldots,r_l$ on J such that $r_j(a) = \alpha_{a,j}$ for $j = 1,\ldots,l$ and $\pi_l(x,\rho) = \prod_{j=1}^{l}(\rho - r_j(x))$. Let $x_0 := \sup J$ and choose $c \in J \cap [x_0 - \delta_{x_0}, x_0]$. Since

$$\prod_{j=1}^{l}(\rho - r_j(c)) = \pi_l(x,\rho) = \prod_{j=1}^{l}(\rho - \alpha_{x_0,j}(c)),$$

we may choose the roots $\alpha_{x_0,1},\ldots,\alpha_{x_0,l}$ in such a way that $r_j(c) = \beta_{x_0,j}(c)$ for $j = 1,\ldots,l$. But then the above considerations yield that

$$\hat{r}_j(x) := \begin{cases} r_j(x) & \text{if } x \in J, \\ \beta_{x_0,j}(x) & \text{if } x \in [x_0-\delta_{x_0}, x_0+\delta_{x_0}] \cap [a,b] \end{cases} \qquad (j = 1,\ldots,l)$$

defines unique functions $\hat{r}_j \in C\big([a, x_0+\delta_{x_0}] \cap [a,b]\big)$ satisfying $\hat{r}_j(a) = \alpha_{a,j}$ and $\pi_l(x,\rho) = \prod_{j=1}^{l}(\rho - \hat{r}_j(x))$ for $x \in [a, x_0+\delta_{x_0}] \cap [a,b]$. This proves that $[a, x_0+\delta_{x_0}] \cap [a,b] \subset J \subset [a,x_0]$. Hence $x_0 = b$ and $J = [a,b]$.

We still have to prove $r_j \in W_p^\kappa(a,b)$ and $r_j^{-1} \in W_p^\kappa(a,b)$ for $j = 1,\dots,l$. For this let $x_0 \in [a,b]$. Set $I := (x_0 - \delta_{x_0}, x_0 + \delta_{x_0}) \cap (a,b)$. From Proposition 2.5.8 we infer for $\rho \in \partial K_{\varepsilon_{x_0}}(r_j(x_0))$ that $\pi_l(\cdot,\rho)|_I^{-1} \in W_p^\kappa(I)$. Since $W_p^\kappa(I)$ is identified with a subspace of $L(W_p^\kappa(I))$ by Proposition 2.3.3 and since the inverse of an element in $W_p^\kappa(I)$ is the inverse of the corresponding multipication operator, $\pi_l(\cdot,\rho)|_I^{-1}$ also depends holomorphically on ρ in a neighbourhood of $\partial K_{\varepsilon_{x_0}}(r_j(x_0))$ by Proposition 1.2.5. Therefore $\frac{\partial \pi_l/\partial \rho}{\pi_l}$ is a continuous mapping from $\partial K_{\varepsilon_{x_0}}(r_j(x_0))$ to $W_p^\kappa(I)$. In view of (7.2.15) this proves that r_j belongs (locally) to $W_p^\kappa(a,b)$. Finally, we have that r_j is continuous and $r_j(x) \neq 0$ for all $x \in [a,b]$. This proves that r_j^{-1} is bounded. An application of Proposition 2.5.8 completes the proof. □

THEOREM 7.2.4. *Let $l \in \{1,\dots,n\}$ be such that $\pi_{l,l} \neq 0$ and $\pi_{i,i} = 0$ for $i = l+1,\dots,n$. Suppose that $\pi_{l,l} \in L_\infty(a,b)$. Then there is a matrix function*

$$C(x,\lambda) = \operatorname{diag}(\lambda^{\nu_1},\dots,\lambda^{\nu_n})C_1(x)$$

with $\nu_1,\dots,\nu_n \in \mathbb{Z}$ and $C_1 \in M_n(W_p^1(a,b))$ such that $\widetilde{A}(\cdot,\lambda)$ given by (7.1.9) *has the form* (7.1.10), *where*

$$A_1 = \operatorname{diag}(0,\dots,0,r_1,\dots,r_l)$$

and $r_j^{-1} \in L_\infty(a,b)$ for $j = 1,\dots,l$, if and only if the following four conditions hold:
i) $\pi_{l,l}^{-1} \in L_\infty(a,b)$;

ii) $$p_i(\cdot,\lambda) = \sum_{j=0}^{l} \lambda^j \pi_{n-i,j} \qquad (i = 0,\dots,n_0-1);$$

iii) $\pi_{i,i} \in W_p^1(a,b)$ *for* $i = 1,\dots,l$
or
$l = 1$ *and* $\dfrac{\pi_{n-i+1,1}}{\pi_{1,1}} \in W_p^1(a,b)$ *for* $i = 1,\dots,n-1$;
iv) *The zeros of $\pi_l(x,\rho)$ are simple and different from zero for all $x \in [a,b]$, where π_l is defined in* (7.2.13).
A. *If* i), ii) *and* iv) *hold and if $l = 1$ or $\pi_{i,l} \in W_p^1(a,b)$ for $i = l+1,\dots,n$, then we can choose $\nu_1 = \dots = \nu_{n_0+1} = 0$, $\nu_i = i - n_0 - 1$ $(i = n_0+2,\dots,n)$ and*

$$C_1 = \left(\begin{array}{ccc|ccc} 1 & & 0 & & & \\ & \ddots & & & 0 & \\ 0 & & 1 & & & \\ \hline -\frac{\pi_{n,l}}{\pi_{l,l}} & \cdots & -\frac{\pi_{l+1,l}}{\pi_{l,l}} & 1 & \cdots & 1 \\ & & & r_1 & \cdots & r_l \\ & 0 & & \vdots & & \vdots \\ & & & r_1^{l-1} & \cdots & r_l^{l-1} \end{array}\right), \tag{7.2.17}$$

where $r_1,\dots,r_l \in W_p^1(a,b)$ *are the roots of* $\pi_l(\cdot,\rho)=0$ *according to Proposition 7.2.3 if* $l>1$.

B. *If* i), ii) *and* iv) *hold and if* $\pi_{i,i} \in W_p^1(a,b)$ *for* $i=1,\dots,l$, *then we can choose* $\nu_1=\dots=\nu_{n_0}=0$, $\nu_i=i-n_0$ $(i=n_0+1,\dots,n)$ *and*

$$C_1 = \left(\begin{array}{ccc|ccc} 1 & & 0 & & & \\ & \ddots & & & 0 & \\ 0 & & 1 & r_1^{-1} & \dots & r_l^{-1} \\ \hline & & & 1 & \dots & 1 \\ & & & r_1 & \dots & r_l \\ & 0 & & \vdots & & \vdots \\ & & & r_1^{l-1} & \dots & r_l^{l-1} \end{array}\right), \tag{7.2.18}$$

where $r_1,\dots,r_l \in W_p^1(a,b)$ *are the roots of* $\pi_l(\cdot,\rho)=0$ *according to Proposition 7.2.3. Note that in case* $l=n$ *the matrix function* C_1 *ist just the lower right block.*

Proof. Assume that we have an asymptotic linearization of the required form. Then i) is clear since (7.2.11) yields $\pi_{l,l}=\prod_{j=1}^{l}(-r_j)$, where the function on the right-hand side is invertible in $L_\infty(a,b)$.

From $\pi_{l,l}\neq 0$ we infer $\deg p_{n_0}=n-n_0$. Hence (7.2.4) yields for $j=n_0$ that

$$1+\sum_{i=n_0+1}^{n-1}(\nu_{i+1}-\nu_i)=l.$$

Since $\nu_{i+1}-\nu_i\le 1$ by (7.2.3), we obtain $\nu_{i+1}=\nu_i+1$ for $i=n_0+1,\dots,n-1$. This proves that the matrix function $\widehat{A}$ defined by (7.2.5) has the form

$$\widehat{A} = \left(\begin{array}{cccccc|ccccc} 0 & \gamma_1 & & & & & & & & & \\ & \cdot & \cdot & 0 & & & & & 0 & & \\ & & \cdot & \cdot & & & & & & & \\ & 0 & & \cdot & \cdot & & & & & & \\ & & & & \cdot & \gamma_{n_0-1} & & & & & \\ & & & & & 0 & \gamma_{n_0} & & & & \\ \hline & & & & & & 0 & 1 & & & \\ & & & & & & & \cdot & \cdot & 0 & \\ & & 0 & & & & & & \cdot & \cdot & \\ & & & & & & & 0 & & \cdot & \cdot \\ & & & & & & & & & 0 & 1 \\ -\pi_{n,\mu_1} & \cdot & \cdot & \cdot & \cdot & -\pi_{l+1,\mu_{n_0}} & -\pi_{l,l} & \cdot & \cdot & \cdot & -\pi_{1,1} \end{array}\right), \tag{7.2.19}$$

where, for $i = 1, \ldots, n_0$, $\gamma_i = 1$ if $\nu_{i+1} = \nu_i + 1$ and $\gamma_i = 0$ if $\nu_{i+1} \le \nu_i$. Furthermore, $\deg p_{i-1} \le 1 + \nu_n - \nu_i =: \mu_i$ for $i = 1, \ldots, n_0$ by (7.2.4).

Now let $n_0 > 0$. Again from $\pi_{l,l}^{-1} \in L_\infty(a,b)$ we infer that the $l \times l$ submatrix of $\widehat{A}$ in the lower right corner has rank l. But since A_1 has n_0 zeros in its diagonal, $\operatorname{rank} \widehat{A} = \operatorname{rank} A_1 = l$. Thus the first n_0 columns of $\widehat{A}$ are linear combinations of the other columns, and we obtain that the entries $\gamma_1, \ldots, \gamma_{n_0-1}$ in the matrix $\widehat{A}$ are zero. This implies $\nu_{n_0} \le \nu_{n_0-1} \le \cdots \le \nu_1$. Hence $\mu_i \le 1 + \nu_n - \nu_{n_0}$ for $i = 1, \ldots, n_0$. Since $\widehat{A}$ has rank l, its n_0-th row must be a linear combination of the last l rows. Hence $\gamma_{n_0} = 0$ or $\gamma_{n_0} \ne 0$ and $\pi_{n,\mu_1} = \cdots = \pi_{l+1,\mu_{n_0}} = 0$. In the first case we obtain $\mu_i \le 1 + \nu_n - \nu_{n_0+1}$, and the second case yields $\deg p_{i-1} < 1 + \nu_n - \nu_{n_0} = 2 + \nu_n - \nu_{n_0+1}$ for $i = 1, \ldots, n_0$. Hence, in both cases, $\deg p_{i-1} \le 1 + \nu_n - \nu_{n_0+1} = l$ for $i = 1, \ldots, n_0$. Thus ii) holds.

Let $l > 1$ or $n_0 \ne 0$ and $\gamma_{n_0} = 1$. Let $j \in \{1, \ldots, l\}$. We have $A_1 e_{n_0+j} = r_j e_{n_0+j}$. Let $q = n_0$ if $n_0 = 0$ or $\gamma_{n_0} = 0$ and $q = n_0 - 1$ if $n_0 > 0$ and $\gamma_{n_0} = 1$. Then $e_k^\mathsf{T} \widehat{A} = 0$ for $k = 1, \ldots, q$ and $e_k^\mathsf{T} \widehat{A} = e_{k+1}^\mathsf{T}$ for $k = q+1, \ldots, n-1$. Hence, in view of $C_1 A_1 = \widehat{A} C_1$, we obtain

$$r_j e_k^\mathsf{T} C_1 e_{n_0+j} = 0 \quad \text{for} \quad k = 1, \ldots, q \tag{7.2.20}$$

and

$$r_j e_k^\mathsf{T} C_1 e_{n_0+j} = e_{k+1}^\mathsf{T} C_1 e_{n_0+j} \quad \text{for} \quad k = q+1, \ldots, n-1. \tag{7.2.21}$$

Since r_j is invertible, (7.2.20) implies that $e_k^\mathsf{T} C_1 e_{n_0+j} = 0$ for $k = 1, \ldots, q$. If $e_n^\mathsf{T} C_1(x) e_{n_0+j} = 0$ for some $x \in [a,b]$, then (7.2.21) and $r_j^{-1} \in L_\infty(a,b)$ would imply that the modulus of $e_k^\mathsf{T} C_1 e_{n_0+j}$ is smaller than a given positive number in a sufficiently small neighbourhood of x for all $k \in \{q+1, \ldots, n\}$. Then the continuity of C_1 would give $e_k^\mathsf{T} C_1(x) e_{n_0+j} = 0$ for these k. But this is impossible since $C_1(x)$ is invertible. Hence $\left(e_n^\mathsf{T} C_1 e_{n_0+j}\right)^{-1} \in W_p^1(a,b)$ by Proposition 2.5.8. From $q \le n-2$, (7.2.21) and Proposition 2.3.1 we infer

$$r_j^{-1} = (e_n^\mathsf{T} C_1 e_{n_0+j})^{-1} (e_{n-1}^\mathsf{T} C_1 e_{n_0+j}) \in W_p^1(a,b).$$

The assumption $\pi_{l,l} \in L_\infty(a,b)$ implies

$$r_j = (-1)^l \pi_{l,l} \prod_{\substack{m=1 \\ m \ne j}}^{l} r_m^{-1} \in L_\infty(a,b).$$

These two properties of r_j yield $r_j \in W_p^1(a,b)$ in view of Proposition 2.5.8. Because of (7.2.11) and Proposition 2.3.1, this proves iii) if $l > 1$ or $n_0 \ne 0$ and $\gamma_{n_0} = 1$.

In case $l = 1$ it follows that $n_0 > 0$ since $n \geq 2$. Hence we still have to consider the case that $l = 1$, $n_0 \neq 0$ and $\gamma_{n_0} = 0$. We have seen in the proof of ii) that $\mu_i \leq 1 + \nu_n - \nu_{n_0+1} = 1$ for $i = 1, \ldots, n-1 (= n_0)$. Note that $A_1 = r_1 e_n e_n^\top$ and $\pi_{1,1} = -r_1$. Let $i \in \{1, \ldots, n\}$. If $\mu_i = 1$, then

$$\pi_{1,1}^{-1} \pi_{n-i+1,1} = r_1^{-1} e_n^\top \widehat{A} e_i = r_1^{-1} e_n^\top C_1 A_1 C_1^{-1} e_i = e_n^\top C_1 e_n e_n^\top C_1^{-1} e_i \in W_p^1(a,b).$$

If $\mu_i \leq 0$, then $\pi_{n-i+1,1} = 0$. This completes the proof of iii).

For the proof of iv) we observe that $r_j^{-1} \in W_p^1(a,b)$ for $j = 1, \ldots, l$ implies that the zeros of $\pi_l(x, \rho)$ are different from zero for all $x \in [a,b]$. And from Proposition 7.2.2 iii) we infer that the roots of $\pi_l(x, \rho)$ are simple for all $x \in [a,b]$.

Conversely, assume that the conditions i)-iv) are fulfilled. If $r_1, \ldots, r_l$ actually occur in the matrices (7.2.17) or (7.2.18), i. e., if $l \geq 2$, then Proposition 7.2.3 is applicable, and we choose $r_1, \ldots, r_l \in W_p^1(a,b)$ such that (7.2.14) holds. Property iv) implies that $r_j - r_i$ is invertible in $L_\infty(a,b)$ if $i \neq j$. By Proposition 7.2.3 we also have $r_1^{-1}, \ldots, r_l^{-1} \in W_p^1(a,b)$. Note that at least one of the cases **A** or **B** is applicable. This proves $C_1 \in M_n(W_p^1(a,b))$ for C_1 given by (7.2.17) or (7.2.18), respectively. Since the lower right $l \times l$ block of C_1 is a Vandermonde matrix, we obtain that

$$\det C_1 = \prod_{1 \leq i < j \leq l} (r_j - r_i)$$

is invertible in $L_\infty(a,b)$. Hence C_1 is invertible in $L_\infty(a,b)$ by CRAMER'S rule. Then Proposition 2.5.8 yields $C_1^{-1} \in M_n(W_p^1(a,b))$. Hence $C(\cdot, \lambda) = C_0(\lambda) C_1$ is invertible in $M_n(W_p^1(a,b))$ if $|\lambda| > 1$, i. e. (7.1.8) is fulfilled.

Next we shall show that

$$C_0^{-1}(\lambda) A(\cdot, \lambda) C_0(\lambda) = \lambda \widehat{A} + O(1) \text{ in } M_n(L_p(a,b)). \tag{7.2.22}$$

Indeed, in case **A** we infer from ii) and (7.2.2) (see also (7.2.18) that (7.2.22) holds with

$$\widehat{A} = \left(\begin{array}{ccc|cccccc} & & & & & & & & \\ & 0 & & & & 0 & & & \\ & & & & & & & & \\ \hline & & & 0 & 1 & & & & \\ & & & & \cdot & \cdot & 0 & & \\ & 0 & & & & \cdot & \cdot & & \\ & & & & 0 & & \cdot & \cdot & \\ & & & & & & 0 & 1 & \\ -\pi_{n,l} & \cdots & -\pi_{l+1,l} & -\pi_{l,l} & \cdot & \cdot & \cdot & \cdot & -\pi_{1,1} \end{array} \right). \tag{7.2.23}$$

And in case **B** we infer from $\nu_n = l$ and (7.2.2) that (7.2.22) holds with

$$
(7.2.24) \qquad \widehat{A} = \left(\begin{array}{c|ccccc}
 & & & & & \\
0 & & & 0 & & \\
 & 1 & & & & \\
\hline
 & 0 & 1 & & & \\
 & & \ddots & \ddots & 0 & \\
0 & & & \ddots & \ddots & \\
 & & 0 & & \ddots & \\
 & & & & 0 & 1 \\
 & -\pi_{l,l} & \cdots & \cdots & \cdots & -\pi_{1,1}
\end{array}\right).
$$

In both cases the representation (7.2.7) yields that (7.1.9) and (7.1.10) hold since the matrix function $\widetilde{A}_0(\cdot,\lambda)$ is a polynomial in λ^{-1}. Also, (7.2.8) holds, i. e.,

$$A_1 = C_1^{-1}\widehat{A}C_1 .$$

We have to prove that

$$
A_1 = \left(\begin{array}{c|cccc}
 & & & & \\
0 & & 0 & & \\
 & & & & \\
\hline
 & r_1 & & & \\
 & & \ddots & 0 & \\
0 & & & \ddots & \\
 & & 0 & & \ddots \\
 & & & & r_l
\end{array}\right).
$$

In case **A** we obtain

$$
\begin{aligned}
(7.2.25) \qquad \widehat{A}C_1e_j &= \widehat{A}\left(e_j - \frac{\pi_{n-j+1,l}}{\pi_{l,l}}e_{n_0+1}\right) \\
&= -\pi_{n-j+1,l}e_n + \pi_{n-j+1,l}e_n = 0 \quad (j = 1,\ldots,n_0).
\end{aligned}
$$

In case **B** we have

$$(7.2.26) \qquad \widehat{A}C_1e_j = \widehat{A}e_j = 0 \quad (j = 1,\ldots,n_0).$$

Let $q := n_0$ in the case **A** or if $n_0 = 0$ and let $q := n_0 - 1$ in the case **B** for $n_0 \neq 0$. Since $C_1(x)e_{n_0+j} \in \operatorname{span}\{e_{q+1},\ldots,e_n\}$ for $j = 1,\ldots,l$ and $x \in [a,b]$, we obtain

$$(7.2.27) \qquad e_k^{\mathsf{T}}\widehat{A}C_1e_{n_0+j} = 0 = e_k^{\mathsf{T}}r_jC_1e_{n_0+j} \quad (k = 1,\ldots,q;\ j = 1,\ldots,l).$$

The definition of C_1 immediately yields

$$e_k^{\mathsf{T}}\widehat{A}C_1 e_{n_0+j} = e_{k+1}^{\mathsf{T}} C_1 e_{n_0+j} = e_k^{\mathsf{T}} r_j C_1 e_{n_0+j} \tag{7.2.28}$$

for $k = q+1,\dots,n-1$ and $j = 1,\dots,l$. Finally, $\pi_l(\cdot, r_j) = 0$ implies

$$e_n^{\mathsf{T}}\widehat{A}C_1 e_{n_0+j} = -\sum_{i=1}^{l} \pi_{i,i} r_j^{l-i} = r_j^l = e_n^{\mathsf{T}} r_j C_1 e_{n_0+j} \quad (j = 1,\dots,l). \tag{7.2.29}$$

The equations (7.2.25) to (7.2.29) prove that

$$C_1 A_1 = \widehat{A}C_1 = C_1 \operatorname{diag}(0,\dots,0,r_1,\dots,r_l). \quad \square$$

We still have to require that the functions $r_1,\dots,r_l$ in the asymptotic linearization obtained in Theorem 7.2.4 satisfy the condition I) in Section 4.1 on page 131. By Proposition 4.1.1 we know that I) holds if and only if one of the conditions II) or III) defined in Section 4.1 holds. It is convenient to introduce the following two conditions:

IV) There is a complex number $\alpha \in \mathbb{C}\setminus\{0\}$ such that for all $x \in [a,b]$ the roots of $\pi_{(\alpha)}(x,\rho) := \alpha^{-l}\pi_l(x,\alpha\rho) = 0$ are real, simple and different from 0.

V) There are a real-valued function $r \in W_p^1(a,b)$ such that $r(x) \neq 0$ for all $x \in [a,b]$ and $\beta_j \in \mathbb{C}$ $(j = 1,\dots,l)$ with $\beta_l \neq 0$ such that

$$\pi_{j,j}(x) = \beta_j r(x)^j \text{ for all } j = 1,\dots,n \text{ and } x \in (a,b), \tag{7.2.30}$$

$$\pi_{(0)}(\rho) := \rho^l + \sum_{j=1}^{l} \beta_j \rho^{l-j} = 0 \text{ has only simple roots.} \tag{7.2.31}$$

PROPOSITION 7.2.5. *Let the assumptions of Proposition* 7.2.3 *be fulfilled. Let* $r_1,\dots,r_l \in W_p^1(a,b)$ *be the roots of* $\pi_l(\cdot,\rho) = 0$ *and set* $r_0 := 0$. *Then*
i) II) $\Leftrightarrow$ IV),
ii) III) $\Leftrightarrow$ V).

Proof. i) Let $\alpha \in \mathbb{C}\setminus\{0\}$. Then

$$\pi_{(\alpha)}(x,\rho) = \prod_{j=1}^{l}\left(\rho - \alpha^{-1} r_j(x)\right).$$

Since the roots of $\pi_l(x,\rho) = 0$ are simple and different from zero by assumption, it is easy to see that II) and IV) are equivalent.
ii) III)$\Rightarrow$V): We have

$$\pi_l(x,\rho) = \prod_{j=1}^{l}(\rho - r_j(x)) = r(x)^l \prod_{j=1}^{l}\left(\frac{\rho}{r(x)} - \alpha_j\right).$$

Hence there are $\beta_j \in \mathbb{C}$ $(j = 1,\dots,l)$ such that

$$\pi_l(x,\rho) = r(x)^l \left(\frac{\rho^l}{r(x)^l} + \sum_{j=1}^{l} \frac{\rho^{l-j}}{r(x)^{l-j}} \beta_j \right).$$

We obtain $\beta_l = \prod_{j=1}^{l} (-\alpha_j) \neq 0$ and $\pi_{j,j}(x) = r(x)^j \beta_j$ for $j = 1,\dots,l$. Finally we infer that $\pi_{(0)}(\rho) = \frac{1}{r(a)^l} \pi_l(a, \rho r(a)) = 0$ has only simple roots.

V)⇒III): Let $\alpha_1,\dots,\alpha_l$ be the zeros of $\pi_{(0)}$. From $\beta_l \neq 0$ we immediately infer that $\alpha_1 \neq 0,\dots,\alpha_l \neq 0$. Then we have for all $x \in [a,b]$

$$\begin{aligned}\prod_{j=1}^{l} (\rho - \alpha_j r(x)) &= r(x)^l \prod_{j=1}^{l} \left(\frac{\rho}{r(x)} - \alpha_j \right) \\ &= r(x)^l \left(\frac{\rho^l}{r(x)^l} + \sum_{j=1}^{l} \frac{\rho^{l-j}}{r(x)^{l-j}} \beta_j \right) \\ &= \rho^l + \sum_{j=1}^{l} \rho^{l-j} \pi_{j,j}(x) = \pi_l(x,\rho).\end{aligned}$$

Hence, for a suitable choice of the indices, $r_\nu = \alpha_\nu r$ holds for $\nu = 0,\dots,l$, where $\alpha_0 = 0$. Since the roots of $\pi_l(a,\rho)$ are simple and different from zero, we obtain $\alpha_\nu \neq \alpha_\mu$ for $\nu, \mu = 0,\dots,l$ and $\nu \neq \mu$. □

COROLLARY 7.2.6. *In Theorem* 7.2.4 *we have that* A_1 *satisfies condition* I) *from Section* 4.1 *if and only if we replace condition* iv) *by*
iv′) $l > 1$ *and there is a number* $\alpha \in \mathbb{C} \setminus \{0\}$ *such that for all* $x \in [a,b]$ *the roots of*

$$\pi_{(\alpha)}(x,\rho) := \rho^l + \sum_{i=1}^{l} \rho^{l-i} \alpha^{-i} \pi_{i,i}(x) = 0$$

are real, simple and different from 0,

or

$l > 1$ *and there are a real-valued function* $r \in W_p^1(a,b)$ *such that* $r(x) \neq 0$ *for all* $x \in [a,b]$ *and* $\beta_j \in \mathbb{C}$ $(j = 1,\dots,l)$ *with* $\beta_l \neq 0$ *such that*

$$\pi_{j,j}(x) = \beta_j r(x)^j \text{ for all } j = 1,\dots,n \text{ and } x \in (a,b),$$

$$\pi_{(0)}(\rho) := \rho^l + \sum_{j=1}^{l} \beta_j \rho^{l-j} = 0 \text{ has only simple roots,}$$

or

$l = 1$ *and there is a number* $\alpha \in \mathbb{C} \setminus \{0\}$ *such that* $\alpha^{-1}\pi_{11}$ *is positive.*

Proof. Let $r_0 = 0$. In case $l = 1$ the equivalence is obvious since $r_1 = -\pi_{1,1}$. In case $l > 1$ it follows from $\pi_l(x,\rho) = \alpha^l \pi_{(\alpha)}(x, \frac{\rho}{\alpha})$ and $\pi_l(x,\rho) = r(x)^l \pi_{(0)}\left(\frac{\rho}{r(x)}\right)$ that iv$'$) implies iv). Hence the assumptions of Proposition 7.2.5 are satisfied, and the result follows from Propositions 7.2.5 and 4.1.1. □

7.3. Birkhoff regular problems

Together with the boundary conditions (7.1.3) and a function $C(x,\lambda)$ satisfying (7.1.8)–(7.1.10) we consider the matrix functions

$$\begin{aligned} W^{(j)}(\lambda) &:= (w_{ki}^{(j)}(\lambda))_{k,i=1}^n C(a_j,\lambda), \\ W(x,\lambda) &:= (w_{ki}(x,\lambda))_{k,i=1}^n C(x,\lambda), \end{aligned} \tag{7.3.1}$$

and set

$$\widehat{T}^R(\lambda)y := \sum_{j=0}^{\infty} W^{(j)}(\lambda)y(a_j) + \int_a^b W(x,\lambda)y(x)\,\mathrm{d}x \quad (y \in (W_p^1(a,b))^n). \tag{7.3.2}$$

DEFINITION 7.3.1. The boundary eigenvalue problem (7.1.2), (7.1.3) is called *Birkhoff regular* if $\pi_{nn} \neq 0$ and if there are matrix functions $C(\cdot,\lambda)$ satisfying (7.1.8)–(7.1.10) and $C_2(\lambda)$ satisfying (5.1.3)–(5.1.5) so that the associated boundary eigenvalue problem $\widetilde{T}^D(\lambda)y = 0$, $C_2(\lambda)^{-1}\widehat{T}^R(\lambda)y = 0$ is Birkhoff regular in the sense of Definition 4.1.2.

We shall assume that $C(\cdot,\lambda) = \{\lambda^{\nu_1},\dots,\lambda^{\nu_n}\}C_1$, where C_1 and $\nu_1,\dots,\nu_n$ are given as in Theorem 7.2.4 **A** or **B**.

The condition $\pi_{nn} \neq 0$ means that $n_0 = 0$. Theorems 5.2.2 and 5.2.3 also hold in this case since the term $A^0(\cdot,\lambda)$ is irrelevant for Birkhoff regularity. For the convenience of the reader we restate these theorems in this section, where we also use Theorem 4.1.3. First let us note that the characteristic function (7.1.4) can be factorized as

$$\pi(\cdot,\rho) = \prod_{\nu=1}^n (\rho - r_\nu(x))$$

according to Proposition 7.2.3. From Theorem 7.2.4 we know that the functions r_ν ($\nu = 1,\dots,n$) are bounded away from zero and mutually different. Furthermore,

$$r_\nu(x) = |r_\nu(x)|e^{i\varphi_\nu}$$

for some φ_ν. As in Section 4.1 we set

$$\delta_\nu(\lambda) := \begin{cases} 0 & \text{if } \Re(\lambda e^{i\varphi_\nu}) < 0, \\ 1 & \text{if } \Re(\lambda e^{i\varphi_\nu}) > 0, \\ 0 & \text{if } \Re(\lambda e^{i\varphi_\nu}) = 0 \text{ and } \Im(\lambda e^{i\varphi_\nu}) > 0, \\ 1 & \text{if } \Re(\lambda e^{i\varphi_\nu}) = 0 \text{ and } \Im(\lambda e^{i\varphi_\nu}) < 0. \end{cases}$$

For $\nu = 1,\dots,n$ we define

$$\Lambda_\nu^1 = \operatorname{diag}(\delta_\nu^1,\dots,\delta_\nu^n),$$

where

$$\delta_\nu^\mu := \begin{cases} 1 & \text{if } \varphi_\mu \in [\varphi_\nu, \varphi_\nu + \pi) \operatorname{mod}(2\pi), \\ 0 & \text{if } \varphi_\mu \notin [\varphi_\nu, \varphi_\nu + \pi) \operatorname{mod}(2\pi). \end{cases}$$

THEOREM 7.3.2. *Suppose that the assumptions of Theorem* 7.2.4 *and property* iv′) *in Corollary* 7.2.6 *are satisfied. Let* $\pi_{nn} \neq 0$. *Assume that* (7.1.2), (7.1.3) *is a two-point boundary eigenvalue problem, i. e., that* (7.1.3) *has the form*

$$\sum_{i=1}^n w_{ki}^{(0)}(\lambda)\eta^{(i-1)}(a) + \sum_{i=1}^n w_{ki}^{(1)}(\lambda)\eta^{(i-1)}(b) = 0 \quad (k = 1,\dots,n).$$

This problem is Birkhoff regular if and only if the following two properties hold:
i) *There is a matrix polynomial* C_2 *whose determinant is not identically zero so that*

$$C_2^{-1}(\lambda)\big(W^{(0)}(\lambda), W^{(1)}(\lambda)\big) = \big(W_0^{(0)}, W_0^{(1)}\big) + O(\lambda^{-1}) \quad \textit{as } \lambda \to \infty,$$

where $\big(W_0^{(0)}, W_0^{(1)}\big)$ *is an* $n \times 2n$ *matrix of rank* n.
ii) *For any matrix polynomial* C_2 *fulfilling* i) *the matrices*

$$W_0^{(0)}\Lambda_\nu^1 + W_0^{(1)}(I_n - \Lambda_\nu^1) \textit{ and } W_0^{(0)}(I_n - \Lambda_\nu^1) + W_0^{(1)}\Lambda_\nu^1 \quad (\nu = 1,\dots,n)$$

are invertible.

Note that it is not necessary to state that $(W_0^{(0)}, W_0^{(1)})$ has rank n since ii) implies this condition.

THEOREM 7.3.3. *Suppose that the assumptions of Theorem* 7.2.4 *and property* iv′) *in Corollary* 7.2.6 *are satisfied. Let* $\pi_{nn} \neq 0$. *The boundary eigenvalue problem* (7.1.2), (7.1.3) *is Birkhoff regular if and only if the following three properties hold:*
i) *There is a matrix polynomial* C_2 *whose determinant is not identically zero so that*

$$C_2^{-1}(\lambda)\big(W^{(0)}(\lambda), W^{(1)}(\lambda)\big) = (W_0^{(0)}, W_0^{(1)}) + O(\lambda^{-1}) \quad \textit{as } \lambda \to \infty,$$

where $(W_0^{(0)}, W_0^{(1)})$ *is an* $n \times 2n$*-matrix of rank* n.
ii) *For any matrix polynomial* C_2 *fulfilling* i) *the matrices*

$$W_0^{(0)}\Lambda_\nu^1 + W_0^{(1)}(I_n - \Lambda_\nu^1) \textit{ and } W_0^{(0)}(I_n - \Lambda_\nu^1) + W_0^{(1)}\Lambda_\nu^1 \quad (\nu = 1,\dots,n)$$

are invertible.
iii) *For any matrix polynomial* C_2 *fulfilling* i) *the estimates*

$$C_2^{-1}(\lambda)W(\cdot,\lambda) = O(1) \quad \textit{in } M_n(L_1(a,b))$$

and, for $j \in \mathbb{N}$,

$$C_2^{-1}(\lambda)W^{(j)}(\lambda) = O(1) \quad \textit{in } M_n(\mathbb{C})$$

hold, and the properties (5.1.4) *and* (5.1.5) *are fulfilled. If* $W^{(j)} \neq 0$ *only for finitely many* j, *then* (5.1.4) *and* (5.1.5) *are automatically satisfied.*

Since $C_2(\lambda)$ and $W(\cdot,\lambda)$ are polynomials, property iii) implies that

$$C_2(\lambda)^{-1}W(\cdot,\lambda) = W_0 + O\Big(\frac{1}{\lambda}\Big) \text{ in } M_n(L_1(a,b)). \tag{7.3.3}$$

7.4. Expansion theorems for Birkhoff regular n-th order differential equations

In this section we shall suppose that $\pi_{n,n} \neq 0$, which is equivalent to $n_0 = 0$. We also suppose that the assumptions of Theorem 7.2.4 and property iv′) in Corollary 7.2.6 are satisfied and that the boundary eigenvalue problem (7.1.2), (7.1.3) is Birkhoff regular, where we assume that the transformation $C(\cdot,\lambda)$ is as in Theorem 7.2.4. Choose the circles Γ_ν according to Theorem 4.3.9 and define

$$Q_\nu f := \frac{1}{2\pi i}\oint_{\Gamma_\nu} \widetilde{J}_1 L^{-1}(\lambda)(\pi_{n,n}\lambda^{n-1}f, 0)\,\mathrm{d}\lambda \quad (f \in L_p(a,b),\ \nu \in \mathbb{N}), \tag{7.4.1}$$

where $\widetilde{J}_1 : W_p^n(a,b) \to L_p(a,b)$ is the canonical embedding. Note that $\pi_{n,n}$ belongs to $L_\infty(a,b)$ by the assumptions made in Theorem 7.2.4. Hence Q_ν is a continuous operator on $L_p(a,b)$.

Let $\widetilde{T}^D(\lambda)$ be given by (7.1.9) and $\widetilde{T}^R(\lambda) = C_2(\lambda)^{-1}\widehat{T}^R(\lambda)$. Since the boundary eigenvalue problem (7.1.2), (7.1.3) is Birkhoff regular, the boundary eigenvalue problem $\widetilde{T}^D(\lambda)y = 0$, $\widetilde{T}^R(\lambda)y = 0$ is Birkhoff regular in the sense of Definition 4.1.2. Together with $\widetilde{T}(\lambda) = (\widetilde{T}^D(\lambda), \widetilde{T}^R(\lambda))$ we consider the operators

$$P_\nu f := -\frac{1}{2\pi i}\oint_{\Gamma_\nu} J\widetilde{T}^{-1}(\lambda)(A_1 f, 0)\,\mathrm{d}\lambda \quad (f \in (L_p(a,b))^n,\ \nu \in \mathbb{N}),$$

see Lemma 4.6.7.

PROPOSITION 7.4.1. *For $f \in L_p(a,b)$ we have*

$$Q_\nu f = e_1^\top C_1 P_\nu C_1^{-1} e_1 f.$$

Proof. We have seen in (6.4.12) that

$$L^{-1}(\lambda)(f_1, f_2) = e_1^\top T^{-1}(\lambda)(e_n f_1, f_2) \tag{7.4.2}$$

for $f_1 \in L_p(a,b)$, $f_2 \in \mathbb{C}^n$, where $T = (T^D, T^R)$ is given by (6.1.3), (6.1.4) and (6.2.1). We have

$$\widetilde{T}(\lambda) = \begin{pmatrix} C(\cdot,\lambda)^{-1} & 0 \\ 0 & C_2(\lambda)^{-1} \end{pmatrix} T(\lambda) C(\cdot,\lambda), \tag{7.4.3}$$

whence

$$\begin{aligned} L^{-1}(\lambda)(f_1, 0) &= e_1^\top C(\cdot,\lambda)\widetilde{T}^{-1}(\lambda)(C(\cdot,\lambda)^{-1}e_n f_1, 0) \\ &= \lambda^{\nu_1-\nu_n} e_1^\top C_1 \widetilde{T}^{-1}(\lambda)(C_1^{-1}e_n f_1, 0). \end{aligned} \tag{7.4.4}$$

From (7.2.19) we obtain that

$$C_1^{-1}\pi_{n,n}e_n = -C_1^{-1}\widehat{A}e_1 = -A_1C_1^{-1}e_1, \tag{7.4.5}$$

whence the proposition is proved if we set $f_1 = \pi_{n,n}f$ and observe that $\nu_1 - \nu_n = 1-n$ by Theorem 7.2.4. $\square$

From (7.2.12) we infer that

$$A_0 = -C_1^{-1}C_1' + C_1^{-1}\begin{pmatrix} & & 0 & \\ & & & \\ -\pi_{n,n-1} & & \cdots & -\pi_{1,0}\end{pmatrix}C_1,$$

and condition (4.1.19) holds if we require in case $p \le \frac{3}{2}$ that there is a number $\hat{p}$ such that $\frac{2}{\hat{p}} < 2 - \frac{1}{p}$ and

$$\pi_{i,i-j} \in W_{\hat{p}}^{1-j}(a,b) \text{ for } i = 1,\dots,n \text{ and } j = 0,1. \tag{7.4.6}$$

As multiplication by C_1 and C_1^{-1} is continuous in $(L_p(a,b))^n$, Theorem 4.6.9 gives

LEMMA 7.4.2. *Let $1 < p < \infty$. Suppose that the boundary eigenvalue problem (7.1.2), (7.1.3) is Birkhoff regular. If $p \le \frac{3}{2}$, then we require that (7.4.6) holds. Suppose that $C(x,\lambda)$ is as in Theorem 7.2.4 and that $W_0 \in M_n(L_q(a,b))$ for some $q > 1$, where W_0 is given by (7.3.3). Assume that a and b are no accumulation points of the set $\{a_j : j \in \mathbb{N}, W_0^{(j)} \ne 0\}$. Then $\lim_{\nu\to\infty} Q_\nu f = f$ for all $f \in L_p(a,b)$.*

Since L is a Fredholm operator function, we can represent the principal parts of L^{-1} in terms of eigen- and associated vectors of L and L^*. Since an eigenvector or associated vector v of L^* belongs to $L_{p'}(a,b) \times \mathbb{C}^n$, we can write $v = (u,d)$ with $u \in L_{p'}(a,b)$ and $d \in \mathbb{C}^n$. For an eigenvalue λ_κ of L let $r(\lambda_\kappa) := \dim N(L(\lambda_\kappa))$ and $m_{\kappa,j}$ $(j = 1,\dots,r_\kappa)$ be the partial multiplicities.

Lemma 7.4.2 and Theorem 1.6.7 lead to

THEOREM 7.4.3. *Let $1 < p < \infty$. Suppose that the boundary eigenvalue problem (7.1.2), (7.1.3) is Birkhoff regular and choose the curves Γ_ν $(\nu \in \mathbb{N})$ with radius ρ_ν according to Theorem 4.3.9. If $p \le \frac{3}{2}$, then we require that (7.4.6) holds. Suppose that $C(x,\lambda)$ is as in Theorem 7.2.4 and that $W_0 \in M_n(L_q(a,b))$ for some $q > 1$, where W_0 is given by (7.3.3). Assume that a and b are no accumulation points of the set $\{a_j : j \in \mathbb{N}, W_0^{(j)} \ne 0\}$. Let $\lambda_0, \lambda_1, \dots$ be the eigenvalues of L and let*

$$\{\eta_{\kappa,l}^{(j)} : j = 1,\dots,r(\lambda_\kappa);\, l = 0,\dots,m_{\kappa,j}-1\}$$

and

$$\{(u_{\kappa,l}^{(j)}, d_{\kappa,l}^{(j)}) : j = 1,\dots,r(\lambda_\kappa);\, l = 0,\dots,m_{\kappa,j}-1\}$$

be biorthogonal CSEAVs of L and L^* at λ_κ, respectively. Then

$$f = \lim_{\nu\to\infty} \sum_{\substack{\kappa\in\mathbb{N}\\ |\lambda_\kappa|<\rho_\nu}} \Bigg(\sum_{j=1}^{r(\lambda_\kappa)} \sum_{l=1}^{\min\{m_{\kappa,j},n\}} \binom{n-1}{l-1} \lambda_\kappa^{n-l} \times$$

$$\times \sum_{h=0}^{m_{\kappa,j}-l} \eta_{\kappa,h}^{(j)} \int_a^b u_{\kappa,m_{\kappa,j}-l-h}^{(j)}(x)\pi_{n,n}(x)f(x)\,\mathrm{d}x \Bigg)$$

in $L_p(a,b)$ holds for all $f\in L_p(a,b)$.

The analog of Lemma 7.4.2 for $p=\infty$ is obtained if we use Theorem 4.7.5. For its formulation we observe that

$$\widehat{W}^{(j)}(\lambda) := \left(w_{ki}^{(j)}(\lambda)\right)_{k,i=1}^n C_0(\lambda),$$
$$\widehat{W}(x,\lambda) := \left(w_{ki}(x,\lambda)\right)_{k,i=1}^n C_0(\lambda),$$

satisfy

$$\widehat{W}^{(j)}(\lambda) = \widehat{W}_0^{(j)} + O(\lambda^{-1}), \quad \widehat{W}(\cdot,\lambda) = \widehat{W}_0 + O(\lambda^{-1}),$$
$$W_0^{(j)} = \widehat{W}_0^{(j)} C_1(a_j), \quad W = \widehat{W} C_1 .$$

THEOREM 7.4.4. *Let $p=\infty$, suppose that the boundary eigenvalue problem given by (7.1.2), (7.1.3) is Birkhoff regular and choose the curves Γ_ν ($\nu\in\mathbb{N}$) with radius ρ_ν according to Theorem 4.3.9. Suppose that $C(x,\lambda)$ is as in Theorem 7.2.4 and that $W_0\in M_n(L_q(a,b))$ for some $q>1$, where W_0 is given by (7.3.3). Assume that a and b are no accumulation points of the set $\{a_j : j\in\mathbb{N}, W_0^{(j)}\neq 0\}$. Let $\lambda_0,\lambda_1,\dots$ be the eigenvalues of L and let*

$$\{\eta_{\kappa,l}^{(j)} : j=1,\dots,r(\lambda_\kappa);\, l=0,\dots,m_{\kappa,j}-1\}$$

and

$$\{(u_{\kappa,l}^{(j)}, d_{\kappa,l}^{(j)}) : j=1,\dots,r(\lambda_\kappa);\, l=0,\dots,m_{\kappa,j}-1\}$$

be biorthogonal CSEAVs of L and L^* at λ_κ, respectively. Then

$$f = \lim_{\nu\to\infty} \sum_{\substack{\kappa\in\mathbb{N}\\ |\lambda_\kappa|<\rho_\nu}} \Bigg(\sum_{j=1}^{r(\lambda_\kappa)} \sum_{l=1}^{\min\{m_{\kappa,j},n\}} \binom{n-1}{l-1} \lambda_\kappa^{n-l} \times$$

$$\times \sum_{h=0}^{m_{\kappa,j}-l} \eta_{\kappa,h}^{(j)} \int_a^b u_{\kappa,m_{\kappa,j}-l-h}^{(j)}(x)\pi_{n,n}(x)f(x)\,\mathrm{d}x \Bigg)$$

holds in $C[a,b]$ for all $f\in C[a,b]\cap BV[a,b]$ satisfying

$$\sum_{j=0}^{\infty} \widehat{W}_0^{(j)} e_1 f(a_j) + \int_a^b \widehat{W}_0(x) e_1 f(x)\,\mathrm{d}x = 0.$$

Proof. We only have to verify condition (4.7.2). Since $n = 0$, that condition is $\widetilde{T}_0^R C_1^{-1} e_1 f = 0$ by Remark 4.7.6. $\square$

7.5. An example for a Birkhoff regular problem with λ-dependent boundary conditions

The following example has been investigated by [HEI1], [SCHM], [TR2]. Consider the boundary eigenvalue problem

$$\eta^{(4)} + K\eta = \lambda\eta, \tag{7.5.1}$$

$$\eta(0) = 0, \quad \eta'(0) = 0, \quad \eta''(1) = 0, \quad \beta(\lambda)\eta'''(1) + \alpha(\lambda)\eta(1) = 0, \tag{7.5.2}$$

where

$$\alpha(\lambda) = \alpha_3\lambda^3 + \alpha_2\lambda^2 + \alpha_1\lambda + \alpha_0, \quad \beta(\lambda) = \beta_2\lambda^2 + \beta_1\lambda + \beta_0,$$

K, α_0, α_1, α_2, α_3, β_0, β_1, β_2 are complex numbers, and $\alpha_3 \neq 0$, $\beta_2 \neq 0$.

If we replace λ by λ^4, then the expansion theorems in Section 7.4 are applicable, and we obtain

THEOREM 7.5.1. *For $1 < p < \infty$, every $f \in L_p(a,b)$ is expandable into eigenfunctions and associated functions of the eigenvalue problem* (7.5.1), (7.5.2) *with λ replaced by λ^4, where the expansion converges in $L_p(a,b)$. If $f \in C[a,b] \cap BV[a,b]$ and satisfies*

$$f(0) = f(1) = 0,$$

then f is expandable into eigenfunctions and associated functions of the eigenvalue problem (7.5.1), (7.5.2) *with λ replaced by λ^4, where the expansion converges in $C[a,b]$.*

Proof. It is easy to check that the assumptions of Theorem 7.2.4 and Corollary 7.2.6 are fulfilled. Here we have $\pi_{4,4} = -1$ and, with $r(x) = 1$,

$$\pi_{(0)}(\rho) = \rho^4 - 1.$$

Then $r_j = \omega_j$, where $\omega_1, \ldots, \omega_4$ are the fourth unit roots. Hence

$$C(x,\lambda) = \begin{pmatrix} 1 & 1 & 1 & 1 \\ \lambda\omega_1 & \lambda\omega_2 & \lambda\omega_3 & \lambda\omega_4 \\ (\lambda\omega_1)^2 & (\lambda\omega_2)^2 & (\lambda\omega_3)^2 & (\lambda\omega_4)^2 \\ (\lambda\omega_1)^3 & (\lambda\omega_2)^3 & (\lambda\omega_3)^3 & (\lambda\omega_4)^3 \end{pmatrix}.$$

The boundary matrices given by (7.3.1) are

$$W^{(0)}(\lambda) = \begin{pmatrix} 1 & 0 & 0 & 0 \\ 0 & 1 & 0 & 0 \\ 0 & 0 & 0 & 0 \\ 0 & 0 & 0 & 0 \end{pmatrix} C(0,\lambda), \quad W^{(1)}(\lambda) = \begin{pmatrix} 0 & 0 & 0 & 0 \\ 0 & 0 & 0 & 0 \\ 0 & 0 & 1 & 0 \\ \alpha(\lambda^4) & 0 & 0 & \beta(\lambda^4) \end{pmatrix} C(1,\lambda),$$

and we immediately infer that we can choose $C_2(\lambda) = \operatorname{diag}(1, \lambda, \lambda^2, \lambda^{12})$. Thus we obtain

$$W_0^{(0)} = \begin{pmatrix} 1 & 1 & 1 & 1 \\ \omega_1 & \omega_2 & \omega_3 & \omega_4 \\ 0 & 0 & 0 & 0 \\ 0 & 0 & 0 & 0 \end{pmatrix}, \quad W_0^{(1)} = \begin{pmatrix} 0 & 0 & 0 & 0 \\ 0 & 0 & 0 & 0 \\ \omega_1^2 & \omega_2^2 & \omega_3^2 & \omega_4^2 \\ \alpha_3 & \alpha_3 & \alpha_3 & \alpha_3 \end{pmatrix}.$$

From Proposition 4.1.7 we infer that any Birkhoff matrix is, up to a permutation of the columns, of the form

$$\begin{pmatrix} 1 & 1 & 0 & 0 \\ \omega_{i_1} & \omega_{i_2} & 0 & 0 \\ 0 & 0 & \omega_{i_3}^2 & \omega_{i_4}^2 \\ 0 & 0 & \alpha_3 & \alpha_3 \end{pmatrix},$$

where $\omega_{i_1} \neq \omega_{i_2}$ and $\omega_{i_3}^2 \neq \omega_{i_4}^2$. This shows that the Birkhoff matrices are invertible. Hence the problem is Birkhoff regular, and the result for $p < \infty$ follows. For the uniform convergence the condition

$$0 = \widehat{W}_0^{(0)} e_1 f(0) + \widehat{W}_0^{(1)} e_1 f(1) = \begin{pmatrix} f(0) \\ 0 \\ 0 \\ \alpha_3 f(1) \end{pmatrix}$$

has to be satisfied. □

The result of Theorem 7.5.1 was obtained in [TR2] in the case of uniform convergence. A slightly weaker result was derived in [HEI1]. In [SCHM] an expansion theorem in $L_2(0,1)$ was obtained, but only under some additional conditions on the functions which are to be expanded.

7.6. Stone regular problems

Let the notations be as at the beginning of Section 7.3.

DEFINITION 7.6.1. The boundary value problem (7.1.2), (7.1.3) is called *strongly s-regular* if $\pi_{n,n} \neq 0$ and there are matrix functions $C(\cdot, \lambda)$ satisfying (7.1.8)–(7.1.11) and $C_2(\lambda)$ satisfying (5.1.3)–(5.1.5) such that the boundary-value problem $\widetilde{T}^D(\lambda)y = 0$, $C_2(\lambda)^{-1}\widehat{T}^R(\lambda)y = 0$ is strongly s-regular in the sense of Definition 5.7.1.

The Stone regularity gives an estimate for the resolvent $\widetilde{T}^{-1}$ of the asymptotically linear system associated with the given problem. Therefore, in view of the results in the proof of Proposition 7.4.1, this gives an estimate for L^{-1}. As for systems, we define auxiliary functions in order to reduce the highest λ-power in the integral over the resolvent $L^{-1}(\lambda)$, where we do not necessarily assume that $\pi_{n,n} \neq 0$. To this end we define the differential operators $\mathscr{H}_i$ for $i = 0, \dots, l$ by

$$\mathscr{H}_i \eta = \begin{cases} \eta^{(n)} + \sum_{k=0}^{n-1} \pi_{n-k,0} \eta^{(k)} & \text{if } i = 0, \\ \sum_{k=0}^{n-i} \pi_{n-k,i} \eta^{(k)} & \text{if } i = 1, \dots, l, \end{cases} \tag{7.6.1}$$

where $l = n - n_0$ and $\eta \in W_p^n(a,b)$. Assume that $\pi_{n-k,i} = 0$ for $k = 0, \dots, n_0 - 1$ and $i = l+1, \dots, n-k$. Then

$$L^D(\lambda) = \sum_{i=0}^{l} \lambda^i \mathscr{H}_i. \tag{7.6.2}$$

PROPOSITION 7.6.2. *Let $\kappa \geq 0$ and $m \geq n - l + 1$. Additionally to the general assumptions of this chapter we also require that $\pi_{n-k,i} \in W_p^{\kappa+i-n+m}(a,b)$ for $i = \max\{0, n - \kappa - m + 1\}, \dots, l-1$, $k = 0, \dots, \min\{n-i, n-1\}$, $\pi_{i,i} \in L_\infty(a,b)$ for $i = 1, \dots, l-1$, $\pi_{n-k,l} \in W_p^{\kappa-n+l-1+m+k}(a,b)$ $(k = 0, \dots, n-l)$, $\pi_{l,l}^{-1} \in L_\infty(a,b)$. Let $f \in W_p^{\kappa+m}(a,b)$. Then there are $f^{[j]} \in W_p^{\kappa-j+m}(a,b)$ $(j = 0, \dots, \kappa+1)$ such that $f^{[0]} = f$ and*

$$\mathscr{H}_l f^{[j]} = -\sum_{i=1}^{\min\{j,l\}} \mathscr{H}_{l-i} f^{[j-i]}$$

for $j = 1, \dots, \kappa+1$.

Proof. The statement is trivial for $j = 0$. Now assume that it holds for $j = 0, \dots, \kappa'$ with $\kappa' \leq \kappa$. Then $f^{[\kappa'+1-i]} \in W_p^{\kappa-\kappa'-1+i+m}(a,b)$ for $i = 1, \dots, \min\{\kappa'+1, l\}$. The assumptions on the $\pi_{n-k,i}$ imply by Proposition 2.3.1 that

$$\sum_{i=1}^{\min\{\kappa'+1,l\}} \mathscr{H}_{l-i} f^{[\kappa'+1-i]} \in W_p^{\kappa-\kappa'-n+l-1+m}(a,b). \tag{7.6.3}$$

Since $\mathscr{H}_l$ can be considered as a differential operator from $W_p^{n-l}(a,b)$ to $L_p(a,b)$, there is a solution $f^{[\kappa'+1]} \in W_p^{n-l}(a,b)$ of the differential equation

$$\mathscr{H}_l f^{[\kappa'+1]} = -\sum_{i=1}^{\min\{\kappa'+1,l\}} \mathscr{H}_{l-i} f^{[\kappa'+1-i]}.$$

If $l = n$, then $\mathscr{H}_l$ is the multiplication by $\pi_{n,n}$, and $f^{[\kappa'+1]} \in W_p^{\kappa-(\kappa'+1)+m}(a,b)$ follows. If $l < n$, then $\pi_{n-k,l} \in W_p^{\kappa-\kappa'-n+l-1+m+k}(a,b)$, and Proposition 2.6.4 yields $f^{[\kappa'+1]} \in W_p^{\kappa-\kappa'-n+l-1+m+n-l}(a,b) = W_p^{\kappa-(\kappa'+1)+m}(a,b)$. □

REMARK 7.6.3. If $m > n - l + 1$, then we do not need that $\pi_{i,i} \in L_\infty(a,b)$. But since we shall apply Proposition 7.6.2 together with Theorem 7.2.4, this is no additional restriction in that case.

PROPOSITION 7.6.4. *Let the assumptions be as in Proposition* 7.6.2 *and, additionally, let* $m \geq n$. *Let the operators* $K_1(\lambda)$ *and* $K_2(\lambda)$ *be as defined in* (6.4.1) *and* (6.4.2). *Then we have for* $f \in W_p^{\kappa+m}(a,b)$ *and* $\kappa' = 0,\ldots,\kappa$ *that*

$$\lambda^{l-1}K_1(\lambda)\mathscr{H}_l f = -\sum_{j=0}^{\kappa'}\left(\sum_{i=0}^{l-2-\kappa'+j}\lambda^{i-j-1}K_1(\lambda)\mathscr{H}_i f^{[j]} - \lambda^{-j-1}f^{[j]}\right) + \lambda^{l-2-\kappa'}K_1(\lambda)\mathscr{H}_l f^{[\kappa'+1]} - \sum_{j=0}^{\kappa'}\lambda^{-j-1}K_2(\lambda)L^R(\lambda)f^{[j]}, \tag{7.6.4}$$

where the functions $f^{[j]}$ *are as in Proposition* 7.6.2.

Proof. For $j = 0,\ldots,\kappa$ we have $f^{[j]} \in W_p^n(a,b)$ and

$$\begin{aligned} f^{[j]} &= L^{-1}(\lambda)L(\lambda)f^{[j]} \\ &= K_1(\lambda)L^D(\lambda)f^{[j]} + K_2(\lambda)L^R(\lambda)f^{[j]} \\ &= \sum_{i=0}^{l}\lambda^i K_1(\lambda)\mathscr{H}_i f^{[j]} + K_2(\lambda)\dot{L}^R(\lambda)f^{[j]}. \end{aligned}$$

From Proposition 7.6.2 we know that

$$-\mathscr{H}_{l-1}f^{[j]} = \mathscr{H}_l f^{[j+1]} + \sum_{i=2}^{\min\{j+1,l\}}\mathscr{H}_{l-i}f^{[j+1-i]} \qquad (j = 0,\ldots,\kappa).$$

Hence we obtain for $j = 0,\ldots,\kappa$ that

$$\lambda^{l-1}K_1(\lambda)\mathscr{H}_l f^{[j]} = \lambda^{-1}f^{[j]} - \sum_{i=0}^{l-2}\lambda^{i-1}K_1(\lambda)\mathscr{H}_i f^{[j]} + \sum_{m=2}^{\min\{j+1,l\}}\lambda^{l-2}K_1(\lambda)\mathscr{H}_{l-m}f^{[j+1-m]} + \lambda^{l-2}K_1(\lambda)\mathscr{H}_l f^{[j+1]} - \lambda^{-1}K_2(\lambda)L^R(\lambda)f^{[j]}. \tag{7.6.5}$$

For $\kappa' = 0$, (7.6.4) coincides with (7.6.5) for $j = 0$. Assume that (7.6.4) holds for some $\kappa' < \kappa$. Then we have with the aid of (7.6.4) and (7.6.5) that

$$\begin{aligned} &\lambda^{l-1}K_1(\lambda)\mathscr{H}_l f \\ &\quad = -\left[\sum_{j=0}^{\kappa'}\left(\sum_{i=0}^{l-2-\kappa'+j}\lambda^{i-j-1}K_1(\lambda)\mathscr{H}_i f^{[j]}\right) - \lambda^{-j-1}f^{[j]}\right] + \lambda^{-\kappa'-2}f^{[\kappa'+1]} \\ &\quad\quad - \sum_{i=0}^{l-2}\lambda^{i-\kappa'-2}K_1(\lambda)\mathscr{H}_i f^{[\kappa'+1]} + \sum_{m=2}^{\min\{\kappa'+2,l\}}\lambda^{l-\kappa'-3}K_1(\lambda)\mathscr{H}_{l-m}f^{[\kappa'+2-m]} \\ &\quad\quad + \lambda^{l-\kappa'-3}K_1(\lambda)\mathscr{H}_l f^{[\kappa'+2]} - \sum_{j=0}^{\kappa'+1}\lambda^{-j-1}K_2(\lambda)L^R(\lambda)f^{[j]} \end{aligned}$$

$$= -\left(\sum_{j=0}^{\kappa'}\sum_{i=0}^{l-3-\kappa'+j}\lambda^{i-j-1}K_1(\lambda)\mathscr{H}_i f^{[j]}\right)$$

$$- \sum_{\substack{j=0\\ l-2-\kappa'+j\geq 0}}^{\kappa'} \lambda^{l-\kappa'-3}K_1(\lambda)\mathscr{H}_{l-2-\kappa'+j}f^{[j]} + \sum_{j=0}^{\kappa'+1}\lambda^{-j-1}f^{[j]}$$

$$- \sum_{i=0}^{l-2}\lambda^{i-\kappa'-2}K_1(\lambda)\mathscr{H}_i f^{[\kappa'+1]} + \sum_{m=2}^{\min\{\kappa'+2,l\}}\lambda^{l-\kappa'-3}K_1(\lambda)\mathscr{H}_{l-m}f^{[\kappa'+2-m]}$$

$$+ \lambda^{l-\kappa'-3}K_1(\lambda)\mathscr{H}_l f^{[\kappa'+2]} - \sum_{j=0}^{\kappa'+1}\lambda^{-j-1}K_2(\lambda)L^R(\lambda)f^{[j]}$$

$$= -\sum_{j=0}^{\kappa'+1}\left(\sum_{i=0}^{l-2-(\kappa'+1)+j}\lambda^{i-j-1}K_1(\lambda)\mathscr{H}_i f^{[j]} - \lambda^{-j-1}f^{[j]}\right)$$

$$+ \lambda^{l-2-(\kappa'+1)}K_1(\lambda)\mathscr{H}_l f^{[(\kappa'+1)+1]} - \sum_{j=0}^{\kappa'+1}\lambda^{-j-1}K_2(\lambda)L^R(\lambda)f^{[j]} \qquad \square$$

THEOREM 7.6.5. *Let* $1 < p < \infty$ *and* $s \geq 1$. *Additionally to the general assumptions of this chapter we require that* $\pi_{n-k,i} \in W_p^{s+i-1}(a,b)$ *provided that* $i = \max\{0, 2-s\}, \dots, n-1$, $k = 0, \dots, \min\{n-i, n-1\}$, $\pi_{n,n} \in W_p^{n+s-2}(a,b)$, $\pi_{n,n}^{-1} \in L_\infty(a,b)$. *If* $p \leq \frac{3}{2}$, *we require that* (7.4.6) *holds. Suppose that the assumptions of Theorem* 7.2.4 *and condition* iv$'$) *in Corollary* 7.2.6 *are satisfied and that the boundary eigenvalue problem* (7.1.2), (7.1.3) *is strongly s-regular. Let*

$$F := \left\{ f \in W_p^{n+s-1}(a,b) : \sum_{j=0}^{s-1}\lambda^{-j}C_2(\lambda)^{-1}L^R(\lambda)f^{[j]} = O(\lambda^{-s-\nu_1}) \right\},$$

where the $f^{[j]}$ *are as in Proposition* 7.6.2 *and* ν_1 *is as in Theorem* 7.2.4. *Let* $\lambda_0, \lambda_1, \dots$ *be the eigenvalues of L and let*

$$\{\eta_{\kappa,l}^{(j)} : j = 1, \dots, r(\lambda_\kappa); l = 0, \dots, m_{\kappa,j}-1\}$$

and

$$\{(u_{\kappa,l}^{(j)}, d_{\kappa,l}^{(j)}) : j = 1, \dots, r(\lambda_\kappa); l = 0, \dots, m_{\kappa,j}-1\}$$

be biorthogonal CSEAVs of L and L^* *at* λ_κ, *respectively. Then*

$$f = \lim_{\nu\to\infty}\sum_{\substack{\kappa\in\mathbb{N}\\ |\lambda_\kappa|<\rho_\nu}}\left(\sum_{j=1}^{r(\lambda_\kappa)}\sum_{l=1}^{\min\{m_{\kappa,j},n\}}\binom{n-1}{l-1}\lambda_\kappa^{n-l} \times \right.$$
$$\left. \times \sum_{h=0}^{m_{\kappa,j}-l}\eta_{\kappa,h}^{(j)}\int_a^b u_{\kappa,m_{\kappa,j}-l-h}^{(j)}(x)\pi_{n,n}(x)f(x)\,\mathrm{d}\right)$$

holds for all $f \in F$, *where the series converges in* $L_p(a,b)$.

Proof. This proof essentially coincides with that of Theorem 5.7.3. The assumptions of Proposition 7.6.2 are satisfied with $l = n$, $\kappa = s-1$, and $m = n$. Let $J_F : F \to L_p(a,b)$ be the canonical inclusion of F into $L_p(a,b)$. Then Proposition 7.6.4 yield for $f \in F$ that

$$\begin{aligned} Q_\nu J_F f =& \frac{1}{2\pi i}\oint_{\Gamma_\nu} \lambda^{n-1}\widetilde{J}_1 K_1(\lambda)\pi_{n,n} f\,d\lambda \\ =& f - \frac{1}{2\pi i}\sum_{j=0}^{s-1}\sum_{i=0}^{n-1-s+j}\oint_{\Gamma_\nu}\lambda^{i-j-1}\widetilde{J}_1 K_1(\lambda)\mathscr{H}_i f^{[j]}\,d\lambda \\ &+ \frac{1}{2\pi i}\oint_{\Gamma_\nu}\lambda^{n-1-s}\widetilde{J}_1 K_1(\lambda)\mathscr{H}_n f^{[s]}\,d\lambda \\ &- \frac{1}{2\pi i}\oint_{\Gamma_\nu}\sum_{j=0}^{s-1}\lambda^{-j-1}\widetilde{J}_1 K_2(\lambda)L^R(\lambda)f^{[j]}\,d\lambda. \end{aligned}$$

Each of the integrals containing K_1 is of the form

$$\oint_{\Gamma_\nu}\lambda^{n-1-\mu}\widetilde{J}_1 K_1(\lambda)g\,d\lambda$$

with $\mu \geq s$. From (7.4.4) we know that

$$\lambda^{n-1}K_1(\lambda)g = e_1^{\mathsf{T}} C_1 \widetilde{T}^{-1}(\lambda)(C_1^{-1}e_n g, 0),$$

whence

$$\left|\oint_{\Gamma_\nu}\lambda^{n-1-\mu}K_1(\lambda)g\,d\lambda\right|_p \leq |C_1|_\infty \left|\oint_{\Gamma_\nu}\lambda^{-\mu}\widetilde{T}^{-1}(\lambda)(C_1^{-1}e_n g, 0)\,d\lambda\right|_p \to 0$$

as $\nu \to \infty$ if $g \in W_p^1(a,b)$ for $\mu = s$ and $g \in L_p(a,b)$ for $\mu > s$. Indeed, if $\mu > s$ this follows from Theorem 4.4.11 i). And if $\mu = s$, then the estimate follows from (5.7.1) and estimates in the proof of Theorem 4.6.9. In particular this holds for all $g = \mathscr{H}_i f^{[j]}$ if we require $f \in W_p^{n+s}(a,b)$. In view of

$$\begin{aligned} K_2(\lambda)c &= e_1^{\mathsf{T}} T^{-1}(\lambda)(0,c) \\ &= e_1^{\mathsf{T}} C(\cdot,\lambda)\widetilde{T}^{-1}(\lambda)(0, C_2(\lambda)^{-1}c) \\ &= e_1^{\mathsf{T}} C_1 \widetilde{T}^{-1}(\lambda)(0, \lambda^{\nu_1} C_2(\lambda)^{-1}c) \end{aligned}$$

for $c \in \mathbb{C}^n$ and Theorem 4.4.11 i) we infer for $f \in F$ that

$$\oint_{\Gamma_\nu}\sum_{j=0}^{s-1}\lambda^{-j-1}K_2(\lambda)L^R(\lambda)f^{[j]}\,d\lambda \to 0$$

as $\nu \to \infty$. The boundedness of $\{P_\nu J_F : \nu \in \mathbb{N}\}$ follows from the above representations as in the proof of Theorem 5.7.3. Here we have used that similarly to the proof of Theorem 5.7.3 we can choose $f^{[j]}$ such that $f \mapsto f^{[j]}$ is a continuous map from $W_p^{n+s}(a,b)$ into $W_p^{n+s-j}(a,b)$. The remaining parts of the proof concerning the convergence are analogous to the corresponding parts in the proof of Theorem 5.7.3. The representation of the residues follows from Theorem 1.6.7, see Theorem 7.4.3. □

For uniform convergence we obtain as in Theorem 5.7.6:

THEOREM 7.6.6. *Let $s \geq 1$. Additionally to the general assumptions of this chapter we require that $\pi_{n-k,i} \in W_\infty^{s+i-1}(a,b)$ for $i = \max\{0, 2-s\}, \ldots, n-1$, $k = 0, \ldots, \min\{n-i, n-1\}$, $\pi_{n,n} \in W_\infty^{n+s-2}$, $\pi_{i,j} \in L_\infty(a,b)$ for the remaining indices, and $\pi_{n,n}^{-1} \in L_\infty(a,b)$. Suppose that the assumptions of Theorem 7.2.4 and condition iv$'$) in Corollary 7.2.6 are satisfied and that the boundary eigenvalue problem (7.1.2), (7.1.3) is strongly s-regular. Let*

$$F := \Big\{ f \in C^{n+s-1}(a,b) : f^{(n+s-1)} \in \mathrm{BV}[a,b],$$

$$\sum_{j=0}^{s} \lambda^{-j} C_2(\lambda)^{-1} L^R(\lambda) f^{[j]} = O(\lambda^{-s-1-\nu_1}) \Big\},$$

where the $f^{[j]}$ are as in Proposition 7.6.2 and ν_1 is as in Theorem 7.2.4. Let $\lambda_0, \lambda_1, \ldots$ be the eigenvalues of L, and let

$$\{\eta_{\kappa,l}^{(j)} : j = 1, \ldots, r(\lambda_\kappa);\, l = 0, \ldots, m_{\kappa,j} - 1\}$$

and

$$\{(u_{\kappa,l}^{(j)}, d_{\kappa,l}^{(j)}) : j = 1, \ldots, r(\lambda_\kappa);\, l = 0, \ldots, m_{\kappa,j} - 1\}$$

be biorthogonal CSEAVs of L and L^* at λ_κ, respectively. Then

$$f = \lim_{\nu \to \infty} \sum_{\substack{\kappa \in \mathbb{N} \\ |\lambda_\kappa| < \rho_\nu}} \Bigg(\sum_{j=1}^{r(\lambda_\kappa)} \sum_{l=1}^{\min\{m_{\kappa,j}, n\}} \binom{n-1}{l-1} \lambda_\kappa^{n-l} \times$$

$$\times \sum_{h=0}^{m_{\kappa,j}-l} \eta_{\kappa,h}^{(j)} \int_a^b u_{\kappa, m_{\kappa,j}-l-h}^{(j)}(x) \pi_{n,n}(x) f(x)\, \mathrm{d} \Bigg)$$

holds for all $f \in F$, where the series is uniformly convergent.

In the two previous expansion theorems we had to require that f belongs to $W_p^{n+s-1}(a,b)$ since the differential operator $\mathscr{H}_0$ occurs already in the very first iteration step. Since there are λ-powers which are smaller than necessary, a substitution for the asymptotically linear differential system would reduce the order of differentiation. However, boundary terms occur which still contain the high derivatives. Therefore, in the next theorems, we shall use an iteration with the asymptotically linear differential system. For this we note that

$$\widetilde{T}^D(\lambda) y = y' - \sum_{i=-1}^{n-1} \lambda^{-i} A_{-i} y \quad (y \in (W_p^1(a,b))^n),$$

where $A_1 = \operatorname{diag}(r_1, \dots, r_n)$,

$$A_0 = -C_1^{-1}(\delta_{i,n}\pi_{n-j+1,n-j})_{i,j=1}^n C_1 - C_1^{-1}C_1',$$
$$A_{-\nu} = -C_1^{-1}(\delta_{i,n}\pi_{n-j+1,n-j-\nu})_{i,j=1}^n C_1 \quad (\nu = 1, \dots, n-1).$$

Here $\pi_{j,m} := 0$ for $m < 0$. Let $\kappa > 0$ and assume that $\pi_{i,j} \in W_p^{\kappa+j-i}(a,b)$ for all $i = 1, \dots, n$ and $j = 0, \dots, i$ for which $\kappa + j - i > 0$ (the remaining $\pi_{i,j}$ belong to $L_p(a,b)$ by our general assumptions). From Proposition 7.2.3 we obtain that $r_i \in W_p^\kappa(a,b)$ for $i = 1, \dots, n$. Therefore C_1, C_1^{-1} and A_1 belong to $M_n(W_p^\kappa(a,b))$. Now it follows that $A_{-\nu} \in M_n(W_p^{\kappa-1-\nu}(a,b))$ for $\nu = 0, \dots, \min\{\kappa-1, n-1\}$, and $A_{-\nu} \in M_n(L_p(a,b))$ for $\nu = \kappa, \dots, n-1$.

For $y \in (L_p(a,b))^n$ and $c \in \mathbb{C}^n$ we set

$$R_1(\lambda)y := \widetilde{T}(\lambda)^{-1}(y,0), \quad R_2(\lambda)y := \widetilde{T}(\lambda)^{-1}(0,c).$$

PROPOSITION 7.6.7. *Let the assumptions be as above. For $f \in W_p^\kappa(a,b)$ we set $y^{[0]} := A_1^{-1}C_1^{-1}e_n\pi_{n,n}f \in (W_p^\kappa(a,b))$ and*

$$y^{[j]} := A_1^{-1}y^{[j-1]\prime} - \sum_{i=0}^{\min\{j-1,n-1\}} A_1^{-1}A_{-i}y^{[j-i-1]} \in (W_p^{\kappa-j}(a,b))^n$$

for $j = 1, \dots, \kappa$. Then

$$\begin{aligned}(7.6.6)\quad R_1(\lambda)C_1^{-1}e_n\pi_{n,n}f = &-\sum_{j=0}^{\kappa'}\lambda^{-j-1}y^{[j]} - \sum_{j=0}^{\kappa'}\sum_{i=\kappa'-j+1}^{n-1}\lambda^{-i-j-1}R_1(\lambda)A_{-i}y^{[j]}\\ &+\lambda^{-1-\kappa'}R_1(\lambda)A_1y^{[\kappa'+1]} + \sum_{j=0}^{\kappa'}\lambda^{-j-1}R_2(\lambda)\widetilde{T}^R(\lambda)y^{[j]}\end{aligned}$$

holds for $\kappa' = 0, \dots, \kappa-1$.

Proof. By definition of $y^{[0]}$ we have $R_1(\lambda)C_1^{-1}e_n\pi_{n,n}f = R_1(\lambda)A_1y^{[0]}$. Then we obtain for $j = 0, \dots, \kappa-1$ that

$$y^{[j]} = R_1(\lambda)y^{[j]\prime} - \sum_{i=-1}^{n-1}\lambda^{-i}R_1(\lambda)A_{-i}y^{[j]} + R_2(\lambda)\widetilde{T}^R(\lambda)y^{[j]},$$

whence

$$\begin{aligned}R_1(\lambda)A_1y^{[j]} = &-\lambda^{-1}y^{[j]} - \sum_{i=1}^{n-1}\lambda^{-i-1}R_1(\lambda)A_{-i}y^{[j]} + \lambda^{-1}R_1(\lambda)A_1y^{[j+1]}\\ &+\sum_{i=1}^{\min\{j,n-1\}}\lambda^{-1}R_1(\lambda)A_{-i}y^{[j-i]} + \lambda^{-1}R_2(\lambda)\widetilde{T}^R(\lambda)y^{[j]}.\end{aligned}$$

This proves (7.6.6) for $\kappa' = 0$. Now assume that (7.6.6) holds for some $\kappa' < \kappa - 1$. With the above equation for $j = \kappa' + 1$ we infer

$$\begin{aligned} R_1(\lambda)A_1 y^{[0]} &= -\sum_{j=0}^{\kappa'} \lambda^{-j-1} y^{[j]} - \sum_{j=0}^{\kappa'} \sum_{i=\kappa'-j+1}^{n-1} \lambda^{-i-j-1} R_1(\lambda)A_{-i} y^{[j]} \\ &\quad - \lambda^{-2-\kappa'} y^{[\kappa'+1]} - \sum_{i=1}^{n-1} \lambda^{-i-2-\kappa'} R_1(\lambda)A_{-i} y^{[\kappa'+1]} + \lambda^{-2-\kappa'} R_1(\lambda)A_1 y^{[\kappa'+2]} \\ &\quad + \sum_{i=1}^{\min\{\kappa'+1,n-1\}} \lambda^{-2-\kappa'} R_1(\lambda)A_{-i} y^{[\kappa'+1-i]} + \sum_{j=0}^{\kappa'+1} \lambda^{-j-1} R_2(\lambda)\widetilde{T}^R(\lambda) y^{[j]} \\ &= -\sum_{j=0}^{\kappa'+1} \lambda^{-j-1} y^{[j]} - \sum_{j=0}^{\kappa'+1} \sum_{i=\kappa'-j+2}^{n-1} \lambda^{-i-j-1} R_1(\lambda)A_{-i} y^{[j]} \\ &\quad + \sum_{j=0}^{\kappa'+1} \lambda^{-j-1} R_2(\lambda)\widetilde{T}^R(\lambda) y^{[j]}. \qquad \square \end{aligned}$$

The definition of the $y^{[j]}$ is suitable for the proof of the extension theorems. However, to calculate the occurring auxiliary boundary conditions, it seems to be better to use the vector functions $C_1 y^{[j]}$:

PROPOSITION 7.6.8. *Let the assumptions be as above. Set*

$$\widetilde{A}_1 := (\delta_{i,j-1} - \delta_{i,n}\pi_{n-j+1,n-j+1})_{i,j=1}^n,$$
$$\widetilde{A}_{-\nu} := -(\delta_{i,n}\pi_{n-j+1,n-j-\nu})_{i,j=1}^n \quad (\nu = 0,\ldots,n-1).$$

For $f \in W_p^\kappa(a,b)$ we set $\tilde{y}^{[0]} := -e_1 f$ and

$$\tilde{y}^{[j]} := \widetilde{A}_1^{-1}\tilde{y}^{[j-1]\prime} - \sum_{i=0}^{\min\{j-1,n-1\}} \widetilde{A}_1^{-1}\widetilde{A}_{-i}\tilde{y}^{[j-i-1]} \in (W_p^{\kappa-j}(a,b))^n$$

for $j = 1,\ldots,\kappa$. Then $\tilde{y}^{[j]} = C_1 y^{[j]}$ for $j = 0,\ldots,\kappa$.

Proof. For $j = 0$ this follows from (7.4.5), whereas a straightforward calculation proves it for $j = 1,\ldots,\kappa$. $\square$

We recall that

$$T^R(\lambda)y = \sum_{j=0}^{\infty} \left(w_{ik}^{(j)}(\lambda)\right)_{i,k=1}^n y(a_j) + \int_a^b \left(w_{ik}(x,\lambda)\right)_{i,k=1}^n y(x)$$

and set $\Xi(\lambda) = \operatorname{diag}(1,\lambda,\ldots,\lambda^{n-1})$.

From the identity (7.6.6) for $\kappa = s+1$ we conclude similarly to Theorems 5.7.3 and 5.7.6:

THEOREM 7.6.9. *Let $1 < p < \infty$ and let s be a positive integer. Suppose that $\pi_{i,j} \in W_p^{s+1+j-i}(a,b)$ for all $i = 1,\dots,n$ and $j = 0,\dots,i$ for which $s+1+j-i > 0$ and that $\pi_{i,j} \in L_p(a,b)$ for the remaining indices. Suppose that the assumptions of Theorem* 7.2.4 *and condition* iv′) *in Corollary* 7.2.6 *are satisfied and that the boundary eigenvalue problem* (7.1.2), (7.1.3) *is strongly s-regular. Let*

$$F := \left\{ f \in W_p^s(a,b) : \sum_{j=0}^{s-1} \lambda^{-j} C_2(\lambda)^{-1} T^R(\lambda) \Xi(\lambda) \tilde{y}^{[j]} = O(\lambda^{-s}) \right\},$$

where the $\tilde{y}^{[j]}$ are defined in Proposition 7.6.8. *Let $\lambda_0, \lambda_1, \dots$ be the eigenvalues of L and let*

$$\{\eta_{\kappa,l}^{(j)} : j = 1,\dots,r(\lambda_\kappa);\, l = 0,\dots,m_{\kappa,j}-1\}$$

and

$$\{(u_{\kappa,l}^{(j)}, d_{\kappa,l}^{(j)}) : j = 1,\dots,r(\lambda_\kappa);\, l = 0,\dots,m_{\kappa,j}-1\}$$

be biorthogonal CSEAVs of L and L^* at λ_κ, respectively. Then

$$f = \lim_{\nu\to\infty} \sum_{\substack{\kappa\in\mathbb{N}\\ |\lambda_\kappa|<\rho_\nu}} \left(\sum_{j=1}^{r(\lambda_\kappa)} \sum_{l=1}^{\min\{m_{\kappa,j},n\}} \binom{n-1}{l-1} \lambda_\kappa^{n-l} \times \right.$$
$$\left. \times \sum_{h=0}^{m_{\kappa,j}-l} \eta_{\kappa,h}^{(j)} \int_a^b u_{\kappa,m_{\kappa,j}-l-h}^{(j)}(x) \pi_{n,n}(x) f(x)\,\mathrm{d} \right)$$

holds for all $f \in F$, where the series converges in $L_p(a,b)$.

THEOREM 7.6.10. *Let s be a positive integer. Suppose that $\pi_{i,j} \in W_\infty^{s+1+j-i}(a,b)$ for all $i = 1,\dots,n$ and $j = 0,\dots,i$ for which $s+1+j-i > 0$ and that $\pi_{i,j}$ belongs to $L_\infty(a,b)$ for the remaining indices. Suppose that the assumptions of Theorem* 7.2.4 *and condition* iv′) *in Corollary* 7.2.6 *are satisfied and that the boundary eigenvalue problem* (7.1.2), (7.1.3) *is strongly s-regular. Let*

$$F := \Big\{ f \in C^s(a,b) : f^{(s)} \in \mathrm{BV}[a,b],$$
$$\sum_{j=0}^{s} \lambda^{-j} C_2(\lambda)^{-1} T^R(\lambda) \Xi(\lambda) \tilde{y}^{[j]} = O(\lambda^{-s-1}) \Big\},$$

where the $\tilde{y}^{[j]}$ are defined in Proposition 7.6.8. *Let $\lambda_0, \lambda_1, \dots$ be the eigenvalues of L and let*

$$\{\eta_{\kappa,l}^{(j)} : j = 1,\dots,r(\lambda_\kappa);\, l = 0,\dots,m_{\kappa,j}-1\}$$

and

$$\{(u_{\kappa,l}^{(j)}, d_{\kappa,l}^{(j)}) : j = 1,\dots,r(\lambda_\kappa);\, l = 0,\dots,m_{\kappa,j}-1\}$$

be biorthogonal CSEAVs of L and L at λ_κ, respectively. Then*

$$f = \lim_{\nu\to\infty} \sum_{\substack{\kappa\in\mathbb{N}\\ |\lambda_\kappa|<\rho_\nu}} \left(\sum_{j=1}^{r(\lambda_\kappa)} \sum_{l=1}^{\min\{m_{\kappa,j},n\}} \binom{n-1}{l-1} \lambda_\kappa^{n-l} \times \right.$$

$$\left. \times \sum_{h=0}^{m_{\kappa,j}-l} \eta_{\kappa,h}^{(j)} \int_a^b u_{\kappa,m_{\kappa,j}-l-h}^{(j)}(x)\pi_{n,n}(x) f(x)\,\mathrm{d}\right)$$

holds for all $f \in F$, where the series converges in $C(a,b)$.

7.7. Boundary eigenvalue problems for $\eta'' + p_1\eta' + p_0\eta = \lambda^2\eta$

We consider the boundary eigenvalue problem

$$\begin{aligned} \eta'' + p_1\eta' + p_0\eta &= \lambda^2\eta, \\ a_0\eta(0) + a_1\eta'(0) + b_0\eta(1) + b_1\eta'(1) &= 0, \\ c_0\eta(0) + c_1\eta'(0) + d_0\eta(1) + d_1\eta'(1) &= 0. \end{aligned}$$

Here we suppose that the boundary conditions are linearly independent and that $p_0, p_1 \in L_p(a,b)$. This problem is well-known and was considered e. g. in Naimark [NA1] and Locker [LO1].

First we want to characterize the Birkhoff regular problems. For this we note that in case $a_1d_1 = b_1c_1$ a suitable linear combination of the boundary conditions yields that we can take $c_1 = d_1 = 0$. Therefore we shall always assume that $c_1 = d_1 = 0$ holds if $a_1d_1 = b_1c_1$.

THEOREM 7.7.1. *In the following three cases the problem is Birkhoff regular:*
CASE *1:* $a_1d_1 \neq b_1c_1$,
CASE *2:* $c_1 = d_1 = 0,\ a_1d_0 + b_1c_0 \neq 0$,
CASE *3:* $a_1 = b_1 = c_1 = d_1 = 0$.

Proof. The differential equation is as considered in Example 4.2.2 with $r_1 = 1$ and $r_2 = -1$. Hence $C(\lambda) = \begin{pmatrix} 1 & 1 \\ \lambda & -\lambda \end{pmatrix}$, and the boundary matrices given by (7.3.1) are

$$W^{(0)}(\lambda) = \begin{pmatrix} a_0 + \lambda a_1 & a_0 - \lambda a_1 \\ c_0 + \lambda c_1 & c_0 - \lambda c_1 \end{pmatrix},$$

$$W^{(1)}(\lambda) = \begin{pmatrix} b_0 + \lambda b_1 & b_0 - \lambda b_1 \\ d_0 + \lambda d_1 & d_0 - \lambda d_1 \end{pmatrix}.$$

We shall apply Theorem 7.3.2 to prove Birkhoff regularity. The matrix polynomial C_2 will be different in the three cases under consideration.

CASE 1: Here we take $C_2(\lambda)=\lambda I_2$ and obtain

$$W_0^{(0)}=\begin{pmatrix} a_1 & -a_1 \\ c_1 & -c_1 \end{pmatrix}, \quad W_0^{(1)}=\begin{pmatrix} b_1 & -b_1 \\ d_1 & -d_1 \end{pmatrix}.$$

Note that the matrices Λ_ν^j occuring in Theorem 4.1.3 are $\operatorname{diag}(1,0)$ and $\operatorname{diag}(0,1)$. Therefore, the Birkhoff matrices are

$$\begin{pmatrix} a_1 & -b_1 \\ c_1 & -d_1 \end{pmatrix} \text{ and } \begin{pmatrix} b_1 & -a_1 \\ d_1 & -c_1 \end{pmatrix},$$

which are invertible. Therefore Theorems 7.3.2 and 4.1.3 yield Birkhoff regularity.

CASE 2: Here we take $C_2(\lambda)=\operatorname{diag}(\lambda,1)$ and obtain

$$W_0^{(0)}=\begin{pmatrix} a_1 & -a_1 \\ c_0 & c_0 \end{pmatrix}, \quad W_0^{(1)}=\begin{pmatrix} b_1 & -b_1 \\ d_0 & d_0 \end{pmatrix}.$$

The Birkhoff matrices are

$$\begin{pmatrix} a_1 & -b_1 \\ c_0 & d_0 \end{pmatrix} \text{ and } \begin{pmatrix} b_1 & -a_1 \\ d_0 & c_0 \end{pmatrix}.$$

Hence the problem is Birkhoff regular by the assumption that $a_1d_0+b_1c_0\neq 0$.

CASE 3: Here we take $C_2(\lambda)=I_2$ and obtain

$$W_0^{(0)}=\begin{pmatrix} a_0 & a_0 \\ c_0 & c_0 \end{pmatrix}, \quad W_0^{(1)}=\begin{pmatrix} b_0 & b_0 \\ d_0 & d_0 \end{pmatrix}.$$

The Birkhoff matrices are

$$\begin{pmatrix} a_0 & b_0 \\ c_0 & d_0 \end{pmatrix} \text{ and } \begin{pmatrix} b_0 & a_0 \\ d_0 & c_0 \end{pmatrix}.$$

They are invertible since the boundary conditions are linearly independent. □

In view of our general asumptions, the considered cases can be written as follows:

CASE 1: $|a_1|+|b_1|>0$, $|c_1|+|d_1|>0$,

CASE 2: $|a_1|+|b_1|>0$, $|c_1|+|d_1|=0$, $a_1d_0+b_1c_0\neq 0$,

CASE 3: $|a_1|+|b_1|=0$, $|c_1|+|d_1|=0$.

Hence we still have to consider the case

$$|a_1|+|b_1|>0,\ c_1=d_1=0,\ a_1d_0+b_1c_0=0.$$

Next we consider a case where no expansion holds.

THEOREM 7.7.2. *The spectrum of the boundary eigenvalue problem is empty in*

CASE 4a: $|a_1|+|b_1|>0$, $c_1=d_1=0$, $a_1d_0+b_1c_0=0$, $b_0=0$, $d_0=0$.

CASE 4b: $|a_1|+|b_1|>0$, $c_1=d_1=0$, $a_1d_0+b_1c_0=0$, $a_0=0$, $c_0=0$.

Proof. We only consider Case 4a; Case 4b is obtained by interchanging the endpoints of the interval $[0,1]$. Under the assumptions of Case 4a we have $b_1 c_0 = 0$. Since $c_0 \neq 0$, $b_1 = 0$ and $a_1 \neq 0$ follow. The second boundary condition reads $c_0 \eta(0) = 0$, and we thus may assume that $a_0 = 0$. Therefore the first boundary condition is $a_1 \eta'(0) = 0$. This is an initial value problem, which does not have nontrivial solutions. □

Therefore we still have to consider the case

$$|a_1| + |b_1| > 0,\; c_1 = d_1 = 0,\; a_1 d_0 + b_1 c_0 = 0,\; (|b_0| + |d_0|)(|a_0| + |c_0|) > 0.$$

Here we shall check for Stone regularity. For this we need additional conditions on the regularity of p_0 and p_1. We start with formal calculations assuming this regularity. Later on we shall state the precise conditions.

According to Corollary 5.5.6 and Proposition 5.5.7 we have to check the coefficients of the polynomial in λ^{-1} for $\kappa = 1, 2$:

$$\det\left(\sum_{r=0}^{k} \lambda^{-r}\left(\widetilde{\widetilde{W}}_r^{(0)} \Lambda_\kappa^1 + \widetilde{\widetilde{W}}_r^{(1)} \Lambda_\kappa^2\right)\right), \tag{7.7.1}$$

where $\Lambda_1^1 = \operatorname{diag}(1,0)$, $\Lambda_2^1 = \operatorname{diag}(0,1)$, $\Lambda_\kappa^2 = I_2 - \Lambda_\kappa^1$,

$$\widetilde{\widetilde{W}}_r^{(j)} = \sum_{\nu=0}^{r} W_\nu^{(j)} \widehat{P}^{[r-\nu]}(a_j).$$

Here the transformation in the proof of Case 2 in Theorem 7.7.1 is taken. Therefore

$$W_0^{(0)} = \begin{pmatrix} a_1 & -a_1 \\ c_0 & c_0 \end{pmatrix}, \quad W_1^{(0)} = \begin{pmatrix} a_0 & a_0 \\ 0 & 0 \end{pmatrix},$$
$$W_0^{(1)} = \begin{pmatrix} b_1 & -b_1 \\ d_0 & d_0 \end{pmatrix}, \quad W_1^{(1)} = \begin{pmatrix} b_0 & b_0 \\ 0 & 0 \end{pmatrix}.$$

Then we have $\widetilde{\widetilde{W}}_0^{(j)} = W_0^{(j)}$ and for $r > 0$

$$\widetilde{\widetilde{W}}_r^{(j)} = W_0^{(j)} \widehat{P}^{[r]}(a_j) + W_1^{(j)} \widehat{P}^{[r-1]}(a_j).$$

From

$$W_0^{(0)} \widehat{P}^{[r]}(0) = \begin{pmatrix} a_1\left(\widehat{P}_{11}^{[r]}(0) - \widehat{P}_{21}^{[r]}(0)\right) & a_1\left(\widehat{P}_{12}^{[r]}(0) - \widehat{P}_{22}^{[r]}(0)\right) \\ c_0\left(\widehat{P}_{11}^{[r]}(0) + \widehat{P}_{21}^{[r]}(0)\right) & c_0\left(\widehat{P}_{12}^{[r]}(0) + \widehat{P}_{22}^{[r]}(0)\right) \end{pmatrix}$$

we infer in view of $a_1 d_0 + b_1 c_0 = 0$ that

$$\det\left(W_0^{(0)} \widehat{P}^{[r]}(0) \Lambda_\kappa^1 + W_0^{(1)} \Lambda_\kappa^2\right) = -2 a_1 d_0 \widehat{P}_{\kappa'\kappa}^{[r]}(0)$$

where $\kappa' = 3 - \kappa$. In the same way we obtain

$$\det\big(W_0^{(0)}\Lambda_\kappa^1 + W_0^{(1)}\widehat{P}^{[r]}(1)\Lambda_\kappa^2\big) = 2a_1 d_0 \widehat{P}^{[r]}_{\kappa\kappa'}(1).$$

Also

$$\det\big(W_0^{(0)}\widehat{P}^{[r]}(0)\Lambda_\kappa^1 + W_1^{(1)}\Lambda_\kappa^2\big) = (-1)^\kappa b_0 c_0\big(\widehat{P}^{[r]}_{1\kappa}(0) + \widehat{P}^{[r]}_{2\kappa}(0)\big),$$
$$\det\big(W_1^{(0)}\Lambda_\kappa^1 + W_0^{(1)}\widehat{P}^{[r]}(1)\Lambda_\kappa^2\big) = (-1)^{\kappa-1} a_0 d_0\big(\widehat{P}^{[r]}_{1\kappa'}(1) + \widehat{P}^{[r]}_{2\kappa'}(1)\big).$$

From

$$W_1^{(0)}\widehat{P}^{[r]}(0) = \begin{pmatrix} a_0\big(\widehat{P}^{[r]}_{11}(0) + \widehat{P}^{[r]}_{21}(0)\big) & a_0\big(\widehat{P}^{[r]}_{12}(0) + \widehat{P}^{[r]}_{22}(0)\big) \\ 0 & 0 \end{pmatrix}$$

we infer

$$\det\big(W_1^{(0)}\widehat{P}^{[r]}(0)\Lambda_\kappa^1 + W_0^{(1)}\Lambda_\kappa^2\big) = (-1)^{\kappa-1} a_0 d_0\big(\widehat{P}^{[r]}_{1\kappa}(0) + \widehat{P}^{[r]}_{2\kappa}(0)\big).$$

In the same way,

$$\det\big(W_0^{(0)}\Lambda_\kappa^1 + W_1^{(1)}\widehat{P}^{[r]}(1)\Lambda_\kappa^2\big) = (-1)^\kappa b_0 c_0\big(\widehat{P}^{[r]}_{1\kappa'}(1) + \widehat{P}^{[r]}_{2\kappa'}(1)\big).$$

From Example 4.2.2 we know that

$$A_0 = \frac{1}{2}p_1\begin{pmatrix} -1 & 1 \\ 1 & -1 \end{pmatrix}, \quad A_{-1} = \frac{1}{2}p_0\begin{pmatrix} -1 & -1 \\ 1 & 1 \end{pmatrix}.$$

THEOREM 7.7.3. *Suppose that $p_1 \in W_p^1(0,1)$ and that*
CASE 5: $|a_1| + |b_1| > 0$, $c_1 = d_1 = 0$, $a_1 d_0 + b_1 c_0 = 0$,
$b_0 c_0 - a_0 d_0 + \frac{1}{2}a_1 d_0\big(p_1(0) + p_1(1)\big) \neq 0$.
Then the problem is 1-regular.

Proof. The assumptions of Theorem 2.8.2 with respect to A_0 and A_{-1} are satisfied with $k = 1$. According to Definition 7.6.1, Corollary 5.5.6, and Proposition 5.5.7 we must show that the coefficient $\alpha_{1,\kappa}$ of λ^{-1} in (7.7.1) for $k = 1$ is different from zero for $\kappa = 1, 2$. We have

$$\begin{aligned}\alpha_{1,\kappa} &= \det\big(W_0^{(0)}\Lambda_\kappa^1 + W_0^{(1)}\widehat{P}^{[1]}(1)\Lambda_\kappa^2\big) \\ &\quad + \det\big(W_0^{(0)}\Lambda_\kappa^1 + W_1^{(1)}\Lambda_\kappa^2\big) \\ &\quad + \det\big(W_0^{(0)}\widehat{P}^{[1]}(0)\Lambda_\kappa^1 + W_0^{(1)}\Lambda_\kappa^2\big) \\ &\quad + \det\big(W_1^{(0)}\Lambda_\kappa^1 + W_0^{(1)}\Lambda_\kappa^2\big) \\ &= 2a_1 d_0 \widehat{P}^{[1]}_{\kappa\kappa'}(1) + (-1)^\kappa b_0 c_0 - 2a_1 d_0 \widehat{P}^{[1]}_{\kappa'\kappa}(0) + (-1)^{\kappa-1} a_0 d_0.\end{aligned}$$

From Proposition 5.5.7 we have

$$\widehat{P}^{[1]}_{\kappa\kappa'} = -(r_\kappa - r_{\kappa'})^{-1} A_{0,\kappa\kappa'} = (-1)^\kappa \frac{1}{4} p_1, \tag{7.7.2}$$

and it follows that

$$\alpha_{1,\kappa} = (-1)^\kappa \Big\{ b_0c_0 - a_0d_0 + \frac{1}{2}a_1d_0\big(p_1(1)+p_1(0)\big)\Big\}. \qquad \square$$

THEOREM 7.7.4. *Suppose that* $p_1 \in W_p^2(0,1)$, $p_0 \in W_p^1(0,1)$, *and that*
CASE 6: $|a_1|+|b_1|>0$, $c_1=d_1=0$, $a_1d_0+b_1c_0=0$,
$b_0c_0-a_0d_0+\frac{1}{2}a_1d_0\big(p_1(0)+p_1(1)\big)=0$, $a_1\neq 0$, $d_0\neq 0$,
$p_0(1)-p_0(0)-\frac{1}{2}\big(p_1'(1)-p_1'(0)\big)+\frac{1}{4}\big(p_1(0)^2-p_1(1)^2\big)\neq 0$.
Then the problem is 2-regular.

Proof. The proof is similar to that of the previous theorem. Here we have to consider the coefficient $\alpha_{2,\kappa}$ of λ^{-2} in (7.7.1). First we calculate

$$\begin{aligned}
&\det\big(W_0^{(0)}\widehat{P}^{[1]}(0)\Lambda_\kappa^1 + W_0^{(1)}\widehat{P}^{[1]}(1)\Lambda_\kappa^2\big)\\
&\quad = (-1)^{\kappa-1}a_1d_0\Big\{\big(\widehat{P}^{[1]}_{1\kappa}(0)-\widehat{P}^{[1]}_{2\kappa}(0)\big)\big(\widehat{P}^{[1]}_{1\kappa'}(1)+\widehat{P}^{[1]}_{2\kappa'}(1)\big)\\
&\qquad + \big(\widehat{P}^{[1]}_{1\kappa}(0)+\widehat{P}^{[1]}_{2\kappa}(0)\big)\big(\widehat{P}^{[1]}_{1\kappa'}(1)-\widehat{P}^{[1]}_{2\kappa'}(1)\big)\Big\}\\
&\quad = (-1)^{\kappa-1}2a_1d_0\big(\widehat{P}^{[1]}_{1\kappa}(0)\widehat{P}^{[1]}_{1\kappa'}(1)-\widehat{P}^{[1]}_{2\kappa}(0)\widehat{P}^{[1]}_{2\kappa'}(1)\big).
\end{aligned}$$

Then we obtain

$$\begin{aligned}
\alpha_{2,\kappa} =& \det\big(W_0^{(0)}\Lambda_\kappa^1 + W_0^{(1)}\widehat{P}^{[2]}(1)\Lambda_\kappa^2\big) + \det\big(W_0^{(0)}\Lambda_\kappa^1 + W_1^{(1)}\widehat{P}^{[1]}(1)\Lambda_\kappa^2\big)\\
&+\det\big(W_0^{(0)}\widehat{P}^{[2]}(0)\Lambda_\kappa^1 + W_0^{(1)}\Lambda_\kappa^2\big) + \det\big(W_1^{(0)}\widehat{P}^{[1]}(0)\Lambda_\kappa^1 + W_0^{(1)}\Lambda_\kappa^2\big)\\
&+\det\big(W_0^{(0)}\widehat{P}^{[1]}(0)\Lambda_\kappa^1 + W_0^{(1)}\widehat{P}^{[1]}(1)\Lambda_\kappa^2\big) + \det\big(W_0^{(0)}\widehat{P}^{[1]}(0)\Lambda_\kappa^1 + W_1^{(1)}\Lambda_\kappa^2\big)\\
&+\det\big(W_1^{(0)}\Lambda_\kappa^1 + W_0^{(1)}\widehat{P}^{[1]}(1)\Lambda_\kappa^2\big) + \det\big(W_1^{(0)}\Lambda_\kappa^1 + W_1^{(1)}\Lambda_\kappa^2\big)\\
=&2a_1d_0\widehat{P}^{[2]}_{\kappa\kappa'}(1) + (-1)^\kappa b_0c_0\big(\widehat{P}^{[1]}_{1\kappa'}(1)+\widehat{P}^{[1]}_{2\kappa'}(1)\big)\\
&-2a_1d_0\widehat{P}^{[2]}_{\kappa'\kappa}(0) + (-1)^{\kappa-1}a_0d_0\big(\widehat{P}^{[1]}_{1\kappa}(0)+\widehat{P}^{[1]}_{2\kappa}(0)\big)\\
&+(-1)^{\kappa-1}2a_1d_0\big(\widehat{P}^{[1]}_{1\kappa}(0)\widehat{P}^{[1]}_{1\kappa'}(1)-\widehat{P}^{[1]}_{2\kappa}(0)\widehat{P}^{[1]}_{2\kappa'}(1)\big)\\
&+(-1)^\kappa b_0c_0\big(\widehat{P}^{[1]}_{1\kappa}(0)+\widehat{P}^{[1]}_{2\kappa}(0)\big) + (-1)^{\kappa-1}a_0d_0\big(\widehat{P}^{[1]}_{1\kappa'}(1)+\widehat{P}^{[1]}_{2\kappa'}(1)\big)\\
=&a_1d_0\big\{2\widehat{P}^{[2]}_{\kappa\kappa'}(1) - 2\widehat{P}^{[2]}_{\kappa'\kappa}(0) + (-1)^{\kappa-1}2\big(\widehat{P}^{[1]}_{1\kappa}(0)\widehat{P}^{[1]}_{1\kappa'}(1)-\widehat{P}^{[1]}_{2\kappa}(0)\widehat{P}^{[1]}_{2\kappa'}(1)\big)\big\}\\
&+(-1)^{\kappa-1}(a_0d_0-b_0c_0)\big(\widehat{P}^{[1]}_{1\kappa}(0)+\widehat{P}^{[1]}_{2\kappa}(0)+\widehat{P}^{[1]}_{1\kappa'}(1)+\widehat{P}^{[1]}_{2\kappa'}(1)\big)\\
=&a_1d_0\big\{2\widehat{P}^{[2]}_{\kappa\kappa'}(1) - 2\widehat{P}^{[2]}_{\kappa'\kappa}(0) + 2\widehat{P}^{[1]}_{\kappa\kappa}(0)\widehat{P}^{[1]}_{\kappa\kappa'}(1) - 2\widehat{P}^{[1]}_{\kappa'\kappa}(0)\widehat{P}^{[1]}_{\kappa'\kappa'}(1)\\
&+\frac{1}{2}(-1)^{\kappa-1}\big(p_1(0)+p_1(1)\big)\big(\widehat{P}^{[1]}_{1\kappa}(0)+\widehat{P}^{[1]}_{2\kappa}(0)+\widehat{P}^{[1]}_{1\kappa'}(1)+\widehat{P}^{[1]}_{2\kappa'}(1)\big)\big\}.
\end{aligned}$$

In view of (7.7.2) it follows that

$$\begin{aligned}\alpha_{2,\kappa} =& 2a_1d_0\{\widehat{P}^{[2]}_{\kappa\kappa'}(1)-\widehat{P}^{[2]}_{\kappa'\kappa}(0)+\widehat{P}^{[1]}_{\kappa\kappa}(0)\widehat{P}^{[1]}_{\kappa\kappa'}(1)-\widehat{P}^{[1]}_{\kappa'\kappa}(0)\widehat{P}^{[1]}_{\kappa'\kappa'}(1)\\ &+(\widehat{P}^{[1]}_{\kappa'\kappa}(0)-\widehat{P}^{[1]}_{\kappa\kappa'}(1))(\widehat{P}^{[1]}_{\kappa\kappa}(0)+\widehat{P}^{[1]}_{\kappa'\kappa}(0)+\widehat{P}^{[1]}_{\kappa\kappa'}(1)+\widehat{P}^{[1]}_{\kappa'\kappa'}(1))\}.\\ =& 2a_1d_0\{\widehat{P}^{[2]}_{\kappa\kappa'}(1)-\widehat{P}^{[2]}_{\kappa'\kappa}(0)+\widehat{P}^{[1]}_{\kappa\kappa}(0)\widehat{P}^{[1]}_{\kappa'\kappa}(0)\\ &+\widehat{P}^{[1]}_{\kappa'\kappa}(0)^2-\widehat{P}^{[1]}_{\kappa\kappa'}(1)^2-\widehat{P}^{[1]}_{\kappa\kappa'}(1)\widehat{P}^{[1]}_{\kappa'\kappa'}(1)\}.\end{aligned}$$

Since $A_{0,\kappa\kappa}=A_{0,\kappa'\kappa'}$, we have in view of (5.5.12) and (7.7.2) that

$$\begin{aligned}\widehat{P}^{[2]}_{\kappa\kappa'} &= (-1)^{\kappa-1}\frac{1}{2}(\widehat{P}^{[1]\,\prime}_{\kappa\kappa'}-A_{0,\kappa\kappa'}\widehat{P}^{[1]}_{\kappa'\kappa'}-A_{-1,\kappa\kappa'})\\ &=\frac{1}{2}(-\frac{1}{4}p_1'+(-1)^{\kappa}\frac{1}{2}p_1\widehat{P}^{[1]}_{\kappa'\kappa'}+\frac{1}{2}p_0)\\ &=-\frac{1}{8}p_1'+\widehat{P}^{[1]}_{\kappa\kappa'}\widehat{P}^{[1]}_{\kappa'\kappa'}+\frac{1}{4}p_0.\end{aligned}$$

Therefore

$$\begin{aligned}\alpha_{2\kappa} &= a_1d_0\left\{\frac{1}{2}(p_0(1)-p_0(0))-\frac{1}{4}(p_1'(1)-p_1'(0))+2\widehat{P}^{[1]}_{\kappa'\kappa}(0)^2-2\widehat{P}^{[1]}_{\kappa\kappa'}(1)^2\right\}\\ &=\frac{1}{2}a_1d_0\left\{p_0(1)-p_0(0)-\frac{1}{2}(p_1'(1)-p_1'(0))+\frac{1}{4}(p_1(0)^2-p_1(1)^2)\right\}. \quad \square\end{aligned}$$

If $p_0=p_1=0$ we take the fundamental system $\{\cosh(\lambda x),\frac{1}{\lambda}\sinh(\lambda x)\}$. For the case not covered by Theorems 7.7.1–7.7.3 the coefficents satisfy $c_1=d_1=0$, $a_1d_0+b_1c_0=0$, and $b_0c_0-a_0d_0=0$. In this case, the characteristic determinant is $-a_1c_0-b_1d_0$. That means, the problem is not Stone regular since the spectrum is either empty or all of $\mathbb{C}$. Therefore we obtain

THEOREM 7.7.5. *Suppose that $p_0=p_1=0$. Then the spectrum of the boundary eigenvalue problem is empty in*
CASE 4c: $|a_1|+|b_1|>0$, $c_1=d_1=0$, $a_1d_0+b_1c_0=0$, $b_0c_0-a_0d_0=0$,
$a_1c_0+b_1d_0\neq 0$,
and the spectrum of the boundary eigenvalue problem is all of $\mathbb{C}$ in
CASE *4d:* $|a_1|+|b_1|>0$, $c_1=d_1=0$, $a_1d_0+b_1c_0=0$, $b_0c_0-a_0d_0=0$,
$a_1c_0+b_1d_0=0$.

Note that Case 4c contains Cases 4a and 4b. In particular, if $p_0=p_1=0$, then every s-regular problem with $s\geq 2$ is 1-regular. However, in case of general p_0 and p_1 it seems to be apparent that s-regular problems which are not $(s-1)$-regular can occur for all $s\in\mathbb{N}\setminus\{0\}$.

7.8. The Regge problem

We consider the Regge problem

$$\eta'' - q\eta - \lambda^2\eta = 0,$$
$$\eta(0) = 0, \quad \eta'(1) + \lambda\eta(1) = 0,$$

where q is a given function. Regularity conditions on this function will be given later. By Example 4.2.2 we have $C(\lambda) = \begin{pmatrix} 1 & 1 \\ \lambda & -\lambda \end{pmatrix}$, the corresponding asymptotically linear system is

$$y' = \lambda \begin{pmatrix} 1 & 0 \\ 0 & -1 \end{pmatrix} y + \frac{q}{2\lambda} \begin{pmatrix} 1 & 1 \\ -1 & -1 \end{pmatrix} y,$$

and the (asymptotically constant) boundary conditions are

$$\begin{pmatrix} 1 & 1 \\ 0 & 0 \end{pmatrix} y(0) + \begin{pmatrix} 0 & 0 \\ 1 & 0 \end{pmatrix} y(1) = 0.$$

Therefore the Birkhoff matrices are

$$\begin{pmatrix} 1 & 0 \\ 0 & 0 \end{pmatrix} \quad \text{and} \quad \begin{pmatrix} 0 & 1 \\ 1 & 0 \end{pmatrix}.$$

Since the first matrix is not invertible, the problem is not Birkhoff regular. In the following, we suppose that the function q is sufficiently smooth. According to Theorem 5.5.5, the problem is s-regular if the coefficient of λ^{-s} of

$$\det\left[\sum_{r=0}^{s} \lambda^{-r} \left(\begin{pmatrix} 1 & 1 \\ 0 & 0 \end{pmatrix} P^{[r]}(0) \begin{pmatrix} 1 & 0 \\ 0 & 0 \end{pmatrix} + \begin{pmatrix} 0 & 0 \\ 1 & 0 \end{pmatrix} P^{[r]}(1) \begin{pmatrix} 0 & 0 \\ 0 & 1 \end{pmatrix} \right)\right]$$

is different from zero, where the $P^{[r]}$ are the matrix functions from Theorem 2.8.2 corresponding to the above system. Since

$$\begin{pmatrix} 0 & 0 \\ 1 & 0 \end{pmatrix} P^{[r]}(1) \begin{pmatrix} 0 & 0 \\ 0 & 1 \end{pmatrix} = \begin{pmatrix} 0 & 0 \\ 0 & P_{12}^{[r]}(1) \end{pmatrix},$$

the coefficients of $\lambda^{-1}, \dots, \lambda^{-j}$ are zero if $P_{12}^{[r]}(1) = 0$ for $r = 1, \dots, j$. Hence, if the problem is s-regular, then $P_{12}^{[r]}(1) \neq 0$ for some $r \in \{1, \dots, s\}$. Conversely, if $P_{12}^{[j]}(1) \neq 0$, but $P_{12}^{[r]}(1) = 0$ for $r = 1, \dots, j-1$, then the coefficient of λ^{-j} of the above determinant is

$$\begin{pmatrix} 1 & 0 \\ 0 & P_{12}^{[j]}(1) \end{pmatrix}.$$

Therefore we have the following result: Suppose that $P_{12}^{[s]}(1) \neq 0$ for some $s \in \mathbb{N}$. Let s be minimal with this property. Then the problem is s-regular but not $s-1$-regular.

By (2.8.19) we have

$$P_{12}^{[r+1]} = \frac{1}{2}\left\{P_{12}^{[r]\prime} - (1-\delta_{r,0})\frac{q}{2}\left(P_{12}^{[r-1]} + P_{22}^{[r-1]}\right)\right\}.$$

Since $P_{12}^{[0]} = 0$, it follows that $P_{12}^{[1]} = 0$. In $P_{22}^{[1]}$ we have an additive constant which can be chosen such that $P_{12}^{[1]}(1) + P_{22}^{[1]}(1) \neq 0$. Thus, for $r = 1, 2, \ldots$, $P_{12}^{[r+1]}(1)$ is a linear combination of $q^{(r-1)}(1), q^{(r-2)}(1), \ldots, q(1)$, where the coefficient of $q^{(r-1)}(1)$ is different from zero. Therefore, suppose that q has a zero of order $s-2$ at 1. Then the Regge problem is s-regular but not $s-1$-regular.

Now let us investigate the regularity conditions on q. Since we need $P_{12}^{[s]}$, we have to take $k = s$ in Theorem 2.8.2 whence $q \in W_p^{s-1}(a,b)$. As in Lemma 5.7.8 we infer that the problem is strongly s-regular. In order to apply Theorem 7.6.9 we must have $q = -\pi_{2,0} \in W_p^{s-1}(a,b)$, i. e., we have the same condition as above. To find the iterates according to Proposition 7.6.8 we note that the nonzero coefficient matrices are $\widetilde{A}_1 = \begin{pmatrix} 0 & 1 \\ 1 & 0 \end{pmatrix}$ and $\widetilde{A}_{-1} = \begin{pmatrix} 0 & 0 \\ q & 0 \end{pmatrix}$. Therefore,

$$\tilde{y}^{[0]} = -\begin{pmatrix} f \\ 0 \end{pmatrix},$$

$$\tilde{y}^{[1]} = -\begin{pmatrix} 0 & 1 \\ 1 & 0 \end{pmatrix}\begin{pmatrix} f' \\ 0 \end{pmatrix} = -\begin{pmatrix} 0 \\ f' \end{pmatrix},$$

$$\tilde{y}^{[j]} = \begin{pmatrix} 0 & 1 \\ 1 & 0 \end{pmatrix}\tilde{y}^{[j-1]\prime} - \begin{pmatrix} 0 & 1 \\ 1 & 0 \end{pmatrix}\begin{pmatrix} 0 & 0 \\ q & 0 \end{pmatrix}\tilde{y}^{[j-2]} = \begin{pmatrix} \tilde{y}_2^{[j-1]\prime} - q\tilde{y}_1^{[j-2]} \\ \tilde{y}_1^{[j-1]\prime} \end{pmatrix} \qquad (j \geq 2),$$

and we obtain

$$\tilde{y}_1^{[j]} = \begin{cases} \tilde{y}_1^{[j-2]\prime\prime} - q\tilde{y}_1^{[j-2]} & \text{if } j \text{ is even and positive,} \\ 0 & \text{if } j \text{ is odd,} \end{cases}$$

$$\tilde{y}_2^{[j]} = \begin{cases} 0 & \text{if } j \text{ is even,} \\ \tilde{y}_1^{[j-1]\prime} & \text{if } j \text{ is odd.} \end{cases}$$

We have

$$C_2(\lambda)^{-1}T^R(\lambda)\Xi(\lambda)\tilde{y}^{[j]} = \begin{pmatrix} 1 & 0 \\ 0 & 0 \end{pmatrix}\tilde{y}^{[j]}(0) + \begin{pmatrix} 0 & 0 \\ 1 & 1 \end{pmatrix}\tilde{y}^{[j]}(1).$$

Then the asymptotic boundary conditions in Theorems 7.6.9 and 7.6.10 for the functions f which are expandable are given by

$$\tilde{y}_1^{[j]}(0) = 0, \quad \tilde{y}_1^{[j]}(1) + \tilde{y}_2^{[j]}(1) = 0,$$

for $j=0,\ldots,\kappa$, where $\kappa=s-1$ if $p<\infty$ and $\kappa=s$ if $p=\infty$. Observing the above recursion formulas we immediately obtain that this is equivalent to

$$\tilde{y}_1^{[2j]}(0)=0,\quad \tilde{y}_1^{[2j]}(1)=0,\quad \tilde{y}_1^{[2k]\prime}(1)=0,$$

for $j=0,\ldots,\left[\frac{\kappa}{2}\right]$ and $k=0,\ldots,\left[\frac{\kappa-1}{2}\right]$. It is easy to see that the latter two conditions are equivalent to $f^{(j)}(1)=0$ for $j=0,\ldots,\kappa$. However, the first condition cannot be expressed this easily. For $j=0$ and $j=1$ it is $f(0)=0$ and $f''(0)=0$, respectively, but for $j=2$ we obtain $f^{(4)}(0)-2q'(0)f'(0)=0$.

We can summarize the above conditions as follows:

Let $s\geq 2$ and $q\in W_p^{s-1}(a,b)$. Suppose that q has a zero of order $s-2$ at 1. Then, for $1<p<\infty$, a function $f\in W_p^s(a,b)$ is expandable in an L_p-convergent series of eigenfunctions and associated functions of the Regge problem if f satisfies the boundary conditions

$$\begin{aligned}\tilde{y}_1^{[2j]}(0)&=0\quad \left(j=0,\ldots,\left[\frac{s-1}{2}\right]\right),\\ f^{(j)}(1)&=0\quad (j=0,\ldots,s-1).\end{aligned}$$

If $p=\infty$, then the series converges uniformly for $f\in C^s(a,b)$ with $f^{(s)}\in \mathrm{BV}[a,b]$ satisfying

$$\begin{aligned}\tilde{y}_1^{[2j]}(0)&=0\quad \left(j=0,\ldots,\left[\frac{s}{2}\right]\right),\\ f^{(j)}(1)&=0\quad (j=0,\ldots,s).\end{aligned}$$

7.9. Almost Birkhoff regular problems

In this section we consider the case $n_0\neq 0$ and shall always assume that the assumptions of Theorem 7.2.4 **A** are satisfied.

DEFINITION 7.9.1. The boundary eigenvalue problem (7.1.2), (7.1.3) is called *almost Birkhoff regular* if the assumptions of Theorem 7.2.4 **A** are fulfilled and if there are matrix functions $C(\cdot,\lambda)$ satisfying (7.1.8)–(7.1.10) and $C_2(\lambda)$ satisfying (5.1.3)–(5.1.5) such that the associated boundary eigenvalue problem $\widetilde{T}^D(\lambda)y=0$, $C_2(\lambda)^{-1}\widehat{T}^R(\lambda)y=0$ is Birkhoff regular in the sense of Definition 4.1.2.

Choose the circles Γ_ν according to Theorem 4.3.9 and define

$$(7.9.1)\quad Q_\nu f:=\frac{1}{2\pi i}\oint_{\Gamma_\nu}\widetilde{J}_1[L^{-1}(\lambda)(\pi_{l,l}\lambda^{l-1}f,0)]^{(n_0)}\,\mathrm{d}\lambda\quad (f\in L_p(a,b),\ \nu\in\mathbb{N}),$$

where $\widetilde{J}_1: W_p^l(a,b)\to L_p(a,b)$ is the canonical embedding. In case $n_0=0$ this definition coincides with (7.4.1). We again have $Q_\nu\in L(L_p(a,b))$. Let P_ν be as considered at the beginning of Section 7.4.

PROPOSITION 7.9.2. *For $f \in L_p(a,b)$ we have*

$$Q_\nu f = e_{n_0+1}^{\mathsf{T}} C_1 P_\nu C_1^{-1} e_{n_0+1} f.$$

Proof. Let $f_1 \in L_p(a,b)$ and $f_2 \in \mathbb{C}^n$. In the proof of Theorem 6.4.1 we have seen that $T^{-1}(\lambda)(e_n f_1, f_2)$ fulfils the assumptions of Proposition 6.1.3, whence

(7.9.2) $$[L^{-1}(\lambda)(f_1,f_2)]^{(j)} = e_{j+1}^{\mathsf{T}} T^{-1}(\lambda)(e_n f_1, f_2)$$

for $j = 0,\dots,n-1$ by (6.4.12) and Proposition 6.1.3. As in (7.4.4) we obtain

(7.9.3) $$[L^{-1}(\lambda)(f_1,0)]^{(n_0)} = \lambda^{\nu_{n_0+1}-\nu_n} e_{n_0+1}^{\mathsf{T}} C_1 \widetilde{T}^{-1}(\lambda)(C_1^{-1} e_n f_1, 0)$$

and

(7.9.4) $$C_1^{-1}\pi_{l,l} e_n f = -C_1^{-1}\widehat{A} e_{n_0+1} f = -A_1 C_1^{-1} e_{n_0+1} f,$$

whence the proposition is proved if we set $f_1 = .\pi_{l,l} f$ and observe that $\nu_{n_0+1} - \nu_n = 1 - l$ by Theorem 7.2.4 **A**. □

Now we obtain as in Lemma 7.4.2 that $P_\nu f \to f$ for all $f \in L_p(a,b)$ if the underlying problem is almost Birkhoff regular. Therefore, every function in $L_p(a,b)$ is expandable into n_0-th derivatives of eigenfunctions and associated functions of the eigenvalue problem. However, we want to expand into eigenfunctions and associated functions themselves. For this we consider two-point boundary value problems in case $1 < p < \infty$. Starting with the classical adjoint problem and considering the family of operators Q_ν ($\nu \in \mathbb{N}$) we obtain $Q_\nu v \to v$ in $L_{p'}(a,b)$ for all $v \in L_{p'}(a,b)$. Therefore,

$$Q_\nu^* f \to f \text{ weakly in } L_p(a,b) \text{ for all } f \in L_p(a,b).$$

But Theorem 6.7.8 immediately yields that this is an expansion into eigenfunctions and associated functions of the given problem.

In order to obtain an expansion into eigenfunctions and associated functions which converges strongly, more use has to be made of the special structure of n-th order differential equations. This problem will be dealt with in the next chapter.

7.10. Notes

The statement of the expansion theorems can be simplified if L only depends on λ^n.

First expansion theorems for non-self-adjoint boundary eigenvalue problems defined by arbitrary n-th order differential equations were proved by BIRKHOFF [BI2] and TAMARKIN [TA3]. Those expansion theorems have been obtained for Birkhoff regular problems with respect to local uniform convergence in the interior of the underlying compact interval. The generalization to Stone regular problems appeared in [ST3]. In [HI2], [HI3], E. HILB independently published some remarkable eigenfunction expansion theorems for second order differential equations, in the latter paper allowing the coefficients to have singularities. In a series of papers W. EBERHARD [EB1]–[EB6] and W. EBERHARD and G. FREILING [EF1]–[EF4] extended the results of BIRKHOFF and TAMARKIN in various directions. In [EB1]–[EB3] EBERHARD investigated irregular boundary eigenvalue problems for n-th order differential equations with separated boundary conditions and in particular falsified an application of the general functional analytic results to such irregular problems by Keldysh in [KE1]; as a consequence, this special application was no longer considered in [KE2]. L_p-expansions have been investigated by Benzinger [BE3], [BE6], [BE7]. Expansions for differential equations with a particular form of λ-dependence have been considered by MÖLLER and USCHOLD in [MU].

Chapter VIII

THE DIFFERENTIAL EQUATION $\mathbf{K}\eta = \lambda\mathbf{H}\eta$

This chapter is concerned with regular two-point boundary eigenvalue problems for n-th order λ-linear differential equations of type $\mathbf{K}\eta = \lambda\mathbf{H}\eta$, where $\mathbf{K}$ and $\mathbf{H}$ are differential operators such that $\mathbf{K}$ is of higher order than $\mathbf{H}$. The boundary conditions are allowed to depend polynomially on the eigenvalue parameter λ. The results on eigenfunction expansions from the foregoing chapter are applicable to boundary eigenvalue problems of this type only if $\mathbf{H}$ is a multiplication operator. In the present chapter expansion statements are established by a specific approach.

To this end, first the proper structure of the asymptotic fundamental system of the differential equation $\mathbf{K}\eta = \lambda\mathbf{H}\eta$ is determined from the asymptotic fundamental matrix which has been constructed in the second chapter for asymptotically λ-linear first order differential systems (Theorems 8.2.1 and 8.2.4, Corollaries 8.3.1 and 8.3.2). This asymptotic fundamental system fully reflects the special structure of the differential equation $\mathbf{K}\eta = \lambda\mathbf{H}\eta$. An asymptotic representation of the inverse of the corresponding fundamental matrix is deduced, which in turn yields an appropriate asymtotic fundamental system of the formally adjoint differential equation $\mathbf{K}^+\zeta = \lambda\mathbf{H}^+\zeta$ (Theorems 8.4.1 and 8.4.2).

These asymptotic fundamental matrices of the original differential equation and of its formally adjoint are most useful for efficient estimates of the GREEN'S function $G(x,\xi,\lambda)$ of the given boundary eigenvalue problem. The essential difference between the estimates of the GREEN'S matrix in the fourth chapter, which have been used for the proofs of the expansion theorems in the previous chapter, and the estimation of the GREEN'S function in this chapter consists in the fact that, in the fourth chapter, the characteristic matrix has been estimated separately from the other terms in the GREEN'S matrix whereas here the asymptotic behavior of the GREEN'S function is investigated as a whole. The condition of almost Birkhoff regularity imposed on the given boundary eigenvalue problem yields that there is no exponential growth of $\mathbf{H}^+G(x,\cdot,\lambda)$ on the regularity circles Γ_ν ($\nu \in \mathbb{N}$). The asymptotic boundary conditions are defined in such a way that those terms in the asymptotic representation of $\int_a^b f(\xi)\mathbf{H}_\xi^+G(x,\xi,\lambda)\,\mathrm{d}\xi$ vanish which would prevent the convergence of the sequence of contour integrals over $\int_a^b f(\xi)\mathbf{H}_\xi^+G(x,\xi,\lambda)\,\mathrm{d}\xi$ along the regularity circles Γ_ν. These asymptotic boundary conditions are given in terms of the coefficients of the differential operators $\mathbf{K}$

and $\mathbf{H}$ as well as of the coefficients in the boundary conditions (Definition 8.5.3). In the special case that $\mathbf{H}$ is a multiplication operator, the asymptotic boundary conditions are determined by the coefficients of the boundary matrices only. In this case the notions Birkhoff regularity and almost Birkhoff regularity coincide.

The expansion theorems state the expandability of sufficiently smooth functions which fulfil the asymptotic boundary conditions up to a certain order. The eigenfunction expansions converge in the topology of some function space $C^s[a,b]$ or $W_p^s(a,b)$, respectively, (Theorems 8.8.2 and 8.8.3). Some examples demonstrate the quality of the achieved results. A boundary eigenvalue problem for a simple fourth order differential equation shows that the asymptotic boundary conditions in general have to include the values of the coefficients of the differential equation at the endpoints of the underlying interval.

8.1. The eigenvalue problem and general assumptions

Let $1 \le p \le \infty$ and p' such that $1/p + 1/p' = 1$. Let $-\infty < a < b < \infty$ and $n \in \mathbb{N}$, $n \ge 2$. In this chapter we consider the differential equation

$$\mathbf{K}\eta = \lambda\mathbf{H}\eta \quad (\eta \in W_1^n(a,b)), \tag{8.1.1}$$

where

$$\mathbf{K}\eta = \eta^{(n)} + \sum_{i=0}^{n-1} k_i \eta^{(i)}, \tag{8.1.2}$$

$$\mathbf{H}\eta = \sum_{i=0}^{n_0} h_i \eta^{(i)}, \tag{8.1.3}$$

with $0 \le n_0 \le n-1$ and $k_i, h_i \in W_{p'}^i(a,b)$. We shall always assume that $h_{n_0} > 0$ and that $h_{n_0}^{-1} \in L_\infty(a,b)$. In view of Proposition 2.5.8 we infer $h_{n_0}^{-1} \in W_{p'}^1(a,b)$ if $n_0 > 0$. If $n_0 = 0$, we suppose that $h_{n_0}^{-1} \in W_{p'}^1(a,b)$.

As usual, we associate the differential operator

$$L^D(\lambda)\eta := \mathbf{K}\eta - \lambda\mathbf{H}\eta \quad (\eta \in W_1^n(a,b)) \tag{8.1.4}$$

with the differential equation (8.1.1). Together with the differential equation (8.1.1) we consider two-point boundary conditions

$$L^R(\lambda)\eta := \left(\sum_{i=0}^{n-1} w_{ki}^{(0)}(\lambda)\eta^{(i-1)}(a) + \sum_{i=0}^{n-1} w_{ki}^{(1)}(\lambda)\eta^{(i-1)}(b) \right)_{k=1}^{n} = 0, \tag{8.1.5}$$

where the $w_{ki}^{(j)}$ are polynomials. For convenience, we define the boundary matrices

$$W^{(j)}(\lambda) = \left(w_{ki}^{(j)}(\lambda)\right)_{k,i=1}^{n} \quad (j = 0, 1). \tag{8.1.6}$$

We shall always assume that the resolvent set of the operator

$$L := (L^D, L^R) \in H(\mathbb{C}, L(W_p^n(a,b), L_p(a,b) \times \mathbb{C}^n)) \tag{8.1.7}$$

is nonempty. Let G denote the GREEN'S function defined in (6.4.5).

The assumptions of Theorem 7.2.4 hold with p' substituted for p if we replace λ by λ^l, where $l = n - n_0$. We have

(8.1.8) $$\pi_{n-i,0} = k_i \ (i = 0, \dots, n-1) \text{ and } \pi_{n-i,l} = -h_i, \ (i = 0, \dots, n_0),$$

whence $\pi_{l,l} = -h_{n_0}$ and $\pi_{i,i} = 0$ if $i \in \{1, \dots, n\}$ and $i \neq l$. According to Theorem 7.2.4, case **B**, we can take $C(x, \lambda) = C_0(\lambda) C_1(x)$, where

(8.1.9) $$C_0(\lambda) = \operatorname{diag}(1, \dots, 1, \lambda, \dots, \lambda^l),$$

(8.1.10) $$C_1 = \left(\begin{array}{ccc|ccc} 1 & & 0 & & & \\ & \ddots & & & 0 & \\ 0 & & 1 & h_{n_0}^{-1}\omega_1^{-1} & \dots & h_{n_0}^{-1}\omega_l^{-1} \\ \hline & & & 1 & \dots & 1 \\ & & & h_{n_0}^{1/l}\omega_1 & \dots & h_{n_0}^{1/l}\omega_l \\ & 0 & & \vdots & & \vdots \\ & & & h_{n_0}^{(l-1)/l}\omega_1^{l-1} & \dots & h_{n_0}^{(l-1)/l}\omega_l^{l-1} \end{array} \right)$$

and

$$\omega_j = \exp\left\{ \frac{2\pi i (j-1)}{l} \right\} \quad (j = 1, \dots, l).$$

We define the formally adjoint $\mathbf{H}^+$ of $\mathbf{H}$ by

(8.1.11) $$\mathbf{H}^+ \eta = \sum_{i=0}^{n_0} (-1)^i (h_i \eta)^{(i)} \quad (\eta \in W_p^{n_0}(a,b)).$$

LEMMA 8.1.1. *If $\lambda \in \rho(L)$, then*

(8.1.12) $$(S(\lambda) f)(x) := \int_a^b (\mathbf{H}^+ G(x, \cdot, \lambda))(\xi) f(\xi) \, d\xi \quad (f \in L_p(a,b), x \in [a,b])$$

is well-defined, and $S \in H(\rho(L), L(L_p(a,b), W_p^{n-n_0-1}(a,b)))$.

Before proving this lemma let us state a consequence.

THEOREM 8.1.2. *We consider circles $\{\gamma_\nu : \nu \in \mathbb{N}\}$ in $\rho(L)$ with centre 0 and radii tending to ∞ as $\nu \to \infty$. The radii will be specified later. For $\nu \in \mathbb{N}$ and $f \in L_p(a,b)$ we define*

(8.1.13) $$(Q_\nu f)(x) := -\frac{1}{2\pi i} \oint_{\gamma_\nu} \int_a^b (\mathbf{H}^+ G(x, \cdot, \lambda))(\xi) f(\xi) \, d\xi \, d\lambda \quad (x \in [a,b]).$$

Then $Q_\nu \in L(L_p(a,b), W_p^{n-n_0-1}(a,b))$.

The operator family $\{Q_\nu : \nu \in \mathbb{N}\}$ leads to the desired expansion theorems.

Let $T(\lambda)$ be the system associated to $L(\lambda^l)$ according to (6.1.3) and let $Y(\cdot,\lambda)$ be a fundamental matrix function of $T^D(\lambda)y = 0$. Let $M(\lambda)$ be the associated characteristic matrix given by

$$M(\lambda) = W^{(0)}(\lambda^l)Y(a,\lambda) + W^{(1)}(\lambda^l)Y(b,\lambda). \tag{8.1.14}$$

In the following lemma, we sometimes omit the eigenvalue parameter λ in order to shorten the formulas.

LEMMA 8.1.3. *We have* $Y^{-1}e_n \in (W^n_{p'}(a,b))^n$ *and*

$$(Y^{-1}e_n)^{(j)} = \sum_{\nu=0}^{j} q_{j,\nu} Y^{-1} e_{n-\nu} \tag{8.1.15}$$

for $j = 0,\dots,n-1$, *where* $q_{j,\nu}(\cdot,\lambda) \in W^{n+\nu-j}_{p'}(a,b)$. *The* $q_{j,\nu}$ *are inductively, with respect to* j, *given by*

$$\begin{aligned}
&q_{j,j} = (-1)^j \quad (j = 0,\dots,n-1),\\
&q_{j,j-1} = -q_{j-1,j-2} \quad (j = 2,\dots,n-1),\\
&q_{j,\nu} = q'_{j-1,\nu} - q_{j-1,\nu-1} \quad (j = 3,\dots,n-1;\ \nu = 1,\dots,j-2),\\
&q_{j,0} = q'_{j-1,0} + \sum_{\nu=0}^{j-1} p_{n-\nu-1} q_{j-1,\nu} \quad (j = 1,\dots,n-1),
\end{aligned}$$

where

$$p_i(\cdot,\lambda) = \begin{cases} k_i & \text{if } n_0 + 1 \le i \le n-1,\\ k_i - \lambda^l h_i & \text{if } 0 \le i \le n_0. \end{cases}$$

Furthermore, the degree of $q_{j,\nu}$ *as a polynomial in* λ *does not exceed* $j-\nu$, *and* $q_{j,\nu}$ *is a polynomial in* λ^l.

Proof. From (2.5.7) and with A given by (6.1.3) and (6.1.4) it follows that

$$(Y^{-1})' = -Y^{-1}Y'Y^{-1} = -Y^{-1}A, \tag{8.1.16}$$

which implies that

$$(Y^{-1})'e_k = -Y^{-1}e_{k-1} + p_{k-1}Y^{-1}e_n \tag{8.1.17}$$

for $k = 2, \ldots, n$. Obviously, (8.1.15) is true if $j = 0$. Assume (8.1.15) holds for some $j \in \{0, \ldots, n-2\}$. Then $(Y^{-1}e_n)^{(j)} \in (W_1^1(a,b))^n$, and we can differentiate (8.1.15). In view of (8.1.17) this leads to

$$\begin{aligned}(Y^{-1}e_n)^{(j+1)} &= \sum_{\nu=0}^{j} (q'_{j,\nu} Y^{-1} e_{n-\nu} - q_{j,\nu} Y^{-1} e_{n-\nu-1} + p_{n-\nu-1} q_{j,\nu} Y^{-1} e_n) \\ &= -q_{j,j} Y^{-1} e_{n-j-1} + \sum_{\nu=1}^{j} (q'_{j,\nu} - q_{j,\nu-1}) Y^{-1} e_{n-\nu} \\ &\quad + \Big(q'_{j,0} + \sum_{\nu=0}^{j} p_{n-\nu-1} q_{j,\nu}\Big) Y^{-1} e_n,\end{aligned}$$

whence (8.1.15) follows for $j+1$. Finally, $(Y^{-1}e_n)^{(n-1)} \in W_{p'}^1(a,b))^n$ implies $Y^{-1}e_n \in W_{p'}^n(a,b)$ by Corollary 2.1.4. $\square$

Proof of Lemma 8.1.1. By (3.2.6) the GREEN'S matrix of the differential operator $T(\lambda)$ is

$$G_T(x,\xi,\lambda) = \begin{cases} Y(x,\lambda)M^{-1}(\lambda)W^{(0)}(\lambda^l)Y(a,\lambda)Y^{-1}(\xi,\lambda) \\ \hfill (a \le \xi \le x \le b), \\ -Y(x,\lambda)M^{-1}(\lambda)W^{(1)}(\lambda^l)Y(b,\lambda)Y^{-1}(\xi,\lambda) \\ \hfill (a \le x < \xi \le b). \end{cases}$$

We have

$$G_T(x,x-0,\lambda) - G_T(x,x+0,\lambda) = G_T(x+0,x,\lambda) - G_T(x-0,x,\lambda) = I_n,$$

which shows that $e_k^{\mathsf{T}} G_T(x,\xi,\lambda) e_j$ is continuous in both variables x and ξ if $k \ne j$. Since the GREEN'S function of L defined in (6.4.5) satisfies

$$G(x,\xi,\lambda^l) = e_1^{\mathsf{T}} G_T(x,\xi,\lambda) e_n,$$

we infer for $k = 0, \ldots, n-n_0-1$ and $j = 0, \ldots, n_0$ in view of Proposition 2.2.2, $(e_1^{\mathsf{T}} Y(\cdot,\lambda))^{(k)} = e_{k+1}^{\mathsf{T}} Y(\cdot,\lambda)$, and Lemma 8.1.3 that

$$\frac{\partial^k}{\partial x^k}\frac{\partial^j}{\partial \xi^j} G(x,\cdot,\lambda) = \begin{cases} e_{k+1}^{\mathsf{T}} Y(x,\lambda)M^{-1}(\lambda)W^{(0)}(\lambda^l)Y(a,\lambda)(Y^{-1}(\cdot,\lambda)e_n)^{(j)} \\ \hfill (a \le \xi < x \le b), \\ -e_{k+1}^{\mathsf{T}} Y(x,\lambda)M^{-1}(\lambda)W^{(1)}(\lambda^l)Y(b,\lambda)(Y^{-1}(\cdot,\lambda)e_n)^{(j)} \\ \hfill (a \le x < \xi \le b). \end{cases}$$

Since $\mathbf{H}^+$ is an n_0-th order differential operator with coefficients in $L_{p'}(a,b)$, $S(\lambda) \in L(L_p(a,b), W_p^{n-n_0-1}(a,b))$ follows. The holomorphy is an immediate consequence of the holomorphic dependence of G on λ. $\square$

8.2. An asymptotic fundamental system for $\mathbf{K}\eta = \lambda^l \mathbf{H}\eta$

In order to obtain the desired estimate of the GREEN'S function we need more information about an asymptotic fundamental system. In this section we suppose that $h_{n_0} = 1$.

THEOREM 8.2.1. *Suppose that* $h_{n_0} = 1$, *set* $l = n - n_0$, *and let* $k \in \mathbb{N}$. *Suppose that* $k \geq \max\{l, n_0 - 1\}$ *if* $n_0 > 0$. *Suppose that*
α) $k_j \in L_{p'}(a,b)$ *for* $j = 0,\dots,n-1-k$ *and* $k_{n-1-j} \in W_{p'}^{k-j}(a,b)$ *for* $j = 0,\dots,\min\{k-1, n-1\}$ *if* $n_0 = 0$,
β) $h_0,\dots,h_{n_0-1} \in W_{p'}^{k}(a,b)$, $k_0,\dots,k_{n_0-1} \in W_{p'}^{k-l}(a,b)$, *and* $k_{n-1-j} \in W_{p'}^{k-j}(a,b)$ *for* $j = 0,\dots,l-1$ *if* $n_0 > 0$. *Let* $\{\pi_1,\dots,\pi_{n_0}\} \subset W_{p'}^{k+n_0}(a,b)$ *be a fundamental system of* $\mathbf{H}\eta = 0$.
For sufficiently large λ *the differential equation* $\mathbf{K}\eta = \lambda^l \mathbf{H}\eta$ *has a fundamental system* $\{\eta_1(\cdot,\lambda),\dots,\eta_n(\cdot,\lambda)\}$ *with the following properties:*
i) *There are functions* $\pi_{\nu r} \in W_{p'}^{k+n_0-lr}(a,b)$ $(1 \leq \nu \leq n_0,\ 1 \leq r \leq [\frac{k}{l}])$ *such that*

$$\eta_\nu^{(\mu)}(\cdot,\lambda) = \pi_\nu^{(\mu)} + \sum_{r=1}^{[\frac{k}{l}]} \lambda^{-lr}\pi_{\nu r}^{(\mu)} + \{o(\lambda^{-k})\}_\infty \qquad (8.2.1)$$

$$(\nu = 1,\dots,n_0;\ \mu = 0,\dots,n_0-1),$$

$$\eta_\nu^{(\mu)}(\cdot,\lambda) = \pi_\nu^{(\mu)} + \sum_{r=1}^{[\frac{k-\mu+n_0-1}{l}]} \lambda^{-lr}\pi_{\nu r}^{(\mu)} + \{o(\lambda^{-k+\mu-n_0+1})\}_\infty \qquad (8.2.2)$$

$$(\nu = 1,\dots,n_0;\ \mu = n_0,\dots,n-1),$$

ii) *Set* $\tilde{k} := \min\{k, k+1-n_0\}$. *Let* $\omega_j = \exp\{\frac{2\pi i(j-1)}{l}\}$ $(j = 1,\dots,l)$. *There are functions* $\varphi_r \in W_{p'}^{k+1-r}(a,b)$, $r = 0,\dots,\tilde{k}$, *such that* φ_0 *is the solution of the initial value problem*

$$\varphi_0' - \frac{1}{l}(h_{n_0-1} - k_{n-1})\varphi_0 = 0, \quad \pi_0(a) = 1, \qquad (8.2.3)$$

and

$$\eta_\nu^{(\mu)}(x,\lambda) = \left[\frac{d^\mu}{dx^\mu}\right]\left\{\sum_{r=0}^{\tilde{k}} (\lambda\omega_{\nu-n_0})^{-r}\varphi_r(x)e^{\lambda\omega_{\nu-n_0}(x-a)}\right\} + \{o(\lambda^{-\tilde{k}+\mu})\}_\infty e^{\lambda\omega_{\nu-n_0}(x-a)} \qquad (8.2.4)$$

$$(\nu = n_0+1,\dots,n;\ \mu = 0,\dots,n-1),$$

where $[\frac{d^\mu}{dx^\mu}]$ *means that we omit those terms of the Leibniz expansion which contain a function* $\varphi_r^{(j)}$ *with* $j > \tilde{k} - r$.

Proof. The regularity assumptions on the coefficients k_i and h_i might be partly weaker than the general assumptions $k_i, h_i \in W_p^i(a,b)$ made at the beginning of this chapter. This means that this theorem holds under the conditions stated here, but we shall only apply it if the general hypotheses of this chapter are satisfied.

We denote the i-th unit vectors in $\mathbb{C}^n$, $\mathbb{C}^{n_0}$, $\mathbb{C}^l$ by e_i, ϵ_i, ε_i. For $i \in \mathbb{Z}\setminus\{1,\dots,n\}$ or $i \in \mathbb{Z}\setminus\{1,\dots,n_0\}$ or $i \in \mathbb{Z}\setminus\{1,\dots,l\}$ we set $e_i := 0$, $\epsilon_i := 0$, $\varepsilon_i := 0$, respectively.

We can write the matrix $A(\cdot,\lambda^l)$ of the corresponding system given by (6.1.4) as

$$A(\cdot,\lambda^l) = \begin{pmatrix} J_{n_0} & \epsilon_{n_0}\varepsilon_1^{\mathsf{T}} \\ \varepsilon_l a_1^{\mathsf{T}}(\lambda) & \varepsilon_l a_2^{\mathsf{T}} + J_l + \lambda^l \varepsilon_l \varepsilon_1^{\mathsf{T}} \end{pmatrix}$$

according to the decomposition $\mathbb{C}^n = \mathbb{C}^{n_0} \oplus \mathbb{C}^l$, where

$$a_1^{\mathsf{T}}(\lambda) := \lambda^l(h_0,\dots,h_{n_0-1}) - (k_0,\dots,k_{n_0-1}) =: \lambda^l a_{11}^{\mathsf{T}} + a_{12}^{\mathsf{T}}, \tag{8.2.5}$$

$$a_2^{\mathsf{T}} := -(k_{n_0},\dots,k_{n-1}), \tag{8.2.6}$$

$$J_r := \begin{pmatrix} 0 & 1 & & & \\ & 0 & \cdot & & 0 \\ & & \cdot & \cdot & \\ & & & \cdot & \cdot \\ & 0 & & \cdot & 1 \\ & & & & 0 \end{pmatrix} \in M_r(\mathbb{C}). \tag{8.2.7}$$

We set

$$\varepsilon^{\mathsf{T}} := \sum_{i=1}^{l} \varepsilon_i^{\mathsf{T}} = (1,\dots,1) \in \mathbb{C}^l, \tag{8.2.8}$$

$$\Omega_l := \operatorname{diag}(\omega_1,\dots,\omega_l), \tag{8.2.9}$$

$$\Xi_r(\lambda) := \operatorname{diag}(1,\lambda,\dots,\lambda^{r-1}) \in M_r(\mathbb{C}), \tag{8.2.10}$$

$$V := \sum_{i=1}^{l} \varepsilon_i \varepsilon^{\mathsf{T}} \Omega_l^{i-1} = \begin{pmatrix} 1 & \cdot & \cdot & \cdot & 1 \\ \omega_1 & \cdot & \cdot & \cdot & \omega_l \\ \vdots & & & & \vdots \\ \omega_1^{l-1} & \cdot & \cdot & \cdot & \omega_l^{l-1} \end{pmatrix}. \tag{8.2.11}$$

If we observe that

$$\varepsilon^{\mathsf{T}} \Omega_l^j \varepsilon = \sum_{i=1}^{l} \omega_i^j = \begin{cases} l & \text{if } j = 0 \operatorname{mod}(l), \\ 0 & \text{if } j \neq 0 \operatorname{mod}(l), \end{cases} \tag{8.2.12}$$

we obtain that V is invertible with

$$V^{-1} = \frac{1}{l} \sum_{i=1}^{l} \Omega_l^{1-i} \varepsilon \varepsilon_i^{\mathsf{T}}. \tag{8.2.13}$$

Also,

$$\varepsilon_i^{\mathsf{T}} V = \varepsilon^{\mathsf{T}} \Omega_l^{i-1} \quad (i = 1, \dots, l) \tag{8.2.14}$$

and

$$(J_l + \varepsilon_l \varepsilon_1^{\mathsf{T}}) V = V \Omega_l. \tag{8.2.15}$$

Then we have in view of (8.1.9) and (8.1.10) that

$$C(\lambda) = C(\cdot, \lambda) = \begin{pmatrix} I_{n_0} & \epsilon_{n_0} \varepsilon^{\mathsf{T}} \Omega_l^{-1} \\ 0 & \lambda \Xi_l(\lambda) V \end{pmatrix}, \tag{8.2.16}$$

and the matrix function $\widetilde{A}(\cdot, \lambda)$ given by (7.1.11) has the form

$$\widetilde{A}(\cdot, \lambda) = C(\lambda)^{-1} A(\cdot, \lambda) C(\lambda) = \sum_{j=-1}^{l} \lambda^{-j} A_{-j}, \tag{8.2.17}$$

where

$$A_1 = \begin{pmatrix} 0 & 0 \\ 0 & \Omega_l \end{pmatrix}, \tag{8.2.18}$$

$$A_0 = \begin{pmatrix} J_{n_0} - \epsilon_{n_0} a_{11}^{\mathsf{T}} & (\epsilon_{n_0 - 1} - (h_{n_0 - 1} - k_{n-1}) \epsilon_{n_0}) \varepsilon^{\mathsf{T}} \Omega_l^{-1} \\ \frac{1}{l} \Omega_l \varepsilon a_{11}^{\mathsf{T}} & \frac{1}{l} (h_{n_0 - 1} - k_{n-1}) \Omega_l \varepsilon \varepsilon^{\mathsf{T}} \Omega_l^{-1} \end{pmatrix}, \tag{8.2.19}$$

$$A_{-j} = \begin{pmatrix} 0 & k_{n-1-j} \epsilon_{n_0} \varepsilon^{\mathsf{T}} \Omega_l^{-1-j} \\ 0 & -\frac{1}{l} k_{n-1-j} \Omega_l \varepsilon \varepsilon^{\mathsf{T}} \Omega_l^{-1-j} \end{pmatrix} \quad (j = 1, \dots, l-1), \tag{8.2.20}$$

$$A_{-l} = \begin{pmatrix} -\epsilon_{n_0} a_{12}^{\mathsf{T}} & k_{n_0 - 1} \epsilon_{n_0} \varepsilon^{\mathsf{T}} \Omega_l^{-1} \\ \frac{1}{l} \Omega_l \varepsilon a_{12}^{\mathsf{T}} & -\frac{1}{l} k_{n_0 - 1} \Omega_l \varepsilon \varepsilon^{\mathsf{T}} \Omega_l^{-1} \end{pmatrix}, \tag{8.2.21}$$

and $h_{n_0 - 1}$ is understood to be zero in case $n_0 = 0$.

For the proof of the representation (8.2.17) of $\widetilde{A}(\cdot, \lambda)$ we first define the auxiliary matrices

$$C_{(1)}(\lambda) = \begin{pmatrix} I_{n_0} & 0 \\ 0 & \lambda \Xi_l(\lambda) V \end{pmatrix}, \quad C_{(2)} = \begin{pmatrix} I_{n_0} & \epsilon_{n_0} \varepsilon^{\mathsf{T}} \Omega_l^{-1} \\ 0 & I_l \end{pmatrix},$$

for which $C(\lambda) = C_{(1)}(\lambda) C_{(2)}$. Then

$$C_{(1)}(\lambda)^{-1} A(\cdot, \lambda) C_{(1)}(\lambda) =$$
$$\begin{pmatrix} J_{n_0} & \lambda \epsilon_{n_0} \varepsilon_1^{\mathsf{T}} \Xi_l(\lambda) V \\ \lambda^{-1} V^{-1} \Xi_l^{-1}(\lambda) \varepsilon_l a_1^{\mathsf{T}}(\lambda) & V^{-1} \Xi_l^{-1}(\lambda) (\varepsilon_l a_2^{\mathsf{T}} + J_l + \lambda^l \varepsilon_l \varepsilon_1^{\mathsf{T}}) \Xi_l(\lambda) V \end{pmatrix}.$$

From (8.2.13), (8.2.14) and (8.2.15) we obtain the identities

$$\varepsilon_1^{\mathsf{T}} \Xi_l(\lambda) V = \varepsilon_1^{\mathsf{T}} V = \varepsilon^{\mathsf{T}}, \quad V^{-1} \varepsilon_l = \frac{1}{l} \Omega_l \varepsilon,$$
$$V^{-1} \Xi_l^{-1}(\lambda) (J_l + \lambda^l \varepsilon_l \varepsilon_1^{\mathsf{T}}) \Xi_l(\lambda) V = \lambda V^{-1} (J_l + \varepsilon_l \varepsilon_1^{\mathsf{T}}) V = \lambda \Omega_l,$$

which yields the representation

$$C_{(1)}(\lambda)^{-1}A(\cdot,\lambda)C_{(1)}(\lambda) =$$
$$\begin{pmatrix} J_{n_0} & 0 \\ \frac{1}{l}\lambda^{-l}\Omega_l \varepsilon a_1^\top(\lambda) & \frac{1}{l}\lambda^{1-l}\Omega_l \varepsilon a_2^\top \Xi_l(\lambda)V \end{pmatrix} + \lambda \begin{pmatrix} 0 & \epsilon_{n_0}\varepsilon^\top \\ 0 & \Omega_l \end{pmatrix}.$$

We already know from Theorem 7.2.4 that A_1 has the representation (8.2.18), which can also be checked here easily. The representations (8.2.19)–(8.2.21) follow from

$$\begin{pmatrix} I_{n_0} & -\epsilon_{n_0}\varepsilon^\top\Omega_l^{-1} \\ 0 & I_l \end{pmatrix} \begin{pmatrix} J_{n_0} & 0 \\ \frac{1}{l}\lambda^{-l}\Omega_l \varepsilon a_1^\top(\lambda) & \frac{1}{l}\lambda^{1-l}\Omega_l \varepsilon a_2^\top \Xi_l(\lambda)V \end{pmatrix} \begin{pmatrix} I_{n_0} & \epsilon_{n_0}\varepsilon^\top\Omega_l^{-1} \\ 0 & I_l \end{pmatrix}$$
$$= \left\{ \begin{matrix} J_{n_0} - \lambda^{-l}\epsilon_{n_0} a_1^\top(\lambda) \\ \frac{1}{l}\lambda^{-l}\Omega_l \varepsilon a_1^\top(\lambda) \end{matrix} \right.$$
$$\left. \begin{matrix} \epsilon_{n_0-1}\varepsilon^\top\Omega_l^{-1} - \lambda^{-l}\epsilon_{n_0} a_1^\top(\lambda)\epsilon_{n_0}\varepsilon^\top\Omega_l^{-1} - \lambda^{1-l}\epsilon_{n_0} a_2^\top \Xi_l(\lambda)V \\ \frac{1}{l}\lambda^{-l}\Omega_l \varepsilon a_1^\top(\lambda)\epsilon_{n_0}\varepsilon^\top\Omega_l^{-1} + \frac{1}{l}\lambda^{1-l}\Omega_l \varepsilon a_2^\top \Xi_l(\lambda)V \end{matrix} \right\},$$

definitions (8.2.5) and (8.2.6), and the identity (8.2.14).

The assumptions α) and β) give $A_{-j} \in M_n(W_{p'}^{k-j}(a,b))$ ($j = 0,\dots,\min\{k,l\}$). According to (8.2.18)–(8.2.21) the coefficient matrix $\widetilde{A}(\cdot,\lambda)$ satisfies the assumptions made in Section 2.8. From Theorem 2.8.2 we obtain that $y' - \widetilde{A}(\cdot,\lambda)y = 0$ has a fundamental system

$$\widetilde{Y}(\cdot,\lambda) = \Big(\sum_{r=0}^{k} \lambda^{-r}P^{[r]} + \{o(\lambda^{-k})\}_\infty\Big)E(\cdot,\lambda) \tag{8.2.22}$$

if λ is sufficiently large, where $P^{[r]} \in M_n(W_{p'}^{k+1-r}(a,b))$ and

$$E(x,\lambda) = \operatorname{diag}(1,\dots,1,e^{\lambda\omega_1(x-a)},\dots,e^{\lambda\omega_l(x-a)}).$$

We infer that

$$\big(\eta_{\mu-1,\nu}(\cdot,\lambda)\big)_{\mu,\nu=1}^{n} := Y(\cdot,\lambda) := C(\lambda)\widetilde{Y}(\cdot,\lambda)\begin{pmatrix} D & 0 \\ 0 & \lambda^{n_0-1}\Omega_l^{n_0} \end{pmatrix} \tag{8.2.23}$$

is a fundamental matrix of $T^D(\lambda)y = 0$ if λ is sufficiently large, where D is an invertible $n_0 \times n_0$ matrix which will be appropriately chosen later. We set

$$\widetilde{Y}(\cdot,\lambda)E(\cdot,\lambda)^{-1} =: \begin{pmatrix} \widetilde{Q}_{11}(\cdot,\lambda) & \widetilde{Q}_{12}(\cdot,\lambda) \\ \widetilde{Q}_{21}(\cdot,\lambda) & \widetilde{Q}_{22}(\cdot,\lambda) \end{pmatrix} \tag{8.2.24}$$

and obtain that

$$Y(\cdot,\lambda) = \begin{pmatrix} \widetilde{Q}_{11}(\cdot,\lambda) + \epsilon_{n_0}\varepsilon^{\mathsf{T}}\Omega_l^{-1}\widetilde{Q}_{21}(\cdot,\lambda) \\ \lambda\Xi_l(\lambda)V\widetilde{Q}_{21}(\cdot,\lambda) \end{pmatrix.$$
$$\begin{matrix} \lambda^{n_0-1}\widetilde{Q}_{12}(\cdot,\lambda)\Omega_l^{n_0} + \lambda^{n_0-1}\epsilon_{n_0}\varepsilon^{\mathsf{T}}\Omega_l^{-1}\widetilde{Q}_{22}(\cdot,\lambda)\Omega_l^{n_0} \\ \lambda^{n_0}\Xi_l(\lambda)V\widetilde{Q}_{22}(\cdot,\lambda)\Omega_l^{n_0} \end{matrix}\Bigg) \begin{pmatrix} D & 0 \\ 0 & I_l \end{pmatrix} E(\cdot,\lambda). \tag{8.2.25}$$

We set $\eta_\nu := \eta_{0,\nu}$. Then $\{\eta_1(\cdot,\lambda),\dots,\eta_n(\cdot,\lambda)\}$ is a fundamental system of $\mathbf{K}\eta = \lambda^l\mathbf{H}\eta$ and $\eta_\nu^{(\mu)} = \eta_{\mu,\nu}$ ($\nu = 1,\dots,n$; $\mu = 0,\dots,n-1$) by Lemma 6.1.5. We shall show that the η_ν have the properties stated in Theorem 8.2.1.

By (8.2.24) and (8.2.22) we have

$$\widetilde{Q}_{ij}(\cdot,\lambda) = \sum_{r=0}^{k}\lambda^{-r}Q_{ij}^{[r]} + \{o(\lambda^{-k})\}_\infty \quad (i,j=1,2), \tag{8.2.26}$$

where the elements of $Q_{ij}^{[r]}$ belong to $W_{p'}^{k+1-r}(a,b)$. We set $Q_{ij}^{[r]} := 0$ for $r<0$.

We infer that (2.8.6)–(2.8.8) are equivalent to the following equations:

$$Q_{11}^{[0]\prime} - (J_{n_0} - \epsilon_{n_0}a_{11}^{\mathsf{T}})Q_{11}^{[0]} = 0, \quad Q_{11}^{[0]}(a) = I_{n_0}, \tag{8.2.27}$$

$$Q_{22}^{[0]\prime} - \frac{1}{l}(h_{n_0-1} - k_{n-1})Q_{22}^{[0]} = 0, \quad Q_{22}^{[0]}(a) = I_l, \tag{8.2.28}$$

$$Q_{12}^{[0]} = 0, \quad Q_{21}^{[0]} = 0, \tag{8.2.29}$$

$$\begin{cases} Q_{11}^{[r]\prime} - (J_{n_0} - \epsilon_{n_0}a_{11}^{\mathsf{T}})Q_{11}^{[r]} = \big(\epsilon_{n_0-1} - (h_{n_0-1} - k_{n-1})\epsilon_{n_0}\big)\varepsilon^{\mathsf{T}}\Omega_l^{-1}Q_{21}^{[r]} \\ \quad + \sum_{j=1}^{l} k_{n-1-j}\epsilon_{n_0}\varepsilon^{\mathsf{T}}\Omega_l^{-1-j}Q_{21}^{[r-j]} - \epsilon_{n_0}a_{12}^{\mathsf{T}}Q_{11}^{[r-l]} \quad (r=1,\dots,k), \end{cases} \tag{8.2.30}$$

$$\begin{cases} Q_{21}^{[r]} = \Omega_l^{-1}Q_{21}^{[r-1]\prime} - \frac{1}{l}\varepsilon a_{11}^{\mathsf{T}}Q_{11}^{[r-1]} - \frac{1}{l}(h_{n_0-1} - k_{n-1})\varepsilon\varepsilon^{\mathsf{T}}\Omega_l^{-1}Q_{21}^{[r-1]} \\ \quad + \frac{1}{l}\sum_{j=1}^{l} k_{n-1-j}\varepsilon\varepsilon^{\mathsf{T}}\Omega_l^{-1-j}Q_{21}^{[r-1-j]} - \frac{1}{l}\varepsilon a_{12}^{\mathsf{T}}Q_{11}^{[r-l-1]} \ (r=1,\dots,k), \end{cases} \tag{8.2.31}$$

$$\begin{cases} Q_{12}^{[r]} = -Q_{12}^{[r-1]\prime}\Omega_l^{-1} + (J_{n_0} - \epsilon_{n_0}a_{11}^{\mathsf{T}})Q_{12}^{[r-1]}\Omega_l^{-1} \\ \quad + \big(\epsilon_{n_0-1} - (h_{n_0-1} - k_{n-1})\epsilon_{n_0}\big)\varepsilon^{\mathsf{T}}\Omega_l^{-1}Q_{22}^{[r-1]}\Omega_l^{-1} \\ \quad + \sum_{j=1}^{l} k_{n-j-1}\epsilon_{n_0}\varepsilon^{\mathsf{T}}\Omega_l^{-1-j}Q_{22}^{[r-1-j]}\Omega_l^{-1} - \epsilon_{n_0}a_{12}^{\mathsf{T}}Q_{12}^{[r-1-l]}\Omega_l^{-1} \\ \qquad (r=1,\dots,k), \end{cases} \tag{8.2.32}$$

(8.2.33)
$$\begin{cases} \Omega_l Q_{22}^{[r]} - Q_{22}^{[r]}\Omega_l \\ = Q_{22}^{[r-1]\prime} - \frac{1}{l}\Omega_l \varepsilon a_{11}^{\mathsf{T}} Q_{12}^{[r-1]} - \frac{1}{l}(h_{n_0-1} - k_{n-1})\Omega_l \varepsilon\varepsilon^{\mathsf{T}}\Omega_l^{-1} Q_{22}^{[r-1]} \\ + \frac{1}{l}\sum_{j=1}^{l} k_{n-1-j}\Omega_l \varepsilon\varepsilon^{\mathsf{T}}\Omega_l^{-1-j} Q_{22}^{[r-1-j]} - \frac{1}{l}\Omega_l \varepsilon a_{12}^{\mathsf{T}} Q_{12}^{[r-1-l]} \\ \qquad (r = 1,\dots,k), \end{cases}$$

(8.2.34)
$$\begin{cases} 0 = \varepsilon_\nu^{\mathsf{T}}\Big\{ Q_{22}^{[k]\prime} - \frac{1}{l}\Omega_l \varepsilon a_{11}^{\mathsf{T}} Q_{12}^{[k]} - \frac{1}{l}(h_{n_0-1} - k_{n-1})\Omega_l \varepsilon\varepsilon^{\mathsf{T}}\Omega_l^{-1} Q_{22}^{[k]} \\ + \frac{1}{l}\sum_{j=1}^{l} k_{n-1-j}\Omega_l \varepsilon\varepsilon^{\mathsf{T}}\Omega_l^{-1-j} Q_{22}^{[k-j]} - \frac{1}{l}\Omega_l \varepsilon a_{12}^{\mathsf{T}} Q_{12}^{[k-l]} \Big\}\varepsilon_\nu \\ \qquad (\nu = 1,\dots,l). \end{cases}$$

Indeed, (2.8.6) is equivalent to the second conditions in (8.2.27) and (8.2.28), (8.2.29) and that $Q_{22}^{[0]}$ is a diagonal matrix function. In view of (8.2.29) we obtain that (2.8.7) and (2.8.8) for $\nu = 0$ are equivalent to the first condition in (8.2.27) and (8.2.30)–(8.2.33). Obviously, (2.8.8) for $\nu = 1,\dots,l$ and (8.2.34) are equivalent. Since (2.8.6) implies that $Q_{22}^{[0]}$ is diagonal, it remains to be shown that the diagonal elements of $Q_{22}^{[0]}$ satisfy the differential equation

$$\eta' - \frac{1}{l}(h_{n_0-1} - k_{n-1})\eta = 0.$$

We conclude this fact from the equation (8.2.33) for $r = 1$. For this purpose we have to observe that the diagonal elements of $\Omega_l Q_{22}^{[1]} - Q_{22}^{[1]}\Omega_l$ are zero, that $Q_{12}^{[0]} = 0$ by (8.2.29), and that the diagonal elements of $\Omega_l \varepsilon\varepsilon^{\mathsf{T}}\Omega_l^{-1}$ have the value 1.

The $Q_{12}^{[r]}$ and $Q_{21}^{[r]}$ are uniquely determined by (8.2.29), (8.2.31) and (8.2.32) (if the terms on the right-hand sides of (8.2.31) and (8.2.32) are considered to be given). But we are completely free to choose $Q_{11}^{[r]}(a)$ for $r = 1,\dots,k$. The representation in part i) will only hold if we make a suitable choice. We require

(8.2.35)
$$Q_{11}^{[r]}(a) = 0 \quad (r = 1,\dots,k).$$

From (8.2.25) and (8.2.26) we infer that there are $\pi_{\nu r\mu} \in W_{p'}^{k+1-r}(a,b)$ for $r = 0,\dots,k$; $\nu = 1,\dots,n_0$ and $\mu = 0,\dots,n_0-1$ such that

(8.2.36)
$$\eta_\nu^{(\mu)}(\cdot,\lambda) = \sum_{r=0}^{k} \lambda^{-r}\pi_{\nu r\mu} + \{o(\lambda^{-k})\}_\infty$$
$$(\nu = 1,\dots n_0;\ \mu = 0,\dots,n_0-1).$$

From (8.2.29) it follows that

$$(\pi_{100}, \dots, \pi_{n_0 00}) = \epsilon_1^{\mathsf{T}} Q_{11}^{[0]} D,$$

and (8.2.27) yields that it is a fundamental system of $\mathbf{H}\eta = 0$. Since two fundamental systems of $\mathbf{H}\eta = 0$ differ by multiplication from the right by an invertible $n_0 \times n_0$ matrix, we can choose the matrix D such that

$$(\pi_1, \dots, \pi_{n_0}) = (\pi_{100}, \dots, \pi_{n_0 00}).$$

By (8.2.25) and (8.2.26) we have

$$\pi_{\nu r \mu} = e_{\mu+1}^{\mathsf{T}} \left(Q_{11}^{[r]} + \epsilon_{n_0} \varepsilon^{\mathsf{T}} \Omega_l^{-1} Q_{21}^{[r]} \right) e_\nu \tag{8.2.37}$$

for $\mu = 0, \dots, n_0 - 1$, $\nu = 1, \dots, n_0$, and $r = 0, \dots, k$. The relations (8.2.27), (8.2.29) and (8.2.30) imply that

$$\epsilon_\mu^{\mathsf{T}} Q_{11}^{[r]\prime} = \epsilon_{\mu+1}^{\mathsf{T}} Q_{11}^{[r]} + \epsilon_{i+1}^{\mathsf{T}} \epsilon_{n_0} \varepsilon^{\mathsf{T}} \Omega_l^{-1} Q_{21}^{[r]}$$

for $\mu = 1, \dots, n_0 - 1$ and $r = 0, \dots, k$. These equations and (8.2.37) show that $\pi_{\nu r \mu} = \pi_{\nu r 0}^{(\mu)}$ for $\mu = 0, \dots, n_0 - 1$; $\nu = 1, \dots, n_0$; $r = 0, \dots, k$. Since $\pi_{\nu r, n_0 - 1} \in W_{p'}^{k+1-r}(a,b)$, it follows that $\pi_{\nu r 0} \in W_{p'}^{k+n_0-r}(a,b)$. From (8.2.25) and (8.2.26) we infer that there are $\pi_{\nu r \mu} \in W_{p'}^{k+n_0-\mu-r}(a,b)$ for $\nu = 1, \dots, n_0$; $\mu = n_0, \dots, n-1$, and $r = n_0 - \mu - 1, \dots, k + n_0 - \mu - 1$ such that

$$\eta_\nu^{(\mu)}(\cdot, \lambda) = \sum_{r=n_0-\mu-1}^{k+n_0-\mu-1} \lambda^{-r} \pi_{\nu r \mu} + \{o(\lambda^{-k-n_0+\mu+1})\}_\infty \tag{8.2.38}$$

$$(\nu = 1, \dots n_0;\ \mu = n_0, \dots, n-1).$$

According to the estimate (2.8.14) in Theorem 2.8.2 differentiation leads to

$$\eta_\nu^{(\mu+1)}(\cdot, \lambda) = \sum_{r=n_0-\mu-1}^{k+n_0-\mu-2} \lambda^{-r} \pi'_{\nu r \mu} + \{o(\lambda^{-k-n_0+\mu+2})\}_{p'} \tag{8.2.39}$$

$$(\nu = 1, \dots n_0;\ \mu = n_0, \dots, n-1).$$

Again by the estimate (2.8.14) in Theorem 2.8.2 we obtain

$$\eta_\nu^{(n_0)}(\cdot, \lambda) = \sum_{r=0}^{k-1} \lambda^{-r} \pi'_{\nu r, n_0 - 1} + \{o(\lambda^{-k-1})\}_{p'} \tag{8.2.40}$$

$$(\nu = 1, \dots n_0;\ \mu = n_0, \dots, n-1)$$

from (8.2.36) for $\mu = n_0 - 1$. From (8.2.38), (8.2.39) and (8.2.40) we deduce $\pi_{\nu r \mu} = 0$ if $r < 0$, and $\pi_{\nu r \mu} = \pi'_{\nu r, \mu - 1}$ if $0 \le r \le k + n_0 - \mu - 1$, where μ runs from n_0 to $n - 1$. The last equation leads to

$$\pi_{\nu r \mu} = \pi_{\nu r, n_0 - 1}^{(\mu - n_0 + 1)} = \pi_{\nu r 0}^{(\mu)}$$

for $\nu = 1,\dots,n_0$; $\mu = n_0,\dots,n-1$ and $r = 0,\dots,k+n_0-\mu-1$. Thus part i) of Theorem 8.2.1 is proved if we show that $\pi_{\nu r0} = 0$ for $r = 1,\dots,k$ if r is not a multiple of l. This is a consequence of the following proposition.

PROPOSITION 8.2.2. *Let $Q_{11}^{[r]}(a) = 0$ for $r = 1,\dots,k$. We assert:*
i) $Q_{11}^{[r]} = 0$ for $r = 1,\dots,k$ if r is not a multiple of l.
ii) For $r = 1,\dots,k$ there are $q^{[r]} \in M_{1,n_0}(W_{p'}^{k+1-r}(a,b))$ such that $Q_{21}^{[r]} = \Omega_l^{1-r}\varepsilon q^{[r]}$.

Proof. The assertion immediately follows from (8.2.30) and (8.2.31) by induction, where we make use of (8.2.12) and the identity $\Omega_l^{-l} = I_l$. □

Now we shall prove assertion ii) of Theorem 8.2.1. First let $\widehat{Q}_{12}^{[r]}$ and $\widehat{Q}_{22}^{[r]}$ ($r = 0,\dots,k$) be arbitrary solutions of (8.2.28), (8.2.29), (8.2.32), (8.2.33) and (8.2.34). We shall show that the matrix functions

$$Q_{12}^{[r]} := \sum_{m=1}^{l} \widehat{Q}_{12}^{[r]}\varepsilon_1\varepsilon_m^{\mathsf{T}}\Omega_l^{-r-1} = \widehat{Q}_{12}^{[r]}\varepsilon_1\varepsilon^{\mathsf{T}}\Omega_l^{-r-1} \tag{8.2.41}$$

and

$$Q_{22}^{[r]} := \sum_{m=1}^{l} (J_l^{\mathsf{T}} + \varepsilon_1\varepsilon_l^{\mathsf{T}})^{m-1}\widehat{Q}_{22}^{[r]}\varepsilon_1\varepsilon_m^{\mathsf{T}}\Omega_l^{-r}, \tag{8.2.42}$$

where $r = 0,\dots,k$, also satisfy (8.2.28), (8.2.29), (8.2.32), (8.2.33) and (8.2.34). The elements of $Q_{12}^{[r]}$ and $Q_{22}^{[r]}$ belong to $W_{p'}^{k+1-r}(a,b)$ because $\widehat{Q}_{12}^{[r]}$ and $\widehat{Q}_{22}^{[r]}$ have this property. $\widehat{Q}_{12}^{[0]} = 0$ implies $Q_{12}^{[0]} = 0$. Since $\widehat{Q}_{22}^{[0]}$ satisfies the differential equation in (8.2.28), also $Q_{22}^{[0]}$ satisfies it. And $Q_{22}^{[0]}(a) = I_l$ follows from $\widehat{Q}_{22}^{[0]}(a) = I_l$ and $(J_l^{\mathsf{T}} + \varepsilon_1\varepsilon_l^{\mathsf{T}})^{m-1}\varepsilon_1 = \varepsilon_m$ for $m = 1,\dots,l$. The identity $\omega_i = \omega_{i-1}\omega_l^{-1}$ $(i = 2,\dots,l)$ leads to

$$\begin{aligned}
\Omega_l^s(J_l^{\mathsf{T}} + \varepsilon_1\varepsilon_l^{\mathsf{T}}) &= \Omega_l^s\Big(\sum_{i=2}^{l}\varepsilon_i\varepsilon_{i-1}^{\mathsf{T}} + \varepsilon_1\varepsilon_l^{\mathsf{T}}\Big) = \sum_{i=2}^{l}\omega_i^s\varepsilon_i\varepsilon_{i-1}^{\mathsf{T}} + \varepsilon_1\varepsilon_l^{\mathsf{T}} \\
&= \omega_l^{-s}\Big(\sum_{i=2}^{l}\omega_{i-1}^s\varepsilon_i\varepsilon_{i-1}^{\mathsf{T}} + \omega_l^s\varepsilon_1\varepsilon_l^{\mathsf{T}}\Big) = \omega_l^{-s}(J_l^{\mathsf{T}} + \varepsilon_1\varepsilon_l^{\mathsf{T}})\Omega_l^s
\end{aligned}$$

for $s \in \mathbb{Z}$ and further to

$$\Omega_l^s(J_l^{\mathsf{T}} + \varepsilon_1\varepsilon_l^{\mathsf{T}})^{m-1} = \omega_m^s(J_l^{\mathsf{T}} + \varepsilon_1\varepsilon_l^{\mathsf{T}})^{m-1}\Omega_l^s \quad (s \in \mathbb{Z}; m = 1,\dots,l) \tag{8.2.43}$$

by induction with respect to m. Since

$$\Omega_l^{-1}\varepsilon_1 = \varepsilon_1,\ \Omega_l^l = I_l,\ \varepsilon_m^{\mathsf{T}}\Omega_l^{-r+j} = \omega_m^{j+1}\varepsilon_m^{\mathsf{T}}\Omega_l^{-r-1},\ \varepsilon^{\mathsf{T}}(J_l^{\mathsf{T}} + \varepsilon_1\varepsilon_l^{\mathsf{T}}) = \varepsilon^{\mathsf{T}},$$

and because of (8.2.43) we obtain

$$Q_{12}^{[r]} = \sum_{m=1}^{l}\Big\{-\widehat{Q}_{12}^{[r-1]\prime}\Omega_l^{-1}\varepsilon_1\varepsilon_m^{\mathsf{T}}\Omega_l^{-r-1} + (J_{n_0} - \epsilon_{n_0}a_{11}^{\mathsf{T}})\widehat{Q}_{12}^{[r-1]}\Omega_l^{-1}\varepsilon_1\varepsilon_m^{\mathsf{T}}\Omega_l^{-r-1}$$

$$
\begin{aligned}
&+ \big(\epsilon_{n_0-1} - (h_{n_0-1} - k_{n-1})\epsilon_{n_0}\big)\,\varepsilon^{\mathsf T}\Omega_l^{-1}\widehat{Q}_{22}^{[r-1]}\Omega_l^{-1}\varepsilon_1\varepsilon_m^{\mathsf T}\Omega_l^{-r-1}\\
&+ \sum_{j=1}^{l} k_{n-1-j}\epsilon_{n_0}\varepsilon^{\mathsf T}\Omega_l^{-1-j}\widehat{Q}_{22}^{[r-1-j]}\Omega_l^{-1}\varepsilon_1\varepsilon_m^{\mathsf T}\Omega_l^{-r-1}\\
&- \epsilon_{n_0} a_{12}^{\mathsf T}\widehat{Q}_{12}^{[r-1-l]}\Omega_l^{-1}\varepsilon_1\varepsilon_m^{\mathsf T}\Omega_l^{-r-1}\Big\}\\
&= \sum_{m=1}^{l}\Big\{-\widehat{Q}_{12}^{[r-1]'}\varepsilon_1\varepsilon_m^{\mathsf T}\Omega_l^{-r-1} + (J_{n_0} - \epsilon_{n_0}a_{11}^{\mathsf T})\widehat{Q}_{12}^{[r-1]}\varepsilon_1\varepsilon_m^{\mathsf T}\Omega_l^{-r-1}\\
&+ \big(\epsilon_{n_0-1} - (h_{n_0-1} - k_{n-1})\epsilon_{n_0}\big)\,\varepsilon^{\mathsf T}\Omega_l^{-1}(J_l^{\mathsf T} + \varepsilon_1\varepsilon_l^{\mathsf T})^{m-1}\widehat{Q}_{22}^{[r-1]}\varepsilon_1\varepsilon_m^{\mathsf T}\Omega_l^{-r}\\
&+ \sum_{j=1}^{l} k_{n-1-j}\epsilon_{n_0}\varepsilon^{\mathsf T}\Omega_l^{-1-j}(J_l^{\mathsf T} + \varepsilon_1\varepsilon_l^{\mathsf T})^{m-1}\widehat{Q}_{22}^{[r-1-j]}\varepsilon_1\varepsilon_m^{\mathsf T}\Omega_l^{-r+j}\\
&- \epsilon_{n_0} a_{12}^{\mathsf T}\widehat{Q}_{12}^{[r-1-l]}\varepsilon_1\varepsilon_m^{\mathsf T}\Omega_l^{-r+l-1}\Big\}
\end{aligned}
$$

for $r = 1,\ldots,k$. We conclude that the matrix functions $Q_{12}^{[r]}$ and $Q_{22}^{[r]}$ satisfy (8.2.32) for $r = 1,\ldots,k$. In a similar way we obtain

$$
\begin{aligned}
&\Omega_l Q_{22}^{[r]} - Q_{22}^{[r]}\Omega_l\\
&= \sum_{m=1}^{l}\Big\{\Omega_l(J_l^{\mathsf T} + \varepsilon_1\varepsilon_l^{\mathsf T})^{m-1}\widehat{Q}_{22}^{[r]}\varepsilon_1\varepsilon_m^{\mathsf T}\Omega_l^{-r} - (J_l^{\mathsf T} + \varepsilon_1\varepsilon_l^{\mathsf T})^{m-1}\widehat{Q}_{22}^{[r]}\varepsilon_1\varepsilon_m^{\mathsf T}\Omega_l^{-r+1}\Big\}\\
&= \sum_{m=1}^{l}(J_l^{\mathsf T} + \varepsilon_1\varepsilon_l^{\mathsf T})^{m-1}(\Omega_l\widehat{Q}_{22}^{[r]} - \widehat{Q}_{22}^{[r]}\Omega_l)\varepsilon_1\varepsilon_m^{\mathsf T}\Omega_l^{-r+1}\\
&= \sum_{m=1}^{l}\Big\{(J_l^{\mathsf T} + \varepsilon_1\varepsilon_l^{\mathsf T})^{m-1}\widehat{Q}_{22}^{[r-1]'}\varepsilon_1\varepsilon_m^{\mathsf T}\Omega_l^{-r+1}\\
&- \frac{1}{l}(J_l^{\mathsf T} + \varepsilon_1\varepsilon_l^{\mathsf T})^{m-1}\Omega_l\varepsilon a_{11}^{\mathsf T}\widehat{Q}_{12}^{[r-1]}\varepsilon_1\varepsilon_m^{\mathsf T}\Omega_l^{-r+1}\\
&- \frac{1}{l}(h_{n_0-1} - k_{n-1})(J_l^{\mathsf T} + \varepsilon_1\varepsilon_l^{\mathsf T})^{m-1}\Omega_l\varepsilon\varepsilon^{\mathsf T}\Omega_l^{-1}\widehat{Q}_{22}^{[r-1]}\varepsilon_1\varepsilon_m^{\mathsf T}\Omega_l^{-r+1}\\
&+ \frac{1}{l}\sum_{j=1}^{l} k_{n-1-j}(J_l^{\mathsf T} + \varepsilon_1\varepsilon_l^{\mathsf T})^{m-1}\Omega_l\varepsilon\varepsilon^{\mathsf T}\Omega_l^{-1-j}\widehat{Q}_{22}^{[r-1-j]}\varepsilon_1\varepsilon_m^{\mathsf T}\Omega_l^{-r+1}\\
&- \frac{1}{l}(J_l^{\mathsf T} + \varepsilon_1\varepsilon_l^{\mathsf T})^{m-1}\Omega_l\varepsilon a_{12}^{\mathsf T}\widehat{Q}_{12}^{[r-1-l]}\varepsilon_1\varepsilon_m^{\mathsf T}\Omega_l^{-r+1}\Big\}\\
&= Q_{22}^{[r-1]'} - \frac{1}{l}\Omega_l\varepsilon a_{11}^{\mathsf T}Q_{12}^{[r-1]} - \frac{1}{l}(h_{n_0-1} - k_{n-1})\Omega_l\varepsilon\varepsilon^{\mathsf T}\Omega_l^{-1}Q_{22}^{[r-1]}\\
&+ \frac{1}{l}\sum_{j=1}^{l} k_{n-1-j}\Omega_l\varepsilon\varepsilon^{\mathsf T}\Omega_l^{-1-j}Q_{22}^{[r-1-j]} - \frac{1}{l}\Omega_l\varepsilon a_{12}^{\mathsf T}Q_{12}^{[r-1-l]}
\end{aligned}
$$

for $r = 1, \ldots, k$. Finally the relationships

$$\varepsilon_\nu^\mathsf{T} \Omega_l \varepsilon = \omega_\nu \varepsilon_1^\mathsf{T} \varepsilon = \omega_\nu \varepsilon_1^\mathsf{T} \Omega_l \varepsilon, \quad \varepsilon_m^\mathsf{T} \Omega_l^{-r} \varepsilon_\nu = \omega_\nu^{-r} \delta_{m\nu}$$

and again (8.2.43) yield

$$\begin{aligned} &\varepsilon_\nu^\mathsf{T} \Big\{ Q_{22}^{[k]\prime} - \frac{1}{l} \Omega_l \varepsilon a_{11}^\mathsf{T} Q_{12}^{[k]} - \frac{1}{l} (h_{n_0-1} - k_{n-1}) \Omega_l \varepsilon \varepsilon^\mathsf{T} \Omega_l^{-1} Q_{22}^{[k]} \\ &+ \frac{1}{l} \sum_{j=1}^{l} k_{n-1-j} \Omega_l \varepsilon \varepsilon^\mathsf{T} \Omega_l^{-1-j} Q_{22}^{[k-j]} - \frac{1}{l} \Omega_l \varepsilon a_{12}^\mathsf{T} Q_{12}^{[k-l]} \Big\} \varepsilon_\nu \\ &= \omega_\nu^{-k} \varepsilon_1^\mathsf{T} \Big\{ \widehat{Q}_{22}^{[k]\prime} - \frac{1}{l} \Omega_l \varepsilon a_{11}^\mathsf{T} \widehat{Q}_{12}^{[k]} - \frac{1}{l} (h_{n_0-1} - k_{n-1}) \Omega_l \varepsilon \varepsilon^\mathsf{T} \Omega_l^{-1} \widehat{Q}_{22}^{[k]} \\ &+ \frac{1}{l} \sum_{j=1}^{l} k_{n-1-j} \Omega_l \varepsilon \varepsilon^\mathsf{T} \Omega_l^{-1-j} \widehat{Q}_{22}^{[k-j]} - \frac{1}{l} \Omega_l \varepsilon a_{12}^\mathsf{T} \widehat{Q}_{12}^{[k-l]} \Big\} \varepsilon_1 = 0. \end{aligned}$$

PROPOSITION 8.2.3. *Let $n_0 \geq 2$. For $r = 0, \ldots, n_0 - 2$ and $i = 1, \ldots, n_0 - r - 1$ we have $\epsilon_i^\mathsf{T} Q_{12}^{[r]} = 0$. For $r = 1, \ldots, n_0 - 1$ we have $\epsilon_{n_0-r}^\mathsf{T} Q_{12}^{[r]} \varepsilon_1 = \varepsilon_1^\mathsf{T} Q_{22}^{[0]} \varepsilon_1$.*

Proof. The first assertion is clear for $r = 0$ by (8.2.29). Assume that it holds for $0 \leq r - 1 < n_0 - 2$. We have to prove the assertion for r and $i = 1, \ldots, n_0 - r - 1$ which is at most $n_0 - 2$. From (8.2.32) and the induction hypothesis we obtain

$$\epsilon_i^\mathsf{T} Q_{12}^{[r]} = -\epsilon_i^\mathsf{T} Q_{12}^{[r-1]\prime} \Omega_l^{-1} + \epsilon_{i+1}^\mathsf{T} Q_{12}^{[r-1]} \Omega_l^{-1} = 0.$$

The second assertion holds for $r = 1$ since, by (8.2.32),

$$\epsilon_{n_0-1}^\mathsf{T} Q_{12}^{[1]} \varepsilon_1 = \varepsilon^\mathsf{T} \Omega_l^{-1} Q_{22}^{[0]} \Omega_l^{-1} \varepsilon_1 = \varepsilon_1^\mathsf{T} Q_{22}^{[0]} \varepsilon_1 .$$

Here we have used that $Q_{22}^{[0]}$ is a diagonal matrix function. Assume that the assertion holds for $1 \leq r - 1 < n_0 - 1$. Then the first assertion and (8.2.32) yield

$$e_{n_0-r}^\mathsf{T} Q_{12}^{[r]} \varepsilon_1 = \epsilon_{n_0-r+1}^\mathsf{T} Q_{12}^{[r-1]} \Omega_l^{-1} \varepsilon_1 = \varepsilon_1^\mathsf{T} Q_{22}^{[0]} \varepsilon_1 . \quad \square$$

Let $n_0 \geq 2$. Proposition 8.2.3 yields $\epsilon_1^\mathsf{T} Q_{12}^{[r]} = 0$ for $r = 0, \ldots, n_0 - 2$. Hence, by (8.2.25) and (8.2.26), there are $\varphi_{\nu r} \in W_{p'}^{k+2-n_0-r}(a,b)$ for $\nu = n_0 + 1, \ldots, n$ and $r = 0, \ldots, k + 1 - n_0$ such that, for $\nu = n_0 + 1, \ldots, n$,

$$\eta_\nu(x, \lambda) = \left\{ \sum_{r=0}^{k+1-n_0} \lambda^{-r} \varphi_{\nu r}(x) + \{o(\lambda^{-k-1+n_0})\}_\infty \right\} e^{\lambda \omega_{\nu-n_0} x}.$$

From (8.2.25) and (8.2.26) we immediately infer that this representation also holds for $n_0 = 1$ and that, for $n_0 = 0$ and $\nu = 1, \ldots, n$,

$$\eta_\nu(x, \lambda) = \left\{ \sum_{r=0}^{k} \lambda^{-r} \varphi_{\nu r}(x) + \{o(\lambda^{-k})\}_\infty \right\} e^{\lambda \omega_{\nu-n_0} x},$$

where $\varphi_{\nu r} \in W_{p'}^{k+1-r}(a,b)$.

If $n_0 > 0$ and $n_0 + 1 \leq \nu < n$, $0 \leq r \leq k+1-n_0$, then we have

$$\begin{aligned}\pi_{\nu r} &= \epsilon_1^{\mathsf{T}}\left\{Q_{12}^{[r+n_0-1]}\Omega_l^{n_0} + \epsilon_{n_0}\varepsilon^{\mathsf{T}}\Omega_l^{-1}Q_{22}^{[r+n_0-1]}\Omega_l^{n_0}\right\}\varepsilon_{\nu-n_0} \\ &= \epsilon_1^{\mathsf{T}}\left\{\widehat{Q}_{12}^{[r+n_0-1]} + \epsilon_{n_0}\varepsilon^{\mathsf{T}}\Omega_l^{-1}\widehat{Q}_{22}^{[r+n_0-1]}\right\}\varepsilon_1\omega_{\nu-n_0}^{-r}\end{aligned} \tag{8.2.44}$$

by (8.2.25), (8.2.26), (8.2.41), (8.2.42) and (8.2.43). If $n_0 = 0$, $1 \leq \nu \leq n$, and $0 \leq r \leq k$, then we have by (8.2.25), (8.2.26), (8.2.42) and (8.2.14) that

$$\varphi_{\nu r} = \varepsilon_1^{\mathsf{T}} V Q_{22}^{[r]}\varepsilon_{\nu-n_0} = \omega_{\nu-n_0}^{-r}\varepsilon^{\mathsf{T}}\widehat{Q}_{22}^{[r]}\varepsilon_1\,. \tag{8.2.45}$$

This leads to

$$\varphi_{\nu r} = \omega_{\nu-n_0}^{-r}\varphi_{n_0+1,r}$$

for $\nu = n_0+1,\ldots,n$ and $r = 0,\ldots,\tilde{k}$. Hence, for $\nu = n_0+1,\ldots,n$,

$$\eta_\nu(x,\lambda) = \left\{\sum_{r=0}^{\tilde{k}}(\lambda\omega_{\nu-n_0})^{-r}\varphi_r(x) + \{o(\lambda^{-\tilde{k}})\}_\infty\right\}e^{\lambda\omega_{\nu-n_0}x}, \tag{8.2.46}$$

where $\varphi_r := \varphi_{n_0+1,r}$.

If $n_0 = 0$, then (8.2.45) yields

$$\varphi_0 = \varphi_{10} = \varepsilon_1^{\mathsf{T}} Q_{22}^{[0]}\varepsilon_1\,.$$

If $n_0 = 1$, then (8.2.44) yields

$$\varphi_0 = \varphi_{20} = \varepsilon^{\mathsf{T}}\Omega_l^{-1}Q_{22}^{[0]}\Omega_l\varepsilon_1 = \varepsilon_1^{\mathsf{T}}Q_{22}^{[0]}\varepsilon_1\,.$$

If $n_0 \geq 2$, then (8.2.44) and the second assertion of Proposition 8.2.3 yield

$$\varphi_0 = \varphi_{n_0+1,0} = \epsilon_1^{\mathsf{T}}Q_{12}^{[n_0-1]}\Omega_l^{n_0}\varepsilon_1 = \varepsilon_1^{\mathsf{T}}Q_{22}^{[0]}\varepsilon_1\,.$$

From (8.2.28) we thus obtain $\varphi_0' - \frac{1}{l}(h_{n_0-1} - k_{n-1})\varphi_0 = 0$ and $\varphi_0(a) = 1$.

If $n_0 = 0$ or $n_0 = 1$, then $\varphi_r = \varphi_{n_0+1,r} \in W_{p'}^{k+1-r}(ha,b)$ for $r = 0,\ldots,k$. Next we prove that this also holds if $n_0 \geq 2$ and $r = 0,\ldots,k+1-n_0$. From (8.2.25) and (8.2.26) we infer for $\nu = n_0+1,\ldots,n$ and $\mu = 0,\ldots,n_0-1$ that

$$\eta_\nu^{(\mu)}(x,\lambda) = \left\{\sum_{r=1-n_0}^{k+1-n_0}\lambda^{-r}\varphi_{\nu r\mu}(x) + \{o(\lambda^{-k-1+n_0})\}_\infty\right\}e^{\lambda\omega_{\nu-n_0}x}, \tag{8.2.47}$$

where

(8.2.48)

$$\varphi_{\nu r\mu} = \omega_{\nu-n_0}^{n_0}\epsilon_{\mu+1}^{\mathsf{T}}\left(Q_{12}^{[r+n_0-1]} + \epsilon_{n_0}\varepsilon^{\mathsf{T}}\Omega_l^{-1}Q_{22}^{[r+n_0-1]}\right)\varepsilon_{\nu-n_0} \in W_{p'}^{k+2-n_0-r}(a,b)$$

for $r=1-n_0,\dots,k+1-n_0$ and $\varphi_{\nu r0}=\varphi_{\nu r}$. For $\nu=n_0+1,\dots,n$, $\mu=1,\dots,n_0-1$ and $r=2-n_0,\dots,k+1-n_0$ the equations (8.2.32) yield

(8.2.49) $$\varphi_{\nu r,\mu-1} = \omega_{\nu-n_0}^{n_0} \epsilon_\mu^{\mathsf{T}} Q_{12}^{[r+n_0-1]} \epsilon_{\nu-n_0}$$

$$= \omega_{\nu-n_0}^{n_0-1}\left(-\epsilon_\mu^{\mathsf{T}} Q_{12}^{[r+n_0-2]'} + \epsilon_{\mu+1}^{\mathsf{T}} Q_{12}^{[r+n_0-2]} + \epsilon_\mu^{\mathsf{T}} \epsilon_{n_0-1} \varepsilon^{\mathsf{T}} \Omega_l^{-1} Q_{22}^{[r+n_0-2]}\right) \varepsilon_{\nu-n_0}$$

$$= \omega_{\nu-n_0}^{-1}\left(-\varphi'_{\nu,r-1,\mu-1} + \varphi_{\nu,r-1,\mu}\right).$$

We shall prove that (8.2.49) leads to

(8.2.50) $$\varphi_{\nu r\mu} \in W_{p'}^{k+1-r-\mu}(a,b)$$

for $\nu=n_0+1,\dots,n$, $\mu=0,\dots,n_0-1$, and $r=1-n_0,\dots,k+1-n_0$. For $\mu=n_0-1$ this is true because of (8.2.48). From (8.2.48) and (8.2.29) we infer $\varphi_{\nu,1-n_0,\mu}=0$ for $\mu=0,\dots,n_0-2$. Suppose that (8.2.50) holds for $r-1$, where $1-n_0<r\le k+1-n_0$. Then (8.2.49) yields $\varphi_{\nu r\mu}\in W_{p'}^{k+1-r-\mu}(a,b)$ for $\mu=0,\dots,n_0-2$.

From (8.2.50) we obtain $\varphi_r = \varphi_{n_0+1,r0} \in W_{p'}^{k+1-r}(a,b)$.

It remains to prove (8.2.4). The equations (8.2.25) and (8.2.26) yield

(8.2.51) $$\eta_\nu^{(\mu)}(x,\lambda) = \left\{\sum_{r=-\mu}^{k-\mu} \lambda^{-r}\varphi_{\nu r\mu}(x) + \{o(\lambda^{-k+\mu})\}_\infty\right\} e^{\lambda\omega_{\nu-n_0}x}$$

for $\nu=n_0+1,\dots,n$ and $\mu=n_0,\dots,n-1$ with $\varphi_{\nu r\mu}\in W_{r'}^{k+1-r-\mu}(a,b)$. According to (8.2.4) we have to prove that

(8.2.52)
$$\eta_\nu^{(\mu)}(x,\lambda) = \left\{\sum_{r=-\mu}^{\tilde{k}-\mu} (\lambda\omega_{\nu-n_0})^{-r} \sum_{j=0}^{\mu} \binom{\mu}{j} \varphi_{r+\mu-j}^{(j)}(x) + \{o(\lambda^{-\tilde{k}+\mu})\}_\infty\right\} e^{\lambda\omega_{\nu-n_0}x}$$

for $\nu=n_0+1,\dots,n$ and $\mu=0,\dots,n-1$, where $\varphi_r:=0$ for $r<0$. This representation is true for $\mu=0$ by (8.2.46). Suppose that it holds for some $\mu<n-1$. Since the right-hand side of (8.2.52) is equal to the right-hand side of (8.2.46), (8.2.47) or (8.2.51), we infer from the estimate (2.8.14) in Theorem 2.8.2 that differentiation is allowed and leads to

$$\eta_\nu^{(\mu+1)}(x,\lambda)$$
$$= \left\{\sum_{r=-\mu}^{\tilde{k}-\mu-1} (\lambda\omega_{\nu-n_0})^{-r} \sum_{j=0}^{\mu} \binom{\mu}{j} \varphi_{r+\mu-j}^{(j+1)}(x) + \{o(\lambda^{-\tilde{k}+\mu+1})\}_{p'}\right\} e^{\lambda\omega_{\nu-n_0}x}$$
$$+ \left\{\sum_{r=-\mu}^{\tilde{k}-\mu} (\lambda\omega_{\nu-n_0})^{-r+1} \sum_{j=0}^{\mu} \binom{\mu}{j} \varphi_{r+\mu-j}^{(j)}(x) + \{o(\lambda^{-\tilde{k}+\mu+1})\}_{p'}\right\} e^{\lambda\omega_{\nu-n_0}x}$$
$$= \left\{\sum_{r=-(\mu+1)}^{\tilde{k}-(\mu+1)} (\lambda\omega_{\nu-n_0})^{-r} \binom{\mu+1}{j} \varphi_{r+(\mu+1)-j}^{(j)}(x) + \{o(\lambda^{-\tilde{k}+(\mu+1)})\}_{p'}\right\} e^{\lambda\omega_{\nu-n_0}x}.$$

Sometimes it is easier to have Theorem 8.2.1 in matricial form.

THEOREM 8.2.4. *Suppose that* $h_{n_0} = 1$, *set* $l = n - n_0$, *and let* $k \in \mathbb{N}$. *Suppose that* $k \geq \max\{l, n_0 - 1\}$ *if* $n_0 > 0$. *Suppose that*
α) $k_j \in L_{p'}(a,b)$ *for* $j = 0, \ldots, n-1-k$ *and* $k_{n-1-j} \in W_{p'}^{k-j}(a,b)$ *for* $j = 0, \ldots, \min\{k-1, n-1\}$ *if* $n_0 = 0$,
β) $h_0, \ldots, h_{n_0-1} \in W_{p'}^k(a,b)$, $k_0, \ldots, k_{n_0-1} \in W_{p'}^{k-l}(a,b)$ *and* $k_{n-1-j} \in W_{p'}^{k-j}(a,b)$ $(j = 0, \ldots, l-1)$ *if* $n_0 > 0$. *Let* $\Phi_{11}^{[0]} \in M_{n_0}(W_{p'}^{k+1}(a,b))$ *be a fundamental matrix of* $\mathbf{H}\eta = 0$.
For sufficiently large λ *the differential equation* $\mathbf{K}\eta = \lambda^l \mathbf{H}\eta$ *has a fundamental matrix function*

$$Y(\cdot,\lambda) = \begin{pmatrix} \Phi_{11}(\cdot,\lambda) & \Phi_{12}(\cdot,\lambda) \\ \Phi_{21}(\cdot,\lambda) & \Phi_{22}(\cdot,\lambda) \end{pmatrix} E(\cdot,\lambda), \tag{8.2.53}$$

where

$$E(x,\lambda) = \operatorname{diag}(1, \ldots, 1, e^{\lambda\omega_1(x-a)}, \ldots, e^{\lambda\omega_l(x-a)})$$

and $\omega_j = \exp\{\frac{2\pi i(j-1)}{l}\}$ $(j = 1, \ldots, l)$, *with the following properties:*
i) *There are* $n_0 \times n_0$ *matrix functions* $\Phi_{11}^{[r]}$ $(r = 1, \ldots, [\frac{k}{l}])$ *with* $\epsilon_{\mu+1}^{\mathsf{T}} \Phi_{11}^{[r]}$ *belonging to* $M_{1,n_0}(W_{p'}^{k+n_0-lr-\mu}(a,b))$ *for* $r = 0, \ldots, [\frac{k}{l}]$ *and* $\mu = 0, \ldots, n_0 - 1$ *such that*

$$\Phi_{11}(\cdot,\lambda) = \sum_{r=0}^{[\frac{k}{l}]} \lambda^{-lr} \Phi_{11}^{[r]} + \{o(\lambda^{-k})\}_\infty, \ \Phi_{11}(\cdot,\lambda\omega_l) = \Phi_{11}(\cdot,\lambda). \tag{8.2.54}$$

ii) *There are* $l \times n_0$ *matrix functions* $\Phi_{21}^{[r]}$ $(r = 0, \ldots, [\frac{k-1}{l}])$ *with* $\varepsilon_j^{\mathsf{T}} \Phi_{21}^{[r]}$ *belonging to* $M_{1,n_0}(W_{p'}^{k-lr-j+1}(a,b))$ *for* $r = 0, \ldots, [\frac{k-1}{l}]$ *and* $j = 1, \ldots, \min\{l, k-lr\}$ *and* $\varepsilon_j^{\mathsf{T}} \Phi_{21}^{[r]} = 0$ *for* $r = 0, \ldots, [\frac{k-1}{l}]$ *and* $j = k - lr + 1, \ldots, n$ *such that*
(8.2.55)

$$\Phi_{21}(\cdot,\lambda) = \sum_{r=0}^{[\frac{k-1}{l}]} \lambda^{-lr} \Phi_{21}^{[r]} + \lambda \Xi_l(\lambda)\{o(\lambda^{-k})\}_\infty, \ \Phi_{21}(\cdot,\lambda\omega_l) = \Phi_{21}(\cdot,\lambda).$$

iii) *Set* $\tilde{k} := \min\{k, k+1-n_0\}$. *For* $r = 0, \ldots, \tilde{k}$ *and* $\mu = 0, \ldots, n-1$ *there are functions* $u_{\mu r} \in W_{p'}^{k+1-r}(a,b)$, *where* $u_{\mu 0} = \varphi_0$ *is the solution of the initial value problem*

$$\varphi_0' - \frac{1}{l}(h_{n_0-1} - k_{n-1})\varphi_0 = 0, \quad \varphi_0(a) = 1,$$

such that, with

$$\Phi_{12}^{[r]} = \operatorname{diag}(u_{0r}, \dots, u_{n_0-1,r})\widehat{V}\Omega_l^{-r},$$

$$\widehat{V} = \begin{pmatrix} 1 & \dots & 1 \\ \omega_1 & \dots & \omega_l \\ \vdots & & \vdots \\ \omega_1^{n_0-1} & \dots & \omega_l^{n_0-1} \end{pmatrix},$$

$$\Phi_{22}^{[r]} = \operatorname{diag}(u_{n_0 r}, \dots, u_{n-1,r})V\Omega_l^{n_0-r}$$

for $r = 0, \dots, \tilde{k}$*, the representations*

(8.2.56) $$\Phi_{12}(\cdot,\lambda) = \Xi_{n_0}(\lambda)\left\{\sum_{r=0}^{\tilde{k}} \lambda^{-r}\Phi_{12}^{[r]} + \{o(\lambda^{-\tilde{k}})\}_\infty\right\},$$

(8.2.57) $$\Phi_{22}(\cdot,\lambda) = \lambda^{n_0}\Xi_l(\lambda)\left\{\sum_{r=0}^{\tilde{k}} \lambda^{-r}\Phi_{22}^{[r]} + \{o(\lambda^{-\tilde{k}})\}_\infty\right\},$$

and

(8.2.58) $$\Phi_{j2}(\cdot,\lambda\omega_l) = \Phi_{j2}(\cdot,\lambda)(J_l + \varepsilon_l\varepsilon_1^{\mathsf{T}}) \quad (j = 1,2)$$

hold. We remind that J_r*,* Ω_l*,* Ξ_r*,* V *are defined in* (8.2.7), (8.2.9), (8.2.10), (8.2.11).

Proof. The statements i) and ii) immediately follow from Theorem 8.2.1 if we observe that the differential equation $\mathbf{K}\eta = \lambda^l \mathbf{H}\eta$ depends on λ^l so that we can take the asymptotic representation for $0 \le \arg\lambda < \frac{2\pi}{l}$ and extend it to arbitrary λ by setting $\eta_j(\cdot,\lambda) = \eta_j(\cdot,\lambda\omega_i)$ for $j = 1, \dots, n_0$ and suitable i (depending on λ).

For the proof of iii) we observe that (8.2.52) yields

$$\eta_\nu^{(\mu)}(x,\lambda) = (\lambda\omega_{\nu-n_0})^\mu \left\{\sum_{r=0}^{\tilde{k}} (\lambda\omega_{\nu-n_0})^{-r} u_{\mu r}(x) + \{o(\lambda^{-\tilde{k}})\}_\infty\right\} e^{\lambda\omega_{\nu-n_0}(x-a)},$$

where the $u_{\mu r}$ have the stated properties. To prove (8.2.56) and (8.2.57) we have to show that

$$\omega_{\nu-n_0}^{\mu-r} u_{\mu r} = e_{\mu+1}^{\mathsf{T}} \begin{pmatrix} \Phi_{12}^{[r]} \\ \Phi_{22}^{[r]} \end{pmatrix} \varepsilon_{\nu-n_0}$$

holds for $\mu = 0, \dots, n-1$, $\nu = n_0, \dots, n$, and $r = 0, \dots, \tilde{k}$. But this follows immediately from

$$\begin{pmatrix} \Phi_{12}^{[r]} \\ \Phi_{22}^{[r]} \end{pmatrix} = \operatorname{diag}(u_{0r}, \dots, u_{n-1,r})\widetilde{V}\Omega_l^{-r},$$

where

$$\widetilde{V} = \begin{pmatrix} \widehat{V} \\ V\Omega_l^{n_0} \end{pmatrix} = \begin{pmatrix} 1 & . & . & . & 1 \\ \omega_1 & . & . & . & \omega_l \\ \vdots & & & & \vdots \\ \omega_1^{n-1} & . & . & . & \omega_l^{n-1} \end{pmatrix}. \tag{8.2.59}$$

Since the differential equation $\mathbf{K}\eta = \lambda^l \mathbf{H}\eta$ depends on λ^l, it is clear that the functions $\hat{\eta}_{n_0+j}(\cdot,\lambda) := \eta_{n_0+1}(\cdot,\lambda\omega_j)$ for $j = 2,\ldots,l$ are solutions of it. But since $\hat{\eta}_{n_0+j}$ and η_{n_0+j} only differ by o-terms in (8.2.4), the above statements also hold if we suppose that $\hat{\eta}_{n_0+j} = \eta_{n_0+j}$ for $j = 2,\ldots,n$. Then (8.2.58) immediately follows. □

Taking only the leading terms, we obtain

COROLLARY 8.2.5. *Let the assumptions of Theorem 8.2.4 be satisfied. Let $\Phi_{11}^{[0]}$ be a fundamental matrix of $\mathbf{H}\eta = 0$ if $n_0 > 0$. Then, for sufficiently large λ, $\mathbf{K}\eta = \lambda^l \mathbf{H}\eta$ has a fundamental matrix function of the form*

$$Y(\cdot,\lambda) = \begin{pmatrix} [\Phi_{11}^{[0]}]_\infty & \pi_0 \Xi_{n_0}(\lambda)[\widehat{V}]_\infty \\ [\Phi_{21}^{[0]}]_\infty & \varphi_0 \lambda^{n_0} \Xi_l(\lambda)[V\Omega_l^{n_0}]_\infty \end{pmatrix} E(\cdot,\lambda),$$

where φ_0 is invertible.

8.3. The asymptotic fundamental system in the general case

In our investigations in Section 8.2 we have assumed that the coefficients of the highest derivatives of $\mathbf{K}$ and $\mathbf{H}$ are 1. In the following we shall suppose that the leading coefficients are bounded and bounded away from zero. Since we can divide the differential equation by the coefficient of the highest derivative of $\mathbf{K}$, it is no restriction to suppose that this coefficient is 1. Now we suppose additionally that the coefficient of the highest derivative of $\mathbf{H}$ is the product of a nonzero complex number α and a positive function. Replacing the eigenvalue parameter λ by $\lambda\alpha^{-1}$ we thus have the assumptions as required at the beginning of Section 8.1, where we impose the regularity conditions stated there; additionally, we require $h_{n_0} \in C^\infty[a,b]$ in order to avoid too complicated regularity considerations.

Now let

$$u(x) = \int_a^x (h_{n_0}(\xi))^{1/l}\, d\xi \quad (x \in [a,b]) \tag{8.3.1}$$

and $\beta = u(b)$. We define the differential operator L_u on $[0,\beta]$ by

$$L_u f = [h_{n_0}^{-n/l} L(f \circ u)] \circ u^{-1} \quad (f \in W_p^n(0,\beta)), \tag{8.3.2}$$

where u^{-1} is the inverse of u. Then

$$L_u = \mathbf{K}_u - \lambda \mathbf{H}_u, \tag{8.3.3}$$

where $\mathbf{K}_u$ is an n-th order differential operator and $\mathbf{H}_u$ is an n_0-th order differential operator. Since

$$(f\circ u)^{(j)} = (u')^j f^{(j)}\circ u + \text{ lower order derivatives in } f$$

and $u' = h_{n_0}^{1/l}$, we infer that the coefficient of the highest derivative is 1 for both $\mathbf{K}$ and $\mathbf{H}$. Also, the coefficients of $\mathbf{K}_u$ and $\mathbf{H}_u$ have the same regularity properties as those of $\mathbf{K}$ and $\mathbf{H}$. Hence L_u satisfies the assumptions of Section 8.2. Then $L_u f = 0$ has a fundamental matrix $Y_u(\cdot,\lambda)$ as stated in Theorem 8.2.1. Since $L_u f = 0$ if and only if $L(f\circ u) = 0$, we infer that $e_1^\mathsf{T} Y_u(\cdot,\lambda)\circ u$ is a fundamental system of $Lf = 0$. Define $Y(\cdot,\lambda)$ by

$$e_\nu^\mathsf{T} Y(\cdot,\lambda) := \left(e_1^\mathsf{T} Y_u(\cdot,\lambda)\circ u\right)^{(\nu-1)} \quad (\nu = 1,\ldots,n).$$

Then $Y(\cdot,\lambda)$ is a fundamental matrix of $\mathbf{K}\eta - \lambda\mathbf{H}\eta = 0$, which satisfies the same asymptotic estimates as $Y_u(\cdot,\lambda)$ in view of Remark 2.8.10 since

$$e_\nu^\mathsf{T} Y(\cdot,\lambda) = \sum_{\mu=1}^{\nu} c_{\nu\mu} e_\mu^\mathsf{T} Y_u(\cdot,\lambda)\circ u \quad (\nu = 1,\ldots,n). \tag{8.3.4}$$

The coefficients $c_{\nu\mu}$ are linear combinations of products of derivatives of u and thus belong to $C^\infty[a,b]$. The elements $c_{\nu\nu}$ are powers of u' and thus are invertible.

In this case (8.2.1) and (8.2.2) are also true if $\mu = 0$ since $\{\pi_1,\ldots,\pi_{n_0}\}$ is a fundamental system of $\mathbf{H}f = 0$ if and only if $\{\pi_1\circ u^{-1},\ldots,\pi_{n_0}\circ u^{-1}\}$ is a fundamental system of $\mathbf{H}_u f = 0$.

In part ii) of Theorem 8.2.1 some changes are necessary. First, the terms $e^{\lambda\omega_{\nu-n_0}(x-a)}$ in (8.2.4) have to be replaced by $e^{\lambda\omega_{\nu-n_0}u(x)}$ (note that a corresponds to the left endpoint of the interval $[0,\beta]$). Since the functions $\eta_\nu^{(\mu)}(\cdot,\lambda)$ have the same asymptotics as in (8.2.1) and (8.2.2), a term-by-term differentiation shows that these representations also hold here. And of course, φ_r corresponds to $\varphi_{u,r}\circ u$, where $\varphi_{u,0}$ is the solution of

$$\varphi_{u,0}' - \frac{1}{l}(h_{u,n_0-1} - k_{u,n-1})\varphi_{u,0} = 0, \quad \varphi_{u,0} = 1,$$

and h_{u,n_0-1}, $k_{u,n-1}$ are the corresponding coefficients of the differential equation $L_u f = 0$. We can write down the corresponding differential equation for φ_0 in terms of the coefficients of the given differential equation. But since we do not need to know φ_0 explicitly, we are not going to find this differential equation.

Therefore, the results of Section 8.2 can be generalized to the case that h_{n_0} is not necessarily 1. This will be summarized in the following three corollaries.

COROLLARY 8.3.1. *Suppose that $h_{n_0}\in C^\infty[a,b]$, $h_{n_0} > 0$, and $h_{n_0}^{-1}\in L_\infty(a,b)$. Set $l = n - n_0$, let $k\in\mathbb{N}$, and suppose that $k\geq \max\{l, n_0-1\}$ if $n_0 > 0$. Suppose that α) $k_j\in L_{p'}(a,b)$ for $j = 0,\ldots,n-1-k$ and $k_{n-1-j}\in W_{p'}^{k-j}(a,b)$ for*

$j=0,\dots,\min\{k-1,n-1\}$ *if* $n_0=0$,
β) $h_0,\dots,h_{n_0-1}\in W^k_{p'}(a,b)$, $k_0,\dots,k_{n_0-1}\in W^{k-l}_{p'}(a,b)$ *and* $k_{n-1-j}\in W^{k-j}_{p'}(a,b)$ $(j=0,\dots,l-1)$ *if* $n_0>0$. *Let* $\{\pi_1,\dots,\pi_{n_0}\}\subset W^{k+n_0}_{p'}(a,b)$ *be a fundamental system of* $\mathbf{H}\eta=0$.
For sufficiently large λ *the differential equation* $\mathbf{K}\eta=\lambda^l\mathbf{H}\eta$ *has a fundamental system* $\{\eta_1(\cdot,\lambda),\dots,\eta_n(\cdot,\lambda)\}$ *with the following properties:*
i) *There are functions* $\pi_{\nu r}\in W^{k+n_0-lr}_{p'}(a,b)$ $(1\le\nu\le n_0,\ 1\le r\le[\frac{k}{l}])$ *such that*

$$\eta_\nu^{(\mu)}(\cdot,\lambda)=\pi_\nu^{(\mu)}+\sum_{r=1}^{[\frac{k}{l}]}\lambda^{-lr}\pi_{\nu r}^{(\mu)}+\{o(\lambda^{-k})\}_\infty \tag{8.3.5}$$
$$(\nu=1,\dots,n_0;\ \mu=0,\dots,n_0-1),$$

$$\eta_\nu^{(\mu)}(\cdot,\lambda)=\pi_\nu^{(\mu)}+\sum_{r=1}^{[\frac{k-\mu+n_0-1}{l}]}\lambda^{-lr}\pi_{\nu r}^{(\mu)}+\{o(\lambda^{-k+\mu-n_0+1})\}_\infty \tag{8.3.6}$$
$$(\nu=1,\dots,n_0;\ \mu=n_0,\dots,n-1),$$

ii) *Set* $\tilde{k}:=\min\{k,k+1-n_0\}$. *Let* $\omega_j=\exp\{\frac{2\pi i(j-1)}{l}\}$ $(j=1,\dots,l)$ *and let* u *be given by* (8.3.1). *For* $r=0,\dots,\tilde{k}$ *there are functions* $\varphi_r\in W^{k+1-r}_{p'}(a,b)$ *such that* $\frac{1}{\varphi_0}\in L_\infty(a,b)$ *and*

$$\eta_\nu^{(\mu)}(x,\lambda)=\left[\frac{d^\mu}{dx^\mu}\right]\left\{\sum_{r=0}^{\tilde{k}}(\lambda\omega_{\nu-n_0})^{-r}\varphi_r(x)e^{\lambda\omega_{\nu-n_0}u(x)}\right\} \tag{8.3.7}$$
$$+\{o(\lambda^{-\tilde{k}+\mu})\}_\infty e^{\lambda\omega_{\nu-n_0}u(x)}$$
$$(\nu=n_0+1,\dots,n;\ \mu=0,\dots,n-1),$$

where $[\frac{d^\mu}{dx^\mu}]$ *means that we omit those terms of the Leibniz expansion which contain a function* $\varphi_r^{(j)}$ *with* $j>\tilde{k}-r$.

COROLLARY 8.3.2. *Suppose that* $h_{n_0}\in C^\infty[a,b]$, $h_{n_0}>0$, *and* $h_{n_0}^{-1}\in L_\infty(a,b)$. *Set* $l=n-n_0$, *let* $k\in\mathbb{N}$, *and suppose that* $k\ge\max\{l,n_0-1\}$ *if* $n_0>0$. *Suppose that* α) $k_j\in L_{p'}(a,b)$ *for* $j=0,\dots,n-1-k$ *and* $k_{n-1-j}\in W^{k-j}_{p'}(a,b)$ *for* $j=0,\dots,\min\{k-1,n-1\}$ *if* $n_0=0$,
β) $h_0,\dots,h_{n_0-1}\in W^k_{p'}(a,b)$, $k_0,\dots,k_{n_0-1}\in W^{k-l}_{p'}(a,b)$ *and* $k_{n-1-j}\in W^{k-j}_{p'}(a,b)$ $(j=0,\dots,l-1)$ *if* $n_0>0$. *Let* $\Phi_{11}^{[0]}\in M_{n_0}(W^{k+1}_{p'}(a,b))$ *be a fundamental matrix of* $\mathbf{H}\eta=0$.
For sufficiently large λ *the differential equation* $\mathbf{K}\eta=\lambda^l\mathbf{H}\eta$ *has a fundamental matrix function*

$$Y(\cdot,\lambda)=\begin{pmatrix}\Phi_{11}(\cdot,\lambda) & \Phi_{12}(\cdot,\lambda)\\ \Phi_{21}(\cdot,\lambda) & \Phi_{22}(\cdot,\lambda)\end{pmatrix}E(\cdot,\lambda), \tag{8.3.8}$$

where

$$E(x,\lambda) = \operatorname{diag}(1,\dots,1,e^{\lambda\omega_1 u(x)},\dots,e^{\lambda\omega_l u(x)}),$$

$\omega_j = \exp\{\frac{2\pi i(j-1)}{l}\}$ $(j = 1,\dots,l)$, *and* u *is given by* (8.3.1), *with the following properties:*

i) *There are* $n_0 \times n_0$ *matrix functions* $\Phi_{11}^{[r]}$ $(r = 0,\dots,[\frac{k}{l}])$ *with* $\epsilon_{\mu+1}^{\mathsf{T}}\Phi_{11}^{[r]}$ *belonging to* $M_{1,n_0}(W_{p'}^{k+n_0-lr-\mu}(a,b))$ *for* $r = 0,\dots,[\frac{k}{l}]$ *and* $\mu = 0,\dots,n_0-1$ *such that*

(8.3.9) $$\Phi_{11}(\cdot,\lambda) = \sum_{r=0}^{[\frac{k}{l}]} \lambda^{-lr}\Phi_{11}^{[r]} + \{o(\lambda^{-k})\}_\infty,\ \Phi_{11}(\cdot,\lambda\omega_l) = \Phi_{11}(\cdot,\lambda).$$

ii) *There are* $l \times n_0$ *matrix functions* $\Phi_{21}^{[r]}$ $(r = 0,\dots,[\frac{k-1}{l}])$ *with* $\varepsilon_j^{\mathsf{T}}\Phi_{21}^{[r]}$ *belonging to* $M_{1,n_0}(W_{p'}^{k-lr-j+1}(a,b))$ *for* $r = 0,\dots,[\frac{k-1}{l}]$ *and* $j = 1,\dots,\min\{l,k-lr\}$ *and* $\varepsilon_j^{\mathsf{T}}\Phi_{21}^{[r]} = 0$ *for* $r = 0,\dots,[\frac{k-1}{l}]$ *and* $j = k-lr+1,\dots,n$ *such that*

(8.3.10)
$$\Phi_{21}(\cdot,\lambda) = \sum_{r=0}^{[\frac{k-1}{l}]} \lambda^{-lr}\Phi_{21}^{[r]} + \lambda\Xi_l(\lambda)\{o(\lambda^{-k})\}_\infty,\ \Phi_{21}(\cdot,\lambda\omega_l) = \Phi_{21}(\cdot,\lambda).$$

iii) *Set* $\tilde{k} := \min\{k,k+1-n_0\}$. *For* $r = 0,\dots,\tilde{k}$ *and* $\mu = 0,\dots,n-1$ *there are functions* $u_{\mu r} \in W_{p'}^{k+1-r}(a,b)$, *where* $\frac{1}{u_{\mu 0}} \in L_\infty(a,b)$ *and* $u_{\mu 0} = h_{n_0}^{\mu/l}u_{00}$ *for* $\mu = 0,\dots,n-1$, *such that, with*

$$\Phi_{12}^{[r]} = \operatorname{diag}(u_{0r},\dots,u_{n_0-1,r})\widehat{V}\Omega_l^{-r},$$

$$\widehat{V} = \begin{pmatrix} 1 & \cdot & \cdot & \cdot & 1 \\ \omega_1 & \cdot & \cdot & \cdot & \omega_l \\ \vdots & & & & \vdots \\ \omega_1^{n_0-1} & \cdot & \cdot & \cdot & \omega_l^{n_0-1} \end{pmatrix},$$

$$\Phi_{22}^{[r]} = \operatorname{diag}(u_{n_0 r},\dots,u_{n-1,r})V\Omega_l^{n_0-r}$$

for $r = 0,\dots,\tilde{k}$, *the representations*

(8.3.11) $$\Phi_{12}(\cdot,\lambda) = \Xi_{n_0}(\lambda)\left\{\sum_{r=0}^{\tilde{k}} \lambda^{-r}\Phi_{12}^{[r]} + \{o(\lambda^{-\tilde{k}})\}_\infty\right\},$$

(8.3.12) $$\Phi_{22}(\cdot,\lambda) = \lambda^{n_0}\Xi_l(\lambda)\left\{\sum_{r=0}^{\tilde{k}} \lambda^{-r}\Phi_{22}^{[r]} + \{o(\lambda^{-\tilde{k}})\}_\infty\right\},$$

and

(8.3.13) $$\Phi_{j2}(\cdot,\lambda\omega_l) = \Phi_{j2}(\cdot,\lambda)(J_l + \varepsilon_l\varepsilon_1^{\mathsf{T}}) \quad (j = 1,2)$$

hold. We recall that J_r, Ω_l, Ξ_r, V *are defined in* (8.2.7), (8.2.9), (8.2.10), (8.2.11).

COROLLARY 8.3.3. *Let the assumptions of Corollary* 8.3.2 *be satisfied. Let* $\Phi_{11}^{[0]}$ *be a fundamental matrix of* $\mathbf{H}\eta = 0$ *if* $n_0 > 0$. *Then, for sufficiently large* λ, $\mathbf{K}\eta = \lambda^l\mathbf{H}\eta$ *has a fundamental matrix function of the form*

$$Y(\cdot,\lambda) = \begin{pmatrix} [\Phi_{11}^{[0]}]_\infty & \pi_0\Xi_{n_0}(\lambda h_{n_0}^{1/l})[\widehat{V}]_\infty \\ [\Phi_{21}^{[0]}]_\infty & \varphi_0\lambda^{n_0}h_{n_0}^{n_0/l}\Xi_l(\lambda h_{n_0}^{1/l})[V\Omega_l^{n_0}]_\infty \end{pmatrix} E(\cdot,\lambda),$$

where φ_0 *is invertible,*

$$E(x,\lambda) = \operatorname{diag}(1,\dots,1,e^{\lambda\omega_1 u(x)},\dots,e^{\lambda\omega_l u(x)}),$$

$\omega_j = \exp\{\frac{2\pi i(j-1)}{l}\}$ $(j = 1,\dots,l)$, *and* u *is given by* (8.3.1).

8.4. The inverse of the asymptotic fundamental matrix

THEOREM 8.4.1. *Suppose that* $h_{n_0} \in C^\infty[a,b]$, $h_{n_0} > 0$, *and* $h_{n_0}^{-1} \in L_\infty(a,b)$. *Set* $l = n - n_0$, *let* $k \in \mathbb{N}$ *and suppose that* $k \geq \max\{l, n_0 - 1\}$ *if* $n_0 > 0$. *Suppose that*
α) $k_j \in L_{p'}(a,b)$ *for* $j = 0,\dots,n-1-k$ *and* $k_{n-1-j} \in W_{p'}^{k-j}(a,b)$ *for* $j = 0,\dots,\min\{k-1,n-1\}$ *if* $n_0 = 0$,
β) $h_0,\dots,h_{n_0-1} \in W_{p'}^k(a,b)$, $k_0,\dots,k_{n_0-1} \in W_{p'}^{k-l}(a,b)$ *and* $k_{n-1-j} \in W_{p'}^{k-j}(a,b)$ $(j = 0,\dots,l-1)$ *if* $n_0 > 0$.
Let $Y(\cdot,\lambda)$ *be the fundamental matrix function of* $\mathbf{K}\eta = \lambda^l\mathbf{H}\eta$ *given as in Corollary* 8.3.2. *We recall that* J_r, Ω_l, Ξ_r, V *are defined in* (8.2.7), (8.2.9), (8.2.10), (8.2.11), *that* ε_j *is the* j*-th unit vector in* $\mathbb{C}^l$, *and* ϵ_j *is the* j*-th unit vector in* $\mathbb{C}^{n_0}$. *Then, for sufficiently large* λ,

$$(8.4.1)\qquad Y(\cdot,\lambda)^{-1} = E(\cdot,-\lambda)\begin{pmatrix} \Psi_{11}(\cdot,\lambda) & \Psi_{12}(\cdot,\lambda) \\ \Psi_{21}(\cdot,\lambda) & \Psi_{22}(\cdot,\lambda) \end{pmatrix},$$

where

$$E(x,\lambda) = \operatorname{diag}(1,\dots,1,e^{\lambda\omega_1 u(x)},\dots,e^{\lambda\omega_l u(x)}),$$

$\omega_j = \exp\{\frac{2\pi i(j-1)}{l}\}$ $(j = 1,\dots,l)$, u *is given by* (8.3.1), *and the* Ψ_{ij} *have the following properties:*
i) *For* $r = 0,\dots,[\frac{k}{l}]$ *there are* $\Psi_{11}^{[r]} \in M_{n_0,n_0}(W_{p'}^{k+1-lr}(a,b))$ *such that*

$$(8.4.2)\qquad \Psi_{11}(\cdot,\lambda) = \sum_{r=0}^{[\frac{k}{l}]}\lambda^{-lr}\Psi_{11}^{[r]} + \{o(\lambda^{-k})\}_\infty.$$

ii) *For* $r = 0,\dots,[\frac{k}{l}]$ *there are* $n_0 \times l$ *matrix functions* $\Psi_{12}^{[r]}$ *with* $\Psi_{12}^{[r]}\varepsilon_j$ *belonging to* $(W_{p'}^{k+1-l(r+1)+j}(a,b))^{n_0}$ *for* $r = 0,\dots,[\frac{k}{l}]$ *and* $j = 1,\dots,l$ *such that*

$$(8.4.3)\qquad \Psi_{12}(\cdot,\lambda) = \lambda^{-l}\left(\sum_{r=0}^{[\frac{k}{l}]}\lambda^{-lr}\Psi_{12}^{[r]} + \{o(\lambda^{-k+l-1})\}_\infty\Xi_l(\lambda)^{-1}\right).$$

iii) *For* $r = 0, \dots, k$ *there are* $\Psi_{21}^{[r]} \in M_{l \times n_0}(W_{p'}^{k+1-r}(a,b))$ *such that*

$$\Psi_{21}(\cdot,\lambda) = \lambda^{-n_0} \sum_{r=0}^{k} \lambda^{-r} \Psi_{21}^{[r]} + \{o(\lambda^{-k+1-n_0})\}_\infty \tag{8.4.4}$$

and

$$\Psi_{21}(\cdot,\lambda\omega_l) = (J_l^\mathsf{T} + \varepsilon_1\varepsilon_l^\mathsf{T})\Psi_{21}(\cdot,\lambda).$$

iv) *For* $r = 0, \dots, k$ *there are* $\Psi_{22}^{[r]} \in M_l(W_{p'}^{k+1-r}(a,b))$ *such that*

$$\Psi_{22}(\cdot,\lambda) = \left\{ \sum_{r=0}^{k} \lambda^{-r} \Psi_{22}^{[r]} + \{o(\lambda^{-k})\}_\infty \right\} \lambda^{-n_0} \Xi_l(\lambda)^{-1} \tag{8.4.5}$$

and

$$\Psi_{22}(\cdot,\lambda\omega_l) = (J_l^\mathsf{T} + \varepsilon_1\varepsilon_l^\mathsf{T})\Psi_{22}(\cdot,\lambda).$$

v) *If* $n_0 > 0$ *then*

$$\Psi_{12}^{[0]}\varepsilon_l = -h_{n_0}^{-1} {\Phi_{11}^{[0]}}^{-1} \epsilon_{n_0}.$$

vi) *We have*

$$\Psi_{22}^{[0]} = {\Phi_{22}^{[0]}}^{-1} = \varphi_0^{-1} h_{n_0}^{-n_0/l} \Omega_l^{-n_0} V^{-1} \Xi_l(h_{n_0}^{-1/l}).$$

Proof. First let us consider the differential operator L_u defined in (8.3.2). Let $\widetilde{Y}_u(\cdot,\lambda)$ be a fundamental matrix of $L_u f = 0$ as given by (8.2.22). According to Theorem 2.8.2 and the Neumann series expansion we have

$$\widetilde{Y}_u(\cdot,\lambda)^{-1} = \widetilde{E}(\cdot,-\lambda) \begin{pmatrix} \widehat{Y}_{11}(\cdot,\lambda) & \widehat{Y}_{12}(\cdot,\lambda) \\ \widehat{Y}_{21}(\cdot,\lambda) & \widehat{Y}_{22}(\cdot,\lambda) \end{pmatrix},$$

where $\widetilde{E}$ corresponds to the matrix function E considered in Section 8.2,

$$\widehat{Y}_{ij}(\cdot,\lambda) = \sum_{r=0}^{k} \lambda^{-r} \widehat{Y}_{ij}^{[r]} + \{o(\lambda^{-k})\}_\infty$$

and the components of $\widehat{Y}_{ij}^{[r]}$ belong to $W_{p'}^{k+1-r}(a,b)$. Also observe that $\widehat{Y}_{21}^{[0]} = 0$ in view of (2.8.17). Writing

$$Y_u(\cdot,\lambda)^{-1} = \widetilde{E}(\cdot,-\lambda) \begin{pmatrix} \Psi_{11}^u(\cdot,\lambda) & \Psi_{12}^u(\cdot,\lambda) \\ \Psi_{21}^u(\cdot,\lambda) & \Psi_{22}^u(\cdot,\lambda) \end{pmatrix},$$

we infer from (8.2.23) and (8.2.16) that

$$\Psi_{11}^u(\cdot,\lambda) = D^{-1}\widehat{Y}_{11}(\cdot,\lambda),$$

$$\Psi_{21}^u(\cdot,\lambda) = \lambda^{1-n_0}\Omega_l^{-n_0}\widehat{Y}_{21}(\cdot,\lambda),$$

$$\Psi_{12}^u(\cdot,\lambda) = D^{-1}\left(-\widehat{Y}_{11}(\cdot,\lambda)\epsilon_{n_0}\varepsilon^\mathsf{T}\Omega_l^{-1} + \widehat{Y}_{12}(\cdot,\lambda)\right)\lambda^{-1}V^{-1}\Xi_l(\lambda)^{-1},$$

$$\Psi_{22}^u(\cdot,\lambda) = \lambda^{1-n_0}\Omega_l^{-n_0}(-\widehat{Y}_{21}(\cdot,\lambda)\epsilon_{n_0}\varepsilon^\mathsf{T}\Omega_l^{-1} + \widehat{Y}_{22}(\cdot,\lambda))\lambda^{-1}V^{-1}\Xi_l(\lambda)^{-1}.$$

In view of (8.3.4) the stated asymptotics and the regularity of the coefficients follow since $Y(\cdot,\lambda)^{-1}$ is obtained from $Y_u(u(\cdot),\lambda)^{-1}$ by multiplication from the right by a lower triangular matrix whose coefficients are in $C^\infty(a,b)$ and do not depend on λ.

We still have to prove the particular shapes of the $\Psi_{ij}^{[r]}$. In view of (8.3.9), (8.3.10), and (8.3.13) we have

$$Y(\cdot,\lambda\omega_l)E(\cdot,\lambda\omega_l)^{-1} = Y(\cdot,\lambda)E(\cdot,\lambda)^{-1}\begin{pmatrix} I_{n_0} & 0 \\ 0 & J_l + \varepsilon_l\varepsilon_1^{\mathsf{T}} \end{pmatrix}.$$

Hence

$$E(\cdot,\lambda\omega_l)Y(\cdot,\lambda\omega_l)^{-1} = \begin{pmatrix} I_{n_0} & 0 \\ 0 & J_l^{\mathsf{T}} + \varepsilon_1\varepsilon_l^{\mathsf{T}} \end{pmatrix} E(\cdot,\lambda)Y(\cdot,\lambda)^{-1}.$$

For Φ_{11} and Φ_{12} we infer that the asymptotic polynomials are invariant under the transformation $\lambda \mapsto \lambda\omega_l$. Hence the asymptotic polynomials are actually asymptotic polynomials in λ^l. Since the highest possible λ-power of $\Psi_{12}(\cdot,\lambda)$ is λ^{-1} by the representation given above, this completes the proof of parts i)–iv).

In order to prove v) and vi) we first observe that $\Phi_{11}(\cdot,\lambda)$ and $\Phi_{22}(\cdot,\lambda)$ are invertible for sufficiently large λ. This follows immediately from Corollary 8.3.3. Using the Schur factorization (1.3.4), the corresponding factorization with indices 1 and 2 interchanged, and the fact that a fundamental matrix is invertible, we also obtain that

$$\Phi_{22} - \Phi_{21}\Phi_{11}^{-1}\Phi_{12} \quad \text{and} \quad \Phi_{11} - \Phi_{12}\Phi_{22}^{-1}\Phi_{21}$$

are invertible. Furthermore, taking the inverses in (1.3.4) and multiplying out yields

$$\Psi_{11} = (\Phi_{11} - \Phi_{12}\Phi_{22}^{-1}\Phi_{21})^{-1}, \tag{8.4.6}$$

$$\Psi_{12} = -\Phi_{11}^{-1}\Phi_{12}(\Phi_{22} - \Phi_{21}\Phi_{11}^{-1}\Phi_{12})^{-1}, \tag{8.4.7}$$

$$\Psi_{21} = -\Phi_{22}^{-1}\Phi_{21}(\Phi_{11} - \Phi_{12}\Phi_{22}^{-1}\Phi_{21})^{-1}, \tag{8.4.8}$$

$$\Psi_{22} = (\Phi_{22} - \Phi_{21}\Phi_{11}^{-1}\Phi_{12})^{-1}. \tag{8.4.9}$$

Here we are interested in Ψ_{12} and Ψ_{22}. From Corollary 8.3.2 we infer

$$\begin{aligned}\Phi_{22}(\cdot,\lambda) &- \Phi_{21}(\cdot,\lambda)\Phi_{11}(\cdot,\lambda)^{-1}\Phi_{12}(\cdot,\lambda) \\ &= \lambda^{n_0}\Xi_l(\lambda)[\Phi_{22}^{[0]}]_\infty - [\Phi_{21}^{[0]}]_\infty[\Phi_{11}^{[0]^{-1}}]_\infty\Xi_{n_0}(\lambda)[\Phi_{12}^{[0]}]_\infty \\ &= \lambda^{n_0}\Xi_l(\lambda)[\Phi_{22}^{[0]}]_\infty.\end{aligned}$$

Hence

$$\Psi_{22}(\cdot,\lambda) = [\Phi_{22}^{[0]^{-1}}]_\infty\lambda^{-n_0}\Xi_l(\lambda)^{-1},$$

which proves vi) in view of Corollary 8.3.3.

We have

$$\Phi_{11}(\cdot,\lambda)^{-1}\Phi_{12}(\cdot,\lambda) = [\Phi_{11}^{[0]^{-1}}]_\infty \varphi_0 \Xi_{n_0}(\lambda h_{n_0}^{1/l})[\widehat{V}]_\infty,$$

which yields

$$\Psi_{12}(\cdot,\lambda) = -[\Phi_{11}^{[0]^{-1}}]_\infty \Xi_{n_0}(\lambda h_{n_0}^{1/l})[\widehat{V}\Omega_l^{-n_0}V^{-1}]_\infty \lambda^{-n_0} h_{n_0}^{-n_0/l} \Xi_l(\lambda h_{n_0}^{1/l})^{-1}.$$

With the aid of (8.2.13) and (8.2.12) we infer

$$\widehat{V}\Omega_l^{-n_0}V^{-1}\varepsilon_l = \frac{1}{l}\sum_{i=1}^{n_0} \epsilon_i \varepsilon^{\mathsf{T}}\Omega_l^{i-n_0}\varepsilon = \sum_{\substack{i=1\\ n_0-i\in l\mathbb{Z}}}^{n_0} \epsilon_i .$$

This implies

$$\Xi_{n_0}(\lambda h_{n_0}^{1/l})[\widehat{V}\Omega_l^{-n_0}V^{-1}\varepsilon_l]_\infty = \lambda^{n_0-1} h_{n_0}^{(n_0-1)/l}[\epsilon_{n_0}]_\infty,$$

whence

$$\Psi_{12}(\cdot,\lambda)\varepsilon_l = -\lambda^{-l} h_{n_0}^{-1}[\Phi_{11}^{[0]^{-1}}\epsilon_{n_0}]_\infty .$$

This proves part v). □

Together with **H** and **K** we consider their formally adjoints $\mathbf{H}^+$ and $\mathbf{K}^+$, respectively, given by

$$\mathbf{H}^+\zeta = \sum_{i=0}^{n_0}(-1)^i(h_i\zeta)^{(i)} \quad (\zeta \in W_{p'}^{n_0}(a,b)),$$

$$\mathbf{K}^+\zeta = (-1)^n\zeta^{(n)} + \sum_{i=0}^{n-1}(-1)^i(k_i\zeta)^{(i)} \quad (\zeta \in W_{p'}^{n}(a,b)),$$

see (8.1.11).

THEOREM 8.4.2. *Suppose that* $h_{n_0} \in C^\infty[a,b]$, $h_{n_0} > 0$, *and* $h_{n_0}^{-1} \in L_\infty(a,b)$. *Let* $k \in \mathbb{N}$ *and suppose that* $k \geq \max\{l, n_0-1\}$ *if* $n_0 > 0$. *Suppose that* $k_j \in W_{p'}^{j}(a,b)$ *for* $j = 0,\ldots,n-1$ *and that*
α) $k_{n-1-j} \in W_{p'}^{k-j}(a,b)$ *for* $j = 0,\ldots,n-1$ *if* $n_0 = 0$ *and* $k \geq n$,
β) $h_0,\ldots,h_{n_0-1} \in W_{p'}^{k}(a,b)$, $k_0,\ldots,k_{n_0-1} \in W_{p'}^{k-l}(a,b)$, *and* $k_{n-1-j} \in W_{p'}^{k-j}(a,b)$ $(j = 0,\ldots,l-1)$ *if* $n_0 > 0$.
Let $\{\eta_1(\cdot,\lambda),\ldots,\eta_n(\cdot,\lambda)\}$ *be the fundamental system of* $\mathbf{K}\eta - \lambda^l\mathbf{H}\eta = 0$ *as considered in Corollary* 8.3.1, *and let* $Y(\cdot,\lambda) = (\eta_\nu^{(\mu-1)}(\cdot,\lambda))_{\nu,\mu=1}^n$. *Set*

$$\zeta_\nu(\cdot,\lambda) = e_\nu^{\mathsf{T}} Y(\cdot,\lambda)^{-1} e_n \quad (\nu = 1,\ldots,n). \tag{8.4.10}$$

Then, for sufficiently large λ, $\{\zeta_1(\cdot,\lambda),\ldots,\zeta_n(\cdot,\lambda)\}$ *is a fundamental system of the differential equation* $\mathbf{K}^+\zeta - \lambda^l\mathbf{H}^+\zeta = 0$ *with the following properties:*

i) *There are a fundamental system* $\{\kappa_1,\dots,\kappa_{n_0}\} \subset W_{p'}^{k+1}(a,b)$ *of* $\mathbf{H}^+\zeta = 0$ *and functions* $\kappa_{\nu r} \in W_{p'}^{k+1-lr}(a,b)$ $(1 \le \nu \le n_0,\ 1 \le r \le [\frac{k}{l}])$ *such that*

$$\zeta_\nu^{(\mu)}(\cdot,\lambda) = \lambda^{-l}\Bigg(\kappa_\nu^{(\mu)} + \sum_{r=1}^{[\frac{k-\mu}{l}]} \lambda^{-lr}\kappa_{\nu r}^{(\mu)} + \{o(\lambda^{-k+\mu})\}_\infty\Bigg) \tag{8.4.11}$$

for $\nu = 1,\dots,n_0$ *and* $\mu = 0,\dots,n-1$.
ii) *For* $r = 0,\dots,k$ *there are functions* $\psi_r \in W_{p'}^{k+1-r}(a,b)$ *such that* ψ_0^{-1} *exists and is bounded, and*

$$\begin{aligned}\zeta_\nu^{(\mu)}(x,\lambda) = (\lambda\omega_{\nu-n_0})^{1-n}\Big\{\Big[\frac{d^\mu}{dx^\mu}\Big]\Big\{\sum_{r=0}^{k}(\lambda\omega_{\nu-n_0})^{-r}\psi_r(x)e^{-\lambda\omega_{\nu-n_0}u(x)}\Big\}\\ + \{o(\lambda^{-k+\mu})\}_\infty e^{-\lambda\omega_{\nu-n_0}u(x)}\Big\}\end{aligned} \tag{8.4.12}$$

for $\nu = n_0+1,\dots,n$ *and* $\mu = 0,\dots,n-1$, *where* u *is given by* (8.3.1) *and* $[\frac{d^\mu}{dx^\mu}]$ *means that we omit those terms of the Leibniz expansion which contain a function* $\psi_r^{(j)}$ *with* $j > k-r$.
If $h_{n_0} = 1$, *then* ψ_0 *is the solution of the initial value problem*

$$\psi_0' + \frac{1}{l}(h_{n_0-1} - k_{n-1})\psi_0 = 0, \quad \psi_0(a) = \frac{1}{l}. \tag{8.4.13}$$

Proof. From (8.1.17) we immediately infer by recursive substitution that

$$\zeta_\nu^{(n)} = (-1)^i(e_\nu^\intercal Y^{-1}e_{n-i})^{(n-i)} + \sum_{j=1}^{i}(-1)^{j-1}(p_{n-j}\zeta_\nu)^{(n-j)}$$

for $i = 0,\dots,n-1$. If we observe that (8.1.17) also holds for $k = 1$ with $e_0 = 0$, then the above identity is also true for $i = n$, which shows that the $\zeta_\nu(\cdot,\lambda)$ satisfy the differential equation $\mathbf{K}^+\zeta - \lambda^l\mathbf{H}^+\zeta = 0$. And if the $\zeta_\nu(\cdot,\lambda)$ would be linearly dependent, then again (8.1.17) would yield that $e_1^\intercal Y(\cdot,\lambda)^{-1},\dots,e_n^\intercal Y(\cdot,\lambda)^{-1}$ were linearly dependent which is impossible. Hence $\{\zeta_1(\cdot,\lambda),\dots,\zeta_n(\cdot,\lambda)\}$ is a fundamental system of the differential equation $\mathbf{K}^+\zeta - \lambda^l\mathbf{H}^+\zeta = 0$.
i) We know from Lemma (8.1.3) and Theorem 8.4.1 ii) that

$$\begin{aligned}\zeta_\nu^{(\mu)}(\cdot,\lambda) &= \sum_{i=0}^{\mu} q_{\mu,i}e_\nu^\intercal Y(\cdot,\lambda)^{-1}e_{n-i}\\ &= \sum_{i=0}^{\mu} q_{\mu,i}\epsilon_\nu^\intercal\Psi_{12}(\cdot,\lambda)\varepsilon_{l-i}\\ &= \lambda^{-l}\Big(\sum_{r=0}^{[\frac{k}{l}]}\lambda^{-lr}\psi_{\nu\mu r} + \lambda^{-k+\mu}\tilde{\psi}_{\nu\mu}(\cdot,\lambda)\Big)\end{aligned} \tag{8.4.14}$$

for $\nu = 1, \ldots, n_0$ and $\mu = 0, \ldots, l-1$ with $\psi_{\nu\mu r} \in W_{p'}^{k+1-lr-\mu}(a,b)$ and $\tilde{\psi}_{\nu\mu}(\cdot,\lambda) = \{o(1)\}_\infty$. Here we have used that the $q_{\mu,i}$ do not depend on λ for $\mu = 0, \ldots, l-1$ and belong to $W_{p'}^{k+1-\mu}(a,b)$.

For $\mu = 0$, the representation (8.4.11) follows. Furthermore, since $\Phi_{11}^{[0]}$ is a fundamental matrix of $\mathbf{H}\eta = 0$ by Corollary 8.3.2, it is also a fundamental matrix of $\frac{1}{h_{n_0}}\mathbf{H}\eta = 0$. Then the same argument as at the beginning of this proof shows that $\left\{\epsilon_\nu^{\mathsf{T}} \Phi_{11}^{[0]^{-1}} \epsilon_{n_0} : \nu = 1, \ldots, n_0\right\}$ is a fundamental system of $\left(\frac{1}{h_{n_0}}\mathbf{H}\right)^+ \zeta = 0$. Since $\kappa_\nu := \psi_{\nu 00} = -h_{n_0}^{-1} \epsilon_\nu^{\mathsf{T}} \Phi_{11}^{[0]^{-1}} \epsilon_{n_0}$ by Theorem 8.4.1 v), it follows that $\{\kappa_1, \ldots, \kappa_{n_0}\}$ is a fundamental system of $\mathbf{H}^+\zeta = 0$.

From the considerations at the beginning of the proof of Theorem 8.4.1 we infer that the $\tilde{\psi}_{\nu\mu}(\cdot,\lambda)$ are linear combinations of products of polynomials in λ^{-1} and o-functions having the properties stated in Theorem 2.8.2. Thus not only $\tilde{\psi}_{\nu\mu}(\cdot,\lambda) = \{o(1)\}_\infty$ but also $\frac{1}{\lambda}\tilde{\psi}'_{\nu\mu}(\cdot,\lambda) \in \{o(1)\}_{p'}$ holds. Hence differentiating and comparing coefficients we obtain

$$\psi'_{\nu\mu r} = \psi_{\nu,\mu+1,r}$$

for $1 \le \nu \le n_0$, $0 \le \mu \le l-2$, $0 \le r \le [\frac{k-\mu-1}{l}]$. This proves i) for $\mu = 0, \ldots, l-1$.

For $\nu = 1, \ldots, n_0$ and $\mu = l, \ldots, n-1$ we have in view of Lemma 8.1.3 and Theorem 8.4.1 that

$$\begin{aligned} \zeta_\nu^{(\mu)}(\cdot,\lambda) &= \sum_{i=0}^{\mu} q_{\mu,i}(\cdot,\lambda) e_\nu^{\mathsf{T}} Y(\cdot,\lambda)^{-1} e_{n-i} \\ &= \sum_{i=0}^{l-1} q_{\mu,i}(\cdot,\lambda) \epsilon_\nu^{\mathsf{T}} \Psi_{12}(\cdot,\lambda) \varepsilon_{l-i} + \sum_{i=l}^{\mu} q_{\mu,i}(\cdot,\lambda) \epsilon_\nu^{\mathsf{T}} \Psi_{11}(\cdot,\lambda) \epsilon_{n-i} \\ &= \lambda^{-l}\Bigg(\sum_{r=-[\frac{\mu}{l}]}^{[\frac{k}{l}]} \lambda^{-lr} \psi_{\nu\mu r} + \lambda^{-k+\mu} \tilde{\psi}_{\nu\mu}(\cdot,\lambda)\Bigg), \end{aligned}$$

where $\tilde{\psi}_{\nu\mu}(\cdot,\lambda) = \{o(1)\}_\infty$ and $\frac{1}{\lambda}\tilde{\psi}'_{\nu\mu}(\cdot,\lambda) = \{o(1)\}_{p'}$. To verify the λ-exponent in the asymptotic representation we have used the statement on the degree of $q_{\nu,i}$ in Lemma 8.1.3. Differentiation of $\zeta_\nu^{(\mu)}$ for $\mu = n_0 - 1, \ldots, n-1$ and comparison with the above representation of $\zeta_\nu^{(\mu)}$ completes the proof of part i).

ii) This proof is similar to the proof of part i). First we consider $\nu = n_0 + 1$. From Lemma 8.1.3 and Theorem 8.4.1 iv) we know that

$$\begin{aligned} \zeta_{n_0+1}^{(\mu)}(\cdot,\lambda) &= \sum_{i=0}^{\mu} q_{\mu,i} e_{n_0+1}^{\mathsf{T}} Y(\cdot,\lambda)^{-1} e_{n-i} = \sum_{i=0}^{\mu} q_{\mu,i} e^{-\lambda u} \varepsilon_1^{\mathsf{T}} \Psi_{22}(\cdot,\lambda) \varepsilon_{l-i} \\ &= \lambda^{-n+1+\mu}\left\{\sum_{r=0}^{k} \lambda^{-r} \psi_{n_0+1,\mu,r} + \{o(\lambda^{-k})\}_\infty\right\} e^{-\lambda u} \end{aligned}$$

for $\mu = 0,\dots,l-1$, where $\psi_{n_0+1,\mu,r} \in W_{p'}^{k+1-r}(a,b)$. With the aid of Theorem 8.4.1 iii) this asymptotic representation can be extended to $\mu = l,\dots,n-1$. Proceeding as in the proof of part i), the representation (8.4.12) for $\nu = n_0+1$ follows.

From Theorem 8.4.1 vi) and (8.2.13) we infer

$$\psi_0 = \varepsilon_1^\mathsf{T}\Psi_{22}^{[0]}\varepsilon_l = \varphi_0^{-1}h_{n_0}^{(1-n)/l}\varepsilon_1^\mathsf{T}\Omega_l^{-n_0}V^{-1}\varepsilon_l = \frac{1}{l}\varphi_0^{-1}h_{n_0}^{(1-n)/l}, \tag{8.4.15}$$

which completes the proof of part ii) for $\nu = n_0+1$ in view of (8.2.3).

Now let $\nu \in \{n_0+2,\dots,n\}$. Then we infer from Theorem 8.4.1 iv) that

$$\begin{aligned}\zeta_\nu(x,\lambda) &= e^{-\lambda\omega_{\nu-n_0}u(x)}\varepsilon_{\nu-n_0}^\mathsf{T}\Psi_{22}(x,\lambda)\varepsilon_l\\ &= e^{-\lambda\omega_{\nu-n_0}u(x)}\varepsilon_{\nu-1-n_0}^\mathsf{T}\Psi_{22}(x,\lambda\omega_l^{-1})\varepsilon_l\\ &= e^{-\lambda\omega_{\nu-n_0}u(x)}\zeta_{\nu-1}(x,\lambda\omega_l^{-1})e^{\lambda\omega_{\nu-1-n_0}u(x)}\\ &= e^{-\lambda\omega_l^{-1}u(x)}\zeta_{\nu-1}(x,\lambda\omega_l^{-1}).\end{aligned}$$

From $\omega_l^{-j} = \omega_{j+1}$ we infer that

$$\zeta_\nu(x,\lambda) = \zeta_{n_0+1}(x,\lambda\omega_{\nu-n_0})$$

for $\nu \in \{n_0+2,\dots,n\}$. Since we have already shown that (8.4.12) holds for $\nu = n_0+1$, it therefore holds for all $\nu = n_0+1,\dots,n$. Finally, (8.4.13) immediately follows from (8.4.15) and (8.2.3). □

REMARK 8.4.3. i) In general, it is not very useful to have estimates where the o-terms dominate all of the other terms. So (8.4.11) is most useful if $-k+\mu \le 0$, i. e., if $\mu \le k$ (note that we suppose that $k \ge l$ if $n_0 > 0$).
ii) If we differentiate in (8.4.12) we obtain

$$\zeta_\nu^{(\mu)}(x,\lambda) = (\lambda\omega_{\nu-n_0})^{1-n+\mu}\left\{\sum_{r=0}^{k}(\lambda\omega_{\nu-n_0})^{-r}\psi_{\mu,r}(x)e^{-\lambda\omega_{\nu-n_0}u(x)} + \{o(\lambda^{-k})\}_\infty e^{-\lambda\omega_{\nu-n_0}u(x)}\right\} \tag{8.4.16}$$

for $\nu = n_0+1,\dots,n$ and $\mu = 0,\dots,n-1$, where $\psi_{\mu,r} \in W_{p'}^{k+1-r}(a,b)$ and $\psi_{\mu,0} = (-h_{n_0}^{1/l})^\mu\psi_0$. Since $\zeta_\nu(\cdot,\lambda)$ is a solution of $(\mathbf{K}^+ - \lambda^l\mathbf{H}^+)\zeta_\nu(\cdot,\lambda) = 0$, it follows that (8.4.16) also holds in case $\mu = n$.
iii) In Theorems 8.2.1, 8.2.4, 8.4.1, 8.4.2 and in Corollaries 8.3.1, 8.3.2 the asymptotic estimates $\{o(\lambda^{-k})\}_\infty$ can be replaced with the estimates $\{O(\lambda^{-k}\tau_{p'}(\lambda))\}_\infty$ or $\{O(\lambda^{-k}\tau_\infty(\lambda))\}_{p'}$, where the latter estimate holds if $k > 0$ or $p < 3$. In case $k = 0$ and $p \ge 3$, which can only occur if $n_0 = 0$, we have $k_{n-1} \in W_{p'}^1(a,b)$ since $n \ge 2$, and then the estimate (2.8.16) holds, see (6.4.10). This follows from the corresponding proofs if we take the estimates (2.8.12) or (2.8.13), respectively,

instead of (2.8.11). Also, these considerations apply to expressions $[\cdot]$ considered below, e. g. we can replace $[c]_\infty$ by $c+\{O(\tau_{p'}(\lambda))\}_\infty$. Furthermore, in the estimates $\{o(\cdot)\}_\infty$ and $\{O(\cdot)\}_\infty$, the functions considered are continuous (even in $W^1_{p'}(a,b)$), so that these estimates hold at each point in the interval $[a,b]$. Finally, by using (2.8.14), we can differentiate $\zeta_\nu^{(\mu)}$ term by term if $\mu \leq n-1$, where $\{o(\lambda^{-k})\}_\infty$ has to be replaced by $\{o(\lambda^{-k+1})\}_{p'}$.

8.5. Almost Birkhoff regular boundary eigenvalue problems

In order to define Birkhoff regular and almost Birkhoff regular problems we first introduce some notations. For $\nu = 1,\ldots,n$ let

$$l_\nu = \deg\left[e_\nu^\mathsf{T}\left(W^{(0)}(\lambda^l)\Xi_n(\lambda), W^{(1)}(\lambda^l)\Xi_n(\lambda)\right)\right], \tag{8.5.1}$$

where $W^{(j)}(\lambda^l)$ and $\Xi_n(\lambda)$ have been defined in (8.1.6) and (8.2.10). Then we have

$$\operatorname{diag}(\lambda^{-l_1},\ldots,\lambda^{-l_n})W^{(j)}(\lambda^l)\Xi_n(\lambda) = W_0^{(j)} + O(\lambda^{-1})$$

for $j=0,1$. We suppose that

$$\operatorname{rank}(W^{(0)}, W^{(1)}) = n.$$

If this condition should not hold, then it might be achieved by taking suitable linear combinations of the boundary conditions. This is similar to the method in Section 5.1. We leave the details to the reader.

Let $\hat{l}_\nu \in \{0,\ldots,l-1\}$ such that

$$l_\nu = \hat{l}_\nu \bmod(l) \quad (\nu = 1,\ldots,n). \tag{8.5.2}$$

We have

$$e_\nu^\mathsf{T} W^{(j)}(\lambda^l)\Xi_n(\lambda) = \lambda^{l_\nu}[e_\nu^\mathsf{T} W_0^{(j)}] \tag{8.5.3}$$

for $\nu = 1,\ldots,n$ and $j=0,1$. Since $W^{(0)}$ and $W^{(1)}$ depend on λ^l, only those entries of $e_\nu^\mathsf{T} W_0^{(j)} e_{\tau+1}$ are different from zero for which $\tau = \hat{l}_\nu \bmod(l)$. We can write these τ as $\tau = \hat{l}_\nu + tl$ for $t = 0,\ldots,L$ where

$$L = \left[\frac{n-1}{l}\right]. \tag{8.5.4}$$

It might be that the τ for $t=L$ is already larger than $n-1$. In this case, all terms corresponding to this t are tacitly understood to be zero. For $\nu = 1,\ldots,n$ and $j=0,1$ let

$$\alpha_{\nu t}^{(j)} := e_\nu^\mathsf{T} W_0^{(j)} e_{\hat{l}_\nu + tl+1} \quad (t = 0,\ldots,L) \tag{8.5.5}$$

and

$$\alpha_\nu^{(j)} = \sum_{t=0}^{L} h_{n_0}^{\hat{l}_\nu/l+t}(a_j)\alpha_{\nu t}^{(j)}, \tag{8.5.6}$$

where $a_0 := a$ and $a_1 := b$.

For $\nu = 1,\ldots,n$ let q'_ν be the degree of $e_\nu^\top(W^{(0)}(\lambda), W^{(1)}(\lambda))$ as a polynomial in λ. For $\nu = 1,\ldots,n$ there are integers $q_\nu \le q'_\nu$ such that

$$e_\nu^\top\big(W^{(0)}(\lambda^l)Y(a,\lambda)(I_n-\Delta_0), W^{(1)}(\lambda^l)Y(b,\lambda)(I_n-\Delta_0)\big) = O(\lambda^{q_\nu l}). \tag{8.5.7}$$

Here we can choose the exponent to be a multiple of l since $Y(\cdot,\lambda)(I_n-\Delta_0)$ is a function of λ^l. We can write

$$e_\nu^\top M(\lambda)e_j =: \lambda^{q_\nu l}[u_{\nu,j}] \tag{8.5.8}$$

for $\nu = 1,\ldots,n$ and $j = 1,\ldots,n_0$, where $M(\lambda)$ is the characteristic matrix given by (8.1.14).

Now let $\vartheta \subset \{1,\ldots,n\}$. Writing $\vartheta = \{\vartheta_1,\ldots,\vartheta_j\}$ we shall use the notations $\vartheta' := \{1,\ldots,n\}\setminus\vartheta =: \{\vartheta'_1,\ldots,\vartheta'_{n-j}\}$. Let $\operatorname{sgn}\vartheta$ be the signum of the permutation given by $\{\vartheta'_1,\ldots,\vartheta'_{n-j},\vartheta_1,\ldots,\vartheta_j\}$. Also, let

$$l_\vartheta = l\sum_{i=1}^{n-j} q_{\vartheta'_i} + \sum_{i=1}^{j} l_{\vartheta_i}. \tag{8.5.9}$$

Let $\Theta_{n,m}$ be the set of all subsets of $\{1,\ldots,n\}$ with m elements. We set

$$\begin{aligned} l^{(0)} &= \max\{l_\vartheta : \vartheta\in\Theta_{n,l}\},\\ l^{(1)} &= \max\{l_\vartheta : \vartheta\in\Theta_{n,l+1}\} \text{ if } n_0>0,\\ l^{(2)} &= \max\{l_\vartheta : \vartheta\in\Theta_{n,l-1}\}. \end{aligned} \tag{8.5.10}$$

For $j = 0,\ldots,l$ we define

$$b_j^{(0)} = \sum_{\substack{\vartheta\in\Theta_{n,l}\\ l_\vartheta = l^{(0)}}} \operatorname{sgn}\vartheta\, u_\vartheta v_{\vartheta,j}, \tag{8.5.11}$$

where

$$u_\vartheta = \begin{vmatrix} u_{\vartheta'_1,1} & \cdots & u_{\vartheta'_1,n_0}\\ \vdots & & \vdots\\ u_{\vartheta'_{n_0},1} & \cdots & u_{\vartheta'_{n_0},n_0}\end{vmatrix} \text{ if } \vartheta \ne \{1,\ldots,n\},\ u_\vartheta = 1 \text{ else}, \tag{8.5.12}$$

$$v_{\vartheta,j} = \begin{vmatrix} \omega_1^{\hat{l}_{\vartheta_1}}\alpha_{\vartheta_1}^{(0)} & \cdots & \omega_j^{\hat{l}_{\vartheta_1}}\alpha_{\vartheta_1}^{(0)} & \omega_{j+1}^{\hat{l}_{\vartheta_1}}\alpha_{\vartheta_1}^{(1)} & \cdots & \omega_l^{\hat{l}_{\vartheta_1}}\alpha_{\vartheta_1}^{(1)}\\ \vdots & & \vdots & \vdots & & \vdots\\ \omega_1^{\hat{l}_{\vartheta_l}}\alpha_{\vartheta_l}^{(0)} & \cdots & \omega_j^{\hat{l}_{\vartheta_l}}\alpha_{\vartheta_l}^{(0)} & \omega_{j+1}^{\hat{l}_{\vartheta_l}}\alpha_{\vartheta_l}^{(1)} & \cdots & \omega_l^{\hat{l}_{\vartheta_l}}\alpha_{\vartheta_l}^{(1)}\end{vmatrix}. \tag{8.5.13}$$

From the definitions of q_ν, q'_ν, and l_ν it immediately follows that $lq_\nu \le l_\nu$, whence $l_\vartheta \le l_{\tilde{\vartheta}}$ if $\vartheta \subset \tilde{\vartheta}$. Therefore

$$l^{(2)} \le l^{(0)} \le l^{(1)}. \tag{8.5.14}$$

DEFINITION 8.5.1. Let $r = l^{(1)} - l^{(0)}$ if $n_0 > 0$. Suppose that $b_j^{(0)} \ne 0$ for $j = \frac{l}{2}$ if l is even and for $j = \frac{l-1}{2}$ and $j = \frac{l+1}{2}$ if l is odd.
i) If $n_0 = 0$ or $n_0 > 0$ and $r = 0$, then the boundary eigenvalue problem (8.1.1), (8.1.5) is called *Birkhoff regular.*
ii) If $n_0 > 0$ and $r > 0$, then the boundary eigenvalue problem (8.1.1), (8.1.5) is called *almost Birkhoff regular of order r.*

REMARK 8.5.2. In order to simplify the notations we shall call the problem (8.1.1), (8.1.5) almost Birkhoff regular of order zero if it is Birkhoff regular.

Together with almost Birkhoff regular problems we have to consider some auxiliary boundary conditions. In order to define them, some further notations are needed. For $\kappa = 1, \ldots, n$ we write

$$\kappa = \hat{l}_\nu + tl - m + 1, \tag{8.5.15}$$

where $0 \le t \le L+1$ and $0 \le m \le l-1$. This representation of κ in terms of l and m is unique. We shall consider numbers κ given by this formula for $0 \le t \le L+1$ and $0 \le m \le l-1$. Several of these numbers may lie outside the admissible set $\{1, \ldots, n\}$ for κ. But we shall avoid imposing unhandy restrictions on m and t. Rather, it is always to be silently understood that all numbers, functions, etc., considered for those t and m for which $\hat{l}_\nu + tl - m + 1 \notin \{1, \ldots, n\}$ are zero. From the definition of l_ν we infer the representation

$$\lambda^{-l_\nu} e_\nu^{\mathsf{T}} W^{(\iota)}(\lambda^l) \Xi_n(\lambda) e_{\hat{l}_\nu + tl - m + 1} = \lambda^{-m}\left(\alpha_{\nu t m}^{(\iota)} + O(\lambda^{-l})\right), \tag{8.5.16}$$

for $\nu = 1, \ldots, n$, $\iota = 0, 1$, t and m as above, where $\alpha_{\nu t m}^{(\iota)} \in \mathbb{C}$.

For $j = 0, \ldots, n-1$ and $\nu = 0, \ldots, j$ define $s_{j,\nu} = [\frac{j-\nu}{l}]$ and $\delta'_\alpha = 1$ if α is not a multiple of l and $\delta'_\alpha = 0$ if α is a multiple of l. Then $q_{j,\nu,0} \in W_1^{n+\nu-j}(a,b)$ is defined recursively with respect to j by

$$\begin{aligned}
q_{j,j,0} &= (-1)^j \quad (j = 0, \ldots, n-1), \\
q_{j,j-1,0} &= -q_{j-1,j-2,0} \quad (j = 2, \ldots, n-1), \\
q_{j,\nu,0} &= \delta'_{j-\nu} q'_{j-1,\nu,0} - q_{j-1,\nu-1,0} \quad (j = 3, \ldots, n-1;\ \nu = 1, \ldots, j-2), \\
q_{j,0,0} &= q'_{j-1,0,0} + \sum_{\nu=0}^{j-1} k_{n-\nu-1} q_{j-1,\nu,0} \quad (j = 1, \ldots, l-1), \\
q_{j,0,0} &= \delta'_j \left(q'_{j-1,0,0} + \sum_{\nu=0}^{j-1-ls_{j,0}} k_{n-\nu-1} q_{j-1,\nu,0} \right) - \sum_{\nu=0}^{j-ls_{j,0}} h_{n_0-\nu} q_{j-1,\nu+l-1,0} \\
&\qquad (j = l, \ldots, n-1).
\end{aligned}$$

For $\kappa = 0,\dots,l-1$ and $t = 1,\dots,L$ such that $\kappa + tl \le n$ we define

(8.5.17)
$$\sigma_{n-\kappa-tl,0} = \sum_{j=n_0-\kappa}^{n_0} \sum_{i=j}^{n_0} (-1)^i \binom{i}{j} h_i^{(i-j)} q_{j,n-\kappa-tl,0}.$$

Furthermore, we set $\sigma_{j,0} := 0$ if $j < 0$. For $m = 0,\dots,l-1$, $i = 0,\dots,m$, and $t = 0,\dots,L$ we define

(8.5.18)
$$\gamma_{i,m}^t = \begin{cases} 1 \text{ if } i = m \text{ and } t = 0, \\ 0 \text{ if } i < m \text{ and } t = 0, \\ \displaystyle\sum_{\tau=0}^{m-i-1} \binom{i+\tau}{\tau} \sigma^{(\tau)}_{n-m+i+\tau-tl,0} + \binom{m}{i} (h^t_{n_0})^{(m-i)} \text{ if } t \ge 1. \end{cases}$$

DEFINITION 8.5.3. For $\nu = 1,\dots,n$ the *asymptotic boundary conditions* are defined by

(8.5.19)
$$U_\nu f := \sum_{i=0}^{\hat{l}_\nu} \sum_{t=0}^{L} \sum_{m=0}^{\hat{l}_\nu - i} \alpha^{(0)}_{\nu tm} \gamma^t_{i,\hat{l}_\nu - m}(a) f^{(i)}(a) + \sum_{i=0}^{\hat{l}_\nu} \sum_{t=0}^{L} \sum_{m=0}^{\hat{l}_\nu - i} \alpha^{(1)}_{\nu tm} \gamma^t_{i,\hat{l}_\nu - m}(b) f^{(i)}(b) = 0.$$

In the above definition, f is a suitable function on $[a,b]$. The actual assumptions on f will be formulated later. Also the regularity conditions on the coefficients will be given later to ensure that all quantities considered above are well-defined.

If $h_{n_0} = 1$ then $h_i^{(i-j)} = 0$ for $i = n_0$ and $i > j$. Hence we obtain

REMARK 8.5.4. If $h_{n_0} = 1$, then we have

$$\sigma_{n-\kappa-tl,0} = (-1)^{n_0} q_{n_0,n-\kappa-tl,0} + \sum_{j=n_0-\kappa}^{n_0-1} \sum_{i=j}^{n_0-1} (-1)^i \binom{i}{j} h_i^{(i-j)} q_{j,n-\kappa-tl,0}.$$

The asymptotic boundary conditions can be calculated explicitly. But this calculation with multiple sums and recursions is very involved. So we are going to consider some special cases in which these conditions are relatively easy.

REMARK 8.5.5. i) In case $L = 0$, i. e., $n_0 = 0$, the asymptotic boundary conditions are

$$U_\nu f = \sum_{i=0}^{\hat{l}_\nu} \alpha^{(0)}_{\nu,0,\hat{l}_\nu - i} f^{(i)}(a) + \sum_{i=0}^{\hat{l}_\nu} \alpha^{(1)}_{\nu,0,\hat{l}_\nu - i} f^{(i)}(b) = 0$$

for $\nu = 1,\dots,n$.

ii) In case $L = 1$, i. e., $1 \le n_0 \le \frac{n}{2}$, we obtain $\gamma^t_{0,0} = \gamma^t_{1,1} = h^t_{n_0}$ for $t = 0, 1$, $\gamma^0_{0,1} = 0$, and, if $l \ge 2$,

$$\gamma^1_{0,1} = h_{n_0-1} - h_{n_0} k_{n-1} + (1-n_0) h'_{n_0}.$$

iii) In case $L=1$ and $\hat{l}_\nu \le 1$ we have

$$U_\nu f = \sum_{i=0}^{\hat{l}_\nu} \delta_{\nu i}^{(0)} f^{(i)}(a) + \sum_{i=0}^{\hat{l}_\nu} \delta_{\nu i}^{(1)} f^{(i)}(b),$$

where

$$\delta_{\nu i}^{(\iota)} = \begin{cases} \alpha_{\nu 00}^{(\iota)} + \alpha_{\nu 10}^{(\iota)} h_{n_0}(a_\iota) & \text{if } i = \hat{l}_\nu, \\ \alpha_{\nu 01}^{(\iota)} + \alpha_{\nu 11}^{(\iota)} h_{n_0}(a_\iota) \\ +\alpha_{\nu 10}^{(\iota)}\{h_{n_0-1}(a_\iota) - h_{n_0}(a_\iota)k_{n-1}(a_\iota) + (1-n_0)h'_{n_0}(a_\iota)\} \\ \qquad\qquad\qquad\qquad \text{if } i = 0 \text{ and } \hat{l}_\nu = 1. \end{cases}$$

Proof. i) is clear by Definition 8.5.3.
ii) All but the formula for $\gamma_{0,1}^1$ immediately follows from Definition 8.5.3. For $\gamma_{0,1}^1$ we calculate

$$\begin{aligned} \gamma_{0,1}^1 &= \sigma_{n_0-1,0} + h'_{n_0} \\ &= \sum_{j=n_0-1}^{n_0} \sum_{i=j}^{n_0} (-1)^i \binom{i}{j} h_i^{(i-j)} q_{j,n_0-1,0} + h'_{n_0} \\ &= (-1)^{n_0-1} h_{n_0-1} q_{n_0-1,n_0-1,0} + (-1)^{n_0} n_0 h'_{n_0} q_{n_0-1,n_0-1,0} \\ &\quad + (-1)^{n_0} h_{n_0} q_{n_0,n_0-1,0} + h'_{n_0} \\ &= h_{n_0-1} - h_{n_0} k_{n-1} + (1-n_0) h'_{n_0}. \end{aligned}$$

iii) This follows from part ii) and

$$\begin{aligned} \delta_{\nu i}^{(\iota)} &= \sum_{t=0}^{1} \sum_{m=0}^{\hat{l}_\nu - i} \alpha_{\nu t m}^{(\iota)} \gamma_{i,\hat{l}_\nu - m}^{t}(a_\iota) \\ &= \alpha_{\nu,0,\hat{l}_\nu - i}^{(\iota)} + \sum_{m=0}^{\hat{l}_\nu - i} \alpha_{\nu 1 m}^{(\iota)} \gamma_{i,\hat{l}_\nu - m}^{1}(a_\iota). \qquad \square \end{aligned}$$

REMARK 8.5.6. Here we suppose that $h_{n_0} = 1$.
i) In case $L=1$, i.e., $1 \le n_0 \le \frac{n}{2}$, we obtain $\gamma_{i,i}^1 = 1$ for $i = 0, \dots, l-1$ and, if $l \ge 2$,

$$\gamma_{i,i+1}^1 = h_{n_0-1} - k_{n-1}$$

for $i = 0, \dots, l-2$.
ii) If $l \ge 3$ (only in this case $\hat{l}_\nu \ge 2$ can occur) and $n_0 \ge 2$, then we have

$$\gamma_{i,i+2}^1 = (n_0 - i - 2)(k'_{n-1} - h'_{n_0-1}) + (k_{n-1} - h_{n_0-1})k_{n-1} + h_{n_0-2} - k_{n-2}$$

for $i = 0, \dots, l-3$.

iii) In case $L = 1$, $n_0 \geq 2$, and $\hat{l}_\nu \leq 2$, the asymptotic boundary conditions are

$$U_\nu f = \sum_{i=0}^{\hat{l}_\nu} \delta_{\nu i}^{(0)} f^{(i)}(a) + \sum_{i=0}^{\hat{l}_\nu} \delta_{\nu i}^{(1)} f^{(i)}(b) = 0,$$

where

$$\delta_{\nu i}^{(\iota)} = \begin{cases} \alpha_{\nu 00}^{(\iota)} + \alpha_{\nu 10}^{(\iota)} & \text{if } i = \hat{l}_\nu, \\ \alpha_{\nu 01}^{(\iota)} + \alpha_{\nu 11}^{(\iota)} + \alpha_{\nu 10}^{(\iota)} (h_{n_0 - 1}(a_\iota) - k_{n-1}(a_\iota)) & \text{if } i = \hat{l}_\nu - 1, \\ \alpha_{\nu 02}^{(\iota)} + \alpha_{\nu 12}^{(\iota)} + \alpha_{\nu 11}^{(\iota)} (h_{n_0 - 1}(a_\iota) - k_{n-1}(a_\iota)) & \\ +\alpha_{\nu 10}^{(\iota)} \{(n_0 - 2)(k_{n-1}'(a_\iota) - h_{n_0-1}'(a_\iota)) & \\ +(k_{n-1}(a_\iota) - h_{n_0-1}(a_\iota))k_{n-1}(a_\iota) + h_{n_0-2}(a_\iota) - k_{n-2}(a_\iota)\} & \\ \qquad \text{if } i = 0 \text{ and } \hat{l}_\nu = 2. & \end{cases}$$

Proof. i) We calculate

$$\begin{aligned} \gamma_{i,i+1}^1 = \sigma_{n_0-1,0} &= (-1)^{n_0-1} h_{n_0-1} q_{n_0-1,n_0-1,0} + (-1)^{n_0} q_{n_0,n_0-1,0} \\ &= h_{n_0-1} - k_{n-1}. \end{aligned}$$

ii) We have

$$\begin{aligned} \gamma_{i,i+2}^1 &= \sigma_{n_0-2,0} + (i+1)\sigma_{n_0-1,0}' \\ &= (-1)^{n_0} q_{n_0,n_0-2,0} + (-1)^{n_0} h_{n_0-2} q_{n_0-2,n_0-2,0} \\ &\quad + (-1)^{n_0-1}(n_0 - 1) h_{n_0-1}' q_{n_0-2,n_0-2,0} \\ &\quad + (-1)^{n_0-1} h_{n_0-1} q_{n_0-1,n_0-2,0} + (i+1)\sigma_{n_0-1,0}'. \end{aligned}$$

A recursive application of the formulæ defining the $q_{j,i,m}$ yields

$$q_{n_0,n_0-2,0} = (-1)^{j-1} j q_{n_0-j,n_0-j-1,0}' + (-1)^j q_{n_0-j,n_0-j-2,0}$$

for $j = 0, \ldots, n_0 - 2$, where we have used $l \geq 3$. Another application of these formulas yields

$$q_{n_0,n_0-2,0} = (-1)^{n_0}(n_0 - 1) q_{1,0,0}' + (-1)^{n_0} k_{n-1} q_{1,0,0} + (-1)^{n_0} k_{n-2} q_{1,1,0}.$$

In the proof of part i) we have already calculated $\sigma_{n_0-1,0}$. Altogether we infer

$$\begin{aligned} \gamma_{i,i+2}^1 &= (n_0 - 1)k_{n-1}' + k_{n-1}^2 - k_{n-2} + h_{n_0-2} \\ &\quad - (n_0 - 1)h_{n_0-1}' - h_{n_0-1} k_{n-1} + (i+1)(h_{n_0-1}' - k_{n-1}') \\ &= (n_0 - i - 2)(k_{n-1}' - h_{n_0-1}') + (k_{n-1} - h_{n_0-1})k_{n-1} + h_{n_0-2} - k_{n-2}. \end{aligned}$$

iii) We have

$$\delta_{\nu i}^{(\iota)} = \sum_{t=0}^{1} \sum_{m=0}^{\hat{l}_\nu - i} \alpha_{\nu t m}^{(\iota)} \gamma_{i,\hat{l}_\nu - m}^{t}(a_\iota)$$

$$= \alpha_{\nu,0,\hat{l}_\nu - i}^{(\iota)} + \sum_{m=0}^{\hat{l}_\nu - i} \alpha_{\nu 1 m}^{(\iota)} \gamma_{i,\hat{l}_\nu - m}^{1}(a_\iota).$$

Now the result follows from parts i) and ii). □

8.6. Estimates of the characteristic determinant

In this section we suppose that the assumptions of Theorem 8.4.2 are satisfied and that additionally $\tilde{k} > 0$ and

(8.6.1) $$k > l \max_{\nu=1}^{n} (q'_\nu - q_\nu + 1) \text{ if } n_0 > 0.$$

Then the estimates $[\cdot]$ which were obtained in the previous section can be replaced by $\cdot + O(\lambda^{-1})$, where Corollary 8.3.2 has been used.

The determinant of the characteristic matrix $M(\lambda)$ given by (8.1.14) is an exponential sum in the sense of Section A.2. Here we use the fundamental matrix $Y(\cdot,\lambda)$ as derived in Corollary 8.3.2. Let Θ be the set of all subsets of $\{1,\dots,l\}$ and set

$$\theta\omega = \sum_{j\in\theta} \omega_j$$

for $\theta \in \Theta$. We consider the set

$$\mathscr{E} := \{\theta\omega : \theta \in \Theta\},$$

for which the set $\widetilde{\mathscr{E}}$ of the vertices of the convex hull of $\mathscr{E}$ has been derived in Theorem A.1.7.

LEMMA 8.6.1. *Suppose that $b_j^{(0)} \neq 0$ for $j = \frac{l}{2}$ if l is even and for $j = \frac{l-1}{2}$ and $j = \frac{l+1}{2}$ if l is odd, where $b_j^{(0)}$ is defined in* (8.5.11). *Then there are a sequence of circles Γ_ν ($\nu \in \mathbb{N}$) with centre 0 and radii tending to infinity, real numbers $\chi_1 < \chi_2 < \cdots < \chi_{t+1} = \chi_1 + 2\pi$, $c_1,\dots,c_t \in \widetilde{\mathscr{E}}$, and $\gamma_1,\dots,\gamma_t \in \mathbb{C}\setminus\{0\}$ such that the estimates*

$$\Re\big((c - c_j)\lambda\big) \le 0$$

and

$$\det M(\lambda)^{-1} = \lambda^{-l^{(0)}} \exp\{-c_j\lambda\}\big(\gamma_j + O(\tau_\infty(\lambda))\big)$$

hold for $j = 1,\dots,t$ and $\lambda \in \bigcup_{\nu\in\mathbb{N}} \Gamma_\nu$ with $\chi_j \le \arg\lambda \le \chi_{j+1}$ and for all c in the convex hull of $\mathscr{E}$, where τ_∞ is defined in (A.3.1).

Proof. We have

$$\det M(\lambda) = \sum_{\theta\in\Theta} b_\theta(\lambda) e^{\theta\omega u(b)},$$

where the coefficient functions $b_\theta(\lambda)$ of $M(\lambda)$ are obtained in the following way: For $\theta \in \Theta$ let

$$\Lambda_\theta^0 := \operatorname{diag}\{0,\dots,0,1-\theta(1),\dots,1-\theta(l)\},$$
$$\Lambda_\theta^1 := \operatorname{diag}\{0,\dots,0,\theta(1),\dots,\theta(l)\},$$

where $\theta(j) = 0$ if $j \notin \theta$ and $\theta(j) = 1$ if $j \in \theta$. Then

$$b_\theta(\lambda) = \det M_\theta(\lambda),$$

where

$$M_\theta(\lambda) = M(\lambda)(I_n - \Delta_0) + W^{(0)}(\lambda^l)P(a,\lambda)\Lambda_\theta^0 + W^{(1)}(\lambda^l)P(b,\lambda)\Lambda_\theta^1$$

and $P(\cdot,\lambda) = Y(\cdot,\lambda)E(\cdot,\lambda)^{-1}$. Note that $\Delta_0 = \Lambda_\theta^0 + \Lambda_\theta^1$.

Together with $\theta \in \Theta$ we shall consider $\theta^s \in \Theta$ given by $\theta^s = (\theta + s)\operatorname{mod}(l)$ for $s \in \mathbb{Z}$. Note that

$$\begin{pmatrix} I_{n_0} & 0 \\ 0 & (J_l + \varepsilon_l\varepsilon_1^\top)^s \end{pmatrix} \Lambda_{\theta^s}^j \begin{pmatrix} I_{n_0} & 0 \\ 0 & (J_l^\top + \varepsilon_1\varepsilon_l^\top)^s \end{pmatrix} = \Lambda_\theta^j$$

for all $\theta \in \Theta$, $s \in \mathbb{Z}$ and $j = 0,1$. Then we infer in view of (8.3.9), (8.3.10), and (8.3.13) that

(8.6.2) $$b_{\theta^s}(\lambda) = (-1)^{(l-1)s} b_\theta(\lambda\omega_l^{-s}).$$

From (8.5.3), (8.5.5), (8.5.6), and Corollary 8.3.3 it follows for $\nu = 1,\dots,n$, $\mu = n_0+1,\dots,n$, and $j = 0,1$ that

$$\lambda^{-l_\nu} e_\nu^\top W^{(j)}(\lambda^l) P(a_j,\lambda) e_\mu = \big(\varphi_0(a_j) + O(\lambda^{-1})\big)\alpha_\nu^{(j)} \omega_{\mu-n_0}^{\hat{l}_\nu}.$$

Therefore we obtain for $j = 0,\dots,l$ that

$$b_{\theta_j}(\lambda) = \lambda^{l(0)} \varphi_0(a)^j \varphi_0(b)^{l-j}\big(b_j^{(0)} + O(\lambda^{-1})\big),$$

where $\theta_j = \{j+1,\dots,l\}$. From Theorem A.1.7 we know that $\theta\omega \in \widetilde{E}$ if and only if $\theta = \theta_j^s$ for $s \in \mathbb{N}$ and $j = \frac{l}{2}$ if l is even or $j = \frac{l-1}{2}$ or $j = \frac{l+1}{2}$ if l is odd. It follows for $\theta\omega \in \widetilde{\mathscr{E}}$ that $b_\theta(\lambda) = \lambda^{l(0)}[b_\theta^{(0)}]$, where $b_\theta^{(0)} \neq 0$. An application of Theorem A.3.1 completes the proof. □

LEMMA 8.6.2. *Let m and k be positive integers. Let $M = (m_1,\dots,m_k) \in M_k(\mathbb{C})$ be invertible and let $A = (a_1,\dots,a_m) \in M_{k,m}(\mathbb{C})$. Then*

$$(\det M)M^{-1}A = (\det M_{\nu\mu}^A)_{\nu,\mu=1}^{k,m},$$

where $M_{\nu\mu}^A := (m_1,\dots,m_{\nu-1},a_\mu,m_{\nu+1},\dots,m_k)$.

Proof. It is well-known that

$$(\det M)M^{-1} = (\det M_{\nu\mu})_{\nu,\mu=1}^{k},$$

where $M_{\nu\mu} := \det(m_1, \dots, m_{\nu-1}, e_\mu, m_{\nu+1}, \dots, m_k)$. Write $A = (a_{\nu\mu})_{\nu,\mu=1}^{k,m}$. Then

$$\begin{aligned}
(\det M)M^{-1}A &= \left(\sum_{j=1}^{k} (\det M_{\nu j}) a_{j\mu}\right)_{\nu,\mu=1}^{k,m} \\
&= \det\left(m_1, \dots, m_{\nu-1}, \sum_{j=1}^{k} a_{j\mu} e_j, m_{\nu+1}, \dots, m_k\right)_{\nu,\mu=1}^{k,m} \\
&= \det(m_1, \dots, m_{\nu-1}, a_\mu, m_{\nu+1}, \dots, m_k)_{\nu,\mu=1}^{k,m}. \qquad \square
\end{aligned}$$

Now we consider $M^{-1}(\lambda)W^{(j)}(\lambda^l)Y(a_j, \lambda)$. In case $n_0 = 0$ we have in view of Corollaries 8.3.2 and 8.3.3 that

$$Y(\cdot, \lambda) = \Xi_n(\lambda)\big(\Phi^{[0]} + \{O(\lambda^{-1}\}_\infty\big)E(\cdot, \lambda),$$

with invertible $\Phi^{[0]}$. This is no longer possible if $n_0 > 0$, and a separate consideration of $M^{-1}(\lambda)$ and $W^{(j)}(\lambda^l)Y(a_j, \lambda)$ as done in the previous chapters is not appropriate.

We recall that the matrices Δ_0 and $\Delta(\lambda)$, $\lambda \neq 0$, defined in Section 4.1 are given by

$$\begin{aligned}
\Delta_0 &= \begin{pmatrix} 0 & 0 \\ 0 & I_l \end{pmatrix} \in M_n(\mathbb{C}), \\
\Delta(\lambda) &= (\delta_1(\lambda), \dots, \delta_1(\lambda), \delta_2(\lambda), \dots, \delta_l(\lambda)) \in M_n(\mathbb{C}),
\end{aligned}$$

where

$$\delta_\nu(\lambda) := \begin{cases} 0 & \text{if } \Re(\lambda e^{i\omega_\nu}) < 0, \\ 1 & \text{if } \Re(\lambda e^{i\omega_\nu}) > 0, \\ 0 & \text{if } \Re(\lambda e^{i\omega_\nu}) = 0 \text{ and } \Im(\lambda e^{i\omega_\nu}) > 0, \\ 1 & \text{if } \Re(\lambda e^{i\omega_\nu}) = 0 \text{ and } \Im(\lambda e^{i\omega_\nu}) < 0. \end{cases}$$

LEMMA 8.6.3. *Let the assumptions be as in Corollary* 8.3.1 *and suppose additionally that* (8.6.1) *holds.*
i) *If the boundary eigenvalue problem* (8.1.1), (8.1.5) *is Birkhoff regular in the sense of Definition* 8.5.1, *then there are circles* Γ_ν $(\nu \in \mathbb{N})$ *with radii tending to infinity such that*

$$\begin{aligned}
&M(\lambda)^{-1}W^{(j)}(\lambda^l)Y(a_j, \lambda)E(a_j, \lambda)^{-1} \\
&\quad = \big(I_n - \Delta(\lambda) + E(b, \lambda)^{-1}\Delta(\lambda)\big)\big(V_{0,j}(\lambda) + O(\tau_\infty(\lambda))\big)
\end{aligned}$$

on $\bigcup_{\nu \in \mathbb{N}} \Gamma_\nu$ *for* $j = 0, 1$, *where* $\mathbb{C} \setminus \{0\}$ *is subdivided into finitely many sectors on each of which* $V_{0,j}$ *is constant.*

ii) *If the boundary eigenvalue problem* (8.1.1), (8.1.5) *is almost Birkhoff regular of order* $r(>0)$, *then there are circles* Γ_ν $(\nu \in \mathbb{N})$ *with radii tending to infinity such that*

$$\begin{aligned}
&M(\lambda)^{-1}W^{(j)}(\lambda^l)Y(a_j,\lambda)(I_n-\Delta_0)\\
&\quad=\big(I_n-\Delta(\lambda)+E(b,\lambda)^{-1}\Delta(\lambda)\big)\big(V_{1,j}(\lambda)+O(\tau_\infty(\lambda))\big),\\
&\Delta_0 M(\lambda)^{-1}W^{(j)}(\lambda^l)Y(a_j,\lambda)E(a_j,\lambda)^{-1}\Delta_0\\
&\quad=\big(I_n-\Delta(\lambda)+E(b,\lambda)^{-1}\Delta(\lambda)\big)\big(V_{2,j}(\lambda)+O(\tau_\infty(\lambda))\big),\\
&(I_n-\Delta_0)M(\lambda)^{-1}W^{(j)}(\lambda^l)Y(a_j,\lambda)E(a_j,\lambda)^{-1}\Delta_0\\
&\quad=\lambda^r\big(I_n-\Delta(\lambda)+E(b,\lambda)^{-1}\Delta(\lambda)\big)\big(V_{3,j}(\lambda)+O(\tau_\infty(\lambda))\big),
\end{aligned}$$

on $\bigcup_{\nu\in\mathbb{N}}\Gamma_\nu$ *for* $j=0,1$, *where* $\mathbb{C}\setminus\{0\}$ *is subdivided into finitely many sectors on each of which* $V_{1,j}$, $V_{2,j}$, $V_{3,j}$ *are constant.*

Proof. We shall prove ii) where also the cases $r=0$ and $n_0=0$ are included. Then i) is a special case of this.

Write $Y(\cdot,\lambda)E(\cdot,\lambda)^{-1}=P(\cdot,\lambda)$. Let $j=0$ or $j=1$ and

$$\begin{aligned}
&\det(M(\lambda))M(\lambda)^{-1}W^{(j)}(\lambda^l)P(a_j,\lambda)=:(c_{\nu\mu}(\lambda))_{\nu,\mu=1}^n,\\
&M(\lambda)^{-1}W^{(j)}(\lambda^l)P(a_j,\lambda)=:(d_{\nu\mu}(\lambda))_{\nu,\mu=1}^n.
\end{aligned}$$

We shall use Lemmas 8.6.1 and 8.6.2 to estimate $c_{\nu\mu}$. Also, λ is always a complex number in $\bigcup_{\nu\in\mathbb{N}}\Gamma_\nu$.

If $\nu\le n_0$ and $\mu\le n_0$, then $c_{\nu\mu}(\lambda)$ has the same form as $\det M(\lambda)$, i. e., the same exponential terms occur, and also the same asymptotics hold. Also note that we have $\exp\{(c-c_j)\lambda\}=O(\lambda^{-1})$ for $c\in\mathscr{E}\setminus\{c_j\}$ and λ in the corresponding sector. Hence

$$(I_n-\Delta_0)M(\lambda)^{-1}W^{(j)}(\lambda^l)Y(a_j,\lambda)(I_n-\Delta_0)=\widetilde{V}_{1,j}(\lambda)+O(\tau_\infty(\lambda))$$

follows in view of Lemma 8.6.1, where $\widetilde{V}_{1,j}(\lambda)$ is constant on sectors.

If $\nu>n_0$ but $\mu\le n_0$, then the column containing the exponential terms $e^{\lambda\omega_{\nu-n_0}u(a_\iota)}$ for $\iota=0,1$ is substituted by a column without exponential terms. Thus, if we multiply $c_{\nu\mu}(\lambda)$ with $e^{\lambda\omega_{\nu-n_0}u(a_i)}$, then the exponentials occuring in this expressions still lie in the convex hull of $\mathscr{E}$ for $i=0,1$. The maximal occuring λ-power in this case is $\lambda^{l^{(2)}}$. Since $l^{(2)}\le l^{(0)}$, we infer that $e^{\lambda\omega_{\nu-n_0}u(a_i)}d_{\nu\mu}(\lambda)=\tilde{d}_{\nu\mu}(\lambda)+O(\tau_\infty(\lambda))$, where i is chosen such that $e^{-\lambda\omega_{\nu-n_0}u(a_i)}$ is the ν-th diagonal element of $I_n-\Delta(\lambda)+E(b,\lambda)^{-1}\Delta(\lambda)$ and where $\tilde{d}_{\nu\mu}(\lambda)$ is constant on sectors.

In case $\nu > n_0$ and $\mu > n_0$, again the exponential terms $e^{\lambda\omega_{\nu-n_0}u(a_\iota)}$ do not occur. Here the maximal occuring λ-power in $c_{\nu\mu}(\lambda)$ is $\lambda^{l^{(0)}}$. Hence

$$e^{\lambda\omega_{\nu-n_0}u(a_i)}d_{\nu\mu}(\lambda) = \tilde{d}_{\nu\mu}(\lambda) + O(\tau_\infty(\lambda)),$$

where a_i is chosen as in the case $\mu \le n_0$ and $\tilde{d}_{\nu\mu}(\lambda)$ is constant on sectors.

In case $\nu \le n_0$ and $\mu > n_0$, a column without exponential terms is substituted. Here the maximal occuring λ-power in $c_{\nu\mu}(\lambda)$ is $\lambda^{l^{(1)}}$. □

LEMMA 8.6.4. *Under the assumptions of Lemma* 8.6.3 *we obtain*

$$\begin{aligned}
&\Delta_0 M(\lambda)^{-1}\operatorname{diag}(\lambda^{l_1},\dots,\lambda^{l_n})\\
&\qquad = \big(I_n - \Delta(\lambda) + E(b,\lambda)^{-1}\Delta(\lambda)\big)\big(V_4(\lambda) + O(\tau_\infty(\lambda))\big),\\
&(I_n - \Delta_0)M(\lambda)^{-1}\operatorname{diag}(\lambda^{l_1},\dots,\lambda^{l_n})\\
&\qquad = \lambda^r\big(I_n - \Delta(\lambda) + E(b,\lambda)^{-1}\Delta(\lambda)\big)\big(V_5(\lambda) + O(\tau_\infty(\lambda))\big),
\end{aligned}$$

on $\bigcup_{\nu\in\mathbb{N}}\Gamma_\nu$, *where* $\mathbb{C}\setminus\{0\}$ *is subdivided into finitely many sectors, on each of which* V_4 *and* V_5 *are constant.*

Proof. The considerations are essentially the same as in Lemma 8.6.3. We only have to note that the case $\mu > n_0$ in Lemma 8.6.3 has to be applied to all μ in Corollary 8.6.4. □

8.7. Asymptotic estimates of the Green's function

In this section we assume that the assumptions of Theorem 8.4.2 and (8.6.1) hold and that $k \ge n$ if $n_0 \ne 0$. Throughout this section we suppose that the boundary eigenvalue problem (8.1.1), (8.1.5) is almost Birkhoff regular of order r.

The operators Q_ν defined in (8.1.13) can be written as

$$\begin{aligned}
(Q_\nu f)(x) &= -\frac{1}{2\pi i}\oint_{\Gamma_\nu}\int_a^b \lambda^{l-1}(\mathbf{H}^+G(x,\cdot,\lambda^l))(\xi)f(\xi)\,\mathrm{d}\xi\,\mathrm{d}\lambda \quad (x\in[a,b])\\
&= -\frac{1}{2\pi i}\oint_{\Gamma_\nu}\lambda^{l-1}(S(\lambda^l)f)(x)\,\mathrm{d}\lambda
\end{aligned}$$

where the Γ_ν are circles centred at 0 and the radius of γ_ν is the l-th power of the radius of Γ_ν. We also have that

$$(Q_\nu f)^{(\mu)}(x) = -\frac{1}{2\pi i}\oint_{\Gamma_\nu}\lambda^{l-1}(S(\lambda^l)f)^{(\mu)}(x)$$

for $\mu = 0,\dots,l-1$, where

$$(S(\lambda^l)f)^{(\mu)}(x) = \int_a^b(\mathbf{H}^+e_{\mu+1}^{\mathsf{T}}G_T(x,\cdot,\lambda)e_n)(\xi)f(\xi)\,\mathrm{d}\xi$$

in view of the derivatives obtained in the proof of Lemma 8.1.1.

For $j = 0, 1$ and $\mu = 0, \ldots, l-1$ we define

$$G_{j,0,\mu}(x,\xi,\lambda) = e_{\mu+1}^{\mathsf{T}} Y(x,\lambda) M^{-1}(\lambda) W^{(j)}(\lambda^l) Y(a_j,\lambda) \times \\ \times (I_n - \Delta_0)(\mathbf{H}^+ Y(\cdot,\lambda)^{-1} e_n)(\xi)$$

$$G_{j,1,\mu}(x,\xi,\lambda) = e_{\mu+1}^{\mathsf{T}} Y(x,\lambda) M^{-1}(\lambda) W^{(j)}(\lambda^l) Y(a_j,\lambda) \times \\ \times \Delta_0(\mathbf{H}^+ Y(\cdot,\lambda)^{-1} e_n)(\xi).$$

Then

$$(S(\lambda^l)f)^{(\mu)}(x) = \int_a^x (G_{0,0,\mu}(x,\xi,\lambda) + G_{0,1,\mu}(x,\xi,\lambda)) f(\xi)\, \mathrm{d}\xi \\ - \int_x^b (G_{1,0,\mu}(x,\xi,\lambda) + G_{1,1,\mu}(x,\xi,\lambda)) f(\xi)\, \mathrm{d}\xi.$$

PROPOSITION 8.7.1. *We have* $G_{j,0,\mu}(x,\xi,\lambda) = O(\lambda^{-l-1})$ *uniformly for* $j = 0, 1$, $\mu = 0, \ldots, l-1$, $x, \xi \in [a,b]$, *and* $\lambda \in \bigcup_{\nu\in\mathbb{N}} \Gamma_\nu$.

Proof. Theorem 8.4.2 yields

$$(I_n - \Delta_0)\mathbf{H}^+(Y(\cdot,\lambda)^{-1} e_n) = \{O(\lambda^{-2l})\}_\infty,$$

and Corollary 8.3.2 yields

$$e_{\mu+1}^{\mathsf{T}} Y(\cdot,\lambda) E(\cdot,\lambda)^{-1} = \{O(\lambda^\mu)\}_\infty.$$

Since

$$E(\cdot,\lambda)(I_n - \Delta(\lambda) + \Delta(\lambda)E(b,\lambda)^{-1}) = \{O(1)\}_\infty$$

by Proposition 4.3.3 i),ii) and since

$$(I_n - \Delta(\lambda) + \Delta(\lambda)E(b,\lambda))M(\lambda)^{-1} W^{(j)}(\lambda^l) Y(a_j,\lambda)(I_n - \Delta_0) = O(1)$$

by Lemma 8.6.3 ii), the statement of the proposition follows. □

To estimate $G_{j,1,\mu}$ for $j = 0, 1$ and $\mu = 0, \ldots, l-1$ we define

$$\widehat{Y}(\cdot,\lambda) = \begin{pmatrix} \Phi_{12}(\cdot,\lambda) \\ \Phi_{22}(\cdot,\lambda) \end{pmatrix} = (\eta_q^{(j-1)}(\cdot,\lambda))_{j=1,q=n_0+1}^{n} \widehat{E}(\cdot,\lambda)^{-1},$$

$$y_\mu(\cdot,\lambda) = e_{\mu+1}^{\mathsf{T}} \widehat{Y}(\cdot,\lambda),$$

and

$$z(\cdot,\lambda) = \widehat{E}(\cdot,\lambda)(\mathbf{H}^+ \zeta_\nu(\cdot,\lambda))_{\nu=n_0+1}^{n},$$

where

$$\widehat{E}(x,\lambda) = \operatorname{diag}(e^{\lambda\omega_1 u(x)}, \ldots, e^{\lambda\omega_l u(x)})$$

and the (matrix) functions Φ_{12}, Φ_{22}, η_ν, ζ_ν, and u are defined in Corollary 8.3.2, Corollary 8.3.1, Theorem 8.4.2, and (8.3.1).

PROPOSITION 8.7.2. *We have*

$$\mathbf{H}^{+}(Y^{-1}e_n) = \sum_{\nu=0}^{n_0} \sigma_\nu Y^{-1} e_{n-\nu},$$

where

$$\sigma_\nu = \sum_{j=\nu}^{n_0}\sum_{i=j}^{n_0}(-1)^i \binom{i}{j} h_i^{(i-j)} q_{j,\nu} \quad (\nu = 0,\dots,n_0)$$

and the $q_{j,\nu}$ *are defined in Lemma* 8.1.3.

Proof. From Lemma 8.1.3 we know that

$$\begin{aligned}\mathbf{H}^{+}(Y^{-1}e_n) &= \sum_{i=0}^{n_0}(-1)^i \sum_{j=0}^{i}\binom{i}{j} h_i^{(i-j)}(Y^{-1}e_n)^{(j)} \\ &= \sum_{i=0}^{n_0}(-1)^i \sum_{j=0}^{i}\binom{i}{j} h_i^{(i-j)} \sum_{\nu=0}^{j} q_{j,\nu} Y^{-1} e_{n-\nu}. \quad \square\end{aligned}$$

If we additionally suppose that $h_i \in W_{p'}^{i+s}(a,b)$ for $i = 0,\dots,n_0-1$ and $s \le l$, then the σ_ν are in $W_{p'}^{s}(a,b)$. Hence the following considerations make sense under this additional assumption. For $j = 0,1$, $\mu = 0,\dots,l-1$, $s = 0,\dots,l$, and $f \in W_p^s(a,b)$ let

$$\begin{aligned}(\widehat{G}_{j,\mu,s}f)(x,\xi,\lambda) =& e_{\mu+1}^{\mathsf{T}} Y(x,\lambda) M^{-1}(\lambda) W^{(j)}(\lambda^l) \widehat{Y}(a_j,\lambda) \widehat{E}(a_j,\lambda) \times \\ & \times \lambda^{-s} \Omega_l^{-s} \widehat{E}(\xi,\lambda)^{-1} D^s(z(\cdot,\lambda) f)(\xi),\end{aligned}$$

where

$$Df := (h_{n_0}^{-1/l} f)'.$$

We suppose that $f \in W_p^s(a,b)$. For $c,d \in [a,b]$ integration by parts leads to

$$\begin{aligned}&\int_c^d \widehat{E}(\xi,\lambda)^{-1} z(\xi,\lambda) f(\xi)\,\mathrm{d}\xi \\ &\qquad = -\sum_{j=0}^{s-1} \lambda^{-j-1}\Omega_l^{-j-1}\widehat{E}(\xi,\lambda)^{-1} h_{n_0}^{-1/l}(\xi) D^j(z(\cdot,\lambda) f)(\xi)\Big|_c^d \\ &\qquad\quad + \int_c^d \lambda^{-s}\Omega_l^{-s}\widehat{E}(\xi,\lambda)^{-1} D^s(z(\cdot,\lambda) f)(\xi)\,\mathrm{d}\xi.\end{aligned}$$

This yields

$$\begin{aligned}
(8.7.1)\qquad & \int_a^x G_{0,1,\mu}(x,\xi,\lambda)f(\xi)\,\mathrm{d}\xi - \int_x^b G_{1,1,\mu}(x,\xi,\lambda)f(\xi)\,\mathrm{d}\xi \\
&= -y_\mu(x,\lambda)\sum_{j=0}^{s-1}\lambda^{-j-1}\Omega_l^{-j-1}h_{n_0}^{-1/l}(x)D^j(z(\cdot,\lambda)f)(x) \\
&+ e_{\mu+1}^{\mathsf{T}}Y(x,\lambda)M^{-1}(\lambda)W^{(0)}(\lambda^l)\widehat{Y}(a,\lambda)\times \\
&\times\sum_{j=0}^{s-1}\lambda^{-j-1}\Omega_l^{-j-1}h_{n_0}^{-1/l}(a)D^j(z(\cdot,\lambda)f)(a) \\
&+ e_{\mu+1}^{\mathsf{T}}Y(x,\lambda)M^{-1}(\lambda)W^{(1)}(\lambda^l)\widehat{Y}(b,\lambda)\times \\
&\times\sum_{j=0}^{s-1}\lambda^{-j-1}\Omega_l^{-j-1}h_{n_0}^{-1/l}(b)D^j(z(\cdot,\lambda)f)(b) \\
&+ \int_a^x(\widehat{G}_{0,\mu,s}f)(x,\xi,\lambda)\,\mathrm{d}\xi - \int_x^b(\widehat{G}_{1,\mu,s}f)(x,\xi,\lambda)\,\mathrm{d}\xi.
\end{aligned}$$

PROPOSITION 8.7.3. *Let $j\in\{0,\dots,l\}$ and suppose that $k\ge j$ and h_i belongs to $W_{p'}^{i+j+1}(a,b)$ for $i=0,\dots,n_0-1$. Then*

$$(8.7.2)\qquad z(\cdot,\lambda)^{(j)} = \lambda^{1-l}\Omega_l\left\{(h_{n_0}^{n/l}\psi_0)^{(j)}\varepsilon + \sum_{r=1}^{k-j}\lambda^{-r}\tilde{\psi}_{j,r} + \{o(\lambda^{-k+j})\}_\infty\right\},$$

where $\tilde{\psi}_{j,r}\in W_{p'}^1(a,b)$.

Proof. From (8.4.16) we immediately infer that

$$z(\cdot,\lambda)^{(\mu)} = \lambda^{1-l}\Omega_l^{1-l}\left\{\sum_{r=-\mu}^{k-\mu}\lambda^{-r}\tilde{\psi}_{\mu,r} + \{o(\lambda^{-k+\mu})\}_\infty\right\}$$

for $\mu=0,\dots,j$. By Remark 8.4.3 we can differentiate term by term if $\mu<j$, and it follows that $\tilde{\psi}_{\mu,r}=0$ if $1\le\mu\le j$ and $r>0$. Finally,

$$\begin{aligned}
z(\cdot,\lambda) &= \left((-1)^{n_0}h_{n_0}(\lambda\omega_{\nu-n_0})^{1-l}\psi_{n_0,0}\right)_{\nu=n_0+1}^{n} + \{o(\lambda^{1-l})\}_\infty \\
&= h_{n_0}^{n/l}\lambda^{1-l}\Omega_l\varepsilon\psi_0 + \{o(\lambda^{1-l})\}_\infty
\end{aligned}$$

gives the stated representation of $\tilde{\psi}_{j,0}$. □

PROPOSITION 8.7.4. *Let $j\in\{0,\dots,l\}$ and suppose that $k\ge j$ and h_i belongs to $W_{p'}^{i+j+1}(a,b)$ for $i=0,\dots,n_0-1$. Then it follows for $\mu=0,\dots,n-1$ that $y_\mu(\cdot,\lambda)\lambda^{-j-1}\Omega_l^{-j-1}h_{n_0}^{-1/l}D^j(z(\cdot,\lambda)f)$ is the sum of a polynomial in λ^l and an asymptotic polynomial in λ^{-l}. The highest possible λ-power has the exponent $([\frac{\mu-j}{l}]-1)l$.*

Proof. From (8.3.13) and Theorem 8.4.1 we infer immediately that the above vector function is invariant under the transformation $\lambda \mapsto \lambda\omega_l$, which proves that it only depends on λ^l. Indeed, this is trivial if $l = 1$, and if $l > 1$, then

$$y_\mu(\cdot,\lambda\omega_l) = y_\mu(\cdot,\lambda)(J_l + \varepsilon_l\varepsilon_1^\mathsf{T}),$$

and

$$(J_l + \varepsilon_l\varepsilon_1^\mathsf{T})\widehat{E}(\cdot,\lambda\omega_l)(J_l^\mathsf{T} + \varepsilon_1\varepsilon_l^\mathsf{T}) = \widehat{E}(\cdot,\lambda)$$

shows that

$$\begin{aligned} z(\cdot,\lambda\omega_l) &= \widehat{E}(\cdot,\lambda\omega_l)(\mathbf{H}^+\zeta_\nu(\cdot,\lambda\omega_l))_{\nu=n_0+1}^n \\ &= \widehat{E}(\cdot,\lambda\omega_l)\mathbf{H}^+\widehat{E}(\cdot,-\lambda\omega_l)\Psi_{22}(\cdot,\lambda\omega_l)\varepsilon_l \\ &= \widehat{E}(\cdot,\lambda\omega_l)\mathbf{H}^+\widehat{E}(\cdot,-\lambda\omega_l)(J_l^\mathsf{T} + \varepsilon_1\varepsilon_l^\mathsf{T})\Psi_{22}(\cdot,\lambda)\varepsilon_l \\ &= (J_l^\mathsf{T} + \varepsilon_1\varepsilon_l^\mathsf{T})\widehat{E}(\cdot,\lambda)\mathbf{H}^+\widehat{E}(\cdot,-\lambda)\Psi_{22}(\cdot,\lambda)\varepsilon_l \\ &= (J_l^\mathsf{T} + \varepsilon_1\varepsilon_l^\mathsf{T})z(\cdot,\lambda). \end{aligned}$$

Finally, (8.2.42) yields

$$(J_l + \varepsilon_l\varepsilon_1^\mathsf{T})(\lambda\omega_l)^{-j-1}\Omega_l^{-j-1}(J_l^\mathsf{T} + \varepsilon_1\varepsilon_l^\mathsf{T}) = \lambda^{-j-1}\Omega_l^{-j-1}$$

since $\omega_l\omega_2 = 1$. The remaining statements follow from Proposition 8.7.3. □

PROPOSITION 8.7.5. *Let $1 \le s \le l$, $k \ge s-1$, and suppose that $\tilde{k} \ge l-1$ and $h_i \in W_{p'}^{i+s}(a,b)$ for $i = 0,\dots,n_0-1$. Then*

$$y_\mu(\cdot,\lambda)\sum_{j=0}^{s-1}\lambda^{-j-1}\Omega_l^{-j-1}h_{n_0}^{-1/l}D^j(z(\cdot,\lambda)f) = \{O(\lambda^{-l})\}_\infty|f|_{p,s-1}$$

for $\mu = 0,\dots,l-1$ and $f \in W_p^s(a,b)$.

Proof. The statement immediately follows from Proposition 8.7.4. □

Now let us consider the two remaining sums. First we have by Proposition 8.7.4 that

$$W^{(\iota)}(\lambda^l)\widehat{Y}(a_\iota,\lambda)\lambda^{-j-1}\Omega_l^{-j-1}h_{n_0}^{-1/l}(a_\iota)D^j(z(\cdot,\lambda)f)(a_\iota)$$

$(j = 0,\dots,s-1;\ \iota = 0,1)$ is a function of λ^l. On the other hand,

(8.7.3)
$$\operatorname{diag}(\lambda^{-l_1},\dots,\lambda^{-l_n})W^{(\iota)}(\lambda^l)\widehat{Y}(a_\iota,\lambda)\lambda^{-j-1}\Omega_l^{-j-1}h_{n_0}^{-1/l}(a_\iota)D^j(z(\cdot,\lambda)f)(a_\iota)$$

satisfies the estimate $O(\lambda^{-j-l})$ in view of (8.3.11), (8.3.12), the definition of l_ν in (8.5.1), and (8.7.2). If we suppose that $\tilde{k} \ge s$, then the above considerations yield

(8.7.4)
$$\begin{aligned} &\operatorname{diag}(\lambda^{-l_1},\dots,\lambda^{-l_n})W^{(\iota)}(\lambda^l)\widehat{Y}(a_\iota,\lambda)\lambda^{-j-1}\Omega_l^{-j-1}h_{n_0}^{-1/l}(a_\iota)D^j(z(\cdot,\lambda)f)(a_\iota) \\ &= \lambda^{-j-l}\big(\beta_j^{(\iota)}(\lambda) + O(\lambda^{j-s})\big), \end{aligned}$$

where $\beta_j^{(\iota)}$ is a polynomial in λ^{-1} of order less than $s-j$. We write

$$\beta^{(\iota)}(\lambda) = \sum_{j=0}^{s-1} \lambda^{-j}\beta_j^{(\iota)}(\lambda).$$

Then $\beta^{(\iota)}$ is a polynomial in λ^{-1} of order less than s. If $s \geq r$ we obtain

$$e_{\mu+1}^{\mathsf{T}} Y(\cdot,\lambda)M^{-1}(\lambda)W^{(\iota)}(\lambda^l)\widehat{Y}(a_\iota,\lambda)\sum_{j=0}^{s-1}\lambda^{-j-1}\Omega_l^{-j-1}h_{n_0}^{-1/l}(a_\iota)D^j(z(\cdot,\lambda)f)(a_\iota)$$
$$= e_{\mu+1}^{\mathsf{T}} Y(\cdot,\lambda)M^{-1}(\lambda)\operatorname{diag}(\lambda^{l_1},\ldots,\lambda^{l_n})\lambda^{-l}\beta^{(\iota)}(\lambda) + \{O(\lambda^{-l})\}_\infty$$

for $\mu = 0,\ldots,\min\{s, l-1\}$, where we have used Lemma 8.6.4 and the fact that $e_{\mu+1}^{\mathsf{T}} Y(\cdot,\lambda)E(\cdot,\lambda)^{-1}\Delta_0 = \{O(\lambda^\mu)\}_\infty$ and $e_{\mu+1}^{\mathsf{T}} Y(\cdot,\lambda)(I_n - \Delta_0) = \{O(1)\}_\infty$ in view of Corollary 8.3.1.

Hence the two remaining sums in (8.7.1) are of order $O(\lambda^{-l})$ if we require

$$\beta^{(0)}(\lambda) + \beta^{(1)}(\lambda) = 0.$$

For later use we note that we can substitute $O(\lambda^{-l})$ by $O(\lambda^{-l-1})$ if $s > \max\{r,\mu\}$. Again from Proposition 8.7.4 and $s \leq l$ we infer that only those terms occur in $\lambda^{-j}\beta_j^{(\iota)}(\lambda)$ for which we take the highest possible λ-power. For the element in the ν-th row this is a number $-l_\nu \operatorname{mod}(l)$, i. e., $-\hat{l}_\nu$. That is,

$$e_\nu^{\mathsf{T}}\beta^{(\iota)}(\lambda) = \lambda^{-\hat{l}_\nu} c_\nu^{(\iota)}.$$

Of course, only terms with $\hat{l}_\nu < s$ must be considered since $e_\nu^{\mathsf{T}}\beta^\iota(\lambda) = 0$ otherwise.

PROPOSITION 8.7.6. *Let* $0 \leq s \leq l$, $k \geq n-1$ *if* $n_0 = 0$, $k \geq n$ *if* $n_0 > 0$, *and* $h_i \in W_{p'}^{i+s}(a,b)$ *for* $i = 0,\ldots,n_0-1$. *For* $\mu = 0,\ldots,\min\{s,l-1\}$ *and* $f \in W_p^s(a,b)$ *we have*

$$e_{\mu+1}^{\mathsf{T}} Y(\cdot,\lambda)M^{-1}(\lambda)W^{(0)}(\lambda^l)\widehat{Y}(a,\lambda)\sum_{j=0}^{s-1}\lambda^{-j-1}\Omega_l^{-j-1}h_{n_0}^{-1/l}(a)D^j(z(\cdot,\lambda)f)(a)$$
$$+ e_{\mu+1}^{\mathsf{T}} Y(\cdot,\lambda)M^{-1}(\lambda)W^{(1)}(\lambda^l)\widehat{Y}(b,\lambda)\sum_{j=0}^{s-1}\lambda^{-j-1}\Omega_l^{-j-1}h_{n_0}^{-1/l}(b)D^j(z(\cdot,\lambda)f)(b)$$
$$= \{O(\lambda^{-l})\}_\infty |f|_{p,s-1}$$

if f *satisfies the boundary conditions* $U_\nu f = 0$ *for* $\nu = 1,\ldots,n$ *with* $\hat{l}_\nu < s$.

To prove Proposition 8.7.6 we have to show that $U_\nu f = c_\nu^{(0)} + c_\nu^{(1)}$. For this we need some preparations.

We see that $c_\nu^{(\iota)}$ is the coefficient of $\lambda^{l_\nu-\hat{l}_\nu-l}$ in

$$e_\nu^\intercal W^{(\iota)}(\lambda^l)\widehat{Y}(a_\iota,\lambda)\sum_{j=0}^{s-1}\lambda^{-j-1}\Omega_l^{-j-1}h_{n_0}^{-1/l}(a_\iota)D^j(z(\cdot,\lambda)f)(a_\iota). \tag{8.7.5}$$

To find this coefficient, we write (8.7.5) as the sum with respect to $\kappa = 1,\dots,n$ of the products of the two terms

$$\lambda^{-l_\nu}e_\nu^\intercal W^{(\iota)}(\lambda^l)\Xi_n(\lambda)e_\kappa \tag{8.7.6}$$

and

$$\lambda^{l_\nu}e_\kappa^\intercal\Xi_n(\lambda)^{-1}\widehat{Y}(a_\iota,\lambda)\sum_{j=0}^{s-1}\lambda^{-j-1}\Omega_l^{-j-1}h_{n_0}^{-1/l}(a_\iota)D^j(z(\cdot,\lambda)f)(a_\iota). \tag{8.7.7}$$

We have already obtained an asymptotic representation of (8.7.6) in (8.5.16), where κ is written as in (8.5.15). To find the coefficient of $\lambda^{l_\nu-\hat{l}_\nu-l}$ in (8.7.5) we thus have to find the coefficient of $\lambda^{l_\nu-\hat{l}_\nu-l+m}$ in (8.7.7) (note that we are looking for the term with the highest possible λ-power in (8.7.5)). That means, we have to find the coefficient $d_{\nu tm}^{(\iota)}$ of $\lambda^{(t-1)l}$ in

$$y_{\hat{l}_\nu+tl-m}(a_\iota,\lambda)\sum_{j=0}^{s-1}\lambda^{-j-1}\Omega_l^{-j-1}h_{n_0}^{-1/l}(a_\iota)D^j(z(\cdot,\lambda)f)(a_\iota). \tag{8.7.8}$$

Then we can write

$$c_\nu^{(\iota)} = \sum_{t=0}^{L+1}\sum_{m=0}^{l-1}\alpha_{\nu tm}^{(\iota)}d_{\nu tm}^{(\iota)}.$$

From Proposition 8.7.4 we infer that only $m+j\le\hat{l}_\nu$ gives a contribution in (8.7.8). In this case, any number $\kappa\in\{1,\dots,n\}$ has a representation (8.5.15) with $l\le L$. So we have that

$$c_\nu^{(\iota)} = \sum_{t=0}^{L}\sum_{m=0}^{\hat{l}_\nu}\alpha_{\nu tm}^{(\iota)}d_{\nu tm}^{(\iota)}. \tag{8.7.9}$$

For $m=0,\dots,l-1$ and $t=0,\dots,L$ such that $m+tl\le n-1$ we define

$$K_m^t(\cdot,\lambda)f = y_{m+tl}(\cdot,\lambda)\sum_{j=0}^{m}\lambda^{-j-1}\Omega_l^{-j-1}h_{n_0}^{-1/l}D^j(z(\cdot,\lambda)f), \tag{8.7.10}$$

where f is a dummy function. For $m=0,\dots,l-1$, $i=0,\dots,m$ and $t=0,\dots,L$ let $\tilde{\gamma}_{i,m}^t$ be the coefficient of $\lambda^{(t-1)l}f^{(i)}$ in $K_m^t(\cdot,\lambda)$. Then

$$d_{\nu tm}^{(\iota)} = \sum_{i=0}^{\hat{l}_\nu-m}\tilde{\gamma}_{i,\hat{l}_\nu-m}^t(a_\iota)f^{(i)}(a_\iota). \tag{8.7.11}$$

Next we shall show that $\gamma_{i,m}^t = \tilde{\gamma}_{i,m}^t$. We extend the definition of $\tilde{\gamma}_{i,m}^t$ to the cases $i=m+1$ and $i=-1$. Let $m\in\{0,\dots,l-1\}$. For $i=m+1$, we take formally the same definition as in (8.7.10), which gives $\tilde{\gamma}^t_{i,m}(\cdot,\lambda)=0$, and we take $\tilde{\gamma}^t_{-1,m}$ to be the coefficient of $\lambda^{(t-1)l}$ in $y_{m+tl}(\cdot,\lambda)z(\cdot,\lambda)$.

PROPOSITION 8.7.7. *For $m=1,\dots,l-1$, $i=0,\dots,m$, and $t=0,\dots,L$ such that $m+tl\le n-1$ we have*

$$\tilde{\gamma}^t_{i,m} = (\tilde{\gamma}^t_{i,m-1})' + \tilde{\gamma}^t_{i-1,m-1},$$

where $\tilde{\gamma}^t_{i,m-1}=0$ if $m=i$.

Proof. We first note that $\widehat{Y}(\cdot,\lambda) = Y(\cdot,\lambda)E(\cdot,\lambda)^{-1}\begin{pmatrix}0\\ I_l\end{pmatrix}$ implies

$$y_{m-1+tl}(\cdot,\lambda)' = y_{m+tl}(\cdot,\lambda) - y_{m-1+tl}(\cdot,\lambda)\lambda\Omega_l h_{n_0}^{1/l}.$$

Since the coefficient of $\lambda^{(t-1)l}$ in $K^t_{m-1}(\cdot,\lambda)$ is $\sum\limits_{i=0}^{m-1}\tilde{\gamma}^t_{i,m-1}f^{(i)}$, it follows that the coefficient of $\lambda^{(t-1)l}f^{(i)}$ in $(K^t_{m-1}(\cdot,\lambda)f)' + y_{m-1+tl}(\cdot,\lambda)z(\cdot,\lambda)f$ is the function $(\tilde{\gamma}^t_{i,m-1})' + \tilde{\gamma}^t_{i-1,m-1}$. We calculate

$$\begin{aligned}
&(K^t_{m-1}(\cdot,\lambda)f)' + y_{m-1+tl}(\cdot,\lambda)z(\cdot,\lambda)f\\
&\quad= y_{m+tl}(\cdot,\lambda)\sum_{j=0}^{m-1}\lambda^{-j-1}\Omega_l^{-j-1}h_{n_0}^{-1/l}D^j(z(\cdot,\lambda)f)\\
&\qquad - y_{m-1+tl}(\cdot,\lambda)\sum_{j=1}^{m-1}\lambda^{-j}\Omega_l^{-j}D^j(z(\cdot,\lambda)f)\\
&\qquad + y_{m-1+tl}(\cdot,\lambda)\sum_{j=0}^{m-1}\lambda^{-j-1}\Omega_l^{-j-1}D^{j+1}(z(\cdot,\lambda)f)\\
&\quad= K^t_m(\cdot,\lambda) - y_{m+tl}(\cdot,\lambda)\lambda^{-m-1}\Omega_l^{-m-1}h_{n_0}^{-1/l}D^m(z(\cdot,\lambda)f)\\
&\qquad + y_{m-1+tl}(\cdot,\lambda)\lambda^{-m}\Omega_l^{-m}D^m(z(\cdot,\lambda)f)\\
&\quad= K^t_m(\cdot,\lambda) - y_{m-1+tl}(\cdot,\lambda)'\lambda^{-m-1}\Omega_l^{-m-1}h_{n_0}^{-1/l}D^m(z(\cdot,\lambda)f).
\end{aligned}$$

Now the result follows if we observe that

$$y_{m-1+tl}(\cdot,\lambda)' = \{O(\lambda^{m-1+tl})\}_\infty \quad\text{and}\quad D^m(z(\cdot,\lambda)f) = \{O(\lambda^{1-l})\}_\infty. \qquad \square$$

PROPOSITION 8.7.8. *For $0\le i\le m\le l-1$ and $t=0,\dots,L$ with $m+tl\le n-1$ we have*

$$\tilde{\gamma}^t_{i,m} = \sum_{\tau=0}^{m-i-1}\binom{i+\tau}{\tau}(\tilde{\gamma}^t_{-1,m-1-i-\tau})^{(\tau)} + \binom{m}{i}(h^t_{n_0})^{(m-i)}.$$

Proof. First let $m=0$. In this case, $\tilde{\gamma}^t_{0,0}$ is the coefficient of $\lambda^{(t-1)l}$ in

$$y_{tl}(\cdot,\lambda)\lambda^{-1}\Omega_l^{-1}h_{n_0}^{-1/l}z(\cdot,\lambda).$$

In view of (8.7.2) we have

$$z(\cdot,\lambda) = \lambda^{1-l}\Omega_l h_{n_0}^{n/l}\psi_0[\varepsilon],$$

and from (8.3.7) we know that

$$y_{tl}(\cdot,\lambda) = \lambda^{tl} h_{n_0}^t \varphi_0[\varepsilon^\top].$$

Then (8.4.15) leads to

$$\begin{aligned}\tilde{\gamma}^t_{0,0} &= h^t_{n_0}\varphi_0\varepsilon^\top h_{n_0}^{-1/l} h_{n_0}^{n/l}\psi_0\varepsilon \\ &= h_{n_0}^{t+(n-1)/l}\, l\varphi_0\psi_0 = h^t_{n_0}.\end{aligned}$$

Now let $m \geq 1$ and suppose that the statement holds for $m-1$. Then, if $i \geq 1$, we infer from Proposition 8.7.7

$$\begin{aligned}\tilde{\gamma}^t_{i,m} =& (\tilde{\gamma}^t_{i,m-1})' + \tilde{\gamma}^t_{i-1,m-1} \\ =& \sum_{\tau=0}^{m-i-2}\binom{i+\tau}{\tau}(\tilde{\gamma}^t_{-1,m-2-i-\tau})^{(\tau+1)} + \binom{m-1}{i}(h^t_{n_0})^{(m-i)} \\ &+ \sum_{\tau=0}^{m-i-1}\binom{i-1+\tau}{\tau}(\tilde{\gamma}^t_{-1,m-1-i-\tau})^{(\tau)} + \binom{m-1}{i-1}(h^t_{n_0})^{(m-i)} \\ =& \sum_{\tau=0}^{m-i-1}\binom{i+\tau}{\tau}(\tilde{\gamma}^t_{-1,m-1-i-\tau})^{(\tau)} + \binom{m}{i}(h^t_{n_0})^{(m-i)}.\end{aligned}$$

With obvious changes in the above calculation, the result also holds for $i = 0$. □

We immediately infer from Proposition 8.7.2

PROPOSITION 8.7.9. *For $\kappa = 0,\ldots,l-1$ and $t = 0,\ldots,L$ such that $1 \leq \kappa + tl \leq n$ we have*

$$e^\top_{\kappa+tl} Y\mathbf{H}^+(Y^{-1}e_n) = \begin{cases} 0 & \text{if } t = 0, \\ \sigma_{n_0-\kappa-(t-1)l} & \text{if } t > 0. \end{cases}$$

The $q_{j,\nu}$ defined in Lemma 8.1.3 are polynomials in λ. Now we shall give an estimate for the order of the highest λ-power and determine the "leading" coefficient. We do not require that the leading coefficient is different from zero.

PROPOSITION 8.7.10. *Let $j \in \{0,\ldots,n-1\}$ and $\nu \in \{0,\ldots,j\}$. Let $s_{j,\nu}$ and $q_{j,\nu,0}$ be definied as in Section* 8.5. *Then*

$$q_{j,\nu}(\cdot,\lambda) = \lambda^{ls_{j,\nu}}[q_{j,\nu,0}].$$

Proof. For $\nu = j$ the statement follows from the definition of $q_{j,j}$ in Lemma 8.1.3. Now let $0 < j \leq n-1$ and suppose that the proposition holds for $j-1$. Then the statement for $0 < \nu \leq j-1$ follows from the definition of $q_{j,\nu}$. If $\nu = 0$ we first observe that $p_{n-i-1}(\cdot,\lambda) = k_{n-i-1}$ if $i \leq l-2$ and $p_{n-i-1}(\cdot,\lambda) = -\lambda^l[h_{n-i-1}]$ if $i \geq l-1$. Now the statement for $\nu = 0$ easily follows. □

PROPOSITION 8.7.11. *For $\kappa = 0,\ldots,l-1$ and $t = 1,\ldots,L$ such that $\kappa + tl \le n$ we have*

$$e^{\mathsf{T}}_{\kappa+tl} Y(\cdot,\lambda)\mathbf{H}^{+}(Y(\cdot,\lambda)^{-1}e_n) = \lambda^{(t-1)l}[\sigma_{n_0-\kappa-(t-1)l,0}],$$

where the $\sigma_{i,0}$ are defined in (8.5.16).

Proof. We have

$$\sigma_{n_0-\kappa-(t-1)l}(\cdot,\lambda) = \sum_{j=n_0-\kappa-(t-1)l}^{n_0} \sum_{i=j}^{n_0} (-1)^i \binom{i}{j} h_i^{(i-j)} q_{j,n_0-\kappa-(t-1)l}(\cdot,\lambda).$$

By Proposition 8.7.10, the largest λ-exponent is obtained for the largest j, and it is $[\frac{n_0-(n_0-\kappa-(t-1)l)}{l}]l = t-1$. And since $[\frac{j-(n_0-\kappa-(t-1)l)}{l}] \ge t-1$ if and only if $j \ge n_0 - \kappa$, an application of Proposition 8.7.9 and 8.7.10 completes the proof. □

PROPOSITION 8.7.12. i) *For $m = 0,\ldots,l-2$ we have $\tilde{\gamma}^0_{-1,m} = 0$.*
ii) *For $m = 0,\ldots,l-2$ and $1 \le t \le L$ we have*

$$\tilde{\gamma}^t_{-1,m} = \sigma_{n_0-m-1-(t-1)l,0}\,.$$

Proof. By definition, $\tilde{\gamma}^t_{-1,m}$ is the coefficient of $\lambda^{(t-1)l}$ in $y_{m+tl}z(\cdot,\lambda)$. Since $Y(\cdot,\lambda)(I_n - \Delta_0)\mathbf{H}^{+}(Y(\cdot,\lambda)^{-1}e_n) = \{O(\lambda^{-2l})\}_\infty$ by Theorem 8.4.2 and Corollary 8.3.2, $\tilde{\gamma}^t_{-1,m}$ is the coefficient of $\lambda^{(t-1)l}$ in $e_{m+1+tl}Y(\cdot,\lambda)\mathbf{H}^{+}(Y(\cdot,\lambda)^{-1}e_n)$. Part i) now immediately follows from Proposition 8.7.9. And part ii) follows from Proposition 8.7.11. □

Proof of Proposition 8.7.6. From Propositions 8.7.8 and 8.7.12 we infer $\gamma^t_{i,m} = \tilde{\gamma}^t_{i,m}$ for $m = 0,\ldots,l-1$, $i = 0,\ldots,m$, and $t = 0,\ldots,L$ such that $m + tl \le n-1$. Now the statement of Proposition 8.7.6 follows in view of (8.7.9) and (8.7.11). □

We write the two remaining integrals on the right hand side of (8.7.1) as in (4.4.4) and obtain

$$\int_a^x (\widehat{G}_{0,\mu,s}f)(x,\xi,\lambda)\,\mathrm{d}\xi - \int_x^b (\widehat{G}_{1,\mu,s}f)(x,\xi,\lambda)\,\mathrm{d}\xi = I_{1,\mu,s}(x,f,\lambda) + I_{2,\mu,s}(x,f,\lambda)$$

where

$$\begin{aligned} I_{1,\mu,s}&(x,f,\lambda) \\ &= \lambda^{-s} y_\mu(x,\lambda)(I_l - \widehat{\Delta}(\lambda))\Omega_l^{-s}\widehat{E}(x,\lambda)\int_a^x \widehat{E}(\xi,\lambda)^{-1}D^s(z(\cdot,\lambda)f)(\xi)\,\mathrm{d}\xi \\ &\quad - \lambda^{-s} y_\mu(x,\lambda)\widehat{\Delta}(\lambda)\Omega_l^{-s}\widehat{E}(x,\lambda)\int_x^b \widehat{E}(\xi,\lambda)^{-1}D^s(z(\cdot,\lambda)f)(\xi)\,\mathrm{d}\xi, \end{aligned}$$

$$I_{2,\mu,s}(x,f,\lambda) = -\lambda^{-s}e^{\mathsf{T}}_{\mu+1}Y(x,\lambda)M^{-1}(\lambda)W^{(1)}(\lambda^l)\widehat{Y}(b,\lambda) \times$$
$$\times \int_a^b (I_l - \widehat{\Delta}(\lambda))\widehat{E}(b,\lambda)\widehat{E}(\xi,\lambda)^{-1}\Omega_l^{-s}D^s(z(\cdot,\lambda)f)(\xi)\,\mathrm{d}\xi$$
$$+\lambda^{-s}e^{\mathsf{T}}_{\mu+1}Y(x,\lambda)M^{-1}(\lambda)W^{(0)}(\lambda^l)\widehat{Y}(a,\lambda) \times$$
$$\times \int_a^b \widehat{\Delta}(\lambda)\widehat{E}(a,\lambda)\widehat{E}(\xi,\lambda)^{-1}\Omega_l^{-s}D^s(z(\cdot,\lambda)f)(\xi)\,\mathrm{d}\xi,$$
$$\widehat{\Delta}(\lambda) = \operatorname{diag}(\delta_1(\lambda),\dots,\delta_l(\lambda)).$$

PROPOSITION 8.7.13. *Let $0 \le s \le l$. Suppose that $h_{n_0} \in C^\infty[a,b]$, $h_{n_0} > 0$ and $h_{n_0}^{-1} \in L_\infty(a,b)$, and that $k_i \in W^i_{p'}(a,b)$ for $i = 0,\dots,n-1$. If $n_0 > 0$, then we additionally suppose that $k_i \in W^{n_0}_{p'}(a,b)$ for $i = 0,\dots,n_0-1$, $k_i \in W^{i+1}_{p'}(a,b)$ for $i = n_0,\dots,n-1$, and $h_i \in W^n_{p'}(a,b)$ for $i = 0,\dots,n_0-1$.*
i) *Let $1 < p < \infty$. For $f \in W^s_p(a,b)$ and $\mu = 0,\dots,\min\{s,l-1\}$ we obtain*

$$\left|\oint_{\Gamma_m} \lambda^{l-1} I_{1,\mu,s}(\cdot,f,\lambda)\,\mathrm{d}\lambda\right|_p = O(1)|f|_{p,s} \text{ as } m \to \infty.$$

ii) *Here let $p < \infty$, i. e. $p' > 1$ if $n_0 = 0$ and $\mu = n-1$, and let $p = \infty$ otherwise. For $f \in C^s[a,b]$ with $f^{(s)} \in BV[a,b]$ and $\mu = 0,\dots,\min\{s,l-1\}$ we have*

$$\left|\oint_{\Gamma_m} \lambda^{l-1} I_{1,\mu,s}(\cdot,f,\lambda)\,\mathrm{d}\lambda\right|_\infty = O(1)|f| \text{ as } m \to \infty,$$

where $|f| = |f|_{(s)} + |f^{(s)}|_{BV[a,b]}$.

Proof. The assumptions of Corollary 8.3.1 and Proposition 8.7.3 are satisfied with $k = n-1$ if $n_0 = 0$ and $k = n$ if $n_0 > 0$. Here we note that Proposition 8.7.3 also holds in case $k = j-1$, where the o-term dominates. We have

$$\lambda^{-\mu} y_\mu(\cdot,\lambda) = \Omega_l^\mu \varepsilon h_{n_0}^{\mu/l}\varphi_0 + \{O(\lambda^{-1})\}_\infty$$

by Corollary 8.3.1 and

$$D^s(z(\cdot,\lambda)f)$$
$$= \lambda^{1-l}\Omega_l\{D^s((h_{n_0}^{n/l}\psi_0)^{(j)}f)\varepsilon + \{O(\lambda^{-1}) + O(\lambda^{-k+s}\tau_{p'}(\lambda))\}_\infty |f|_{p,s}\}$$

by Proposition 8.7.3, where we have made use of Remark 8.4.3. If $\mu < s < n$ or $\mu < s-1$ and $s = n$, then the above representations and Proposition 4.3.3 yield $I_{1,\mu,s}(\cdot,f,\lambda) = \{O(\lambda^{-l})\}_\infty$, and the result follows. If $\mu = s < n$ or $\mu + 1 = s = n$, the part coming from $\{O(\lambda^{-1})\}_\infty$ in $\lambda^{-\mu}y_\mu(\cdot,\lambda)$ can be estimated as above. Therefore we still have to estimate

(8.7.12)
$$\tilde{y}(x)(I_l - \widehat{\Delta}(\lambda))\widehat{E}(x,\lambda) \int_a^x \widehat{E}(\xi,\lambda)^{-1} D^s(\tilde{z}f)(\xi)\,\mathrm{d}\xi$$
$$- \tilde{y}(x)\widehat{\Delta}(\lambda)\widehat{E}(x,\lambda) \int_x^b \widehat{E}(\xi,\lambda)^{-1} D^s(\tilde{z}f)(\xi)\,\mathrm{d}\xi$$

and

(8.7.13)
$$\tilde{y}(x)(I_l - \widehat{\Delta}(\lambda))\widehat{E}(x,\lambda) \int_a^x \widehat{E}(\xi,\lambda)^{-1} \{O(\tau_{p'}(\lambda)\}_\infty(\xi)\,\mathrm{d}\xi\, |f|_{p,s}$$
$$- \tilde{y}(x)\widehat{\Delta}(\lambda)\widehat{E}(x,\lambda) \int_x^b \widehat{E}(\xi,\lambda)^{-1} \{O(\tau_{p'}(\lambda)\}_\infty(\xi)\,\mathrm{d}\xi\, |f|_{p,s},$$

where $\tilde{y}$ and $\tilde{z}$ are vector functions in suitable Sobolev spaces (in case $s = n$ we have $\tilde{z} = 0$). Now the desired estimate of (8.7.12) follows from Propositions 4.6.5 and 4.7.2, respectively, whereas the estimate of (8.7.13) follows from Propositions 4.3.5 and 4.6.4. For part ii) we have to observe that $\{O(\tau_{p'}(\lambda))\}_\infty$ can be replaced by $\{O(\lambda^{-1})\}_\infty$ if $\mu < k$. In the remaining case, i. e. $n_0 = 0$ and $\mu = n - 1$, we have $p' > 1$, so that the above proof also holds here. □

PROPOSITION 8.7.14. *Suppose that the assumptions of Proposition* 8.7.13 *are satisfied and that* (8.6.1) *holds. Suppose that the boundary eigenvalue problem* (8.1.1), (8.1.5) *is almost Birkhoff regular of order* $r < l$.

ii) *Let* $r < s \leq l$. *In case* $n_0 = 0$ *we can even take* $s = r$. *Let* $1 < p < \infty$. *For* $f \in W_p^s(a,b)$ *and* $\mu = 0, \ldots, \min\{s, l-1\}$ *we obtain*

$$\left| \oint_{\Gamma_m} \lambda^{l-1} I_{2,\mu,s}(\cdot, f, \lambda)\,\mathrm{d}\lambda \right|_p = O(1)|f|_{p,s} \text{ as } m \to \infty.$$

ii) *Let* $r \leq s \leq l$. *For* $f \in C^s[a,b]$ *with* $f^{(s)} \in BV[a,b]$ *and* $\mu = 0, \ldots, s$ *we have*

$$\left| \oint_{\Gamma_m} \lambda^{l-1} I_{2,\mu,s}(\cdot, f, \lambda)\,\mathrm{d}\lambda \right|_\infty = O(1)|f| \text{ as } m \to \infty,$$

where $|f| = |f|_{(s)} + |f^{(s)}|_{BV[a,b]}$.

Proof. In view of Lemma 8.6.3 we have that

$$\lambda^{-s} e_{\mu+1}^{\mathsf{T}} Y(x,\lambda) M^{-1}(\lambda) W^{(j)}(\lambda) \widehat{Y}(a_j,\lambda)$$
$$= \lambda^{-s} y_\mu(x,\lambda) \widehat{E}(x,\lambda) \left(I_n - \widehat{\Delta}(\lambda) + \widehat{E}(b,\lambda)^{-1}\widehat{\Delta}(\lambda)\right) \left(v_{\mu,j}(\lambda) + O(\tau_\infty(\lambda))\right)$$
$$+ \lambda^{r-s} e_{\mu+1}^{\mathsf{T}} Y(x,\lambda)(I_n - \Delta_0)\left(\tilde{v}_{\mu,j}(\lambda) + O(\tau_\infty(\lambda))\right),$$

where $\lambda^{-s} y_\mu(\cdot,\lambda)$ and $\lambda^{r-s} e_{\mu+1}^{\mathsf{T}} Y(\cdot,\lambda)$ are of the form $v + \{O(\lambda^{-1})\}_\infty$, the coefficients of v belong to $L_\infty(a,b)$, and $v_{\mu,j}(\lambda)$ and $\tilde{v}_{\mu,j}(\lambda)$ are constant on sectors. Now the terms containing $O(\tau_\infty(\lambda))$, λ^{r-s} if $r < s$, or $\{O(\tau_{p'}(\lambda))\}_\infty$ (from $D^s(z(\cdot,\lambda)f)$) and terms with $\mu < s$ are easily estimated as in Proposition 8.7.13. Since terms with λ^{r-s} do not occur if $n_0 = 0$, we do not need any restriction with respect to r in this case. The remaining terms coming from $y_\mu(\cdot,\lambda)$ are of the form a considered in Propositions 4.6.3 and 4.7.1, respectively. In part ii) we still have to estimate terms where the exponential terms in the variable x are missing. But this corresponds to $x = a$ in the case with full exponential terms. Since the estimates in Proposition 4.7.1 are uniformly on $[a,b]$, they extend to this case. □

From Propositions 8.7.1, 8.7.5, 8.7.6, 8.7.13, and 8.7.14 we obtain

LEMMA 8.7.15. *Let the assumptions be as in Proposition* 8.7.14. *Suppose that the boundary eigenvalue problem* (8.1.1), (8.1.5) *is almost Birkhoff regular of order* $r < l$.
i) *Let* $1 < p < \infty$, $s \in \{r,\dots,l-1\}$ *if* $n_0 = 0$, $s \in \{r+1,\dots,l-1\}$ *if* $n_0 > 0$, *and let* F *be the Banach space*

$$F = \{f \in W_p^s(a,b) : U_\nu f = 0,\ \nu = 1,\dots,n,\ \hat{l}_\nu < s\}.$$

Then $\{Q_m J_F : m \in \mathbb{N}\}$ *is bounded in* $L(F, W_p^s(a,b))$, *where* J_F *is the canonical inclusion map from* F *into* $W_p^s(a,b)$.
ii) *Let* $s \in \{r,\dots,l-1\}$ *and let* F *be the Banach space*

$$F = \{f \in C^s[a,b] : f^{(s)} \in BV[a,b],\ U_\nu f = 0,\ \nu = 1,\dots,n,\ \hat{l}_\nu < s\}.$$

Then $\{Q_m J_F : m \in \mathbb{N}\}$ *is bounded in* $L(F, C^s[a,b])$, *where* J_F *is the canonical inclusion map from* F *into* $C^s[a,b]$.

8.8. Expansion theorems

PROPOSITION 8.8.1. *Suppose that the assumptions of Proposition* 8.7.14 *are satisfied. Suppose that the boundary eigenvalue problem* (8.1.1), (8.1.5) *is almost Birkhoff regular of order* $r < l$.
i) *Let* $1 < p < \infty$, $s \in \{r,\dots,l-1\}$ *if* $n_0 = 0$, $s \in \{r+1,\dots,l-1\}$ *if* $n_0 > 0$, *and let* H *be the set*

$$H = \{f \in W_p^{s+1}(a,b) : U_\nu f = 0,\ \nu = 1,\dots,n,\ \hat{l}_\nu < s\}.$$

Then $Q_m f \to f$ *in* $W_p^s(a,b)$ *as* $m \to \infty$ *for all* $f \in H$.
ii) *Let* $s \in \{r,\dots,l-1\}$ *and let* H *be the set*

$$H = \{f \in C^{s+1}[a,b] : f^{(s+1)} \in BV[a,b],\ U_\nu f = 0,\ \nu = 1,\dots,n,\ \hat{l}_\nu \le s\}.$$

Then $Q_m f \to f$ *in* $C^s[a,b]$ *as* $m \to \infty$ *for all* $f \in H$.

Proof. We use the representation of Q_m as obtained in Section 8.7 and follow the steps which were used to prove Lemma 8.7.15. Let $f \in W_p^{s+1}(a,b)$ with s as in i) or ii). Then we can iterate one step further in (8.7.1), i.e., we replace s by $s+1$. Then Proposition 8.7.3 holds with $s+1$ instead of s since $i+s+1 \le n$ for $i \le n_0 - 1$ shows that its assumptions are satisfied. Therefore we have for $\mu = 0,\dots,s$ that

$$\frac{1}{2\pi i}\oint_{\Gamma_m} \lambda^{l-1} y_\mu(\cdot,\lambda) \sum_{j=0}^{s} \lambda^{-j-1}\Omega_l^{-j-1} h_{n_0}^{-1/l} D^j(z(\cdot,\lambda)f)\,\mathrm{d}\lambda$$

is the sum of the coefficient of λ^{-l} of the asymptotic polynomial

$$y_\mu(\cdot,\lambda)\sum_{j=0}^{s}\lambda^{-j-1}\Omega_l^{-j-1}h_{n_0}^{-1/l}D^j(z(\cdot,\lambda)f) \tag{8.8.1}$$

and an o-term. In view of Proposition 8.7.4 the coefficient of λ^{-l} of (8.8.1) coincides with the coefficient of λ^{-l} on the right-hand side of (8.7.10) with $\mu = m$ and $t = 0$. By definition of $\tilde{\gamma}^0_{i,m}$ and the fact that $\gamma^0_{i,m} = \tilde{\gamma}^0_{i,m}$, the above integral can be written as

$$\sum_{i=0}^{\mu}\gamma^0_{i,\mu}f^{(i)} + \{o(1)\}_\infty = f^{(\mu)} + \{o(1)\}_\infty.$$

Now we want to show that the integral over Γ_m of the second and third summands in (8.7.1) multiplied by λ^{l-1} (with $s-1$ replaced by s) tends to 0 in $L_p(a,b)$ as $m\to\infty$ for $\mu = 1,\dots,s$, where $p=\infty$ in part ii). In part ii) this immediately follows from Proposition 8.7.6 and the considerations preceding it since $U_\nu f = 0$ for $\nu = 1,\dots,n$ with $\hat{l}_\nu \le s$. In part i), the coefficient of λ^{-s} in $\beta^{(0)}(\lambda)+\beta^{(1)}(\lambda)$ will be different from 0 in general, and therefore we still have to show that

$$\oint_{\Gamma_m} e^{\mathsf{T}}_{\mu+1}Y(\cdot,\lambda)M^{-1}(\lambda)\operatorname{diag}(\lambda^{l_1},\dots,\lambda^{l_n})\lambda^{-s-1}\,\mathrm{d}\lambda$$

tends to 0 in $L_\infty(a,b)$ as $m\to\infty$ for $\mu = 1,\dots,s$. In view of the assumption $s > r$ in case $n_0 \ne 0$ we infer from Lemma 8.6.4 and $Y(\cdot,\lambda)(I_n-\Delta_0) = O(1)$ that only the integral

$$\oint_{\Gamma_m}\lambda^{-s-1}e^{\mathsf{T}}_{\mu+1}Y(\cdot,\lambda)\Delta_0M^{-1}(\lambda)\operatorname{diag}(\lambda^{l_1},\dots,\lambda^{l_n})\,\mathrm{d}\lambda$$

has to be considered. From Corollary 8.3.2 we infer that

$$\lambda^{-s}e^{\mathsf{T}}_{\mu+1}Y(\cdot,\lambda)\Delta_0 = \lambda^{\mu-s}(h_{n_0}^{\mu/l}\varphi_0\varepsilon^{\mathsf{T}}\Omega_l^{\mu} + \{O(\lambda^{-1})\}_\infty)E(\cdot,\lambda)\Delta_0.$$

This representation and Lemmas 8.6.4 and 2.7.2 ii) show that there is a constant $C>0$ such that, for $h\in(L_{p'}(a,b))^n$,

$$\Big|\int_a^b\oint_{\Gamma_m}\lambda^{-s-1}e^{\mathsf{T}}_{\mu+1}Y(x,\lambda)\Delta_0M^{-1}(\lambda)\operatorname{diag}(\lambda^{l_1},\dots,\lambda^{l_n})\,\mathrm{d}\lambda\,h(x)\,\mathrm{d}x\Big|$$
$$\le C|h|_{p'}\oint_{\Gamma_m}|\lambda|^{-1}(1+|\Re(\lambda)|)^{-1/p}|d\lambda| = C|h|_{p'}\oint_{|\lambda|=1}(1+\rho_m|\Re(\lambda)|)^{-1/p}|d\lambda|,$$

where ρ_m is the radius of Γ_m. By LEBESGUE'S dominated convergence theorem, the integral tends to 0 as $m\to\infty$, and

$$\Big|\oint_{\Gamma_m}\lambda^{-s-1}e^{\mathsf{T}}_{\mu+1}Y(x,\lambda)\Delta_0M^{-1}(\lambda)\operatorname{diag}(\lambda^{l_1},\dots,\lambda^{l_n})\,\mathrm{d}\lambda\Big|_p \to 0$$

as $m\to\infty$ is shown.

Finally, also the two remaining integrals are of the form $\{o(1)\}_\infty$. To see this we observe that we know from the proof of Proposition 8.7.14 that

$$\lambda^{-s} e^{\mathsf{T}}_{\mu+1} Y(\cdot,\lambda) M^{-1}(\lambda) W^{(j)}(\lambda) \widehat{Y}(a_j,\lambda) = \{O(1)\}_\infty$$

on $\bigcup_{m\in\mathbb{N}} \Gamma_m$ and that Lemma 2.7.2 gives

$$\int_a^b (I_l - \widehat{\Delta}_l(\lambda)) \widehat{E}(b-\xi,\lambda) \Omega_l^{-s-1} D^{s+1}(z(\cdot,\lambda)f)(\xi)\,\mathrm{d}\xi = o(1),$$

$$\int_a^b \widehat{\Delta}_l(\lambda) \widehat{E}(a-\xi,\lambda) \Omega_l^{-s-1} D^{s+1}(z(\cdot,\lambda)f)(\xi)\,\mathrm{d}\xi = o(1). \quad \square$$

Since the linear functionals U_ν for $\nu = 1,\dots,n$ with $l_\nu \le s$ are continuous on the space F considered in Proposition 8.7.15 ii) and since it follows as in the proof of Theorem 4.7.5 that H considered in Proposition 8.8.1 ii) is dense in F defined in Lemma 8.7.15 ii), Lemma 8.7.15 ii), Proposition 8.8.1 ii) and Lemma 4.6.8 yield

THEOREM 8.8.2. *Suppose that the boundary eigenvalue problem* (8.1.1), (8.1.5) *is almost Birkhoff regular of order $r < l$. Let $s \in \{r,\dots,l-1\}$ and suppose that $h_{n_0} \in C^\infty[a,b]$, $h_{n_0} > 0$, $h_{n_0}^{-1} \in L_\infty(a,b)$, and $k_i \in W_1^i(a,b)$ for $i = 0,\dots,n-1$. If $s = n-1$, then we suppose that $k_i \in W_p^i(a,b)$ for some $p > 1$ and $i = 0,\dots,n-1$. If $n_0 > 0$, then we additionally suppose that $k_i \in W_1^{n_0}(a,b)$ for $i = 0,\dots,n_0-1$, $k_i \in W_1^{i+1}(a,b)$ for $i = n_0,\dots,n-1$, and $h_i \in W_1^n(a,b)$ for $i = 0,\dots,n_0-1$. Then $Q_m f \to f$ as $m \to \infty$ in $C^s[a,b]$ for each $f \in C^s[a,b]$ with $f^{(s)} \in BV[a,b]$ which satisfies the boundary conditions $U_\nu f = 0$ for $\nu = 1,\dots,n$ with $\hat{l}_\nu \le s$.*

Similarly, we obtain

THEOREM 8.8.3. *Suppose that the boundary eigenvalue problem* (8.1.1), (8.1.5) *is almost Birkhoff regular of order $r < l$. Let $1 < p < \infty$, $s \in \{r,\dots,l-1\}$ if $n_0 = 0$, $s \in \{r+1,\dots,l-1\}$ if $n_0 > 0$, and suppose that $h_{n_0} \in C^\infty[a,b]$, $h_{n_0} > 0$, $h_{n_0}^{-1} \in L_\infty(a,b)$, and $k_i \in W_{p'}^i(a,b)$ for $i = 0,\dots,n-1$. If $n_0 > 0$, then we additionally suppose that $k_i \in W_{p'}^{n_0}(a,b)$ for $i = 0,\dots,n_0-1$, $k_i \in W_{p'}^{i+1}(a,b)$ for $i = n_0,\dots,n-1$, and $h_i \in W_{p'}^n(a,b)$ for $i = 0,\dots,n_0-1$. Then $Q_m f \to f$ as $m \to \infty$ in $W_p^s(a,b)$ for each $f \in W_p^s(a,b)$ which satisfies the boundary conditions $U_\nu f = 0$ for $\nu = 1,\dots,n$ with $\hat{l}_\nu < s$.*

REMARK 8.8.4. The convergence in the previous theorems can be written as the convergence of a series in eigenvectors and associated vectors as in Section 7.4. For this let L be the differential operator associated with the boundary eigenvalue problem (8.1.1), (8.1.5), and let L^* be its adjoint. Let $\lambda_0, \lambda_1, \dots$ be the eigenvalues of L and let

$$\{\eta_{\kappa,h}^{(j)} : j = 1,\dots,r(\lambda_\kappa);\, h = 0,\dots,m_{\kappa,j}-1\}$$

and

$$\left\{\left(u_{\kappa,h}^{(j)}, d_{\kappa,h}^{(j)}\right) : j = 1, \ldots, r(\lambda_\kappa); h = 0, \ldots, m_{\kappa,j} - 1\right\}$$

be biorthogonal CSEAVs of L and L^* at λ_κ, respectively. Then the statement $Q_m f \to f$ in the above theorems can be written as

$$f^{(\mu)} = \lim_{m\to\infty} \sum_{\substack{\kappa\in\mathbb{N} \\ \lambda_\kappa < \rho_m}} \left(\sum_{j=1}^{r(\lambda_\kappa)} \sum_{h=0}^{m_{\kappa,j}-1} (\eta_{\kappa,h}^{(j)})^{(\mu)} \int_a^b (\mathbf{H}^+ u_{\kappa,m_{\kappa,j}-1-h}^{(j)})(x) f(x)\,\mathrm{d}x \right)$$

in $L_p(a,b)$ for $\mu = 0, \ldots, s$ for a suitable sequence of positive numbers $\rho_m \to \infty$, where we have used the representation of the Green's function given in Theorem 6.7.8.

8.9. The differential equation $\eta^{(4)} - \alpha\eta''' = \lambda\eta''$

In this section we consider a boundary eigenvalue problem for a differential equation of the form (8.1.1) with $h_{n_0} = 1$ where the boundary conditions which must be satisfied for the expansion theorems depend on the coefficients of the differential equation. For this we must have $\hat{l}_\nu$ at least 1, i. e., l at least 2, and a corresponding $\alpha_{\nu 10}^{(\iota)} \neq 0$. By (8.5.16) this means that e_4 must be defined, and we therefore need that n is at least 4. Here we consider a differential equation were these numbers are minimal, namely

$$\eta^{(4)} - \alpha\eta''' - \lambda\eta'' = 0 \quad (\alpha \in \mathbb{C})$$

together with the boundary conditions

$$\eta(1) = 0,\ \eta'(0) = 0,\ \eta''(1) = 0,\ \eta'''(0) = 0.$$

Therefore $\mathbf{H}\eta = \eta''$, and a fundamental system of $\mathbf{H}\eta = 0$ is $\{1, x\}$. Hence the fundamental matrix considered in Theorem 8.2.1 satisfies

$$Y(x,\lambda)(I_n - \Delta_0) = \begin{pmatrix} 1 & x & 0 & 0 \\ 0 & 1 & 0 & 0 \\ 0 & 0 & 0 & 0 \\ 0 & 0 & 0 & 0 \end{pmatrix} + \{O(\lambda^{-2})\}_\infty.$$

For the boundary matrices defined in (8.1.6) we infer

$$(8.9.1) \qquad W^{(0)}(\lambda^2)\begin{pmatrix} 1 & 0 \\ 0 & 1 \\ 0 & 0 \\ 0 & 0 \end{pmatrix} = \begin{pmatrix} 0 & 0 \\ 0 & 1 \\ 0 & 0 \\ 0 & 0 \end{pmatrix}, \quad W^{(1)}(\lambda^2)\begin{pmatrix} 1 & 1 \\ 0 & 1 \\ 0 & 0 \\ 0 & 0 \end{pmatrix} = \begin{pmatrix} 1 & 1 \\ 0 & 0 \\ 0 & 0 \\ 0 & 0 \end{pmatrix}.$$

According to the requirement (8.5.7) we can take

$$q_1 = q_2 = 0,\ q_3 = q_4 = -1.$$

Also, we obviously have that the numbers defined in (8.5.1) are

$$l_1 = 0,\ l_2 = 1,\ l_3 = 2,\ l_4 = 3.$$

We infer

$$l^{(0)} = 5,\ l^{(1)} = 6,$$

see (8.5.10). By Definition 8.5.1 the problem is almost Birkhoff regular of order 1 if we show that $u_{\{3,4\}} \neq 0$ and $v_{\{3,4\},1} \neq 0$ for the numbers defined in (8.5.12) and (8.5.13). By definition, $u_{\{3,4\}}$ is the determinant of the matrix obtained by summing up the two matrices in (8.9.1) and taking their first two columns. So $u_{\{3,4\}} \neq 0$. Also, by definition (note that $\omega_1 = 1$, $\omega_2 = -1$),

$$v_{\{3,4\},1} = \begin{vmatrix} \alpha_3^{(0)} & \alpha_3^{(1)} \\ \alpha_4^{(0)} & -\alpha_4^{(1)} \end{vmatrix},$$

where the $\alpha_\nu^{(j)}$ are defined in (8.5.6). Now $\alpha_3^{(0)} = 0$ and $\alpha_4^{(1)} = 0$ since the third boundary condition only contains terms at 1 and the fourth boundary condition only contains terms at 0. Obviously,

$$\alpha_3^{(1)} = \sum_{t=0}^{1} \alpha_{3t}^{(1)} = 1,\ \alpha_4^{(0)} = \sum_{t=0}^{1} \alpha_{4t}^{(0)} = 1,$$

which shows that $u_{\{3,4\}} \neq 0$.

Hence we can apply Theorem 8.8.2 for $s = 1$. For the numbers defined in (8.5.16) we have $\alpha_{100}^{(1)} = 1$, $\alpha_{200}^{(0)} = 1$, $\alpha_{310}^{(1)} = 1$, $\alpha_{410}^{(0)} = 1$, and all other $\alpha_{\nu tm}^{(\iota)}$ are zero. Hence, in view of Remark 8.5.6 iii), the boundary terms $U_\nu f = 0$ are

$$U_1 f = f(1),\ U_2 f = f'(0),\ U_3 f = f(1),\ U_4 f = \alpha f(0) + f'(0).$$

If $\alpha \neq 0$, we thus have the three asymptotic boundary conditions

$$f(1) = 0,\ f(0) = 0,\ f'(0) = 0.$$

If $\alpha = 0$, we only have two asymptotic boundary conditions

$$f(1) = 0,\ f'(0) = 0.$$

8.10. The differential equation $\eta^{(4)} + K\eta = \lambda \mathbf{H}\eta$

In this section we consider the differential equation

$$\eta^{(4)} + K\eta - \lambda \mathbf{H}\eta = 0$$

with the boundary conditions

$$\eta(0) = 0,\ \eta'(0) = 0,\ \eta''(1) = 0,\ \beta(\lambda)\eta'''(1) + \alpha(\lambda)\eta(1) = 0,$$

where K is a function,

$$\alpha(\lambda) = \alpha_3\lambda^3 + \alpha_2\lambda^2 + \alpha_1\lambda,$$
$$\beta(\lambda) = \beta_2\lambda^2 + \beta_1\lambda,$$

with $\alpha_3 > 0$ and $\beta_2 > 0$, and

A. $\mathbf{H}\eta = \eta$,

B. $\mathbf{H}\eta = \eta' + G_1\eta$,

C. $\mathbf{H}\eta = \eta'' + G_1\eta' + G_2\eta$,

with functions G_1 and G_2. We suppose that the functions K, G_1, G_2 satisfy the assumptions of Theorems 8.8.2 or 8.8.3. A sufficient condition is that these three functions belong to $W_\infty^4(0,1)$. The differential equation is of the form (8.1.1).

Case **A.** Here $n_0 = 0$ and

$$l_1 = 0,\ l_2 = 1,\ l_3 = 2,\ l_4 = 12.$$

By Definition 8.5.1 the problem is Birkhoff regular if $\nu_{\{1,2,3,4\},2} \neq 0$, where this number is defined in (8.5.13). We have, if we observe that $\alpha_\nu^{(0)} = 0$ if $\nu = 3,4$ and $\alpha_\nu^{(1)} = 0$ if $\nu = 1,2$,

$$\begin{aligned}\nu_{\{1,2,3,4\},2} &= \begin{vmatrix} \alpha_1^{(0)} & \alpha_1^{(0)} & 0 & 0 \\ \alpha_2^{(0)} & i\alpha_2^{(0)} & 0 & 0 \\ 0 & 0 & \alpha_3^{(1)} & -\alpha_3^{(1)} \\ 0 & 0 & \alpha_4^{(1)} & \alpha_4^{(1)} \end{vmatrix} \\ &= 2(i-1)\alpha_1^{(0)}\alpha_2^{(0)}\alpha_3^{(1)}\alpha_4^{(1)} \\ &= 2(i-1)\alpha_{100}^{(0)}\alpha_{200}^{(0)}\alpha_{300}^{(1)}\alpha_{400}^{(1)} \neq 0\end{aligned}$$

since

$$\alpha_{100}^{(0)} = 1,\ \alpha_{200}^{(0)} = 1,\ \alpha_{300}^{(1)} = 1,\ \alpha_{400}^{(1)} = \alpha_3.$$

For the boundary conditions we need further

$$\alpha_{201}^{(0)} = 0,\ \alpha_{301}^{(0)} = 0,\ \alpha_{302}^{(1)} = 0.$$

Hence we have

$$U_1 f = f(0),\ U_2 f = f'(0),\ U_3 f = f''(1),\ U_4 f = f(1).$$

Therefore the asymptotic boundary conditions are

$$f(0) = 0,\ f'(0) = 0,\ f(1) = 0,\ f''(1) = 0.$$

Which of these conditions are actually needed in the corresponding expansion theorems in Section 8.8 depends on s and p. Of course, this also holds for the cases **B** and **C** below.

Case **B.** Here $n_0 = 1$ and

$$l_1 = 0,\ l_2 = 1,\ l_3 = 2,\ l_4 = 9.$$

Choosing $q_\nu = q'_\nu$, we have

$$q_1 = 0,\ q_2 = 0,\ q_3 = 0,\ q_4 = 3.$$

Then

$$l^{(0)} = 12,\ l^{(1)} = 12.$$

Hence the problem is Birkhoff regular if

$$b_1^{(0)} \neq 0 \quad \text{and} \quad b_2^{(0)} \neq 0.$$

We have

$$b_j^{(0)} = -u_{\{1,2,3\}} v_{\{1,2,3\},j} + u_{\{2,3,4\}} v_{\{2,3,4\},j}.$$

Since the first two boundary conditions are taken at 0 and the last two at 1, we infer

$$v_{\{1,2,3\},1} = \begin{vmatrix} * & 0 & 0 \\ * & 0 & 0 \\ 0 & * & * \end{vmatrix} = 0,$$

$$v_{\{2,3,4\},2} = \begin{vmatrix} * & * & 0 \\ 0 & 0 & * \\ 0 & 0 & * \end{vmatrix} = 0.$$

Also,

$$u_{\{1,2,3\}} = \alpha_3 \pi_1(1),\ u_{\{2,3,4\}} = \pi_1(0),$$

where π_1 is a nontrivial solution of $\mathbf{H}\eta = 0$. Since $\mathbf{H}\eta = 0$ is a first order differential equation, the nontrivial solution π_1 does not have zeros, and

$$u_{\{1,2,3\}} \neq 0 \quad \text{and} \quad u_{\{2,3,4\}} \neq 0$$

follows. So we must show that

$$v_{\{1,2,3\},2} \neq 0 \quad \text{and} \quad v_{\{2,3,4\},1} \neq 0.$$

We have

$$v_{\{1,2,3\},2} = \begin{vmatrix} \alpha_1^{(0)} & \alpha_1^{(0)} & 0 \\ \omega_1 \alpha_2^{(0)} & \omega_2 \alpha_2^{(0)} & 0 \\ 0 & 0 & \omega_3^2 \alpha_3^{(1)} \end{vmatrix} = (\omega_2 - \omega_1)\omega_3^2 \alpha_1^{(0)} \alpha_2^{(0)} \alpha_3^{(1)},$$

$$v_{\{2,3,4\},1} = \begin{vmatrix} \omega_1 \alpha_2^{(0)} & 0 & 0 \\ 0 & \omega_2^2 \alpha_3^{(1)} & \omega_3^2 \alpha_3^{(1)} \\ 0 & \alpha_4^{(1)} & \alpha_4^{(1)} \end{vmatrix} = \omega_1(\omega_2^2 - \omega_3^2)\alpha_2^{(0)} \alpha_3^{(1)} \alpha_4^{(1)},$$

where $\omega_1, \omega_2, \omega_3$ are the distinct third unit roots. In view of (8.5.6) we must show that

$$\alpha_{\nu 00}^{(j)} + \alpha_{\nu 10}^{(j)} \neq 0 \text{ for } \nu = 1,2,\ j = 0 \text{ and } \nu = 3,4,\ j = 1,$$

where we have used that $\alpha_{\nu t}^{(j)} = \alpha_{\nu t 0}^{(j)}$. We have

$$\alpha_{100}^{(0)} = 1,\ \alpha_{200}^{(0)} = 1,\ \alpha_{300}^{(1)} = 1,\ \alpha_{400}^{(1)} = \alpha_3, \alpha_{410}^{(1)} = \beta_2,$$

and all the other $\alpha_{\nu t m}^{(j)}$ are zero. Since α_3 and β_2 are positive, the Birkhoff regularity of the problem follows. From Remark 8.5.5 iii) we infer

$$\delta_{10}^{(0)} = 1,$$
$$\delta_{20}^{(0)} = 0,\ \delta_{21}^{(0)} = 1,$$
$$\delta_{40}^{(1)} = \alpha_3 + \beta_2,$$

which implies

$$U_1 f = f(0),\ U_2 f = f'(0),\ U_4 f = (\alpha_3 + \beta_2) f(1).$$

For the remaining boundary condition a direct calculation yields

$$U_3 f = \sum_{i=0}^{2} \gamma_{i,2}^{0}(1) f^{(i)}(1) = f''(1).$$

This leads to the asymptotic boundary conditions

$$f(0) = 0,\ f'(0) = 0,\ f(1) = 0,\ f''(1) = 0.$$

Case **C**. Here $n_0 = 2$ and

$$l_1 = 0,\ l_2 = 1,\ l_3 = 2,\ l_4 = 7.$$

Choosing $q_\nu = q'_\nu$, we have

$$q_1 = 0,\ q_2 = 0,\ q_3 = 0,\ q_4 = 3.$$

Then

$$l^{(0)} = 9,\ l^{(1)} = 10,$$

and

$$b_1^{(0)} = u_{\{3,4\}} v_{\{3,4\},1} + u_{\{2,3\}} v_{\{2,3\},1} = u_{\{2,3\}} v_{\{2,3\},1}$$

since $v_{\{3,4\},1} = 0$ as the third and fourth boundary conditions do not contain terms at 0. First observe that

$$u_{\{2,3\}} = \alpha_3 \begin{vmatrix} \pi_1(0) & \pi_2(0) \\ \pi_1(1) & \pi_2(1) \end{vmatrix},$$

where $\{\pi_1, \pi_2\}$ is a fundamental system of $\mathbf{H}\eta = 0$. Since the condition $u_{\{2,3\}} \neq 0$ is independent of the choice of the fundamental system, we may choose $\pi_1(0) = 0$. Since $\pi_2(0) \neq 0$ in this case, $u_{\{2,3\}} \neq 0$ if and only if the boundary value problem

(8.10.1) $$\mathbf{H}\eta = 0,\ \eta(0) = 0,\ \eta(1) = 0$$

has only the trivial solution.

We have

$$v_{\{2,3\},1} = \begin{vmatrix} \alpha_2^{(0)} & -\alpha_2^{(1)} \\ \alpha_3^{(0)} & \alpha_3^{(1)} \end{vmatrix} = \alpha_2^{(0)}\alpha_3^{(1)} = 1$$

since, in this case,

$$\alpha_{100}^{(0)} = 1,\ \alpha_{200}^{(0)} = 1,\ \alpha_{310}^{(1)} = 1,\ \alpha_{410}^{(1)} = \beta_2, \alpha_{401}^{(1)} = \alpha_3,$$

and all the other $\alpha_{v\iota r}^{(\iota)}$ are zero. Hence, if the boundary eigenvalue problem (8.10.1) has only the trivial solution, then the problem is almost Birkhoff regular of order 1. From Remark 8.5.5 iii) we infer

$$\begin{aligned}
&\delta_{10}^{(0)} = 1,\\
&\delta_{20}^{(0)} = 0,\ \delta_{21}^{(0)} = 1,\\
&\delta_{30}^{(1)} = 1,\\
&\delta_{40}^{(1)} = \alpha_3 + \beta_2 G_1(1),\ \delta_{41}^{(1)} = \beta_2.
\end{aligned}$$

It follows that

$$U_1 f = f(0),\ U_2 f = f'(0),\ U_3 f = f(1),\ U_4 f = (\alpha_3 + \beta_2 G_1(1)) f(1) + \beta_2 f'(1),$$

and the asymptotic boundary conditions are

$$f(0) = 0,\ f'(0) = 0,\ f(1) = 0,\ f'(1) = 0.$$

8.11. A boundary eigenvalue problem with associated functions at each eigenvalue

Here we consider the boundary eigenvalue problem

$$y^{(4)} - \lambda y'' = 0; \quad y(1) = 0,\ y'(0) = 0,\ y'(1) = 0,\ y''(1) = 0.$$

This is an example from [KF, p. 79]. Kaufmann calculated the eigenvectors of the given problem and its adjoint. He realized that they are not biorthogonal and could not get an expansion theorem. According to our results, we immediately know that to every eigenfunction there must be an associated function.

We shall investigate the regularity of the problem and calculate its eigenfunctions and associated functions as well as those of the adjoint problem.

First let us consider regularity. We use the notations of Section 8.5. Obviously, $n_0 = 2$ and

$$\begin{aligned}
&l_1 = 0,\ l_2 = 1,\ l_3 = 1,\ l_4 = 2,\\
&q_1 = q_2 = q_3 = q_4 = 0.
\end{aligned}$$

This gives

$$l^{(0)} = 3,\ l^{(1)} = 4,$$

and

$$b_1^{(0)} = u_{\{3,4\}}v_{\{3,4\},1} - u_{\{2,4\}}v_{\{2,4\},1}.$$

As the third and the fourth boundary conditions only have terms at 1, $v_{\{3,4\},1} = 0$. Also, $\{1,x\}$ is a fundamental system of $y'' = 0$ whence

$$u_{\{2,4\}} = \begin{vmatrix} 1 & 1 \\ 0 & 1 \end{vmatrix} \neq 0.$$

Finally,

$$v_{\{2,4\},1} = \begin{vmatrix} \alpha_2^{(0)} & -\alpha_2^{(1)} \\ \alpha_4^{(0)} & \alpha_4^{(1)} \end{vmatrix} = \alpha_2^{(0)}\alpha_4^{(1)} = 1.$$

Hence the problem is almost Birkhoff regular of order 1.

It is easy to see that

$$\alpha_{100}^{(1)} = 1,\ \alpha_{200}^{(0)} = 1,\ \alpha_{300}^{(1)} = 1,\ \alpha_{410}^{(1)} = 0,$$

and that all other $\alpha_{\nu t m}^{(t)}$ are zero. By Remark 8.5.5 iii) we obtain

$$U_1 f = f(1),\ U_2 f = f'(0),\ U_3(f) = f'(1),\ U_4(f) = f(1).$$

Therefore the asymptotic boundary conditions are

$$f(1) = 0,\ f'(0) = 0,\ f'(1) = 0.$$

In order to find the eigenfunctions and associated functions, it is convenient to introduce the new eigenvalue parameter $\rho = -i\sqrt{\lambda}$, $\Re(\rho) \geq 0$, where we also note for later use that $\frac{d\rho}{d\lambda} = -\frac{1}{2\rho}$. Then a fundamental matrix of the given differential equation for $\lambda \neq 0$ is

$$Y(x,\lambda) = \begin{pmatrix} 1 & x & \cos(\rho x) & \sin(\rho x) \\ 0 & 1 & -\rho\sin(\rho x) & \rho\cos(\rho x) \\ 0 & 0 & -\rho^2\cos(\rho x) & -\rho^2\sin(\rho x) \\ 0 & 0 & \rho^3\sin(\rho x) & -\rho^3\cos(\rho x) \end{pmatrix}.$$

The characteristic matrix is

$$M(\lambda) = \begin{pmatrix} 1 & 1 & \cos\rho & \sin\rho \\ 0 & 1 & 0 & \rho \\ 0 & 1 & -\rho\sin\rho & \rho\cos\rho \\ 0 & 0 & -\rho^2\cos\rho & -\rho^2\sin\rho \end{pmatrix},$$

and its determinant is

$$\det M(\lambda) = \begin{vmatrix} 1 & 0 & \rho \\ 1 & -\rho\sin\rho & \rho\cos\rho \\ 0 & -\rho^2\cos\rho & -\rho^2\sin\rho \end{vmatrix} = \begin{vmatrix} -\rho\sin\rho & \rho(\cos\rho - 1) \\ -\rho^2\cos\rho & -\rho^2\sin\rho \end{vmatrix}$$
$$= \rho^3(1 - \cos\rho).$$

Clearly, $M(0)$ is invertible if we take the fundamental system $\{1,x,x^2,x^3\}$ at 0. Hence the eigenvalues with respect to ρ are the numbers $2k\pi$, $k\in\mathbb{Z}\setminus\{0\}$, and the spectrum of the given problem consists of the numbers $\lambda_k=-4k^2\pi^2$, $k\in\mathbb{N}\setminus\{0\}$. Since $\rho\mapsto 1-\cos\rho$ has a double zero at these numbers, the total multiplicity of each eigenvectors is 2.

Now let us calculate the eigenvectors of $M(\lambda_k)$. We have

$$M(\lambda_k)=\begin{pmatrix}1&1&1&0\\0&1&0&2k\pi\\0&1&0&2k\pi\\0&0&-4k^2\pi^2&0\end{pmatrix},$$

which has rank 3. Hence the dimension of the eigenspace is 1, and there must be an associated vector.

So each eigenvalue λ_k of M has a root function c_k of order 1. To find it we have to solve

$$M(\lambda_k)c_k(\lambda_k)=0,\ M(\lambda_k)c_k'(\lambda_k)=-M'(\lambda_k)c_k(\lambda_k).$$

Obviously, we can choose

$$c_k(\lambda_k)=\begin{pmatrix}2k\pi\\-2k\pi\\0\\1\end{pmatrix}.$$

In view of Theorem 6.3.4 this means that an eigenfunction of the boundary eigenvalue problem at λ_k is given by

$$\eta_k(x)=\sin(2k\pi x)-2k\pi x+2k\pi.$$

With

$$M'(\lambda_k)=-\frac{1}{4k\pi}\begin{pmatrix}0&0&0&1\\0&0&0&1\\0&0&-2k\pi&1\\0&0&-4k\pi&-4k^2\pi^2\end{pmatrix}$$

we infer that $c_k'(\lambda_k)$ is found from

$$\begin{pmatrix}1&1&1&0\\0&1&0&2k\pi\\0&1&0&2k\pi\\0&0&-4k^2\pi^2&0\end{pmatrix}c_k'(\lambda_k)=\frac{1}{4k\pi}\begin{pmatrix}1\\1\\1\\-4k^2\pi^2\end{pmatrix}.$$

This shows that we can take

$$c_k'(\lambda_k)=\frac{1}{4k\pi}\begin{pmatrix}-1\\1\\1\\0\end{pmatrix}.$$

Since $e_1^\top Y c_k$ is a root function of the eigenvalue problem at λ_k, we have that

$$\eta_{k,1}(x) = e_1^\top \frac{\partial}{\partial\lambda} Y(x,\lambda_k)c_k(\lambda_k) + e_1^\top Y(x,\lambda_k)c_k'(\lambda_k)$$

is an associated function of the eigenvalue problem at λ_k. We calculate

$$\eta_{k,1}(x) = \left(0, 0, \tfrac{x}{4k\pi}\sin(2k\pi x), -\tfrac{x}{4k\pi}\cos(2k\pi x)\right)\begin{pmatrix} 2k\pi \\ -2k\pi \\ 0 \\ 1 \end{pmatrix}$$

$$+\frac{1}{4k\pi}(1, x, \cos(2k\pi x), \sin(2k\pi x))\begin{pmatrix} -1 \\ 1 \\ 1 \\ 0 \end{pmatrix}$$

$$= \frac{1-x}{4k\pi}\left(\cos(2k\pi x) - 1\right).$$

Now we shall find the root functions of M^* at λ_k. We have

$$M^*(\lambda_k) = \begin{pmatrix} 1 & 0 & 0 & 0 \\ 1 & 1 & 1 & 0 \\ 1 & 0 & 0 & -4k^2\pi^2 \\ 0 & 2k\pi & 2k\pi & 0 \end{pmatrix},$$

and an eigenvector is

$$\hat{d}_k = \begin{pmatrix} 0 \\ 1 \\ -1 \\ 0 \end{pmatrix}.$$

We have

$$M'^*(\lambda_k) = -\frac{1}{4k\pi}\begin{pmatrix} 0 & 0 & 0 & 0 \\ 0 & 0 & 0 & 0 \\ 0 & 0 & -2k\pi & -4k\pi \\ 1 & 1 & 1 & -4k^2\pi^2 \end{pmatrix},$$

and an associated vector $\hat{d}_{k,1}$ is given by

$$\begin{pmatrix} 1 & 0 & 0 & 0 \\ 1 & 1 & 1 & 0 \\ 1 & 0 & 0 & -4k^2\pi^2 \\ 0 & 2k\pi & 2k\pi & 0 \end{pmatrix}\hat{d}_{k,1} = \frac{1}{4k\pi}\begin{pmatrix} 0 & 0 & 0 & 0 \\ 0 & 0 & 0 & 0 \\ 0 & 0 & -2k\pi & -4k\pi \\ 1 & 1 & 1 & -4k^2\pi^2 \end{pmatrix}\begin{pmatrix} 0 \\ 1 \\ -1 \\ 0 \end{pmatrix}$$

$$= \begin{pmatrix} 0 \\ 0 \\ \frac{1}{2} \\ 0 \end{pmatrix}.$$

Hence we can choose

$$\hat{d}_{k,1} = \begin{pmatrix} 0 \\ 0 \\ 0 \\ -\frac{1}{8k^2\pi^2} \end{pmatrix}.$$

To establish biorthogonal root functions, we must find $\alpha \neq 0$ and β such that, with

$$d_k(\lambda) = (\alpha + \beta(\lambda - \lambda_k))\hat{d}_k + \alpha(\lambda - \lambda_k)\hat{d}_{k,1},$$

the relation

$$d_k^*(\lambda)M(\lambda)c_k(\lambda) = (\lambda - \lambda_k)^2\left(1 + O(\lambda - \lambda_k)^2\right)$$

holds, where

$$c_k(\lambda) = \begin{pmatrix} 2k\pi - \frac{1}{4k\pi}(\lambda - \lambda_k) \\ -2k\pi + \frac{1}{4k\pi}(\lambda - \lambda_k) \\ \frac{1}{4k\pi}(\lambda - \lambda_k) \\ 1 \end{pmatrix}.$$

A straightforward calculation yields

$$\begin{aligned} d_k^*(\lambda)M(\lambda)c_k(\lambda) &= (\alpha + \beta(\lambda - \lambda_k))\frac{1}{4k\pi}(\lambda - \lambda_k)\rho\sin\rho \\ &\quad + \frac{\alpha}{32k^3\pi^3}(\lambda - \lambda_k)^2\rho^2\cos\rho \\ &\quad + (\alpha + \beta(\lambda - \lambda_k))\rho(1 - \cos\rho) + \frac{\alpha}{8k^2\pi^2}(\lambda - \lambda_k)\rho^2\sin\rho. \end{aligned}$$

Using the Taylor expansions

$$\begin{aligned} \rho\sin\rho &= -\frac{1}{2}(\lambda - \lambda_k) + \frac{1}{32k^2\pi^2}(\lambda - \lambda_k)^2 + O((\lambda - \lambda_k)^3), \\ \rho^2\cos\rho &= 4k^2\pi^2 - (\lambda - \lambda_k) + O((\lambda - \lambda_k)^2), \\ \rho(1 - \cos\rho) &= \frac{1}{16k\pi}(\lambda - \lambda_k)^2 + O((\lambda - \lambda_k)^4), \\ \rho^2\sin\rho &= -k\pi(\lambda - \lambda_k) + \frac{3}{16k\pi}(\lambda - \lambda_k)^2 + O((\lambda - \lambda_k)^3) \end{aligned}$$

we infer

$$\begin{aligned} d_k^*(\lambda)M(\lambda)c_k(\lambda) &= -\frac{\alpha}{8k\pi}(\lambda - \lambda_k)^2 + \frac{\alpha}{128k^3\pi^3}(\lambda - \lambda_k)^3 - \frac{\beta}{8k\pi}(\lambda - \lambda_k)^3 \\ &+ \frac{\alpha}{8k\pi}(\lambda - \lambda_k)^2 - \frac{\alpha}{32k^3\pi^3}(\lambda - \lambda_k)^3 + \frac{\alpha}{16k\pi}(\lambda - \lambda_k)^2 \\ &+ \frac{\beta}{16k\pi}(\lambda - \lambda_k)^3 - \frac{\alpha}{8k\pi}(\lambda - \lambda_k)^2 + \frac{3\alpha}{128k^3\pi^3}(\lambda - \lambda_k)^3 + O((\lambda - \lambda_k)^4) \\ &= -\frac{\alpha}{16k\pi}(\lambda - \lambda_k)^2 - \frac{\beta}{16k\pi}(\lambda - \lambda_k)^3 + O((\lambda - \lambda_k)^4). \end{aligned}$$

This yields $\alpha = -16k\pi$ and $\beta = 0$, whence

$$d_k(\lambda) = \begin{pmatrix} 0 \\ -16k\pi \\ 16k\pi \\ -\frac{2}{k\pi}(\lambda - \lambda_k) \end{pmatrix}.$$

To find the eigenfunctions and associated functions of the adjoint problem we must calculate $L^R U_L$ according to Theorem 6.3.4. In the definition (6.3.3) of U_L we can replace the number a by any other element of the interval $[a,b]$. In our case, it is best to take this value to be 1, i. e.

$$(U_L(\lambda)f)(x) = e_1^\mathsf{T} Y(x,\lambda) \int_1^x Y(t,\lambda)^{-1} e_4 f(t)\,\mathrm{d}t.$$

Then

$$\begin{aligned}(U_L(\lambda)f)'(x) &= e_1^\mathsf{T} Y'(x,\lambda) \int_1^x Y(t,\lambda)^{-1} e_4 f(t)\,\mathrm{d}t \\ &= e_2^\mathsf{T} Y(x,\lambda) \int_1^x Y(t,\lambda)^{-1} e_4 f(t)\,\mathrm{d}t.\end{aligned}$$

Repeating the differentiation we infer for $j = 0, \ldots, 3$ that

$$(U_L(\lambda)f)^{(j)}(x) = e_{j+1}^\mathsf{T} Y(x,\lambda) \int_1^x Y(t,\lambda)^{-1} e_4 f(t)\,\mathrm{d}t.$$

Now all boundary terms at 1 are zero, and hence

$$L^R U_L(\lambda) f = -e_2^\mathsf{T} Y(0,\lambda) \int_0^1 Y(t,\lambda)^{-1} e_4 f(t)\,\mathrm{d}t \begin{pmatrix} 0 \\ 1 \\ 0 \\ 0 \end{pmatrix}.$$

We calculate

$$\begin{aligned}\langle (L^R U_L(\lambda))^* d_k(\lambda), f\rangle &= \langle d_k(\lambda), L^R U_L(\lambda) f\rangle \\ &= 16k\pi e_2^\mathsf{T} Y(0,\lambda) \int_0^1 Y(t,\lambda)^{-1} e_4 f(t)\,\mathrm{d}t\end{aligned}$$

whence

$$((L^R U_L(\lambda))^* d_k(\lambda))(t) = 16k\pi e_2^\mathsf{T} Y(0,\lambda) Y(t,\lambda)^{-1} e_4.$$

Since

$$e_2^\mathsf{T} Y(0,\lambda) = (0,1,0,\rho),$$

we must calculate $e_2^\mathsf{T} Y(t,\lambda)^{-1} e_4$ and $e_4^\mathsf{T} Y(t,\lambda)^{-1} e_4$. From

$$\det Y(x,\lambda) = \rho^5$$

and CRAMER'S rule we obtain

$$e_2^\mathsf{T} Y(x,\lambda)^{-1} e_4 = \rho^{-5} \begin{vmatrix} 1 & \cos(\rho x) & \sin(\rho x) \\ 0 & -\rho\sin(\rho x) & \rho\cos(\rho x) \\ 0 & -\rho^2\cos(\rho x) & -\rho^2\sin(\rho x) \end{vmatrix} = \rho^{-2}$$

and

$$e_4^\top Y(x,\lambda)^{-1}e_4=\rho^{-5}\begin{vmatrix}1 & x & \cos(\rho x)\\ 0 & 1 & -\rho\sin(\rho x)\\ 0 & 0 & -\rho^2\cos(\rho x)\end{vmatrix}=-\rho^{-3}\cos(\rho x).$$

So we have

$$((L^R U_L(\lambda))^* d_k)(t)=16k\pi\rho^{-2}(1-\cos(\rho t)).$$

From

$$\rho^{-2}(1-\cos(\rho t))=\frac{1}{4k^2\pi^2}(1-\cos(2k\pi t))$$
$$+\frac{1}{16k^4\pi^4}[1-\cos(2k\pi t)-k\pi t\sin(2k\pi t)](\lambda-\lambda_k)+O((\lambda-\lambda_k)^2)$$

we infer

$$((L^R U_L(\lambda))^* d_k)(t)=\frac{4}{k\pi}(1-\cos(2k\pi t))$$
$$+\frac{1}{k^3\pi^3}[1-\cos(2k\pi t)-k\pi t\sin(2k\pi t)](\lambda-\lambda_k)+O((\lambda-\lambda_k).$$

Altogether we obtain from Theorem 6.3.4 that

$$\left\{\sin(2k\pi x)+2k\pi(1-x),\frac{1-x}{4k\pi}(\cos(2k\pi x)-1)\right\},$$
$$\left\{\frac{4}{k\pi}(\cos(2k\pi x)-1),\frac{1}{k^3\pi^3}(\cos(2k\pi x)-1+k\pi x\sin(2k\pi x)\right\}$$

are biorthogonal canonical systems of eigenfunctions and associated functions of the given boundary eigenvalue problem and its formally adjoint at $-4k^2\pi^2$.

From Theorem 6.7.8 and the definition of Q_m we therefore obtain the expansion

$$f(x)=4\sum_{k=1}^{\infty}\left\{[\sin(2k\pi x)+2k\pi(1-x)]\int_0^1 t\sin(2k\pi t)f(t)\,dt\right.$$
$$\left.+(1-x)(\cos(2k\pi x)-1)\int_0^1\cos(2k\pi t)f(t)\,dt\right\},$$

where $f\in C^1[a,b]$ such that $f'\in BV[a,b]$ and $f(1)=0$, $f'(0)=0$, $f'(1)=0$, and the series converges in $C^1[a,b]$, see Theorem 8.8.2.

8.12. Notes

The results in this chapter are essentially due to TRETTER [TR2], [TR3]. Whereas TRETTER uses the classical asymptotic expansion of the fundamental system, we use first order systems and matrix representation as much as possible, which, in our opinion, shortens and clarifies proofs. Also, we have given special attention to the case $h_{n_0}\neq 1$. Although it is in principle easy to transform this problem to the case $h_{n_0} = 1$, this transformation is cumbersome and timeconsuming, so that we feel that it is very helpful to have formulas which allow us to check directly for Birkhoff or almost Birkhoff regularity.

Further results on boundary eigenvalue problems for the differential equation $\mathbf{K}\eta = \lambda \mathbf{H}\eta$ with λ-independent boundary conditions, in particular concerning completeness, minimality and basisness of the corresponding eigenfunctions and associated functions can be found in the publications [SHTR1], [SHTR2] of SHKALIKOV and TRETTER. Only very recently Tretter succeeded to extend these results to the case of λ-polynomial boundary conditions in [TR6], [TR10]. As in the case of first order systems, see the notes in Section 5.8, the proofs of the minimality and basisness properties are based on TRETTER'S new linearization method in [TR7], TRETTER'S theory of linear pencils $A - \lambda B$ in [TR9] and sharp asymptotic estimates of the Green's function as published in [MM5] and newly presented in Section 8.7 of this monograph.

KAMKE was the first who systematically studied boundary eigenvalue problems for the differential equation $\mathbf{K}\eta = \lambda \mathbf{H}\eta$ with λ-independent boundary conditions in the self-adjoint case, see the monograph [KK2]. In a series of papers EBERHARD [EB4], [EB5], [EB6] and EBERHARD and FREILING [EF2], [EF3] treated the non-self-adjoint case for the differential eqaution $\mathbf{K}\eta = \lambda \mathbf{H}\eta$ together with λ-polynomial boundary conditions. According to their expansion theorems, the class of the expandable functions is relatively small: apart from fulfilling the usual smoothness conditions these functions and their derivatives up to a certain order have to vanish at both endpoints of the underlying interval.

The Orr-Sommerfeld equation is an important special case of a differential equation of type $\mathbf{K}\eta = \lambda \mathbf{H}\eta$ and will be paid a particular consideration in Chapter X.

Chapter IX

n-TH ORDER DIFFERENTIAL EQUATIONS AND n-FOLD EXPANSIONS

In general, for n-th order differential equations there are too many eigenvectors and associated vectors in the sense that expansions are not unique. This non-uniqueness is due to the fact that the eigenvalue parameter occurs nonlinearly. In this chapter we study boundary eigenvalue problems for n-th order scalar differential equations together with two-point boundary conditions, both with λ-polynomial coefficients, which have been investigated by SHKALIKOV in [SH5]. Following his approach the corresponding operator function $L(\lambda) = (L^D(\lambda), L^R(\lambda))$ is linearized with respect to λ in a product of Sobolev spaces (Theorem 9.1.3). It is shown that an n-fold eigenfunction expansion holds if the original boundary eigenvalue problem is Birkhoff regular (Theorem 9.3.3).

9.1. Shkalikov's linearization

Let $n \in \mathbb{N}$, $n \geq 2$, and $1 \leq p \leq \infty$. Consider the differential operator

(9.1.1) $$L^D(\lambda) = \sum_{i=0}^{n} \lambda^i \mathscr{H}_i,$$

where $\lambda \in \mathbb{C}$,

$$\mathscr{H}_i \eta = \begin{cases} \eta^{(n)} + \sum\limits_{k=0}^{n-1} \pi_{n-k,0} \eta^{(k)} & \text{if } i = 0, \\ \sum\limits_{k=0}^{n-i} \pi_{n-k,i} \eta^{(k)} & \text{if } i = 1, \dots, n, \end{cases}$$

all the coefficient functions $\pi_{n-k,i}$ belong to $L_p(a,b)$, $\pi_{i,i}$ for $i = 1, \dots, n$ and $\pi_{n,n}^{-1}$ belong to $L_\infty(a,b)$, and $\eta \in W_p^n(a,b)$, see (7.6.1). With this differential equation, λ-polynomial two-point boundary conditions are associated:

(9.1.2) $$L^R(\lambda)\eta = \sum_{\nu=0}^{m} \lambda^\nu U^\nu \eta = 0.$$

Then

(9.1.3) $$L(\lambda) = (L^D(\lambda), L^R(\lambda)) : W_p^n(a,b) \to L_p(a,b) \times \mathbb{C}^n$$

is a continuous operator and depends holomorphically on λ. We write

$$U^{\nu}\eta = \sum_{j=0}^{n-1} U^{\nu,j}\eta^{(j)},$$

where

$$e_k^{\mathsf{T}} U^{\nu,j}\eta^{(j)} = \alpha_{\nu,k,j}\eta^{(j)}(a) + \beta_{\nu,k,j}\eta^{(j)}(b) \quad (k = 1,\ldots,n)$$

with complex numbers $\alpha_{\nu,k,j}$ and $\beta_{\nu,k,j}$.

By the substitution

(9.1.4) $$y_1 := \eta,\ y_2 := \lambda y_1 = \lambda\eta,\ \ldots,\ y_n := \lambda y_{n-1} = \lambda^{n-1}\eta,$$

we obtain the identity

$$L^D(\lambda)\eta = \lambda\pi_{n,n}y_n + \sum_{i=0}^{n-1}\mathscr{H}_i y_{i+1},$$

whence the equation $L^D(\lambda)\eta = 0$ becomes equivalent to

$$H^D y = \lambda y,$$

where

$$H^D y = \begin{pmatrix} y_2 \\ \vdots \\ y_n \\ -\pi_{n,n}^{-1}\sum_{i=0}^{n-1}\mathscr{H}_i y_{i+1} \end{pmatrix},\quad y = \begin{pmatrix} y_1 \\ \vdots \\ y_n \end{pmatrix}.$$

To make H^D a proper operator, we have to define its domain and range spaces. To this end let $r,k \in \mathbb{N}$ and set

(9.1.5) $$\mathscr{W}_p^{r,k}(a,b) := W_p^{r+k-1}(a,b) \oplus W_p^{r+k-2}(a,b) \oplus \cdots \oplus W_p^{r}(a,b).$$

Then $H^D : \mathscr{W}_p^{1,n}(a,b) \to \mathscr{W}_p^{0,n}(a,b)$ is a continuous operator and can be written in the matrix form

(9.1.6) $$H^D = \begin{pmatrix} 0 & 1 & & \\ & & \ddots & \\ & & & 1 \\ -\pi_{n,n}^{-1}\mathscr{H}_0 & \cdot & \cdot & -\pi_{n,n}^{-1}\mathscr{H}_{n-1} \end{pmatrix}.$$

The substitution also yields

$$\begin{cases} \lambda^{\nu}U^{\nu,j}\eta^{(j)} = U^{\nu,j}y_{\nu+1}^{(j)} & \text{if } j+\nu < n, \\ \lambda^{\nu}U^{\nu,j}\eta^{(j)} = \lambda^{\nu-n+j+1}U^{\nu,j}y_{n-j}^{(j)} & \text{if } j+\nu \geq n, \end{cases}$$

whence

$$L^R(\lambda)\eta = \sum_{j=0}^{m} \lambda^j U_j \tilde{y}, \tag{9.1.7}$$

where

$$\begin{aligned} U_0 &= (U^{0,0} \dots U^{0,n-1}, U^{1,0} \dots U^{1,n-2}, \dots, U^{n-1,0}),\\ U_j &= (0 \dots 0 U^{j,n-1}, 0 \dots 0 U^{j+1,n-2}, 0 \dots 0, U^{j+n-1,0}) \end{aligned} \tag{9.1.8}$$

for $j = 1, \dots, m$, $U^{k,r} = 0$ if $k > m$,

$$\begin{aligned} \tilde{y}^{\mathsf{T}} &= (\tilde{y}_1^{\mathsf{T}}, \dots, \tilde{y}_n^{\mathsf{T}}),\\ \tilde{y}_j^{\mathsf{T}} &= (y_j, \dots, y_j^{(n-j)}) \quad (j = 1, \dots, n). \end{aligned}$$

Here, in U_j for $j > 0$, $U^{j+k,n-1-k}$ is applied to the function $y_{k+1}^{(n-1-k)}$.

Define the boundary operator

$$\widetilde{H}^R(\lambda) : \mathscr{W}_p^{1,n}(a,b) \to \mathbb{C}^n$$

by

$$\widetilde{H}^R(\lambda)y := \sum_{j=0}^{m} \lambda^j U_j \tilde{y}, \tag{9.1.9}$$

and the operator

$$\widetilde{H}(\lambda) : \mathscr{W}_p^{1,n}(a,b) \to \mathscr{W}_p^{0,n}(a,b) \oplus \mathbb{C}^n$$

by

$$\widetilde{H}(\lambda) = \begin{pmatrix} H^D - \lambda \\ \widetilde{H}^R(\lambda) \end{pmatrix}. \tag{9.1.10}$$

PROPOSITION 9.1.1. *The operator function $\widetilde{H}$ is globally equivalent to a canonical extension of L.*

Proof. Define the multiplication operator $\widetilde{D}(\lambda) : \mathscr{W}_p^{1,n}(a,b) \to \mathscr{W}_p^{1,n}(a,b)$ by

$$\widetilde{D}(\lambda) := \begin{pmatrix} 1 & & & \\ \lambda & 1 & & 0 \\ \vdots & & \ddots & \\ \lambda^{n-1} & 0 & & 1 \end{pmatrix}. \tag{9.1.11}$$

Obviously, $\widetilde{D}$ is holomorphic, and $\widetilde{D}(\lambda)$ is invertible for each $\lambda \in \mathbb{C}$. Define $\widetilde{C}(\lambda) : \mathscr{W}_p^{0,n}(a,b) \oplus \mathbb{C}^n \to (L_p(a,b) \oplus \mathscr{W}_p^{1,n-1}(a,b)) \oplus \mathbb{C}^n$ by

$$\widetilde{C}(\lambda) := \begin{pmatrix} \widetilde{C}_{11}(\lambda) & 0 \\ \widetilde{C}_{21}(\lambda) & I_n \end{pmatrix},$$

where

$$\widetilde{C}_{11}(\lambda) := \begin{pmatrix} \mathscr{H}_{1,n-1}(\lambda)B(\lambda)^{-1} & -\pi_{n,n} \\ -B(\lambda)^{-1} & 0 \end{pmatrix} \tag{9.1.12}$$

with the invertible $(n-1)\times(n-1)$ matrix

$$B(\lambda) := \begin{pmatrix} -1 & & & \\ \lambda & -1 & & 0 \\ & \cdot & \cdot & \\ 0 & & \lambda & -1 \end{pmatrix}, \tag{9.1.13}$$

$$\mathscr{H}_{1,n-1}(\lambda) := (\mathscr{H}_1, \dots, \mathscr{H}_{n-2}, \mathscr{H}_{n-1} + \lambda\pi_{n,n}), \tag{9.1.14}$$

and

$$\widetilde{C}_{21}(\lambda) := \widetilde{H}^R(\lambda)\begin{pmatrix} 0 & 0 \\ B(\lambda)^{-1} & 0 \end{pmatrix}. \tag{9.1.15}$$

It is easy to see that $\widetilde{C}$ depends holomorphically on λ and that $\widetilde{C}(\lambda)$ is invertible for each $\lambda \in \mathbb{C}$. With

$$v(\lambda) = \begin{pmatrix} \lambda \\ \vdots \\ \lambda^{n-1} \end{pmatrix}$$

we have

$$\begin{pmatrix} \lambda \\ 0 \\ \vdots \\ 0 \end{pmatrix} = -B(\lambda)v(\lambda),$$

and therefore

$$\begin{aligned} (H^D - \lambda)\widetilde{D}(\lambda) - \begin{pmatrix} -B(\lambda)v(\lambda) & B(\lambda) \\ \pi_{n,n}^{-1}\mathscr{H}_0 & \pi_{n,n}^{-1}\mathscr{H}_{1,n-1} \end{pmatrix} \begin{pmatrix} 1 & 0 \\ v(\lambda) & I_{n-1} \end{pmatrix} \\ = -\begin{pmatrix} 0 & B(\lambda) \\ \pi_{n,n}^{-1}L^D(\lambda) & \pi_{n,n}^{-1}\mathscr{H}_{1,n-1}(\lambda) \end{pmatrix}, \end{aligned}$$

whence

$$\widetilde{C}_{11}(\lambda)(H^D - \lambda)\widetilde{D}(\lambda) = \begin{pmatrix} L^D(\lambda) & 0 \\ 0 & \mathrm{id}_{\mathscr{W}_p^{1,n-1}(a,b)} \end{pmatrix}.$$

Since

$$\widetilde{C}_{21}(\lambda) = -\widetilde{H}^R(\lambda)\begin{pmatrix} 0 & 0 \\ 0 & I_{n-1} \end{pmatrix}\widetilde{C}_{11}(\lambda),$$

we infer that

$$\widetilde{C}_{21}(\lambda)(H^D - \lambda)\widetilde{D}(\lambda) = -\widetilde{H}^R(\lambda)\begin{pmatrix} 0 & 0 \\ 0 & \mathrm{id}_{\mathscr{W}_p^{1,n-1}(a,b)} \end{pmatrix}.$$

Therefore, in view of (9.1.4), (9.1.7), and (9.1.9),

$$(\widetilde{C}_{21}(\lambda)(H^D-\lambda)\widetilde{D}(\lambda)+\widetilde{H}^R(\lambda)\widetilde{D}(\lambda))y=\widetilde{H}^R(\lambda)\begin{pmatrix}1\\ \lambda\\ \vdots\\ \lambda^{n-1}\end{pmatrix}y_1=L^R(\lambda)y_1,$$

and it follows that

$$\widetilde{C}(\lambda)\widetilde{H}(\lambda)\widetilde{D}(\lambda)=\begin{pmatrix}L^D(\lambda) & 0\\ 0 & \mathrm{id}_{\mathscr{W}_p^{1,n-1}(a,b)}\\ L^R(\lambda) & 0\end{pmatrix}. \qquad \square \tag{9.1.16}$$

If $\widetilde{H}^R(\lambda)$ is linear in λ, then the above proposition yields a linearization of L. Otherwise we still have to linearize the boundary operator. In [LM] it was shown that a canonical extension of $\widetilde{H}^R(\lambda)$ is equivalent on $\mathbb{C}\setminus\{0\}$ to a colligation (linear system). Since we want equivalence on all of $\mathbb{C}$, we have to modify the construction. Essentially, we have to introduce an additional permutation, which makes the result slightly more complicated than that in [LM].

For $i=1,\dots,n$ let ν_i be the degree of the operator polynomial $e_i^{\mathsf{T}}\sum\limits_{j=0}^{m}\lambda^j U_j$. Let

$$N:=\sum_{i=1}^{n}\nu_i. \tag{9.1.17}$$

Without loss of generality we may assume that $\nu_1\ge\nu_2\ge\dots\ge\nu_n$. Let $q:=0$ if $\nu_1=0$ and $q:=\max\{j:\nu_j>0\}$ otherwise. Define

$$H^{R,0}:\mathscr{W}_p^{1,n}(a,b)\to\mathbb{C}^n$$

by

$$H^{R,0}y:=\begin{pmatrix}e_1^{\mathsf{T}}U_{\nu_1}\\ \vdots\\ e_n^{\mathsf{T}}U_{\nu_n}\end{pmatrix}\tilde{y},$$

and

$$H^{R,1}:\mathscr{W}_p^{1,n}(a,b)\to\mathbb{C}^N$$

by

$$H^{R,1}y = \begin{pmatrix} e_1^{\mathsf{T}} U_{\nu_1-1} \\ \vdots \\ e_1^{\mathsf{T}} U_0 \\ e_2^{\mathsf{T}} U_{\nu_2-1} \\ \vdots \\ e_2^{\mathsf{T}} U_0 \\ \vdots \\ e_q^{\mathsf{T}} U_{\nu_q-1} \\ \vdots \\ e_q^{\mathsf{T}} U_0 \end{pmatrix} \tilde{y}.$$

Let $A = \operatorname{diag}(A^{(1)}, \ldots, A^{(q)})$, where $A^{(j)}$ is the $\nu_j \times \nu_j$ matrix

$$A^{(j)} = (\delta_{i+1,k})_{i,k=1}^{\nu_j} = \begin{pmatrix} 0 & 1 & 0 & \\ & & \ddots & \\ & & & 1 \\ 0 & . & . & 0 \end{pmatrix}.$$

For $j = 1, \ldots, q+1$ let $k_j := \sum_{l=1}^{j-1} \nu_l$. Let G be the $n \times N$ matrix whose entries in the j-th row and the $k_j + 1$-th column are 1 for $j = 1, \ldots, q$ and whose other entries are zero. Define

$$H^R : \mathscr{W}_p^{1,n}(a,b) \oplus \mathbb{C}^N \to \mathbb{C}^n \oplus \mathbb{C}^N$$

by

$$H^R = \begin{pmatrix} H^{R,0} & G \\ H^{R,1} & A \end{pmatrix}$$

and set

$$\widehat{H}^R(\lambda) = \begin{pmatrix} H^{R,0} & G \\ H^{R,1} & A - \lambda I_N \end{pmatrix}.$$

Define an $n+N$ dimensional permutation matrix J which permutes the rows of $\widehat{H}^R(\lambda)$ as follows: For $j = 1, \ldots, q$, the j-th row becomes the $n + k_j + 1$-th row. The rows numbered $n + k_j + 1$ to $n + k_j + \nu_j - 1$ are shifted one row down, and the $n + k_j + \nu_j$-th row becomes the j-th row. Then we have

$$\text{(9.1.18)} \qquad J\widehat{H}^R(\lambda) = \begin{pmatrix} \widehat{H}^{R,0} & -\lambda \widehat{G} \\ \widehat{H}^{R,1} & I_N - \lambda A^{\mathsf{T}} \end{pmatrix},$$

where

$$\widehat{H}^{R,0}y = U_0\tilde{y},$$

$$\widehat{H}^{R,1}y = \begin{pmatrix} e_1^\intercal U_{\nu_1} \\ \vdots \\ e_1^\intercal U_1 \\ e_2^\intercal U_{\nu_2} \\ \vdots \\ e_2^\intercal U_1 \\ \vdots \\ e_q^\intercal U_{\nu_q} \\ \vdots \\ e_q^\intercal U_1 \end{pmatrix} \tilde{y},$$

and $\widehat{G}$ is the $n \times N$ matrix whose entries in the j-th row and k_{j+1}-th column are 1 for $j = 1, \dots, q$ and whose other entries are zero. Since $I_N - \lambda A^\intercal$ is invertible for all $\lambda \in \mathbb{C}$, we have

(9.1.19)
$$\begin{pmatrix} \widehat{H}^{R,0} + \lambda \widehat{G}(I_N - \lambda A^\intercal)^{-1}\widehat{H}^{R,1} & 0 \\ 0 & I_N \end{pmatrix} = \begin{pmatrix} I_n & \lambda \widehat{G}(I_N - \lambda A^\intercal)^{-1} \\ 0 & I_N \end{pmatrix} \times$$
$$\times \begin{pmatrix} \widehat{H}^{R,0} & -\lambda \widehat{G} \\ \widehat{H}^{R,1} & I_N - \lambda A^\intercal \end{pmatrix} \begin{pmatrix} I_{\mathscr{W}_p^{1,n}(a,b)} & 0 \\ -(I_N - \lambda A^\intercal)^{-1}\widehat{H}^{R,1} & (I_N - \lambda A^\intercal)^{-1} \end{pmatrix}.$$

PROPOSITION 9.1.2. *The $\mathbb{C}^N$-extension of $\widetilde{H}^R$ is globally equivalent to $\widehat{H}^R$.*

Proof. In view of the preceding observations it it sufficient to show that

(9.1.20)
$$\widehat{\widehat{H}}^R(\lambda) := \widehat{H}^{R,0} + \lambda \widehat{G}(I_N - \lambda A^\intercal)^{-1}\widehat{H}^{R,1} = \widetilde{H}^R(\lambda).$$

For $j > q$ we have $e_j^\intercal \widehat{G} = 0$ whence

$$e_j^\intercal \widehat{\widehat{H}}^R(\lambda) = e_j^\intercal \widehat{H}^{R,0} = e_j^\intercal \widetilde{H}^R(\lambda).$$

For $j \le q$ we have

$$e_j^\intercal \widehat{\widehat{H}}^R(\lambda)y = e_j^\intercal U_0\tilde{y} + \lambda e_{\nu_j}^\intercal \sum_{k=0}^{\nu_j - 1} \lambda^k \left(A^{(j)\intercal}\right)^k \begin{pmatrix} e_j^\intercal U_{\nu_j} \\ \vdots \\ e_j^\intercal U_1 \end{pmatrix} \tilde{y}$$
$$= e_j^\intercal U_0\tilde{y} + \sum_{k=0}^{\nu_j - 1} \lambda^{k+1} e_j^\intercal U_{k+1}\tilde{y} = e_j^\intercal \widetilde{H}^R(\lambda)y,$$

where the last identity follows from (9.1.7) and (9.1.9). □

We define

$$\mathscr{H} : \mathscr{W}_p^{1,n}(a,b) \oplus \mathbb{C}^N \to \mathscr{W}_p^{0,n}(a,b) \oplus \mathbb{C}^n \oplus \mathbb{C}^N$$

by the block operator representation

$$\mathscr{H} = \begin{pmatrix} H^D & 0 \\ H^{R,0} & G \\ H^{R,1} & A \end{pmatrix}. \tag{9.1.21}$$

Let $\widetilde{I} : \mathscr{W}_p^{1,n}(a,b) \oplus \mathbb{C}^N \to \mathscr{W}_p^{0,n}(a,b) \oplus \mathbb{C}^n \oplus \mathbb{C}^N$ be the canonical embedding. From Propositions 9.1.1 and 9.1.2 we immediately infer

THEOREM 9.1.3. *The canonical $\mathscr{W}_p^{1,n-1}(a,b) \oplus \mathbb{C}^N$ extension of $L(\lambda)$ is globally equivalent on $\mathbb{C}$ to $\mathscr{H} - \lambda\widetilde{I}$.*

Propositions 9.1.1 and 9.1.2 also give this equivalence explicitly. But the permutation matrix J makes the representation cumbersome. Therefore, we consider instead

$$\widehat{\mathscr{H}}(\lambda) := \begin{pmatrix} \mathrm{id}_{\mathscr{W}_p^{0,n}(a,b)} & 0 \\ 0 & J \end{pmatrix} (\mathscr{H} - \lambda\widetilde{I}). \tag{9.1.22}$$

COROLLARY 9.1.4. *Let*

$$C(\lambda) : \mathscr{W}_p^{0,n}(a,b) \oplus \mathbb{C}^n \oplus \mathbb{C}^N \to (L_p(a,b) \oplus \mathscr{W}_p^{1,n-1}(a,b)) \oplus \mathbb{C}^n \oplus \mathbb{C}^N$$

be given by

$$C(\lambda) := \begin{pmatrix} \widetilde{C}_{11}(\lambda) & 0 & 0 \\ \widetilde{C}_{21}(\lambda) & I_n & \lambda\widehat{G}(I_N - \lambda A^{\mathsf{T}})^{-1} \\ 0 & 0 & I_N \end{pmatrix}$$

and $D(\lambda) : \mathscr{W}_p^{1,n}(a,b) \oplus \mathbb{C}^N \to \mathscr{W}_p^{1,n}(a,b) \oplus \mathbb{C}^N$ be given by

$$D(\lambda) := \begin{pmatrix} \widetilde{D}(\lambda) & 0 \\ -(I_N - \lambda A^{\mathsf{T}})^{-1}\widehat{H}^{R,1}\widetilde{D}(\lambda) & (I_N - \lambda A^{\mathsf{T}})^{-1} \end{pmatrix},$$

where $\widetilde{D}(\lambda)$, $\widetilde{C}_{11}(\lambda)$ and $\widetilde{C}_{21}(\lambda)$ are given by (9.1.11), (9.1.12) *and* (9.1.15), *respectively. Then*

$$\begin{pmatrix} L^D(\lambda) & 0 & 0 \\ 0 & \mathrm{id}_{\mathscr{W}_p^{1,n-1}(a,b)} & 0 \\ L^R(\lambda) & 0 & 0 \\ 0 & 0 & I_N \end{pmatrix} = C(\lambda)\widehat{\mathscr{H}}(\lambda)D(\lambda).$$

Proof. In view of Propositions 9.1.1 and 9.1.2 and the representations (9.1.16), (9.1.19), and (9.1.20) this follows from

$$C(\lambda) = \begin{pmatrix} \widetilde{C}_{11}(\lambda) & 0 & 0 \\ \widetilde{C}_{21}(\lambda) & I_n & 0 \\ 0 & 0 & I_N \end{pmatrix} \begin{pmatrix} \mathrm{id}_{\mathscr{W}_p^{0,n}(a,b)} & 0 & 0 \\ 0 & I_n & \lambda\widehat{G}(I_N - \lambda A^{\mathsf{T}})^{-1} \\ 0 & 0 & I_N \end{pmatrix}$$

and

$$D(\lambda) = \begin{pmatrix} \mathrm{id}_{\mathscr{W}_p^{1,n}(a,b)} & 0 \\ -(I_N - \lambda A^{\mathsf{T}})^{-1}\widehat{H}^{R,1} & (I_N - \lambda A^{\mathsf{T}})^{-1} \end{pmatrix} \begin{pmatrix} \widetilde{D}(\lambda) & 0 \\ 0 & I_N \end{pmatrix}. \qquad \square$$

The linearization $\mathscr{H} - \lambda\widetilde{I}$ essentially coincides with the linearization obtained in [SH5, p. 1321]; Shkalikov considers a restriction of this operator to a domain of finite codimension, see Section 9.3.

9.2. A first convergence result

In this section let $1 < p < \infty$. Furthermore, we suppose that the assumptions of Theorem 7.2.4 are satisfied, that $\pi_{1,0} \in L_{p'}(a,b)$, $1/p + 1/p' = 1$, and that one of the conditions II), III), IV) or V) considered in Proposition 7.2.5 holds. We set

$$\widetilde{C}_{21}^{0}(\lambda) = \begin{pmatrix} 0 \\ B(\lambda)^{-1} \end{pmatrix} \quad \text{and} \quad H_j^{R,1}y = \begin{pmatrix} e_j^{\mathsf{T}}U_{\nu_j-1} \\ \vdots \\ e_j^{\mathsf{T}}U_0 \end{pmatrix}\tilde{y} \quad (j = 1,\dots,q).$$

PROPOSITION 9.2.1. *Suppose that the problem* (9.1.1), (9.1.2), *i. e.* $L(\lambda)\eta = 0$, *is Birkhoff regular. Let* $\begin{pmatrix} f \\ c \end{pmatrix} \in (W_p^n(a,b))^n \oplus \mathbb{C}^N$. *We write*

$$c = \begin{pmatrix} c_{11} \cdots c_{1\nu_1} \cdots c_{q1} \cdots c_{q\nu_q} \end{pmatrix}^{\mathsf{T}}.$$

Suppose that

$$C_2(\lambda)^{-1}\Bigg(\sum_{k=1}^{n}\sum_{j=0}^{k-1}\lambda^{j-k}\Bigg\{\sum_{l=0}^{n-1-j} U^{j,l}(e_k^{\mathsf{T}}f)^{(l)} + \sum_{l=1}^{m}\lambda^l U^{l+j,n-1-j}(e_k^{\mathsf{T}}f)^{(n-1-j)}\Bigg\} + \sum_{j=1}^{q}\sum_{r=1}^{\nu_j}\lambda^{\nu_j-r}e_jc_{jr}\Bigg) = O(\lambda^{-n}). \tag{9.2.1}$$

Then there is a sequence of positive numbers $(\rho_\nu)_{\nu\in\mathbb{N}}$ *with* $\rho_\nu \to 0$ *as* $\nu \to \infty$ *such that, for* $\Gamma_\nu = \{\lambda : |\lambda| = \rho_\nu\}$,

$$-\frac{1}{2\pi i}\oint_{\Gamma_\nu}(\mathscr{H} - \lambda\widetilde{I})^{-1}\widetilde{I}\begin{pmatrix} f \\ c \end{pmatrix}\,\mathrm{d}\lambda \to \begin{pmatrix} f \\ c \end{pmatrix} \tag{9.2.2}$$

in $\mathscr{W}_p^{0,n}(a,b) \oplus \mathbb{C}^N$ *as* $\nu \to \infty$.

Proof. From the definition of $\mathscr{H}$ in (9.1.21) it follows that

$$\mathscr{H} - \lambda\widetilde{I} =: \begin{pmatrix} \widetilde{\mathscr{H}^0}(\lambda) & \widetilde{G} \\ H^{R,1} & A - \lambda I_N \end{pmatrix}$$

with

$$\widetilde{\mathscr{H}^0}(\lambda) = \begin{pmatrix} H^D - \lambda\widehat{I} \\ H^{R,0} \end{pmatrix}, \quad \widetilde{G} = \begin{pmatrix} 0 \\ G \end{pmatrix},$$

where $\widehat{I} : \mathscr{W}_p^{1,n}(a,b) \to \mathscr{W}_p^{0,n}(a,b)$ is the canonical embedding. In an analogous manner to formula (1.3.4) one can see that the inverse operator

$$\widetilde{\mathscr{H}_0^0}(\lambda) := (\widetilde{\mathscr{H}^0}(\lambda) - \widetilde{G}(A - \lambda I_N)^{-1} H^{R,1})^{-1}$$

exists for $\lambda \in \rho(A) \cap \rho(\mathscr{H})$. With the notation

$$\widetilde{\mathscr{H}_1^0}(\lambda) := \begin{pmatrix} \widetilde{\mathscr{H}_0^0}(\lambda) & -\widetilde{\mathscr{H}_0^0}(\lambda)\widetilde{G}(A - \lambda I_N)^{-1} \end{pmatrix} \tag{9.2.3}$$

we obtain that

$$(\mathscr{H} - \lambda\widetilde{I})^{-1} = \begin{pmatrix} \widetilde{\mathscr{H}_1^0}(\lambda) \\ -(A - \lambda I_N)^{-1} H^{R,1} \widetilde{\mathscr{H}_1^0}(\lambda) \end{pmatrix} + \begin{pmatrix} 0 & 0 \\ 0 & (A - \lambda I_N)^{-1} \end{pmatrix} \tag{9.2.4}$$

for $\lambda \in \rho(A) \cap \rho(\mathscr{H})$. With

$$\widetilde{\mathscr{H}_2^0}(\lambda) := \widetilde{\mathscr{H}_1^0}(\lambda) + \begin{pmatrix} \widetilde{C}_{21}^0(\lambda) & 0 & 0 & 0 \end{pmatrix}$$

we first shall prove

$$-\frac{1}{2\pi i} \oint_{\Gamma_\nu} \widetilde{\mathscr{H}_2^0}(\lambda)\widetilde{I} \begin{pmatrix} f \\ c \end{pmatrix} \, \mathrm{d}\lambda \to f \tag{9.2.5}$$

in $\mathscr{W}_p^{0,n}(a,b)$ as $\nu \to \infty$.

We shall use $\widehat{\mathscr{H}}(\lambda)^{-1}$ to find a representation of $\widetilde{\mathscr{H}_2^0}(\lambda)$. For this we write $L^{-1} = (K_1, K_2)$, see Section 6.4. Then

$$\begin{pmatrix} L^D(\lambda) & 0 & 0 \\ 0 & \mathrm{id}_{\mathscr{W}_p^{1,n-1}(a,b)} & 0 \\ L^R(\lambda) & 0 & 0 \\ 0 & 0 & I_N \end{pmatrix}^{-1} = \begin{pmatrix} K_1(\lambda) & 0 & K_2(\lambda) & 0 \\ 0 & \mathrm{id}_{\mathscr{W}_p^{1,n-1}(a,b)} & 0 & 0 \\ 0 & 0 & 0 & I_N \end{pmatrix}.$$

With the aid of (9.2.4), (9.1.22) and Corollary 9.1.4, a straightforward calculation shows that

$$\widetilde{\mathscr{H}_2^0}(\lambda) \begin{pmatrix} \mathrm{id}_{\mathscr{W}_p^{0,n}} & 0 \\ 0 & J^{-1} \end{pmatrix} = \begin{pmatrix} 1 \\ \lambda \\ \vdots \\ \lambda^{n-1} \end{pmatrix} \begin{pmatrix} K_1(\lambda) & K_2(\lambda) \end{pmatrix} \times \tag{9.2.6}$$

$$\times \begin{pmatrix} \mathscr{H}_{1,n-1}(\lambda)B(\lambda)^{-1} & -\pi_{n,n} & 0 & 0 \\ \widetilde{H}^R(\lambda)\widetilde{C}_{21}^0(\lambda) & 0 & I_n & \lambda\widehat{G}(I_N - \lambda A^\mathsf{T})^{-1} \end{pmatrix}.$$

Writing

$$\widetilde{\mathscr{H}_2^0}(\lambda) = \begin{pmatrix} \widetilde{\mathscr{H}_{21}^0}(\lambda) & \widetilde{\mathscr{H}_{22}^0}(\lambda) \end{pmatrix} \tag{9.2.7}$$

with $\widetilde{\mathscr{H}_{21}^0}(\lambda) : \mathscr{W}_p^{0,n}(a,b) \to \mathscr{W}_p^{1,n}(a,b)$ and $\widetilde{\mathscr{H}_{22}^0}(\lambda) : \mathbb{C}^{n+N} \to \mathscr{W}_p^{1,n}(a,b)$ gives

$$\widetilde{\mathscr{H}_{21}^0}(\lambda) = \begin{pmatrix} 1 \\ \lambda \\ \vdots \\ \lambda^{n-1} \end{pmatrix} \begin{pmatrix} K_1(\lambda) & K_2(\lambda) \end{pmatrix} \begin{pmatrix} \mathscr{H}_{1,n-1}(\lambda)B(\lambda)^{-1} & -\pi_{n,n} \\ \widetilde{H}^R(\lambda)\widetilde{C}_{21}^0(\lambda) & 0 \end{pmatrix}, \tag{9.2.8}$$

$$\widetilde{\mathscr{H}_{22}^0}(\lambda) = \begin{pmatrix} 1 \\ \lambda \\ \vdots \\ \lambda^{n-1} \end{pmatrix} K_2(\lambda) \begin{pmatrix} I_n & \lambda\widehat{G}(I_N - \lambda A^{\mathsf{T}})^{-1} \end{pmatrix} J.$$

In view of

$$J = \sum_{j=1}^{q} e_{n+k_j+1} e_j^{\mathsf{T}} + \sum_{j=q+1}^{n} e_j e_j^{\mathsf{T}} + \sum_{j=1}^{q} \sum_{k=n+k_j+1}^{n+k_{j+1}-1} e_{k+1} e_k^{\mathsf{T}} + \sum_{j=1}^{q} e_j e_{n+k_{j+1}}^{\mathsf{T}},$$

a straightforward calculation yields

$$\widetilde{\mathscr{H}_{22}^0}(\lambda) = \begin{pmatrix} 1 \\ \lambda \\ \vdots \\ \lambda^{n-1} \end{pmatrix} K_2(\lambda) \left\{ \sum_{j=1}^{n} \lambda^{\nu_j} e_j e_j^{\mathsf{T}} + \sum_{j=1}^{q} \sum_{r=1}^{\nu_j} \lambda^{\nu_j - r} e_j e_{n+k_j+r}^{\mathsf{T}} \right\}. \tag{9.2.9}$$

Observe that we obtain similar to (7.4.4) that

$$(K_1(\lambda)h)^{(\mu)} = \lambda^{\mu-n+1} e_{\mu+1}^{\mathsf{T}} C_1 \widetilde{T}^{-1}(\lambda)(C_1^{-1} e_n h, 0) \quad (\mu = 0, \dots, n-1)$$

for $h \in L_p(a,b)$, where $\widetilde{T}$ is given by (7.4.3) and C_1 is as in Theorem 7.2.4. Here $n_0 = 0$, and in both cases (7.2.17) and (7.2.18) we have $C_1 = (r_k^{i-1})_{i,k=1}^n$, where $r_1, \dots, r_n$ are the zeros of the characteristic function (7.1.4). Since we assume that the problem (9.1.1), (9.1.2) is Birkhoff regular, the problem $\widetilde{T}(\lambda)y = 0$ is Birkhoff regular, see Definition 7.3.1. Let ρ_ν be the radii as assigned for these Birkhoff regular problems. Then

$$\oint_{\Gamma_\nu} (K_1(\lambda)h(\lambda))^{(\mu)} \, \mathrm{d}\lambda \to 0 \quad \text{in } L_p(a,b) \text{ as } \nu \to \infty \tag{9.2.10}$$

follows for $\mu = 0, \dots, n-1$ and $h(\lambda) = O(\lambda^{n-2-\mu})$ in $L_p(a,b)$ as $\lambda \to \infty$, see Theorem 4.4.11 ii). And for $h \in L_p(a,b)$ and $\mu = 0, \dots, n-1$ we infer in view of (7.4.5) and Theorem 4.6.9 that

$$\frac{1}{2\pi i} \oint_{\Gamma_\nu} \lambda^{n-1-\mu} (K_1(\lambda)h)^{(\mu)} \, \mathrm{d}\lambda = e_{\mu+1}^{\mathsf{T}} C_1 P_\nu C_1^{-1} e_1 \pi_{n,n}^{-1} h \to \delta_{\mu,0} \pi_{n,n}^{-1} h \tag{9.2.11}$$

in $L_p(a,b)$ as $\nu \to \infty$. From (7.4.2) and (7.4.3) it follows that

$$(K_2(\lambda)c)^{(\mu)} = \lambda^{\mu} e_{\mu+1}^{\mathsf{T}} C_1 \widetilde{T}(\lambda)^{-1}(0, C_2(\lambda)^{-1}c),$$

and Theorem 4.4.11 ii) yields that

$$\oint_{\Gamma_\nu} (K_2(\lambda)c(\lambda))^{(\mu)}\, d\lambda \to 0 \quad \text{in } L_p(a,b) \text{ as } \nu \to \infty \tag{9.2.12}$$

holds for $\mu = 0,\dots,n-1$ and each vector function c with coefficients in $\mathbb{C}^n$ such that $C_2(\lambda)^{-1}c(\lambda) = O(\lambda^{-\mu-1})$.

For $k = 1,\dots,n-1$ we have

$$B(\lambda)^{-1}e_k = -\sum_{j=k}^{n-1} \lambda^{j-k} e_j, \tag{9.2.13}$$

and therefore

$$\mathscr{H}_{1,n-1}(\lambda)B(\lambda)^{-1}e_k = -\sum_{j=k}^{n} \lambda^{j-k}\mathscr{H}_j.$$

Here and in the following, e_k is the k-th unit vector in $\mathbb{C}^{n-1}$ or $\mathbb{C}^n$, where it is clear from the context which space is taken. For $f =: \begin{pmatrix} f_1 \\ f_2 \end{pmatrix} \in (W_p^n(a,b))^{n-1} \oplus W_p^n(a,b)$ we obtain

$$\begin{aligned} &-K_1(\lambda)\mathscr{H}_{1,n-1}(\lambda)B(\lambda)^{-1}e_k e_k^{\mathsf{T}} f_1 \\ &\qquad = K_1(\lambda)L^D(\lambda)\lambda^{-k}e_k^{\mathsf{T}} f - \sum_{j=0}^{k-1} \lambda^{j-k} K_1(\lambda)\mathscr{H}_j e_k^{\mathsf{T}} f \end{aligned} \tag{9.2.14}$$

and

$$K_1(\lambda)\pi_{n,n}f_2 = K_1(\lambda)L^D(\lambda)\lambda^{-n}e_n^{\mathsf{T}} f - \sum_{j=0}^{n-1} \lambda^{j-n} K_1(\lambda)\mathscr{H}_j e_n^{\mathsf{T}} f. \tag{9.2.15}$$

For $s+\mu \le n-1$ and $k = 1,\dots,n-1$ it follows from (9.2.10) that

$$\begin{aligned} &-\frac{1}{2\pi i}\left(\oint_{\Gamma_\nu} \lambda^s K_1(\lambda)\mathscr{H}_{1,n-1}(\lambda)B(\lambda)^{-1}e_k e_k^{\mathsf{T}} f_1 \, d\lambda\right)^{(\mu)} \\ &\qquad = \frac{1}{2\pi i}\left(\oint_{\Gamma_\nu} \lambda^{-k+s} K_1(\lambda)L^D(\lambda)e_k^{\mathsf{T}} f \, d\lambda\right)^{(\mu)} + o(1) \end{aligned} \tag{9.2.16}$$

and

$$\begin{aligned} &\frac{1}{2\pi i}\left(\oint_{\Gamma_\nu} \lambda^s K_1(\lambda)\pi_{n,n} f_2 \, d\lambda\right)^{(\mu)} \\ &\qquad = \frac{1}{2\pi i}\left(\oint_{\Gamma_\nu} \lambda^{-n+s} K_1(\lambda)L^D(\lambda)e_n^{\mathsf{T}} f \, d\lambda\right)^{(\mu)} + o(1) \end{aligned} \tag{9.2.17}$$

in $L_p(a,b)$ as $\nu \to \infty$. For $h \in W_p^n(a,b)$ we have

$$K_1(\lambda)L^D(\lambda)h = h - K_2(\lambda)L^R(\lambda)h$$

by definition of $K_1(\lambda)$ and $K_2(\lambda)$. Therefore, for $k = 1, \ldots, n$,

$$\begin{aligned}(9.2.18)\qquad &\frac{1}{2\pi i}\oint_{\Gamma_\nu} \lambda^{-k+s}K_1(\lambda)L^D(\lambda)e_k^\mathsf{T} f\,\mathrm{d}\lambda \\ &= \delta_{k,s+1}e_k^\mathsf{T} f - \frac{1}{2\pi i}\oint_{\Gamma_\nu} \lambda^{-k+s}K_2(\lambda)L^R(\lambda)e_k^\mathsf{T} f\,\mathrm{d}\lambda.\end{aligned}$$

Altogether, we obtain for $s = 0, \ldots, n-1$ and $\mu = 0, \ldots, n-1-s$ in view of (9.2.6), (9.2.7), (9.2.8), (9.2.9), (9.2.16), (9.2.17), and (9.2.18) that

$$\begin{aligned}(9.2.19)\qquad &-\frac{1}{2\pi i}\left(\oint_{\Gamma_\nu} e_{s+1}^\mathsf{T}\widetilde{\mathscr{H}}_2^0(\lambda)\widetilde{I}\begin{pmatrix} f \\ c\end{pmatrix}\mathrm{d}\lambda\right)^{(\mu)} \\ &= -\frac{1}{2\pi i}\left(\oint_{\Gamma_\nu} \lambda^s K_1(\lambda)(\mathscr{H}_{1,n-1}(\lambda)B(\lambda)^{-1}f_1 - \pi_{n,n}f_2)\,\mathrm{d}\lambda\right)^{(\mu)}\mathrm{d}\lambda \\ &\quad - \frac{1}{2\pi i}\left(\oint_{\Gamma_\nu} \lambda^s K_2(\lambda)\widetilde{H}^R(\lambda)\widetilde{C}_{21}^{(0)}(\lambda)f_1\,\mathrm{d}\lambda\right)^{(\mu)} \\ &\quad - \frac{1}{2\pi i}\left(\oint_{\Gamma_\nu} \lambda^s K_2(\lambda)\sum_{j=1}^{q}\sum_{r=1}^{\nu_j}\lambda^{\nu_j - r}e_j c_{jr}\,\mathrm{d}\lambda\right)^{(\mu)} \\ &= e_{s+1}^\mathsf{T} f^{(\mu)} - \frac{1}{2\pi i}\left(\oint_{\Gamma_\nu} \lambda^s K_2(\lambda)\left\{\sum_{k=1}^{n}\lambda^{-k}L^R(\lambda)e_k^\mathsf{T} f\right.\right. \\ &\quad \left.\left. + \widetilde{H}^R(\lambda)\widetilde{C}_{21}^{(0)}(\lambda)f_1 + \sum_{j=1}^{q}\sum_{r=1}^{\nu_j}\lambda^{\nu_j - r}e_j c_{jr}\right\}\mathrm{d}\lambda\right)^{(\mu)} + o(1)\end{aligned}$$

in $L_p(a,b)$ as $\nu \to \infty$. From (9.2.13) we infer that

$$(9.2.20)\qquad \widetilde{C}_{21}^0(\lambda)f_1 = -\sum_{k=1}^{n-1}\sum_{j=k}^{n-1}\lambda^{j-k}e_{j+1}e_k^\mathsf{T} f$$

and therefore, in view of (9.1.7), (9.1.9), and (9.1.8),

$$\begin{aligned}&\sum_{k=1}^{n}\lambda^{-k}L^R(\lambda)e_k^\mathsf{T} f + \widetilde{H}^R(\lambda)\widetilde{C}_{21}^0(\lambda)f_1 \\ &= \sum_{k=1}^{n}\lambda^{-k}\left(L^R(\lambda) - \sum_{j=k}^{n-1}\widetilde{H}^R(\lambda)\lambda^j e_{j+1}\right)e_k^\mathsf{T} f \\ &= \sum_{k=1}^{n}\sum_{j=0}^{k-1}\lambda^{j-k}\widetilde{H}^R(\lambda)e_{j+1}e_k^\mathsf{T} f \\ &= \sum_{k=1}^{n}\sum_{j=0}^{k-1}\lambda^{j-k}\left\{\sum_{l=0}^{n-1-j}U^{j,l}(e_k^\mathsf{T} f)^{(l)} + \sum_{l=1}^{m}\lambda^l U^{l+j,n-1-j}(e_k^\mathsf{T} f)^{(n-1-j)}\right\}.\end{aligned}$$

Hence (9.2.5) follows in view of (9.2.1) and (9.2.12).

In view of (9.2.4) this proves the convergence for the components belonging to $\mathscr{W}_p^{0,n}(a,b)$ in (9.2.2). Here we have to note that in (9.2.5) the operator function $\widetilde{\mathscr{H}_2^0}(\lambda)$ can be replaced by $\widetilde{\mathscr{H}_1^0}(\lambda)$ since $\widetilde{C}_{21}^0(\lambda)$ depends polynomially on λ. Since

$$-\frac{1}{2\pi i}\oint_{\Gamma_\nu}(A-\lambda I_N)^{-1}c\,\mathrm{d}\lambda = c,$$

the proof of (9.2.2) will be complete if we show that

$$\frac{1}{2\pi i}\oint_{\Gamma_\nu}(A-\lambda I_N)^{-1}H^{R,1}\widetilde{\mathscr{H}_1^0}(\lambda)\widetilde{I}\begin{pmatrix} f\\ c\end{pmatrix}\,\mathrm{d}\lambda$$
$$= -\sum_{s=0}^{\nu_1-1}A^sH^{R,1}\frac{1}{2\pi i}\oint_{\Gamma_\nu}\lambda^{-s-1}\widetilde{\mathscr{H}_1^0}(\lambda)\widetilde{I}\begin{pmatrix} f\\ c\end{pmatrix}\,\mathrm{d}\lambda$$

tends to 0 as $\nu\to\infty$ for all $\begin{pmatrix} f\\ c\end{pmatrix}$ satisfying (9.2.1). For this it is sufficient to prove for $s\in\mathbb{N}$ that

$$\frac{1}{2\pi i}\oint_{\Gamma_\nu}\lambda^{-s-1}\widetilde{\mathscr{H}_1^0}(\lambda)\widetilde{I}\begin{pmatrix} f\\ c\end{pmatrix}\,\mathrm{d}\lambda\to 0 \tag{9.2.21}$$

in $\mathscr{W}_1^{1,n}(a,b)$ as $\nu\to\infty$.

We have by (9.2.20) that

$$-\frac{1}{2\pi i}\oint_{\Gamma_\nu}\lambda^{-s-1}\begin{pmatrix}\widetilde{C}_{21}^0(\lambda) & 0 & 0 & 0\end{pmatrix}\widetilde{I}\begin{pmatrix} f\\ c\end{pmatrix}\,\mathrm{d}\lambda$$
$$=\sum_{k=1}^{n-1}\sum_{j=k}^{n-1}\frac{1}{2\pi i}\oint_{\Gamma_\nu}\lambda^{-s-1+j-k}e_{j+1}e_k^{\mathsf{T}}f\,\mathrm{d}\lambda = \sum_{k=1}^{n-1-s}e_{s+k+1}e_k^{\mathsf{T}}f.$$

On the other hand,

$$\frac{1}{2\pi i}\oint_{\Gamma_\nu}\lambda^{-s-1}\widetilde{\mathscr{H}_2^0}(\lambda)\widetilde{I}\begin{pmatrix} f\\ c\end{pmatrix}\,\mathrm{d}\lambda$$
$$=\sum_{j=0}^{\min\{n-1,s\}}\frac{1}{2\pi i}\oint_{\Gamma_\nu}\lambda^{-s-1}e_{j+1}e_{j+1}^{\mathsf{T}}\widetilde{\mathscr{H}_2^0}(\lambda)\widetilde{I}\begin{pmatrix} f\\ c\end{pmatrix}\,\mathrm{d}\lambda$$
$$+\sum_{j=s+1}^{n-1}e_{j+1}\frac{1}{2\pi i}\oint_{\Gamma_\nu}e_{j-s}^{\mathsf{T}}\widetilde{\mathscr{H}_2^0}(\lambda)\widetilde{I}\begin{pmatrix} f\\ c\end{pmatrix}\,\mathrm{d}\lambda,$$

where we have used that the definition of $\widetilde{\mathscr{H}_2^0}(\lambda)$ immediately gives the identity $\lambda^{-s-1}e_{j+1}^{\mathsf{T}}\widetilde{\mathscr{H}_2^0}(\lambda) = e_{j-s}^{\mathsf{T}}\widetilde{\mathscr{H}_2^0}(\lambda)$. From (9.2.5) we know that

$$\frac{1}{2\pi i}\oint_{\Gamma_\nu}e_{j-s}^{\mathsf{T}}\widetilde{\mathscr{H}_2^0}(\lambda)\widetilde{I}\begin{pmatrix} f\\ c\end{pmatrix}\,\mathrm{d}\lambda\to -e_{j-s}^{\mathsf{T}}f$$

in $W_p^{n-(j-s)+1}(a,b)$ for $j = s+1, \dots, n-1$. Therefore we still have to show that

$$\oint_{\Gamma_\nu} \lambda^{-s-1} e_{j+1}^\top \widetilde{\mathscr{H}_2^0}(\lambda) \tilde{I} \begin{pmatrix} f \\ c \end{pmatrix} \, d\lambda \to 0$$

in $W_1^{n-j}(a,b)$ as $\nu \to \infty$ for $s \in \mathbb{N}$ and $j = 0, \dots, \min\{n-1, s\}$. From (9.2.1) and an estimate as in (9.2.19) we infer for $j = 1, \dots, \min\{n-1, s\}$ and $\mu = 0, \dots, n-j$ that

$$\begin{aligned} (9.2.22) \qquad & \left(\oint_{\Gamma_\nu} \lambda^{-s-1} e_{j+1}^\top \widetilde{\mathscr{H}_2^0}(\lambda) \tilde{I} \begin{pmatrix} f \\ c \end{pmatrix} \, d\lambda \right)^{(\mu)} \\ & = \left(\oint_{\Gamma_\nu} \lambda^{-s} e_j^\top \widetilde{\mathscr{H}_2^0}(\lambda) \tilde{I} \begin{pmatrix} f \\ c \end{pmatrix} \, d\lambda \right)^{(\mu)} = o(1) \end{aligned}$$

holds in $L_p(a,b)$ as $\nu \to \infty$. In the same way, the estimate

$$(9.2.23) \qquad \left(\oint_{\Gamma_\nu} \lambda^{-s-1} e_1^\top \widetilde{\mathscr{H}_2^0}(\lambda) \tilde{I} \begin{pmatrix} f \\ c \end{pmatrix} \, d\lambda \right)^{(\mu)} = o(1)$$

in $L_p(a,b)$ as $\nu \to \infty$ holds for $\mu = 0, \dots, n-1$. We still have to consider the latter integral for $\mu = n$. For this we note that

$$\widetilde{\mathscr{H}_2^0}(\lambda) \tilde{I} \begin{pmatrix} f \\ c \end{pmatrix} = \begin{pmatrix} 1 \\ \lambda \\ \vdots \\ \lambda^{n-1} \end{pmatrix} L^{-1}(\lambda) g(\lambda),$$

where $g(\lambda)$ is a certain polynomial vector function in λ. We have

$$\left(L^{-1}(\lambda) g(\lambda) \right)^{(n)} = \left(L^D(\lambda) - \mathscr{L}^D(\lambda) \right) L^{-1}(\lambda) g(\lambda),$$

where

$$\mathscr{L}^D(\lambda)\eta = \sum_{k=0}^{n-1} \pi_{n-k,0} \eta^{(k)} + \sum_{i=1}^{n} \lambda^i \mathscr{H}_i \eta.$$

Since $g(\lambda)$ depends polynomially on λ, we infer that

$$\begin{aligned} & \left(\oint_{\Gamma_\nu} \lambda^{-s-1} e_1^\top \widetilde{\mathscr{H}_2^0}(\lambda) \tilde{I} \begin{pmatrix} f \\ c \end{pmatrix} \, d\lambda \right)^{(n)} = - \oint_{\Gamma_\nu} \lambda^{-s-1} \mathscr{L}^D(\lambda) e_1^\top \widetilde{\mathscr{H}_2^0}(\lambda) \tilde{I} \begin{pmatrix} f \\ c \end{pmatrix} \, d\lambda \\ & = - \sum_{k=0}^{n-1} \pi_{n-k,0} \left(\oint_{\Gamma_\nu} \lambda^{-s-1} e_1^\top \widetilde{\mathscr{H}_2^0}(\lambda) \tilde{I} \begin{pmatrix} f \\ c \end{pmatrix} \, d\lambda \right)^{(k)} \\ & \quad - \sum_{i=1}^{n} \mathscr{H}_i \left(\oint_{\Gamma_\nu} \lambda^{-s} e_i^\top \widetilde{\mathscr{H}_2^0}(\lambda) \tilde{I} \begin{pmatrix} f \\ c \end{pmatrix} \, d\lambda \right), \end{aligned}$$

which is of the form $o(1)$ in $L_1(a,b)$ as $\nu \to \infty$ by (9.2.22), (9.2.23) and $\pi_{1,0} \in L_{p'}(a,b)$. This completes the proof of (9.2.21). $\square$

9.3. The expansion theorem

In this section we shall show that the convergence holds for a larger class of functions than established in Proposition 9.2.1. For this we suppose that the conditions posed at the beginning of the previous section are satisfied. Additionally, we assume that the boundary conditions are normalized. Thus we can take $C_2(\lambda) = \operatorname{diag}(\lambda^{\gamma_1}, \dots, \lambda^{\gamma_n})$, where γ_i is the degree of the operator polynomial

$$e_i^{\mathsf{T}} \sum_{\nu=0}^{m} \lambda^{\nu} (U^{\nu,0}, \lambda U^{\nu,1}, \dots, \lambda^{n-1} U^{\nu,n-1}).$$

Then $\gamma_i = \max\{l + j : e_i^{\mathsf{T}} U^{j,l} \neq 0\}$, and therefore $\gamma_i = \nu_i + n - 1$ for $i = 1, \dots, q$ and

$$\gamma_i = \max\{l + j \le n - 1 : e_i^{\mathsf{T}} U^{j,l} \neq 0\} \quad (i = q+1, \dots, n).$$

PROPOSITION 9.3.1. *Assume that the boundary conditions are normalized. Then* (9.2.1) *is equivalent to*

$$\sum_{k=0}^{\gamma_i} e_i^{\mathsf{T}} U^k \big(e_{k+j}^{\mathsf{T}} f\big) = 0 \quad (i = q+1, \dots, n; j = 1, \dots, n-1-\gamma_i). \tag{9.3.1}$$

Proof. For $i = 1, \dots, q$, the i-th component of the left-hand side in (9.2.1) satisfies the estimate $O(\lambda^{-n})$. For $i = q+1, \dots, n$, the i-th component of the left-hand side of (9.2.1) is

$$\lambda^{-\gamma_i} \sum_{j=1}^{n} \lambda^{-j} \sum_{k=j}^{n} \sum_{l=0}^{n-1-k+j} e_i^{\mathsf{T}} U^{k-j,l} \big(e_k^{\mathsf{T}} f\big)^{(l)}.$$

Condition (9.2.1) means that for $j = 1, \dots, n-1-\gamma_i$ the coefficient of $\lambda^{-\gamma_i - j}$ is zero. Finally, we replace $k - j$ by k and observe that $e_i^{\mathsf{T}} U^{k,l} = 0$ if $l > \gamma_i - k$. □

We define the operators $\tilde{Q}_\nu : \mathscr{W}_p^{0,n}(a,b) \oplus \mathbb{C}^N \to \mathscr{W}_p^{0,n}(a,b) \oplus \mathbb{C}^N$ by

$$\tilde{Q}_\nu := -\frac{1}{2\pi i} \oint_{\Gamma_\nu} (\mathscr{H} - \lambda \tilde{I})^{-1} \tilde{I} \, \mathrm{d}\lambda \quad (\nu \in \mathbb{N}),$$

where, in this case, $\tilde{I}$ is the canonical embedding from $\mathscr{W}_p^{0,n}(a,b) \oplus \mathbb{C}^N$ into $\mathscr{W}_p^{0,n}(a,b) \oplus \mathbb{C}^n \oplus \mathbb{C}^N$.

PROPOSITION 9.3.2. *Assume that the boundary conditions are normalized. Let F_0 be the set of all $\begin{pmatrix} f \\ c \end{pmatrix} \in (W_p^n(a,b))^n \oplus \mathbb{C}^N$ satisfying the conditions* (9.3.1), *and let F be the closure of F_0 in $\mathscr{W}_p^{0,n}(a,b) \oplus \mathbb{C}^N$. Then $\{\tilde{Q}_\nu|_F : \nu \in \mathbb{N}\}$ is a bounded subset of $L(F, \mathscr{W}_p^{0,n}(a,b) \oplus \mathbb{C}^N)$.*

Proof. Since the operators $\tilde{Q}_\nu$ are continuous, it is sufficient to show that $\{\tilde{Q}_\nu|_{F_0}\}$ is a bounded subset of $L(F_0, \mathscr{W}_p^{0,n}(a,b) \oplus \mathbb{C}^N)$, where F_0 is equipped with the relative topology induced by $\mathscr{W}_p^{0,n}(a,b) \oplus \mathbb{C}^N$.

In view of Theorem 4.4.11 ii) we have that the integrals (9.2.10) and (9.2.12) satisfy the estimates $o(1)$ in $L_p(a,b)$ as $\nu \to \infty$ uniformly for $h(\lambda) = O(\lambda^{n-2-\mu})$ in $L_p(a,b)$ as $\lambda \to \infty$ and $C_2(\lambda)^{-1}c(\lambda) = O(\lambda^{-\mu-1})$, respectively. Using Lemma 4.6.7 in (9.2.11), that integral satisfies $O(1)|h|_p$. Redoing the proof of Proposition 9.2.1 with these estimates completes the proof of Proposition 9.3.2. □

THEOREM 9.3.3. *Let $1 < p < \infty$. Consider the boundary eigenvalue problem* (9.1.1), (9.1.2), *and assume that it is Birkhoff regular and that the boundary conditions are normalized. For the coefficients of the differential equation* (9.1.1) *we suppose that they belong to $L_p(a,b)$ and that, additionally, $\pi_{i,i} \in W_p^1(a,b)$ for $i = 1,\ldots,n$, $\pi_{n,n}^{-1} \in L_\infty(a,b)$, $\pi_{1,0} \in L_{p'}(a,b)$, $1/p + 1/p' = 1$. Let*

$$\pi(\cdot,\rho) = \rho^n + \sum_{i=0}^{n-1} \rho^i \pi_{n-i,n-i}$$

be the characteristic function of the differential equation (9.1.1) *and assume that either*
i) *there is a number $\alpha \in \mathbb{C}\setminus\{0\}$ such that for all $x \in [a,b]$ the roots of $\pi(x,\alpha\rho) = 0$ are real, simple and different from* 0,
or
ii) *there are a real-valued function $r \in W_p^1(a,b)$ such that $r(x) \neq 0$ for all $x \in [a,b]$ and $\beta_j \in \mathbb{C}$ $(j = 1,\ldots,n)$ such that $\pi_{j,j}(x) = \beta_j r(x)^j$ for $j = 1,\ldots,n$ and $x \in (a,b)$ and $\rho^n + \sum_{j=1}^{n} \beta_j \rho^{n-j} = 0$ has only simple roots.*
Let F be the finite-codimensional closed subspace of $\mathscr{W}_p^{0,n}(a,b) \oplus \mathbb{C}^N$ which consists of those elements $\begin{pmatrix} f \\ c \end{pmatrix}$ satisfying

$$\sum_{k=0}^{\gamma_i} e_i^{\mathsf{T}} U^k \left(e_{k+j}^{\mathsf{T}} f\right) = 0 \quad (i = q+1,\ldots,n; j = 1,\ldots,n-1-\gamma_i). \tag{9.3.2}$$

where

$$\gamma_i = \max\{l + j \leq n-1 : e_i^{\mathsf{T}} U^{j,l} \neq 0\} \quad (i = q+1,\ldots,n). \tag{9.3.3}$$

Then

$$\begin{pmatrix} f \\ c \end{pmatrix} = -\lim_{\nu\to\infty} \oint_{\Gamma_\nu} (\mathscr{H} - \lambda\widetilde{I})^{-1}\widetilde{I}\begin{pmatrix} f \\ c \end{pmatrix} \mathrm{d}\lambda \tag{9.3.4}$$

for $\begin{pmatrix} f \\ c \end{pmatrix} \in F$, i. e., the elements $\begin{pmatrix} f \\ c \end{pmatrix}$ are expandable into a series of eigenvectors and associated vectors of $\mathscr{H} - \lambda\widetilde{I}$ which converges in $\mathscr{W}_p^{0,n}(a,b) \oplus \mathbb{C}^N$. Here we have used the same symbol $\widetilde{I}$ for the canonical embeddings from $\mathscr{W}_p^{1,n}(a,b) \oplus \mathbb{C}^N$ and $\mathscr{W}_p^{0,n}(a,b) \oplus \mathbb{C}^N$ into $\mathscr{W}_p^{0,n}(a,b) \oplus \mathbb{C}^n \oplus \mathbb{C}^N$.

Proof. From Propositions 9.3.2, 9.3.1 and 9.2.1 and Lemma 4.6.8 the stated result follows for F as defined in Proposition 9.3.2. We still have to show that the space F as defined in Theorem 9.3.3 is the closure of F_0 defined in Proposition 9.3.2. $F_0 \subset F$ is obvious, and F is closed since the linear functionals on the left-hand side of (9.3.2) are continuous on $\mathscr{W}_p^{0,n}(a,b)$. To prove that F_0 is dense in F let u be a continuous linear functional on F such that $u|_{F_0} = 0$. We must show that $u = 0$. Let v be a continuous linear extension of u to $\mathscr{W}_p^{0,n}(a,b) \oplus \mathbb{C}^N$ and w the restriction of v to $(W_p^n(a,b))^n \oplus \mathbb{C}^N$. Since $w|_{F_0} = u|_{F_0} = 0$, w is a linear combination of the linear functionals given by (9.3.2). Since $(W_p^n(a,b))^n \oplus \mathbb{C}^N$ is dense in $\mathscr{W}_p^{0,n}(a,b) \oplus \mathbb{C}^N$, also v can be written as this linear combination of the linear functionals given by (9.3.2). But this implies $u = v|_F = 0$. □

Now let us compare our results with those of Shkalikov. He considers minimality, completeness and basis property for the eigenvectors and associated vectors for $p = 2$. A system of vectors in a Banach space E is said to be *complete* if their linear span is dense in E. Shkalikov only requires that the coefficients belong to $L_1(a,b)$, but he always assumes that the $\pi_{i,i}$ are constant. In [SH5, Theorem 3.1], it is shown that the eigenvectors and associated vectors of the operator [SH5, (1.36), (1.37)] form a Riesz basis in parenthesis. The operator [SH5, (1.36)] is

$$\begin{pmatrix} H^D & 0 \\ -H^{R,1} & A \end{pmatrix}$$

and its domain is the set of all $\begin{pmatrix} y \\ c \end{pmatrix} \in \mathscr{W}_p^{1,n}(a,b) \oplus \mathbb{C}^N$ which satisfy (9.3.1) and

$$H^{R,0} y = Gc.$$

Of course, we can replace L^R by $-L^R$ without changing the problem under consideration. Then this operator coincides with a restriction of the operator $\mathscr{H}$ defined in (9.1.21).

The space for which Shkalikov shows that the Riesz basis property holds coincides with the space for which we obtained expandability in Theorem 9.3.3.

We note that Shkalikov also proved the basis property in certain subspaces of $\mathscr{W}_p^{r,n}(a,b) \oplus \mathbb{C}^N$ for $r > 0$, where the coefficients have to be in $W_1^r(a,b)$. Since we did not consider expansions of this form in our general exposition, we shall not dwell on this.

Finally, we are going to show that our expansion result in Theorem 9.3.3 yields completeness and minimality. Expandability and completeness imply basisness but not Riesz basisness.

Let $\mathscr{W}_{p,U}^{j,n}(a,b)$ denote the set of all elements in $\mathscr{W}_p^{j,n}(a,b)$ satisfying the boundary conditions (9.3.2), where $j \in \mathbb{N}$.

PROPOSITION 9.3.4. *Assume that* $\rho(\mathscr{H}) \neq \emptyset$ *and let* $\begin{pmatrix} f \\ c \end{pmatrix} \in \mathscr{W}_p^{1,n}(a,b) \oplus \mathbb{C}^N$ *be an eigenvector or associated vector of* $\mathscr{H} - \lambda\widetilde{I}$ *at* $\mu \in \sigma(\mathscr{H})$. *Then* f *belongs to* $\mathscr{W}_{p,U}^{1,n}(a,b)$.

Proof. In view of Proposition 1.10.2 we have

$$(\mathscr{H} - \mu\widetilde{I}) \begin{pmatrix} f \\ c \end{pmatrix} = \begin{pmatrix} g \\ 0 \\ c \end{pmatrix},$$

where $\begin{pmatrix} g \\ c \end{pmatrix}$ is either zero or an eigenvector or associated vector of $\mathscr{H} - \lambda\widetilde{I}$ at μ. Using a proof by induction, we may assume that $g \in \mathscr{W}_{p,U}^{1,n}(a,b)$. Then, by definition of $\mathscr{H}$ in (9.1.21),

$$H^D f - \mu f = g,$$

and therefore

$$f_{l+1} - \mu f_l = g_l \tag{9.3.5}$$

for $l = 1, \dots, n-1$. From $e_i^{\mathsf{T}} G = 0$ for $i = q+1, \dots, n$ we infer

$$0 = e_i^{\mathsf{T}} (H^{R,0} f + Gc) = \sum_{k=0}^{\gamma_i} e_i^{\mathsf{T}} U^k e_{k+1}^{\mathsf{T}} f,$$

which proves (9.3.2) for $i = q+1, \dots, n$ and $j+1$. For $j = 2, \dots, n-1-\gamma_i$ we use induction. Assume (9.3.2) holds for $j-1$. Then (9.3.5) and $g \in \mathscr{W}_{p,U}^{1,n}(a,b)$ imply

$$\sum_{k=0}^{\gamma_i} e_i^{\mathsf{T}} U^k e_{k+j}^{\mathsf{T}} f = \mu \sum_{k=0}^{\gamma_i} e_i^{\mathsf{T}} U^k e_{k+j-1}^{\mathsf{T}} f + \sum_{k=0}^{\gamma_i} e_i^{\mathsf{T}} U^k e_{k+j-1}^{\mathsf{T}} g = 0. \qquad \square$$

THEOREM 9.3.5. *Let the assumptions of Theorem* 9.3.3 *be satisfied. Then every element* $\begin{pmatrix} f \\ c \end{pmatrix} \in \mathscr{W}_{p,U}^{0,n}(a,b) \oplus \mathbb{C}^N$ *is expandable into a series of eigenvectors and associated vectors of the operator function* $\mathscr{H} - \lambda\widetilde{I}$ *which converges in* $\mathscr{W}_{p,U}^{0,n}(a,b) \oplus \mathbb{C}^N$, *and the set of eigenvectors and associated vectors of* $\mathscr{H} - \lambda\widetilde{I}$ *is complete and minimal in* $\mathscr{W}_{p,U}^{0,n}(a,b) \oplus \mathbb{C}^N$.

Proof. Since the problem is Birkhoff regular, we have $\rho(\mathscr{H}) \neq \emptyset$. From Theorem 9.3.3 and Proposition 9.3.4 we know that every element $\begin{pmatrix} f \\ c \end{pmatrix} \in \mathscr{W}_{p,U}^{0,n}(a,b) \oplus \mathbb{C}^N$ is expandable into a series of eigenvectors and associated vectors of the operator function $\mathscr{H} - \lambda\widetilde{I}$ which converges in $\mathscr{W}_{p,U}^{0,n}(a,b) \oplus \mathbb{C}^N$ since all the terms of the expansion belong to $\mathscr{W}^{1,n}_{p,U}(a,b) \oplus \mathbb{C}^N$. This also shows that the system of eigenvectors and associated vectors of $\mathscr{H} - \lambda\widetilde{I}$ is complete in $\mathscr{W}^{0,n}_{p,U}(a,b) \oplus \mathbb{C}^N$. We shall use Proposition 1.10.6 to show that the system of eigenvectors and associated vectors of $\mathscr{H} - \lambda\widetilde{I}$ is minimal. The operator $\widetilde{I}$ is the product of the embeddings $W_p^{k+1}(a,b) \hookrightarrow W_p^k(a,b)$ $(k = 0,\ldots,n-1)$, which are compact by Theorem 2.4.2. Hence $\widetilde{I}$ is compact. Since the embedding $\mathscr{W}^{1,n}_{p,U}(a,b) \oplus \mathbb{C}^N \hookrightarrow \mathscr{W}^{0,n}_{p}(a,b) \oplus \mathbb{C}^N$ is continuous, the system of eigenvectors and associated vectors of $\mathscr{H} - \lambda\widetilde{I}$ is minimal in $\mathscr{W}^{0,n}_{p}(a,b) \oplus \mathbb{C}^N$ and therefore also in $\mathscr{W}^{0,n}_{p,U}(a,b) \oplus \mathbb{C}^N$. □

REMARK 9.3.6. From Theorem 9.3.5 and Proposition 9.3.4 it is immediately clear that $\begin{pmatrix} f \\ c \end{pmatrix} \in \mathscr{W}^{0,n}_{p}(a,b) \oplus \mathbb{C}^N$ is expandable into a series of eigenvectors and associated vectors of the operator function $\mathscr{H} - \lambda\widetilde{I}$ which converges in $\mathscr{W}^{0,n}_{p}(a,b) \oplus \mathbb{C}^N$ if and only if $f \in \mathscr{W}^{0,n}_{p,U}(a,b)$.

9.4. Notes

In this chapter we followed the linearization procedure $y_1 = \eta$, $y_2 = \lambda\eta, \ldots,$ $y_n = \lambda^{n-1}\eta$. This type of linearization has been frequently used in connection with the investigation of λ-polynomial operator matrices, see e. g. GOHBERG and KREIN [GK, Section V.9], MARKUS [MA4, Chapter 2] and the references in the monographs. A different linearization method can be used for the differential operator $L^D(\lambda)$ by simultaneously transforming this operator to a first order system and linearizing it with respect to λ, see [KRI], [LMM1], [LMM2]. A rather particular type of linearization method was used in [LMMS] and [MÖ5].

Chapter X

APPLICATIONS

In this chapter we apply our results to some spectral problems which have been considered in the literature. In particular, we investigate whether these problems are Birkhoff or Stone regular. Sections 10.1–10.4 deal with problems occurring in mechanics. In Section 10.5, a problem from meteorology is discussed. The well-known Orr-Sommerfeld equation is studied in Section 10.6. Finally, Sections 10.7–10.9 deal with problems from hydrodynamics and magnetohydrodynamics.

10.1. The clamped-free elastic bar

HAUGER and LEONHARD [HL1, HL2] have investigated the equation of motion of a clamped-free elastic bar. Separation of variables leads to the eigenvalue problem

$$\eta^{(4)} - \lambda \frac{(1+\kappa)^2}{(1+\kappa\cdot)^2}\eta'' - \Omega^2\eta = 0, \tag{10.1.1}$$

$$\eta(0) = 0,\ \eta'(0) = 0,\ \eta''(1) = 0,\ \eta'''(1) - \lambda(1-\gamma)\eta'(1) = 0, \tag{10.1.2}$$

where κ, Ω, γ are constants and $\kappa > -1$, $\Omega > 0$. The differential equation is of the form (8.1.1) with $n = 4$, $n_0 = 2$,

$$h_{n_0}(x) = \frac{(1+\kappa)^2}{(1+\kappa x)^2},$$

$\mathbf{H}\eta = h_{n_0}\eta''$, and a fundamental system of $\mathbf{H}\eta = 0$ is $\{1, x\}$. By Theorem 8.2.1 there is a fundamental matrix $Y(\cdot,\lambda)$ of (10.1.1) such that

$$Y(x,\lambda)(I_n - \Delta_0) = \begin{pmatrix} 1 & x & 0 & 0 \\ 0 & 1 & 0 & 0 \\ 0 & 0 & 0 & 0 \\ 0 & 0 & 0 & 0 \end{pmatrix} + \{O(\lambda^{-1})\}_\infty,$$

where $\Lambda_0 = \operatorname{diag}(0,0,1,1)$. For the boundary matrices defined in (8.1.6) we obtain

$$W^{(0)}(\lambda) = \begin{pmatrix} 1 & 0 & 0 & 0 \\ 0 & 1 & 0 & 0 \\ 0 & 0 & 0 & 0 \\ 0 & 0 & 0 & 0 \end{pmatrix}, \quad W^{(1)}(\lambda) = \begin{pmatrix} 0 & 0 & 0 & 0 \\ 0 & 0 & 0 & 0 \\ 0 & 0 & 1 & 0 \\ 0 & -\lambda(1-\gamma) & 0 & 1 \end{pmatrix}. \tag{10.1.3}$$

According to the requirement (8.5.7) we can take

$$q_1 = q_2 = 0,\ q_3 = -1,\ q_4 = 1.$$

Also, we obviously have that the numbers defined in (8.5.1) are

$$l_1 = 0,\ l_2 = 1,\ l_3 = 2,\ l_4 = 3.$$

We infer

$$l^{(0)} = 5,\ l^{(1)} = 6,$$

see (8.5.10). We have

$$\alpha_{100}^{(0)} = 1,\ \alpha_{200}^{(0)} = 1,\ \alpha_{310}^{(1)} = 0,\ \alpha_{400}^{(1)} = \gamma - 1,\ \alpha_{410}^{(1)} = 1,$$

see (8.5.16), and all the other $\alpha_{\nu tm}^{(j)}$ are zero. Note that $\alpha_{\nu t0}^{(j)} = \alpha_{\nu t}^{(j)}$. We want to show that

$$b_1^{(0)} = u_{\{2,3\}} v_{\{2,3\},1} + u_{\{3,4\}} v_{\{3,4\},1} \neq 0,$$

see (8.5.12) and (8.5.13) for the definition of these numbers. Then the problem is almost Birkhoff regular of order 1 by Definition 8.5.1. Here $u_{\{2,3\}}$ and $u_{\{3,4\}}$ are the determinants of the matrices built by taking the rows number 1,4 and 1,2, respectively, of

$$\begin{pmatrix} 1 & 0 \\ 0 & 1 \\ * & * \\ 0 & \gamma - 1 \end{pmatrix}.$$

Hence

$$u_{\{2,3\}} = \gamma - 1 \neq 0,\ u_{\{3,4\}} = 1 \neq 0,$$

and

$$v_{\{2,3\},1} = \begin{vmatrix} \alpha_2^{(0)} & \alpha_2^{(1)} \\ \alpha_3^{(0)} & -\alpha_3^{(1)} \end{vmatrix} = -\alpha_2^{(0)}\alpha_3^{(1)} = -h_{n_0}^{1/2}(0)h_{n_0}(1) \neq 0,$$

$$v_{\{3,4\},1} = \begin{vmatrix} \alpha_3^{(0)} & \alpha_3^{(1)} \\ \omega_1\alpha_4^{(0)} & \omega_2\alpha_4^{(1)} \end{vmatrix} = 0$$

since $\alpha_3^{(0)} = \alpha_4^{(0)} = 0$. This shows that $b_1^{(0)} \neq 0$.

To calculate the boundary conditions $U_\nu f = 0$ we use Remark 8.5.5 iii). We need only $\delta_{10}^{(0)}, \delta_{20}^{(0)}, \delta_{21}^{(0)}, \delta_{30}^{(1)}, \delta_{40}^{(1)}, \delta_{41}^{(1)}$, since $\delta_{\nu i}^{(0)} = 0$ if $\nu = 3,4$ and $\delta_{\nu i}^{(1)} = 0$ if $\nu = 1,2$. With

$$h_{n_0 - 1} = 0,\ k_{n-1} = 0,$$

$$h_{n_0}(1) = 1,\ h'_{n_0}(x) = -2\kappa\frac{(1+\kappa)^2}{(1+\kappa x)^3},\ h'_{n_0}(1) = -\frac{2\kappa}{1+\kappa}$$

we infer

$$\delta_{10}^{(0)} = 1,\ \delta_{20}^{(0)} = 0,\ \delta_{21}^{(0)} = 1,\ \delta_{30}^{(1)} = 1,\ \delta_{40}^{(1)} = \frac{2\kappa}{1+\kappa},\ \delta_{41}^{(1)} = \gamma - 1 + 1 = \gamma.$$

Hence the boundary conditions $U_\nu f = 0$ are given by

$$U_1 f = f(0),\ U_2 f = f'(0),\ U_3 f = f(1),\ U_4 f = \frac{2\kappa}{1+\kappa} f(1) + \gamma f'(1),$$

which can be rewritten as

$$f(0) = 0,\ f'(0) = 0,\ f(1) = 0,\ \gamma f'(1) = 0.$$

By Theorem 8.8.2 we obtain an expansion into eigenfunctions and associated functions in the space $C^1[0,1]$ of the problem (10.1.1), (10.1.2) for all $f \in C^1[0,1]$ with $f' \in BV[0,1]$ which satisfy these boundary conditions. This expansion theorem was obtained by TRETTER in [TR2], [TR3].

10.2. Control of beams

In this section we consider one beam (N=1) or $N(>1)$ beams connected by joints, see [CDKP] for more details. The problem is governed by the partial differential equation

$$m_j \frac{\partial^2 y}{\partial t^2} + E_j I_j \frac{\partial^4 y}{\partial x^4} = 0, \quad \text{on } (a_{j-1}, a_j) \text{ for } j = 1, \dots, N,$$

initial conditions, and the boundary conditions

$$y(0,t) = 0, \quad \frac{\partial y}{\partial x}(0,t) = 0,$$

$$\left.\begin{aligned} y(a_j^-,t) = y(a_j^+,t), \quad \frac{\partial y}{\partial x}(a_j^-,t) &= \frac{\partial y}{\partial x}(a_j^+,t), \\ E_j I_j \frac{\partial^3 y}{\partial x^3}(a_j^-,t) - E_{j+1} I_{j+1} \frac{\partial^3 y}{\partial x^3}(a_j^+,t) &= u_{0j}(t), \\ -[E_j I_j \frac{\partial^2 y}{\partial x^2}(a_j^-,t) - E_{j+1} I_{j+1} \frac{\partial^2 y}{\partial x^2}(a_j^+,t)] &= u_{1j}(t), \end{aligned}\right\} \quad (1 \le j \le N-1)$$

$$E_N I_N \frac{\partial^3 y}{\partial x^3}(L,t) = u_{ON}(t),$$

$$-E_N I_N \frac{\partial^2 y}{\partial x^2}(L,t) = u_{ON}(t).$$

Here the joints, if any, are at $a_1 < a_2 < \dots < a_{N-1}$. For convenience, we have set $a_0 := 0$ and $a_N := L$. In the control problem, the functions u_{lj} $(l = 1,2,\ 1 \le j \le N)$ are connected with the unknown function. Here we consider the case

$$u_{0j}(t) = k_{0j} \frac{\partial y}{\partial t}(a_j,t), \quad u_{1j}(t) = k_{1j} \frac{\partial^2 y}{\partial x \partial t}(a_j,t),$$

where k_{0j}, k_{1j} are positive constants (for $j = 1,\dots,N-1$ we might also allow them to be zero). Using the separation of variables technique,

$$y(x,t) = \psi(t)\eta(x),$$

this leads to the differential equations

$$\psi'' + \lambda^2\psi = 0$$

$$E_jI_j\eta^{(4)} - \lambda^2 m_j\eta = 0 \quad \text{on } (a_{j-1},a_j) \text{ for } j = 1,\dots,N, \tag{10.2.1}$$

where λ is the eigenvalue parameter coming from the separation. Taking the solution $\psi(t) = e^{i\lambda t}$ (the case $\psi(t) = e^{-i\lambda t}$ is similar), the boundary conditions can be written as

$$\eta(0) = 0, \quad \eta'(0) = 0,$$

$$\left.\begin{aligned} \eta(a_j^-) = \eta(a_j^+), \quad \eta'(a_j^-) = \eta'(a_j^+), \\ E_jI_j\eta'''(a_j^-) - E_{j+1}I_{j+1}\eta'''(a_j^+) = k_{0j}i\lambda\eta(a_j), \\ -[E_jI_j\eta''(a_j^-) - E_{j+1}I_{j+1}\eta''(a_j^+)] = k_{1j}i\lambda\eta'(a_j), \end{aligned}\right\} \quad (j = 1,\dots,N-1),$$

$$E_NI_N\eta'''(L) = k_{0N}i\lambda\eta(L),$$

$$-E_NI_N\eta''(L) = k_{1N}i\lambda\eta'(L).$$

Of course, in case $N > 1$, the solutions of the eigenvalue problem do not belong to $H^4(0,L)$ in general. But we can consider the differential equation on the intervals (a_{j-1},a_j) separately. Therefore we arrive at a system of ordinary differential equations. In Section 10.3 we shall consider the case $N = 1$ and in Section 10.4 the case $N > 1$.

10.3. Control of one beam

In [KR12] A. M. KRALL considered the case $N = 1$. He mentions that by results of COLE and LANGER an eigenfunction expansion holds. Here we shall show that this is indeed true by showing that the problem is Birkhoff regular. Replacing λ by λ^2 we have according to Theorem 7.2.4 that the transformation matrix is $C(\lambda) = \operatorname{diag}(1,\lambda,\lambda^2,\lambda^3)V$ where

$$V = \begin{pmatrix} 1 & 1 & 1 & 1 \\ \beta & i\beta & -\beta & -i\beta \\ \beta^2 & -\beta^2 & \beta^2 & -\beta^2 \\ \beta^3 & -i\beta^3 & -\beta^3 & i\beta^3 \end{pmatrix}$$

and $\beta = \sqrt[4]{m_1(E_1I_1)^{-1}}$. Therefore, the boundary matrices defined in (7.3.1) are

$$W^{(0)}(\lambda) = \begin{pmatrix} 1 & 0 & 0 & 0 \\ 0 & \lambda & 0 & 0 \\ 0 & 0 & 0 & 0 \\ 0 & 0 & 0 & 0 \end{pmatrix} V,$$

$$W^{(1)}(\lambda) = \begin{pmatrix} 0 & 0 & 0 & 0 \\ 0 & 0 & 0 & 0 \\ -k_{0N}i\lambda^2 & 0 & 0 & E_N I_N \lambda^3 \\ 0 & -k_{1N}i\lambda^3 & -E_N I_N \lambda^2 & 0 \end{pmatrix} V,$$

and we obtain according to Theorem 7.3.2 that

$$W_0^{(0)} = \begin{pmatrix} 1 & 0 & 0 & 0 \\ 0 & 1 & 0 & 0 \\ 0 & 0 & 0 & 0 \\ 0 & 0 & 0 & 0 \end{pmatrix} V, \quad W_0^{(1)} = \begin{pmatrix} 0 & 0 & 0 & 0 \\ 0 & 0 & 0 & 0 \\ 0 & 0 & 0 & E_N I_N \\ 0 & -k_{1N}i & 0 & 0 \end{pmatrix} V.$$

Altogether, there are 4 Birkhoff matrices, the first one being

$$W_0^{(0)} \begin{pmatrix} 1 & 0 & 0 & 0 \\ 0 & 1 & 0 & 0 \\ 0 & 0 & 0 & 0 \\ 0 & 0 & 0 & 0 \end{pmatrix} + W_0^{(1)} \begin{pmatrix} 0 & 0 & 0 & 0 \\ 0 & 0 & 0 & 0 \\ 0 & 0 & 1 & 0 \\ 0 & 0 & 0 & 1 \end{pmatrix}, \tag{10.3.1}$$

and the other ones are obtained by a cyclic permutations of the zeros and ones in the diagonals of the diagonal matrices in the previous formula, see Proposition 4.1.7. The Birkhoff matrix (10.3.1) is invertible,

$$\begin{pmatrix} 1 & 1 & 0 & 0 \\ \beta & i\beta & 0 & 0 \\ 0 & 0 & -E_N I_N \beta^3 & iE_N I_N \beta^3 \\ 0 & 0 & k_{1N}i\beta & -k_{1N}\beta \end{pmatrix},$$

and also the other three Birkhoff matrices are invertible. Therefore the problem is Birkhoff regular, and by Theorem 7.4.3 every function $f \in L_p(a,b)$ $(1 < p < \infty)$ can be expanded into a series of eigenfunctions and associated functions of (10.2.1) for $N = 1$ with the corresponding boundary conditions.

10.4. Control of multiple beams

This problem has to be transformed to one for which our results apply. We define

$$\eta_j(x) := \eta(a_{j-1} + (a_j - a_{j-1})x) \quad (x \in (0,1),\ j = 1, \ldots, N).$$

We abbreviate $\alpha_{j,r} := (a_j - a_{j-1})^r E_j I_j$. Then the differential equation (10.2.1) can be written as a system

$$\alpha_{j,4}\eta_j^{(4)} - \lambda^2 m_j \eta_j = 0 \text{ on } (0,1) \text{ for } j = 1, \ldots, N,$$

and the boundary conditions are

$$\eta_1(0) = 0, \quad \eta_1'(0) = 0,$$

$$\left.\begin{aligned} \eta_j(1) &= \eta_{j+1}(0), \\ (a_j - a_{j-1})\eta_j'(1) &= (a_{j+1} - a_j)\eta_{j+1}'(0), \\ \alpha_{j,3}\eta_j'''(1) - \alpha_{j+1,3}\eta_{j+1}'''(0) &= k_{0j} i\lambda \eta_j(1), \\ -[\alpha_{j,2}\eta_j''(1) - \alpha_{j+1,2}\eta_{j+1}''(0)] &= \tilde{k}_{1j} i\lambda \eta_j'(1), \end{aligned}\right\} \quad (j = 1, \dots, N-1),$$

$$\alpha_{N,3}\eta_N'''(1) = k_{0N} i\lambda \eta_N(1),$$

$$-\alpha_{N,2}\eta_N''(1) = \tilde{k}_{1N} i\lambda \eta_N'(1),$$

where $\tilde{k}_{1j} = (a_j - a_{j-1})k_{1j}$.

We again replace λ by λ^2 and set $\beta_j := \sqrt[4]{m_j\alpha_{j,4}^{-1}}$. Defining V_j as we defined V in Section 10.3 with β replaced by β_j it follows that

$$\operatorname{diag}(1, \lambda, \lambda^2, \lambda^3)V_j y_j = (\eta_j, \eta_j', \eta_j'', \eta_j''')^{\mathsf{T}}$$

transforms the differential equations into the first order systems

$$y_j' = \lambda \beta_j \Omega y_j,$$

where $\Omega = \operatorname{diag}(1, i, -1, -i)$. Therefore, the system of fourth order differential equations is transformed into a $4N \times 4N$ system of first order differential equations. The boundary matrices are, see (7.3.1),

$$W^{(0)}(\lambda) = \begin{pmatrix} B_1^{(0)}(\lambda) & & & \\ & B_2^{(0)}(\lambda) & & 0 \\ & & \ddots & \\ & 0 & & B_N^{(0)}(\lambda) \\ & & & D \end{pmatrix},$$

$$W^{(1)}(\lambda) = \begin{pmatrix} D & & & \\ B_1^{(1)}(\lambda) & & & 0 \\ & B_2^{(1)}(\lambda) & & \\ & & \ddots & \\ & 0 & & B_N^{(1)}(\lambda) \end{pmatrix},$$

where D is the 2×4 zero matrix, the matrices $B_1^{(0)}(\lambda)$ and $B_N^{(1)}(\lambda)$ are 2×4 matrices, and all the other matrices $B_j^{(k)}(\lambda)$ are 4×4 matrices. The $B_j^{(k)}(\lambda)$ are given by

$$B_1^{(0)}(\lambda) = \begin{pmatrix} 1 & 0 & 0 & 0 \\ 0 & \lambda & 0 & 0 \end{pmatrix} V_1,$$

$$B_N^{(1)}(\lambda) = \begin{pmatrix} k_{0N} i\lambda^2 & 0 & 0 & -\alpha_{N,3}\lambda^3 \\ 0 & -\tilde{k}_{1N} i\lambda^3 & -\alpha_{N,2}\lambda^2 & 0 \end{pmatrix} V_N,$$

$$B_j^{(0)}(\lambda) = \begin{pmatrix} 1 & 0 & 0 & 0 \\ 0 & (a_j - a_{j-1})\lambda & 0 & 0 \\ 0 & 0 & 0 & \alpha_{j,3}\lambda^3 \\ 0 & 0 & \alpha_{j,2}\lambda^2 & 0 \end{pmatrix} V_j \quad (j = 2, \ldots, N),$$

$$B_j^{(1)}(\lambda) = \begin{pmatrix} -1 & 0 & 0 & 0 \\ 0 & -(a_j - a_{j-1})\lambda & 0 & 0 \\ k_{0j} i\lambda^2 & 0 & 0 & -\alpha_{j,3}\lambda^3 \\ 0 & -\tilde{k}_{1j} i\lambda^3 & -\alpha_{j,2}\lambda^2 & 0 \end{pmatrix} V_j$$
$$(j = 1, \ldots, N-1).$$

Therefore we have

$$W_0^{(0)} = \begin{pmatrix} B_{1,0}^{(0)} & & & \\ & B_{2,0}^{(0)} & & 0 \\ & & \ddots & \\ & 0 & & B_{N,0}^{(0)} \\ & & & D \end{pmatrix},$$

$$W_0^{(1)} = \begin{pmatrix} D & & & \\ B_{1,0}^{(1)} & & & 0 \\ & B_{2,0}^{(1)}(\lambda) & & \\ & & \ddots & \\ & 0 & & B_{N,0}^{(1)} \end{pmatrix},$$

where in case all k_{1j} are different from zero we have

$$B_{1,0}^{(0)} = \begin{pmatrix} 1 & 0 & 0 & 0 \\ 0 & 1 & 0 & 0 \end{pmatrix} V_1, \quad B_{N,0}^{(1)} = \begin{pmatrix} 0 & 0 & 0 & -\alpha_{N,3} \\ 0 & -\tilde{k}_{1N} i & 0 & 0 \end{pmatrix} V_N,$$

$$B_{j,0}^{(0)} = \begin{pmatrix} 1 & 0 & 0 & 0 \\ 0 & a_j - a_{j-1} & 0 & 0 \\ 0 & 0 & 0 & \alpha_{j,3} \\ 0 & 0 & 0 & 0 \end{pmatrix} V_j \quad (j = 2, \ldots, N),$$

$$B_{j,0}^{(1)} = \begin{pmatrix} -1 & 0 & 0 & 0 \\ 0 & -(a_j - a_{j-1}) & 0 & 0 \\ 0 & 0 & 0 & -\alpha_{j,3} \\ 0 & -\tilde{k}_{1j} i & 0 & 0 \end{pmatrix} V_j \quad (j = 1, \ldots, N-1).$$

As in the case of one beam we have four Birkhoff matrices. The first Birkhoff matrix is obtained by taking the first and second column from $W_0^{(0)}$, the third and fourth column from $W_0^{(1)}$, and so on. This matrix has the form

$$\begin{pmatrix} C_1^{(0)} & & & & & & & \\ & C_1^{(1)} & C_2^{(0)} & & & & 0 & \\ & & & C_2^{(1)} & C_3^{(0)} & & & \\ & & & & & \ddots & & \\ & & 0 & & & & C_{N-1}^{(1)} & C_N^{(0)} \\ & & & & & & & & C_N^{(1)} \end{pmatrix},$$

where $C_1^{(0)}$ and $C_N^{(1)}$ are 2×2 matrices and the other $C_j^{(k)}$ are 4×2 matrices. These matrices are

$$C_1^{(0)} = \begin{pmatrix} 1 & 1 \\ \beta_1 & i\beta_1 \end{pmatrix}, \quad C_N^{(1)} = \begin{pmatrix} \alpha_{N,3}\beta_N^3 & -\alpha_{N,3}i\beta_N^3 \\ \tilde{k}_{1N}i\beta_N & -\tilde{k}_{1N}\beta_N \end{pmatrix},$$

$$C_j^{(0)} = \begin{pmatrix} 1 & 1 \\ (a_j - a_{j-1})\beta_j & i(a_j - a_{j-1})\beta_j \\ \alpha_{j,3}\beta_j^3 & -i\alpha_{j,3}\beta_j^3 \\ 0 & 0 \end{pmatrix} \quad (j = 2,\dots,N),$$

$$C_j^{(1)} = \begin{pmatrix} -1 & -1 \\ (a_j - a_{j-1})\beta_j & i(a_j - a_{j-1})\beta_j \\ \alpha_{j,3}\beta_j^3 & -\alpha_{j,3}i\beta_j^3 \\ \tilde{k}_{1j}i\beta_j & -\tilde{k}_{1j}\beta_j \end{pmatrix} \quad (j = 1,\dots,N-1).$$

Since the matrices $C_1^{(0)}$ and $C_N^{(1)}$ are invertible if $k_{1N} \neq 0$, the problem is Birkhoff regular if and only if the matrices $\begin{pmatrix} C_j^{(1)} & C_{j+1}^{(0)} \end{pmatrix}$ $(j = 1,\dots,N-1)$ are invertible. But

$$\det\begin{pmatrix} C_j^{(1)} & C_{j+1}^{(0)} \end{pmatrix} = 2(1-i)(a_{j+1} - a_j)\beta_{j+1}\beta_j\tilde{k}_{1j}(\alpha_{j+1,3}\beta_{j+1}^3 + \alpha_{j,3}\beta_j^3) \neq 0$$

since the $k_{1,j} \neq 0$, $\alpha_{j,3}$ and β_j are positive.

The other Birkhoff matrices have—after a permutation of their columns—the same shape as the above one, where now the matrices $C_j^{(k)}$ are formed by different rows of the $B_{j,0}^{(k)}$. In the same way as above we see that these Birkhoff matrices are invertible.

This shows that the first order system is Birkhoff regular if the constants $k_{1,j}$ are nonzero for all $j = 1,\dots,N$. Also in case not all of the $k_{1,j}$ are zero it can be shown that the problem is Birkhoff regular.

Now it is easy to see that one can apply the methods of Section 7.4 to show that every function in $L_p(0,L)$ $(1 < p < \infty)$ can be expanded into a series of eigenfunctions and associated functions of (10.2.1) for $N > 1$ with the corresponding boundary conditions.

10.5. An example from meteorology

PROUDMAN and DOODSON [PRD] investigated the effect of atmospheric conditions to the sea-level. This led to an eigenvalue problem. The corresponding expansion in eigenvectors and associated vectors has been discussed by PROUDMAN [PR].

The boundary eigenvalue problem is given by

(10.5.1) $$(\eta v')' + \lambda v = -1,$$

(10.5.2) $$v'(0) = 0,\ v(1) = 0,\ \int_0^1 v(x)\,dx - \alpha\lambda = 0,$$

where η is a real continuously differentiable function on $(0,1)$ greater than a positive number, and α is a real nonzero constant.

Proudman showed that there are numbers A_s such that

(10.5.3) $$\sum_s A_s v_s = 0,$$

(10.5.4) $$\sum_s A_s = 1$$

holds, where $\{v_s\}$ is the sequence of the eigenvectors. Furthermore, he proved an expansion

(10.5.5) $$\sum_s B_s v_s = f,$$

(10.5.6) $$\sum_s B_s = 0$$

for functions having Sturm-Liouville expansions. The proof is done by considering the initial value problem

$$(\eta V')' + \lambda V = 0,$$
$$V(0) = 1,\ V'(0) = 0.$$

and the homogeneous boundary eigenvalue problem

$$(\eta V')' + \lambda V = 0,$$
$$V'(0) = 0,\ V(1) = 0.$$

Here we use a different approach to transform the problem into a homogeneous one. For this we observe that we may consider the value -1 on the right-hand side of the differential equation as a constant function ($\neq 0$); the special value only means a normalization of the eigenfunction. Thus we can replace -1 by a function $-y_3$ with $y_3' = 0$. We set $y_1 := v$ and $y_2 := \eta v'$. Then the boundary value problem (10.5.1), (10.5.2) can be written as

$$y_2' + \lambda y_1 + y_3 = 0, \quad y_1' - \frac{1}{\eta} y_2 = 0, \quad y_3' = 0,$$

$$y_2(0) = 0, \quad y_1(1) = 0, \quad \int_0^1 y_1(x)\,dx - \alpha\lambda y_3(0) = 0,$$

where the case $y_3 = 0$ has to be excluded; indeed, below we shall show that $y_3 \neq 0$ for every nontrivial solution of this boundary eigenvalue problem.

With $y := \begin{pmatrix} y_1 \\ y_2 \\ y_3 \end{pmatrix}$ this boundary eigenvalue problem has the form

$$y' - \begin{pmatrix} 0 & \eta^{-1} & 0 \\ -\lambda & 0 & -1 \\ 0 & 0 & 0 \end{pmatrix} y = 0, \tag{10.5.7}$$

(10.5.8)
$$\begin{pmatrix} 0 & 1 & 0 \\ 0 & 0 & 0 \\ 0 & 0 & -\lambda\alpha \end{pmatrix} y(0) + \begin{pmatrix} 0 & 0 & 0 \\ 1 & 0 & 0 \\ 0 & 0 & 0 \end{pmatrix} y(1) + \int_0^1 \begin{pmatrix} 0 & 0 & 0 \\ 0 & 0 & 0 \\ 1 & 0 & 0 \end{pmatrix} y(x)\,dx = 0.$$

In the following, we make the substitution $\lambda = \rho^2$.

We consider the transformation matrix

$$C(\cdot,\rho) = \begin{pmatrix} i & -i & 1 \\ \rho\sqrt{\eta} & \rho\sqrt{\eta} & 0 \\ 0 & 0 & -\rho^2 \end{pmatrix},$$

which has the inverse

$$C^{-1}(\cdot,\rho) = \frac{1}{2} \begin{pmatrix} -i & (\rho\sqrt{\eta})^{-1} & -i\rho^{-2} \\ i & (\rho\sqrt{\eta})^{-1} & i\rho^{-2} \\ 0 & 0 & -2\rho^{-2} \end{pmatrix}.$$

Then we obtain the following differential system

$$0 = C^{-1}(\cdot,\rho) \left\{ (C(\cdot,\rho)y)' - \begin{pmatrix} 0 & \eta^{-1} & 0 \\ -\rho^2 & 0 & -1 \\ 0 & 0 & 0 \end{pmatrix} C(\cdot,\rho)y \right\} \tag{10.5.9}$$

$$= y' + C^{-1}(\cdot,\rho)\left\{ C'(\cdot,\rho)y - \begin{pmatrix} 0 & \eta^{-1} & 0 \\ -\rho^2 & 0 & -1 \\ 0 & 0 & 0 \end{pmatrix} C(\cdot,\rho) \right\} y$$

$$= y' - \frac{1}{2}\begin{pmatrix} -i & (\rho\sqrt{\eta})^{-1} & -i\rho^{-2} \\ i & (\rho\sqrt{\eta})^{-1} & i\rho^{-2} \\ 0 & 0 & -2\rho^{-2} \end{pmatrix} \begin{pmatrix} \frac{\rho}{\sqrt{\eta}} & \frac{\rho}{\sqrt{\eta}} & 0 \\ -i\rho^2 - \frac{1}{2}\frac{\rho\eta'}{\sqrt{\eta}} & i\rho^2 - \frac{1}{2}\frac{\rho\eta'}{\sqrt{\eta}} & 0 \\ 0 & 0 & 0 \end{pmatrix} y$$

$$= y' - \rho \begin{pmatrix} -\frac{i}{\sqrt{\eta}} & 0 & 0 \\ 0 & \frac{i}{\sqrt{\eta}} & 0 \\ 0 & 0 & 0 \end{pmatrix} y + \frac{1}{4}\begin{pmatrix} \frac{\eta'}{\eta} & \frac{\eta'}{\eta} & 0 \\ \frac{\eta'}{\eta} & \frac{\eta'}{\eta} & 0 \\ 0 & 0 & 0 \end{pmatrix} y.$$

We set

$$C_2(\rho) := \begin{pmatrix} \rho\sqrt{\eta(0)} & 0 & 0 \\ 0 & 1 & 0 \\ 0 & 0 & \rho^4 \end{pmatrix}.$$

Denoting the boundary operator given in (10.5.8) by $T^R(\lambda)$ we obtain the following boundary conditions:

$$0 = C_2^{-1}(\rho)T^R(\rho^2)(C(\cdot,\rho)y) = \begin{pmatrix} 1 & 1 & 0 \\ 0 & 0 & 0 \\ 0 & 0 & \alpha \end{pmatrix} y(0) \tag{10.5.10}$$

$$+ \begin{pmatrix} 0 & 0 & 0 \\ i & -i & 1 \\ 0 & 0 & 0 \end{pmatrix} y(1) + \rho^{-4}\int_0^1 \begin{pmatrix} 0 & 0 & 0 \\ 0 & 0 & 0 \\ i & -i & 1 \end{pmatrix} y(x)\,dx.$$

Now we shall prove that the boundary value problem is Birkhoff regular in the sense of Definition 4.1.2. For this we calculate

$$\Delta_0 = \begin{pmatrix} 1 & 0 & 0 \\ 0 & 1 & 0 \\ 0 & 0 & 0 \end{pmatrix},$$

$$W_0^{(0)} = \begin{pmatrix} 1 & 1 & 0 \\ 0 & 0 & 0 \\ 0 & 0 & \alpha \end{pmatrix}, \qquad W_0^{(1)} \begin{pmatrix} 0 & 0 & 0 \\ i & -i & 1 \\ 0 & 0 & 0 \end{pmatrix},$$

$$\Delta(\lambda) = \begin{pmatrix} 1 & 0 & 0 \\ 0 & 0 & 0 \\ 0 & 0 & 0 \end{pmatrix} \quad \text{or} \quad \Delta(\lambda) = \begin{pmatrix} 0 & 0 & 0 \\ 0 & 1 & 0 \\ 0 & 0 & 0 \end{pmatrix}.$$

From $P^{[0]} = \operatorname{diag}\left(\eta(0)^{\frac{1}{4}}\eta^{-\frac{1}{4}}, \eta(0)^{\frac{1}{4}}\eta^{-\frac{1}{4}}, 1\right)$, see (2.8.17), (2.8.18), we infer

$$\widetilde{M}_2(I_n - \Delta_0) = \left(W_0^{(0)} + W_0^{(1)}P^{[0]}(1)\right)\begin{pmatrix} 0 & 0 & 0 \\ 0 & 0 & 0 \\ 0 & 0 & 1 \end{pmatrix} = \begin{pmatrix} 0 & 0 & 0 \\ 0 & 0 & 1 \\ 0 & 0 & \alpha \end{pmatrix}.$$

Now the two matrices

$$\begin{pmatrix} 1 & 0 & 0 \\ 0 & -i & 1 \\ 0 & 0 & \alpha \end{pmatrix}, \qquad \begin{pmatrix} 0 & 1 & 0 \\ i & 0 & 1 \\ 0 & 0 & \alpha \end{pmatrix}$$

are invertible which proves that the problem is Birkhoff regular.

We are interested in expansions into eigenfunctions of the problem (10.5.1), (10.5.2). By construction, an eigenfunction of (10.5.1), (10.5.2) is the first component of an eigenfunction of (10.5.7), (10.5.8) whose third component is 1. We shall show that $y_3 \neq 0$ for each eigenfunction y of (10.5.7), (10.5.8); hence, after normalization, the first component of each eigenfunction of (10.5.7), (10.5.8) is an eigenfunction of (10.5.1), (10.5.2). Indeed, assume that $\begin{pmatrix} y_1 \\ y_2 \\ 0 \end{pmatrix}$ is an eigenfunction of (10.5.7), (10.5.8) for some eigenvalue λ. Using $y_2' + \lambda y_1 = 0$ and the boundary conditions, we infer

$$y_2(1) = -\lambda \int_0^1 y_1(x)\,\mathrm{d}x = 0.$$

Therefore, $\begin{pmatrix} y_1 \\ y_2 \end{pmatrix}$ is a solution of a first order system satisfying the initial conditions $y_1(1) = 0$, $y_2(1) = 0$, and $y_1 = 0$, $y_2 = 0$ follows.

Note that the eigenspaces are one-dimensional. Indeed, if there were two linearly independent eigenfunctions of (10.5.1), (10.5.2) for some eigenvalue, then the problem (10.5.7), (10.5.8) would have the same property. Taking a suitable linear combination, we would obtain an eigenfunction of (10.5.7), (10.5.8) whose third component is zero. But this is impossible as we have seen in the previous paragraph.

Also observe that 0 is no eigenvalue of (10.5.1), (10.5.2). Indeed, if 0 were an eigenvalue and $\begin{pmatrix} y_1 \\ y_2 \\ 1 \end{pmatrix}$ an eigenfunction of (10.5.7), (10.5.8), then $y_2(x) = -x$ and

$$y_1(x) = \int_x^1 \frac{t}{\eta(t)}\,\mathrm{d}t.$$

But since $\eta > 0$, this contradicts the boundary condition $\int_0^1 y_1(x)\,\mathrm{d}x = 0$.

We denote the eigenvalues of (10.5.9), (10.5.10) by ρ_j ($j \in \mathbb{N}$) and the corresponding chains of eigenfunctions and associated functions by $w_{j,0}, \ldots, w_{j,r_j-1}$. Here we multiply the last boundary condition in (10.5.10) by ρ^4 in order to have polynomial dependence on ρ. Let $\widehat{P}:\mathbb{C}^3\to\mathbb{C}^2$ be the projection onto the first two components. Let $1<p<\infty$. From Theorem 5.3.2 we know that the expansion

$$\begin{pmatrix} f_1 \\ f_2 \end{pmatrix} = \sum_{j\in\mathbb{N}}\sum_{k=0}^{r_j-1} c_{j,k}(f_1,f_2)\widehat{P}w_{j,k} \tag{10.5.11}$$

holds for all $f_1,f_2\in L_p(0,1)$, where the series converges (in parenthesis) in the space $(L_p(0,1))^2$.

Also, if $f_1,f_2\in C[a,b]\cap \mathrm{BV}[a,b]$ and $f_1(0)+f_2(0)=0$, $f_1(1)=f_2(1)$, then the expansion (10.5.11) holds in $(L_\infty(a,b))^2$. This follows from Theorem 5.3.3. Here we have to verify the boundary conditions $B(\lambda)(f_1,f_2,0)=0$. By Remark 4.7.7 it is sufficient to consider $B_0(\lambda)(f_1,f_2,0)$. Since $P^{[0]}$ is a diagonal matrix function, it is easy to see that $(I_n-\Delta_0)P^{[0]^{-1}}A_0=0$. Hence the condition is

$$W_0^{(0)}\begin{pmatrix} f_1(0) \\ f_2(0) \\ 0 \end{pmatrix} + W_0^{(1)}\begin{pmatrix} f_1(1) \\ f_2(1) \\ 0 \end{pmatrix} = 0,$$

and the above conditions follow.

Finally let us note that 0 is an eigenvalue. A corresponding eigenfunction is $\begin{pmatrix} -1 \\ 1 \\ 2i \end{pmatrix}$. And there is also an associated function.

Now we consider (10.5.7), (10.5.8). Let λ_j $(j\in\mathbb{N})$ be its eigenvalues. Obviously, $\{\lambda_j : j\in\mathbb{N}\}\cup\{0\}=\{\rho_j^2 : j\in\mathbb{N}\}$. Let ρ be a nonzero eigenvalue of (10.5.9), (10.5.10). Then we we have an eigenfunction

$$w=C(\cdot,\rho)^{-1}\begin{pmatrix} y_1 \\ y_2 \\ 1 \end{pmatrix},$$

where $\begin{pmatrix} y_1 \\ y_2 \\ 1 \end{pmatrix}$ is the (unique normalized) eigenfunction of (10.5.7), (10.5.8) corresponding to the eigenvalue $\lambda=\rho^2$. Then

$$w^-:=C(\cdot,-\rho)^{-1}\begin{pmatrix} y_1 \\ y_2 \\ 1 \end{pmatrix}$$

is an eigenfunction of (10.5.9), (10.5.10) for the eigenvalue $-\rho$. This shows that to each eigenvalue λ_j there correspond two eigenvalues ρ_k,ρ_l.

Assuming that (10.5.7), (10.5.8) has no associated vectors, the expansion (10.5.11) can be written as

$$\begin{pmatrix} f_1 \\ f_2 \end{pmatrix} = \sum_{j\in\mathbb{N}} \left(c_j^+(f_1,f_2)\widehat{P}C(\cdot,\sqrt{\lambda_j})^{-1} + c_j^-(f_1,f_2)\widehat{P}C(\cdot,-\sqrt{\lambda_j})^{-1} \right) \begin{pmatrix} y_{1,j} \\ y_{2,j} \\ 1 \end{pmatrix} + w,$$

where $\begin{pmatrix} y_{1,j} \\ y_{2,j} \\ 1 \end{pmatrix}$ is an eigenfunction of (10.5.7), (10.5.8) for the eigenvalue λ_j, say $0 \leq \arg\sqrt{\lambda_j} < \pi$ for definiteness, and w is the term corresponding to the eigenvalue 0 of (10.5.9), (10.5.10).

We consider the contour integral leading to the above expansion. Let us denote the operator associated with the problem (10.5.7), (10.5.8) by $T(\lambda)$, and the operator associated with (10.5.9), (10.5.10) by $\widetilde{T}(\rho)$. Since

$$\widetilde{T}(\rho) = \begin{pmatrix} C^{-1}(\cdot,\rho) & 0 \\ 0 & C_2^{-1}(\rho) \end{pmatrix} T(\rho^2)C(\cdot,\rho),$$

we can write (10.5.11) as

$$\begin{aligned} \begin{pmatrix} f_1 \\ f_2 \end{pmatrix} &= -\lim_{\nu\to\infty} \frac{1}{2\pi i} \oint_{\Gamma_\nu} \widehat{P}\widetilde{T}^{-1}(\rho) \left(A \begin{pmatrix} f_1 \\ f_2 \\ 0 \end{pmatrix}, 0 \right) d\rho \\ &= -\lim_{\nu\to\infty} \frac{1}{2\pi i} \oint_{\Gamma_\nu} \widehat{P}C(\cdot,\rho)^{-1}T^{-1}(\rho^2) \left(C(\cdot,\rho)A \begin{pmatrix} f_1 \\ f_2 \\ 0 \end{pmatrix}, 0 \right) d\rho, \end{aligned}$$

where A is the coefficient of ρ in (10.5.9) and Γ_ν is a circle with centre 0 and radius r_ν. We have

$$\begin{aligned} C(\cdot,\rho)A \begin{pmatrix} f_1 \\ f_2 \\ 0 \end{pmatrix} &= \begin{pmatrix} i & -i & 1 \\ \rho\sqrt{\eta} & \rho\sqrt{\eta} & 0 \\ 0 & 0 & -\rho^2 \end{pmatrix} \begin{pmatrix} \frac{-i}{\sqrt{\eta}} & 0 & 0 \\ 0 & \frac{i}{\sqrt{\eta}} & 0 \\ 0 & 0 & 0 \end{pmatrix} \begin{pmatrix} f_1 \\ f_2 \\ 0 \end{pmatrix} \\ &= \begin{pmatrix} \frac{f_1+f_2}{\sqrt{\eta}} \\ -i\rho(f_1-f_2) \\ 0 \end{pmatrix}. \end{aligned}$$

If we set $f_2 = -f_1$ and subtract the second component from the first one in the above expansion, then we obtain, with $f := f_1$,

$$f = \lim_{\nu\to\infty} \frac{1}{2\pi i} \oint_{\Gamma_\nu} \begin{pmatrix} 1 & 0 & \rho^{-2} \end{pmatrix} T^{-1}(\rho^2) \left(\begin{pmatrix} 0 \\ \rho f \\ 0 \end{pmatrix}, 0 \right) d\rho.$$

If γ_ν denotes the circle with centre 0 and radius r_ν^2, then we can write

$$f = \lim_{\nu\to\infty} \frac{1}{2\pi i} \oint_{\gamma_\nu} \begin{pmatrix} 1 & 0 & \lambda^{-1} \end{pmatrix} T^{-1}(\lambda) \left(\begin{pmatrix} 0 \\ f \\ 0 \end{pmatrix}, 0 \right) \mathrm{d}\lambda.$$

Hence we obtain

$$f = \begin{pmatrix} 0 & 0 & 1 \end{pmatrix} T^{-1}(0) \left(\begin{pmatrix} 0 \\ f \\ 0 \end{pmatrix}, 0 \right) + \sum_{j\in\mathbb{N}} A_j(f)(y_{1,j} + \lambda_j^{-1}).$$

It is easy to check that

$$T^{-1}(0) \left(\begin{pmatrix} 0 \\ f \\ 0 \end{pmatrix}, 0 \right) = \begin{pmatrix} c\int_x^1 \frac{t}{\eta(t)}\,\mathrm{d}t - \int_x^1 \frac{1}{\eta(\tau)} \int_0^\tau f(t)\,\mathrm{d}t\,\mathrm{d}\tau \\ \int_0^x f(t)\,\mathrm{d}t - cx \\ c \end{pmatrix},$$

where

$$c = \int_0^1 \frac{1}{\eta(\tau)} \int_0^\tau f(t)\,\mathrm{d}t\,\mathrm{d}\tau \Big/ \int_0^1 \frac{t}{\eta(t)}\,\mathrm{d}t.$$

Hence we obtain the expansion

$$f = \sum_{j\in\mathbb{N}} A_j(f) y_{1,j} + \sum_{j\in\mathbb{N}} A_j(f)\lambda_j^{-1} + \int_0^1 \frac{1}{\eta(\tau)} \int_0^\tau f(t)\,\mathrm{d}t\,\mathrm{d}\tau \Big/ \int_0^1 \frac{t}{\eta(t)}\,\mathrm{d}t.$$

This holds for all $f \in L_p(0,1)$ if $1 < p < \infty$. If $p = \infty$, we have to impose the restriction $f \in C[a,b] \cap \mathrm{BV}[a,b]$ and $f(1) = 0$. Since $y_{1,j}$ is an eigenfunction of (10.5.1), (10.5.2), we have $y_{1,j}(1) = 0$. Therefore, for uniform convergence, the above expansion splits into two parts:

$$f = \sum_{j\in\mathbb{N}} A_j(f) y_{1,j} \tag{10.5.12}$$

and

$$\sum_{j\in\mathbb{N}} A_j(f)\lambda_j^{-1} = -\int_0^1 \frac{1}{\eta(\tau)} \int_0^\tau f(t)\,\mathrm{d}t\,\mathrm{d}\tau \Big/ \int_0^1 \frac{t}{\eta(t)}\,\mathrm{d}t. \tag{10.5.13}$$

This is not the expansion with the coefficients as obtained by PROUDMAN. So let us consider a different transformation. We take the transformation matrix

$$C(\cdot,\rho) = \begin{pmatrix} i & -i & 0 \\ \rho\sqrt{\eta} & \rho\sqrt{\eta} & 0 \\ 0 & 0 & -1 \end{pmatrix}. \tag{10.5.14}$$

Then

$$C^{-1}(\cdot,\rho) = \frac{1}{2} \begin{pmatrix} -i & (\rho\sqrt{\eta})^{-1} & 0 \\ i & (\rho\sqrt{\eta})^{-1} & 0 \\ 0 & 0 & -2 \end{pmatrix},$$

and we obtain the differential system

$$0 = C^{-1}(\cdot,\rho)\left\{(C(\cdot,\rho)y)' - \begin{pmatrix} 0 & \eta^{-1} & 0 \\ -\rho^2 & 0 & -1 \\ 0 & 0 & 0 \end{pmatrix} C(\cdot,\rho)y\right\}$$

$$= y' - \rho\begin{pmatrix} -\frac{i}{\sqrt{\eta}} & 0 & 0 \\ 0 & \frac{i}{\sqrt{\eta}} & 0 \\ 0 & 0 & 0 \end{pmatrix} y + \frac{1}{4}\begin{pmatrix} \frac{\eta'}{\eta} & \frac{\eta'}{\eta} & 0 \\ \frac{\eta'}{\eta} & \frac{\eta'}{\eta} & 0 \\ 0 & 0 & 0 \end{pmatrix} y - \frac{1}{\rho}\begin{pmatrix} 0 & 0 & \frac{1}{2\sqrt{\eta}} \\ 0 & 0 & \frac{1}{2\sqrt{\eta}} \\ 0 & 0 & 0 \end{pmatrix} y.$$

We set

$$C_2(\rho) := \begin{pmatrix} \rho\sqrt{\eta(0)} & 0 & 0 \\ 0 & i & 0 \\ 0 & 0 & \rho^2 \end{pmatrix} \tag{10.5.15}$$

and obtain the following boundary conditions:

$$0 = C_2^{-1}(\rho)T^R(\rho^2)(C(\cdot,\rho)y) = \begin{pmatrix} 1 & 1 & 0 \\ 0 & 0 & 0 \\ 0 & 0 & \alpha \end{pmatrix} y(0)$$

$$+ \begin{pmatrix} 0 & 0 & 0 \\ 1 & -1 & 0 \\ 0 & 0 & 0 \end{pmatrix} y(1) + \rho^{-2}\int_0^1 \begin{pmatrix} 0 & 0 & 0 \\ 0 & 0 & 0 \\ 1 & -1 & 0 \end{pmatrix} y(x)\,dx.$$

Here

$$\widetilde{M}_2(I_n - \Delta_0) = \begin{pmatrix} 0 & 0 & 0 \\ 0 & 0 & 0 \\ 0 & 0 & \alpha \end{pmatrix}$$

and the two Birkhoff matrice are

$$\begin{pmatrix} 1 & 0 & 0 \\ 0 & -1 & 0 \\ 0 & 0 & \alpha \end{pmatrix}, \qquad \begin{pmatrix} 0 & 1 & 0 \\ 1 & 0 & 0 \\ 0 & 0 & \alpha \end{pmatrix}.$$

Since they are invertible, also this problem is Birkhoff regular in the sense of Definition 4.1.2.

Since the differential system is not holomorphic on $\mathbb{C}$, we do not obtain an expansion in terms of eigenfunctions of this system. But we are not interested in eigenfunction expansions of this system, only in the first component of it. Most important is that the Birkhoff regularity yields as in the previous case that

$$\begin{pmatrix} f_1 \\ f_2 \end{pmatrix} = -\lim_{\nu\to\infty}\frac{1}{2\pi i}\oint_{\Gamma_\nu} \widehat{P}C(\cdot,\rho)^{-1}T^{-1}(\rho)\left(C(\cdot,\rho)A\begin{pmatrix} f_1 \\ f_2 \\ 0 \end{pmatrix}, 0\right) d\rho$$

with the same assumptions on f_1, f_2 as above, and where A is the coefficient of ρ in the differential system. We have

$$C(\cdot,\rho)A\begin{pmatrix} f_1 \\ f_2 \\ 0 \end{pmatrix} = \begin{pmatrix} \frac{f_1+f_2}{\sqrt{\eta}} \\ -i\rho(f_1-f_2) \\ 0 \end{pmatrix}.$$

If we set $f_2 = -f_1$ and subtract the second component from the first one in the above expansion, then we obtain, with $f := f_1$,

$$f = \lim_{\nu\to\infty}\frac{1}{2\pi i}\oint_{\Gamma_\nu}\begin{pmatrix} 1 & 0 & 0 \end{pmatrix} T^{-1}(\rho^2)\left(\begin{pmatrix} 0 \\ \rho f \\ 0 \end{pmatrix}, 0\right) \mathrm{d}\rho.$$

If γ_ν denotes the circle with centre 0 and radius r_ν^2, then we can write

$$f = \lim_{\nu\to\infty}\frac{1}{2\pi i}\oint_{\gamma_\nu}\begin{pmatrix} & 0 & 0 \end{pmatrix} T^{-1}(\lambda)\left(\begin{pmatrix} 0 \\ f \\ 0 \end{pmatrix}, 0\right) \mathrm{d}\lambda.$$

Hence we obtain

$$f = \sum_{j\in\mathbb{N}} A_j(f)y_{1,j}.$$

This holds for all $f \in L_p(0,1)$ if $1 < p < \infty$. If $p = \infty$, we have to impose the restriction $f \in C[a,b] \cap \mathrm{BV}[a,b]$ and $f(1) = 0$. This coincides with the first expansion (10.5.12), but we do not obtain the identity (10.5.13) here.

Now we shall use still another method. We consider the operator T associated with (10.5.7), (10.5.8). As we have done for Stone regular problems, we write for $f \in (W_p^1(0,1))^3$

$$\frac{1}{\lambda}f = T^{-1}(\lambda)T_1 f + \frac{1}{\lambda}T^{-1}(\lambda)T_0 f, \tag{10.5.16}$$

where $T(\lambda) =: \lambda T_1 + T_0$. Now we are looking for $f^{[1]} \in (W_p^1(0,1))^3$ such that

$$T_0 f = T_1 f^{[1]}. \tag{10.5.17}$$

That is, we have to satisfy

$$\begin{pmatrix} f_1' - \eta^{-1}f_2 \\ f_2' - f_3 \\ f_3' \end{pmatrix} = \begin{pmatrix} 0 \\ f_2^{[1]} \\ 0 \end{pmatrix},$$

and, for the boundary part,

$$f_1(0) = f_2(1) = 0, \quad -\alpha f_3^{[1]}(0) = \int_0^1 f_1(x)\,\mathrm{d}x.$$

That means, we have to choose f_3 to be constant. Since the third components of the eigenfunctions of T are constant, this is also necessary for an eigenfunction expansion. Furthermore, we have to require that $\eta f_1' = f_2 \in W_p^1(0,1)$ and also $(\eta f_1')' = f_2' = f_3 - f_2^{[1]} \in W_p^1(0,1)$, and, for the boundary part, $f_1(0) = f_2(1) = 0$. The last boundary condition can always be satisfied by a suitable choice of $f_3^{[1]}$. Inserting (10.5.17) into (10.5.16) yields

$$\frac{1}{\lambda} f = T^{-1}(\lambda) T_1 f + \frac{1}{\lambda^2} f^{[1]} - \frac{1}{\lambda^2} T^{-1}(\lambda) T_0 f^{[1]}.$$

Integrating along γ_ν as defined above we obtain

$$f = \frac{1}{2\pi i} \oint_{\gamma_\nu} T^{-1}(\lambda) T_1 f \, d\lambda - \frac{1}{2\pi i} \oint_{\Gamma_\nu} \frac{1}{\rho^3} T^{-1}(\rho^2) T_0 f^{[1]} \, d\rho.$$

With the second transformation, i. e., (10.5.14) and (10.5.15), we obtain

$$\begin{aligned} &\frac{1}{2\pi i} \oint_{\Gamma_\nu} \frac{1}{\rho^3} T^{-1}(\rho^2) T_0 f^{[1]} \, d\rho \\ &= \frac{1}{2\pi i} \oint_{\Gamma_\nu} \frac{1}{\rho^2} \rho^{-1} C(\cdot,\rho) \widetilde{T}^{-1}(\rho) \begin{pmatrix} C^{-1}(\cdot,\rho) T_0^D f^{[1]} \\ C_2^{-1}(\rho) T_0^R f^{[1]} \end{pmatrix} d\rho. \end{aligned}$$

Since $\rho^{-1} C(\cdot,\rho) = O(1)$, $C^{-1}(\cdot,\rho) = O(1)$, and $C_2^{-1}(\rho) = O(1)$, we obtain in view of Theorem 4.4.9 ii) that

$$f = \lim_{\nu \to \infty} \frac{1}{2\pi i} \oint_{\gamma_\nu} T^{-1}(\lambda) T_1 f \, d\lambda$$

holds in $L_p(a,b)$ for $f = (f_1, f_2, f_3)^{\mathsf{T}}$ satisfying the above conditions.

Now assume that the problem has only simple eigenvalues. Taking $f_1 = f_2 = 0$ and $f_3 = 1$, we obtain the expansion (10.5.3), (10.5.4) of PROUDMAN. And taking $f_3 = 0$, we obtain the expansion (10.5.5), (10.5.6) of PROUDMAN, albeit under the stronger assumption that $(\eta f_1')' \in W_p^1(0,1)$ and $f_1(0) = f_1'(1) = 0$. As we have seen for Stone regular problems in Section 5.7, the condition $(\eta f_1')' \in W_p^1(0,1)$ might be weakened.

The expansions obtained by PROUDMAN show that the coefficients of an expansion in eigenvectors of the problem (10.5.1), (10.5.2) are not unique. Taking an "intermediate" substitution

$$\widehat{T}^D \begin{pmatrix} y_1 \\ y_2 \end{pmatrix} := \begin{pmatrix} (\eta y_1')' + \lambda y_1 + y_2 \\ y_2' \end{pmatrix},$$

$$\widehat{T}^R \begin{pmatrix} y_1 \\ y_2 \end{pmatrix} := \left(y_1'(0), y_1(1), -\lambda \alpha y_2(0) + \int_0^1 y_1(x) \, dx \right)^{\mathsf{T}},$$

we obtain an operator

$$\widehat{T}(\lambda) := (\widehat{T}^D(\lambda), \widehat{T}^R(\lambda)) : W_p^2(0,1) \times W_p^1(0,1) \to (L_p(0,1))^2 \times \mathbb{C}^3 .$$

As for first order systems of differential equations and n-th order scalar differential equations, the operator $\widehat{T}(\lambda)$ is a Fredholm operator of index 0. Also, writing

$\widehat{T}(\lambda) = \widehat{T}_0 + \lambda \widehat{T}_1$, we see that $\widehat{T}_1$ is compact in view of Theorem 2.4.2. Since the eigenvalues of $\widehat{T}$ are the eigenvalues of (10.5.1), (10.5.2), we infer that $\rho(\widehat{T}) \neq \emptyset$. Therefore, Proposition 1.10.6 shows that the coefficients of an expansion

$$\begin{pmatrix} f \\ c \end{pmatrix} = \sum c_j \begin{pmatrix} y_{1,j} \\ 1 \end{pmatrix}$$

are unique for all $f \in L_p(0,1)$ and $c \in \mathbb{C}$ for which such an expansion holds.

In the previous results we assumed for simplicity that there are no associated functions. We have seen that this means that all eigenvalues are simple. Of course, our general results about the principal parts of inverses of Fredholm operator functions immediately give expansion theorems in case there are associated functions.

Let us show that associated functions can occur. Let c be a positive constant and consider the case $\eta = c$. For $\lambda \neq 0$ and $\cos\sqrt{\frac{\lambda}{c}} \neq 0$,

$$v(x) := \frac{\cos\left(\sqrt{\frac{\lambda}{c}}x\right)}{\lambda \cos\sqrt{\frac{\lambda}{c}}} - \frac{1}{\lambda}$$

is the unique function satisfying (10.5.1) and the first two boundary conditions in (10.5.2). In order to satisfy the third boundary condition, we have to satisfy

$$\sqrt{\frac{c}{\lambda}} \tan\sqrt{\frac{\lambda}{c}} - 1 = \alpha\lambda^2. \tag{10.5.18}$$

Then λ is a double eigenvalue if it satisfies (10.5.18) and

$$-\frac{1}{2}\sqrt{c}\lambda^{-\frac{3}{2}} \tan\sqrt{\frac{\lambda}{c}} + \frac{1}{2\lambda}\left(1 + \tan^2\sqrt{\frac{\lambda}{c}}\right) = 2\alpha\lambda.$$

These two equations hold if and only if (10.5.18) and

$$-\frac{1}{4}\sqrt{\frac{c}{\lambda}} \tan\sqrt{\frac{\lambda}{c}} + \frac{1}{4}\left(1 + \tan^2\sqrt{\frac{\lambda}{c}}\right) = \sqrt{\frac{c}{\lambda}} \tan\sqrt{\frac{\lambda}{c}} - 1$$

are satisfied. Setting $t = \sqrt{\frac{\lambda}{c}}$ the latter equation means that we are looking for a zero of

$$h(t) = \frac{5}{t}\tan t - 5 - \tan^2 t$$

with $t \neq 0$. In the interval $(0, \frac{\pi}{2})$ we have $h(t) = \frac{2}{3}t^2 + O(t^4)$ near zero, i. e. $h(t) > 0$ for small positive t. But $h(t) \to -\infty$ as $t \to \frac{\pi}{2}$, and so there must be a number $t_0 \in (0, \frac{\pi}{2})$ such that $h(t_0) = 0$. With $\lambda = ct_0^2$ and α such that (10.5.18) is satisfied we obtain a multiple eigenvalue. Note that $\alpha > 0$.

10.6. The Orr-Sommerfeld equation

We consider the Orr-Sommerfeld equation

$$\phi^{(4)} - 2\alpha^2\phi'' + \alpha^4\phi - iR\alpha\{(u-\lambda)(\phi'' - \alpha^2\phi) + 2\phi\} = 0 \tag{10.6.1}$$

on the interval $[a,b]$ subject to the boundary conditions

$$\phi(a) = \phi(b) = \phi'(a) = \phi'(b) = 0, \tag{10.6.2}$$

see [DR1, p. 156]. This differential equation results if one considers the perturbation of a given plane flow. Here the Reynolds number R and the wave number α are given nonzero real numbers and u is a given function describing an unperturbed flow. Examples for u are the plane Couette flow where $u(x) = x$ and the plane Poiseuille flow where $u(x) = 1 - x^2$ on $[-1,1]$. Expansion theorems for this problem where obtained by SCHENSTEDT [SS1] and DIPRIMA and HABETLER [DH1]. The differential equation is as considered in Chapter VIII. Here we have $l = n - l = 2$, and the numbers given by (8.5.1) and (8.5.7) are $l_1 = l_2 = 0$, $l_3 = l_4 = 1$, $q_\nu = 0$ $(\nu = 1,\dots,4)$. Therefore, $l^{(0)} = l^{(1)} = 2$. We have

$$b_1^{(0)} = u_{\{3,4\}} v_{\{3,4\},1},$$

where

$$u_{\{3,4\}} = \left|\begin{pmatrix} u_{1,1} & u_{1,2} \\ u_{2,1} & u_{2,2} \end{pmatrix}\right|,$$

$$v_{\{3,4\},1} = \left|\begin{pmatrix} \omega_1\alpha_3^{(0)} & \omega_2\alpha_3^{(1)} \\ \omega_1\alpha_4^{(0)} & \omega_2\alpha_4^{(1)} \end{pmatrix}\right| = -\alpha_3^{(0)}\alpha_4^{(1)},$$

see (8.5.11), (8.5.12), and (8.5.13). Since $\mathbf{H}\phi = -iR\alpha(\phi'' - \alpha^2\phi)$, a fundamental system of $\mathbf{H}\phi = 0$ is given by $\{\cosh(\alpha(x-a)), \sinh(\alpha(x-a))\}$. In view of Theorem 8.2.1 and (8.5.8) it follows that

$$\begin{pmatrix} u_{1,1} & u_{1,2} \\ u_{2,1} & u_{2,2} \end{pmatrix} = \begin{pmatrix} I_2 & 0 \end{pmatrix} [M(\lambda)] \begin{pmatrix} I_2 \\ 0 \end{pmatrix}$$
$$= \begin{pmatrix} 1 & 0 \\ \cosh(\alpha(b-a)) & \sinh(\alpha(b-a)) \end{pmatrix},$$

and therefore $u_{\{3,4\}} \neq 0$. From (8.5.5) and (8.5.6) we infer $\alpha_{30}^{(0)} = 1$, $\alpha_{31}^{(0)} = 0$, $\alpha_{40}^{(1)} = 1$, $\alpha_{41}^{(1)} = 0$, and thus $\alpha_3^{(0)} \neq 0$, $\alpha_4^{(1)} \neq 0$. Altogether, $b_1^{(0)} \neq 0$, and hence the boundary eigenvalue problem is Birkhoff regular.

Since $\alpha_{100}^{(0)} = 1$, $\alpha_{200}^{(1)} = 1$, $\alpha_{300}^{(0)} = 1$, $\alpha_{400}^{(1)} = 1$, and all the other $\alpha_{\nu t m}^{(\iota)}$ are zero, we obtain for the asymptotic boundary conditions by Remark 8.5.5 iii):

$$U_1 f = f(a),\ U_2(f) = f(b),\ U_3(f) = f'(a),\ U_4(f) = f'(b).$$

Then Theorems 8.8.2 and 8.8.3 yield that a function f is expandable into a series of eigenfunctions and associated functions of the eigenvalue problem (10.6.1), (10.6.2) if
i) $f \in C[a,b] \cap BV[a,b]$ such that $f(a) = f(b) = 0$, and the series converges in $C[a,b]$,
or
ii) $f \in C^1[a,b]$ such that $f' \in BV[a,b]$ and $f(a) = f'(a) = f(b) = f'(b) = 0$, and the series converges in $C^1[a,b]$,
or
iii) $f \in L_p(a,b)$, $1 < p < \infty$, and the series converges in $L_p(a,b)$,
or
iii) $f \in W_p^1(a,b)$, $1 < p < \infty$, such that $f(a) = f(b) = 0$, and the series converges in $W_p^1(a,b)$.

10.7. A system of differential equations in the theory of viscous fluids

In the theory of viscous fluids, the following system of differential equations occurs, see [DR, p. 155]:

$$\begin{aligned} \{D^2 - \alpha^2 - i\alpha R(U-c)\}u &= R\tilde{U}w + i\alpha Rp, \\ \{D^2 - \alpha^2 - i\alpha R(U-c)\}w &= RDp, \\ i\alpha u + Dw &= 0, \end{aligned}$$

where $D = d/dx$, U and $\tilde{U}$ are given functions, c is the eigenvalue parameter, u, w, p are the unknown functions, and α and R are nonzero real constants. Boundary conditions are given by

$$u(a) = u(b) = w(a) = w(b) = 0.$$

Eliminating the unknowns u and p this leads to the Orr-Sommerfeld equation as considered in Section 10.6. Hence we obtain expansions in terms of the eigenfunction components corresponding to w and u. For the w-component we obtain exactly the same expansions as in Section 10.6, whereas for the u-component the corresponding expansions in the spaces $L_p(a,b)$ and $C[a,b]$ hold, where the functions f which are expanded additionally satisfy $\int_a^b f(x)\,dx = 0$.

10.8. Heat-conducting viscous fluid

After linearization and separation of variables, the differential equation of a heat-conducting viscous fluid leads to the differential equation

$$(D^2 - a^2)(D^2 - a^2 - s)(D^2 - a^2 - s/P)\eta + a^2R\eta = 0,$$

where $D = d/dx$ and R and P are the Rayleigh and the Prandtl number, respectively, and the boundary conditions

$$\eta(c) = \eta'(c) = \eta^{(4)} - (2a^2 + s/P)\eta''(c) = 0 \qquad \text{(rigid boundary)}$$

$$\eta(c) = \eta''(c) = \eta^{(4)}(c) = 0 \qquad \text{(free boundary)}$$

at the endpoints $c = 0, 1$, see [DR1, p. 43]. Here s is a certain "eigenvalue" parameter, and a is a parameter occuring in the separation of variables. Therefore, we can also consider a as an eigenvalue parameter. Hence we shall investigate the expansion problem with respect to the eigenvalues s and a. Of course, this should be considered as a two-parameter problem, but we ask here what happens if we fix one of the parameters. We write the differential equation in the form

$$\eta^{(6)} + P_4(a,s)\eta^{(4)} + P_2(a,s)\eta'' + P_0(a,s)\eta = 0$$

where

$$\begin{aligned} P_4(a,s) &= -\{3a^2 + s(1 + 1/P)\}, \\ P_2(a,s) &= a^2(a^2 + s) + (2a^2 + s)(a^2 + s/P), \\ P_0(a,s) &= -a^2(a^2 + s)(a^2 + s/P) + a^2 R. \end{aligned}$$

Let us first take the eigenvalue parameter $\lambda = a$. Then

$$\pi_{2,2} = -3,\ \pi_{4,4} = 3,\ \pi_{6,6} = -1,$$

and $\pi_{j,j} = 0$ for $j = 1, 3, 5$. In this case we have $n_0 = 0$. The characteristic function is

$$\rho^6 - 3\rho^4 + 3\rho^2 - 1 = (\rho^2 - 1)^3.$$

The roots are not simple, and hence our expansion theorems are not applicable.

Now consider the eigenvalue parameter $\lambda^2 = s$. Then we obtain

$$\pi_{2,2} = -(1 + 1/P),\ \pi_{4,4} = 1/P,$$

and $\pi_{j,j} = 0$ for $j = 1, 3, 5, 6$. The results of Chapter VIII are not applicable since the differential equation contains two different powers of the eigenvalue parameter. Therefore we shall transform the boundary eigenvalue problem into one for a first order system. We are going to consider three different transformations.

Writing the differential equation in the usual way as a system for the vector function $y = (\eta, \eta', \ldots, \eta^{(5)})^{\mathsf{T}}$, the transformation

$$y = \begin{pmatrix} 1 & 0 & 0 & 0 & 0 & 0 \\ \alpha & 1 & 0 & 0 & 0 & 0 \\ a^2 & 0 & 1 & 0 & 0 & 0 \\ \alpha a^2 & a^2 & \lambda & 1 & 0 & 0 \\ a^4 & 0 & \lambda^2 + 2a^2 & 0 & 1 & 0 \\ \alpha a^4 & a^4 & \lambda^3 + 2a^2\lambda & \lambda^2 + 2a^2 & \frac{\lambda}{\sqrt{P}} & 1 \end{pmatrix} \tilde{y}$$

yields a full linearization of the resulting system with the coefficient matrix of λ being diagonal:

$$\tilde{y}' = \begin{pmatrix} \alpha & 1 & 0 & 0 & 0 & 0 \\ a^2-\alpha^2 & -\alpha & 1 & 0 & 0 & 0 \\ 0 & 0 & \lambda & 1 & 0 & 0 \\ 0 & 0 & a^2 & -\lambda & 1 & 0 \\ 0 & 0 & 0 & 0 & \frac{\lambda}{\sqrt{P}} & 1 \\ -a^2R & 0 & 0 & 0 & a^2 & -\frac{\lambda}{\sqrt{P}} \end{pmatrix} \tilde{y}.$$

Here α is a free parameter. Note that this transformation even holds in case $P = 1$ when 1 and -1 are double roots of the characteristic equation. However, due to the fact that the transformation matrix is lower triangular, the last column of each boundary matrix is zero. Indeed, it is easy to see that for the asymptotic boundary matrices, the last three columns of the leading (λ-independent) matrix are always zero. Then this also holds for the Birkhoff matrices, and therefore this λ-linearized problem is not Birkhoff regular.

Of course, one might ask for Stone regularity. We leave this to the interested reader.

Now we are asking if the asymptotic linearizations in Theorem 7.2.4 lead to Birkhoff regular problems. Here we must suppose that $P \neq 1$. Taking first Theorem 7.2.4 A we obtain the transformation matrix

$$C(\lambda) = \begin{pmatrix} 1 & 0 & 0 & 0 & 0 & 0 \\ 0 & 1 & 0 & 0 & 0 & 0 \\ a^2 & 0 & 1 & 1 & 1 & 1 \\ 0 & 0 & \lambda & -\lambda & \frac{\lambda}{\sqrt{P}} & -\frac{\lambda}{\sqrt{P}} \\ 0 & 0 & \lambda^2 & \lambda^2 & \frac{\lambda^2}{P} & \frac{\lambda^2}{P} \\ 0 & 0 & \lambda^3 & -\lambda^3 & \frac{\lambda^3}{P^{3/2}} & -\frac{\lambda^3}{P^{3/2}} \end{pmatrix}. \tag{10.8.1}$$

With Theorem 7.2.4 B we have the transformation matrix

$$C(\lambda) = \begin{pmatrix} 1 & 0 & 0 & 0 & 0 & 0 \\ 0 & 1 & 1 & -1 & \sqrt{P} & -\sqrt{P} \\ 0 & 0 & \lambda & \lambda & \lambda & \lambda \\ 0 & 0 & \lambda^2 & -\lambda^2 & \frac{\lambda^2}{\sqrt{P}} & -\frac{\lambda^2}{\sqrt{P}} \\ 0 & 0 & \lambda^3 & \lambda^3 & \frac{\lambda^3}{P} & \frac{\lambda^3}{P} \\ 0 & 0 & \lambda^4 & -\lambda^4 & \frac{\lambda^4}{P^{3/2}} & -\frac{\lambda^4}{P^{3/2}} \end{pmatrix}. \tag{10.8.2}$$

In both cases, the coefficient matrix of λ in the transformed system is the same as in the fully linearized system. For the transformed system with the transformation

(10.8.1) we obtain that the upper left-hand 2×2 block of the λ-independent matrix is

$$A_{0,00} = \begin{pmatrix} 0 & 1 \\ a^2 & 0 \end{pmatrix}.$$

The fundamental matrix of $y_0' = A_{0,00}y$ which is the identity matrix at 0 is given by

$$\begin{pmatrix} \cosh(ax) & \frac{1}{a}\sinh(ax) \\ a\sinh(ax) & \cosh(ax) \end{pmatrix}.$$

A straightforward calculation shows that the problem is Birkhoff regular only in the case when both boundaries are free. For the transformed system with the transformation (10.8.2) we obtain that the upper left-hand 2×2 block of the λ-independent matrix is

$$A_{0,00} = \begin{pmatrix} 0 & 1 \\ 0 & 0 \end{pmatrix}.$$

The fundamental matrix of $y_0' = A_{0,00}y$ which is the identity matrix at 0 is given by

$$\begin{pmatrix} 1 & x \\ 0 & 0 \end{pmatrix}.$$

Here a straightforward calculation shows that the problem is Birkhoff regular with any combination of rigid and free boundaries.

In case we use the transformation (10.8.1), the results of Section 7.9 are applicable, and we obtain that every function in $L_p(a,b)$ $(1\le p\le\infty)$ is expandable into a series of second order derivatives of eigenfunctions and associated functions of the given eigenvalue problem—where the series converges in paranthesis in $L_p(a,b)$—if both boundaries are free. Also, one can show that the adjoint problem is of the same form. Therefore, the eigenfunctions and associated functions of the given eigenvalue problem with free boundaries are complete in the reflexive space $L_p(a,b)$.

In case we use the transformation (10.8.2), the results of Section 7.9 are not applicable. Therefore we use a modification which works in this case. We shall first prove an abstract result.

Suppose that the general assumptions of Chapter VI are satisfied. For $j = 1,\dots,n-1$ we define

$$L_j^D(\lambda)f := f^{(n-j)} + \sum_{i=0}^{n-1-j} p_{j+i}(\cdot,\lambda)f^{(i)} \quad (f\in W_p^{n-j}(a,b)).$$

Note that

$$L_j^D(\lambda)f = e_n^{\mathsf{T}}T^D(\lambda)(0,\dots,0,f,f',\dots,f^{(n-j-1)})^{\mathsf{T}}.$$

Similarly, we define

$$L_j^R(\lambda)f = T^R(\lambda)(0,\dots,0,f,f',\dots,f^{(n-j-1)})^\top \quad (f \in W_p^{n-j}(a,b)).$$

PROPOSITION 10.8.1. *Let* $j \in \{1,\dots,n-1\}$, $\lambda \in \rho(L)$, $f_1 \in W_p^{n-j}(a,b)$, *and* $f_2 \in \mathbb{C}^n$. *Then*

$$[L^{-1}(\lambda)(L_j^D(\lambda)f_1, f_2 + L_j^R(\lambda)f_1)]^{(k)} = e_{k+1}^\top T^{-1}(\lambda)(e_j f_1, f_2) \tag{10.8.3}$$

for $k = 0\dots,j$.

Proof. Let $y := T^{-1}(\lambda)(e_j f_1, f_2)$. Then $T^D(\lambda)y = e_j f_1$, and it follows that

$$y_i' = y_{i+1} \ (i = 1,\dots,n-1, i \neq j), \quad y_j' = y_{j+1} + f_1.$$

Therefore

$$y_i = y_1^{(i-1)} \ (i = 2,\dots,j), \quad y_i = y_1^{(i-1)} - f_1^{(i-j-1)} \ (i = j+1,\dots,n). \tag{10.8.4}$$

Hence the last component of $T^D(\lambda)y = e_j f$ leads to

$$L^D(\lambda)y_1 - L_j^D(\lambda)f_1 = 0.$$

Then

$$f_2 = T^R(\lambda)y = L^R(\lambda)y_1 - L_j^R(\lambda)f_1,$$

which proves altogether

$$(L_j^D(\lambda)f_1, f_2 + L_j^R(\lambda)f_1) = L(\lambda)y_1,$$

and with the aid of (10.8.4) it follows that

$$[L^{-1}(\lambda)(L_j^D(\lambda)f_1, f_2 + L_j^R(\lambda)f_1)]^{(k)} = y_1^{(k)} = y_{k+1} = e_{k+1}^\top T^{-1}(\lambda)(e_j f_1, f_2). \quad \square$$

Now we return to the transformation (10.8.2). In this case, the transformation $\widehat{A}$ is given by (7.2.24), and thus $\widehat{A}e_4 = e_3$ since $\pi_{3,3} = 0$. As in (7.9.4) this leads to

$$C_1^{-1}e_3 = C_1^{-1}\widehat{A}e_4 = A_1 C_1^{-1}e_4,$$

and therefore, in view of Proposition 10.8.1 and with $\widetilde{T}$ as defined in Section 7.4,

$$[L^{-1}(\lambda)(L_3^D(\lambda)f, L_3^R(\lambda)f)]^{(3)} = \lambda e_4^\top C_1 \widetilde{T}^{-1}(\lambda)(A_1 C_1^{-1}e_4 f, 0)$$

for $f \in W_p^2(0,1)$. This yields that

$$-\frac{1}{2\pi i}\oint_{\Gamma_\nu} \lambda^{-1}[L^{-1}(\lambda)(L_3^D(\lambda)f, L_3^R(\lambda)f)]^{(3)}\,d\lambda = e_4^\top C_1 P_\nu C_1^{-1} e_4 f \to f$$

as $\nu \to \infty$ in $L_p(a,b)$ for all $f \in W_p^2(0,1)$, see Theorem 4.6.9. The left hand side can be expressed in terms of third derivatives of eigenfunctions and associated functions of the given problem and an additional residue at 0. In view of the structure of $\widehat{A}$ it is impossible to avoid this additional residue, i. e., for $c \in \mathbb{C}^n$ such that $e_i^\top \widehat{A}c = 0$ for $i < j$ and $e_j^\top \widehat{A}c \neq 0$ for $i = j$ where $j \geq 3$ we have $e_i^\top c \neq 0$ for some $i > j$.

Taking $a = 0$ we obtain that a fundamental system of the differential equation at $\lambda = 0$ is given by the functions $\frac{x^j}{j!}$ $(j = 0, \ldots, 5)$, and with any combination of rigid and free boundary conditions, the characteristic matrix is different from zero. Since the boundary eigenvalue problem depends polynomially on a and R, it follows that the characteristic matrix depends holomorphically on a and R. Therefore it is nonzero for almost all a and R (which have a physical meaning).

Therefore let us assume that $0 \in \rho(L)$. Then it follows that for each $f \in W_p^3(0,1)$ the function

$$f + [L^{-1}(0)(L_3^D(0)f, L_3^R(0)f)]^{(3)} \tag{10.8.5}$$

is expandable into a series of third order derivatives of eigenfunctions and associated functions of the given eigenvalue problem.

We want to determine those functions which have the representation (10.8.5). Obviously,

$$f \mapsto f + [L^{-1}(0)(L_3^D(0)f, L_3^R(0)f)]^{(3)}$$

is a continuous operator from $W_p^3(0,1)$ to $W_p^3(0,1)$. In order to find an estimate for the defect of this operator we first consider the case $a = 0$. Let $h \in W_p^3(0,1)$. Above we have seen that the eigenvalue problem

$$g^{(6)} = \frac{1}{2}h^{(3)},$$

$$g(0) = g(1) = 0,\ g^{(4)}(0) = \frac{1}{2}h'(0),\ g^{(4)}(1) = \frac{1}{2}h'(1),\ g^{(\iota_c)}(c) = 0,$$

for $c = 0, 1$, where $\iota_c = 1$ if the boundary is rigid and $\iota_c = 2$ if the boundary is free, has a (unique) solution $g \in W_p^6(0,1)$. With $f = h - g^{(3)} \in W_p^3(0,1)$ and

$$v := f + [L^{-1}(0)(L_3^D(0)f, L_3^R(0)f)]^{(3)} - h$$

it follows in view of $L^D(0)\eta = \eta^{(6)}$ that

$$\begin{aligned} v^{(3)} &= f^{(3)} + L^D(0)L^{-1}(0)(L_3^D(0)f, L_3^R(0)f) - h^{(3)} \\ &= f^{(3)} + L_3^D(0)f - h^{(3)} = 2f^{(3)} - h^{(3)} \end{aligned}$$

since

$$L_3^D(0)f = f^{(3)} - 3a^2 f' = f^{(3)}.$$

But

$$2f^{(3)} - h^{(3)} = 2(h^{(3)} - g^{(6)}) - h^{(3)} = 0$$

shows that $v^{(3)} = 0$. With

$$w := L^{-1}(0)(L_3^D(0)f, L_3^R(0)f) - g$$

we have

$$v = f + g^{(3)} + w^{(3)} - h = w^{(3)},$$

and therefore, for $c = 0, 1$,

$$v'(c) = w^{(4)}(c) = f'(c) - g^{(4)}(c) = h'(c) - 2g^{(4)}(c) = 0,$$

which gives altogether that v is constant. And since the third derivative does not occur in the boundary part in $L(0)$, we do not know if we can find f such that $v = 0$. This shows that for $a = 0$ the operator given by (10.8.5) has defect at most 1. Also the nullity is at most 1. Indeed, if

$$f + [L^{-1}(0)(L_3^D(0)f, L_3^R(0)f)]^{(3)} = 0,$$

then we set

$$g := L^{-1}(0)(L_3^D(0)f, L_3^R(0)f)$$

and obtain

$$2f^{(3)} = f^{(3)} + L_3^D(0)f = f^{(3)} + g^{(6)} = 0$$

and $2f'(c) = f'(c) + g^{(4)}(c) = 0$ for $c = 0, 1$. Therefore f must be constant. Then we can find an extension of this operator which is at most one-dimensional in the domain and range spaces such that this extension is invertible. Since this operator depends holomorphically on a and R (for which $L(0)$ is invertible), it follows for almost all a and R that each function f which belong to a subspace of $W_p^{(3)}(0,1)$ of codimension at most 1 can be expanded into a series of third order derivatives of eigenfunctions and associated functions of the given eigenvalue problem, which converges in $L_p(0,1)$.

Of course, if both boundaries are free, the result which we obtained from the transformation (10.8.1) is better. However, we can improve the result considerably if one boundary is rigid (say at 0) and the other boundary is free (say at 1). An obvious generalization of Proposition 6.6.8 shows that all eigenfunctions and associated functions of the boundary eigenvalue problem satisfy those boundary conditions which are independent of the eigenvalue parameter. Therefore, for each function $f \in W_p^6(a,b)$ which satisfies $f(0) = 0$, $f'(0) = 0$, and $f''(1) = 0$ such that $f^{(3)}$ can be expanded into a series of third order derivatives of eigenfunctions and associated functions of the given eigenvalue problem which converges in $L_p(0,1)$, it follows by repeated integration that f can be expanded into a series of eigenfunctions and associated functions of the given eigenvalue problem which converges in $W_p^3(0,1)$. Since these functions must also satisfy the boundary condition $f(1) = 0$, the set of all $f^{(3)}$ such that f satisfies these four boundary conditions has codimension 1 in $W_p^{(3)}(0,1)$ and must therefore coincide with the set for which we obtained the expansion into a series of third order derivatives of eigenfunctions and associated functions of the given eigenvalue problem. Hence we obtain that each function $f \in W_p^6(a,b)$ which satisfies $f(0) = f(1) = f'(0) = f''(1) = 0$ can be expanded into a series of eigenfunctions and associated functions of the given eigenvalue problem which converges in $W_p^3(0,1)$.

The expansions into second and third order derivatives is not really what we want. Even in the last case, we have to take the functions in $W_p^6(0,1)$ in order to obtain an expansion in $W_p^3(0,1)$. Appropriate expansions theorems should be obtained by generalizing the results of Chapter VIII to the case when the differential equation depends polynomially on λ.

10.9. Motions of an incompressible magnetized plasma

In [LI] LIFSCHITZ considered, among other problems, the Lundquist equations describing motions of an incompressible magnetized plasma of unit density. After separation of variables, this leads to the following system of ordinary differential equations

$$\begin{pmatrix} \sigma+(\alpha-\beta)\frac{\mathrm{d}}{\mathrm{d}\psi}+(\alpha-\beta)\mathscr{N} & (\alpha+\beta)\mathscr{N} \\ (\alpha-\beta)\mathscr{N} & \sigma+(\alpha+\beta)\frac{\mathrm{d}}{\mathrm{d}\psi}+(\alpha+\beta)\mathscr{N} \end{pmatrix} \begin{pmatrix} \tilde{\mathbf{v}}_{\perp,+} \\ \tilde{\mathbf{v}}_{\perp,-} \end{pmatrix} = 0,$$

where σ is the eigenvalue parameter, α and β are real constants characterizing the relative magnitude of the velocity and magnetic field. The 2-vector functions $\tilde{\mathbf{v}}_{\perp,\pm}$ are the horizontal components of $\tilde{\mathbf{v}}_{\pm} = \mathbf{v} \pm \mathbf{b}$, where $\mathbf{v}$ is the plasma velocity and $\mathbf{b}$ is the magnetic field. The angular variable ψ is the independent variable, and therefore the periodic boundary conditions

$$\begin{pmatrix} \tilde{\mathbf{v}}_{\perp,+} \\ \tilde{\mathbf{v}}_{\perp,-} \end{pmatrix}(0) = \begin{pmatrix} \tilde{\mathbf{v}}_{\perp,+} \\ \tilde{\mathbf{v}}_{\perp,-} \end{pmatrix}(2\pi)$$

have to be imposed. Finally,

$$\mathscr{N} = \begin{pmatrix} 0 & -\dfrac{1-\delta^2}{\rho^2(1-\delta\cos 2\psi)+1-\delta^2} \\ 1 & \dfrac{\delta\rho^2\sin 2\psi}{\rho^2(1-\delta\cos 2\psi)+1-\delta^2} \end{pmatrix},$$

where $\rho \in (0,\infty)$ is the radial variable, which is fixed here, and $\delta \in [0,1)$ is the ellipticity parameter of the elliptic flow. Lifschitz states that for $|\alpha| \neq |\beta|$ the spectrum is discrete but that the classical arguments of Birkhoff and Langer are not applicable to this problem since the eigenvalues are not asymptotically simple. Hence the author was not able to prove completeness of the system of corresponding eigenfunctions. Here, as with any non-self-adjoint problem, associated functions can occur, and it is very difficult to show that the eigenvalues are algebraically simple or at least that the algebraic multiplicity coincides with the geometric multiplicity, which would guarantee that there are no associated functions.

However, we shall show that the system of eigenfunctions and associated functions is complete. To this end, we introduce the vector function

$$y = \begin{pmatrix} (\alpha-\beta)\tilde{\mathbf{v}}_{\perp,+} \\ (\alpha+\beta)\tilde{\mathbf{v}}_{\perp,-} \end{pmatrix},$$

and the eigenvalue problem becomes

$$y' - (\sigma A_1 + A_0)y = 0, \quad y(0) = y(2\pi), \tag{10.9.1}$$

where

$$A_1 = -\begin{pmatrix} \frac{1}{\alpha-\beta} I_2 & 0 \\ 0 & \frac{1}{\alpha+\beta} I_2 \end{pmatrix}, \quad A_0 = -\begin{pmatrix} \mathcal{N} & \mathcal{N} \\ \mathcal{N} & \mathcal{N} \end{pmatrix},$$

and A_0 depends continuously on ψ.

The matrices $\Delta(\sigma)$ are of the form $\operatorname{diag}(I_2, 0)$ and $\operatorname{diag}(0, I_2)$ if $|\beta| > |\alpha|$ and I_4 and 0 if $|\beta| < |\alpha|$. Since $W^{(0)} = -W^{(2\pi)} = I_4$ if writing the boundary conditions in the form $W^{0)}y(0) + W^{2\pi}(2\pi) = 0$, it is immediately clear that all Birkhoff matrices are invertible. Hence this problem is Birkhoff regular, and it follows from Theorem 5.3.2 that every function in $(L_2(0,2\pi))^4$ is expandable into eigenfunctions and associated functions of this problem.

The above result shows that the system of eigenvectors and associated vectors is complete, and from Proposition 1.10.5 we know that it is also minimal. Hence the existence of at least one associated function would imply that the system of eigenvectors would not be complete.

Now we are going to investigate the case of circular flow, i. e., $\delta = 0$, where a fundamental system and the eigenvalues can be calculated explicitly. Writing

$$a := \frac{1}{\alpha-\beta}, \quad b := \frac{1}{\alpha+\beta}, \quad c := \frac{1}{\sqrt{\rho^2+1}},$$

a straightforward calculation shows that

$$p(\sigma,\lambda) = \det(\sigma A_1 + A_0 - \lambda) = (\sigma a + \lambda)^2(\sigma b + \lambda)^2 + c^2(\sigma(a+b) + 2\lambda)^2.$$

The matrix $\sigma A_1 + A_0$ is similar to an upper triangular matrix, and therefore the fundamental matrix $Y(\psi)$ of (10.9.1) is similar to an upper triangular matrix with diagonal elements $\exp(\lambda_1\psi), \ldots, \exp(\lambda_4\psi)$, where $\lambda_j = \lambda_j(\sigma)$, $j = 1, \ldots, 4$, are the four zeros of $p(\sigma,\lambda)$. Hence σ is an eigenvalue of (10.9.1) if and only if $\lambda_j(\sigma) \in i\mathbb{Z}$ for some j, i. e., $p(\sigma, ik) = 0$ for at least one $k \in \mathbb{Z}$.

We want to investigate if and when there exist eigenvalues with associated functions. So let σ be an eigenvalue of (10.9.1). Then the above considerations shows that the geometric multiplicity of σ, i. e., the defect of $Y(\sigma,2\pi) - Y(\sigma,0)$, is at least as large as the number of different integers k such that $p(\sigma, ik) = 0$.

On the other hand, the algebraic multiplicity of the eigenvalue σ is the multiplicity of the zero of

$$\det(Y(\tau,2\pi)-Y(\tau,0))=\prod_{j=1}^{4}(\exp(2\pi\lambda_j(\tau))-1)$$

at $\tau=\sigma$. If $\exp(2\pi\lambda_j(\sigma))=1$ for some j, then $\lambda_j(\sigma)=ik$ for some integer k, and thus

$$\frac{\lambda_j(\tau)-ik}{\exp(2\pi\lambda_j(\tau))-1}\to\frac{1}{2\pi}\text{ as }\tau\to\sigma.$$

This shows that the algebraic multiplicity of the eigenvalue σ equals the sum of the multiplicities of the zeros of

$$\prod_{j=1}^{4}(\lambda_j(\tau)-ik)=p(\tau,ik)$$

at $\tau=\sigma$, the summation being over all integers k.

Therefore, the geometric multiplicity of the eigenvalue σ equals its algebraic multiplicity if for any k for which $p(\sigma,ik)=0$, σ is a simple zero. In other words, if there exist associated functions for the problem (10.9.1), there must be an integer k and a complex number σ which is a multiple zero of $p(\sigma,ik)$.

Obviously, $p(\sigma,\lambda)=0$ if and only if

(10.9.2) $$(\sigma a+\lambda)(\sigma b+\lambda)=i\varepsilon_1 c(\sigma(a+b)+2\lambda),\ \varepsilon_1=-1,1,$$

i. e., with $\lambda=ik$,

(10.9.3) $$\sigma=-i\frac{a+b}{2ab}(k-\varepsilon_1 c)\pm\frac{i}{2ab}\sqrt{(a-b)^2(k^2-2\varepsilon_1 ck)+c^2(a+b)^2}.$$

Hence $p(\sigma,ik)$ has a multiple zero if and only if σ satisfies (10.9.2) for $\varepsilon_1=-1$ and $\varepsilon_1=1$ or if $(a-b)^2(k^2-2\varepsilon_1 ck)+c^2(a+b)^2=0$.

The first case occurs if and only if both sides of (10.9.2) are zero, which is satisfied if and only if $\sigma=0$ and $k=0$ or $a=b$ and $\sigma=-i\frac{k}{a}$, $k\in\mathbb{Z}$. We observe that

$$\begin{aligned}p(\sigma,0)&=a^2b^2\sigma^4+c^2\sigma^2(a+b)^2\\&=\sigma^2(a^2b^2\sigma^2+c^2(a+b)^2).\end{aligned}$$

Hence, if $a+b\neq 0$, then $\sigma=0$ is a double zero for $k=0$. But

$$A_0=-\begin{pmatrix}\mathcal{N}&\mathcal{N}\\\mathcal{N}&\mathcal{N}\end{pmatrix}=-\frac{1}{2}\begin{pmatrix}I_2&I_2\\-I_2&I_2\end{pmatrix}\begin{pmatrix}0&0\\0&2\mathcal{N}\end{pmatrix}\begin{pmatrix}I_2&-I_2\\I_2&I_2\end{pmatrix}$$

shows that A_0 is diagonalizable since $\mathcal{N}$ is diagonalizable, and 0 is a double eigenvalue of A_0. Thus $Y(2\pi)-Y(0)$ for $\sigma=0$ and $k=0$ has defect 2, and therefore $\sigma=0$ has no associated functions if $a+b\neq 0$.

If $a+b=0$, then $\sigma=0$ has algebraic multiplicity 4, but $Y(2\pi)-Y(0)$ is similar to

$$\operatorname{diag}(0,0,\exp(4\pi ic)-1,\exp(4\pi ic)-1),$$

which has geometric multiplicity 4 if $c=\frac{1}{2}$, i. e., $\rho=\sqrt{3}$, and geometric multiplicity 2 for all other $\rho>0$. As $a+b=0$ occurs if and only if $\alpha=0$, it follows that for $\alpha=0$ and $\rho\neq\sqrt{3}$ there are associated functions.

Now let $a=b$, i. e., $\beta=0$. Then

$$\begin{aligned} p(\sigma,ik) &= (\sigma a+ik)^4+4c^2(\sigma a+ik)^2 \\ &= (\sigma a+ik)^2((\sigma a+ik)^2+4c^2), \end{aligned}$$

i. e., $\sigma=-i\frac{k}{a}$ is a double zero. But, for $a=b$,

$$\begin{aligned} -i\frac{k}{a}A_1+A_0 &= -\begin{pmatrix} -ikI_2+\mathcal{N} & \mathcal{N} \\ \mathcal{N} & -ikI_2+\mathcal{N} \end{pmatrix} \\ &= -\frac{1}{2}\begin{pmatrix} I_2 & I_2 \\ -I_2 & I_2 \end{pmatrix}\begin{pmatrix} -ikI_2 & 0 \\ 0 & -ikI_2+2\mathcal{N} \end{pmatrix}\begin{pmatrix} I_2 & -I_2 \\ I_2 & I_2, \end{pmatrix} \end{aligned}$$

and it clearly follows that $Y(2\pi)-Y(0)$ has defect (at least) 2. Hence the algebraic and geometric multiplicities coincide.

If $(a-b)^2(k^2-2\varepsilon_1 ck)+c^2(a+b)^2=0$, we have already covered the case $a+b=0$ and $k=0$. For $a+b=0$, it remains the case $k=2\epsilon_1 c$, i. e., $c=\frac{1}{2}$ and $k=\varepsilon_1$. Then

$$\begin{aligned} p(\sigma,ik)=p(\sigma,i\varepsilon_1) &= (\sigma a+ie_1)^2(-\sigma a+i\varepsilon_1)^2-1 \\ &= (a^2\sigma^2-1)^2-1 \\ &= a^2\sigma^2(a^2\sigma^2-2). \end{aligned}$$

Again, $\sigma=0$ is a double eigenvalue, and we have already seen above that in this case the geometric multiplicity is at least 2. Hence there are no associated functions in this case.

Finally, for $a+b\neq 0$, $(a-b)^2(k^2-2\varepsilon_1 ck)+c^2(a+b)^2=0$ can only be satisfied if $k=\varepsilon_1$, i. e.,

$$(a-b)^2(1-2c)+c^2(a+b)^2=0$$

or

$$c^2+(1-2c)\left(\frac{\beta}{\alpha}\right)^2=0. \tag{10.9.4}$$

Since $0 < c < 1$, it is easy to see that this case occurs if and only if $|\beta| > |\alpha|$. From (10.9.3) we obtain that $\sigma = -i\frac{a+b}{2ab}\varepsilon_1(1-c)$, and therefore, after some straightforward calculations,

$$p(\sigma,\lambda) = \left(\lambda^2 - 2i\lambda\varepsilon_1\frac{1-2c}{1-c} - 1 + 2c\right)^2 + 4c^2\left(-i\varepsilon_1\frac{1-2c}{1-c} + \lambda\right)^2.$$

In case $\varepsilon_1 = 1$ the four solutions λ of $p(\sigma,\lambda) = 0$ are i, $i(1-\frac{2c^2}{1-c})$, $i(1-\frac{2c}{1-c})$, $i(2c-1)$, and only one of them belongs to $i\mathbb{Z}$ (note that $\frac{1}{2} < c < 1$ due to (10.9.4)). Hence σ is an eigenvalue of algebraic multiplicity 2 and geometric multiplicity 1, and associated vectors occur.

To summarize, we have shown that in case $\delta = 0$ there are associated functions if and only if $|\alpha| < |\beta|$.

Appendix A

EXPONENTIAL SUMS

A.1. The convex hull of sums of complex numbers

Let $n \in \mathbb{N} \setminus \{0\}$. We consider sets $\mathscr{P}_1, \ldots, \mathscr{P}_n \subset \mathbb{C}$ with the property that, for each $j \in \{1, \ldots, n\}$, $0 \in \mathscr{P}_j$ and there is a number $c_j \in \mathscr{P}_j$ such that $c_j \neq 0$ and $\mathscr{P}_j \subset \overline{0, c_j}$. We set

$$\mathscr{E} := \left\{ \sum_{j=1}^{n} z_j : z_j \in \mathscr{P}_j,\ j = 1, \ldots, n \right\}. \tag{A.1.1}$$

Let $\mathscr{P}$ be the convex hull of $\mathscr{E}$.

Obviously, there is a natural number m with $1 \leq m \leq n$ such that the points of $\mathscr{P}_1, \ldots, \mathscr{P}_n$ lie on m different lines $g_1, \ldots, g_m$ with $0 \in g_j$ $(j = 1, \ldots, m)$. We have $g_j = \mathbb{R}e^{i\varphi_j}$, where we may assume without loss of generality that $0 \leq \varphi_1 < \varphi_2 < \cdots < \varphi_m < \pi$. For $l \in \mathbb{Z} \setminus \{0\}$ and $j \in \{1, \ldots, m\}$ we set $\varphi_{lm+j} := \varphi_j + l\pi$ and $g_{lm+j} := g_j$. Then $\varphi_k < \varphi_{k+1}$ and $\varphi_{k+m} = \varphi_k + \pi$ hold for all $k \in \mathbb{Z}$. For $j \in \mathbb{Z}$ we set

$$a_j := \sum_{\substack{k=1 \\ c_k \in \mathbb{R}_+ e^{i\varphi_j}}}^{n} c_k.$$

Then

$$e^{-i\varphi_j} a_{j+m} \leq 0 \leq e^{-i\varphi_j} a_j.$$

Since there is at least one $k \in \{1, \ldots, n\}$ with $c_k \in g_j \setminus \{0\}$, we have $a_{j+m} \neq 0$ or $a_j \neq 0$. Hence $a_{j+m} - a_j \neq 0$.

For $j \in \mathbb{Z}$ we set

$$\mathscr{E}^j := \left\{ \sum_{\substack{k=1 \\ c_k \in g_j}}^{n} z_k : z_k \in \mathscr{P}_k,\ k = 1, \ldots, n \right\}.$$

Note that $a_{j+2m} = a_j$ and $\mathscr{E}^{j+m} = \mathscr{E}^j$ for all $j \in \mathbb{Z}$.

PROPOSITION A.1.1. *For each $j \in \mathbb{Z}$ we have $\mathscr{E}^j \subset \mathscr{E}$, and the line segment $\overline{a_j, a_{j+m}}$ is the convex hull of $\mathscr{E}^j$. Let $z_k \in \overline{0, c_k}$ $(k = 1, \ldots, n;\ c_k \in g_j)$ and*

$$\sum_{\substack{k=1\\ c_k \in g_j}}^{n} z_k = a_j.$$

Then

$$z_k = \begin{cases} c_k & \text{if } c_k \in \mathbb{R}_+ e^{i\varphi_j},\\ 0 & \text{if } c_k \in \mathbb{R}_- e^{i\varphi_j}. \end{cases} \tag{A.1.2}$$

Proof. Let $j \in \mathbb{Z}$. The statement $\mathscr{E}^j \subset \mathscr{E}$ is obvious since $0 \in \mathscr{P}_k$ for all $k \in \{1, \ldots, n\}$.

From $g_j = \mathbb{R}e^{i\varphi_j}$ we immediately infer

$$a_j = \sum_{\substack{k=1\\ c_k \in \mathbb{R}_+ e^{i\varphi_j}}}^{n} c_k \in \mathscr{E}^j \quad \text{and} \quad a_{j+m} = \sum_{\substack{k=1\\ c_k \in \mathbb{R}_- e^{i\varphi_j}}}^{n} c_k \in \mathscr{E}^j.$$

Hence $\overline{a_j, a_{j+m}}$ is a subset of the convex hull of $\mathscr{E}^j$.

Now let $z \in \mathscr{E}^j$. Then there are $z_k \in \mathscr{P}_k$ $(k = 1, \ldots, n;\ c_k \in g_j)$ such that

$$z = \sum_{\substack{k=1\\ c_k \in g_j}}^{n} z_k.$$

For $c_k \in \mathbb{R}_+ e^{i\varphi_j}$ we have $e^{-i\varphi_j} z_k \in [0, e^{-i\varphi_j} c_k]$ and for $c_k \in \mathbb{R}_- e^{i\varphi_j} = \mathbb{R}_+ e^{i\varphi_{j+m}}$ we have $e^{-i\varphi_j} z_k \in [e^{-i\varphi_j} c_k, 0]$ since $\mathscr{P}_k \subset \overline{0, c_k}$. Hence

$$e^{-i\varphi_j} a_{j+m} \le \sum_{\substack{k=1\\ c_k \in \mathbb{R}_- e^{i\varphi_j}}}^{n} e^{-i\varphi_j} z_k \le e^{-i\varphi_j} z \le \sum_{\substack{k=1\\ c_k \in \mathbb{R}_+ e^{i\varphi_j}}}^{n} e^{-i\varphi_j} z_k \le e^{-i\varphi_j} a_j,$$

which proves $z \in \overline{a_j, a_{j+m}}$ and $z = a_j$ if and only if (A.1.2) holds. □

PROPOSITION A.1.2. *The convex hull $\mathscr{P}$ of $\mathscr{E}$ is*

$$\mathscr{P} = \sum_{j=1}^{m} \overline{a_j, a_{j+m}}.$$

Proof. The definitions of $\mathscr{E}$ and $\mathscr{E}^j$ immediately yield

$$\mathscr{E} = \sum_{j=1}^{m} \mathscr{E}^j.$$

Hence

$$\operatorname{cv} \mathscr{E} = \sum_{j=1}^{m} \operatorname{cv} \mathscr{E}^j,$$

where cv denotes the convex hull. An application of Proposition A.1.1 completes the proof. □

For $j \in \mathbb{Z}$ we set

$$b_j := \sum_{k=j}^{j+m-1} a_k \in \mathscr{E}.$$

Note that $b_{j+1} - b_j = a_{j+m} - a_j \neq 0$. The definition of the a_k yields

$$b_j = \sum_{k=j}^{j+m-1} \sum_{\substack{\nu=1 \\ c_\nu \in \mathbb{R}_+ e^{i\varphi_k}}}^{n} c_\nu .$$

Since $c_\nu \in \mathbb{R}_+ e^{i\varphi_k}$ holds for some $k \in \{j, \dots, j+m-1\}$ if and only if $\arg c_\nu - \varphi_j \in [0, \pi) \operatorname{mod}(2\pi)$, we obtain the representation

$$b_j = \sum_{\substack{\nu=1 \\ \varphi_j \le \arg c_\nu < \varphi_j + \pi}}^{n} c_\nu . \tag{A.1.3}$$

THEOREM A.1.3. i) *$\mathscr{P}$ is a convex polygon with $2m$ vertices, the set of the vertices of $\mathscr{P}$ is*

$$\widetilde{\mathscr{E}} = \{b_j : j = 1, \dots, 2m\},$$

and the boundary of $\mathscr{P}$ is

$$\partial \mathscr{P} = \bigcup_{j=1}^{2m} \overline{b_j, b_{j+1}} .$$

ii) *Let $j \in \{1, \dots, 2m\}$, $z_k \in \overline{0, c_k}$ $(k = 1, \dots, n)$ and $z = \sum_{k=1}^{n} z_k$. Then $z \in \overline{b_j, b_{j+1}}$ if and only if for all $l \in \{j+1, \dots, j+m-1\}$ and all $k \in \{1, \dots, n\}$ with $c_k \in g_l$*

$$z_k = \begin{cases} c_k & \text{if } c_k \in \mathbb{R}_+ e^{i\varphi_l}, \\ 0 & \text{if } c_k \in \mathbb{R}_- e^{i\varphi_l}. \end{cases}$$

iii) *The representation of b_j $(j = 1, \dots, 2m)$ as an element of $\mathscr{E}$ is unique.*

Proof. For $m = 1$ we have $\mathscr{E} = \mathscr{E}^1$, $b_1 = a_1$ and $b_2 = a_2$. In this case the assertion of Theorem A.1.3 follows from Proposition A.1.1.

Now let $m \ge 2$. i) We set

$$H_j := \{z \in \mathbb{C} : \Im\big((z - b_j)\overline{(b_{j+1} - b_j)}\big) \ge 0\} \quad (j = 1, \dots, 2m)$$

and

$$H := \bigcap_{j=1}^{2m} H_j .$$

We assert that

$$\mathscr{P} = H \tag{A.1.4}$$

and

$$\partial \mathscr{P} = \bigcup_{j=1}^{2m} \overline{b_j, b_{j+1}}. \tag{A.1.5}$$

First we shall prove that

$$\mathscr{E} \subset H_j \tag{A.1.6}$$

holds for $j = 1, \ldots, 2m$. For this let $j \in \{1, \ldots, 2m\}$ and $z \in \mathscr{E}$. Then there are $z_k \in \mathscr{P}_k$ $(k = 1, \ldots, n)$ such that

$$z = \sum_{k=1}^{n} z_k = \sum_{l=j}^{j+m-1} \sum_{\substack{k=1 \\ c_k \in g_l}}^{n} z_k.$$

For $l \in \mathbb{Z}$ we set

$$d_l := \sum_{\substack{k=1 \\ c_k \in g_l}}^{n} z_k.$$

Then $d_l \in \mathscr{E}^l$ and

$$z = \sum_{l=j}^{j+m-1} d_l. \tag{A.1.7}$$

Since $d_l \in \overline{a_l, a_{l+m}}$ by Proposition A.1.1, $(d_l - a_l)e^{-i\varphi_{l+m}} \geq 0$. Hence

$$\begin{aligned} (z - b_j)\overline{(b_{j+1} - b_j)} &= \sum_{l=j}^{j+m-1} (d_l - a_l)\overline{(a_{j+m} - a_j)} \\ &= \sum_{l=j}^{j+m-1} |d_l - a_l| e^{i\varphi_{l+m}} |a_{j+m} - a_j| e^{-i\varphi_{j+m}}. \end{aligned} \tag{A.1.8}$$

From $\varphi_{j+m} < \varphi_{l+m} < \varphi_{j+m} + \pi$ for $l = j+1, \ldots, j+m-1$ we infer that $\Im\left((z - b_j)\overline{(b_{j+1} - b_j)}\right) \geq 0$ and

$$\Im\left((z - b_j)\overline{(b_{j+1} - b_j)}\right) = 0 \Leftrightarrow \forall l \in \{j+1, \ldots, j+m-1\}\ d_l = a_l. \tag{A.1.9}$$

This proves (A.1.6). Thus we obtain

$$\mathscr{P} \subset H \tag{A.1.10}$$

since H_j is a convex set for each $j = 1, \ldots, 2m$.

Now we shall prove that

$$\partial H = \bigcup_{j=1}^{2m} \overline{b_j, b_{j+1}}. \tag{A.1.11}$$

Since H is closed, $z \in \partial H$ holds if and only if $z \in H$ and $z \in \partial H_j$ for at least one $j \in \{1, \dots, 2m\}$. From (A.1.10) we infer $b_j \in H$ for $j \in \mathbb{Z}$. The convexity of H implies $\overline{b_j, b_{j+1}} \subset H$ for $j = 1, \dots, 2m$. Since the boundary of H_j is the straight line consisting of those complex numbers z for which $\Im\left((z - b_j)\overline{(b_{j+1} - b_j)}\right) = 0$, we obtain $\overline{b_j, b_{j+1}} \subset \partial H_j$ for $j = 1, \dots, 2m$. This proves

$$\partial H \supset \bigcup_{j=1}^{2m} \overline{b_j, b_{j+1}}.$$

Conversely, let $z \in \partial H$. Then there is a $j \in \{1, \dots, 2m\}$ such that $z \in \partial H_j$. Since ∂H_j is a straight line with $b_j, b_{j+1} \in \partial H_j$, there is a $t \in \mathbb{R}$ with $z = (1-t)b_j + tb_{j+1}$. Then

$$\begin{aligned}(z - b_{j+1})\overline{(b_{j+2} - b_{j+1})} &= (1-t)(b_j - b_{j+1})\overline{(b_{j+2} - b_{j+1})} \\ &= (1-t)|a_j - a_{j+m}|e^{i\varphi_j}|a_{j+1} - a_{j+m+1}|e^{-i\varphi_{j+m+1}}.\end{aligned}$$

We have $0 < \varphi_{j+1} - \varphi_j < \pi$ because of $m \geq 2$. Then

$$\varphi_j - \varphi_{j+m+1} = -(\varphi_{j+1} - \varphi_j) - \pi$$

implies

$$-2\pi < \varphi_j - \varphi_{j+m+1} < -\pi.$$

We have $a_j \neq a_{j+m}$, $a_{j+1} \neq a_{j+m+1}$, and $\Im\left((z - b_{j+1})\overline{(b_{j+2} - b_{j+1})}\right) \geq 0$ since $z \in H_{j+1}$. A comparison of the imaginary parts of the above equation yields $1 - t \geq 0$. Analogously we consider

$$\begin{aligned}(z - b_j)\overline{(b_j - b_{j-1})} &= t(b_{j+1} - b_j)\overline{(b_j - b_{j-1})} \\ &= t|a_j - a_{j+m}|e^{i\varphi_{j+m}}|a_{j-1} - a_{j+m-1}|e^{-i\varphi_{j+m-1}}.\end{aligned}$$

Because of

$$(z - b_j)\overline{(b_j - b_{j-1})} = (z - b_{j-1})\overline{(b_j - b_{j-1})} - (b_j - b_{j-1})\overline{(b_j - b_{j-1})}$$

we have

$$\Im\left((z - b_j)\overline{(b_j - b_{j-1})}\right) = \Im\left((z - b_{j-1})\overline{(b_j - b_{j-1})}\right) \geq 0$$

since $z \in H_{j-1}$. From $0 < \varphi_{j+m} - \varphi_{j+m-1} < \pi$, $a_j \neq a_{j+m}$ and $a_{j-1} \neq a_{j+m-1}$ we infer $t \geq 0$. Thus $t \in [0, 1]$ which implies $z \in \overline{b_j, b_{j+1}}$. This proves (A.1.11).

Let $r := \max\{|b_j| : j = 1, \dots, 2m\}$ and $U := \{z \in \mathbb{C} : |z| > r\}$. Since H_1 is a half-plane, we have $U \not\subset H_1$ and hence $U \not\subset H$. From (A.1.11) we know $U \cap \partial H = \emptyset$. Hence the set $U \cap H$ has no boundary point in U. Since U is connected, $U \cap H = \emptyset$. This proves that H is bounded and hence compact.

Now we shall prove that

$$H \subset \mathscr{P}. \tag{A.1.12}$$

Assume that (A.1.12) is false. Then there is a number $z_0 \in H \setminus \mathscr{P}$. For $t \in \mathbb{R}$ we set $\alpha(t) = tz_0 + (1-t)c_1$. Since $\alpha(0) = c_1 \in \mathscr{P}$ and $\alpha(1) = z_0 \notin \mathscr{P}$, the convexity of $\mathscr{P}$ implies that

$$\{tz_0 + (1-t)c_1 : t \geq 1\} \cap \mathscr{P} = \emptyset. \tag{A.1.13}$$

On the other hand the compactness of H implies

$$z_0 \in \{tz_0 + (1-t)c_1 : t \geq 1\} \not\subset H.$$

Hence there is a $t \geq 1$ with $z := tz_0 + (1-t)c_1 \in \partial H$. According to (A.1.13), $z \notin \mathscr{P}$. This is a contradiction since (A.1.11) yields $\partial H \subset \mathscr{P}$. Now (A.1.10), (A.1.11) and (A.1.12) prove (A.1.4) and (A.1.5).

Since $\mathscr{P} = H$ is the intersection of $2m$ half-spaces, $\mathscr{P}$ is a convex polygon. The representation (A.1.5) of the boundary of $\mathscr{P}$ immediately implies that $\widetilde{\mathscr{E}} \subset \{b_j : j = 1, \dots, 2m\}$. For $j \in \{1, \dots, 2m\}$ and $k \in \{j+2, \dots, j+m\}$ we have

$$b_k = \sum_{l=k}^{k+m-1} a_l = \sum_{l=j}^{k-1} a_{l+m} + \sum_{l=k}^{j+m-1} a_l.$$

The summand for $l = j+1$ on the right hand side is a_{j+m+1}. As $a_{j+m+1} \neq a_{j+1}$, we infer from (A.1.9) that b_k does not lie on the straight line through the points b_j and b_{j+1}. For $k \in \{j+m+1, \dots, j+2m-1\}$ we have

$$b_k = \sum_{l=k}^{k+m-1} a_l = \sum_{l=k-m}^{k-1} a_{l+m} = \sum_{l=j}^{k-m-1} a_l + \sum_{l=k-m}^{j+m-1} a_{l+m}.$$

The summand for $l = j+m-1$ on the right hand side is a_{j-1}. As $a_{j-1} \neq a_{j+m-1}$, we infer from (A.1.9) that b_k does not lie on the straight line through the points b_j and b_{j+1}. We have proved that no three points of $\{b_1, \dots, b_{2m}\}$ lie on a straight line. This proves that each b_j is a vertex of $\mathscr{P}$.

ii) Let $z_k \in \overline{0,c_k}$ and $z = \sum_{k=1}^{n} z_k$. Then, by (A.1.9), $z \in \overline{b_j,b_{j+1}}$ if and only if

$$\sum_{\substack{k=1\\ c_k \in g_l}}^{n} z_k = a_l \tag{A.1.14}$$

for $l = j+1,\ldots,j+m-1$. The application of the last statement of Proposition (A.1.1) completes the proof of ii).
iii) If in ii) additionally $z = b_j$, then (A.1.8) and (A.1.9) yield that (A.1.14) also holds for $l = j$. Again by Proposition A.1.1 we infer that all z_k $(k = 1,\ldots,n)$ are uniquely determined by $z = b_j$. □

COROLLARY A.1.4. *For all $j \in \{1,\ldots,2m\}$ we have*

$$\overline{b_j,b_{j+1}} \cap \mathscr{E} = b_j - a_j + \mathscr{E}^j .$$

Proof. Let $z \in \mathscr{E}$. Then we have the representation $z = \sum_{k=1}^{n} z_k$, where $z_k \in \mathscr{P}_k \subset \overline{0,c_k}$. From Theorem A.1.3 ii) we infer that $z \in \overline{b_j,b_{j+1}} \cap \mathscr{E}$ if and only if

$$\begin{aligned} z &= \sum_{l=j+1}^{j+m-1} \sum_{\substack{k=1\\ c_k \in \mathbb{R}_+ e^{i\varphi_l}}}^{n} c_k + \sum_{\substack{k=1\\ c_k \in g_j}}^{n} z_k \\ &= b_j - a_j + \sum_{\substack{k=1\\ c_k \in g_j}}^{n} z_k . \end{aligned}$$

From the definition of $\mathscr{E}^j$ we immediately infer that $z \in \overline{b_j,b_{j+1}} \cap \mathscr{E}$ if and only if $z \in b_j - a_j + \mathscr{E}^j$. □

REMARK A.1.5. Since $b_{j+1} - b_j = a_{j+m} - a_j$ and $b_{j+m-1} - b_{j+m} = a_j - a_{j+m}$ for $j = 1,\ldots,m$, the line segments $\overline{b_j,b_{j+1}}$ and $\overline{b_{j+m},b_{j+m+1}}$ are parallel to g_j and have the length $|a_{j+m} - a_j|$.

PROPOSITION A.1.6. *Assume in addition that for each $j \in \{1,\ldots,n\}$ the points 0 and c_j are no accumulation points of $\mathscr{P}_j$. Then for each $j \in \{1,\ldots,2m\}$ the set $\mathscr{E} \setminus \overline{b_j,b_{j+1}}$ has no accumulation point in $\overline{b_j,b_{j+1}}$.*

Proof. First we shall prove for each $j \in \{1,\ldots,n\}$ that the point a_j is no accumulation point of $\mathscr{E}^j$. For this let $(z_k)_{k\in\mathbb{N}}$ be a sequence in $\mathscr{E}^j$ such that $z_k \to a_j$ as $k \to \infty$. Then there are $z_{k,l} \in \mathscr{P}_l$ such that

$$z_k = \sum_{\substack{l=1\\ c_l \in g_j}}^{n} z_{k,l} .$$

Since every term on the right hand side of

$$e^{-i\varphi_j}(z_k - a_j) = \sum_{\substack{l=1 \\ c_l \in \mathbb{R}_+ e^{i\varphi_j}}}^{n} \left(e^{-i\varphi_j} z_{k,l} - e^{-i\varphi_j} c_l\right) + \sum_{\substack{l=1 \\ c_l \in \mathbb{R}_- e^{i\varphi_j}}}^{n} e^{-i\varphi_j} z_{k,l}$$

is nonpositive, $z_k \to a_j$ as $k \to \infty$ implies $z_{k,l} \to c_l$ if $c_l \in \mathbb{R}_+ e^{i\varphi_j}$ and $z_{k,l} \to 0$ if $c_l \in \mathbb{R}_- e^{i\varphi_j}$ as $k \to \infty$. Since 0 and c_l are no accumulation points of $\mathscr{P}_l$, we have $z_{k,l} = c_l$ if $c_l \in \mathbb{R}_+ e^{i\varphi_j}$ and $z_{k,l} = 0$ if $c_l \in \mathbb{R}_- e^{i\varphi_j}$ for sufficiently large k. Hence $z_k = a_j$ if k is sufficiently large. This proves that a_j is an isolated point of $\mathscr{E}^j$.

Now let $(z_k)_{k\in\mathbb{N}}$ be a sequence in $\mathscr{E}$ such that $z_k \to z$ as $k \to \infty$ for some $z \in \overline{b_j, b_{j+1}}$. According to (A.1.7) we write

$$z_k = \sum_{l=j}^{j+m-1} d_{k,l}\,.$$

Since $\Im\left((z - b_j)\overline{(b_{j+1} - b_j)}\right) = 0$, we have $\Im\left((z_k - b_j)\overline{(b_{j+1} - b_j)}\right) \to 0$ as $k \to \infty$. Each term on the right hand side of (A.1.8) has a non-negative imaginary part. Thus (A.1.8) yields $d_{k,l} \to a_l$ for $k \to \infty$ and $l = j+1, \ldots, j+m-1$. The first part of the proof gives $d_{k,l} = a_l$ for $l = j+1, \ldots, j+m-1$ if k is sufficiently large. Hence $z_k \in \overline{b_j, b_{j+1}}$ if k is sufficiently large by Theorem A.1.3 ii). This proves that $\mathscr{E} \setminus \overline{b_j, b_{j+1}}$ has no accumulation point in $\overline{b_j, b_{j+1}}$. $\square$

Now we consider a special case. Let

$$\omega_j := \exp(2\pi i \tfrac{j-1}{n}) \quad (j \in \mathbb{Z}),$$

$c_j := \omega_j$, $\mathscr{P}_j := \{0, c_j\}$ $(j = 1, \ldots, n)$. For a finite subset θ of $\mathbb{Z}$ let

$$\theta\omega := \sum_{j\in\theta} \omega_j.$$

Then

$$\mathscr{E} = \{\theta\omega : \theta \subset \{1, \ldots, n\}\}.$$

For $j \in \mathbb{Z}$ and $r \in \mathbb{N}$ we set

$$\theta_r^j := \{j, j+1, \ldots, j+r-1\}.$$

For $\theta \subset \{1, \ldots, n\}$ and $\tilde\theta \subset \mathbb{Z}$ we write $\theta \sim \tilde\theta$ if $\#\theta = \#\tilde\theta$ and if for each $\vartheta \in \theta$ there is a $\tilde\vartheta \in \tilde\theta$ such that $\tilde\vartheta - \vartheta \in n\mathbb{Z}$.

THEOREM A.1.7. i) *Let n be even. Then the set of the vertices of $\mathscr{P}$ is*

$$\widetilde{\mathscr{E}} = \left\{ \theta^j_{\frac{n}{2}} \omega : j \in \{1,\dots,n\} \right\}.$$

Let $\theta \subset \{1,\dots,n\}$. Then $\theta\omega \in \partial\mathscr{P}$ if and only if $\theta \sim \theta^j_r$ for some numbers $r \in \{\frac{n}{2}-1, \frac{n}{2}, \frac{n}{2}+1\}$ and $j \in \{1,\dots,n\}$.
ii) *Let n be odd. Then the set of the vertices of $\mathscr{P}$ is*

$$\widetilde{\mathscr{E}} = \left\{ \theta^j_r \omega : r \in \{\tfrac{n-1}{2}, \tfrac{n+1}{2}\},\ j \in \{1,\dots,n\} \right\}.$$

Let $\theta \subset \{1,\dots,n\}$. Then $\theta\omega \in \partial\mathscr{P}$ if and only if $\theta \sim \theta^j_r$ for some $r \in \{\frac{n-1}{2}, \frac{n+1}{2}\}$ and $j \in \{1,\dots,n\}$.

Proof. i) In this case, $m = \frac{n}{2}$ and $\varphi_j = 2\pi\frac{j-1}{n}$ for $j \in \mathbb{Z}$. Obviously,

$$a_j = \omega_j \qquad (j = 1,\dots,2m).$$

Since

$$b_j = \sum_{k\in\theta^j_m} \omega_k = \theta^j_{\frac{n}{2}}\omega \qquad (j = 1,\dots,2m),$$

Theorem A.1.3 yields the representation of $\widetilde{\mathscr{E}}$.

Let $\theta \subset \{1,\dots,n\}$ and set $\vartheta_k = 1$ if $k \in \theta$ and $\vartheta_k = 0$ if $k \in \{1,\dots,n\} \setminus \theta$. Then

$$\theta\omega = \sum_{k=1}^{n} \vartheta_k \omega_k .$$

According to Theorem A.1.3, $\theta\omega \in \partial P$ if and only if there is a number $j \in \{1,\dots,n\}$ such that for all $l \in \{j+1,\dots,j+m-1\}$ we have $\vartheta_k = 1$ if $k \in l+2m\mathbb{Z}$ and $\vartheta_k = 0$ if $k \in l+m+2m\mathbb{Z}$. This holds if and only if there is a set $\tilde{\theta} \subset \{j,\dots,j+2m-1\}$ with $\theta^{j+1}_{m-1} \subset \tilde{\theta} \subset \theta^j_{m+1}$ and $\theta\omega = \tilde{\theta}\omega$. This proves part i) because of

$$\theta^{j+1}_{m-1} \subset \tilde{\theta} \subset \theta^j_{m+1} \Leftrightarrow \tilde{\theta} \in \left\{ \theta^{j+1}_{\frac{n}{2}-1}, \theta^j_{\frac{n}{2}}, \theta^{j+1}_{\frac{n}{2}}, \theta^j_{\frac{n}{2}+1} \right\},$$

$\theta^{n+1}_{\frac{n}{2}} \sim \theta^1_{\frac{n}{2}}$ and $\theta^{n+1}_{\frac{n}{2}-1} \sim \theta^1_{\frac{n}{2}-1}$.
ii) In this case, $m = n$ and $\varphi_j = \pi\frac{j-1}{n}$ for $j \in \mathbb{Z}$. Obviously,

$$a_{2j-1} = \omega_j \qquad (j = 1,\dots,m),$$

$$a_{2j} = 0 \qquad (j = 1,\dots,m),$$

and

$$\mathscr{E}^{2j-1} = \{\omega_j, 0\} \qquad (j = 1,\dots,\tfrac{m+1}{2}),$$

$$\mathscr{E}^{2j} = \{\omega_{j+\frac{m+1}{2}}, 0\} \qquad (j = 1,\dots,\tfrac{m-1}{2}).$$

For $j=1,\dots,m$ we have

$$b_{2j-1}=\sum_{k=2j-1}^{2j+m-2}a_k=\sum_{k=j}^{j+\frac{m+1}{2}-1}\omega_k=\theta^j_{\frac{m+1}{2}}\omega$$

and

$$b_{2j}=\sum_{k=2j}^{2j+m-1}a_k=\sum_{k=j+1}^{j+\frac{m+1}{2}-1}\omega_k=\theta^{j+1}_{\frac{m-1}{2}}\omega.$$

Theorem A.1.3 yields the representation of $\widetilde{\mathscr{E}}$.

By Corollary A.1.4 we have $\overline{b_j,b_{j+1}}\cap\mathscr{E}=b_j-a_j+\mathscr{E}^j$ for $j\in\{1,\dots,2m\}$. Since $\#\mathscr{E}^j=2$, we obtain $\overline{b_j,b_{j+1}}\cap\mathscr{E}=\{b_j,b_{j+1}\}$. Hence $\theta\omega\in\partial\mathscr{P}\cap\mathscr{E}$ for $\theta\subset\{1,\dots,n\}$ if and only if $\theta\omega=b_j$ for some $j\in\{1,\dots,2m\}$. Theorem A.1.3 iii) and the above representation of b_j complete the proof. □

A.2. Estimates of exponential sums

Let Ω be an unbounded subset of $\mathbb{C}$. Since we are interested in estimates for large $\lambda\in\Omega$, we may assume for simplicity that $|\lambda|\geq 1$ for $\lambda\in\Omega$. For $\tilde{a}:\Omega\to\mathbb{C}$ and $a\in\mathbb{C}$ we write as in Section 2.7

$$\tilde{a}(\lambda)=[a]$$

if

$$\tilde{a}(\lambda)-a\to 0\quad\text{as}\quad\lambda\to\infty.$$

For $\nu\in\mathbb{R}$ and $\lambda\in\mathbb{C}\setminus\{0\}$ let $\lambda^\nu=\exp(\nu\log\lambda)$, where log is the principal value of the logarithm, i.e. the inverse of $\exp:\mathbb{R}+i(-\pi,\pi]\to\mathbb{C}\setminus\{0\}$. For $\nu\in\mathbb{N}$, λ^ν is the ν-th power of λ. For $\lambda\in\mathbb{C}\setminus\{0\}$ we have

$$\begin{aligned}\arg(-\lambda):=\Im(\log(-\lambda))&=\pi+\arg\lambda\quad\text{if }\arg\lambda\leq 0,\\ \arg(-\lambda)&=-\pi+\arg\lambda\quad\text{if }\arg\lambda>0.\end{aligned}$$

Hence

$$\begin{aligned}(-\lambda)^\nu&=(-1)^\nu\lambda^\nu\quad\text{if }\arg\lambda\leq 0,\\ (-\lambda)^\nu&=(-1)^{-\nu}\lambda^\nu\quad\text{if }\arg\lambda>0.\end{aligned}$$

Let $\mathscr{N}$ be a countable set with at least two elements, $c_j\in\mathbb{C}$ be pairwise different and $b_j:\Omega\to\mathbb{C}$ for $j\in\mathscr{N}$. Suppose that

(A.2.1) $$\sup\{|c_j|:j\in\mathscr{N}\}<\infty.$$

We assume that for all $j\in\mathscr{N}$ there are $a_j\in\mathbb{C}$ and $\nu_j\in\mathbb{R}$ such that

(A.2.2) $$b_j(\lambda)=\lambda^{\nu_j}[a_j],$$

$\sup\{\nu_j : j \in \mathscr{N}\} < \infty$,

(A.2.3)
$$\sum_{j\in\mathscr{N}} |a_j| < \infty,$$

and

(A.2.4)
$$\varepsilon(\lambda) := \sum_{j\in\mathscr{N}} |\varepsilon_j(\lambda)| < \infty, \quad \varepsilon(\lambda) \to 0 \quad (\lambda \to \infty),$$

where

(A.2.5)
$$\varepsilon_j(\lambda) := \lambda^{-\nu_j} b_j(\lambda) - a_j \qquad (j \in \mathscr{N}).$$

The function $D : \Omega \to \mathbb{C}$ defined by

(A.2.6)
$$D(\lambda) := \sum_{j\in\mathscr{N}} b_j(\lambda) \exp(c_j\lambda) \qquad (\lambda \in \Omega)$$

is called an *exponential sum*. The estimate

$$|b_j(\lambda) \exp(c_j\lambda)| \leq |\lambda|^{\nu_j} \left(|a_j| + |\varepsilon_j(\lambda)|\right) \exp\{|c_j\lambda|\},$$

the boundedness of the set of the c_j and the set of the ν_j, and the assumptions (A.2.3) and (A.2.4) prove that the exponential sum (A.2.6) is absolutely convergent.

In this section we shall estimate exponential sums. We start with a special case and shall generalize it step by step.

Let $\mathbb{K} = \mathbb{R}$ or $\mathbb{K} = \mathbb{C}$. We equip $\mathbb{K}^k$ with the Euclidean norm. For $x \in \mathbb{K}^k$ and $\delta > 0$,

$$\overset{(-)}{K}_\delta(x) := \{y \in \mathbb{K}^k : |y - x| \underset{(=)}{<} \delta\}$$

denotes the open (closed) ball with centre x and radius δ. Then the boundary $\partial K_\delta(x)$ of $\overset{(-)}{K}_\delta(x)$ is given by

$$\partial K_\delta(x) = \{y \in \mathbb{K}^k : |y - x| = \delta\}.$$

For two subsets A and B of $\mathbb{K}^k$ we define

$$d(A,B) := \inf\{|x - y| : x \in A, y \in B\},$$

where $d(A,B) = \infty$ if $A = \emptyset$ or $B = \emptyset$. For $A = \{x\}$ we write $d(x,B)$ instead of $d(\{x\},B)$.

For a countable nonempty set $\mathscr{N}$ let

$$l_1(\mathscr{N}) := \left\{ (x_j)_{j\in\mathscr{N}} \in \mathbb{C}^{\mathscr{N}} : \sum_{j\in\mathscr{N}} |x_j| < \infty \right\}.$$

It is well-known that $l_1(\mathscr{N})$ is a Banach space if it is equipped with the norm $|x|_1 := \sum_{j\in\mathscr{N}} |x_j| < \infty$ $(x = (x_j)_{j\in\mathscr{N}} \in l_1(\mathscr{N}))$.

For $B > 0$ let

$$\Pi_B(\mathcal{N}) := \{(x_j)_{j\in\mathcal{N}} \in \mathbb{C}^{\mathcal{N}} : \forall j \in \mathcal{N}\ |x_j| \le B\}.$$

We equip $\Pi_B(\mathcal{N})$ with the product topology on $\mathbb{C}^{\mathcal{N}}$. Since $\mathcal{N}$ is countable, this topology is metrizable, see e.g. [HO, p. 118]. For $x = (x_j)_{j\in\mathcal{N}} \in l_1(\mathcal{N})$ and $y = (y_j)_{j\in\mathcal{N}} \in \Pi_B(\mathcal{N})$ the complex number

(A.2.7) $$m(x,y) := \sum_{j\in\mathcal{N}} x_j y_j$$

is well-defined.

PROPOSITION A.2.1. *$m : l_1(\mathcal{N}) \times \Pi_B(\mathcal{N}) \to \mathbb{C}$ is continuous.*

Proof. Let $x^n, x \in l_1(\mathcal{N})$, $y^n, y \in \Pi_B(\mathcal{N})$ $(n \in \mathbb{N})$ with $\lim_{n\to\infty} x^n = x$ and $\lim_{n\to\infty} y^n = y$. Let $\mathcal{N}_2 \subset \mathcal{N}$ be finite and $\mathcal{N}_1 := \mathcal{N} \setminus \mathcal{N}_2$. The estimate

$$\begin{aligned}|m(x^n,y^n) - m(x,y)| &\le |m(x^n - x, y^n)| + |m(x, y^n - y)| \\ &\le B|x^n - x|_1 + 2B \sum_{j\in\mathcal{N}_1} |x_j| + \sum_{j\in\mathcal{N}_2} |x_j||y_j^n - y_j|\end{aligned}$$

yields

$$\limsup_{n\to\infty} |m(x^n,y^n) - m(x,y)| \le 2B \sum_{j\in\mathcal{N}_1} |x_j|.$$

From

$$\inf\Big\{ \sum_{j\in\mathcal{N}_1} |x_j| : \mathcal{N}_1 \subset \mathcal{N}, \mathcal{N} \setminus \mathcal{N}_1 \text{ finite} \Big\} = 0$$

we infer $\lim_{n\to\infty} m(x^n,y^n) = m(x,y)$. Since $l_1(\mathcal{N}) \times \Pi_B(\mathcal{N})$ is metrizable as a product of metrizable spaces, the continuity of m is proved. □

PROPOSITION A.2.2. *Let $U \subset \mathbb{K}$ be unbounded, $\mathcal{N}$ a countable nonempty set, $f = (f_j)_{j\in\mathcal{N}} : U \to l_1(\mathcal{N})$ and $g = (g_j)_{j\in\mathcal{N}} : U \to \mathbb{C}^{\mathcal{N}}$. Suppose that g satisfies $\sup\{|g_j(t)| : t \in U,\ j \in \mathcal{N}\} < \infty$ and one of the following two conditions holds:*
i) *there is an element $f(\infty) \in l_1(\mathcal{N})$ such that $f(t) \to f(\infty)$ in $l_1(\mathcal{N})$ as $|t| \to \infty$, and $g_j(t) \to 0$ as $|t| \to \infty$ for all $j \in \mathcal{N}$,*
ii) *$|f(t)|_1 \to 0$ as $|t| \to \infty$.*
Then

(A.2.8) $$\sum_{j\in\mathcal{N}} f_j(t) g_j(t) \to 0 \quad \text{as } |t| \to \infty.$$

Proof. We set $B := \sup\{|g_j(t)| : t \in U,\ j \in \mathcal{N}\}$. Then $g(t) \in \Pi_B(\mathcal{N})$ for all $t \in U$. If i) is fulfilled, then $g(t) \to 0$ as $|t| \to \infty$ in $\Pi_B(\mathcal{N})$. In this case, (A.2.8) follows from Proposition A.2.1. If ii) is fulfilled, then (A.2.8) follows from

$$\Big| \sum_{j\in\mathcal{N}} f_j(t) g_j(t) \Big| \le B|f(t)|_1. \quad \square$$

PROPOSITION A.2.3. *We consider the exponential sum* (A.2.6)

$$D(\lambda) = \sum_{j\in\mathcal{N}} b_j(\lambda)\exp(c_j\lambda),$$

where $c_j \in \mathbb{R}$ $(j \in \mathcal{N})$. *Assume that there are* $\alpha, \beta \in \mathcal{N}$ *such that* $c_\alpha \le 0$, $c_\beta \ge 0$, $c_\alpha < c_j < c_\beta$ $(j \in \mathcal{N} \setminus \{\alpha, \beta\})$, $a_\alpha \ne 0$, $a_\beta \ne 0$. *Assume that there is a number* $d \in \mathbb{R}$ *such that* $\nu_j = dc_j$ *for all* $j \in \mathcal{N}$. *Then there are numbers* $M > 0$, $K_0 \ge 0$, *and* $g_0 > 0$ *such that for all* $\lambda \in \Omega$ *satisfying* $|\lambda| > K_0$ *and* $|\Re(\lambda) + d\log|\lambda|| \ge M$ *the estimate*

$$|D(\lambda)| \ge g_0$$

holds.

Proof. The definition (A.2.5) of the ε_j yields

$$\begin{aligned}
&|a_\alpha^{-1}\lambda^{-\nu_\alpha}\exp(-c_\alpha\lambda)D(\lambda) - 1| \\
&\le \left|\frac{\varepsilon_\alpha(\lambda)}{a_\alpha}\right| + \sum_{j\in\mathcal{N}\setminus\{\alpha\}} \left|\frac{a_j + \varepsilon_j(\lambda)}{a_\alpha}\lambda^{\nu_j-\nu_\alpha}\exp\{(c_j - c_\alpha)\lambda\}\right| \\
&\le \frac{|\varepsilon_\alpha(\lambda)|}{|a_\alpha|} + |a_\alpha|^{-1} \sum_{j\in\mathcal{N}\setminus\{\alpha\}} \big(|a_j| + |\varepsilon_j(\lambda)|\big)\exp\{(c_j - c_\alpha)(\Re(\lambda) + d\log|\lambda|)\}.
\end{aligned}$$

Applying Proposition A.2.2 i) to $\sum_{j\in\mathcal{N}\setminus\{\alpha\}} |a_j|\exp\{(c_j - c_\alpha)t\}$ $(t \le 0)$ and Proposition A.2.2 ii) to $\sum_{j\in\mathcal{N}\setminus\{\alpha\}} |\varepsilon_j(\lambda)|\exp\{(c_j - c_\alpha)(\Re(\lambda) + d\log|\lambda|)\}$ we obtain that there are numbers $M_1 > 0$ and $K_1 \ge 0$ such that for all $\lambda \in \Omega$ with $|\lambda| \ge K_1$ and $\Re(\lambda) + d\log|\lambda| \le -M_1$ the estimate

$$\left|a_\alpha^{-1}\lambda^{-\nu_\alpha}\exp(-c_\alpha\lambda)D(\lambda) - 1\right| \le \frac{1}{2}$$

holds. This yields

$$\begin{aligned}
|D(\lambda)| &\ge |a_\alpha|\exp\{c_\alpha(\Re(\lambda) + d\log|\lambda|)\}\big(1 - |a_\alpha^{-1}\lambda^{-\nu_\alpha}\exp(-c_\alpha\lambda)D(\lambda) - 1|\big) \\
&\ge \frac{|a_\alpha|}{2}.
\end{aligned}$$

We apply this result to

$$\begin{aligned}
D(-\lambda) &= \sum_{j\in\mathcal{N}} \big((-1)^{\nu_j}(a_j + \varepsilon_j(-\lambda))\big)\lambda^{\nu_j}\exp(-c_j\lambda) \quad (\lambda \in \Omega, \arg\lambda \le 0), \\
D(-\lambda) &= \sum_{j\in\mathcal{N}} \big((-1)^{-\nu_j}(a_j + \varepsilon_j(-\lambda))\big)\lambda^{\nu_j}\exp(-c_j\lambda) \quad (\lambda \in \Omega, \arg\lambda > 0),
\end{aligned}$$

and obtain that there are $M_2 > 0$ and $K_2 \geq 0$ such that for all $\lambda \in \Omega$ with $|\lambda| \geq K_2$ and $-\Re(\lambda) - d\log|\lambda| \leq -M_2$ the estimate

$$|D(\lambda)| \geq \frac{|a_\beta|}{2}$$

holds. Then the statement of the proposition follows with $M := \max\{M_1, M_2\}$, $K_0 := \max\{K_1, K_2\}$ and $g_0 := \min\left\{\frac{|a_\alpha|}{2}, \frac{|a_\beta|}{2}\right\}$. □

PROPOSITION A.2.4. *Let $S \subset \mathbb{C}$ be compact and V be a compact topological space. Let $f : S \times V \to \mathbb{C}$ be continuous. Suppose that for all $x \in V$ the function $f(\cdot, x)$ is holomorphic in the interior $\mathring{S}$ of S and not identically zero in any nonempty open subset of $\mathring{S}$. For $x \in V$ and $\delta > 0$ we set*

$$N(x,\delta) := \{z \in S : f(z,x) = 0,\, d(z,\partial S) \geq \delta\}, \tag{A.2.9}$$

$$S(x,\delta) := \{z \in S : d(z, N(x,\delta)) \geq \delta,\, d(z,\partial S) \geq 2\delta\}. \tag{A.2.10}$$

Then for each $\delta > 0$ there are numbers $g_0(\delta) > 0$ and $l(\delta) > 0$ such that
i) *$\#N(x,\delta) \leq l(\delta)$ for all $x \in V$;*
ii) *$|f(z,x)| \geq g_0(\delta)$ for all $x \in V$ and $z \in S(x,\delta)$.*

Proof. i) Let $\delta > 0$ and set $\widetilde{S} := \{z \in S : d(z,\partial S) \geq \delta\}$. Let $x \in V$. The set

$$\widehat{N}(x) := \{z \in \mathring{S} : f(z,x) = 0\}$$

is a discrete subset of $\mathring{S}$ since $f(\cdot,x)$ is holomorphic in $\mathring{S}$ and not identically zero in any nonempty open subset of $\mathring{S}$. Thus for each $z \in \widetilde{S} \subset \mathring{S}$ there is a number $\varepsilon_z > 0$ such that $\overline{K}_{\varepsilon_z}(z) \subset \mathring{S}$ and $f(\zeta, x) \neq 0$ for all $\zeta \in \overline{K}_{\varepsilon_z} \setminus \{z\}$. $\widetilde{S}$ is a closed subset of the compact set S and hence $\widetilde{S}$ is compact. Thus there is a finite number of elements $z_1, \ldots, z_j$ in $\widetilde{S}$ such that

$$\widetilde{S} \subset \bigcup_{\nu=1}^{j} \overline{K}_{\varepsilon_{z_\nu}}(z_\nu).$$

Since the sets $\partial K_{\varepsilon_{z_\nu}}(z_\nu)$ are compact,

$$g_x(y) := \max\left\{ |f(z,y) - f(z,x)| : z \in \bigcup_{\nu=1}^{j} \partial K_{\varepsilon_{z_\nu}}(z_\nu) \right\} \qquad (y \in V)$$

defines a continuous function $g_x : V \to \mathbb{R}_+$. From $g_x(x) = 0$ and $f(z,x) \neq 0$ for all $z \in \partial K_{\varepsilon_{z_\nu}}(z_\nu)$ $(\nu = 1, \ldots, j)$ we infer that there is a neighbourhood $V_x \subset V$ of x such that

$$g_x(y) < \min\left\{ |f(z,x)| : z \in \bigcup_{\nu=1}^{j} \partial K_{\varepsilon_{z_\nu}}(z_\nu) \right\}$$

for all $y \in V_x$. ROUCHÉ's theorem yields that for each $\nu = 1, \dots, j$ the number of the zeros of $f(\cdot, y)$ in $\overline{K}_{\varepsilon_{z_\nu}}(z_\nu)$ counted according to their multiplicities does not depend on y. Hence there is a natural number $n(x, \delta)$ such that $\# N(y, \delta) \leq n(x, \delta)$ for all $y \in V_x$. Since the set V is compact, it can be covered by a finite number of neighbourhoods V_{x_i} $(i = 1, \dots, s)$ having this property. Now i) holds with $l(\delta) := \max\{n(x_i, \delta) : i = 1 \dots, s\}$.
ii) Let $\delta > 0$. Set

$$S_\delta := \bigcup_{x \in V} S(x, \delta) \times \{x\}.$$

We are going to show that $S \times V \setminus S_\delta$ is an open subset of $S \times V$. Then S_δ is a closed subset of the compact set $S \times V$. Hence S_δ is compact. Since f is continuous on $S \times V \supset S_\delta$ and nonzero on S_δ, the number

$$g_0(\delta) := \min\{|f(z,x)| : (z,x) \in S_\delta\}$$

is well-defined and positive if $S_\delta \neq \emptyset$.

To prove the openness of $S \times V \setminus S_\delta$ choose some $(z,x) \in S \times V \setminus S_\delta$. Then $z \notin S(x, \delta)$. If $d(z, \partial S) < 2\delta$, then

$$d(z', \partial S) \leq |z' - z| + d(z, \partial S) < 2\delta$$

holds for all $(z', x') \in S \times V$ such that $|z' - z| < 2\delta - d(z, \partial S)$. Hence (z,x) belongs to the interior of $S \times V \setminus S_\delta$.

If $d(z, \partial S) \geq 2\delta$, then $d(z, N(x, \delta)) < \delta$. There exists $z_0 \in N(x, \delta)$ such that $|z - z_0| < \delta$. Let $0 < \delta' < \delta - |z - z_0|$. From $d(z, \partial S) \geq 2\delta$ we obtain $K_{\delta'}(z_0) \subset \mathring{S}$ and from i) we know that z_0 is an isolated zero of $f(\cdot, x)$. Hence there are numbers $0 < \delta_1 < \delta'$ and $\varepsilon > 0$ such that $|f(\zeta, x)| \geq \varepsilon$ holds for all $\zeta \in \partial K_{\delta_1}(z_0)$. From the continuity of f and the compactness of $\partial K_{\delta_1}(z_0) \subset S$ we infer that there is a neighbourhood V_x of x in V such that $|f(\zeta, y) - f(\zeta, x)| < \varepsilon$ for all $y \in \dot{V}_x$ and $\zeta \in \partial K_{\delta_1}(z_0)$. ROUCHÉ's theorem yields that for all $y \in V_x$ the number of the zeros of $f(\cdot, y)$ in $K_{\delta_1}(z_0)$, counted according to their multiplicities, is equal to the number of the zeros of $f(\cdot, x)$ in $K_{\delta_1}(z_0)$. Since $f(z_0, x) = 0$, for each $y \in V_x$ there is a number $z_y \in K_{\delta_1}(z_0)$ such that $f(z_y, y) = 0$. The estimate

$$d(z_y, \partial S) \geq d(z, \partial S) - |z_y - z_0| - |z_0 - z| > 2\delta - \delta_1 - |z_0 - z| > \delta$$

yields $z_y \in N(y, \delta)$. For $(\zeta, y) \in K_{\delta' - \delta_1}(z) \times V_x$ we infer

$$|\zeta - z_y| \leq |\zeta - z| + |z - z_0| + |z_y - z_0| < \delta' - \delta_1 + |z - z_0| + \delta_1 < \delta.$$

This proves $(K_{\delta' - \delta_1}(z) \times V_x) \cap S_\delta = \emptyset$. Hence (z,x) belongs to the interior of $S \times V \setminus S_\delta$. □

COROLLARY A.2.5. *Let* $M, \widetilde{M} > 0$, $\mathcal{N}$ *be a countable nonempty set,* $a_j \in \mathbb{C}$, $c_j \in \mathbb{R}$ $(j \in \mathcal{N})$, $\sum_{j\in\mathcal{N}} |a_j| < \infty$, $\alpha, \beta \in \mathcal{N}$, $a_\alpha \neq 0$, $a_\beta \neq 0$, $c_\alpha \leq 0$, $c_\beta \geq 0$, $c_\alpha < c_j < c_\beta$ $(j \in \mathcal{N} \setminus \{\alpha, \beta\})$,

$$\widehat{S} := \{z \in \mathbb{C} : |\Re(z)| \leq M, |\Im(z)| \leq \widetilde{M}\},$$
$$S := \{z \in \mathbb{C} : |\Re(z)| \leq M+1, |\Im(z)| \leq \widetilde{M}+1\},$$
$$V := [0, 2\pi]^{\mathcal{N}}.$$

Define $f : S \times V \to \mathbb{C}$ *by*

$$f(z,x) := \sum_{j\in\mathcal{N}} a_j \exp(ix_j) \exp(c_j z),$$

where $(z,x) \in S \times V$ *and* $x = (x_j)_{j\in\mathcal{N}}$. *Then there are a natural number* l_1, *for each* $\delta > 0$ *a number* $g_0(\delta) > 0$, *and, for each* $x \in V$, l_1 *balls in* S *with radius* δ *such that for all* $z \in \widehat{S}$ *outside of these balls the estimate*

$$|f(z,x)| \geq g_0(\delta) \tag{A.2.11}$$

holds.

Proof. The sets $S \subset \mathbb{C}$ and, by TIHONOV'S theorem, V are compact, where V is endowed with the product topology. For each $x \in V$, $f(\cdot, x)$ is the restriction to S of an exponential sum of the form (A.2.6). In this case, the exponential sum is an entire function since, on compact sets, it is the uniform limit of entire functions. Because of $a_\alpha \exp(ix_\alpha) \neq 0$ and $a_\beta \exp(ix_\beta) \neq 0$ this entire function is not identically zero by Proposition A.2.3. For $j \in \mathcal{N}$ we define the function $g_j : S \times V \to \mathbb{C}$ by $g_j(z,x) = \exp(ix_j)\exp(c_j z)$. Then the function g_j is continuous and $|g_j(z,x)| \leq \exp\{\max\{c_\beta, -c_\alpha\}(M+1)\} =: B$. Thus $g := (g_j)_{j\in\mathcal{N}} : S \times V \to \Pi_B(\mathcal{N})$ is continuous. Obviously, $f(z,x) = m(a, g(z,x))$, where m is given by (A.2.7) and $a := (a_j)_{j\in\mathcal{N}}$. The continuity of m proved in Proposition A.2.1 yields that f is continuous. Hence the assumptions of Proposition A.2.4 are fulfilled. Take $l(\frac{1}{2})$ from Proposition A.2.4 and set $l_1 := l(\frac{1}{2})$.

Now let $0 < \delta \leq \frac{1}{2}$ and $x \in V$. Let $N(x,\delta)$ and $S(x,\delta)$ be given by (A.2.9) and (A.2.10). We assert

$$S(x,\delta) \cap \widehat{S} = \{z \in \widehat{S} : d(z, N(x, \tfrac{1}{2})) \geq \delta\}. \tag{A.2.12}$$

For the proof of (A.2.12) let $z \in S(x,\delta) \cap \widehat{S}$. The relation $N(x,\delta) \supset N(x,\frac{1}{2})$ is obvious and $d(z, N(x,\delta)) \geq \delta$ is clear from the definition of $S(x,\delta)$. Hence $d(z, N(x,\frac{1}{2})) \geq \delta$. Conversely let $z \in \widehat{S}$ with $d(z, N(x,\frac{1}{2})) \geq \delta$. From $|\Re(z)| \leq M$ and $|\Im(z)| \leq \widetilde{M}$ we infer $d(z, \partial S) \geq 1 \geq 2\delta$. For $\zeta \in N(x,\delta) \setminus N(x,\frac{1}{2})$ we have $d(\zeta, \partial S) < \frac{1}{2}$ and hence $|z - \zeta| \geq d(z,\partial S) - d(\zeta, \partial S) > \frac{1}{2}$. This proves

$$d(z, N(x,\delta)) = \min\{d(z, N(x,\tfrac{1}{2})), d(z, N(x,\delta) \setminus N(x,\tfrac{1}{2}))\} \geq \delta.$$

According to (A.2.12), $S(x,\delta)\cap\widehat{S}$ is the complement in $\widehat{S}$ of at most l_1 balls with radius δ since $\#N(x,\frac{1}{2})\le l_1$ by definition of l_1 and Propsition A.2.4. For $\delta\le\frac{1}{2}$ the corollary now follows from Proposition A.2.4. For $\delta>\frac{1}{2}$ it obviously holds with $g_0(\delta)=g_0(\frac{1}{2})$. □

PROPOSITION A.2.6. *We consider the exponential sum* (A.2.6)

$$D(\lambda)=\sum_{j\in\mathscr{N}} b_j(\lambda)\exp(c_j\lambda),$$

where $b_j(\lambda)=\lambda^{\nu_j}[a_j]$ *according to* (A.2.2)–(A.2.5), $\nu_j\in\mathbb{R}$, $c_j\in\mathbb{R}$ $(j\in\mathscr{N})$, $\alpha,\beta\in\mathscr{N}$, $a_\alpha\neq 0$, $a_\beta\neq 0$, $0=c_\alpha<c_j<c_\beta$ $(j\in\mathscr{N}\setminus\{\alpha,\beta\})$. *Assume that there is a number* $d\in\mathbb{R}$ *such that* $\nu_j=dc_j$ *for all* $j\in\mathscr{N}$. *Then there are a natural number l, for each* $\delta>0$ *numbers* $K(\delta)>0$ *and* $g(\delta)>0$, *and for all* $R>K(\delta)$ *there are l balls of radius* δ *such that for all* $\lambda\in\Omega$ *with* $R\le|\lambda|\le R+1$ *outside of these balls the estimate*

$$|D(\lambda)|\ge g(\delta)$$

holds.

Proof. From Proposition A.2.3 we know that there are numbers $K_0>0$, $M>0$, and $g_1>0$ such that the estimate $|D(\lambda)|\ge g_1$ holds for $\lambda\in\Omega$ satisfying $|\lambda|\ge K_0$ and $|\Re(\lambda)+d\log|\lambda||\ge M$.

We set $\widetilde{M}:=|d|\pi+1$ and define S, $\widehat{S}$, V, and f as in Corollary A.2.5. Then the assumptions of Corollary A.2.5 are fulfilled. We take l_1 from Corollary A.2.5 and set $l:=2l_1$. Because of $\frac{(\log(2r))^2}{r}\to 0$ $(r\to\infty)$ there is a number $K_1\ge K_0$ such that for all $R>K_1$ the estimates

$$2R-3-\big(M+1+|d|\log(R+2)\big)^2\ge 0 \tag{A.2.13}$$

and

$$R>8\big(M+3+|d|(\log(2R)+\pi)\big) \tag{A.2.14}$$

hold.

Now let $\delta>0$ and choose $g_0(\frac{\delta}{2})$ according to Corollary A.2.5. Because of (A.2.4) there is a $K(\delta)\ge\max\{K_1,2\}$ such that for all $\lambda\in\Omega$ with $|\lambda|\ge K(\delta)$ the estimate

$$\sum_{j\in\mathscr{N}}|\varepsilon_j(\lambda)|\exp(c_jM)\le\tfrac{1}{2}g_0(\tfrac{\delta}{2}) \tag{A.2.15}$$

holds. We set $g(\delta):=\min\{\frac{1}{2}g_0(\frac{\delta}{2}),g_1\}$ and

$$D_0(\lambda)=\sum_{j\in\mathscr{N}} a_j\lambda^{\nu_j}\exp(c_j\lambda).$$

We shall show that for all $R \geq K(\delta)$ there are at most l balls of radius δ such that for all λ in

$$W := \{\lambda \in \Omega : R \leq |\lambda| \leq R+1, |\Re(\lambda) + d \log|\lambda|| \leq M\}$$

outside of these balls the estimate

$$|D_0(\lambda)| \geq g_0(\tfrac{\delta}{2}) \tag{A.2.16}$$

holds. Then for these λ the estimates (A.2.15) and (A.2.16) imply that

$$\begin{aligned}\left|\sum_{j\in\mathcal{N}} b_j(\lambda)\exp(c_j\lambda)\right| &\geq |D_0(\lambda)| - \sum_{j\in\mathcal{N}} |\varepsilon_j(\lambda)\exp\{c_j(\lambda + d\log\lambda)\}| \\ &\geq |D_0(\lambda)| - \sum_{j\in\mathcal{N}} |\varepsilon_j(\lambda)|\exp(c_jM) \geq \tfrac{1}{2}g_0(\tfrac{\delta}{2}),\end{aligned}$$

which proves the statement of the proposition.

For the proof of (A.2.16) we set

$$W_0^\sigma := \{\lambda \in W : \sigma\Im(\lambda) \geq 0\},$$

where $\sigma \in \{1,-1\}$. Because of $W = W_0^1 \cup W_0^{-1}$ it is sufficient to prove that for all $R \geq K(\delta)$ and $\sigma \in \{1,-1\}$ there are at most l_1 balls of radius δ such that for all $\lambda \in W_0^\sigma$ outside of these balls the estimate $|D_0(\lambda)| \geq g_0(\frac{\delta}{2})$ holds. We set

$$W_1^\sigma := \{\lambda \in \Omega : R-1 \leq |\lambda| \leq R+2, |\Re(\lambda) + d\log|\lambda|| \leq M+1, \sigma\Im(\lambda) \geq 0\}.$$

For $\lambda \in \mathbb{C} \setminus \{0\}$ we set

$$h^\sigma(\lambda) := \lambda + d\log\lambda - i\sigma R$$

and assert

$$h^\sigma(W_0^\sigma) \subset \widehat{S}, \quad h^\sigma(W_1^\sigma) \subset S, \tag{A.2.17}$$

$$W_1^\sigma \subset K_{\frac{R}{4}}(i\sigma R), \tag{A.2.18}$$

$$h^\sigma \text{ is injective on } K_{\frac{R}{4}}(i\sigma R), \tag{A.2.19}$$

$$S \subset h^\sigma(K_{\frac{R}{4}}(i\sigma R)). \tag{A.2.20}$$

For the proof of (A.2.17) let $\lambda \in W_r^\sigma$ $(r = 0,1)$. From

$$|\Re(h^\sigma(\lambda))| = |\Re(\lambda) + d\log|\lambda|| \leq M + r \tag{A.2.21}$$

and (A.2.13) we obtain

$$\begin{aligned}|\Im(\lambda)|^2 = |\lambda|^2 - |\Re(\lambda)|^2 &\geq (R-r)^2 - \big(M + r + |d|\log(R+r+1)\big)^2 \\ &= (R-r-1)^2 + \big(2R - 2r - 1 - (M+r+|d|\log(R+r+1))^2\big) \\ &\geq (R-r-1)^2 + \big(2R - 3 - (M+r+|d|\log(R+2))^2\big) \\ &\geq (R-r-1)^2.\end{aligned}$$

This estimate, $\sigma\Im(\lambda) \geq 0$ and $\sigma\Im(\lambda) \leq |\lambda| \leq R+r+1$ imply

$$-r-1-|d|\pi \leq \sigma\Im(\lambda) - |d|\pi - R \leq \sigma\Im(h^\sigma(\lambda)) \leq \sigma\Im(\lambda) - R + |d|\pi \leq r+1+|d|\pi. \tag{A.2.22}$$

The estimates (A.2.21) and (A.2.22) prove (A.2.17).

For the proof of (A.2.18) let $\lambda \in W_1^\sigma$. Then

$$|\Re(\lambda)| \leq |\Re(\lambda) + d\log|\lambda|| + |d|\log|\lambda| \leq M+1+|d|\log(R+2)$$

and, by (A.2.22),

$$-2 \leq \sigma\Im(\lambda) - R \leq 2.$$

These estimates and (A.2.14) yield

$$|\lambda - i\sigma R| \leq M+3+|d|\log(R+2) < \frac{R}{4}.$$

For the proof of (A.2.19) let $\lambda_0 \in K_{\frac{R}{4}}(i\sigma R)$. The function $\lambda \mapsto \lambda - \lambda_0$ has exactly one simple zero in $K_{\frac{R}{2}}(i\sigma R)$. For $\lambda \in \partial K_{\frac{R}{2}}(i\sigma R)$ we obtain because of $|\lambda|, |\lambda_0| \geq \frac{R}{2} \geq 1$, $\sigma\Im(\lambda) > 0$, $\sigma\Im(\lambda_0) > 0$, and (A.2.14) that

$$\begin{aligned}|d\log\lambda - d\log\lambda_0| &\leq |d|\big(\big|\log|\lambda| - \log|\lambda_0|\big| + \pi\big)\\ &\leq |d|(\log(2R)+\pi) < \frac{R}{4} < |\lambda - i\sigma R| - |\lambda_0 - i\sigma R|\\ &\leq |\lambda - \lambda_0|.\end{aligned}$$

According to ROUCHÉ's theorem, the mapping

$$\lambda \mapsto \lambda + d\log\lambda - (\lambda_0 + d\log\lambda_0)$$

has exactly one zero in $K_{\frac{R}{2}}(i\sigma R)$. Hence there is exactly one $\lambda \in K_{\frac{R}{2}}(i\sigma R)$ such that

$$h^\sigma(\lambda) = \lambda_0 + d\log\lambda_0 - i\sigma R = h^\sigma(\lambda_0).$$

This proves (A.2.19).

For the proof of (A.2.20) let $z \in S$. The definition of S, $\widetilde{M} = |d|\pi + 1$ and (A.2.14) yield

$$|z| \leq |\Re(z)| + |\Im(z)| \leq M+3+|d|\pi < \frac{R}{8}. \tag{A.2.23}$$

Hence the function $\lambda \mapsto \lambda - z - i\sigma R$ has exactly one zero in $K_{\frac{R}{4}}(i\sigma R)$. For $\lambda \in \partial K_{\frac{R}{4}}(i\sigma R)$ we obtain from $2R \geq |\lambda| \geq \frac{R}{2} \geq 1$, (A.2.14), and (A.2.23) that

$$\begin{aligned}|d\log\lambda| &\leq |d|(\log|\lambda| + \pi) \leq |d|(\log(2R)+\pi) < \frac{R}{8}\\ &< |\lambda - i\sigma R| - |z| \leq |\lambda - z - i\sigma R|.\end{aligned}$$

An application of ROUCHÉ's theorem yields an element $\lambda \in K_{\frac{R}{4}}(i\sigma R)$ such that $\lambda + d\log\lambda - z - i\sigma R = 0$. Hence

$$z = \lambda + d\log\lambda - i\sigma R = h^{\sigma}(\lambda) \in h^{\sigma}(K_{\frac{R}{4}}(i\sigma R)).$$

This completes the proof of (A.2.17)–(A.2.20).

By (A.2.19), the inverse function h_1^{σ} of $h^{\sigma}|_{K_{\frac{R}{4}}(i\sigma R)}$ is well-defined. For $j \in \mathcal{N}$ we define $k_j \in \mathbb{Z}$ by

$$2\pi k_j \le Rc_j\sigma < 2\pi(k_j+1).$$

We set

$$x_j := Rc_j\sigma - 2\pi k_j,\ x := (x_j)_{j\in\mathcal{N}}.$$

For $\lambda \in W_1^{\sigma}$ we obtain

$$\begin{aligned} D_0(\lambda) &= \sum_{j\in\mathcal{N}} a_j \exp\{c_j(\lambda + d\log\lambda)\} \\ &= \sum_{j\in\mathcal{N}} a_j \exp(i\sigma c_j R)\exp\{c_j h^{\sigma}(\lambda)\} \\ &= \sum_{j\in\mathcal{N}} a_j \exp(ix_j)\exp\{c_j h^{\sigma}(\lambda)\} \\ &= f(h^{\sigma}(\lambda), x), \end{aligned}$$

where f is the function defined in Corollary A.2.5. By Corollary A.2.5 there are l_1 balls $K_{\frac{\delta}{2}}(z_i)$ $(z_i \in S; i = 1,\ldots,l_1)$ such that the estimate $|f(z,x)| \ge g_0(\frac{\delta}{2})$ holds for $z \in \widehat{S} \setminus \bigcup_{i=1}^{l_1} K_{\frac{\delta}{2}}(z_i)$. Let $\lambda \in W_0^{\sigma} \setminus \bigcup_{i=1}^{l_1} K_{\delta}(h_1^{\sigma}(z_i))$. Then $\lambda = h_1^{\sigma}(h^{\sigma}(\lambda))$ because of (A.2.17) and (A.2.20), and (A.2.18) yields $\lambda \in K_{\frac{R}{4}}(i\sigma R)$. Since S is convex, we obtain for $i = 1,\ldots,l_1$ with the aid of the mean value theorem, (A.2.20), and (A.2.14) that

$$\begin{aligned} |\lambda - h_1^{\sigma}(z_i)| &= |h_1^{\sigma}(h^{\sigma}(\lambda)) - h_1^{\sigma}(z_i)| \le \sup_{\zeta\in S} |h_1^{\sigma\prime}(\zeta)|\,|h^{\sigma}(\lambda) - z_i| \\ &\le \sup_{\xi\in K_{\frac{R}{4}}(i\sigma R)} \left|\frac{1}{h^{\sigma\prime}(\xi)}\right| |h^{\sigma}(\lambda) - z_i| \le \frac{1}{1 - \frac{2|d|}{R}} |h^{\sigma}(\lambda) - z_i| \\ &\le 2|h^{\sigma}(\lambda) - z_i|. \end{aligned}$$

Hence

$$|h^{\sigma}(\lambda) - z_i| \ge \frac{1}{2}|\lambda - h_1^{\sigma}(z_i)| \ge \frac{\delta}{2}.$$

From (A.2.17) we infer $h^{\sigma}(\lambda) \in \widehat{S} \setminus \bigcup_{i=1}^{l_1} K_{\frac{\delta}{2}}(z_i)$. Then Corollary A.2.5 yields

$$|D_0(\lambda)| = |f(h^{\sigma}(\lambda), x)| \ge g_0(\tfrac{\delta}{2}). \quad \square$$

DEFINITION A.2.7. We consider an exponential sum of the form (A.2.6) with the representation (A.2.2) of the b_j. Let $\mathscr{M}$, $\mathscr{M}' \subset \mathscr{N}$, where $\#\mathscr{M} \geq 2$ and $\mathscr{M}'$ is a finite subset of $\mathscr{M}$. The pair $(\mathscr{M}, \mathscr{M}')$ is called *weakly regular* if the following properties hold:
i) there are $\alpha, \beta \in \mathscr{M}$ such that for all $j \in \mathscr{M}$ there is a number $\tau \in [0,1]$ such that $c_j = \tau c_\alpha + (1-\tau)c_\beta$, i.e., the line segment $\overline{c_\alpha, c_\beta}$ is the convex hull of the set $\{c_j : j \in \mathscr{M}\}$;
ii) for all $j \in \mathscr{M}$ there are $j_1, j_2 \in \mathscr{M}'$ and a number $\tau \in [0,1]$ such that

$$c_j = \tau c_{j_1} + (1-\tau)c_{j_2} \quad \text{and} \quad v_j \leq \tau v_{j_1} + (1-\tau)v_{j_2};$$

iii) if $j \in \mathscr{M}'$ for a triple $\{j, j_1, j_2\}$ which fulfils ii), then $j \in \{j_1, j_2\}$.

If $(\mathscr{M}, \mathscr{M}')$ is weakly regular, then condition i) implies that the convex hull of $\mathscr{M}$ is a line segment whose endpoints belong to $\mathscr{M}'$ by ii).

Condition ii) says that those $j \in \mathscr{M}$ which do not have a representation $c_j = \tau c_{j_1} + (1-\tau)c_{j_2}$ with $j_1, j_2 \in \mathscr{M}'$, $\tau \in (0,1)$ and $v_j \leq \tau v_{j_1} + (1-\tau)v_{j_2}$ must belong to $\mathscr{M}'$. For in this case we have $\tau = 0$ or $\tau = 1$, and the only representation fulfilling ii) is the trivial one with $c_j = c_{j_1}$ or $c_j = c_{j_2}$ which implies $j = j_1 \in \mathscr{M}'$ or $j = j_2 \in \mathscr{M}'$.

REMARK A.2.8. For $j, j_1, j_2 \in \mathscr{M}$ such that $c_j \in \overline{c_{j_1}, c_{j_2}}$ and $j \notin \{j_1, j_2\}$ the following conditions are equivalent:
i) There is a number $\tau \in (0,1)$ such that

$$c_j = \tau c_{j_1} + (1-\tau)c_{j_2} \quad \text{and} \quad v_j \leq \tau v_{j_1} + (1-\tau)v_{j_2};$$

ii)
$$\frac{v_j - v_{j_1}}{|c_j - c_{j_1}|} \leq \frac{v_{j_2} - v_{j_1}}{|c_{j_2} - c_{j_1}|};$$

iii)
$$\frac{v_j - v_{j_1}}{|c_j - c_{j_1}|} \leq \frac{v_{j_2} - v_j}{|c_{j_2} - c_j|}.$$

Proof. The equivalence of i) and ii) immediately follows from

$$c_j = \tau c_{j_1} + (1-\tau)c_{j_2} \Leftrightarrow c_j - c_{j_1} = (1-\tau)(c_{j_2} - c_{j_1})$$

and

$$v_j \leq \tau v_{j_1} + (1-\tau)v_{j_2} \Leftrightarrow v_j - v_{j_1} \leq (1-\tau)(v_{j_2} - v_{j_1}).$$

In the same way, the equivalence of i) and iii) immediately follows from

$$c_j = \tau c_{j_1} + (1-\tau)c_{j_2} \Leftrightarrow \tau(c_j - c_{j_1}) = (1-\tau)(c_{j_2} - c_j)$$

and

$$v_j \leq \tau v_{j_1} + (1-\tau)v_{j_2} \Leftrightarrow \tau(v_j - v_{j_1}) \leq (1-\tau)(v_{j_2} - v_j). \quad \square$$

Now we deduce a method how to construct a set $\mathscr{M}'$ such that $(\mathscr{M},\mathscr{M}')$ is weakly regular. By Definition A.2.7 i) the existence of $\alpha,\beta\in\mathscr{M}$ such that $\{c_j : j\in\mathscr{M}\}\subset\overline{c_\alpha,c_\beta}$ is necessary. We set $\gamma_0:=\alpha$ and $\mathscr{M}_0:=\mathscr{M}$ and assume that the set of real numbers

$$\left\{\frac{v_j-v_\alpha}{|c_j-c_\alpha|} : j\in\mathscr{M}_0\setminus\{\gamma_0\}\right\}$$

has a maximum d. Further we assume that the set

$$\left\{\frac{|c_j-c_\alpha|}{|c_\beta-c_\alpha|} : j\in\mathscr{M}_0\setminus\{\gamma_0\},\ \frac{v_j-v_\alpha}{|c_j-c_\alpha|}=d\right\}$$

has a maximum t_1. Then there is a (unique) $\gamma_1\in\mathscr{M}$ so that $|c_{\gamma_1}-c_\alpha|=t_1|c_\beta-c_\alpha|$ and $v_{\gamma_1}-v_\alpha=d|c_{\gamma_1}-c_\alpha|$. Obviously, each element $j\in\mathscr{M}$ with $|c_j-c_\alpha|<|c_{\gamma_1}-c_\alpha|$ has a representation of the form ii) with $j_1=\alpha$ and $j_2=\gamma_1$. If $t_1<1$, then we repeat this procedure with the set

$$\mathscr{M}_1:=\{j\in\mathscr{M}_0 : |c_j-c_\alpha|\geq t_1|c_\beta-c_\alpha|\}.$$

Then we obtain an element $\gamma_2\in\mathscr{M}$ if the corresponding maxima exist. We proceed in this way and assume that we can construct $\gamma_1,\gamma_2,\dots$ Further we assume that $\gamma_n=\beta$ for some $n\in\mathbb{N}$. Then $\mathscr{M}':=\{\gamma_0,\dots,\gamma_n\}$ fulfils the conditions i) and ii). Furthermore, it is easy to see that iii) holds. For if ii) holds with $\{\gamma_{i_1},\gamma_{i_2},\gamma_{i_3}\}\subset\mathscr{M}'$ and $i_2<i_1<i_3$, we would have

$$\frac{v_{\gamma_{i_1}}-v_{\gamma_{i_2}}}{|c_{\gamma_{i_1}}-c_{\gamma_{i_2}}|}\leq\frac{v_{\gamma_{i_3}}-v_{\gamma_{i_1}}}{|c_{\gamma_{i_3}}-c_{\gamma_{i_1}}|}$$

by Remark A.2.8. By the definition of γ_{i_2+1} we have

$$\frac{v_{\gamma_{i_1}}-v_{\gamma_{i_2}}}{|c_{\gamma_{i_1}}-c_{\gamma_{i_2}}|}\leq\frac{v_{\gamma_{i_2+1}}-v_{\gamma_{i_2}}}{|c_{\gamma_{i_2+1}}-c_{\gamma_{i_2}}|}.$$

From $|c_{\gamma_{i_1}}-c_{\gamma_{i_2}}|=|c_{\gamma_{i_1}}-c_{\gamma_{i_2+1}}|+|c_{\gamma_{i_2+1}}-c_{\gamma_{i_2}}|$ we then immediately infer

$$\begin{aligned}&(v_{\gamma_{i_1}}-v_{\gamma_{i_2+1}})|c_{\gamma_{i_1}}-c_{\gamma_{i_2}}|-(v_{\gamma_{i_1}}-v_{\gamma_{i_2}})|c_{\gamma_{i_1}}-c_{\gamma_{i_2+1}}|\\&=(v_{\gamma_{i_1}}-v_{\gamma_{i_2}})|c_{\gamma_{i_2+1}}-c_{\gamma_{i_2}}|-(v_{\gamma_{i_2+1}}-v_{\gamma_{i_2}})|c_{\gamma_{i_1}}-c_{\gamma_{i_2}}|\leq 0,\end{aligned}$$

which proves

$$\frac{v_{\gamma_{i_1}}-v_{\gamma_{i_2+1}}}{|c_{\gamma_{i_1}}-c_{\gamma_{i_2+1}}|}\leq\frac{v_{\gamma_{i_1}}-v_{\gamma_{i_2}}}{|c_{\gamma_{i_1}}-c_{\gamma_{i_2}}|}.$$

By induction we obtain

$$\frac{v_{\gamma_{i_1}}-v_{\gamma_{i_1-1}}}{|c_{\gamma_{i_1}}-c_{\gamma_{i_1-1}}|}\leq\frac{v_{\gamma_{i_1}}-v_{\gamma_{i_2}}}{|c_{\gamma_{i_1}}-c_{\gamma_{i_2}}|}\leq\frac{v_{\gamma_{i_3}}-v_{\gamma_{i_1}}}{|c_{\gamma_{i_3}}-c_{\gamma_{i_1}}|}.$$

From Remark A.2.8 we infer

$$\frac{\nu_{\gamma_{i_1}} - \nu_{\gamma_{i_1-1}}}{|c_{\gamma_{i_1}} - c_{\gamma_{i_1-1}}|} \le \frac{\nu_{\gamma_{i_3}} - \nu_{\gamma_{i_1-1}}}{|c_{\gamma_{i_3}} - c_{\gamma_{i_1-1}}|},$$

which contradicts the definition of γ_{i_1}. □

One can show that the existence of the maxima and the finiteness of the procedure are necessary for the existence of a weakly regular pair $(\mathscr{M}, \mathscr{M}')$. Furthermore, $\mathscr{M}'$ is unique if it exists. Obviously, the above method is possible if $\mathscr{M}$ is finite. Hence we obtain

REMARK A.2.9. Assume that $\mathscr{M}$ is finite and that the points $\{c_j : j \in \mathscr{M}\}$ lie on a straight line. Then there is a subset $\mathscr{M}'$ of $\mathscr{M}$ such that $(\mathscr{M}, \mathscr{M}')$ is weakly regular.

REMARK A.2.10. Let $D(\lambda)$ be an exponential sum with the representation (A.2.2) of the b_j. Let $\mathscr{M} \subset \mathscr{N}$ be such that the convex hull of $\mathscr{M}$ is a line segment, the endpoints of which we denote by c_α and c_β. Assume that on $\mathscr{M}$ the ν_j do not depend on j, i.e., there is a $\nu \in \mathbb{R}$ such that $\nu_j = \nu$ for all $j \in \mathscr{M}$. Then $(\mathscr{M}, \{\alpha, \beta\})$ is weakly regular.

Proof. Since $\nu_j = \tau\nu_{j_1} + (1-\tau)\nu_{j_2}$ holds for all $j, j_1, j_2 \in \mathscr{M}$ and $\tau \in [0,1]$, we can take $j_1 = \alpha$ and $j_2 = \beta$ in the representation ii) of Definition A.2.7. □

PROPOSITION A.2.11. *We consider the exponential sum* (A.2.6)

$$D(\lambda) = \sum_{j \in \mathscr{N}} b_j(\lambda) \exp(c_j \lambda),$$

where $b_j(\lambda) = \lambda^{\nu_j}[a_j]$ according to (A.2.2)–(A.2.5), $\nu_j \in \mathbb{R}$, $c_j \in \mathbb{R}_+$ $(j \in \mathscr{N})$, $c_{j_0} = 0$, $\nu_{j_0} = 0$ *for some $j_0 \in \mathscr{N}$. Assume that there is a finite subset $\mathscr{N}'$ of $\mathscr{N}$ such that $(\mathscr{N}, \mathscr{N}')$ is weakly regular and that $a_j \ne 0$ for all $j \in \mathscr{N}'$. Then there are a natural number $l > 0$, for each $\delta > 0$ numbers* $K(\delta) > 0$ *and* $g(\delta) > 0$, *and for all $R > K(\delta)$ there are l balls of radius δ such that for all $\lambda \in \Omega$ with $R \le |\lambda| \le R+1$ outside of these balls we have the estimate*

$$|D(\lambda)| \ge g(\delta).$$

Proof. Since $\mathscr{N}'$ is finite, we may assume without loss of generality that there is some $k \in \mathbb{N} \setminus \{0\}$ such that $\mathscr{N}' = \{0, \ldots, k\}$ and $0 = c_0 < c_1 < \cdots < c_k$. For $\kappa = 1, \ldots, k$ we set

$$d_\kappa := \frac{\nu_\kappa - \nu_{\kappa-1}}{c_\kappa - c_{\kappa-1}}.$$

By condition iii) of Definition A.2.7 we obtain for each $\kappa = 1, \ldots, k-1$ that there is a $\tau \in (0,1)$ such that

$$c_\kappa = \tau c_{\kappa-1} + (1-\tau)c_{\kappa+1} \quad \text{and} \quad \nu_\kappa > \tau\nu_{\kappa-1} + (1-\tau)\nu_{\kappa+1}.$$

Hence, by Remark A.2.8,

$$d_\kappa = \frac{v_\kappa - v_{\kappa-1}}{c_\kappa - c_{\kappa-1}} > \frac{v_{\kappa+1} - v_\kappa}{c_{\kappa+1} - c_\kappa} = d_{\kappa+1} \qquad (\kappa = 1, \ldots, k-1). \tag{A.2.24}$$

The definition of d_κ yields for $\kappa_1, \kappa_2 \in \{0, \ldots, k\}$ such that $\kappa_1 < \kappa_2$ that

$$v_{\kappa_2} - v_{\kappa_1} = \sum_{\kappa=\kappa_1+1}^{\kappa_2} (v_\kappa - v_{\kappa-1}) = \sum_{\kappa=\kappa_1+1}^{\kappa_2} d_\kappa (c_\kappa - c_{\kappa-1}).$$

With the aid of (A.2.24) we infer for $\kappa_1, \kappa_2 \in \{0, \ldots, k\}$ such that $\kappa_1 \leq \kappa_2$ that

$$d_{\kappa_2}(c_{\kappa_2} - c_{\kappa_1}) \leq v_{\kappa_2} - v_{\kappa_1} \leq d_{\kappa_1+1}(c_{\kappa_2} - c_{\kappa_1}). \tag{A.2.25}$$

Let $j \in \mathscr{N} \setminus \mathscr{N}'$. Then there are $\kappa \in \{1, \ldots, k\}$, $\kappa_1, \kappa_2 \in \{0, \ldots, k\}$, and $\tau, t \in (0,1)$ such that

$$c_j = \tau c_{\kappa-1} + (1-\tau) c_\kappa, \tag{A.2.26}$$

$$c_j = t c_{\kappa_1} + (1-t) c_{\kappa_2}, \tag{A.2.27}$$

$$v_j \leq t v_{\kappa_1} + (1-t) v_{\kappa_2}. \tag{A.2.28}$$

Here we may assume that $\kappa_1 < \kappa_2$. Obviously, $\kappa_1 \leq \kappa - 1$ and $\kappa \leq \kappa_2$. We shall show that

$$v_j \leq \tau v_{\kappa-1} + (1-\tau) v_\kappa =: v'_j. \tag{A.2.29}$$

The number v'_j is unique since κ and τ are uniquely determined by c_j. From (A.2.26) and (A.2.27) we infer

$$\begin{aligned} \tau &= \frac{c_j - c_\kappa}{c_{\kappa-1} - c_\kappa} = \frac{t c_{\kappa_1} + (1-t) c_{\kappa_2} - c_\kappa}{c_{\kappa-1} - c_\kappa} \\ &= \frac{t(c_\kappa - c_{\kappa_1}) + (1-t)(c_\kappa - c_{\kappa_2})}{c_\kappa - c_{\kappa-1}}. \end{aligned}$$

With the aid of (A.2.28), (A.2.25) and (A.2.24) we calculate

$$\begin{aligned} v_j &\leq t v_{\kappa_1} + (1-t) v_{\kappa_2} \\ &= t(v_{\kappa_1} - v_\kappa) + (1-t)(v_{\kappa_2} - v_\kappa) + v_\kappa \\ &\leq t d_\kappa (c_{\kappa_1} - c_\kappa) + (1-t) d_{\kappa+1} (c_{\kappa_2} - c_\kappa) + v_\kappa \\ &\leq \frac{v_\kappa - v_{\kappa-1}}{c_\kappa - c_{\kappa-1}} \{ t(c_{\kappa_1} - c_\kappa) + (1-t)(c_{\kappa_2} - c_\kappa) \} + v_\kappa \\ &= \tau v_{\kappa-1} + (1-\tau) v_\kappa. \end{aligned}$$

If $v_j < v'_j$, then

$$b_j(\lambda) = \lambda^{v'_j} \{ \lambda^{v_j - v'_j} [a_j] \} = \lambda^{v'_j} [0].$$

For each $j \in \mathcal{N}$ we have the representation

$$b_j(\lambda) = \lambda^{\nu'_j}(a'_j + \varepsilon'_j(\lambda)),$$

where $a'_j = a_j$ and $\varepsilon'_j = \varepsilon_j$ if $\nu_j = \nu'_j$, and $a'_j = 0$ and $\varepsilon'_j(\lambda) = \lambda^{\nu_j - \nu'_j}(a_j + \varepsilon_j(\lambda))$ if $\nu_j < \nu'_j$. The estimate (A.2.3) obviously remains true with a'_j instead of a_j. We have

$$\varepsilon'(\lambda) := \sum_{j \in \mathcal{N}} |\varepsilon'_j(\lambda)| \le \sum_{j \in \mathcal{N}} \left(|a_j| + |\varepsilon_j(\lambda)|\right) < \infty$$

since $|\lambda^{\nu_j - \nu'_j}| = |\lambda|^{\nu_j - \nu'_j} \le 1$ for $|\lambda| \ge 1$. We set $x_j(\lambda) := |\lambda|^{\nu_j - \nu'_j}$ if $\nu < \nu'_j$ and $x_j(\lambda) = 0$ if $\nu_j = \nu'_j$. Then $x_j(\lambda) \to 0$ as $|\lambda| \to \infty$ for all $j \in \mathcal{N}$. Proposition A.2.2 i) yields

$$\sum_{j \in \mathcal{N}} |\varepsilon'_j(\lambda)| \le \sum_{j \in \mathcal{N}} |a_j| x_j(\lambda) + \varepsilon(\lambda) \to 0 \quad \text{as } \lambda \to \infty.$$

Hence we can suppose that $\nu_j = \nu'_j$.

If $k = 1$, then the assumptions of Proposition A.2.6 hold with $d = \frac{\nu_1}{c_1}$. Thus, for $k = 1$, Proposition A.2.11 is an immediate consequence of Proposition A.2.6.

Now let $k \ge 2$. Let $\kappa \in \{1, \dots, k\}$ and $j \in \mathcal{N}$ such that $c_j \in [c_{\kappa-1}, c_\kappa]$, i. e., $c_j = \tau c_{\kappa-1} + (1 - \tau) c_\kappa$ for some $\tau \in [0, 1]$. From $\nu_j = \nu'_j$ we infer

$$\nu_j - \nu_{\kappa-1} = d_\kappa (c_j - c_{\kappa-1}) \quad \text{if} \quad c_j \in [c_{\kappa-1}, c_\kappa]. \tag{A.2.30}$$

For $\kappa = 1, \dots, k-1$ we choose $t_\kappa \in \mathbb{R}$ with $d_\kappa > t_\kappa > d_{\kappa+1}$ and set

$$\begin{aligned}
V_\kappa^+ &:= \{\lambda \in \Omega : |\lambda| \ge 1,\ \Re(\lambda) + t_\kappa \log|\lambda| \ge 0\} \quad (\kappa = 1, \dots, k-1), \\
V_\kappa^- &:= \{\lambda \in \Omega : |\lambda| \ge 1,\ \Re(\lambda) + t_\kappa \log|\lambda| < 0\} \quad (\kappa = 1, \dots, k-1), \\
V_1 &:= V_1^-, \\
V_\kappa &:= V_{\kappa-1}^+ \cap V_\kappa^- \quad (\kappa = 2, \dots, k-1), \\
V_k &:= V_{k-1}^+.
\end{aligned}$$

For $\kappa \in \{1, \dots, k\}$ let

$$\begin{aligned}
\mathcal{N}_\kappa &:= \{j \in \mathcal{N} : c_{\kappa-1} \le c_j \le c_\kappa\}, \\
\mathcal{N}_\kappa^- &:= \{j \in \mathcal{N} : c_j < c_{\kappa-1}\}, \\
\mathcal{N}_\kappa^+ &:= \{j \in \mathcal{N} : c_j > c_\kappa\}.
\end{aligned}$$

For $\lambda \in \Omega$ and $j \in \mathcal{N}_\kappa$ such that $j \notin \{\kappa, \kappa - 1\}$ or $\lambda \notin V_\kappa$ let $\varepsilon_j^\kappa(\lambda) := \varepsilon_j(\lambda)$. For $\lambda \in V_\kappa$ we set

$$\varepsilon_{\kappa-1}^\kappa(\lambda) := \varepsilon_{\kappa-1}(\lambda) + \sum_{j \in \mathcal{N}_\kappa^-} b_j(\lambda) \lambda^{-\nu_{\kappa-1}} \exp\{c_j \lambda - c_{\kappa-1} \lambda\}$$

and

$$\varepsilon_\kappa^\kappa(\lambda) := \varepsilon_\kappa(\lambda) + \sum_{j\in\mathcal{N}_\kappa^+} b_j(\lambda)\lambda^{-\nu_\kappa}\exp\{c_j\lambda - c_\kappa\lambda\}.$$

We shall prove that for $\kappa = 1,\dots,k$ the functions $D^\kappa : V_\kappa \to \mathbb{C}$ defined by

$$D^\kappa(\lambda) := \sum_{j\in\mathcal{N}_\kappa} (a_j + \varepsilon_j^\kappa(\lambda))\lambda^{\nu_j-\nu_{\kappa-1}}\exp\{(c_j - c_{\kappa-1})\lambda\} \quad (\lambda \in V_\kappa) \tag{A.2.31}$$

fulfil the assumptions of Proposition A.2.6. By (A.2.30), the equation $\nu_j - \nu_{\kappa-1} = d_\kappa(c_j - c_{\kappa-1})$ holds for each $j \in \mathcal{N}_\kappa$. Since, by assumption, $a_{\kappa-1} \neq 0$ and $a_\kappa \neq 0$, we have to prove that

$$\sum_{j\in\mathcal{N}_\kappa} |\varepsilon_j^\kappa(\lambda)| \to 0 \quad (\lambda \in V_\kappa, \lambda \to \infty),$$

which obviously holds if

$$\sum_{j\in\mathcal{N}_\kappa^-} |b_j(\lambda)\lambda^{-\nu_{\kappa-1}}\exp\{c_j\lambda - c_{\kappa-1}\lambda\}| \to 0 \quad (\lambda \in V_\kappa, \lambda \to \infty)$$

and

$$\sum_{j\in\mathcal{N}_\kappa^+} |b_j(\lambda)\lambda^{-\nu_\kappa}\exp\{c_j\lambda - c_\kappa\lambda\}| \to 0 \quad (\lambda \in V_\kappa, \lambda \to \infty)$$

are fulfilled. As, by (A.2.3)–(A.2.5), $(b_j(\lambda)\lambda^{-\nu_j})_{j\in\mathcal{N}_\kappa^\pm} \to (a_j)_{j\in\mathcal{N}_\kappa^\pm}$ in $l_1(\mathcal{N}_\kappa^\pm)$ as $\lambda \to \infty$, this follows from Proposition A.2.2 i) and

$$\lambda^{\nu_j-\nu_{\kappa-1}}\exp\{c_j\lambda - c_{\kappa-1}\lambda\} = o(1) \qquad (j \in \mathcal{N}_\kappa^-, \lambda \in V_{\kappa-1}^+) \tag{A.2.32}$$

$$\lambda^{\nu_j-\nu_\kappa}\exp\{c_j\lambda - c_\kappa\lambda\} = o(1) \qquad (j \in \mathcal{N}_\kappa^+, \lambda \in V_\kappa^-) \tag{A.2.33}$$

for those $\kappa = 2,\dots,k$ in (A.2.32) and $\kappa = 1,\dots,k-1$ in (A.2.33), respectively, for which $V_{\kappa-1}^+$ or V_κ^- are unbounded. We have to prove (A.2.32) and (A.2.33).

First we consider (A.2.32). Let $j \in \mathcal{N}_\kappa^-$. Then there is a number $\kappa' \le \kappa - 1$ such that $c_j \in [c_{\kappa'-1}, c_{\kappa'})$. From (A.2.30), the definition of $d_{\kappa-1}$, and (A.2.24) we infer

$$\begin{aligned} \nu_j - \nu_{\kappa-1} &= \nu_j - \nu_{\kappa'-1} + \sum_{\iota=\kappa'}^{\kappa-1}(\nu_{\iota-1} - \nu_\iota) \\ &= d_{\kappa'}(c_j - c_{\kappa'-1}) + \sum_{\iota=\kappa'}^{\kappa-1} d_\iota(c_{\iota-1} - c_\iota) \\ &\le d_{\kappa-1}(c_j - c_{\kappa'}) + \sum_{\iota=\kappa'+1}^{\kappa-1} d_{\kappa-1}(c_{\iota-1} - c_\iota) \\ &= d_{\kappa-1}(c_j - c_{\kappa-1}). \end{aligned} \tag{A.2.34}$$

For $j \in \mathcal{N}_\kappa^-$ and $\lambda \in V_{\kappa-1}^+$ we thus obtain

$$\begin{aligned}
&|\lambda^{\nu_j-\nu_{\kappa-1}} \exp\{c_j\lambda - c_{\kappa-1}\lambda\}| = \exp\{(c_j - c_{\kappa-1})\Re(\lambda) + (\nu_j - \nu_{\kappa-1}) \log|\lambda|\} \\
&= \exp\{(c_j - c_{\kappa-1})(\Re(\lambda) + t_{\kappa-1} \log|\lambda|)\} \times \\
&\quad \times \exp\{((\nu_j - \nu_{\kappa-1}) - t_{\kappa-1}(c_j - c_{\kappa-1})) \log|\lambda|\} \\
&\le \exp\{(d_{\kappa-1} - t_{\kappa-1})(c_j - c_{\kappa-1}) \log|\lambda|\}.
\end{aligned}$$

Since $(d_{\kappa-1} - t_{\kappa-1})(c_j - c_{\kappa-1}) < 0$, (A.2.32) is proved.

Now we consider (A.2.33). Let $j \in \mathcal{N}_\kappa^+$. Then there is a number $\kappa' > \kappa$ such that $c_j \in (c_{\kappa'-1}, c_{\kappa'}]$. From (A.2.30), the definition of d_κ, and (A.2.24) we infer

$$\begin{aligned}
\nu_j - \nu_\kappa &= \nu_j - \nu_{\kappa'-1} + \sum_{\iota=\kappa+1}^{\kappa'-1} (\nu_\iota - \nu_{\iota-1}) \\
&= d_{\kappa'}(c_j - c_{\kappa'-1}) + \sum_{\iota=\kappa+1}^{\kappa'-1} d_\iota(c_\iota - c_{\iota-1}) \\
&\le d_{\kappa+1}(c_j - c_{\kappa'-1}) + \sum_{\iota=\kappa+1}^{\kappa'-1} d_{\kappa+1}(c_\iota - c_{\iota-1}) \\
&= d_{\kappa+1}(c_j - c_\kappa).
\end{aligned}$$

For $j \in \mathcal{N}_\kappa^+$ and $\lambda \in V_\kappa^-$ we thus obtain

$$\begin{aligned}
&|\lambda^{\nu_j-\nu_\kappa} \exp\{c_j\lambda - c_\kappa\lambda\}| = \exp\{(c_j - c_\kappa)\Re(\lambda) + (\nu_j - \nu_\kappa) \log|\lambda|\} \\
&= \exp\{(c_j - c_\kappa)(\Re(\lambda) + t_\kappa \log|\lambda|)\} \exp\left\{((\nu_j - \nu_\kappa) - t_\kappa(c_j - c_\kappa)) \log|\lambda|\right\} \\
&\le \exp\{(d_{\kappa+1} - t_\kappa)(c_j - c_\kappa) \log|\lambda|\}.
\end{aligned}$$

Since $(d_{\kappa+1} - t_\kappa)(c_j - c_\kappa) < 0$, (A.2.33) is proved.

Hence we can apply Proposition A.2.6 to D^κ if V_κ is unbounded. For these κ let l_κ, $K_\kappa(\delta) > 1$ and $g_\kappa(\delta)$ be the numbers from the assertion of Proposition A.2.6 for D^κ. For κ such that V_κ is bounded let $l_\kappa := 0$, $K_\kappa := \sup\{|\lambda| : \lambda \in V_\kappa\}$ and $g_\kappa(\delta) := 1$. We set

$$l := \sum_{\kappa=1}^{k} l_\kappa, \quad K(\delta) = \max_{\kappa=1}^{k} K_\kappa(\delta), \quad g(\delta) := \min_{\kappa=1}^{k} g_\kappa(\delta).$$

Let $\delta > 0$ and $R \ge K(\delta)$. Then there are l balls of radius δ such that for all $\kappa = \{1, \dots, k\}$ and all $\lambda \in V_\kappa$ with $R \le |\lambda| \le R+1$ outside of these balls the estimate

$$|D^\kappa(\lambda)| \ge g(\delta) \tag{A.2.35}$$

holds. Let $\lambda \in \Omega$ with $R \le |\lambda| \le R+1$ outside of these balls. Then $|\lambda| > 1$ and one and only one of the following inequalities holds:

$$\frac{\Re(\lambda)}{\log|\lambda|} < -t_1,$$

$$-t_{\kappa-1} \le \frac{\Re(\lambda)}{\log|\lambda|} < -t_\kappa \qquad (\kappa = 2,\dots,k-1),$$

$$-t_{k-1} \le \frac{\Re(\lambda)}{\log|\lambda|}.$$

Hence there is a unique $\kappa \in \{1,\dots,k\}$ with $\lambda \in V_\kappa$. The definition (A.2.31) yields

$$\begin{aligned} D^\kappa(\lambda) &= \sum_{j\in\mathcal{N}_\kappa} b_j(\lambda)\lambda^{-\nu_{\kappa-1}}\exp\{(c_j - c_{\kappa-1})\lambda\} \\ &\quad + \sum_{j\in\mathcal{N}_\kappa^-} b_j(\lambda)\lambda^{-\nu_{\kappa-1}}\exp\{(c_j - c_{\kappa-1})\lambda\} \\ &\quad + \sum_{j\in\mathcal{N}_\kappa^+} b_j(\lambda)\lambda^{-\nu_\kappa}\exp\{(c_j - c_\kappa)\lambda\}\lambda^{\nu_\kappa-\nu_{\kappa-1}}\exp\{(c_\kappa - c_{\kappa-1})\lambda\} \\ &= \lambda^{-\nu_{\kappa-1}}\exp\{-c_{\kappa-1}\lambda\}D(\lambda). \end{aligned}$$

If $\kappa \ge 2$ and $j = 0$, the estimate (A.2.34) yields $0 \le \nu_{\kappa-1} - d_{\kappa-1}c_{\kappa-1}$. Because of $V_\kappa \subset V_{\kappa-1}^+$ we conclude

$$\begin{aligned} c_{\kappa-1}\Re(\lambda) &+ \nu_{\kappa-1}\log|\lambda| \\ &= c_{\kappa-1}(\Re(\lambda) + d_{\kappa-1}\log|\lambda|) + (\nu_{\kappa-1} - d_{\kappa-1}c_{\kappa-1})\log|\lambda| \\ &\ge c_{\kappa-1}(\Re(\lambda) + t_{\kappa-1}\log|\lambda|) \ge 0. \end{aligned}$$

Hence $|\lambda^{\nu_{\kappa-1}}\exp\{c_{\kappa-1}\lambda\}| \ge 1$ holds since it is trivial in case $\kappa = 1$ because of $\nu_0 = c_0 = 0$. Finally the estimate (A.2.35) yields

$$|D(\lambda)| = |\lambda^{\nu_{\kappa-1}}\exp\{c_{\kappa-1}\lambda\}||D^\kappa(\lambda)| \ge g(\delta). \quad \square$$

We now consider a general exponential sum (A.2.6) and set

$$\mathcal{E} := \{c_j : j \in \mathcal{N}\}.$$

Let $\mathcal{P}$ be the convex hull of $\mathcal{E}$. If $\mathcal{P}$ is a convex polygon, then there are a number S and line segments P_s $(s = 1,\dots,S)$ such that $\partial\mathcal{P} = \bigcup_{s=1}^{S} P_s$, where the endpoints of P_s are the vertices of $\mathcal{P}$. Let

$$\mathcal{N}_s := \{j \in \mathcal{N} : c_j \in P_s\} \qquad (s = 1,\dots,S).$$

DEFINITION A.2.12. The exponential sum (A.2.6) with the representation (A.2.2) of the b_j is called *weakly regular* if the following three conditions hold:
i) $\mathcal{P}$ is a convex polygon;
ii) for each $s \in \{1,\dots,S\}$ there is a finite subset $\mathcal{M}_s$ of $\mathcal{N}_s$ such that $(\mathcal{N}_s, \mathcal{M}_s)$ is

weakly regular in the sense of Definition A.2.7;
iii) For all $s \in \{1,\ldots,S\}$ the set

$$\mathcal{E}_s := \{c_j : j \in \mathcal{N} \setminus \mathcal{N}_s : \nu_j > \min\{\nu_\kappa : \kappa \in \mathcal{N}, c_\kappa \in P_s \cap \mathcal{E}\}\}$$

has no accumulation point in P_s.

We denote the set of the vertices of $\mathcal{P}$ by $\widetilde{\mathcal{E}}$ and define

$$\mathcal{M} := \bigcup_{s=1}^{S} \mathcal{M}_s$$

and

$$\widehat{\mathcal{E}} := \{c_j : j \in \mathcal{M}\}. \tag{A.2.36}$$

Since the endpoints of P_s belong to $\mathcal{M}_s$, we have $\widetilde{\mathcal{E}} \subset \widehat{\mathcal{E}}$.

REMARK A.2.13. Let $D(\lambda)$ be an exponential sum with the representation (A.2.2) of the b_j. Assume that the ν_j do not depend on j, i. e., there is a number $\nu \in \mathbb{R}$ such that $\nu_j = \nu$ for all $j \in \mathcal{N}$.
Then the exponential sum is weakly regular if $\mathcal{P}$ is a convex polygon. In this case $\widetilde{\mathcal{E}} = \widehat{\mathcal{E}}$.

Proof. We have to show that the conditions ii) and iii) of Definition A.2.12 are fulfilled. For each $s \in \{1,\ldots,S\}$ the line segment P_s is the convex hull of $\mathcal{N}_s$. The condition ii) of Definition A.2.12 is fulfilled in view of Remark A.2.10, where $\mathcal{M}_s$ is the set of the endpoints of the line segment P_s. The condition iii) of Definition A.2.12 holds trivially since the sets $\mathcal{E}_s$ are empty. Finally, since the endpoints of the line segments P_s are vertices of $\mathcal{P}$, we obtain $\widehat{\mathcal{E}} \subset \widetilde{\mathcal{E}}$. □

THEOREM A.2.14. *On an unbounded subset* Ω *of* $\mathbb{C}$ *we consider the exponential sum* (A.2.6)

$$D(\lambda) = \sum_{j \in \mathcal{N}} b_j(\lambda) \exp(c_j \lambda) \qquad (\lambda \in \Omega)$$

fulfilling (A.2.1)–(A.2.5), *where* $b_j(\lambda) = \lambda^{\nu_j}[a_j]$, $\nu_j \in \mathbb{R}$, $a_j \in \mathbb{C}$, $c_j \in \mathbb{C}$. *Assume that the exponential sum is weakly regular in the sense of Definition* A.2.12 *and that* $a_j \neq 0$ *for all* $j \in \mathcal{M}$, *i. e.,* $a_j \neq 0$ *if* $c_j \in \widehat{\mathcal{E}}$. *Then the following assertions hold:*
i) *For all* $\lambda \in \mathbb{C} \setminus \{0\}$ *there is a number* $c(\lambda) \in \widetilde{\mathcal{E}}$ *such that for all* $c \in \mathcal{P}$ *the estimate* $\Re\big((c - c(\lambda))\lambda\big) \leq 0$ *holds.*
ii) *There are a positive integer* l *and for each* $\delta > 0$ *numbers* $K(\delta) > 0$ *and* $g(\delta) > 0$ *satisfying the following property: for each* $R > K(\delta)$ *there are* l *balls of radius* δ *such that for all* $\lambda \in \Omega$ *with* $R \leq |\lambda| \leq R+1$ *outside of these balls the estimate*

$$|\lambda^{-\nu(\lambda)} D(\lambda) \exp\{-c(\lambda)\lambda\}| \geq g(\delta)$$

holds, where $\nu(\lambda) := \nu_j$ *if* $c(\lambda) = c_j$.

Proof. For simplicity of notation we may assume that $\widetilde{\mathscr{E}} = \{c_1, \dots, c_S\}$, where $P_s = \overline{c_{s-1}, c_s}$ $(s = 1, \dots, S;\, c_0 := c_S)$. Note that in case $\mathscr{P}$ is a line segment, i. e., $S = 2$, the sets $P_1 = \overline{c_2, c_1}$ and $P_2 = \overline{c_1, c_2}$ coincide. For $\lambda \in \mathbb{C} \setminus \{0\}$ we set

$$d(\lambda) := \max\{\Re(c\lambda) : c \in \mathscr{P}\}, \tag{A.2.37}$$

which exists since $\mathscr{P}$ is compact. Since $\mathscr{P}$ is the convex hull of the set of its vertices, there is a number $s(\lambda) \in \{1, \dots, S\}$ such that $\Re(c_{s(\lambda)}\lambda) = d(\lambda)$. Hence i) follows from (A.2.37) with $c(\lambda) = c_{s(\lambda)}$.

ii) Let $(s, \alpha) \in \bigcup_{s=1}^{S} \{s\} \times (P_s \cap \widetilde{\mathscr{E}}) =: Q$. Then there are $\beta(s, \alpha), \gamma(s, \alpha) \in \widetilde{\mathscr{E}}$ and $t(s, \alpha) \in \{1, \dots, S\} \setminus \{s\}$ such that $P_s = \overline{\alpha, \beta(s, \alpha)}$ and $P_{t(s,\alpha)} = \overline{\alpha, \gamma(s, \alpha)}$. Clearly, $t(s, \alpha) = s + 1$ or $t(s, \alpha) = s - 1$, where $P_0 := P_S$ and $P_{S+1} := P_1$. For $(s, \alpha) \in Q$ set

$$V_{s,\alpha} := \{\lambda \in \mathbb{C} \setminus \{0\} : \Re(\alpha\lambda) = d(\lambda),\, \Re\big((\beta(s, \alpha) - \alpha)\lambda\big) \geq \Re\big((\gamma(s, \alpha) - \alpha)\lambda\big)\}.$$

For $(s, \alpha) \in Q$ and $\lambda \in V_{s,\alpha}$ we assert

$$\Re((c - \alpha)\lambda) \leq 0 \qquad (c \in P_s), \tag{A.2.38}$$

$$\Re((c - \alpha)\lambda) < 0 \qquad (c \in \mathscr{P} \setminus P_s). \tag{A.2.39}$$

Let $(s, \alpha) \in Q$ and $\lambda \in V_{s,\alpha}$. For each $c \in \mathscr{P}$ we have $\Re((c - \alpha)\lambda) \leq 0$ by definition of $V_{s,\alpha}$ and $d(\lambda)$. This proves (A.2.38). Now let $c \in \mathscr{P} \setminus P_s$. In this case, $S > 2$. Assume that (A.2.39) does not hold for this c. Then $\Re((c - \alpha)\lambda) = 0$. By (A.2.37), $\mathscr{P}$ lies on one side of the line through the points c and α. Hence $\overline{c, \alpha} \subset \partial\mathscr{P}$, which proves $c \in P_s \cup P_{t(s,\alpha)}$. We infer $c \in P_{t(s,\alpha)} \setminus \{\alpha\}$ because of $c \in \mathscr{P} \setminus P_s$. Then $\gamma(s, \alpha)$ lies on the line through c and α, and $\Re((c - \alpha)\lambda) = 0$ implies $\Re\big((\gamma(s, \alpha) - \alpha)\lambda\big) = 0$. Hence, by definition of $V_{(s,\alpha)}$,

$$0 \geq \Re\big((\beta(s, \alpha) - \alpha)\lambda\big) \geq \Re\big((\gamma(s, \alpha) - \alpha)\lambda\big) = 0.$$

This implies that α lies on the line through $\beta(s, \alpha)$ and $\gamma(s, \alpha)$, which contradicts the fact that α is a vertex of $\mathscr{P}$. Here we have to note that $\beta(s, \alpha) \neq \gamma(s, \alpha)$ since $S > 2$. Hence the assumption is false and (A.2.39) is proved.

For $(s, \alpha) \in Q$ we set

$$\Omega_{s,\alpha} := \Omega \cap V_{s,\alpha}.$$

Let $(s, \alpha) \in Q$. Since α is an endpoint of P_s, there is a $\varphi_{s,\alpha} \in [0, 2\pi)$ such that $e^{i\varphi_{s,\alpha}}(c_j - \alpha) > 0$ for $j \in \mathscr{N}_s \setminus \{j(s, \alpha)\}$, where $\alpha = c_{j(s,\alpha)}$. For $j \in \mathscr{N}_s \setminus \{j(s, \alpha)\}$ and $z \in W_{s,\alpha} := e^{-i\varphi_{s,\alpha}} \cdot \Omega_{s,\alpha}$ we set $\varepsilon_j^{s,\alpha}(z) := \varepsilon_j(e^{i\varphi_{s,\alpha}} z)$. For $z \in W_{s,\alpha}$ we set

$$\varepsilon_{j(s,\alpha)}^{s,\alpha}(z) := \varepsilon_{j(s,\alpha)}(e^{i\varphi_{s,\alpha}} z) + \sum_{j \in \mathscr{N} \setminus \mathscr{N}_s} b_j(e^{i\varphi_{s,\alpha}} z)(e^{i\varphi_{s,\alpha}} z)^{-\nu_{j(s,\alpha)}} \exp\{(c_j - \alpha)e^{i\varphi_{s,\alpha}} z\}$$

and

$$D_{s,\alpha}(z) := \sum_{j \in \mathscr{N}_s} (a_j + \varepsilon_j^{s,\alpha}(z))(e^{i\varphi_{s,\alpha}} z)^{\nu_j - \nu_{j(s,\alpha)}} \exp\{(c_j - \alpha)e^{i\varphi_{s,\alpha}} z\}.$$

We assert that

$$\varepsilon_{j(s,\alpha)}^{s,\alpha}(z) \to 0 \quad \text{as } z \to \infty \text{ in } W_{s,\alpha}. \tag{A.2.40}$$

Since $(a_j + \varepsilon_j(\lambda))_{j \in \mathcal{N} \setminus \mathcal{N}_s} \to (a_j)_{j \in \mathcal{N} \setminus \mathcal{N}_s}$ in $l_1(\mathcal{N} \setminus \mathcal{N}_s)$ as $|\lambda| \to \infty$ by (A.2.3) and (A.2.4), this will follow from Proposition A.2.2 if we show that

$$\sup\{|\lambda^{\nu_j - \nu_{j(s,\alpha)}} \exp\{(c_j - \alpha)\lambda\}| : \lambda \in \Omega_{s,\alpha}, |\lambda| \geq 1, j \in \mathcal{N} \setminus \mathcal{N}_s\} < \infty \tag{A.2.41}$$

and, for all $j \in \mathcal{N} \setminus \mathcal{N}_s$,

$$\lambda^{\nu_j - \nu_{j(s,\alpha)}} \exp\{(c_j - \alpha)\lambda\} \to 0 \text{ as } \lambda \in \Omega_{s,\alpha} \text{ and } |\lambda| \to \infty. \tag{A.2.42}$$

Since the function d given by (A.2.37) is continuous, the set $\{\lambda \in V_{s,\alpha} : |\lambda| = 1\}$ is compact. From condition iii) of Definition A.2.12 we infer that the closure $\overline{\mathcal{E}_s}$ of $\mathcal{E}_s$ is a compact subset of $\mathcal{P} \setminus P_s$. Hence (A.2.39) implies

$$\sup\{\Re((c-\alpha)\lambda) : \lambda \in V_{s,\alpha}, |\lambda| = 1\} =: \eta_{s,\alpha,c} < 0 \quad (c \in \mathcal{P} \setminus P_s), \tag{A.2.43}$$

$$\sup\{\Re((c-\alpha)\lambda) : \lambda \in V_{s,\alpha}, |\lambda| = 1, c \in \overline{\mathcal{E}_s}\} =: \eta_{s,\alpha} < 0. \tag{A.2.44}$$

Set $\nu := \sup\{\nu_j : j \in \mathcal{N}\}$. From (A.2.39) and (A.2.44) we infer that

$$\begin{aligned} &\sup\{|\lambda^{\nu_j - \nu_{j(s,\alpha)}} \exp\{(c_j - \alpha)\lambda\}| : \lambda \in \Omega_{s,\alpha}, |\lambda| = r, r \geq 1, j \in \mathcal{N} \setminus \mathcal{N}_s\} \\ &\quad \leq \sup\{1, r^{\nu - \nu_{j(s,\alpha)}} \exp(\eta_{s,\alpha} r) : r \geq 1\} < \infty. \end{aligned}$$

This proves (A.2.41). Finally, (A.2.43) implies

$$|\lambda^{\nu_j - \nu_{j(s,\alpha)}} \exp\{(c_j - \alpha)\lambda\}| \leq |\lambda|^{\nu_j - \nu_{j(s,\alpha)}} \exp\{\eta_{s,\alpha,c_j} |\lambda|\}$$

for all $j \in \mathcal{N} \setminus \mathcal{N}_s$ and $\lambda \in V_{s,\alpha}$, which proves (A.2.42).

Hence all mappings $D_{s,\alpha}$ $((s,\alpha) \in Q)$ fulfil the assumptions of Proposition A.2.11 if $\Omega_{s,\alpha}$ is unbounded. For these (s,α) choose the numbers $l_{s,\alpha}$, $K_{s,\alpha}(\delta)$ and $g_{s,\alpha}(\delta)$ according to Proposition A.2.11. If $\Omega_{s,\alpha}$ is bounded, we set $l_{s,\alpha} := 0$, $K_{s,\alpha}(\delta) := \sup\{|\lambda| : \lambda \in \Omega_{s,\alpha}\}$ and $g_{s,\alpha}(\delta) := 1$. We set

$$l := \sum_{(s,\alpha) \in Q} l_{s,\alpha},$$

$$K(\delta) := \max\{K_{s,\alpha}(\delta) : (s,\alpha) \in Q\},$$

$$g(\delta) := \min\{g_{s,\alpha}(\delta) : (s,\alpha) \in Q\}.$$

Now let $\delta > 0$ and $R > K(\delta)$. Then for each $(s,\alpha) \in Q$ there are numbers $z_{s,\alpha}^1, \ldots, z_{s,\alpha}^{l_{s,\alpha}} \in \mathbb{C}$ so that for all $z \in W_{s,\alpha}$ with $R \leq |z| \leq R+1$ and $\min_{j=1}^{l_{s,\alpha}} |z - z_{s,\alpha}^j| \geq \delta$ the estimate

$$|D_{s,\alpha}(z)| \geq g_{s,\alpha}(\delta) \tag{A.2.45}$$

holds according to Proposition A.2.11.

Now let $\lambda \in \Omega$ with $R \le |\lambda| \le R+1$ and $|\lambda - e^{i\varphi_{s,\alpha}} z^j_{s,\alpha}| \ge \delta$ for all $(s,\alpha) \in Q$ and $1 \le j \le l_{s,\alpha}$. Let $\alpha = c(\lambda)$. Then there are $s_1, s_2 \in \{1,\dots,S\}$ and $\alpha_1, \alpha_2 \in \widetilde{E}$ such that $P_{s_1} = \overline{\alpha,\alpha_1}$ and $P_{s_2} = \overline{\alpha,\alpha_2}$. We may assume that $\Re((\alpha_1-\alpha)\lambda) \ge \Re((\alpha_2-\alpha)\lambda)$ and set $s := s_1$. Then $\beta(s,\alpha) = \alpha_1$, $\gamma(s,\alpha) = \alpha_2$, $\lambda \in V_{s,\alpha}$, and $\alpha = c_s$ or $\alpha = c_{s-1}$. Hence $e^{-i\varphi_{s,\alpha}}\lambda \in W_{s,\alpha}$,

$$|e^{-i\varphi_{s,\alpha}}\lambda - z^j_{s,\alpha}| = |\lambda - e^{i\varphi_{s,\alpha}} z^j_{s,\alpha}| \ge \delta$$

for $j = 1,\dots,l_{s,\alpha}$, and

$$\begin{aligned} D_{s,\alpha}(e^{-i\varphi_{s,\alpha}}\lambda) &= \sum_{j\in\mathcal{N}_s} (a_j + \varepsilon_j(\lambda))\lambda^{\nu_j-\nu(\lambda)} \exp\{(c_j - c(\lambda))\lambda\} \\ &\quad + \sum_{j\in\mathcal{N}\setminus\mathcal{N}_s} b_j(\lambda)\lambda^{-\nu(\lambda)} \exp\{(c_j - c(\lambda))\lambda\} \\ &= \lambda^{-\nu(\lambda)} \exp\{-c(\lambda)\lambda\} D(\lambda). \end{aligned}$$

From (A.2.45) we infer

$$|\lambda^{-\nu(\lambda)} \exp\{-c(\lambda)\lambda\} D(\lambda)| = |D_{s,\alpha}(e^{-i\varphi_{s,\alpha}}\lambda)| \ge g(\delta). \quad \square$$

THEOREM A.2.15. *On an unbounded subset Ω of $\mathbb{C}$ we consider the exponential sum* (A.2.6)

$$D(\lambda) = \sum_{j\in\mathcal{N}} b_j(\lambda) \exp(c_j\lambda) \qquad (\lambda \in \Omega)$$

fulfilling (A.2.1)–(A.2.5), *where $b_j(\lambda) = \lambda^{\nu_j}[a_j]$, $\nu_j \in \mathbb{R}$, $a_j \in \mathbb{C}$, $c_j \in \mathbb{C}$. Assume that the exponential sum is weakly regular in the sense of Definition* A.2.12 *and that $a_j \ne 0$ for all $j \in \mathcal{M}$, i.e., $a_j \ne 0$ if $c_j \in \widehat{\mathcal{E}}$. Then there is an increasing sequence $(\rho_\nu)_{\nu=1}^\infty$ of positive real numbers with $\rho_\nu \to \infty$ as $\nu \to \infty$ and a number $\varepsilon > 0$ such that for all $\nu \in \mathbb{N}$ and all $\lambda \in \Omega$ with $|\lambda| = \rho_\nu$ there is a number $c(\lambda) \in \widetilde{\mathcal{E}}$ such that*

$$\Re\big((c - c(\lambda))\lambda\big) \le 0 \quad \textit{for all} \quad c \in \mathcal{P}$$

and

$$|\lambda^{-\nu(\lambda)} D(\lambda) \exp\{-c(\lambda)\lambda\}| \ge \varepsilon,$$

where $\nu(\lambda) := \nu_j$ if $c(\lambda) = c_j$.

Proof. The assumptions of this theorem are the same as in Theorem A.2.14. Hence the assertion of Theorem A.2.14 holds. Take l from Theorem A.2.14, choose $\delta > 0$ such that $2l\delta < 1$ and take $K(\delta)$ and $g(\delta) > 0$ from Theorem A.2.14. Let $R_\nu = K(\delta) + \nu$. Since $2l\delta < 1$, there is a number $\rho_\nu \in (R_\nu, R_\nu + 1)$ such that $\{\lambda \in \mathbb{C} : |\lambda| = \rho_\nu\}$ does not intersect any of the l balls of radius δ from the assertion of Theorem A.2.14. Hence the desired estimate follows from Theorem A.2.14 with $\varepsilon = g(\delta)$. $\square$

REMARK A.2.16. The two estimates in Theorem A.2.15 yield

$$|\lambda^{-\nu}D(\lambda)\exp\{-c\lambda\}| \geq \varepsilon$$

for all $c \in \mathscr{P}$, where $\nu := \min\{\nu_j : j \in \mathscr{N}, c_j \in \widetilde{\mathscr{E}}\}$.

A.3. Improved estimates for exponential sums

In this section we consider an exponential sum as in Section A.2 such that there is a number ν such that $\nu_j = \nu$ for all $j \in \mathscr{N}$, and such that

$$\varepsilon_j(\lambda) = O(\tau_p(\lambda))$$

for all $j \in \mathscr{N}$, where $1 < p \leq \infty$ and τ_p is a function of the form

$$\tau_p(\lambda) = \max_{q=1}^{t}(1 + |\Re(\lambda e^{i\chi_q})|))^{-1+1/p}, \tag{A.3.1}$$

for some numbers $0 \leq \chi_1 < \chi_2 < \cdots < \chi_t < 2\pi$. Here the numbers t and χ_q might be different in different formulas. We also assume that $\mathscr{P}$ is a convex polygon, that the vertices of $\mathscr{P}$, i. e., the elements of $\widetilde{\mathscr{E}}$, are no accumulation points of $\mathscr{E}$, and that the points in $\partial\mathscr{P}$ are no accumulation points of $\mathscr{E} \setminus \partial\mathscr{P}$.

Then, in Proposition A.2.3, $d = 0$, and we obtain the estimates

$$D(\lambda)^{-1} = \begin{cases} (a_\alpha^{-1} + O(\tau_p(\lambda)))\exp(-c_\alpha\lambda) & \text{if } \Re(\lambda) \leq -M, \\ (a_\beta^{-1} + O(\tau_p(\lambda)))\exp(-c_\beta\lambda) & \text{if } \Re(\lambda) \geq M, \end{cases} \tag{A.3.2}$$

for some $M > 0$.

Indeed, as in the proof of Proposition A.2.3, we have

$$\begin{aligned} &|a_\alpha^{-1}\exp(-c_\alpha\lambda)D(\lambda) - 1| \\ &\quad\leq \frac{|\varepsilon_\alpha(\lambda)|}{|a_\alpha|} + |a_\alpha|^{-1}\sum_{j\in\mathscr{N}\setminus\{\alpha\}}(|a_j| + |\varepsilon_j(\lambda)|)\exp\{(c_j - c_\alpha)\Re(\lambda)\} \\ &\quad= O(\tau_p(\lambda)) \quad \text{if } \Re(\lambda) \leq 0 \end{aligned}$$

since $\sup_{j\in\mathscr{N}\setminus\{\alpha\}}(c_j - c_\alpha) > 0$. Also, from the proof of Proposition A.2.3 we know that this term is less or equal to $\frac{1}{2}$ if $\Re(\lambda) \leq -M_1$. Then the stated estimate follows for λ with negative real part. The proof for λ with positive real part is analogous.

We can also substitute the estimate of $D(\lambda)$ in Proposition A.2.6 by the estimate

$$D(\lambda)^{-1} = \begin{cases} (a_\alpha^{-1} + O(\tau_p(\lambda)))\exp(-c_\alpha\lambda) & \text{if } \Re(\lambda) < 0, \\ (a_\beta^{-1} + O(\tau_p(\lambda)))\exp(-c_\beta\lambda) & \text{if } \Re(\lambda) > 0, \end{cases} \tag{A.3.3}$$

for all λ outside the balls as considered in Proposition A.2.6. In view of (A.3.2) it suffices to consider the values of λ for which $|\Re(\lambda)| \leq M$. Since $D(\lambda)\exp(-c_\alpha\lambda)$

and $D(-\lambda)\exp(c_\beta\lambda)$ also satisfy the assumptions of Proposition A.2.6, we immediately infer

$$D(\lambda)^{-1} = O(1)\exp(-c_\alpha\lambda) \quad \text{and} \quad D(\lambda)^{-1} = O(1)\exp(-c_\beta\lambda)$$

outside the balls considered in Proposition A.2.6. And for λ so that $|\Re(\lambda)| \le M$, the estimate $O(1)$ can be written as $O(\tau_p(\lambda))$.

In this case, Proposition A.2.11 is a special case of Proposition A.2.6.

Now we are going to consider Theorem A.2.14. The set $V_{s,\alpha}$ defined in the proof of that theorem is a sector. Since

$$|\exp\{(c_j-\alpha)e^{i\varphi_{s,\alpha}}z\}| \le \exp\{-\tilde{c}_{s,\alpha}\Re(z)\}$$

for $z \in W_{s,\alpha}$ and $j \in \mathscr{N} \setminus \mathscr{N}_s$, where $\tilde{c}_{s,\alpha} > 0$, we have $\varepsilon^{s,\alpha}_{j(s,\alpha)} = o(1)$ and $O(\tau_p(\lambda))$ in $W_{s,\alpha}$. Then $D_{s,\alpha}$ defined in the proof of Theorem A.2.14 satisfies the assumptions of Proposition A.2.11. Hence we infer

$$D(\lambda)^{-1}\lambda^{\nu}\exp\{\alpha\lambda\} = a_{s,\alpha}^{-1} + O(\tau_p(\lambda))$$

on each sector $\Omega_{s,\alpha}$. Thus we can formulate Theorem A.2.15 in the following way:

THEOREM A.3.1. *On an unbounded subset Ω of $\mathbb{C}$ we consider the exponential sum* (A.2.6)

$$D(\lambda) = \sum_{j\in\mathscr{N}} b_j(\lambda)\exp(c_j\lambda) \qquad (\lambda \in \Omega)$$

fulfilling (A.2.1)–(A.2.5), *where $b_j(\lambda) = \lambda^{\nu_0}[a_j]$, $\nu_0 \in \mathbb{R}$, $a_j \in \mathbb{C}$, $c_j \in \mathbb{C}$, and $\varepsilon_j(\lambda) = O(\tau_p(\lambda))$ for $j \in \mathscr{N}$, where $1 < p \le \infty$. Assume that $\mathscr{P}$ is a convex polygon, that $\widetilde{\mathscr{E}}$ is isolated in $\mathscr{E}$, that the points of $\partial\mathscr{P}$ are no accumulation points of $\mathscr{E} \setminus \partial\mathscr{P}$, and that $a_j \ne 0$ if $c_j \in \widehat{\mathscr{E}}$. Then there are $0 \le \chi_1 < \chi_2 < \cdots < \chi_{t+1} = \chi_1 + 2\pi$, $\gamma_1,\ldots,\gamma_t \in \mathbb{C}\setminus\{0\}$, $c_1,\ldots,c_t \in \widetilde{\mathscr{E}}$, and an increasing sequence $(\rho_\nu)_{\nu=1}^\infty$ of positive real numbers with $\rho_\nu \to \infty$ as $\nu \to \infty$ such that*

$$\Re\big((c-c_j)\lambda\big) \le 0 \quad \textit{for all} \quad c \in \mathscr{P}$$

and

$$D(\lambda)^{-1} = \lambda^{-\nu_0}\exp\{-c_j\lambda\}\big(\gamma_j + O(\tau_p(\lambda))\big)$$

hold for all $\nu \in \mathbb{N}$ and all $\lambda \in \Omega$ such that $|\lambda| = \rho_\nu$ and $\chi_j \le \arg\lambda \le \chi_{j+1}$ for $j = 1,\ldots,t$.

In the above theorem we have that $\{c_1,\ldots,c_t\} = \widetilde{\mathscr{E}}$.

Bibliography

[AD] ADAMS, R. A.: Sobolev spaces. Academic Press, New York – San Francisco – London 1975.

[AL] ALIEV, N. A.: Expansion of matrix functions in terms of solutions of spectral problems (Russian). Differentsial'nye Uravneniya 2 (1966), 847–852.

[ALR] ALIEV, N. A., RASULOV, M. L.: A formula for the expansion of an arbitrary matrix-function in terms of the solution of a spectral problem (Russian). Diffentsial'nye Uravneniya 3 (1967), 941–947.

[ALM] ALIMOV, S. A.: On expandability of continuous functions from Sobolev classes into eigenfunctions of the Laplace operator (Russian). Sibirsk. Mat. Zh. 19 (1978), 721–734, Engl. transl.: Siberian Math. J. 19 (1979), 507–517.

[AN] ANNABERDYEV, CH. O.: On multiple basicity of eigenfunctions of a certain boundary value problem (Russian). Uspekhi Mat. Nauk 37 (1982), no. 5(227), 163–164, Engl. transl.: Russian Math. Surveys 37 (1982), no. 5, 155–156.

[ASH] ASHUROV, R. R.: On the asymptotic estimate in L_2 of spectral functions of nonselfadjoint operators (Russian). Dokl. Akad. Nauk UzSSR (1990), no. 10, 10–12.

[AT1] ATKINSON, F. V.: A spectral problem for completely continuous operators. Acta Math. Sci. Hungar. 3 (1952), 53–60.

[AT2] ATKINSON, F. V.: Discrete and continuous boundary problems. Mathematics in Science and Engineering, Vol. 8, Academic Press, New York – London 1964.

[ALMS] ATKINSON, F. V., LANGER, H., MENNICKEN, R., SHKALIKOV, A. A.: The essential spectrum of some matrix operators. Math. Nachr. 167(1994), 5-22.

[AV] AVAKYAN, V. A.: The Sturm-Liouville problem with spectral parameter in the boundary condition (Russian). Izv. Akad. Nauk Armyan. SSR Ser. Mat. 14 (1979), 91–106.

[BAP] BABAKHANOVA, S. SH., PASHAEVA, EH. EH.: On a certain multiple eigen-function expansion of a differential operator with integral boundary conditions (Russian). Dokl. Akad. Nauk AzSSR 38 (1982), no. 5, 26–29.

[BCH] BANERJEE, S., CHAKRAVARTY, N. K.: An eigenvalue problem associated with an $n \times n$ matrix differential operator. Prog. Math. Allahabad 16 (1982), 67–75.

[BA1] BART, H.: Meromorphic operator valued functions. Thesis, Vrije Universiteit, Amsterdam, Mathematical Centrwe Tracts, 44, 1973.

[BA2] BART, H.: Holomorphic relative inverses of operator valued functions. Math. Ann. 208 (1974), 179–194.

[BM] BAUER, G., MENNICKEN, R.: Störungstheorie für diskrete Spektraloperatoren. Regensburger Math. Schriften 10, 1983.

[BG] BAUMGÄRTEL, H.: Endlichdimensionale analytische Störungstheorie. Berlin 1972.

[BEL] BELOW, J. VON: Sturm-Liouville eigenvalue problems on networks. Math. Methods Appl. Sci. 10 (1988), 383–395.

[BGP] BENEDEK, A., GÜICHAL, E., PANZONE, R.: On certain non-harmonic Fourier expansions of non-regular Sturm-Liouville systems. Notas de Algebra y Analisis, N° 4, Instituto de Matemática, Universidad Nacional del Sur, Bahia Blanca 1974.

[BP1] BENEDEK, A., PANZONE, R.: Remarks on a problem in the flow of heat for a solid in contact with a fluid. Rev. Un. Mat. Argentina 29 (1980), 120–130.

[BP2] BENEDEK, A., PANZONE, R.: On Sturm-Liouville problems with the square-root of the eigenvalue parameter contained in the boundary conditions. Notas de Algebra y Analiysis, N° 10, Instituto de Matemática, Universidad Nacional del Sur, Bahia Blanca 1981.

[BP3] BENEDEK, A., PANZONE, R.: On the structure of the set of eigenfunctions of certain irregular boundary problems. In: E. B. Christoffel, eds. P. L. Butzer, F. Féher, Birkhäuser, Basel 1981, 401–410.

[BP4] BENEDEK, A., PANZONE, R.: Problemas de contorno para ecuaciones diferenciales ordinarias de segundo orden con condiciones de borde dependientes del parametro espectral. Trabajos de Matemática, Buenos Aires 1982, 1–21.

[BE1] BENZINGER, H.: Eigenfunction expansions for n-th order non-regular boundary value problems. Ph.D. thesis, Syracuse, New York, 1967.

[BE2] BENZINGER, H.: Green's function for ordinary differential operators. J. Differential Equations 1970, 478–496.

[BE3] BENZINGER, H.: The L^p behavior of eigenfunction expansions. Trans. Amer. Math. Soc. 174 (1972), 333–344.

[BE4] BENZINGER, H.: Completeness of eigenvectors in Banach spaces. Proc. Amer. Math. Soc. 38 (1973), 319–324.

[BE5] BENZINGER, H.: A biorthogonal expansion related to the zeros of an entire function. Proc. Amer. Math. Soc. 49 (1975), 135–142.

[BE6] BENZINGER, H.: The L^2 behavior of eigenfunction expansions. Indiana Univ. Math. J. 25 (1976), 691–702.

[BE7] BENZINGER, H.: The spectral analysis of Stone regular differential operators. In: Ordinary and partial differential equations, eds.: W. N. Everitt and B. D. Sleeman, Lecture Notes in Math. 564 (1976), 34–40.

[BE8] BENZINGER, H.: Eigenvalues of regular differential operators. Proc. Roy. Soc. Edinburgh 79A (1977), 299–305.

[BE9] BENZINGER, H.: Pointwise and norm convergence of a class of biorthogonal expansions. Trans. Amer. Math. Soc. 231 (1977), 259–271.

[BE10] BENZINGER, H.: Perturbation of the heat equation. J. Differential Equations 32 (1979), 398–419.

[BEZ] BEZVERKHNIJ, V. A.: Expansion into eigenfunctions of a certain differential operators in the space of vector functions (Russian). Uspekhi Mat. Nauk 30, No. 1(181) (1975), 225–226.

[BBS] BINDING, P. A., BROWNE, P. J., SEDDIGHI, K.: Sturm-Liouville problems with eigenparameter dependent boundary conditions. Proc. Edinburgh Math. Soc. (2) 37 (1994), 57–72.

[BI1] BIRKHOFF, G. D.: On the asymptotic character of the solutions of certain linear differential equations containing a parameter. Trans. Amer. Math. Soc. 9 (1908), 219–231.

[BI2] BIRKHOFF, G. D.: Boundary value and expansion problems of ordinary linear differential equations. Trans. Amer. Math. Soc 9 (1908), 373–395.

[BIL] BIRKHOFF, G. D., LANGER, R. E.: The boundary problems and developments associated with a system of ordinary differential equations of the first order. Amer. Acad. Proc. 58 (1923), 49–128.

[BLI] BLISS, G. A.: A boundary value problem for a system of ordinary linear differential equations of the first order. Trans. Amer. Math. Soc. 28 (1926), 561–584.

[BL1] BLOSHANSKAYA, S. K.: On the basis property of the system of eigenfunctions and associated functions of a nonselfadjoint differential operator (Russian). Dokl. Akad. Nauk SSSR 252 (1980), 17–19, Engl. transl.: Soviet Math. Dokl. 21 (1980), 653–655.

[BL2] BLOSHANSKAYA, S. K.: On the basis property of systems of eigen- and associated functions of a differential operator of the second order. In: Functional methods in equations of mathematical physics, Moskva 1980, 12–15.

[BL3] BLOSHANSKAYA, S. K.: On a necessary condition for being a basis (Russian). Differentsial'nye Uravneniya 17 (1981), no. 5, 778–788, 954, Engl. transl.: Differential Equations 17 (1981), 508–516.

[BL4] BLOSHANSKAYA, S. K.: On the existence of associated functions of a second-order operator (Russian). Dokl. Akad. Nauk SSSR 262 (1982), 1036-1039, Engl. transl.: Soviet Math. Dokl. 25 (1982), 149–152.

[BOC1] BÔCHER, M.: Boundary problems in one dimension. Proceedings of the fifth international congress of mathematicians, Cambridge 1912, Cambridge University Press 1913, 163–195.

[BOC2] BÔCHER, M.: Applications and generalizations of the conception of adjoint systems. Trans. Amer. Math. Soc. 14 (1913), 403–420.

[BOU] BOUNITZKY, E.,: Sur la fonction de Green des équations différentielles linéaires ordinaires. J. Math. Pures Appl. VI, 5 (1909), 65–125.

[BO1] BOZHINOV, N. S.: On uniqueness and completeness theorems for expansion in eigen- and associated functions of a nonlocal Sturm-Liouville operator on a finite interval (Russian). Differentsialnye Uravneniya 26 (1990), No. 5, 741–753, Engl. transl.: Differential Equations 26 (1990), No. 5, 521–530.

[BO2] BOZHINOV, N. S.: Sufficient conditions for the basis property, convergence and summability of the root expansion of a nonlocal Sturm-Liouville problem (Russian). Dokl. Akad. Nauk SSSR 319 (1991), No. 5, 1042–1047, Engl. transl.: Soviet Math. Dokl. 44 (1992), No. 1, 257–262.

[BO3] BOZHINOV, N. S.: On the root expansion of a nonlocal Sturm-Liouville integro-differential operator with an integral part of Volterra type (Russian). Dokl. Akad. Nauk 335 (1994), No. 5, 549–552.

[BRA] BRAAKSMA, B. L. J.: Global reduction of linear differential systems involving a small parameter. SIAM J. Math. Anal. 2 (1971), 149–165.

[BR1] BROWN, R. C.: The existence of the adjoint in linear differential systems with discontinuous boundary conditions. Ann. Mat. Pura Appl. (IV) 94 (1972) 269–274.

[BR2] BROWN, R. C.: Adjoint domains and generalized splines. Czechoslovak Math. J. 25 (1975), 134–147.

[BR3] BROWN, R. C.: Duality theory for nth order differential operators under Stieltjes boundary conditions. SIAM J. Math. Anal. 6 (1975), 882–900.

[BR4] BROWN, R. C.: Duality theory for N-th order differential operators under Stieltjes boundary conditions. II: Nonsmooth coefficients and nonsingular measures. Ann. Mat. Pura Appl. (IV) 105 (1975), 141–170.

[BR5] BROWN, R. C.: Differential operators with abstract boundary conditions. Canad. J. Math. 30 (1978), 262–288.

[BR6] BROWN, R. C.: Generalized boundary value problems with abstract side conditions and their adjoints I. Czechoslovak Math. J. 30 (1980), 7–27.

[BRGK] BROWN, R. C., GREEN, G. B., KRALL, A. M.: Eigenfunction expansions under multipoint-integral boundary conditions. Ann. Mat. Pura Appl. (IV) 95 (1973), 231–243.

[BRK1] BROWN, R. C., KRALL, A. M.: Ordinary differential operators under Stieltjes boundary conditions. Trans. Amer. Math. Soc. 198 (1974), 73–92.

[BRK2] BROWN, R. C., KRALL, A. M.: N-th order ordinary differential systems under Stieltjes boundary conditions. Czechoslovak Math. J. 27 (1977), 119–131.

[BRT] BROWN, R. C., TVRDÝ, M.: Generalized boundary value problems with abstract side conditions and their adjoints II. Czechoslovak Math. J. 31 (1981), 501–509.

[BRTV] BROWN, R. C., TVRDÝ, M., VEJVODA, O.: Duality theory for linear n-th order integro-differential operators with domain in L^p_m determined by interface side conditions. Czechoslovak Math. J. 32 (1982), 183–196.

[BUD1] BUDAK, A. B.: Expansions in eigenfunctions of a fourth-order differential operator with a piecewise-constant leading coefficient (Russian). Differentsial'nye Uravneniya 16 (1980), 1545–1558, Engl. transl.: Differential Equations 16 (1981), 982–992.

[BUD2] BUDAK, A. B.: Expansions in eigenfunctions of a fourth order differential operator with discontinuous coefficients (Russian). Differentsial'nya Uravneniya 16 (1980), 2115–2126, Engl. transl.: Differential Equations 16 (1981), 1337–1344.

[BU] BURCKEL, R. B.: An introduction to classical complex analysis, I. Academic Press, New York 1979.

[CA] CARLSON, R.: Expansions associated with non-self-adjoint boundary-value problems. Proc. Amer. Math. Soc. 73(1979), 173–179.

[CH1] CHAKRAVARTY, N. K.: Some problems in eigenfunction expansions (I). Quart. J. Math. 16 (1965), 135–150.

[CH2] CHAKRAVARTY, N. K.: Some problems in eigenfunction expansions (II). Quart. J. Math. 19 (1968), 213–224.

[CDKP] CHEN, G., DELFOUR, M., KRALL, A. M., PAYRE,G.: Modelling, stabilization and control of serially connected beams. SIAM J. Control Optim. 25 (1987), 526–546.

[COD] CODDINGTON, E. A.: Spectral theory of ordinary differential operators. Lecture Notes in Math., 448 (1975), 1–24.

[COL] CODDINGTON, E. A., LEVINSON, N.: Theory of differential equations. McGraw-Hill, New York – Toronto – London 1955.

[CO1] COLE, R. H.: Reduction of an n-th order differential equation and m-point boundary conditions to an equivalent matrix system. Amer. J. Math. 68 (1946), 179–184.

[CO2] COLE, R. H.: The expansion problem with boundary conditions at a finite set of points. Canad. J. Math. 13 (1961), 462–479.

[CO3] COLE, R. H.: General boundary conditions for an ordinary linear differential system. Trans. Amer. Math. Soc. 111 (1964), 521–550.

[CO4] COLE, R. H.: The two-point boundary problem. Amer. Math. Monthly 72 (1965), 701–711.

[CO5] COLE, R. H.: Theory of ordinary differential equations. Meredith Corp., New York 1968.

[CL] COLLATZ, L.: Eigenwertaufgaben mit technischen Anwendungen. Leipzig 1949.

[CON] CONWAY, J. B.: A course in functional analysis. Springer-Verlag, New York 1985.

[DAS] DAS, J., SENGUPTA, P. K.: On the pointwise convergence of the eigenfunction expansion associated with some iterated boundary value problems. Proc. Amer. Math. Soc. 98 (1986), 593–600.

[DE1] DEGUTIS, YU. EH.: Expansion in eigenfunctions and associated functions (Russian). Litovsk. Mat. Sb. 19 (1979), no. 2, 81–86, Engl. transl.: Lithuanian Math. J. 19 (1980), 215–218.

[DE2] DEGUTIS, YU. EH.: Expansion in eigenfunctions of a differential operator with irregular boundary conditions (Russian). Litovsk. Mat. Sb. 21 (1981), no. 4, 75–81, Engl. transl.: Lithuanian Math. J. 21 (1981), no. 4, 313–316.

[DI] DIEUDONNÉ, J.: Foundations of modern analysis. I. Academic Press, New York – London 1960.

[DIJ] DIJKSMA, A.: Eigenfunction expansions for a class of J-selfadjoint ordinary differential operators witrh boundary conditions containing the eigenvalue parameter. Proc. Roy. Soc. Edinburgh 86A (1980), 1–27.

[DLS1] DIJKSMA, A., LANGER, H., DE SNOO, H. S. V.: Selfadjoint Π_κ-extensions of symmetric subspaces: an abstract approach to boundary problems with spectral parameter in the boundary conditions. Integral Equations Operator Theory 7 (1984), 459–515.

[DLS2] DIJKSMA, A., LANGER, H., DE SNOO, H. S. V.: Symmetric Sturm-Liouville operators with eigenvalue-depending boundary conditions. Can. Math. Soc. Conference Proc. 8 (1987), 87–116.

[DLS3] DIJKSMA, A., LANGER, H., DE SNOO, H. S. V.: Hamiltonian systems with eigenvalue depending boundary conditions. Operator Theory: Adv. Appl. 35 (1988), 37–83.

[DIN] DINCULEANU, N.: Vector measures. Pergamon Press, Oxford 1967.

[DH] DIPRIMA, R. C., HABETLER, G. J.: A completeness theorem for non-selfadjoint eigenvalue problems in hydrodynamic stability. Arch. Rational Mech. Anal. 34 (1969), 218–227.

[DO] DOLGANOVA, S. M.: The summation of principal vector expansions in eigenvectors of certain nonselfadjoint differential operators (Russian). Differentsial'nye Uravneniya 4 (1968), 518–525, Engl. transl.: Differential Equations 4 (1968), 272–275.

[DR] DRAZIN, P. G., REID, W. H.: Hydrodynamic stability. Cambridge University Press, Cambridge 1981.

[EB1] EBERHARD, W.: Das asymptotische Verhalten der Greenschen Funktion irregulärer Eigenwertprobleme mit zerfallenden Randbedingungen. Math. Z. 86 (1964), 45–53.

[EB2] EBERHARD, W.: Die Entwicklung nach Eigenfunktionen irregulärer Eigenwertprobleme mit zerfallenden Randbedingungen. Math. Z. 86 (1964), 205–214.

[EB3] EBERHARD, W.: Die Entwicklung nach Eigenfunktionen irregulärer Eigenwertprobleme mit zerfallenden Randbedingungen. II. Math. Z. 90 (1965), 126–137.

[EB4] EBERHARD, W.: Über nichtselbstadjungierte Eigenwertprobleme bei gewöhnlichen linearen Differentialgleichungen vom Typ $My = \lambda Ny$. Habilitationsschrift 1969.

[EB5] EBERHARD, W.: Über das asymptotische Verhalten von Lösungen der linearen Differentialgleichung $M[y] = \lambda N[y]$ für große Werte von $|\lambda|$. Math. Z. 119 (1971), 160–170.

[EB6] EBERHARD, W.: Über Eigenwerte und Eigenfunktionen einer Klasse von nichtselbstadjungierten Randwertproblemen. Math. Z. 119 (1971), 171–178.

[EF1] EBERHARD, W., FREILING, G.: Nicht S-hermitische Rand- und Eigenwertprobleme. Math. Z. 133 (1973), 187–202.

[EF2] EBERHARD, W., FREILING, G.: Das Verhalten der Greenschen Matrix und der Entwicklungen nach Eigenfunktionen N-regulärer Eigenwertprobleme. Math. Z. 136 (1974), 13–30.

[EF3] EBERHARD, W., FREILING, G.: Stone-reguläre Eigenwertprobleme. Math. Z. 160 (1978), 139–161.

[EF4] EBERHARD, W., FREILING, G.: An expansion theorem for eigenvalue problems with several turning points. Analysis 13 (1993), 301–308.

[EE] EDMUNDS, D. E., EVANS, W. D.: Spectral theory and differential operators. Clarendon Press, Oxford 1987.

[EH1] EHTKIN, A. E.: Some boundary value problems with a spectral parameter in the boundary conditions (Russian). Funktsional. Anal. i Prilozhen. 18 (1982), 138–146. Engl. transl.: Amer. Math. Soc. Transl. Ser. 2, 136 (1987), 35–41.

[EH2] EHTKIN, A. E.: On a first order boundary value problem with a spectral parameter in the boundary condition (Russian). Funktsional. Anal. i Prilozhen. 19 (1982), 177-190.

[EH3] EHTKIN, A. E.: On a second order boundary value problem with a spectral parameter in the boundary condition (Russian). Funktsional. Anal. i Prilozhen. 23 (1984), 142–153.

[EV] EVANS, W. D.: A non-self-adjoint differential operator in $L^2[a,b)$. Quart. J. Math. (2) 21 (1970), 371–383.

[FMMM] FAIERMAN, M., MARKUS, A., MATSAEV, V., MÖLLER, M.: On n-fold expansions for ordinary differential operators. Math. Nachr. 238 (2002), 62–77.

[FO] FOLLAND, G. B.: Spectral analysis of nonself-adjoint differential operators. J. Differential Equations 39 (1981), 151–185.

[FR1] FREILING, G.: Ein Äquikonvergenzsatz für eine Klasse von singulären Differentialoperatoren. Math. Z. 153 (1977), 255–265.

[FR2] FREILING, G.: Reguläre Eigenwertprobleme mit Mehrpunkt-Integral-Randbedingungen. Math. Z. 171 (1980), 113–131.

[FR3] FREILING, G.: Zur Vollständigkeit des Systems der Eigenfunktionen und Hauptfunktionen irregulärer Operatorbüschel. Math. Z. 188 (1984), 55-68.

[FR4] FREILING, G.: Über die mehrfache Vollständigkeit des Systems der Eigenfunktionen und assoziierten Funktionen irregulärer Operatorbüschel in $L_2[0,1]$. Z. Angew. Math. Mech. 65 (1985), 336–338.

[FR5] FREILING, G.: Boundary value problems associated with multiple roots of the characteristic equation. Resultate Math. 11 (1987), 44–62.

[FR6] FREILING, G.: On the completeness and minimality of the derived chains of eigen- and associated functions of boundary eigenvalue problems nonlinearly dependent on the parameter. Resultate Math. 14 (1988), 64–83.

[FR7] FREILING, G.: General boundary eigenvalue problems for systems of differential equations with multiple roots of the characteristic equation. Proc. Roy. Soc. Edinburgh 108A (1988), 45–56.

[FR8] FREILING, G.: Multiple expansions in series of eigen- and associated functions of boundary eigenvalue problems nonlinearly dependent on the spectral parameter. II. Resultate Math. 17 (1990), 241-255.

[FRK] FREILING, G., KAUFMANN, F.-J.: On uniform and L_p-convergence of eigenfunction expansions for indefinite eigenvalue problems. Integral Equations Operator Theory 13 (1990), 193-215.

[FRR1] FREILING, G., RYKHLOV, V.: On a general class of Birkhoff-regular eigenvalue problems. Differ. Integral Equ. 8 (1995), 2157–2176.

[FRR2] FREILING, G., RYKHLOV, V.: Pointwise convergence of eigenfunction expansions for a general class of regular eigenvalue problems. Methods Funct. Anal. Topology 3 (1997), 27–45.

[FRT] FREILING, G., TROOSHIN, I. YU.: Abel-summability of eigenfunction expansions of three-point boundary value problems. Math. Nachr. 190 (1998), 129–148.

[FRY] FREILING, G., YURKO, V.: Expansion theorems for boundary value problems with a singularity. Analysis 18 (1998), 25–39.

[FT] FREITAS, P.: A nonlocal Sturm-Liouville eigenvalue problem. Proc. Roy. Soc. Edinburgh 124A (1994), 169–188.

[FRO] FROBENIUS, G.: Über adjungierte lineare Differentialausdrücke. J. reine angew. Math. 85 (1878), 185–213.

[FU] FULTON, C. T.: Two-point boundary value problems with eigenvalue parameter contained in the boundary conditions. Proc. Roy. Soc. Edinburgh 77A (1977), 293–308.

[FUN1] FUNTAKOV, V. N.: Expansions in eigenfunctions of non-self-adjoint second-order differential equations (Russian). Differentsial'nye Uravneniya 6 (1970), 2011–2020, Engl. transl.: Differential Equations 6 (1970), 1528–1535.

[FUN2] FUNTAKOV, V. N.: Eigenfunction expansion of a not self-adjoint Dirac operator (Russian). Spektr. Teor. Oper. 4 (1982), 199–208.

[GAK] GADIROVA, T. A., KERIMOV, T. G.: Multiple expansion by eigenfunctions of differential operator with integral boundary conditions (Russian). Izv. Akad. Nauk Azerbaĭdzhan. SSR Ser. Fiz.-Tekh. Mat. Nauk 1977 (1977), No. 1, 52–56.

[GAD] GADZHIEV, M. K.: Convergence of the Fourier series with respect to eigenfunctions of a four-point spectral problem (Russian). In: Spectral theory of differential operators, Baku 1984, 37–39.

[GAL] GALOYAN, R. S.: On completeness of a certain system of eigen and associated functions (Russian). Izv. Akad. Nauk Armyan. SSR Ser. Mat. 15 (1980), 310–322.

[GDZ] GASYMOV, M. G., DZAVADOV, M. G.: Multiple completeness of a part of the set of eigenfunctions and adjoint functions of differential operator bundles (Russian). Dokl. Akad. Nauk SSSR 203 (1972), 521–524, Engl. transl.: Soviet Math. Dokl 13 (1972), 521–524.

[GM1] GASYMOV, M. G., MAGERRAMOV, A. M.: On multiple completeness of systems of eigen- and associated functions of a class of differential operators (Russian). Dokl. Akad. Nauk AzSSR 30 (1974), No. 3, 9–12.

[GM2] GASYMOV, M. G., MAGERRAMOV, A. M.: Investigation of a class of differential operator pencils of even order (Russian). Dokl. Akad. Nauk SSSR 265 (1982), N°2 (Russian), English transl.: Soviet Math. Dokl. 26 (1982), No. 1, 55–58.

[GKH] GERDZHIKOV, V. S., KHRISTOV, E. KH.: On expansions in products of solutions of two Dirac systems (Russian). Mat. Zametki 28 (1980), 501–512.

[GOH] GOHBERG, I. C.: On linear operators depending analytically on a parameter (Russian). Dokl. Akad. Nauk SSSR (N.S.) 78 (1951), 629–632.

[GGK] GOHBERG, I. C., GOLDBERG, S., KAASHOEK, M. A.: Classes of linear operators, Vol. I. Oper. Theory: Adv. Appl., Vol 49, Birkhäuser-Verlag Basel Boston Berlin 1990.

[GKL] GOHBERG, I. C., KAASHOEK, M. A., LAY, D. C.: Spectral classification of operators and operator functions. Bull. Amer. Math. Soc. 82 (1976), 587–589.

[GK] GOHBERG, I. C., KREIN, M. G.: Introduction to the theory of linear nonselfadjoint operators. Translations of mathematical monographs 18, American Mathematical Society, Providence 1969.

[GS] GOHBERG, I. C., SIGAL, E. I.: An operator generalization of the logarithmic residue theorem and the theorem of Rouché (Russian). Mat. Sb. (N.S.) 84(126) (1971), 607–629, Engl. transl.: Math. USSR-Sb. 13 (1971), 603–625.

[GO] GOLDBERG, S.: Unbounded linear operators. McGraw-Hill, New York 1966.

[GH1] GOLDBERG, W., HOCHSTADT, H.: On a periodic boundary value problem with only two simple eigenvalues. J. Math. Anal. Appl. 80 (1981), 1–6.

[GH2] GOLDBERG, W., HOCHSTADT, H.: On a periodic boundary value problem with only a finite number of simple eigenvalues. J. Math. Anal. Appl. 91 (1983), 340–351.

[GOR] GORBACHUK, M. L.: Completeness of the system of eigenfunctions and associated functions of a nonself-adjoint boundary value problem for a differential operator equation of second order (Russian). Funktsional. Anal. i Prilozhen. 7 (1973), no. 1, 68–69, Engl. transl.: Functional Anal. Appl. 7 (1973), 58–59.

[GG] GORBACHUK, V. I., GORBACHUK, M. L.: On boundary-value problems for a first-order differential equation with operator coefficients and the expansion in eigenfunctions of this equation (Russian). Dokl. Akad. Nauk SSSR 208 (1973) 1286–1271, Engl. transl.: Soviet Math. Dokl. 14 (1973), 244–248.

[GRK] GREEN, G. B., KRALL, A. M.: Linear differential systems with infinitely many boundary points. Ann. Mat. Pura Appl. (4) 91 (1972), 53–67.

[GRO] GROMOV, V. P.: On expansions in eigenelements of a linear operator (Russian). Dokl. Akad. Nauk SSSR 265 (1982), 281–284, Engl. transl.: Soviet Math. Dokl. 26 (1982), 59–62.

[GUK] GUBREEV, G. M., KOVALENKO, A. I.: Tests for completeness of root subspaces of a differentiation operator with abstract boundary conditions (Russian). Mat. Zametki 30 (1981), 543–552, Engl. transl.: Math. Notes 30 (1982), 765–770.

[GP] GUPTA, P. K. S.: On the convergence of eigenfunction expansions associated with a vector-matrix differential equation. J. Indian Inst. Sci. 60 (1978), no. 3, 89–110.

[GUR] GUREVIC, A. P.: On the expansion of a differential operator in eigenfunctions and associated functions (Russian). Trudy Molod. Ucenyh Mat. Meh. (1969), 22–27.

[GUS] GUSEJNOV, I. M.: Spectral analysis of a certain class of a system of first order differential equations (Russian). Dokl. Akad. Nauk AzSSR 38 (1982), no. 10, 13–17.

[GSE] GUSEJNOV, M. M.: Double expansions in eigen- and associated functions of a non-selfadjoint problem (Russian). Izv. Akad. Nauk Azerbaĭdzhan. SSR Ser. Fiz.-Tekhn. Mat. Nauk 1974 (1974), no. 6, 50–54.

[HA] HALTINER, G. J.: The theory of linear differential systems based upon a new definition of the adjoint. Duke Math. J. 15 (1948), 893–920.

[HLP] HARDY, G., LITTLEWOOD, J. E., PÓLYA, G.: Inequalities. 2nd ed., Cambridge University Press, Cambridge 1952.

[HL1] HAUGER, W., LEONHARD, M.: Influence of the end-supports on the stability of bars. J. Sound Vibr. 55(1) (1977), 153–156.

[HL2] HAUGER, W., LEONHARD, M.: Untersuchung zur Genauigkeit des Galerkinschen Verfahrens anhand exakterKnicklasten elastischer Stäbe unter nichtkonservativer Belastung. Der Stahlbau 6/1978, 169–174.

[HEI1] HEISECKE, G.: Rand-Eigenwertprobleme $N(y) = \lambda P(y)$ bei λ-abhängigen Randbedingungen. Mitt. Math. Sem. Gießen 145 (1980), 1–74.

[HEI2] HEISECKE, G.: Rand-Eigenwertprobleme $N(y) = \lambda P(y)$ bei λ-abhängigen Randbedingungen. Z. Angew. Math. Mech. 61 (1981), T242–T244.

[HS] HEWITT, E., STROMBERG, K.: Real and abstract analysis. Springer-Verlag, New York – Heidelberg – Berlin 1965.

[HI1] HILB, E.: Über Integraldarstellungen willkürlicher Funktionen. Math. Ann. 66 (1909), 1–66.

[HI2] HILB, E.: Über Reihenentwicklungen nach den Eigenfunktionen linearer Differentialgleichungen 2ter Ordnung. Math. Ann. 71 (1911), 76–87.

[HI3] HILB, E.: Über gewöhnliche Differentialgleichungen mit Singularitäten und die dazugehörigen Entwicklungen willkürlicher Funktionen. Math. Ann. 76 (1915), 333-339.

[HT] HILLE, E., TAMARKIN, J. D.: On the summability of Fourier series. I. Trans. Amer. Math. Soc. 34 (1932), 757–783.

[HIN] HINTON, D. B.: An expansion theorem for an eigenvalue problem with eigenvalue parameter in the boundary condition. Quart. J. Math. Oxford Ser. (2) 30 (1979), 33–42.

[HÖ1] HÖRMANDER, L.: Linear partial differential operators. Springer-Verlag, Berlin – Heidelberg – New York 1963.

[HÖ2] HÖRMANDER, L.: The analysis of linear partial differential operators I. Springer-Verlag, Berlin – Heidelberg – New York – Tokyo 1983.

[HO] HORVÁTH, J.: Topological vector spaces and distributions. Reading Mass. 1966.

[HPY] HUTSON, V., PYM, J. S.: Applications of functional analysis and operator theory. Academic Press, London 1980.

[IL1] IL'IN, V. A.: On uniform convergence of expansions in eigen and associated functions of a non-selfadjoint ordinary differential operator and in trigonometric Fourier series (Russian). Dokl. Akad. Nauk SSSR, 223 (1975), 548–551, Engl. transl.: Soviet Math. Dokl. 16 (1975), 932–936.

[IL2] IL'IN, V. A.: On the equiconvergence of an expansion in a trigonometric Fourier series and the eigenfunctions of the Keldyš pencil of ordinary nonselfadjoint differential operators (Russian). Dokl. Akad. Nauk SSSR 225 (1975), No. 3, Engl. transl.: Soviet Math. Dokl. 16 (1975), 1494–1496.

[IL3] IL'IN, V. A.: Necessary and sufficient conditions for a subsystem of the eigen- and associated functions of a Keldys bundle of ordinary differential operators to be a basis (Russian). Dokl. Akad. Nauk SSSR 227 (1976), 796–799, Engl. transl.: Soviet Math. Dokl. 17 (1976), 513–516.

[IL4] IL'IN, V. A.: On the properties of a reduced system of eigenfunctions and associated functions of a Keldys bundle of ordinary differential operators (Russian). Dokl. Akad. Nauk SSSR 230 (1976), 30–33, Engl. transl.: Soviet Math. Dokl. 17 (1976), 1247–1250.

[IL5] IL'IN, V. A.: On the existence of a reduced system of eigenfunctions and generalized eigenfunctions for a nonselfadjoint ordinary differential operator (Russian). Trudy Mat. Inst. Steklov 142 (1976), 148–155, Engl. transl.: Proc. Steklov Inst. Math. 142 (1979), 157–164.

[IL6] IL'IN, V. A.: Approximation of functions by biorthogonal series in the eigen- and associated functions of differential operators. In: The theory of the approximation of functions, Proc. Int. Conf. Collect. Artic. Kaluga 1975, 1977, 206–213.

[IL7] IL'IN, V. A.: Convergence of eigenfunction expansions at points of discontinuity of the coefficients of a differential operator (Russian). Mat. Zametki, 22 (1977), no. 5, 679-698, Engl. transl.: Math. Notes 22 (1977), 870–882.

[IL8] IL'IN, V. A.: Necessary and sufficient conditions for the basis property in L_p, and equiconvergence with a trigonometric series, of spectral decompositions and expansions in systems of exponential functions (Russian). Dokl. Akad. Nauk SSSR 273 (1983), N^0 4, 789–793, Engl. transl.: Soviet Math. Dokl. 28 (1983), 701–705.

[IL9] IL'IN, V. A.: On the unconditional basis property for systems of eigenfunctions and associated functions of a second-order differential operator on a closed interval (Russian). Dokl. Akad. Nauk SSSR 273 (1983), 1048–1053, Engl. transl.: Soviet Math. Dokl. 28 (1983), 743–747.

[IL10] IL'IN, V. A.: Necessary and sufficient conditions for the eigenvectors and associated vectors of discontinuous second order operators to be Riesz bases (Russian). Differentsial'nye Uravneniya 22 (1986), no. 12, 2059–2071, Engl. transl.: Differential Equations 22 (1986), 1409–1419.

[IT] IL'IN, V. A., TIKHOMIROV, V. V.: Basis property of Riesz means of spectral expansions corresponding to a nonself-adjoint ordinary differential operator of order n (Russian). Differentsial'nye Uravneniya 18 (1982), no. 12, 2098–2126, Engl. transl.: Differential Equations 18 (1983), 1486–1511.

[IV] IVANOV, S. A.: Regge problem for vector functions (Russian). Teor. Oper. Teor. Funkts. 1 (1983), 68–86.

[JW] JEGGLE, H., WENDLAND, W.: On the discrete approximation of eigenvalue problems with holomorphic parameter dependence. Proc. Royal Soc. Edinb. 78A (1977), 1–29.

[JO] JONES, C. A.: Multiple eigenvalues and mode classification in plane Poiseuille flow. Quart. J. Mech. Appl. Math. 41 (1988), 363–382.

[JON] JONES, W. R.: Differential systems with integral boundary conditions. J. Differential Equations 3 (1967), 191–202.

[KAS] KAASHOEK, M. A.: Analytic equivalence of the boundary eigenvalue operator function and its characteristic matrix function. Integral Equations Operator Theory 9 (1986), 275–285.

[KAV] KAASHOEK, M. A., VERDUYN LUNEL, S. M.: Characteristic matrices and spectral properties of evolutionary systems. Trans. Amer. Math. Soc. 334 (1992), 479–517.

[KC] KAC, I. S.: The multiplicity of the spectrum of a differential operator of second order and the expansion into eigenfunctions (Russian). Izv. Akad. Nauk SSSR Ser. Mat. 27 (1963), 1081-1112, 28 (1964), 951–952.

[KK1] KAMKE, E.: Differentialgleichungen. Lösungsmethoden und Lösungen. I. Gewöhnliche Differentialgleichungen. 5th ed., Leipzig 1956.

[KK2] KAMKE, E.: Differentialgleichungen. I. Gewöhnliche Differentialgleichungen. 4th ed., Leipzig 1962.

[KA] KATO, T.: Perturbation theory for linear operators. 2nd ed., Springer-Verlag, Berlin – Heidelberg – New York 1966.

[KFF] KAUFFMAN, R.: Non-normalizable eigenfunction expansions for ordinary differential operators. In: Ordinary and partial differential equations, Proc. 7th Conf., Dundee/Scotl., Lecture Notes in Math. 964 (1982), 362–367.

[KF] KAUFMANN, F.-J.: Abgeleitete Birkhoff-Reihen bei Randeigenwertproblemen zu $N(Y) = \lambda P(Y)$ mit λ-abhängigen Randbedingungen. Thesis, Math.-Naturw. Fak. RWTH Aachen 1988.

[KL] KAUFMANN, F.-J., LUTHER, W. J.: Degree of convergence of Birkhoff series, direct and inverse theorems. J. Math. Anal. Appl. 187 (1994), 156–168.

[KE1] KELDYSH, M. V.: On the eigenvalues and eigenfunctions of certain classes of nonselfadjoint equations (Russian). Dokl. Akad. Nauk SSSR 77 (1951), 11–14.

[KE2] KELDYSH, M. V.: On the completeness of eigenfunctions of some classes of nonselfadjoint linear equations (Russian). Uspekhi Mat. Nauk 26 (1971), 15–41, Engl. transl.: Russian Math. Surveys 26 (1971), 15–44.

[KER] KERIMOV, N. B.: On the unconditional basis property of the system of eigenfunctions and associated functions of a fourth order differential operator (Russian). Dokl. Akad. Nauk SSSR 286 (1986), 803–808, Engl. transl.: Soviet Math. Dokl. 33 (1986), 185–189.

[KRM] KERIMOV, YA. R.: Expansion in eigenfunctions of a nonselfadjoint differential operator. In: Spectral theory of operators, Baku 1977, 208–241.

[KES] KESEL'MAN, G. M.: Spectral property of a perturbed Sturm-Liouville operator with nonlocal boundary conditions (Russian). Differentsial'nye Uravneniya 21 (1985), 494–499, Engl. transl.: Differential Equations 21 (1985), 336–341.

[KH1] KHROMOV, A. P.: On the summation of the expansions in eigenfunctions of the boundary value problem for an ordinary differential equation with splitting boundary conditions and on an analog of the Weierstrass theorem (Russian). Obyknov. Diff. Uravn. Razlozen. Rjady Fur'e 1 (1968), 29–41.

[KH2] KHROMOV, A. P.: On uniform convergence in eigenfunctions of the expansions of differential operators of second order (Russian). Differ. Uravn. vycisl. Mat. 5 (1975), no. 2, 3–20.

[KS] KHRUSHCHEV, S. V.: The Regge problem for strings. Unconditionally convergent eigenfunction expansions, and unconditional bases of exponentials in $L_2(-T,T)$. J. Operator Theory 14 (1985), 67–85.

[KY] KHRYPTUN, V. G.: Expansion in eigenfunctions of a linear operator (Russian). Differentsial'nye Uravnenija 7 (1971), 373–375, Engl. transl.: Differential Equations 7 (1971), 291–293.

[KN] KNESER, A.: Die Theorie der Integralgleichungen und die Darstellung willkürlicher Funktionen in der mathematischen Physik. Math. Ann. 63 (1907), 477–524.

[KOB] KOBAYASHI, M.: Eigenfunction expansions: a discontinuous version. SIAM J. Appl. Math. 50 (1990), 910-917.

[KOC] KOCHUBEJ, A. N.: Expansions with respect to root vectors of abstract differential operators (Russian). Differentsial'nye Uravnenija 11(1975), 1341–1342, Engl. transl. Differential Equations 11 (1975), 1009–1010.

[KOG] KOGAN, B. L.: Double completeness of the system of eigenfunctions and associated functions of the Regge problem (Russian). Funktsional. Analiz i Prilozhen. 5 (1971), no. 3, 70–74, Engl. transl.: Functional Anal. Appl. 5 (1971), 229–232.

[KOM] KOMORNIK, V.: On the equiconvergence of expansions by Riesz bases formed by eigenfunctions of a linear differential operator of order 2. Acta Math. Hungar. 44 (1984), 311-325.

[KSH] KOSTYUCHENKO, A. G., SHKALIKOV, A. A.: Summability of eigenfunction expansions of differential operators and convolution operators (Russian). Funktsional. Anal. i Prilozhen. 12 (1978), N° 4, 24–40, Engl. transl.: Functional Anal. Appl. 12 (1978), 262–276.

[KM] KOZLOV, V., MAZ'YA, V.: Differential equations with operator coefficients. Springer, Berlin 1999.

[KR1] KRALL, A. M.: Nonhomogeneous differential operators. Michigan Math. J. 12 (1965), 247–255.

[KR2] KRALL, A. M.: Differential operators and their adjoints under integral and multiple point boundary conditions. J. Differential Equations 4 (1968), 327–336.

[KR3] KRALL, A. M.: Boundary value problems with interior point boundary conditions. Pacific J. Math. 29 (1969), 161–166.

[KR4] KRALL, A. M.: Adjoint differential systems in inner product spaces. SIAM J. Appl. Math. 20 (1971), 40–43.

[KR5] KRALL, A. M.: Differential-boundary operators. Trans. Amer. Math. Soc. 154 (1971), 429–458.

[KR6] KRALL, A. M.: An eigenfunction expansion for a nonselfadjoint, interior point boundary value problem. Trans. Amer. Math. Soc. 170 (1972), 137–147.

[KR7] KRALL, A. M.: Adjoints of multipoint-integral boundary value problems. Proc. Amer. Math. Soc. 37 (1973), 213–216.

[KR8] KRALL, A. M.: Stieltjes differential-boundary operators. Proc. Amer. Math. Soc. 41 (1973), 80–86.

[KR9] KRALL, A. M.: Stieltjes differential-boundary operators. II. Pacific J. Math. 55 (1974), 207–218.

[KR10] KRALL, A. M.: The development of general differential and general differential-boundary systems. Rocky Mountain J. Math. 5 (1975), 493–542.

[KR11] KRALL, A. M.: n-th order Stieltjes differential boundary operators and Stieltjes differential boundary systems. J. Differential Equations 24 (1977), 253–267.

[KR12] KRALL, A. M.: Asymptotic stability of the Euler-Bernoulli beam with boundary control. J. Math. Anal. Appl. 137 (1989), 288–295.

[KRR] KRALL, A. M., RACE, D.: Self-adjointness for the Weyl problem under an energy norm. Quaestiones Math. 18 (1995), 407–426.

[KRA1] KRAVICKIĬ, A. O.: Expansion in a series of eigenfunctions of a nonself-adjoint boundary value problem (Russian). Dokl. Akad. Nauk SSSR 170 (1966), No. 6; Engl. transl.: Soviet Math. Dokl. 7 (1966), 1361–1365.

[KRA2] KRAVICKIĬ, A. O.: The two-fold expansion of a certain nonself-adjoint boundary value problem in series of eigenfunctions (Russian). Differential nye Uravnenija 4 (1968), 165–177. Engl. transl.: Differential Equations 4 (1968), 86–92.

[KT] KREĬN, S. G., TROFIMOV, V. P.: On holomorphic operator-functions of several complex variables (Russian). Funktsional. Anal. i Prilozhen. 3 (1969), N° 4, 85–86; Engl. transl.: Functional Anal. Appl. 3 (1969), 330–331.

[KRI] KRIMMER, W.: Im Parameter nichtlineare Eigenwertprobleme bei Differentialgleichungen im Komplexen und Entwicklungen analytischer Funktionen nach Speziellen Funktionen. Thesis, Regensburg 1984.

[LL1] LANG, P., LOCKER, J.: Spectral decomposition of a Hilbert space by a Fredholm operator. J. Funct. Anal. 79 (1988), 9–17.

[LL2] LANG, P., LOCKER, J.: Spectral representation of the resolvent of a discrete operator. J. Funct. Anal. 79 (1988), 18–31.

[LL3] LANG, P., LOCKER, J.: Denseness of the generalized eigenvectors of an H–S discrete operator. J. Funct. Anal. 82 (1989), 316–329.

[LL4] LANG, P., LOCKER, J.: Spectral theory for a differential operator: Characteristic determinant and Green's function. J. Math. Anal. Appl. 141 (1989), 405–423.

[LL5] LANG, P., LOCKER, J.: Spectral theory of two-point differential operators determined by $-D^2$. I. Spectral properties. J. Math. Anal. Appl. 141 (1989), 538–558.

[LL6] LANG, P., LOCKER, J.: Spectral properties of two-point differential operators determined by $-D^2$. II. Analysis of cases. J. Math. Anal. Appl. 146 (1990), 148–191.

[LMM1] LANGER, H., MENNICKEN, R., MÖLLER, M.: On Floquet eigenvalue problems for first-order differential systems in the complex domain. J. reine angew. Math. 425 (1992), 87–121.

[LMM2] LANGER, H., MENNICKEN, R., MÖLLER, M.: Expansions of analytic functions in series of Floquet solutions of first order differential systems. Math. Nachr. 162 (1993), 279–314.

[LMMS] LANGER, H., MENNICKEN, R., MÖLLER, M., SATTLER, A.: Expansions of analytic functions in products of Bessel functions. Resultate Math. 24 (1993), 129–146.

[LMT] LANGER, H., MENNICKEN, R., TRETTER, C.: A self-adjoint linear pencil $Q - \lambda P$ of ordinary differential operators. Methods Funct. Anal. Topology 2 (1996), 38–54.

[LM] LANGER, H., MÖLLER, M.: Linearization of boundary eigenvalue problems. Integral Equations Operator Theory 14 (1991), 105–119.

[LA1] LANGER, R. E.: Developments associated with a boundary problem not linear in the parameter. Trans. Amer. Math. Soc. 25 (1923), 155–172.

[LA2] LANGER, R. E.: On the theory of integral equations with discontinuous kernels. Trans. Amer. Math. Soc. 28 (1926), 585–639.

[LA3] LANGER, R. E.: The boundary problem associated with a differential equation in which the coefficient of the parameter changes sign. Trans. Amer. Math. Soc. 31 (1929), 1–24.

[LA4] LANGER, R. E.: The asymptotic location of the roots of a certain transcendental equation. Trans. Amer. Math. Soc. 31 (1929), 837–844.

[LA5] LANGER, R. E.: The expansion problem in the theory of ordinary linear differential systems of the second order. Trans. Amer. Math. Soc. 31 (1929), 868–906.

[LA6] LANGER, R. E.: The boundary problem associated with a differential system rational in the parameter. Trans. Amer. Math. Soc. 32 (1930), 238–250.

[LA7] LANGER, R. E.: On the zeros of exponential sums and integrals. Bull. Amer. Math. Soc. 37 (1931), 213-239.

[LA8] LANGER, R. E.: The asymptotic solutions of certain linear ordinary differential equations of the second order. Trans. Amer. Math. Soc. 36 (1934), 90–106.

[LA9] LANGER, R. E.: The boundary problem of an ordinary linear differential system in the complex domain. Trans. Amer. Math. Soc. 46 (1939), 151–190.

[LA10] LANGER, R. E.: A theory for ordinary differential boundary problems of the second order and of the highly irregular type. Trans. Amer. Math. Soc. 53 (1943), 292–361.

[LB] LANGER, R. E., BROWN, E. P.: On a class of integral equations with discontinuous kernels. Trans. Amer. Math. Soc. 29 (1927), 683–715.

[LBO] LANGINESTRA, A. V., BOYCE, W. E.: Convergence and evaluation of sums of reciprocal powers of eigenvalues of boundary value problems nonlinear in the eigenvalue parameter. SIAM J. Math. Anal. 5 (1974), 64–90.

[LAZ1] LAŽETIĆ, N.: On the derivatives of partial sums of expansions in eigenfunctions and associated functions of nonselfadjoint operators of Sturm-Liouville type (Russian). Dokl. Akad. Nauk SSSR 260 (1981), 22–26, Engl. transl.: Soviet Math. Dokl. 24 (1981), 212–215.

[LAZ2] LAŽETIĆ, N.: On the convergence of expansions with respect to eigen- and adjoint functions of a non-self-adjoint Sturm-Liouville type operator with discontinuous coefficients (Russian). Publ. Inst. Math. (Beograd) (N. S.) 30(44) (1981), 89–109.

[LAZ3] LAŽETIĆ, N.: Necessary and sufficient conditions for a system of eigenfunctions and associated functions of a Sturm-Liouville operator with discontinuous coefficients to possess the basis property. Publ. Inst. Math. (Beograd) (N. S.) 38(52) (1985), 119–147.

[LE] LEVIN, B. JA.: Distribution of zeros of entire functions. 2nd ed., Providence, Rhode Island 1980.

[LV] LEVITAN, B. M.: On the completeness of products of solutions of two Sturm-Liouville equations. Differential Integral Equations 7 (1994), 1–14.

[LS] LEVITAN, B. M., SARGSJAN, I. S.: Introduction to spectral theory (Russian). Moscow 1970, Engl. transl.: Transl. Math. Monographs 39, AMS, Providence, Rhode Island, 1975.

[LI] LIFSCHITZ, A.: Exact description of the spectrum of elliptical vortices in hydrodynamics and magnetohydrodynamics. Phys. Fluids 7 (1995), No. 7, 1626–1636.

[LO1] LOCKER, J.: Functional analysis and two-point differential operators. Longman,, Pitman Research Notes in Mathematics 144, Harlow 1986.

[LO2] LOCKER, J.: The nonspectral Birkhoff-regular differential operators determined by $-D^2$. J. Math. Anal. Appl. 154 (1991), 243–254.

[LO3] LOCKER, J.: Eigenfunction expansions for the nonspectral differential operators determined by $-D^2$. J. Differential Equations 96 (1992), 318–339.

[LO4] LOCKER, J.: The spectral theory of second order two-point differential operators: I. A priori estimates for the eigenvalues and completeness. Proc. Roy. Soc. Edinburgh 121A (1992), 279–301.

[LO5] LOCKER, J.: The spectral theory of second order two-point differential operators. II. Asymptotic expansions and the characteristic determinant. J. Differ. Equations 114 (1994), 272–287.

[LO6] LOCKER, J.: The spectral theory of second order two-point differential operators. III. The eigenvalues and their asymptotic formulas. Rocky Mt. J. Math. 26 (1996), 679–706.

[LO7] LOCKER, J.: The spectral theory of second order two-point differential operators. IV. The associated projections and the subspace $S_\infty(L)$. Rocky Mt. J. Math. 26 (1996), 1473–1498.

[LO8] LOCKER, J.: Spectral theory of non-self-adjoint two-point differential operators. AMS, Providence, Rhode Island, 2000.

[LOLA] LOCKER, J., LANG, P.: Eigenfunction expansions for the nonspectral differential operators determined by $-D^2$. J. Differential Equations 96 (1992), 318–339.

[MAK] MAKHMUDOV, A. P.: Nonlinear perturbations of regular spectral problems with a parameter in the boundary conditions and the completeness of their eigenfunctions (Russian). Dokl. Akad. Nauk. SSSR 277 (1984), 1318–1323, Engl. transl.: Soviet Math. Dokl. 30 (1984), 276–281.

[MP] MAKSUDOV, F. G., PASHAEVA, EH. EH.: On a multiple expansion in eigenfunctions (Russian). Dokl. Akad. Nauk SSSR 249 (1979), 544–547, Engl. transl.: Soviet Math. Dokl. 20 (1979), 1310–1313.

[MV] MAKSUDOV, F. G., VELIEV, O. A.: Nonselfadjoint differential operators in the space of vector-valued functions with periodic coefficients (Russian). Dokl. Akad. Nauk SSSR 258 (1981), 26–30, Engl. transl.: Soviet Math. Dokl. 23 (1981), 475–478.

[MAM1] MAMEDOV, K. S.: A theorem on multiple completeness of eigen- and adjoint vector functions of polynomial differential operator bundles. Spectral theory of operators, Baku 1975, 133-139.

[MAM2] MAMEDOV, K. S.: On the convergence of multiple expansions by the systems of eigen- and adjoined vector-functions of a polynomial differential-operator bundle (Russian). Izv. Akad. Nauk Azerbaĭdzhan. SSR Ser. Fiz.-Tekh. Mat. Nauk 1977 (1977), no. 1, 36–40.

[MA1] MARKUS, A. S.: A basis of root vectors of a dissipative operator (Russian). Dokl. Akad. Nauk SSSR 132 (1960), 524–527, Engl. transl.: Soviet Math. Dokl. 1 (1960), 599–602.

[MA2] MARKUS, A. S.: Expansion in root vectors of a slightly perturbed selfadjoint operator (Russian). Dokl. Akad. Nauk SSSR 142 (1962), 538–541, Engl. transl.: Soviet Math. Dokl. 3 (1962), 104–108.

[MA3] MARKUS, A. S.: Some criteria for the completeness of a system of root vectors of a linear operator in Banach spaces (Russian). Mat. Sb. (N. S.) 70 (1966), 526–561, Engl. transl.: Amer. Math. Soc. Transl. (2) 85 (1970), 51–91.

[MA4] MARKUS, A. S.: Introduction to the spectral theory of polynomial operator pencils. Transl. Math. Monographs 71, AMS, Providence, Rhode Island, 1988.

[MAS] MARKUS, A. S., SIGAL, E. I.: The multiplicity of the characteristic number of an analytic operator function (Russian). Mat. Issled. 5 (1970), no. 3(17), 129–147.

[MS] MCSHANE, E. J.: Integration. Princeton University Press, Princeton 1944.

[ME] MENNICKEN, R.: Spectral theory for certain operator polynomials. In: Functional Analysis, Holomorphy and Approximation Theory II, ed.: G. Zapata, North-Holland Math. Studies 86 (Notas de Matemática 92) (1984), 203–243.

[MM1] MENNICKEN, R., MÖLLER, M.: Root functions, eigenvectors, associated vectors and the inverse of a holomorphic operator function. Arch. Math. (Basel) 42 (1984), 455–463.

[MM2] MENNICKEN, R., MÖLLER, M.: A generalization of a theorem of Keldyš. In: Functional Analysis: surveys and recent results, III, Paderborn, 1983, 275–290, Noth-Holland, Amsterdam, 1984.

[MM3] MENNICKEN, R., MÖLLER, M.: Root functions of boundary eigenvalue operator functions. Integral Equations Operator Theory 9 (1986), 237–265.

[MM4] MENNICKEN, R., MÖLLER, M.: Boundary eigenvalue problems. In: Operator Theory: advances and applications, Vol. 19, eds.: H. Bart, I. Gohberg and M. A. Kaashoek, Birkhäuser, Basel – Boston – Stuttgart 1986, 279–331.

[MM5] MENNICKEN, R., MÖLLER, M.: Boundary eigenvalue problems. Notas de Algebra y Analisis, N° 14, Instituto de Matemática, Universidad Nacional del Sur, Bahia Blanca 1986.

[MM6] MENNICKEN, R., MÖLLER, M.: Equivalence of boundary eigenvalue operator functions and their characteristic matrix functions. Math. Ann. 279 (1987), 267–275.

[MIK1] MIKHAJLETS, V. A.: The solvability and completeness of the system of eigenfunctions and associated functions of nonselfadjoint boundary-value problems for a Sturm-Liouville operator equation (Russian). Dokl. Akad. Nauk SSSR 218 (1974), 284–286, Engl. transl.: Soviet Math. Dokl. 15 (1974), 1327–1330.

[MIK2] MIKHAJLETS, V. A.: Boundary-value problems for an operational Sturm-Liouville equation with a complete system of eigen- and associated functions (Russian). Differentsial'nye Uravnenijya 11 (1975), 1595–1600, 1715, Engl. transl.: Differential Equations 11 (1975), 1189–1192.

[MIR] MIRZOEV, S. S.: Multiple completeness of a part of the root vectors of fourth order polynomial operator pencils with the normal smooth part (Russian). Spektr. Teor. Oper. 4 (1982), 148–161.

[MOH] MOHR, E.: Gewöhnliche Eigenwertaufgaben zweiter Ordnung, bei denen der Parameter in einer Randbedingung auftritt. Ann. Mat. Pura Appl. IV. Ser. 138 (1984), 157–189.

[MÖ1] MÖLLER, M.: Equivalence of differential-boundary eigenvalue operators and their characteristic matrices. Math. Nachr. 141 (1989), 299–312.

[MÖ2] MÖLLER, M.: The inverse and the adjoint of differential-boundary operators.

[MÖ3] MÖLLER, M.: Expansion theorems for Birkhoff-regular differential-boundary operators. Proc. Roy. Soc. Edinburgh 107A (1987), 349–374.

[MÖ4] MÖLLER, M.: An expansion theorem for Birkhoff-regular n-th order differential-boundary operators. In: Differential Equations and Applications: Proceedings of the International Conference on Theory and Applications of Differential Equations. Columbus, Ohio 1988. Ohio University Press, 1989, Vol. II, 244 –253.

[MÖ5] MÖLLER, M.: On the Carlitz expansion in products of Bessel functions. Resultate Math. 24 (1993), 147–152.

[MU] MÖLLER, M., USCHOLD, C.: Über Randeigenwertprobleme für Differentialgleichungen mit der charakteristischen Matrix $\lambda^p(\lambda^l - 1) = 0$. Regensburger Math. Schriften 17, 1988.

[MOL] MOLODENKOV, V. A.: Expansion in eigenfunctions of a boundary value problem (Russian). Differ. Integral. Uravn., Saratov 1972, 9–17.

[MOK] MOLODENKOV, V. A., KHROMOV, A. P.: Expansion in eigenfunctions of a boundary value problem for the differential operator (Russian). Differ. Uravn. vycisl. Mat. 1 (1972), 17–26.

[MO] MOYER, R. D.: The adjoints of ordinary differential operators. J. Differential Equations 4 (1968), 337–349.

[MUS] MUSTAFAEV, G. B.: An expansion formula connected with the boundary value problem with a complex parameter for systems of differential equations of the first order (Russian). Differentsial'nye Uravneniya 1 (1965), 1246–1259.

[MUT] MUSTAFIN, M. A.: A condition for the basis of summation by the Abel method of series in root vectors of the nonselfadjoint Sturm-Liouville operator (Russian). Differentsialnye Uravneniya 29 (1993), No. 1, 116–117, 181, Engl. transl.: Differential Equations 29 (1993), no. 1, 98–99.

[NA1] NAIMARK, M. A.: Linear differential operators, part I: Elementary theory of linear differential operators. Ungar, New York 1967.

[NA2] NAIMARK, N. A.: Linear differential operators, part II: Linear differential operators in Hilbert space. Ungar, New York 1968.

[NI] NIESSEN, H.-D.: Singuläre S-Hermitesche Rand-Eigenwertprobleme. Manuscripta Math. 3 (1970), 35–68.

[OR1] ORAZOV, M. B.: On the completeness of systems of elementary solutions for some operator differential equations on a half-line and on an interval (Russian). Dokl. Akad. Nauk SSSR 245 (1979), No 4, 788–792; Engl. transl.: Soviet Math. Dokl. 20 (1979), 347–352.

[OR2] ORAZOV, M. B.: Completeness of elementary solutions in the problems of steady-state vibrations of an elastic rectangular plate (Russian). Funktsional Anal. Prilozhen. 14 (1980), no. 1, 77-78; Engl. transl.: Functional Anal. Appl. 14 (1980), 63–64.

[OR3] ORAZOV, M. B.: On the completeness of the elementary solutions of the problem of stationary oscillations of a finite cylinder (Russian). Uspekhi Mat. Nauk 35 (1980), no. 5, 237–238; Engl. transl.: Russian Math. Surveys 35 (1980), no. 5, 267–268.

[OS] ORAZOV, M. B., SHKALIKOV, A. A.: The n-fold basis property of the characteristic functions of certain regular boundary value problems (Russian). Sibirsk. Mat. J. 17 (1976), N^o,3, 627–639, Engl. transl.: Siberian Math. J. 17 (1976), 483–492.

[ORU] ORUDZHEV, E. G.: Boundary value problems for a fourth order differential equation that depends polynomially on the spectral parameter (Russian). Ḋokl. Akad. Nauk. AzSSR 45 (1989), no. 10, 7–12.

[PE1] PECHENTSOV, A. S.: Asymptotic expansions of solutions of linear differential equations containing a parameter (Russian). Diffentsial'nye Uravneniya 17 (1981), 1611–1620; Engl. transl.: Differential Equations 17 (1981), 1023–1031.

[PE2] PECHENTSOV, A. S.: Boundary-value problems for differential equations with a parameter, having characteristic equations with multiple roots (Russian). Differentsial'nye Uravneniya 20 (1984), no. 2, 263–273, Engl. transl.: Differential Equations 20 (1984), 206–215.

[PER1] PERSSON, J.: The vibrating rod with point masses. Boll. Unione Mat. Ital. B(7) 1 (1987), 185–195.

[PER2] PERSSON, J.: Distribution differential equations. Regularization and bending of a rod. Math. Scand. 66 (1990), 291–300.

[PIT] PITTS, C. G. C.: An equiconvergence theorem for a class of eigenfunction expansions. Trans. Amer. Math. Soc. 189 (1974), 337–350.

[PL] PLIEV, V. T.: On the completeness of derived chains corresponding to half the eigenelements of a pencil of ordinary differential operators (Russian). Dokl. Akad. Nauk 288 (1986), no. 3, 552–556; Engl. transl.: Soviet Math. Dokl. 33 (1986), 721–724.

[PO] POINCARÉ, H.: Sur les équations de la physique mathématique. Rend. Circ. Mat. Palermo 8 (1894), 57–156.

[PS] POPOV, I. YU., STREPETOV, A. V.: On the completeness of a system of eigenfunctions in the two-sided Regge problem (Russian). Vestnik Leningrad. Univ. 1983, no. 13, Mat. Mekh. Astronom. no. 3 (1983), 25–31, Engl. transl.: Vestnik Leningrad Univ. Math. 16 (1984), 169–176.

[PR] PROUDMAN, J.: On a class of expansions. Proc. London Math. Soc. 2(24) 1926, 131–139.

[PRD] PROUDMAN, J., DOODSON, A. T.: Time-relations in metereological effects on the sea. Proc. London Math. Soc. 2(24) 1926, 140–149.

[RAD1] RADZIEVSKIĬ, G. V.: On completeness of derived chains corresponding to boundary value problems on a finite interval (Russian). Ukrain. Mat. Zh. 31 (1979), 279–288; Engl. transl.: Ukrainian Math. J. 31 (1980), 215–222.

[RAD2] RADZIEVSKIĬ, G. V.: The problem of the completeness of root vectors in the spectral theory of operator-valued functions (Russian). Uspekhi. Mat. Nauk 37 (1982), no. 2, 81–145, Engl. transl.: Russian Math. Surv. 37 (1982), no. 2, 91–164.

[RAK] RAKHIMOV, M.: The unconditional basis property, on a closed interval, of systems of eigenfunctions and associated functions of a quadratic bundle of nonself-adjoint differential operators (Russian). Differentsial'nye Uravneniya, 22 (1986), no. 1, 94–103, Engl. transl.: Differential Equations 22 (1986), 75–81.

[RAO] RAO, B. S. R.: Eigenfunction expansions for a fourth order differential equation. Indian J. Pure Appl. Math. 3 (1972), 1141–1149.

[RA] RASULOV, M. L.: Methods of contour integration. North-Holland, Amsterdam 1967.

[RO] ROACH, G. F.: Green's functions. Cambridge 1982.

[RR] ROBERTSON, A. P., ROBERTSON, W.: Topological vector spaces. Cambridge 1966.

[RU] RUSSAKOVSKII, E. M.: The matrix Sturm-Liouville problem with a spectral parameter in the boundary conditions (Russian). Funktsional. Anal. i Prilozhen. 27 (1993). No. 1, 86–88, Engl. transl.: Functional Anal. Appl. 27 (1993), 73–74.

[SA] SALAFF, S.: Regular boundary conditions for ordinary differential operators. Trans. Amer. Math. Soc. 134 (1968), 355-373.

[SCHSCH1] SCHÄFKE, F. W., SCHNEIDER, A.: S-hermitesche Rand-Eigenwertprobleme. I. Math. Ann. 162 (1965/1966), 9–26.

[SCHSCH2] SCHÄFKE, F. W., SCHNEIDER, A.: S-hermitesche Rand-Eigenwertprobleme. II. Math. Ann. 165 (1966), 236–260.

[SCHSCH3] SCHÄFKE, F. W., SCHNEIDER, A.: S-hermitesche Rand-Eigenwertprobleme. III. Math. Ann. 177 (1968), 67–94.

[SS] SCHENSTEDT, I. V.: Contributions to the theory of hydrodynamic stability. Ph. D. thesis, University of Michigan 1960.

[SCHM] SCHMIDT, P.: Zur Spektraltheorie einiger Schwingungsprobleme der Technischen Mechanik. Master's thesis, Regensburg 1990.

[SCHN1] SCHNEIDER, A.: Untersuchungen über singuläre reelle S-hermitesche Differentialgleichungssysteme im Normalfall. Math. Z. 107 (1968), 271–296.

[SCHN2] SCHNEIDER, A.: A note on eigenvalue problems with eigenvalue parameter in the boundary conditions. Math. Z. 136 (1974), 163–167.

[SCHNV] SCHNEIDER, A., VONHOFF, R.: Indefinite boundary eigenvalue problems in a Pontryagin space setting. Results Math. 35 (1999), 325–354.

[SCHU] SCHULTZE, B.: On the definition of Stone-regularity. J. Differential Equations 31 (1979), 224–229.

[SR] SCHUR, A.: Zur Entwicklung willkürlicher Funktionen nach Lösungen von Systemen linearer Differentialgleichungen. Math. Ann. 82 (1921), 211–236.

[STV] SCHWABIK, Š., TVRDÝ, M., VEJVODA, O.: Differential and integral equations. Boundary value problems and adjoints. Dordrecht 1979.

[SHI1] SHILOVSKAYA, O. K.: Expansion of linear differential operators in eigenfunctions (Russian). Differ. Integral. Uravn. (1972), 53–79.

[SHI2] SHILOVSKAYA, O. K.: An expansion in eigenfunctions of a differential operator of second order in case of non-regular boundary conditions (Russian). Issled. Differ. Uravn. Teor. Funkcii (1973), 58–74.

[SH1] SHKALIKOV, A. A.: The completeness of eigenfunctions and associated functions of an ordinary differential operator with irregular-separated boundary conditions (Russian). Funktional. Anal. i Prilozhen. 10 (1976), no. 4, 69–80, Engl. transl.: Functional Anal. Appl. 10 (1976), 305–316.

[SH2] SHKALIKOV, A. A.: On the basis problem of the eigenfunctions of an ordinary differential operator (Russian). Uspekhi Mat. Nauk 34, N° 5 (1979), 235–236, Engl. Transl.: Russian Math. Surveys 34, N° 5 (1979), 249–250.

[SH3] SHKALIKOV, A. A.: The basis problem of the eigenfunctions of ordinary differential operators with integral boundary conditions (Russian). Vestnik Moskov. Univ. Ser. I Mat. Mekh. 37 (1982), N° 6, 12–21, Engl. transl.: Moscow Univ. Math. Bull. 37 (1982), N° 6, 10–20.

[SH4] SHKALIKOV, A. A.: Boundary-value problems for ordinary differential equations with a parameter in the boundary conditions (Russian). Funktsional. Anal. i Prilozhen. 16 (1982), no. 4, 92–93, Engl. transl.: Functional Anal. Appl. 16 (1982), no. 4, 92–93.

[SH5] SHKALIKOV, A. A.: Boundary problems for ordinary differential equations with parameter in the boundary conditions (Russian). Trudy Sem. Petrovsk. 9 (1983), 190–229, Engl. transl.: J. Soviet Math. 1311–1342.

[SH6] SHKALIKOV, A. A.: On estimates of meromorphic functions and summation of series in the root vectors of nonselfadjoint operators (Russian). Dokl. Akad. Nauk 268 (1983), no. 6, 1310–1314; Engl. transl.: Soviet Math. Dokl. 27 (1983), 259–263.

[SH7] SHKALIKOV, A. A.: On minimality and completeness of systems constructed from some of the eigen- and associated elements of quadratic operator pencils (Russian). Dokl. Akad. Nauk 285 (1985), no. 6, 1334–1339; Engl. transl.: Soviet Math. Dokl. 32 (1985), 902–907.

[SHTR1] SHKALIKOV, A. A., TRETTER, C.: Kamke problems. Properties of the eigenfunctions. Math. Nachr. 170 (1994), 251–275.

[SHTR2] SHKALIKOV, A. A., TRETTER, C.: Spectral analysis for linear pencils $N - \lambda P$ of ordinary differential operator. Math. Nachr. 179 (1996), 275–305.

[SU] SHUBIN, M. A.: Holomorphic families of subspaces of a Banach space (Russian). Mat. Issled. 5 (1970), vyp. 4(18), 153–165.

[SIN] SIN'KO, G. I.: Expansion in eigenfunctions of a boundary value problem with a spectral parameter entering into the boundary conditions (Russian). Funktsional. Anal. i Prilozhen. 19 (1982), 99-106.

[SN] SNOO, H. S. V. DE: Pointwise convergence of eigenfunctions expansions associated with ordinary differential operators. J. Math. Anal. Appl. 92 (1983), 172–179.

[SO] SOBOLEV, S. I.: Einige Anwendungen der Funktionalanalysis auf Gleichungen der Mathematischen Physik. Berlin 1964.

[STA] STALLARD, F. W.: Differential systems with interface conditions. Oak Ridge, Tennessee, Oak Ridge National Laboratory, ORNL 1876, 1955.

[STO] STOEBER, T.: Entwicklungssätze für eine Klasse indefiniter Eigenwertprobleme mit stückweise stetiger Gewichtsfunktion. Mitt. Math. Sem. Gießen 206 (1992).

[ST1] STONE, M. H.: An unusual type of expansion problem. Trans. Amer. Math. Soc. 26 (1924), 335–355.

[ST2] STONE, M. H.: A comparison of the series of Fourier and Birkhoff. Trans. Amer. Math. Soc. 28 (1926), 695–761.

[ST3] STONE, M. H.: Irregular differential systems of order two and the related expansion problem. Trans. Amer. Math. Soc. 29 (1927), 23–53.

[STRA] STRELITZ, SH., ABRAMOVICH, S.: On the existence of eigenvalues of differential operators dependent on a parameter. Trans. Amer. Math. Soc. 258 (1980), 407–429.

[STP] STREPETOV, A. V.: Completeness and basis property for a system of eigenfunctions of the two-sided Regge problem for a polar operator (Russian). Vestnik Leningrad Univ. 1983, no. 19, Mat. Mekh. Astronom. no. 4 (1983), 44–50, Engl. transl.: Vestnik Leningrad Univ. Math. 16 (1984), 255–262.

[SZN] SZ.-NAGY, B.: On a spectral problem of Atkinson. Acta Math. Sci. Hungar. 3 (1952), 61–66

[TA1] TAMARKIN, J. D.: Sur quelques points de la théorie des équations différentielles linèaires ordinaires et sur la généralisation des séries de Fourier. Ren. Circ. Mat. Palermo 34 (1912), 345–382.

[TA2] TAMARKIN, J. D.: Sur un probléme de la théorie des équations différentielles linéaires ordinaires. Rend. Circ. Mat. Palermo 37 (1914), 376–378.

[TA3] TAMARKIN, J. D.: Some general problems of the theory of ordinary linear differential equations and expansion of an arbitrary function in series of fundamental functions (Russian). Petrograd 1917.

[TA4] TAMARKIN, J. D.: On Fredholm's integral equations whose kernels are analytic in a parameter. Ann. Math. (2) 28 (1926–1927), 127–152.

[TA5] TAMARKIN, J. D.: Some general problems of the theory of ordinary linear differential equations and expansion of an arbitrary function in series of fundamental functions. Math. Z. 27 (1928), 1–54.

[TAL] TAMARKIN, J. D., LANGER, R. E.: On integral equations with discontinuous kernels. Trans. Amer. Math. Soc. 30 (1928), 453–471.

[TAY] TAYLOR, A. E.: Linear operations which depend analytically on a parameter. Ann. Math. (2) 39 (1938), 574–593.

[TL] TAYLOR, A. E., LAY, D. C.: Introduction to functional analysis. New York 1980.

[TIK1] TIKHOMIROV, V. V.: On the Riesz means of expansions in a trigonometric Fourier series and in the eigenfunctions of a Keldyš pencil of ordinary nonselfadjoint differential operators (Russian). Dokl. Akad. Nauk SSSR 226 (1976), 1015–1017, Engl. transl.: Soviet Math. Dokl. 17 (1976), 241–243.

[TIK2] TIKHOMIROV, V. V.: On Riesz' means of expansions into eigenfunctions and associated functions of a non-selfadjoint ordinary differential operator (Russian). Mat. Sb. N. S. 102(144) (1977), 33-55.

[TI] TITCHMARSH, E. C.: Introduction to the theory of Fourier integrals. 2nd ed., Oxford 1948.

[TIW] TIWARI, S.: On eigenvector expansions associated with a matrix differential operator. Indian J. Math. 23 (1981), 41–55.

[TR1] TRETTER, C.: Zum Eigenwertproblem der Bewegungsgleichung des einseitig eingespannten elastischen Stabes. Master's thesis, Regensburg 1989.

[TR2] TRETTER, C.: Asymptotische Randbedingungen für Entwicklungssätze bei Randeigenwertproblemen zu $N(y) = \lambda P(y)$ mit λ-abhängigen Randbedingungen. Thesis, Regensburg 1992.

[TR3] TRETTER, C.: On λ-nonlinear boundary eigenvalue problems. Akademie Verlag, Berlin 1993.

[TR4] TRETTER, C.: Asymptotic boundary conditions for expansion theorems for boundary eigenvalue problems of the form $N(y) = \lambda P(y)$ with λ-depending boundary conditions.. Z. Angew. Math. Mech. 73 (1993), T769–T773.

[TR5] TRETTER, C.: On fundamental systems for differential equations of Kamke type. Math. Z. 219 (1995), 609–629.

[TR6] TRETTER, C.: Nonselfadjoint spectral problems for linear pencils $N - \lambda P$ of ordinary differential operators with λ-linear boundary conditions: Completeness results. Integral Equations Operator Theory 26 (1996), 222–248.

[TR7] TRETTER, C.: A linearization for a class of λ-nonlinear boundary eigenvalue problems. J. Math. Anal. Appl. 247 (2000), 331–355.

[TR8] TRETTER, C.: Spectral problems for systems of differential equations $y' + A_0 y = \lambda A_1 y$ with λ-polynomial boundary conditions. Math. Nachr. 214 (2000), 129–172.

[TR9] TRETTER, C.: Linear operator pencils $A - \lambda B$ with discrete spectrum. Integral Equations Operator Theory 37 (2000), 357–373.

[TR10] TRETTER, C.: Boundary eigenvalue problems for differential equations $N\eta = \lambda P\eta$ and λ-polynomial boundary conditions. J. Differential Equations 170 (2001), 408–471.

[TRO] TROFIMOV, V. P.: The root subspaces of operators that depend analytically on a parameter (Russian). Mat. Issled. 3 (1968), No. 3, 117–125.

[TU] TURRITTIN, H. L.: Asymptotic solutions of certain ordinary differential equations associated with multiple roots of the characteristic equation. Amer. J. Math. 58 (1936), 364–376.

[TY] TURYN, L.: Perturbation of two-point boundary value problems with eigenvalue parameter in the boundary conditions. Proc. Roy. Soc. Edinburgh 94A (1983), 213–219.

[VA1] VAGABOV, A. I.: On the question of completeness of the system of eigenfunctions of a nonregular ordinary differential pencil (Russian). Dokl. Akad. Nauk SSSR 257 (1981), no. 1, 15–18; Engl. transl.: Soviet Math. Dokl. 23 (1981), 207–210.

[VA2] VAGABOV, A. I.: Multiple completeness of eigenfunctions of a type of ordinary differential bundle (Russian). Differential'nye Uravneniya 18 (1982), no. 2, 194–197, 363; Engl. transl.: Differential Equations 18 (1982), 152-155.

[VA3] VAGABOV, A. I.: On a case of violation of the multiple completeness property of the eigenfunctions for an ordinary differential pencil (Russian). Dokl. Akad. Nauk SSSR 263 (1982), 11-14, Engl. transl.: Soviet Math. Dokl. 25 (1982), 263–266.

[VA4] VAGABOV, A. I.: Convergence of expansions in Fourier series for ordinary differential operators (Russian). Mat. Zametki 32 (1982), no. 3, 303–307; Engl. transl.: Math. Notes 32 (1982), 631–634.

[VA5] VAGABOV, A. I.: On the incompleteness of the system of eigenelements of differential operators in a space of vector-valued functions (Russian). Dokl. Akad. Nauk SSSR 271 (1983), no. 1, 17–19; Engl. transl.: Soviet Math. Dokl. 28 (1983), 8–11.

[VA6] VAGABOV, A. I.: On equiconvergence of expansions in trigonometric Fourier series and in principal functions of ordinary differential operators (Russian). Izv. Akad. Nauk SSSR Ser. Mat. 48 (1984), no. 3, 614–630; Engl. transl.: Math. USSR-Izv. 24 (1983), 567–582.

[VA7] VAGABOV, A. I.: Completeness of the eigenfunction system of an ordinary differential bundle not of the Keldysh type (Russian). Differentsial'nye Uravneniya 20 (1984), no. 3, 375–382, Engl. transl.: Differential Equations 20 (1984), 281–287.

[VA8] VAGABOV, A. I.: Completeness of eigenfunctions of irregular differential operators in a space of vector-functions (Russian). Mat. Zametki 40 (1986), N° 2, 197–202, 286; Engl. transl.: Math. Notes 40 (1986), N° 1–2, 611–615.

[VA9] VAGABOV, A. I.: Conditions of multiple completeness of the eigenelements of an ordinary differential bundle (Russian). Izv. Vyssh. Uchebn. Zaved. Mat. 1986 (1986), no. 4(287), 13–20, Engl. transl.: Soviet Math. 30 (1986), no. 4, 16–24.

[VA10] VAGABOV, A. I.: n-fold summability of series in principal functions of pencils of ordinary differential operators (Russian). Dokl. Akad. Nauk SSR 299 (1988), N° 3, 534–537; Engl. transl.: Soviet Math. Dokl. 37 (1988), N° 2, 394–398.

[VAS] VASS, J. I.: A class of boundary problems of highly irregular type. Duke Math. J. 2 (1936), 151–165.

[VEG1] VEGA, K. E.: On the completeness of the eigenfunctions of a boundary value problem (Russian). Uspekhi Mat. Nauk 27 (1972), no. 4(166), 205–206.

[VEG2] VEGA, K. E.: On the completeness of a subsystem of eigenfunctions of one class of boundary value problems (Russian). Funktsional. Anal. i Prilozhen. 6 (1972), no. 3, 65–67, Engl. transl.: Funktional Anal. Appl. 6 (1972), 228–230.

[VEL] VELIEV, O. A.: Spectral expansions of differential operators with periodic complex-valued coefficients (Russian). Dokl. Akad. Nauk AzSSR 36 (1980), no. 6, 13–17.

[VE] VERBLUNSKY, S.: A uniqueness theorem for series of generalized eigenfunctions. Proc. London Math. Soc. (3) 37 (1978), 1–31.

[VM] VIZITEĬ, V. N., MARKUS, A. S.: On convergence of multiple expansions in the eigenvectors and associated vectors of an operator pencil. Mat. Sb. 66 (108) (1965), 287–320; Engl. transl.: Amer. Math. Soc. Transl. (2) 87 (1970).

[VOR] VORONINA, S. K.: Necessary and sufficient conditions for the eigenfunctions and associated functions of an ordinary differential operator to form a basis in $L(0,1)$ (Russian). Differentsial'nye Uravneniya 14 (1978), 407–417, Engl. transl.: Differential Equations 14 (1978), 282–289.

[WAL] WALKER, P. W.: An expansion theory for non-self-adjoint boundary-value problems. Proc. Roy. Soc. Edinburgh 108A (1988), 11–26.

[WT] WALTER, J.: Regular eigenvalue problems with eigenvalue parameter in the boundary conditions. Math. Z. 133 (1973), 301–312.

[WA] WASOW, W.: Asymptotic expansions for ordinary differential equations. New York – London – Sydney 1965.

[WE] WERMUTH, E.: Konvergenzuntersuchungen bei Eigenfunktionsentwicklungen zu Randeigenwertproblemen n-ter Ordnung mit parameterabhängigen Randbedingungen. Thesis, Math.-Naturw. Fak. RWTH Aachen 1984.

[WEX] WEXLER, D.: On boundary value problems for an ordinary linear differential system. Ann. Mat. Pura Appl. (4) 80 (1968), 123–134.

[WEY1] WEYL, H.: Über gewöhnliche Differentialgleichungen mit Singularitäten und die zugehörigen Entwicklungen willkürlicher Funktionen. Math. Ann. 68 (1910), 220–269.

[WEY2] WEYL, H.: Ramifications old and new of the eigenvalue problem. Bull. Amer. Math. Soc. 56 (1950), 115–139.

[WHY1] WHYBURN, W. M.: Differential equations with general boundary conditions. Bull. Amer. Math. Soc. 48 (1942), 692–704.

[WHY2] WHYBURN, W. M.: Differential systems with boundary conditions at more than two points. Proceedings of the Conference on Differential Equations, University of Maryland,, 1955, 1-21, College Park, Md., 1956.

[WI1] WILDER, C. E.: Expansion problems of ordinary linear differential equations with auxiliary conditions at more than two points. Trans. Amer. Math. Soc. 18 (1917), 415–442.

[WI2] WILDER, C. E.: Problems in the theory of ordinary linear differential equations with auxiliary conditions at more than one point. Trans. Amer. Math. Soc. 19 (1918), 157–166.

[WOL] WOLTER, M.: Das asymptotische Verhalten der Greenschen Funktion N-irregulärer Eigenwertprobleme mit zerfallenden Randbedingungen. Math. Methods Appl. Sci. 5 (1983), 331–345.

[WOO] WOOD, A. D.: Eigenfunction expansions by the delta function and contour integral method. J. Inst. Math. Appl. 14 (1974), 57–67.

[WR] WRAY, S. D.: Absolutely convergent expansions associated with a boundary value problem with the eigenvalue parameter contained in one boundary condition. Czechoslovak Math. J. 32 (1982), 608–622.

[WU] WU, J. J.: On adjoint operators associated with boundary problems. J. Sound Vibration 39 (1975), 195–206.

[XI] XIA, N. M.: The eigenfunction expansion of the solution for the nonhomogeneous Sturm-Liouville problem containing white noise. J. Math. anal. Appl. 159 (1991), 88–109.

[YA] YAKUBOV, S. YA.: Multiple completeness of root vectors of an unbounded polynomial operator pencil. Rev. Roumaine Math. Pures Appl. 31 (1986), 423-438.

[YM1] YAKUBOV, S. YA., MAMEDOV, K. S.: Multiple completeness of a system of eigen and associated elements of a polynomial operator bundle and multiple expansions with respect to this system (Russian). Funktsional. Anal. i Prilozhen. 9 (1975), N° 1, 91–93; Engl. transl.: Functional Anal. Appl. 9 (1975), N° 1, 90-92.

[YM2] YAKUBOV, S. YA., MAMEDOV, K. S.: Completeness of the eigen- and associated functions of certain nonregular boundary-value problems for ordinary differential equations (Russian). Funktional. Anal. i Prilozhen. 14 (1980), no. 4, 93–94; Engl. transl.: Functional Anal. Appl. 14 (1981), 327–328.

[ZA] ZAYED, E. M. E.: Regular eigenvalue problem with eigenvalue parameter in the boundary conditions. Proc. Math. Phys. Soc. Egypt 58 (1984), 55–62 (1986).

[ZIM1] ZIMMERBERG, H. J.: Two-point boundary problems involving a parameter linearly. Illinois J. Math. 4 (1960), 593–608.

[ZIM2] ZIMMERBERG, H. J.: Two-point boundary conditions linear in a parameter. Pacific J. Math. 12 (1962), 385–393.

[ZU] ZUBOV, V. M.: On a theory of a generalized Green's function (Russian). Izv. Vyssh. Uchebn. Zaved. Mat. 1981 (1981), no. 9(232), 69–73, Engl. transl.: Soviet Math. 25 (1981), no. 9, 86–91.

Notations

$\mathbb{C}$	the complex numbers	$\mathbb{N}$	the nonnegative integers
$\mathbb{R}$	the real numbers	$\Re(z)$	the real part of z
$\mathbb{R}_+$	the nonnegative real numbers	$\Im(z)$	the imaginary part of z
$\mathbb{R}_-$	the nonpositive real numbers	#	the cardinality of a (finite) set
$\mathbb{Z}$	the integers		

e_i, ε_i, ϵ_i unit vectors in $\mathbb{C}^n$, $\mathbb{C}^l$, $\mathbb{C}^{n_0}$

Index